PHYSICS
A Textbook for Advanced Level Students

PHYSICS

A Textbook for Advanced Level Students

T. Duncan BSc

Formerly Senior Lecturer in Education, University of Liverpool

John Murray

First published 1982

Text set in 9/11pt Linotron 202 Times, printed and bound
in Great Britain at The Pitman Press, Bath

British Library Cataloguing in Publication Data

Duncan, Tom
    Physics: a textbook for Advanced Level students.
    1. Physics
    I. Title
    530   QL23

ISBN 0-7195-3889-0

Preface

This book is a combined volume of the second editions of the two sixth form textbooks, *Advanced Physics: Materials and Mechanics* and *Advanced Physics: Fields, Waves and Atoms*. These are substantially the same as the original editions but have been updated throughout to be in line with recent syllabus revisions, particularly the chapter on electronics.

Once again I would like to express my gratitude to those who helped with the original manuscripts, Dr J. W. Warren, Mr J. Dawber, Professor J. Stringer, Dr J. C. Gibbings, Mr B. Baker, Drs B. L. N. and H. M. Kennett. Dr M. G. Pellatt of B.D.H. Chemicals Ltd. very kindly gave advice on the new section dealing with liquid crystals. I am also grateful to those students and teachers, particularly Mr J. V. Thornton of St. Mary's College, Crosby, who have written to me with suggestions for improving the earlier books.

Thanks are also due to my wife not only for typing the revision but for continuing to suffer uncomplainingly the inevitable domestic inconvenience that writing or revising a textbook involves.

For permission to use questions from recent examinations grateful acknowledgement is made to the various examining boards, indicated by the following abbreviations: *A.E.B.* (Associated Examining Board); *C* (Cambridge Local Examination Syndicate); *J.M.B.* (Joint Matriculation Board); *L* (University of London); *O* (Oxford Local Examinations); *O and C* (Oxford and Cambridge Schools Examination Board); *S* (Southern Universities Joint Board); *W* (Welsh Joint Education Committee).

T.D.

Acknowledgements

Thanks are due to the following who have kindly permitted the reproduction of copyright photographs:

Figs. 1.1*a*, *b*, Cambridge Scientific Instruments Ltd; 1.1*c*, 1968 The Plessey Co Ltd; 1.1*d*, A. Dinsdale, The British Ceramic Research Association; 1.7, Dr R. T. Southin; 1.9, Dr B. Ralph, Cambridge University; 1.10, J. W. Martin (from *Elementary Science of Metals*, Wykeham Publications (London) Ltd); 1.16*b*, *c*, *d*, *e*, The Royal Society (*Proceedings*, 1947, A, 190); 1.19*a*, G. Bell & Sons Ltd (from *An Approach to Modern Physics* by E. Andrade); 1.19*b*, Dr C. Henderson, Aberdeen University; 2.1, 2.2, Avery-Denison Ltd; 2.13, V. A. Phillips and J. A. Hugo; 2.14, The Royal Society (*Proceedings*, 1947, A, 190); 2.15, Joseph V. Laukonis, Research Laboratories, General Motors Corporation; 2.17*b*, Penguin Books Ltd (from *Revolution in Optics* by S. Tolansky); 2.22, British Engine Boiler & Electrical Insurance Company Ltd; 2.24, Vosper Thornycroft, Southampton; 2.25*a*, Central Office of Information, from *Project*; 2.25*c*, Fulmer Research Institute Ltd; 2.28*a*, Professor E. H. Andrews, Queen Mary College; 2.28*b*, Malayan Rubber Fund Board (London) Inc.; 2.20, Eidenbenz and Eglin, used in *Science*, ed. J. Bronowski (Aldus Books); 3.20, Avo Limited, Dover; 3.22, Central Electricity Generating Board; 3.24, Royal Aircraft Establishment, Farnborough; 3.25, Mullard Ltd; 3.30, Educational Measurements Ltd; 4.4*a*, *b*, *c*, U.S.I.S.; 5.37*a*, *b*, Penguin Books Ltd (from *Revolution in Optics* by S. Tolansky); 5.37*c*, Rank Precision Industries; 5.76*a*, Penguin Books Ltd (from *Revolution in Optics* by S. Tolansky); 5.87, Hale Observatories; 5.97, Times Newspapers Ltd; 6.1, Shell Chemicals U.K. Ltd; 6.17, reprinted by permission of the publishers, D. C. Heath & Co, Lexington (from *PSSC Physics*); 6.25*a*, *b*, 6.34*a*, *b*, J. T. Jardine, Moray House College of Education; 6.35*b*, Lord Blackett; 7.9, Popperfoto; 7.29, NASA; 7.30, AP Laserphoto; 8.1*a*, *b*, Mrs F. B. Farquharson; 8.25, Royal Aircraft Establishment, Farnborough; 9.1, Popperfoto; 9.3*b*, British Leyland (Austin-Morris) Ltd; 9.3*c*, Pressure Dynamics Ltd; 9.14*a*, L. H. Newman (photo by W. J. C. Murray); 9.14*b*, Focal Press (photo by Oskar Kreisal from *Focal Encyclopaedia of Photography*); 10.9, Dr J. C. Gibbings, Department of Mechanical Engineering, University of Liverpool, and the Editor, *Physics Bulletin* – © J. C. Gibbings; Figs. 11.3, Aerofilms Ltd; 11.8*a*, *b*, Education Development Center, Inc; 12.25, 21.27*a*, 21.47, and 21.63*a*, Unilab Ltd; 12.37, United Kingdom Atomic Energy Authority; 13.23*a*, *b*, Leybold-Heraeus; 13.30*b*, National Physical Laboratory; 13.35*c*, Walden Precision Apparatus Ltd; 14.2*b*, Royal Institution; 14.14*b*, Mullard Ltd; 14.16, Science Museum, London; 14.23*a*, GEC Turbine Generators Ltd; 14.32*a*, Central Electricity Generating Board; 14.32*b*, Electricity Council; 14.35, Central Electricity Generating Board; 14.48*a*, *b*, Mullard Ltd; 14.51, McGraw-Hill Book Co.; 16.3*a*, A. M. Lock & Co. Ltd; 16.8*b*, Griffin and George Ltd; 16.10*b*, 16.15*b* and 16.20*a*, *b*, *c*, W. Llowarch, *Ripple Tank Studies of Wave Motion*, (Clarendon Press, Oxford); 16.19, Decca Navigator Company Ltd; 16.21*a*, *b*, from *PSSC Physics* (D. C. Heath and Company); 16.34, Decca Radar Ltd; 17.2*a*, *b*, United States Information Service; 17.4, 17.17, GLC Architect's Department; 18.1*a*, Philip Harris Ltd; 18.3*b*, D. G. A. Dyson; 18.12, Bausch and Lomb Optical Company Ltd; 18.19, C. B. Daish; 18.21, 18.24*a*, Addison-Wesley Publishing Company; 18.53, 18.54, Open University; 18.57, Barnes Engineering Company; 19.8, BBC Publications; 19.20, Rolls-Royce (1971) Ltd; 20.10*c*, 21.18*a*, RS Components Ltd; 20.16*a*, *b*, from *Nuffield Advanced Physics, Teachers' Guide, Unit 1* (Longman Group Ltd); 20.23, AEI Scientific Apparatus Ltd; 20.31*a*, Professor H. Hill; 20.31*b*, G. Mollenstedt and H. Duker; 21.3*b*, *c*, Tektronix UK Ltd; 21.9, 21.59*a*, *d*, Mullard Ltd; 21.59*e*, *f*, Department of Industry; 21.67, Thandar Electronics Ltd; 22.7*a*, *b*, Panax Equipment Ltd; 22.11*a*, *b*, C. T. R. Wilson; 22.20*a*, 22.28*a*, *b*, and 22.29, United Kingdom Atomic Energy Authority; 22.20*b*, UKAEA (Courtesy National Hospital); 22.22 AERE Harwell; 22.34*a*, *b*, Photo CERN; 22.35*a*, *b*, Lord Blackett's estate; A16.1*b* A. G. Gaydon.

Page 1, Professor S. Tolansky, Royal Holloway College; Page 131, British Aerospace; Page 219, MIT Museum and Historical Collections; Page 337, W. Llowarch; Page 425, Philips Electronic Components and Materials.

The following photographs, taken by the author, show equipment or experiments from these sources:

Figs 13.16 and 17.30*a*, *b*, Philip Harris Ltd; 16.3*b*, A. M. Lock & Co Ltd; 16.4, 16.28, Unilab Ltd; 22.11*c*, University of Liverpool.

Contents

Part 1 MATERIALS

1 Structure of materials

Materials science

Advances in technology depend increasingly on the development of better materials. This is especially true of those industries engaged in aircraft production, space projects, telecommunications, computer manufacture and nuclear power engineering. Structural materials are required to be stronger, stiffer and lighter than existing ones. In some cases they may have to withstand high temperatures or exposure to intense radioactivity. Materials with very precise electrical, magnetic, thermal, optical or chemical properties are also demanded.

A great deal has been known for many years about materials that are useful in everyday life and industry. For example, the metallurgist has long appreciated that alloys can be made by adding one metal to another or that heating, cooling or hammering metals changes their mechanical behaviour. Materials *technology* is a long-established subject. The comparatively new subject of materials *science* is concerned with the study of materials as a whole and not just with their physical, chemical or engineering properties. As well as asking *how* materials behave, the materials scientist also wants to know *why* they behave as they do. Why is steel strong, glass brittle and rubber extensible? To begin to find answers to such questions has required the drawing together of ideas from physics, chemistry, metallurgy and other disciplines.

The deeper understanding of materials which we now have has come from realizing that the properties of matter in bulk depend largely on the way the atoms are arranged when they are close together. Progress has been possible because of the invention of instruments for 'seeing' finer and finer details. The electron microscope, which uses beams of electrons instead of beams of light as in the optical microscope, reveals structure just above the atomic level. The field ion microscope and X-ray apparatus allow investigation at that level.

The scanning electron microscope, Fig. 1.1a, is a development from the electron microscope and 'scans' a surface with electrons in the way that a television screen is scanned. It gives higher magnifications and much greater depth of focus than optical microscopes using reflected light. It is useful for examining the surfaces of semiconductors, the hairlike fibres and 'whiskers' that are so important in the manufacture of the new generation of composite materials, man-made fibres, and corroded and fractured surfaces. A view of the end of a torn wire ($\times$ 75) is shown in Fig. 1.1b and of lead–tin telluride crystals ($\times$ 30) in Fig. 1.1c.

Materials science is a rapidly advancing subject with exciting prospects for the future. Its importance lies in the help it can give with the selection of materials for particular applications, with the design of new materials and with the improvement of existing ones. The strength of even a tea cup has been improved by research into ceramics, as Fig. 1.1d shows.

Atoms, molecules and Brownian motion

The modern atomic theory was proposed in 1803 by John Dalton, an English schoolmaster. He thought of atoms as tiny, indivisible particles, all the atoms of a given element being exactly alike and different from those of other elements in behaviour and mass. By making simple assumptions he explained the gravimetric (i.e. by weight) laws of chemical combination but failed to account satisfactorily for the volume relationships which exist between combining gases. This required the introduction in 1811 by the Italian scientist, Amedeo Avogadro, of the molecule as the smallest particle of an element or compound capable of existing independently and consisting of two or more atoms, not necessarily identical. Thus, whilst we could only have atoms of elements, molecules of both elements and compounds were possible.

(b)

(a)

(c)

(d)

Fig. 1.1

At the end of the nineteenth century some scientists felt that evidence, more direct than that provided by the chemist, was needed to justify the basic assumption that atoms and molecules exist. In 1827 the Scottish botanist, Robert Brown, discovered that fine pollen grains suspended in water were in a state of constant movement, describing small, irregular paths but never stopping. The effect, which has been observed with many kinds of small particles suspended in both liquids and gases, is called *Brownian motion*. It is now considered to be due to the unequal bombardment of the suspended particles by the molecules of the surrounding medium.

Very small particles are essential. If the particle is

fairly large, the impacts, occurring on every side and irregularly, will cancel out and there will be no average resultant force on the particle. Only if the particle is small will it suffer impacts with a few hundred molecules at any instant and the chances of these cancelling out are proportionately less. It is then likely that for a short time most of the impacts will be in one direction; shortly afterwards the direction will have changed. The phenomenon can be observed in smoke in a small glass cell which is illuminated strongly from one side and viewed from above with a low-power microscope, Fig. 1.2. How would the random motion be affected by (*i*) cooling the air to a low temperature, (*ii*) using smaller smoke particles?

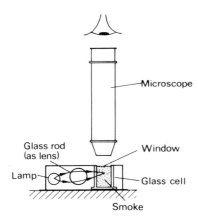

Fig. 1.2

The effect, on its own, does not offer conclusive proof for molecules but it clearly reveals that on the microscopic scale there is great activity in matter which macroscopically (on a large scale) appears to be at rest. The theory of the motion was worked out by Einstein and is found to correspond closely with observation. His basic assumption was that the suspended particles have the same mean kinetic energy as the molecules of the fluid and so behave just like very large molecules. Their motions should therefore be similar to those of the fluid molecules.

The Avogadro constant: mole

Atomic and *molecular masses* (previously called *atomic* and *molecular weights*) give the masses of atoms and molecules compared with the mass of another kind of atom. Originally the hydrogen atom was taken as the standard with atomic mass 1 since it has the smallest mass. In 1960 it was agreed internationally, for various reasons, to base atomic and molecular masses on the atom of carbon (more precisely, on the carbon 12 isotope $^{12}_{6}C$). On the carbon scale the atomic mass of carbon 12 is taken as exactly 12 making that of hydrogen 1.008 and of oxygen 16.00. Nowadays atomic masses are found very accurately using a *mass spectrometer*.

It follows from the definition of atomic mass that any number of atoms of carbon will have, near enough, 12 times the mass of the *same* number of atoms of hydrogen. Therefore any mass of hydrogen, say 1 g, will contain the same number of atoms as 12 g of carbon. In general, the atomic mass of any element expressed in grams, contains the same number of atoms as 12 g of carbon. This number is thus, by definition, a constant. It is called the *Avogadro constant* and is denoted by *L*. Its accepted experimental value is 6.02×10^{23}.

The number of molecules in the molecular mass in grams of a substance is also (because of the way molecular masses are defined) the same for all substances and equal to the Avogadro constant. There are, therefore, 6.02×10^{23} molecules in 2 g of hydrogen (molecular mass 2) and in 18 g of water (molecular mass 18). In fact, the Avogadro constant is useful when dealing with other particles besides atoms and molecules and a quantity which contains 6.02×10^{23} particles is called, especially by chemists, a *mole*. We can thus have a mole of atoms, a mole of molecules, a mole of ions, a mole of electrons, etc.—all contain 6.02×10^{23} particles. We must always have a mole of some kind of particle and so

$$L = 6.02 \times 10^{23} \text{ particles per mole}$$

It should be noted that the mole (abbreviation mol) is based on the gram and not the kilogram, which makes it an anomaly in the SI system of units. Sometimes, however, it is expressed in terms of the number of particles per kilogram-mole and its value then is 6.02×10^{26}.

The Avogadro constant has been measured in various ways. In an early method alpha particles emitted by a radioactive source were counted by allowing those within a small known angle to strike a fluorescent screen. Each particle produced one scintillation on the screen and if it is assumed that one particle is emitted by each radioactive atom an approximate value for *L* can be obtained (see question 6, p. 18). Other methods give more reliable results—one involves X-ray crystallography.

Size of a molecule

(a) *Monolayer experiments.* An experimental determination of the size of a molecule was made by Lord Rayleigh in 1899. He used the fact that certain organic substances, such as olive oil, spread out over a clean water surface to form very thin films.

A simple procedure for performing the experiment is to obtain a drop of olive oil by dipping the end of a loop of thin wire, mounted on a card, into olive oil, quickly withdrawing it and then estimating the diameter of the drop by holding it against a $\frac{1}{2}$ mm scale and viewing the drop and scale through a lens, Fig. 1.3a. If the drop is then transferred to the centre of a waxed tray overbrimming with water, the surface of which has been previously cleaned by drawing two waxed booms across it and then lightly dusted with lycopodium powder, Fig. 1.3b, it spreads out into a circular film pushing the powder before it. Assuming the drop is spherical, the thickness of the film can be calculated if its diameter is measured. It is found to be about 2×10^{-9} metre, i.e. 2 nanometres (2 nm).

Oil-film experiments do not necessarily prove that matter is particulate but from them we can infer that if molecules exist and if the film is one molecule thick, i.e. a monolayer, then in the case of olive oil one dimension of its molecule is 2 nm.

(b) *Predictions from kinetic theory of gases.* Information about the molecular world can sometimes be obtained from observations of the behaviour of matter in bulk, i.e. from macroscopic observations. Thus with the help of the kinetic theory of gases, expressions can be derived relating such properties as rate of diffusion with the size of the gas molecules involved.

(c) *Using the Avogadro constant.* Consider copper which has atomic mass 64 and density 9.0 g cm^{-3}. One mole of copper atoms, therefore, has mass 64 g and volume 64/9 cm^3; it contains 6.0×10^{23} atoms. The volume available to each atom is $64/(9 \times 6 \times 10^{23})$ cm^3 and the radius r of a sphere having this volume is given by

$$\frac{4}{3}\pi r^3 = \frac{64}{9 \times 6 \times 10^{23}}$$

$$\therefore \quad r = 0.14 \times 10^{-7} \text{ cm}$$

$$= 0.14 \times 10^{-9} \text{ m}$$

$$= 0.14 \text{ nm}$$

If copper atoms are spherical, would their radius be larger or smaller than this even if they were packed tightly? Why? A more accurate way of calculating the size of a copper atom is indicated in questions 10 to 13 on p. 18.

A word of caution is necessary regarding atomic dimensions. Nowadays atoms and molecules are no longer pictured as having hard, definite surfaces like a ball and there is, therefore, little point in trying to give their diameters too exact values; most are within the range 0.1 to 0.5 nm. Also, although we shall usually treat atoms and molecules as spheres, it is necessary on occasion to consider them as having other shapes.

Periodic table

With the passage of time the early nineteenth-century picture of an indivisible atom came to be doubted in the light of fresh information. During the 1860s chemical knowledge increased sufficiently for it to be clear that there were elements with similar chemical properties. Moreover, atomic masses were being established with greater certainty and attempts were made to relate properties and atomic masses.

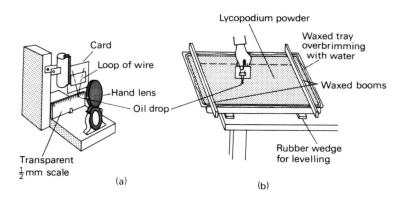

Fig. 1.3

Table 1.1

Group 1	Group 2	Group 3	Group 4	Group 5	Group 6	Group 7	Group 0
1 Hydrogen							2 Helium
3 Lithium	4 Beryllium	5 Boron	6 Carbon	7 Nitrogen	8 Oxygen	9 Fluorine	10 Neon
11 Sodium	12 Magnesium	13 Aluminium	14 Silicon	15 Phosphorus	16 Sulphur	17 Chlorine	18 Argon

It was found that if the elements were arranged in order of increasing atomic masses then, at certain repeating intervals, elements occurred with similar chemical properties. Sometimes it was necessary to place an element of larger atomic mass before one of slightly smaller atomic mass to preserve the pattern. The first eighteen elements of this arrangement, called the Periodic table, are shown in Table 1.1. The third and eleventh (3 + 8) elements are the alkali metals lithium and sodium; the ninth and seventeenth (9 + 8) are the halogens fluorine and chlorine—here the repeating interval is eight. The serial number of an element in the table is called its *atomic number*.

The Periodic table suggests that the atoms of the elements may not be simple entities but are somehow related. There must be similarities between the atoms of similar elements and it would seem that the similarity might be due to the way they are built up.

We now believe that atoms are composed of three types of particles—protons, neutrons and electrons. (Many other subatomic particles, such as positrons, mesons and antiprotons, are known but most are short-lived and are not primary components.) Protons and neutrons are packed together into a very small nucleus which is surrounded by a cloud of electrons, the diameter of the atom as a whole being at least 10 000 times greater than that of the nucleus. The comparative masses and charges of the three basic particles are given in Table 1.2. The nucleus is positively charged and the electron cloud negatively charged but the number of protons equals the number of electrons so that the atom is electrically neutral.

The number of protons in the nucleus of an atom has been found to be the same as its atomic number which therefore means that each element in the Periodic table has one more proton and one more electron in its atom than the previous element. Hydrogen, the first element, has one proton and one electron. Helium, the second element, has two protons and two electrons.

Lithium, with atomic number three, has three protons and three electrons. Neutrons are present in all nuclei except that of hydrogen.

Table 1.2

Particle	Mass	Charge
electron	1	$-e$
proton	1836	$+e$
neutron	1839	0

e = electronic charge

Interatomic bonds

Materials consist of atoms held together by the attractive forces they exert on each other. These forces are electrical in nature and create interatomic bonds of various types. The type formed in any case depends on the outer electrons in the electron clouds of the atoms involved.

(a) *Ionic bond.* This is formed between the atoms of elements at opposite sides of the Periodic table, for example between sodium (Group 1) and chlorine (Group 7) when they are brought together to form common salt. A sodium atom has a loosely held outer electron which is readily accepted by a chlorine atom. The sodium atom becomes a positive ion, i.e. an atom deficient of an electron, and the chlorine atom becomes a negative ion, i.e. an atom with a surplus electron. The two ions are then bonded by the electrostatic attraction between their unlike charges.

A sodium ion attracts *all* neighbouring chloride ions in other pairs of bonded ions and vice versa. Each ion becomes surrounded by ions of opposite sign and the resulting structure depends among other things on the relative sizes of the two kinds of ion.

The ionic bond is strong. Ionic compounds are usually solid at room temperature and have high melting points. They are good electrical insulators in the solid state since the electrons are nearly all firmly bound to particular ions and few are available for conduction.

(*b*) *Covalent bond.* In ionic bonding electron *transfer* occurs from one atom to another. In covalent bonding electron *sharing* occurs between two or more atoms. Thus the atoms of carbon can form covalent bonds with other carbon atoms. Each carbon atom has four outer electrons, Fig. 1.4*a*, and all can be shared with four other carbon atoms to make four bonds, Fig. 1.4*b*, each consisting of two interlocking electron clouds.

Covalent bonds are also strong and many covalent compounds have similar mechanical properties to ionic compounds. However, unlike the latter, they do not conduct electricity when molten.

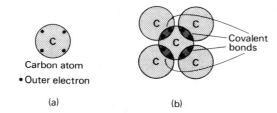

Fig. 1.4

(*c*) *Metallic bond.* Metal atoms have one or two outer electrons that are in general loosely held and are readily lost. In a metal we picture many free electrons drifting around randomly, not attached to any particular atom as they are in covalent bonding. All atoms share *all* the free electrons. The atoms thus exist as positive ions in a 'sea' of free electrons, Fig. 1.5; the strong attraction between the ions and electrons constitutes the metallic bond.

The nature of the metallic bond has a profound influence on the various properties of metals, as we shall see later.

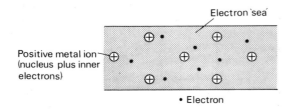

Fig. 1.5

(*d*) *Van der Waals bond.* Van der Waals forces are very weak and are present in all atoms and molecules. They arise because, although the centres of negative and positive charges in an atom coincide over a period of time, they do not coincide at any instant—for reasons too advanced to be considered here. There is a little more of the electron cloud on one side of the nucleus than the other. A weak electric 'dipole' is produced giving rise to an attractive force between opposite ends of such dipoles in neighbouring atoms.

The condensation and solidification at low temperatures of oxygen, hydrogen and other gases is caused by van der Waals forces binding their molecules together. (In the molecules of such gases the atoms are held together covalently.) Van der Waals forces are also important when considering polymers (p. 16).

Two further points: first, sometimes more than one of the previous four types of bonding is involved in a given case; second, information about the strength of interatomic bonds in solids is obtained from heat of sublimation measurements in which solid is converted directly to vapour and all atoms separated from their neighbours (see question 8, p. 18) and for liquids latent heat of vaporization measurements provide the information (see page 72).

States of matter

The existence of three states or phases of matter is due to a struggle between interatomic (intermolecular) forces and the motion which atoms (molecules) have because of their internal energy (see p. 65).

(*a*) *Solids.* In the four types of interatomic force just discussed only attractions were considered but there must also be interatomic repulsions, otherwise matter would collapse. Evidence, both theoretical and experimental, suggests that at distances greater than one atomic diameter the attractive force exceeds the repulsive one, whilst for small distances, i.e. less than one atomic diameter, the reverse is true. In Fig. 1.6*a* the dotted graphs show how the *short-range attractive* force and the *very short-range repulsive* force between two atoms vary with the separation of the atoms; the total or resultant force is shown by the continuous graph.

It can be seen that for one value of the separation r_0, the resultant interatomic force is zero. This is the situation that normally exists in a solid, but if the atoms come closer together—for example, when the solid is compressed—they repel each other; they attract when they are pulled farther apart. We have only considered two atoms whereas in a solid each atom has interactions with many of its close neighbours. The conclusion

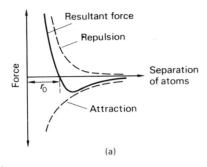

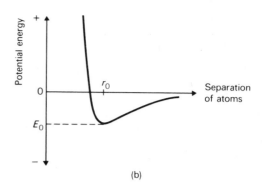

Fig. 1.6

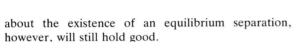

about the existence of an equilibrium separation, however, will still hold good.

In an ionic bond the short-range attractive part of the interatomic force arises from the attraction between positive and negative ions which pulls them together until their electron clouds start to overlap, thus creating a very short-range repulsive force. The attractive and repulsive forces in the other types of bond also arise from the electric charges in atoms.

Now consider the motion of the atoms, the other contestant. In a solid the atoms vibrate about their equilibrium positions alternately attracting and repelling one another but the interatomic forces have the upper hand. The atoms are more or less locked in position and so solids have shape and appreciable stiffness.

The corresponding potential energy–separation curve for two atoms (or molecules) is shown in Fig. 1.6b. At the equilibrium separation r_0 when the resultant force is zero, the p.e. must have its minimum value E_0. This is so because any attempt to change the separation involves overcoming an opposing force—an attractive one if the separation increases and a repulsive one if it decreases. E_0 is called the *bonding energy;* it is the energy needed to pull the atoms apart so that their p.e. increases to zero. They are then quite free from one another's influence.

Bonding energy will be considered later in connection with latent heat (p. 72) and surface energy (p. 204).

(*b*) *Liquids.* As the temperature is increased the atoms have larger amplitudes of vibration and eventually they are able partly to overcome the interatomic forces of their immediate neighbours. For short spells they are within range of the forces exerted by other atoms not quite so near. There is less order and the solid melts. The atoms and molecules of a liquid are not much farther apart than in a solid but they have greater

speeds, due to the increased temperature, and move randomly in the liquid while continuing to vibrate. The difference between solids and liquids involves a difference of structure rather than a difference of distance between atoms and molecules.

Although the forces between the molecules in a liquid do not enable it to have a definite shape, they must still exist otherwise the liquid would not hold together or exhibit surface tension (i.e. behave as if it had a skin on its surface) and viscosity nor would it have latent heat of vaporization.

(*c*) *Gases.* In a gas or vapour the atoms and molecules move randomly with high speeds through all the space available and are now comparatively far apart. On average their spacing at s.t.p. is about 10 molecular diameters and their mean free path (i.e. the distance travelled between collisions) is roughly 300 molecular diameters. Molecular interaction only occurs for those brief spells when molecules collide and large repulsive forces operate between them.

Conditions in gases and solids are, by comparison, simpler than those in liquids and in general their behaviour is better understood.

Types of solids

There are two main types of solids.

(*a*) *Crystalline.* Most solids, including all metals and many minerals, are crystalline. In substances such as sugar the crystal form is evident but less so in the case of metals, although large crystals of zinc are often visible on a freshly galvanized iron surface.

The crystalline structure of a metal can be revealed by polishing the surface, treating it with an etching chemical, sometimes a dilute acid, and then viewing it under an optical microscope. The metal is seen to consist of a mass of tiny crystals, called grains, at various angles to one another; it is said to be *polycrystalline*. Grain sizes are generally small, often about 0.25 mm across. Fig. 1.7 shows crystal grains in a cross-section of an aluminium-copper casting. The grains show up on the surface after etching because 'steps' are formed on each grain due to the rate of chemical action differing with different grain orientations. Light is then reflected in various directions by the different grains so that some appear light and others dark, Fig. 1.8.

Fig. 1.7

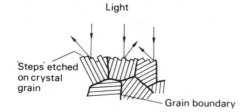

Fig. 1.8

Grain boundaries are revealed at the atomic level by the field ion microscope, Fig. 1.9, which uses beams of helium ions to 'illuminate' the object rather than

electrons or light. The field ion micrograph in Fig. 1.10 shows the tip of an iridium needle, viewed from the point and magnified 2 500 000 times. Each bright spot represents an iridium atom and the abrupt pattern change is clearly visible at the grain boundary.

The essence of the structure of a crystal, whether it be a large single crystal or a tiny grain in a polycrystalline specimen, is that the arrangement of atoms, ions or molecules repeats itself regularly many times, i.e. there is a long-range order.

(*b*) *Amorphous.* Here the particles are assembled in a more disordered way and only show order over short distances; there is no long-range order. The structure of an amorphous solid has been likened to that of an instantaneous photograph of a liquid. It is much more difficult to unravel but is the subject of considerable research. The many types of glass are the commonest of the amorphous solids; we can think of them as having a structure of groups of atoms (e.g. of silicon and oxygen) that would have been crystalline had it not been distorted.

Crystal structures

The structure adopted by a crystalline solid depends on various factors including the kind of bond(s) formed and the size and shape of the particles involved. For example, in metals where all the positive ions attract all electrons (p. 8), the bonding pulls equally in all directions, i.e. is non-directional, and every ion tends to surround itself by as many other ions as is geometrically possible. A close-packed structure results. On the other hand, in covalent solids the bonding is directional, i.e. every shared electron is localized between only two atoms. This does not encourage close-packing since the number of atoms immediately surrounding each atom is limited to the number of covalent bonds it forms. Would you describe the ionic bond as directional or non-directional?

Some common crystal structures will now be described.

(*a*) *Face-centred cubic* (FCC) packing is shown in Fig. 1.11*a*; there is a particle at the centre of each of the six faces of the cube in addition to the eight at the corners. Copper and aluminium have this structure. The sodium crystal can be regarded as two interpenetrating FCC structures, one of sodium ions and the other of chloride ions, Fig. 1.11*b*; each sodium ion is surrounded by six chloride ions and vice versa.

(*b*) *Hexagonal close-packing* (HCP) is represented in Fig. 1.12; it is built up from layers of hexagons. Zinc and magnesium form HCP crystals.

Fig. 1.9

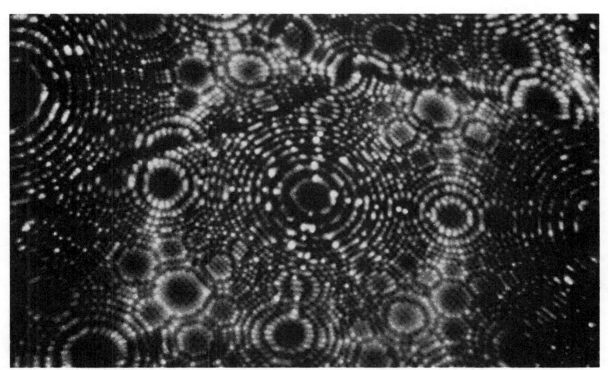

Fig. 1.10

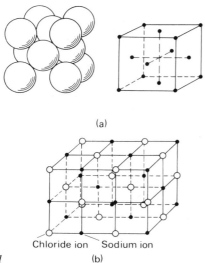

(a)

Chloride ion   Sodium ion

Fig. 1.11   (b)

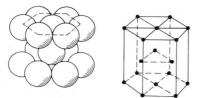

Fig. 1.12

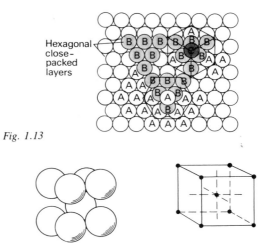

Hexagonal close-packed layers

Fig. 1.13

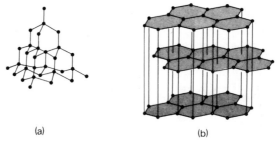

Fig. 1.14

These two structures give the closest possible packing and account for 60% of all metals. They are not very dissimilar if we consider how they can be assembled from successive layers. Fig. 1.13 shows a layer A of hexagonal close-packed spheres in which each sphere touches six others—this is the closest packing possible for spheres. A second hexagonal close-packed layer B can be placed on top and the packing of these two layers will be closest when the spheres of B sit in the hollows formed by three neighbouring spheres of A. A third hexagonal close-packed layer can be placed on top of B in two ways. If it rests in the hollows of B so that its spheres are directly above the *spheres* in A, then an HCP crystal results, Fig. 1.13 (bottom), and the layer stacking is ABAB. However, if the third layer rests in other hollows in B, its spheres can be above *hollows* in A and the structure is FCC with layer stacking ABCABC.

(*c*) *Body-centred cubic* (BCC) packing has a particle at the centre of the cube and one at each corner, Fig. 1.14. Alkali metals have this less closely packed structure.

(*d*) *Tetrahedral* structures have a particle at the centre of a regular tetrahedron and one at each of the four corners, Fig. 1.15*a*. This more open arrangement is found in carbon (as diamond), silicon and germanium—all substances which form covalent bonds. The hardness of diamond is partly due to the fact that its atoms are not in layers and so cannot slide over each other as they can in graphite, the other crystalline form of carbon. Graphite forms layers of six-membered rings of carbon atoms that are about two-and-a-half times farther apart than are the carbon atoms in the layers, Fig. 1.15*b*. The forces between the layers are weak, thus explaining why graphite flakes easily and is soft and suitable for use in pencils and as a lubricant. Graphite and diamond provide a good example of the importance of structure in determining properties.

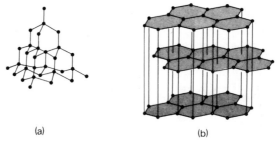

(a)   (b)

Fig. 1.15

Two further points: first, there is in every crystal structure a typical cell, called the *unit cell*, which is repeated over and over again—Figs. 1.11*a*, 1.12 and 1.14 are examples of unit cells; second, the structures described are those of perfect crystals. In practice there are imperfections in crystals and these are important in determining the properties of a material, as we shall see later.

Bubble raft

Soap bubbles pack together in an orderly manner and provide a good representation, in two dimensions, of how atoms are packed in a crystal.

A bubble raft is made by attaching a glass jet (about 1 mm bore) or a 25 gauge hypodermic needle on a 1 cm³ syringe barrel, to the gas tap, via a length of rubber tubing and a screw clip. The jet is held below the surface of a 'soap' solution (1 Teepol, 8 glycerol and 32 water is satisfactory) in a shallow glass dish, at a constant depth which gives bubbles of about 2 mm diameter, Fig. 1.16a. If the dish is placed on an overhead projector a magnified image of the raft can be viewed. What pulls the bubbles together and what keeps them from getting too close?

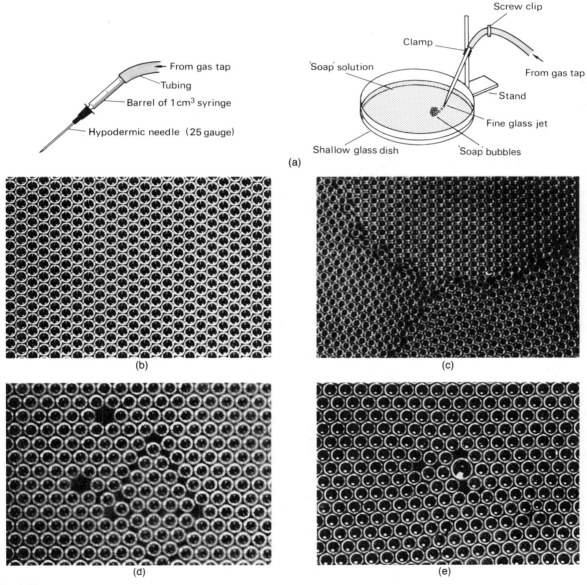

Fig. 1.16

A perfect, hexagonally close-packed array is shown in Fig. 1.16b. Grain boundaries are readily seen in Fig. 1.16c, 'vacancies' in Fig. 1.16d and a bubble of a different size in Fig. 1.16e. What might (d) and (e) represent in a real crystal and what effect do they have on the structure?

X-ray crystallography

Before the discovery which enabled the arrangement of atoms in a crystal to be determined experimentally, crystallographers had simply assumed that the regular external shapes of crystals were due to the atoms being arranged in regular, repeating patterns.

In 1912 three German physicists, Max von Laue, W. Friedrich and P. Knipping, found that a beam of X-rays, on passing through a crystal, formed a pattern of spots on a photographic plate (see Fig. 1.19b). Shortly afterwards W. L. (later Sir Lawrence) Bragg and his father Sir William Bragg showed how the pattern could be used to reveal the positions of the atoms in a crystal. Together they proceeded to unravel the atomic structures of many substances and started the science of X-ray crystallography. In recent years the structures of many complex organic molecules, including some like DNA (deoxyribonucleic acid) that play a vital part in the life process, have been discovered by this technique.

The analysis of crystal structures by X-rays depends on the fact that X-rays, like light, have a wave-like nature and when they fall on a crystal they are scattered by the atoms so that in some directions the scattered beams reinforce each other while in others they cancel each other. X-rays are used because their wavelengths are of the same order as the atomic spacings in crystals—about 10 000 times less than those of light.

The simplest way of regarding what occurs when X-rays fall on a crystal is to consider the crystal as made up of regularly spaced layers of atoms each of which produces a weak 'reflected' beam such that the angle of incidence equals the angle of reflection, as for the reflection of light by a mirror. In Fig. 1.17a reflection by a single layer of atoms is shown: most of the beam passes through. Fig. 1.17b shows a beam of X-rays falling on a set of parallel layers of atoms. If all the reflected waves are to combine to produce a strong reflected beam (and give an intense spot on a photograph) then they must all emerge in step. For this to happen the path difference between successive layers must be a whole number of wavelengths of the X-rays—as they are in this case. Otherwise crests of one wave may coincide with troughs from another and the two tend to cancel out.

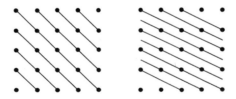

Fig. 1.18

The atoms in a crystal can be considered as arranged in several different sets of parallel planes, from all of which strong reflections may be obtained to give a pattern of spots characteristic of the particular structure. Two other possible sets of planes are shown for the array in Fig. 1.18. In a polycrystalline sample many planes are involved at once and thousands of spots are produced resulting in circles or circular arcs on the photograph. Fig. 1.19a is due to a polycrystalline sample of gold and Fig. 1.19b to a single crystal of the same material.

Microwave analogue

Microwaves are very short radio waves with wavelengths extending from 1 cm or so to about 1 m and are used for radar and satellite communication. A large-scale demonstration with microwaves shows that

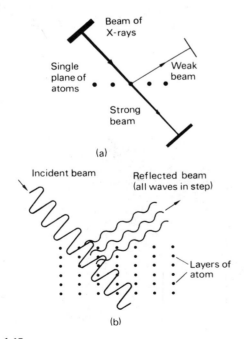

Fig. 1.17

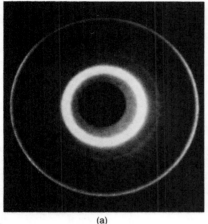

Fig. 1.19   (a)   (b)

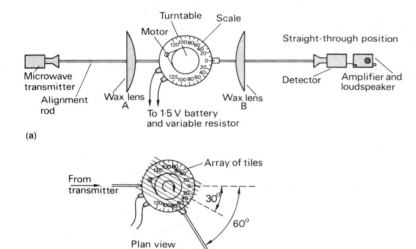

Microwave transmitter · Alignment rod · Wax lens A · Turntable · Motor · Scale · Straight-through position · Detector · Amplifier and loudspeaker · To 1·5 V battery and variable resistor · Wax lens B

(a)

From transmitter · Array of tiles · 30° · 60° · Plan view · To detector

(c)

(b)

Fig. 1.20   (d)

regularly spaced polystyrene tiles or layers of spheres give strong reflections at certain angles, depending on the tile or layer separation and that this can be explained as being due to interference between waves reflected from successive layers—as it can with X-rays and layers of atoms.

The apparatus is shown in Fig. 1.20a; wax lens A produces a parallel beam from the 3 cm microwave transmitter and B focuses it on the detector. First it should be shown, with the transmitter and detector in line, that microwaves can largely penetrate a polystyrene ceiling tile but not a metal sheet.

(a) *One tile*. A single tile held vertically on the turntable reflects (partially) the beam at *any* angle. Through what angle from the straight-through position must the detector be turned when the *glancing* angle (i.e. the angle between the tile and the incident beam) is θ (see p. 84)?

(b) *Two tiles*. When a second tile is brought up behind and parallel to the first (already positioned for the reflected beam to be detected), the intensity of the reflection rises and falls as the extra path to the second tile varies between an even and an odd number of half-wavelengths. Interference is occurring between the waves reflected from each tile when they come together. Will this occur for all glancing angles?

(c) *Ten tiles at 3* cm *centre-to-centre spacing*, Fig. 1.20b. If the detector is swung round as the array of tiles revolves on the turntable, a strong reflection is obtained only when the detector makes an angle of 60° with the straight-through position. The tiles then bisect the angle between the detector and the straight-through position, i.e. they make an angle of 30° with the straight-through position, and give a glancing angle of 30°, Fig. 1.20c.

(d) *Polystyrene ball 'crystal'*. This is made from seven hexagonal close-packed layers of 5 cm diameter spheres glued together to give a face-centred cubic (FCC) structure (see Appendix 6). As the crystal rotates on the turntable, Fig. 1.20d, *two pairs* of strong reflections are obtained per revolution when the detector is at 44° from the straight-through position ($\theta = 22°$), the time between each pair being greater than that between the two signals in each. At 50° ($\theta = 25°$) there are *two single* strong reflections per revolution and similarly at 74° ($\theta = 37°$).

The first pair of 44° reflections are produced when two easily identifiable sets of vertical 'hexagonal' layers (i.e. with the packing of spheres in each layer the closest possible) bisect *in turn* the angle between the straight-through position and the detector; the second pair arises half a revolution later when the 'backs' of the two sets of the same layers are in the bisecting position. At 50°, a set of 'square' layers (i.e. with less closely packed spheres—see question 9, p. 18) is responsible for the strong signals when, twice in each revolution, it bisects the angle between the beam and the detector.

Polymers

Polymers are materials with giant molecules each containing anything from 1000 to 100 000 atoms and are usually carbon (organic) compounds. An example of a natural polymer is cellulose whose long, tough fibres give strength and stiffness to the roots, stems and leaves of plants and trees. Rubber, wool, proteins, resins and silk are others. Man-made polymers include plastics such as polythene, Perspex and polystyrene, fibres like nylon and Terylene, synthetic rubbers and the epoxy resins which are well known for their strong bonding properties and toughness.

The unravelling of the intricacies of nature's polymers required X-ray apparatus, the electron microscope and other instruments. Their molecules were each found to consist of a large number of repeating units, called *monomers*, arranged in a long flexible chain. Thus every molecule of cellulose comprises a long chain of from a few hundred to several thousand glucose sugar ($C_6H_{12}O_6$) molecules.

Artificial polymers are made by a chemical reaction known as *polymerization*, in which large numbers of small molecules join together to form a large one. Polythene or polyethylene (to give it its full name) is made by polymerizing ethylene (C_2H_4), a gas obtained when petroleum is 'cracked'. In one process the ethylene molecules, heated to 100 °C–300 °C under a pressure several thousand times greater than atmospheric, link with one another to give the long chain molecules of polythene, Fig. 1.21.

If the chains run parallel to each other, like wires in a cable, the structure shows a certain amount of order and is said to be 'crystalline', Fig. 1.22a. This contrasts with the disorder of tangled chains in an 'amorphous' structure, Fig. 1.22b. Many polymers have both crystalline and amorphous regions, Fig. 1.22c. If crystallinity predominates an X-ray photograph shows sharp spots (but the pattern is never as sharp as for wholly crystalline materials) and the polymer is fairly strong and rigid. A polymer with a largely amorphous structure is soft and flexible and gives diffuse rings in an X-ray photograph. The proportion of crystalline to amorphous regions in a polymer depends on its chemical composition, molecular arrangement and how it has

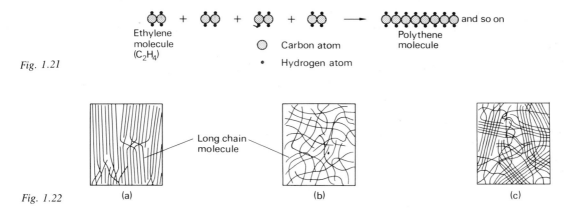

Ethylene
molecule
(C₂H₄)

◯ Carbon atom

• Hydrogen atom

Polythene
molecule

Fig. 1.21

Long chain molecule

Fig. 1.22   (a)   (b)   (c)

been processed. The intermolecular forces between chains are of the weak van der Waals type, but in crystalline structures the chains are close together over comparatively large distances and so the total effect of these forces is to produce a stiff material.

Crystallization is one of two principles that have been used to produce strong, stiff polymers (e.g. polythene, nylon); the other is the formation of strong covalent bonds between chains—a process known as cross-linking. In vulcanizing raw rubber, i.e. heating it with a controlled amount of sulphur, a certain number of sulphur atoms form cross-links between adjacent rubber molecules to give a more solid material than raw rubber which is too soft for use, Fig. 1.23. As more cross-links are added to rubber it stiffens and ultimately becomes the hard material called ebonite.

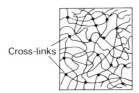

Cross-links

Fig. 1.23

Polymers such as ebonite and bakelite (the first plastic to be made) with many strong cross-links do not soften with increased temperature but set once and for all after their initial moulding. They are called *thermosetting plastics* and remain comparatively strong until excessive heating leads to breakdown of the cross-links and chemical decomposition. By contrast in *thermoplastic* polymers only the weak van der Waals forces hold the chains together and these materials can be softened by heating and if necessary remoulded. On cooling they recover their original properties and retain

any new shape. This treatment can be repeated almost indefinitely so long as temperatures are below those causing decomposition, i.e. breakdown of the covalent bonds that hold together the atoms in the long chain.

The possibility of using man-made polymers in the future as load-bearing structural materials for houses, buildings, cars, boats and aircraft depends largely on how far their strength and stiffness can be increased. As we have seen, two methods are used to do this at present. One is by having 'crystallized' long chains—a physical feature—and the other requires cross-links to be formed between chains—a chemical feature. Current research is directed towards producing molecular chains which are themselves stiff (most existing man-made polymer chains are inherently flexible) by polymerizing monomers which have a ring-shaped structure. It is then hoped to achieve the desired strength and stiffness by crystallizing and cross-linking those chains.

QUESTIONS

1. Experiment shows that 3 g of carbon combine with 8 g of oxygen to form 11 g of carbon dioxide. If 1 atom of carbon reacts with 2 atoms of oxygen to give 1 molecule of carbon dioxide (i.e. $C + O_2 = CO_2$)

(a) compare the mass of an oxygen atom with that of a carbon atom. (*Hint:* start by supposing that 1 g of C contains x atoms.)

(b) what is the atomic mass of oxygen on the carbon 12 scale?

(c) what mass of oxygen contains the same number of atoms as 12 g of C?

2. (a) If the atomic mass of nitrogen is 14, what mass of nitrogen contains the same number of atoms as 12 g of carbon?

(b) What mass of chlorine contains the same number of atoms as 32 g of oxygen? (Atomic masses of chlorine and oxygen are 35.5 and 16 respectively.)

3. Taking the value of the Avogadro constant as 6.0×10^{23} how many atoms are there in (a) 14 g of iron (at. mass 56), (b) 81 g of aluminium (at. mass 27), (c) 6.0 g of carbon (at. mass 12)?

4. What is the mass of (a) one atom of magnesium (at. mass 24), (b) three atoms of uranium (at. mass 238), (c) one molecule of water (mol. mass 18)? Take the value of the Avogadro constant as 6.0×10^{23}.

5. (a) A *mole* is the name given to the *quantity of substance* which contains a certain number of particles. What is this number?
   (b) What is the mass of 1 mole of hydrogen molecules?
   (c) If the density of hydrogen at s.t.p. is 9.0×10^{-5} g cm^{-3} what volume does 1 mole of hydrogen molecules occupy at s.t.p.?
   (d) How many molecules are there in 1 cm^3 of hydrogen at s.t.p.?

6. By counting scintillations it is found that 1.00 mg of polonium in decaying completely emits approximately 2.90×10^{18} alpha particles. If one particle is emitted by each atom and the atomic mass of polonium is 210, what is the Avogadro constant?

7. Estimate (a) the mass and (b) the diameter of a water molecule (assumed spherical) if water has molecular mass 18 and the Avogadro constant is 6.0×10^{23} per mole.

8. (a) Suggest an *approximate* but reasonable value for the heat of sublimation of copper (in J g^{-1}) from the following data.

   Specific latent heat of fusion = 2.0×10^2 J g^{-1}
   Specific latent heat of vaporization = 4.8×10^3 J g^{-1}

What additional information would enable a better estimate to be made?
   (b) If the Avogadro constant is 6.0×10^{23} per mole and the atomic mass of copper is 64, what is the heat of sublimation of copper in J/atom? Why is a knowledge of this quantity useful?

Square packing
(a)

Hexagonal packing
(b)

Pyramid
(c)

Fig. 1.24

9. (a) 'Square' and 'hexagonal' methods of packing spheres are shown in Figs. 1.24a and b respectively. How many other spheres are touched by (i) A, (ii) B? In which arrangement is the packing closest?
   (b) Fig. 1.24c is a pyramid of spheres in which the second and successive layers are formed by placing balls in the hollows of the layer below it. How are the balls packed in (i) the sloping sides of the pyramid, (ii) the horizontal layers?

10. One face of the unit cell of an FCC crystal is shown in Fig. 1.25, atoms being represented by spheres. If r is the atomic radius in cm, calculate (a) the length of a side of the unit cell, (b) the volume of a unit cell, (c) the number of unit cells in 1 cm^3.

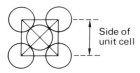

Side of unit cell

Fig. 1.25

11. In a crystal built up from a large number of similar unit cells the atoms at the corners and on the faces of individual cells are shared with neighbouring cells. In an FCC unit cell
   (a) how many corner atoms are there?
   (b) how many neighbouring cells share each corner atom?
   (c) what is the effective number of corner atoms per cell?
   (d) how many *face* atoms are there?
   (e) how many neighbouring cells share each face atom?
   (f) what is the effective number of face atoms per cell?
   (g) what is the total effective number of atoms per cell?
   (h) what is the total effective number of atoms in 1 cm^3 of unit cells (use your answer from 10(c)).

12. X-ray diffraction shows that copper, atomic mass 64 and density 9.0 g cm^{-3}, has an FCC structure. If the Avogadro constant is 6.0×10^{23} per mole, how many atoms are there in 1.0 cm^3 of copper?

13. Using your answers to 11(h) and 12, calculate the atomic radius of copper.

14. For crystalline sodium chloride, draw the unit cell which is repeated throughout the lattice. Label precisely the two kinds of particle at the lattice sites. What are the forces maintaining them in their relative positions?
   Calculate the distance between adjacent particles in crystalline sodium chloride, given that its formula weight is 58.5 and its density is 2.16 g cm^{-3} (2.16×10^3 kg m^{-3}). (The Avogadro constant = 6.03×10^{23} mole^{-1}.)
   Discuss the effect of a small stress on a crystalline lattice.
   (C. Phys. Sc.)

2 Mechanical properties

Stress and strain

The mechanical properties of a material are concerned with its behaviour under the action of external forces—a matter of importance to engineers when selecting a material for a particular job. Four important mechanical properties are strength, stiffness, ductility and toughness.

Strength deals with how great an applied force a material can withstand before breaking. *Stiffness* tells us about the opposition a material sets up to being distorted by having its shape or size, or both, changed. A stiff material is not very flexible. There is no such thing as a perfectly stiff or rigid (unyielding) material; all 'give' in some degree although the deformation may often be very small. *Ductility* or workability relates to the ability of the material to be hammered, pressed, bent, rolled, cut or stretched into useful shapes. A *tough* material is one which is not brittle, i.e. it does not crack readily. Steel has all four properties, putty has none of them. Glass is strong and stiff but not tough or ductile. Which properties would you ascribe to rubber, nylon and diamond?

Information about mechanical properties may be obtained by observing the behaviour of a wire or strip of material when it is stretched. The stretching of short rods or 'test-pieces' is done using a machine like that in Figs. 2.1 and 2.2.

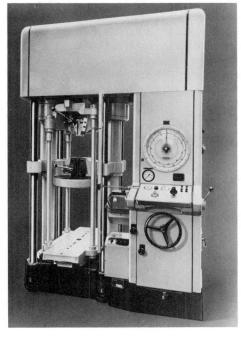

Fig. 2.1

Fig. 2.2

19

The extension produced in a sample of material depends on (*i*) the nature of the material, (*ii*) the stretching force, (*iii*) the cross-section area of the sample and (*iv*) its original length. What effect would you expect (*iii*) and (*iv*) to have? To enable fair comparisons to be made between samples having different sizes the terms stress and strain are used when referring to the deforming force and the deformation it produces. *Stress is the force acting on unit cross-section area* and for a force F and area A it equals F/A, Fig. 2.3. The unit of stress is the *pascal* (Pa) which equals one newton per square metre (N m^{-2}). *Strain is the extension of unit length.* If e = extension and l = original length, the strain is e/l. Strain is a ratio and has no unit. A stress which causes an increase of length puts the sample in tension, and so we talk about a tensile stress and a tensile strain.

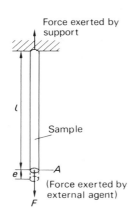

Fig. 2.3

The shape of the stress-strain graph for the stretching of a sample (e.g. a wire) depends not only on the material but also on its previous treatment and method of manufacture. For a ductile material, i.e. a metal, it has the *general* form shown by OEPAD in Fig. 2.4. There are two main parts.

(*a*) *Elastic deformation.* The first part of the graph from O to E is a straight line through the origin showing that strain is directly proportional to stress, i.e. doubling the stress doubles the strain. Over this range the material suffers elastic deformation, i.e. it returns to its original length when the stress is removed and none of the extension remains.

(*b*) *Plastic deformation.* As the stress is increased the graph becomes non-linear but the deformation remains elastic until at a certain stress corresponding to point P, and called the *yield point*, permanent or plastic deformation starts. Henceforth the material behaves

rather like Plasticine and retains some of its extension if the stress is removed. Recovery is no longer complete and on reducing the stress at A, for example, the specimen recovers along AO' where AO' is almost parallel to OE. OO' is the permanent plastic extension produced. If the stress is reapplied, the curve O'AD is followed. At D the specimen develops one or more 'waists' and *ductile fracture* occurs at one of them. The stress at D is the greatest the material can bear and is called the *breaking stress* or *ultimate tensile strength*; it is a useful measure of the strength of a material.

The specimen appears to 'give' at P, and over the plastic region a given stress increase produces a greater increase of strain than previously. None the less it still opposes deformation and any increase of strain requires increased stress. Beyond P the material is said to *work-harden* or *strain-harden*.

Few ductile materials behave elastically for strains as great as $\frac{1}{2}\%$ (i.e. for an extension of $\frac{1}{2}$ cm in a 100 cm long wire) but they may bear large plastic strains, up to 50%, before fracture. The breaking stress for steels may occur at stresses half as great again as that at the yield point, while for very ductile metals with low yield points it may be several times greater than the stress at the yield point. It is desirable that metals used in engineering structures should carry loads which only deform them elastically. On the other hand the fabrication of metals into objects of various shapes requires them to withstand considerable plastic deformation before fracture, i.e. be very ductile. The dominant position of metals in modern technology arises from their strength and ductility.

A brittle material such as a glass may give a curve like OB in Fig. 2.4 and fractures almost immediately after the elastic stage; little or no plastic deformation occurs and the glass is non-ductile.

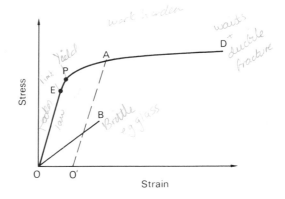

Fig. 2.4

The Young modulus

The stress–strain curve for the stretching of metals and some other materials (e.g. glasses), over almost all the elastic region, is a straight line through the origin. That is, *tensile strain is directly proportional to tensile stress during elastic deformation*. This statement is known as Hooke's law and in more elementary work it is often stated in the form: extension varies as the load. In mathematical terms it can be written

$$tensile\ strain \varpropto tensile\ stress$$

or $$\frac{tensile\ stress}{tensile\ strain} = a\ constant$$

This constant is called the *Young modulus* and is denoted by E. Its value depends on the nature of the material and *not* on the dimensions of the sample. If a material has large E, it resists elastic deformation strongly and a large stress is required to produce a small strain. E is thus a measure of the opposition of a material to change of length strains such as occur when a wire or rod is stretched elastically, i.e. it measures *elastic stiffness*.

The Young modulus is of great importance in engineering. In the early days of iron railway bridge construction, engineers relied heavily on 'rule of thumb' methods. It required a series of disasters like that of the Tay Bridge in Scotland in 1879 and a reputed collapse rate of twenty-five bridges per year at about the same time in the U.S.A. before it was accepted that reliable strength calculations were necessary for safety and the economical use of materials. The value of E is one of the pieces of information which must be known to calculate accurately the deformation (deflections) that will occur in a loaded structure and its parts. When a beam bends, one surface is compressed and the other stretched, Fig. 2.5, so that E is involved; 'beam theory' is one of the foundation stones of engineering.

If a stretching force F acting on a wire of cross-section area A and original length l causes an extension e we can write

$$E = \frac{tensile\ stress}{tensile\ strain} = \frac{F/A}{e/l} = \frac{Fl}{Ae}$$

Like stress, E is expressed in pascals (Pa) since strain is a ratio. Suppose a load of 1.5 kg attached to the end of a wire 3.0 m long of diameter 0.46 mm stretches it by 2.0 mm then $F = ma = mg = 1.5 \times 9.8$ N ($g = 9.8$ m s^{-2})

$$E = \frac{Fl}{Ae}$$

$$= \frac{(1.5 \times 9.8\ \text{N})(3.0\ \text{m})}{(\pi \times 0.23^2 \times 10^{-6}\ \text{m}^2)(2.0 \times 10^{-3}\ \text{m})}$$

$$= \frac{1.5 \times 9.8 \times 3.0}{\pi \times 0.23^2 \times 10^{-6} \times 2.0 \times 10^{-3}}\ \frac{\text{Nm}}{\text{m}^2\text{m}}$$

$$= 1.3 \times 10^{11}\ \text{Pa}$$

Approximate values of E for some common materials are given in Table 2.1.

Table 2.1

Material	The Young modulus $E/10^{10}$ Pa
steel	21
copper	13
glasses	7
polythene	about 0.5
rubber	about 0.005

Glasses are surprisingly stiff (and strong). The high elasticity of rubber (i.e. its ability to regain its original shape after a very large deformation) is not to be confused with its low elastic modulus. For steel a large stress gives a small strain while the same stress applied to rubber will give a very much larger strain. This

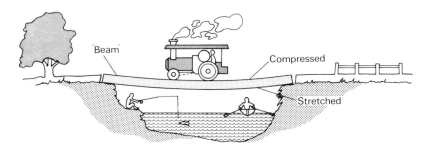

Fig. 2.5

means that by our scientific definition, steel has a greater modulus of elasticity than rubber.

Stretching experiments

1. Copper. Using an arrangement like that in Fig. 2.6 the extensions produced in a 2 metre length of copper wire (SWG 32) are found as it is loaded to breaking with 100 gram slotted weights.

A load–extension graph is plotted to see if Hooke's law is obeyed and to find the percentage strain copper can withstand before its elastic limit is exceeded.

The breaking stress of copper can be determined and, if wire of another gauge is used (say SWG 26), whether the breaking stress depends on wire thickness. The percentage plastic strain borne by copper before it breaks may also be found.

2. Steel. If *1* is repeated for 2 metres of steel wire (SWG 44) the same kind of information can be obtained.

3. Rubber. A load–extension graph can be plotted for a strip of rubber (5 cm long and 2 mm wide cut from a rubber band), suspended vertically and loaded with 100 g slotted weights and then unloaded. Information may be obtained about Hooke's law, and also the number of times its original length that rubber can be extended. The breaking stress of rubber should be found.

4. Polythene. A strip 15 cm long and 1 cm wide, cut *cleanly* from a piece of polythene, can be investigated as in *3*.

5. Glass. The breaking stress of glass may be found by hanging weights from a glass thread which has been freshly drawn from a length of 3 mm diameter soda glass rod. If parts of the rod are left at the top and bottom, the thread can be supported by clamping one end and a hook made at the other end for the weights.

6. The Young modulus for a wire. Using the apparatus of Fig. 2.7*a* or *b* the extensions of a wire can be measured with greater accuracy.

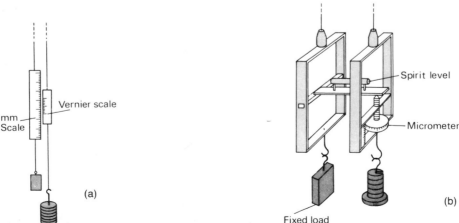

Fig. 2.6

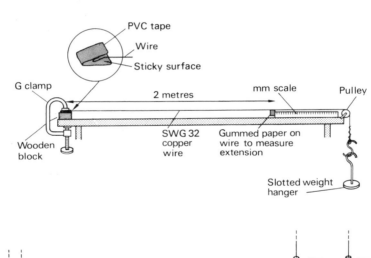

Fig. 2.7

In Fig. 2.7a the right-hand wire is under test and carries a vernier scale (see Appendix 3) which, when the right-hand wire is loaded, moves over a millimetre scale attached to the left-hand wire and enables the extension to be measured. The alternative and more accurate arrangement in Fig. 2.7b is known as Searle's apparatus. In this case the micrometer screw (see Appendix 3) is adjusted, after the addition of a load to the right-hand wire, so that the bubble of the spirit level is centralized. The extension is then found from the scale readings. By having two wires of the same material suspended from the same support, errors are eliminated if there is a change of temperature or if the support yields, since both wires will be affected equally.

Initially both wires should have loads that keep them taut and free from kinks. Readings are then taken as the load on the right-hand wire is increased by equal steps, without exceeding the elastic limit. The strain should therefore not be more than 0.1%, i.e. the wire should not be stretched much beyond 1/1000th of its original length. The length l of the wire to the top of the vernier or micrometer is measured with a metre rule and the diameter $(2r)$ found at various points along its length with a micrometer screw gauge.

From a graph of 'load' against 'extension', an average value of 'load'/'extension' in kg mm^{-1} is given by PQ/OQ, Fig. 2.8. The Young modulus can then be calculated from

$$E = \frac{Fl}{Ae}$$

where F/e is expressed in N m^{-1}, i.e. $F/e = $ PQ $\times$ $g/(\text{OQ} \times 10^{-3})$, $A(= \pi r^2)$ in m^2 and l in m.

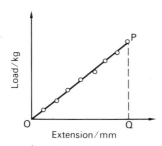

Fig. 2.8

Deformation and dislocations

The deformation behaviour of materials can be explained at the atomic level.

(a) *Elastic strain.* This is due to the stretching of the interatomic bonds that hold atoms together. The atoms

are pulled apart very slightly; each is displaced a tiny distance from its equilibrium position and the material lengthens. Hooke's law is a result of the fact that the 'interatomic force–separation' graph, Fig. 2.9a, is a straight line for atomic separations close to the equilibrium separation r_0, Fig. 2.9b.

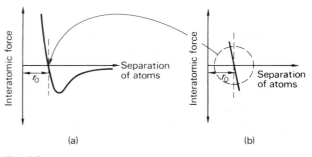

Fig. 2.9

The Young modulus E, the measure of elastic stiffness (i.e. resistance to elastic deformation), is high for materials with strong interatomic bonds. Covalent and ionic solids, and to a lesser extent metals, are in this category. Diamond (pure carbon), the hardest known natural substance, has a large number of very strong covalent bonds per unit volume and a very high value of E.

As well as determining the stiffness of a material, the Young modulus also governs ultimately, in theory, its strength since this too depends on the forces between atoms. However, we shall see later (pp. 26–7) that there are other factors which prevent solids displaying their theoretical strengths.

(b) *Plastic strain.* The ability to undergo plastic strain (and be ductile) is a property of crystalline materials. The yielding which occurs could therefore be attributed to the slipping of layers of atoms (or ions) over one another. With close-packed layers like those in Fig. 2.10a, the atoms would have to be moved farther apart, Fig. 2.10b. This would be resisted by the interatomic bonds, many of which will have to be broken simultaneously.

Fig. 2.10

Calculations based on the known strength of bonds show that the stresses needed to produce slip in this

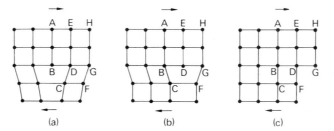

Fig. 2.11   (a)   (b)   (c)

way are many times greater than those which actually cause plastic strain. The problem is, therefore, not so much to explain the strength of metals as their weakness. This led to a search for defects in crystal structures and in 1934 G. I. Taylor of Cambridge University proposed the *dislocation* as one such defect.

The idea is that occasionally, due perhaps to growth faults during crystallization, there is an incomplete plane of atoms (or ions) in the crystal lattice, for example AB in Fig. 2.11a. We shall now see how the movement of the dislocation produces the same effect as a plane of atoms slipping over other planes, but much more easily.

If a stress is applied as shown by the arrows, atom B, whose bonds have already been weakened by the distortion of the structure, moves a small distance to the right and forms a bond with atom C, Fig. 2.11b. Plane of atoms DE is now incomplete. Then D flicks over and joins with F leaving GH as the incomplete plane, Fig. 2.11c. The result is just the same as if half-plane AB had slipped over the planes below BDG to the surface of the crystal. This process would have involved breaking a great many bonds at the same time. Instead, the dislocation, by moving a single line at a time has broken many fewer bonds and required a much smaller stress to do it. No atom has in fact moved more than a small fraction of the atomic spacing. Plastic deformation by this mechanism is clearly only possible in the well-ordered structure of a crystalline material.

The passage of a dislocation in a crystal is like the movement of a ruck in a carpet. A greater force is needed to drag one carpet over another by pulling one end of it, than to make a ruck in the carpet and kick it along, Fig. 2.12. Calculations confirm that the stresses to make dislocations move in metals are in good agreement with their measured plastic flow stresses.

The first and most direct evidence for the existence of dislocations was obtained in 1956 by J. W. Menter, also of Cambridge University, using an electron microscope. An electron micrograph (an electron microscope photograph) is shown in Fig. 2.13 of an aluminium–copper alloy in which the planes of atoms are spaced

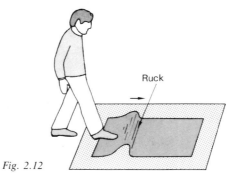

Ruck

Fig. 2.12

about 0.20 nm apart. A dislocation can be seen; the extra plane of atoms ends in the white circle and distorts the arrangement of the surrounding planes.

Dislocations can be obtained in a bubble raft (p. 13) and made to move if the raft is squeezed between two glass slides dipping into the 'soap' solution. There is one to be seen in Fig. 2.14.

Strengthening metals

Pure metals produced commercially are generally too weak or soft to be of much mechanical use—a rod of pure copper the thickness of a pencil is easily bent by hand. Their weakness can be attributed to the fact that they contain a moderate number of dislocations which can move about easily in the orderly crystal structure, thus allowing deformation under relatively small stresses. The traditional methods of making metals stronger and stiffer all involve obstructing dislocation movement by 'barriers', i.e. pockets of disorder in the lattice. Three barriers will be considered.

(*a*) *'Foreign' atoms*. In an alloy such as steel 'foreign' atoms (e.g. carbon) are introduced into the lattice of iron, disturbing its perfection and opposing dislocation motion. This makes for greater strength and stiffness.

(*b*) *Other dislocations*. One problem with the dislocation model is that when the dislocations have slipped out of the crystal, as in Fig. 2.11c, the crystal is

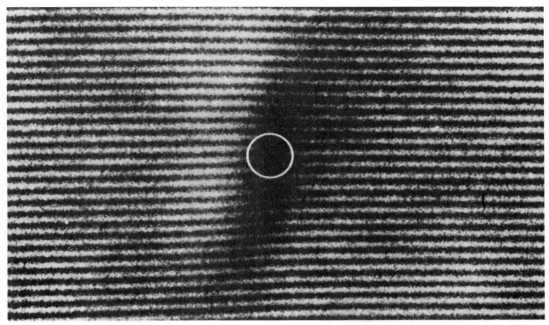

Fig. 2.13

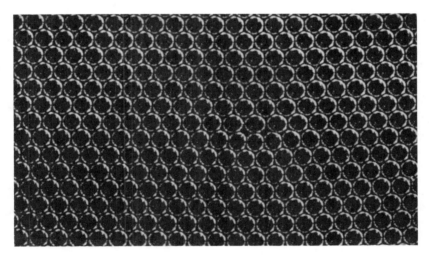

Fig. 2.14

then perfect and should have its theoretical strength. In general this is not observed and it would seem that further dislocations are generated whilst slip is occurring. This view can be justified in a more advanced treatment but here it is sufficient to say that as the metal is submitted to further stress, dislocations are created, move, meet and thereby obstruct each other's progress. A 'traffic jam' of dislocations builds up.

(*c*) *Grain boundaries.* In practice most metal samples are polycrystalline, i.e. consist of many small crystals or grains at different angles to each other (see Fig. 1.7, p. 10). The boundary between two grains is imperfect and can act as an obstacle to dislocation movement. In general, the smaller the grains the more difficult is it to deform the metal. Why?

An obvious way of strengthening metals would be to eliminate dislocations altogether and produce in effect perfect crystals. So far this has only been possible for tiny, hairlike single-crystal specimens called 'whiskers' that are only a few micrometres thick and are seldom more than a few millimetres long. Their strength, however, approaches the theoretical value and they can withstand elastic strains of 4 or 5% (compared with $\frac{1}{2}$% or less for most common engineering materials). Unfortunately, perhaps due to surface oxidation, dislocations soon develop and the 'whisker' weakens. At present they are the subject of much research. Fig. 2.15 shows 'whiskers' of pure iron magnified four times, each one virtually a perfect crystal, and having about fifty times the tensile strength of the same thickness of ordinary soft iron.

Fig. 2.15

Cracks and fracture

Cracks, both external and internal and however small, play an important part in the fracture of a material and prevent it displaying its theoretical strength. Different types of fracture usually occur in brittle and ductile materials.

(a) *Brittle fracture.* This happens after little or no plastic deformation (i.e. during elastic deformation) by the very rapid propagation of a crack. It takes place, for example, when a glass rod is cut by making a small but sharp notch on it with a glass knife or file and then 'bending' it, as in Fig. 2.16—with the notch on the far side of the rod. Why?

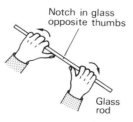

Fig. 2.16

Round a scratch, notch or crack there is a concentration of stress which in general is greater the smaller the radius of curvature of the tip of the crack (i.e. the sharper the crack). Such stress concentrations may be seen by viewing a lamp through two 'crossed' Polaroid squares (i.e. one square is rotated to cut off most of the light coming from the other to give a dark field of view) having a strip of polythene between them, Fig. 2.17a. Pulling the strip causes colours to appear and is an indication that the polythene is under stress. (The phenomenon is called *photoelasticity* and is used to study stresses in plastic models of engineering structures. Fig. 2.17b shows these round a triangular hole in a plastic block under pressure.) When the strip is cut halfway across and again pulled, colours are seen at the tip of the cut showing the stress is high there. If it is sufficiently high, interatomic bonds are broken, the cut spreads and breaks a few more bonds at the new tip. Eventually complete fracture occurs.

Fig. 2.17 (b)

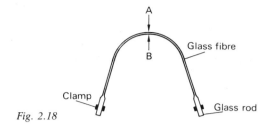

Fig. 2.18

Even tiny surface scratches, and they seem to arise inevitably on all materials, can lead to fracture. For example, a freshly-drawn ½-metre-long glass fibre can be bent into an arc but it fractures if 'scratched' at A, Fig. 2.18, by gently stroking a few times with another glass fibre. It is less likely to break if scratched at B. Why?

In brittle materials like concrete and glass, cracks spread more readily when the specimen is stretched or bent, i.e. is in tension. Crack propagation is much more difficult if such materials are used in compression (i.e. squeezed) so that any cracks close up, Fig. 2.19. Thus prestressed concrete contains steel rods that are in tension because they were stretched whilst the concrete was poured on them and set. As well as providing extra tensile strength these keep the concrete in compression even if the whole prestressed structure is in tension.

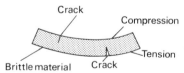

Fig. 2.19

Glass can also be prestressed and its surface compressed, making it more resistant to crack propagation. In thermal toughening, jets of air are used to cool the hot glass and cause the outside to harden and contract whilst the inside is still soft. Later the inside contracts, pulls on the now reluctant-to-yield surface which is thus compressed. As we saw previously stresses in transparent materials are revealed by polarized light. The pattern of the air jets used for cooling the prestressed glass of a car windscreen can be seen through polarizing spectacles or sometimes by sunlight that has been partially polarized by reflection from the car bonnet. Prestressing glass in this way can increase its toughness 3000 times.

(b) *Ductile fracture.* In this case fracture follows appreciable plastic deformation, by *slow* crack propagation. After thinning uniformly along its length during the plastic stage, the specimen develops a 'waist' or 'neck' in which cavities form. These join up into a crack and this travels out to the surface of the specimen, Fig. 2.20a, to give 'cup and cone' shaped, dull fracture surfaces, like those in Fig. 2.20b for an aluminium rod. It is possible that the internal cavities are formed during the later stages of plastic strain when stress concentrations arise in regions having a large number of interlocking dislocations.

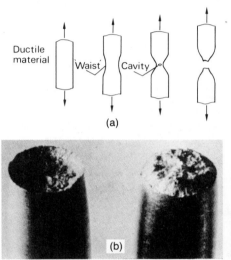

Fig. 2.20

Cracks propagate with much greater difficulty in metals due to the action of dislocations. These can move to places of high stress such as the tip of a crack, thus reducing the effective stress by causing it to be shared among a greater number of interatomic bonds, Fig. 2.21. The crack tip is thereby deformed plastically, blunted by the dislocation and further cracking possibly stopped. For the same reason surface scratches on metals have practically no effect. In brittle materials dislocation movement is impossible (why?) and high local stresses can build up at cracks under an applied force.

Fatigue and creep

These are two other important aspects of the mechanical behaviour of metals that are conveniently considered here.

(a) *Fatigue.* This may cause fracture, often with little or no warning and happens when a metal is subjected to a large number of cycles of *varying* stress even if the maximum value of the stress could be applied *steadily* with complete safety. It is estimated that about 90% of

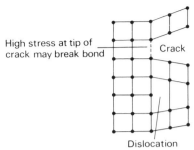
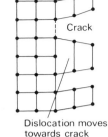
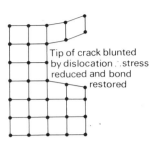

High stress at tip of crack may break bond — Crack

Dislocation

Crack

Dislocation moves towards crack

Tip of crack blunted by dislocation ∴stress reduced and bond restored

Fig. 2.21

Fig. 2.22

all metal failures are due to fatigue; it occurs in aircraft parts, in engine connecting rods, axles, etc.

A typical fatigue fracture in a steel shaft is shown in Fig. 2.22; starting as a fine crack, probably at a point of high stress, it has spread slowly, producing a smooth surface (as on right of photograph), until it was half-way across the shaft which then broke suddenly. Stress concentrations may be due to bad design (e.g. rapid changes of diameter), bad workmanship (e.g. a tool mark) and are common at holes—the Comet aircraft disasters of 1954 were caused by fatigue failure started round small rivet holes.

For many ferrous metals there is a safe stress varia-tion below which failure will not occur even for an infinite number of cycles. With other materials 'limited-life' design only is possible. As yet no fully comprehen-sive theory of fatigue exists.

(b) *Creep*. In general this occurs at high temperature and results in the metal continuing to deform as time passes, even under constant stress. The effect is thought to be associated with dislocation motion due to the vibratory motion of atoms and in the creep resistant alloys that have been developed this motion is re-stricted. Such alloys are used, for example, to make the turbine blades of jet engines where high stress at high temperature has to be withstood without change of dimensions.

Some low melting-point metals can creep at room temperature, thus unsupported lead pipes gradually sag and the lead sheeting on church roofs has to be replaced periodically.

Composite materials

Composites are produced by combining materials so that the combination has the most desirable features of the components. The idea is not new. Wattle and daub (interlaced twigs and mud) have been used to build homes for a long time; straw and clay are the ingred-ients of bricks; Eskimos freeze moss into ice to give a less brittle material for igloo construction; reinforced concrete contains steel rods or steel mesh. In all cases the composite has better mechanical properties than any of its components.

The production of composites is man's attempt to copy nature. Wood is a composite of cellulose fibres cemented together with lignin. Bone is another compos-ite material. Many modern technological applications require materials that are strong and stiff but light and heat resistant. The development of composites to meet these requirements is at present a major concern of materials scientists throughout the world and offers exciting possibilities to engineers in the future.

The highest *strength-to-weight* and *stiffness-to-weight* ratios are possessed not by metals but by materials such as glass, carbon and boron whose atoms are linked by many strong covalent bonds. (The strength of covalent bonds and their number per unit volume in these covalent solids accounts for the high strength and stiffness; the directional nature of the bond explains the non-close-packed structure and consequently small density, see p. 12). Unfortunately these materials are brittle, partly because their structures make dislocation motion difficult under an applied stress and partly because they usually have small surface scratches that develop into cracks.

In modern composite materials the desirable properties of covalent solids are exploited by incorporating them as *fibres* in a weaker, yielding material called the *matrix*. Freshly drawn fibres are fairly scratch-free and are therefore strong. The matrix has three functions: first, it has to bond with and hold the fibres together so that the applied load is transmitted to them; second, it must protect the surface of the fibres from scratches; third, if cracks do appear it should prevent them from spreading from one fibre to another—it can do this by acting as a barrier to the crack and deflecting it harmlessly along the interface it forms with the fibre, Fig. 2.23. A plastic resin or a ductile metal makes a suitable matrix. Fibre-reinforced composites are strong to stresses applied along the fibres.

Fig. 2.23

Fibre-glass was the first of the successful modern composites. It consists of high-strength glass fibres in a plastic resin (called glass-reinforced plastic—GRP) which is widely used where the stiffness and heat resistance required are not too great, e.g. for making boat hulls, storage tanks, pipes, car components. Fig. 2.24 shows the launching at Southampton in 1972 of the minesweeper H.M.S. *Wilton*, believed to be the world's first true ship (as distinct from boat) to have a hull made of GRP. Lightweight lift jet engines for vertical take-off also use GRP extensively for low-temperature parts and produce a thrust sixteen times their own weight. The best engines in commercial service today produce a thrust less than five times their weight.

Fig. 2.24

Carbon-fibre-reinforced plastics (CFRP) are similar but carbon fibres (about 6×10^5 per cm^2 of cross-section) of greater strength and stiffness replace glass fibres. These are stronger and stiffer than steel and much lighter (see question 7, p. 35). At present they are costly to produce; nevertheless CFRP are particularly attractive to the aircraft industry. They are not subject to fatigue failure or to high-stress concentrations round holes and cracks as metals are. They do not, however, flow plastically like metals but are elastic until failure, when extensive damage may occur. Good design is therefore necessary to avoid overloading. Their resistance to corrosion is also poor and prohibits certain applications. Fig. 2.25a shows a bundle of

(a)

(b)

Fig. 2.25   (c)

carbon fibres being assembled for tensile testing; Fig. 2.25b is a section of a piece of CFRP.

Much research is taking place on composites. The interface between a fibre and its matrix holds the key to the formation of a successful composite; at present a great deal is not understood about the properties of the interface. If the fibres are not covered uniformly with the matrix, small holes form at the interface. In service these cause high-stress concentrations and premature failure. They may also allow liquids and gases to penetrate the composite and attack the fibres. The effect of coating the fibre with some other material to protect its surface from scratches, corrosion, etc., before combining it with the matrix is being studied. Fig. 2.25c is a scanning electron microscope photograph of a tungsten-coated carbon fibre (× 3000).

The fibre-reinforcement of metals and concrete is also being investigated.

Strain energy

Energy has to be supplied to stretch a wire. If the stretching force is provided by hanging weights, there is a loss of potential energy which is stored in the stretched wire as *strain energy*. Provided the elastic limit is not exceeded this energy can usually be recovered completely (rubber is a notable exception as we shall see in the next section). If it is exceeded, the part of the energy used to cause crystal slip (i.e. plastic strain) is retained by the wire.

We require to determine the strain energy stored in a wire stretched by a known amount. Consider a material with a force–extension graph like that in Fig. 2.26. (This is of the same form as its stress-strain graph.)

Suppose the wire is already extended by e_1 and then suffers a further extension δe_1 which is so infinitesimally small that the shaded area is near enough a rectangle. If F_1 is the average but *nearly* constant value of the stretching force during the extension δe_1, then the energy transferred to the wire, i.e. the work done δW_1, is given by

$$\delta W_1 = \text{force} \times \text{distance}$$
$$= F_1 \times \delta e_1$$
$$= \text{area of shaded strip}$$

∴ Total work done during whole extension e
$$= \text{area OAB}$$

i.e. Strain energy stored in wire = area OAB

If Hooke's law is obeyed, OA is a straight line (as shown) and OAB is a triangle.

Therefore strain energy in wire for whole extension e and final stretching force F, is given by

$$\text{strain energy} = \text{area} \triangle \text{OAB}$$
$$= \tfrac{1}{2}\text{AB} \times \text{OB}$$
$$= \tfrac{1}{2}Fe$$

The expression $\tfrac{1}{2}Fe$ gives the strain energy in joules if F is in newtons and e in metres.

If l is the original length of the wire and A its cross-section area, then volume of wire = Al.

∴ Strain energy per unit volume = $\tfrac{1}{2}Fe/(Al)$
$$= \frac{1}{2}\left(\frac{F}{A} \times \frac{e}{l}\right)$$

But F/A = stress and e/l = strain.

∴ Strain energy per unit volume = $\tfrac{1}{2}$(stress × strain)

If Hooke's law is not obeyed so that OA is not a straight line, what is the value of the strain energy in the wire for extension e? If the wire suffers plastic deformation, e.g. an extension OD, what represents the total strain energy stored?

Rubber

The two most striking mechanical properties of rubber are (a) its range of elasticity is great—some rubbers can be stretched to more than ten times their original length (i.e. 1000% strain) before the elastic limit is reached and (b) its value of the Young modulus is about 10^4 times smaller than most solids and *increases* as the temperature rises, an effect not shown by any other material.

Rubber is a polymer consisting of up to 10^4 isoprene molecules (C_5H_8) joined end-to-end into a long chain of carbon atoms, Fig. 2.27. The enormous extensibility and low value of E cannot be due to the stretching of

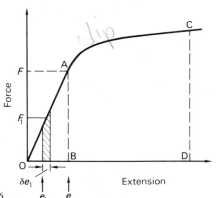

Fig. 2.26

Strain energy per unit volume = $\frac{1}{2}\sigma\epsilon$

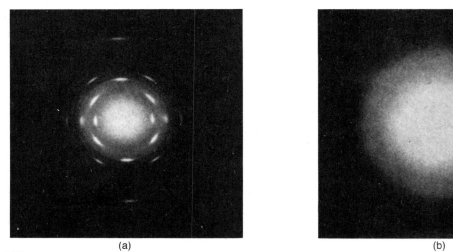

Isoprene monomer
(C_5H_8)

Fig. 2.27

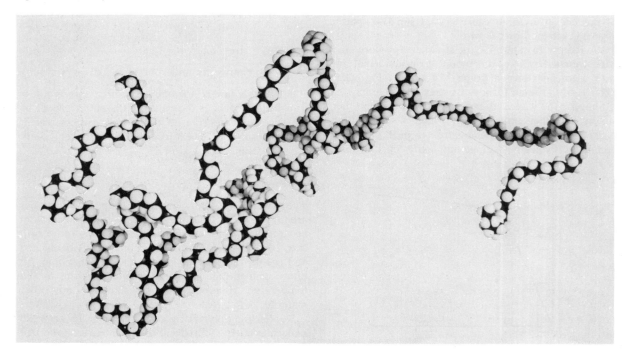

(a)   (b)

Fig. 2.28

Fig. 2.29

the strong covalent bonds between atoms in the carbon chain.

If a sample of stretched rubber is 'photographed' by a beam of high-energy electrons, sharp spots are obtained, Fig. 2.28a, like those produced by X-rays and a crystal. This suggests there is some order among the molecules in such a sample. Fig. 2.28b is a similar photograph of unstretched rubber. A plausible explanation of the behaviour of rubber might be that its long-chain molecules are intertwined and jumbled up like cooked spaghetti. Fig. 2.29 is a model of one rubber molecule. A stretching force would tend to make the chains uncoil and straighten out into more or less orderly lines alongside each other. When the force is removed they coil up again. There is also some cross-bonding between chains, achieved during manufacture by vulcanizing raw rubber (see p. 17); this cross-linking as well as causing stiffening also greatly increases the reversible strain possible by anchoring together the long molecules. Fully extended rubber is strong because the bonds are then stretched directly.

The rise in value of E with temperature can be attributed to the greater disorder among the chains when the material is heated; their resistance to alignment by a stretching force therefore increases.

If a stress–strain curve is plotted for the loading and unloading of a piece of rubber the two parts do not coincide, Fig. 2.30. OABC is for stretching and CDEO for contracting. The strain for a given stress is greater when unloading than when loading. The unloading strain can be considered to 'lag behind' the loading strain; the effect is called *elastic hysteresis*. It occurs with other substances, noticeably with polythene and glasses and to a small extent with metals.

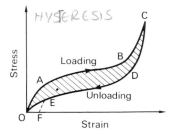

Fig. 2.30

We note that Hooke's law is not obeyed. This is typical of non-crystalline polymers and contrasts with the usually linear behaviour of crystalline materials, e.g. metals which obey the law during elastic strain. It is also evident that rubber stretches easily at first but is stiffer at large extensions. It has been shown (p. 31) that the area enclosed by OABC and the strain axis,

represents the energy supplied to cause stretching; similarly the area under CDEO represents the energy given up by the rubber during contraction. The shaded area is called a *hysteresis loop*; it is a measure of the energy 'lost' as heat during one expansion–contraction cycle. The changes in temperature can be felt by placing a wide rubber band on the lips; if it is stretched quickly the temperature rises due to the transfer of mechanical energy to the rubber. When released under control the temperature falls. (The temperature change on free contraction is small.)

Rubber with a hysteresis loop of small area is said to have *resilience*. This is an important property where the rubber undergoes continual compression and relaxation as does each part of a car tyre when it touches the road and rotates on. If the rubber used in tyres does not have high resilience there is appreciable loss of energy resulting in increased petrol consumption or lower maximum speed. Should the heat build-up be large the tyre may disintegrate.

When rubber is stretched and released there may be a small permanent set as shown by the dotted line EF in Fig. 2.30.

Elastic moduli

All deformations of a body whether stretches, compressions, bends or twists can be regarded as consisting of one or more of three basic types of strain. For many materials experiment shows that *provided the elastic limit is not exceeded*

$$\frac{stress}{strain} = \text{a constant}$$

This is a more general statement of Hooke's law. The constant is called an *elastic modulus* of the material for the type of strain under consideration. There are three moduli, one for each kind of strain.

(*a*) *The Young modulus* (E). This has already been considered (p. 21) and is concerned with change of length strains. It is defined by

$$E = \frac{tensile\ stress}{tensile\ strain}$$

where stress is force per unit area (F/A) and strain is change of length per unit length (e/l).

(*b*) *Shear modulus* (G). In this case the strain involves a change of shape without change of volume. Thus if a tangential force F is applied along the top surface of area A of a rectangular block of material fixed to the bench, the block suffers a change of shape

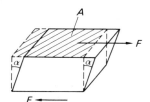

Fig. 2.31

and is deformed so that the front and rear faces become parallelograms, Fig. 2.31. The shear stress is F/A and angle α is taken as a measure of the strain produced. (The force F on the bottom surface of the block is exerted by the bench.) The shear modulus is defined as

$$G = \frac{shear\ stress}{shear\ strain} = \frac{F}{A\alpha}$$

When a wire is twisted, a small square on the surface becomes a rhombus, Fig. 2.32, and is an example of a shear strain. G can be found from experiments on the twisting of wires. If a spiral spring is stretched, the wire itself is not extended but is twisted, i.e. sheared. The extension thus depends on the shear modulus of the material as well as on the dimensions of the spring.

Wire

Fig. 2.32

(c) *Bulk modulus* (K). If a body of volume V is subjected to an *increase* of external pressure δp which changes its volume by δV, Fig. 2.33, the deformation is a change of volume without a change of shape. The bulk stress is δp, i.e. increase in force per unit area, and the bulk strain $\delta V/V$, i.e. change of volume/original volume; the bulk modulus K is defined by

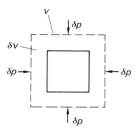

Fig. 2.33

$$K = \frac{bulk\ stress}{bulk\ strain} = \frac{-\delta p}{\delta V/V}$$

$$= -V\frac{\delta p}{\delta V}$$

The negative sign is introduced to make K positive since δV, being a decrease, is negative.

Note. δ (pronounced 'delta') is the Greek letter small d and when used as a prefix to the symbol for a quantity it indicates a change in that quantity is being considered.

Solids have all three moduli, liquids and gases only K. All moduli have the same units—pascals.

QUESTIONS

1. (a) Why are stresses and strains rather than forces and extensions generally considered when describing the deformation behaviour of solids?

(b) A length of copper of square cross-section measuring 1.0 mm by 1.0 mm is stretched by a tension of 40 N. What is the tensile stress in Pa?

(c) If the breaking stress of steel is 1.0×10^9 Pa will a wire of this material of cross-section area 4.0×10^{-4} cm^2 break when a 10 kg mass is hung from it? ($g = 10$ m s^{-2})

(d) A strip of rubber 6 cm long is stretched until it is 9 cm long. What is the tensile strain in the rubber as (i) a ratio, (ii) a percentage?

(e) A wire originally 2 m long suffers a 0.1% strain. What is its stretched length?

2. (a) The Young modulus for steel is greater than that for brass. Which would stretch more easily? Which is stiffer?

(b) How does a deformed body behave when the deforming force is removed if the strain is (i) elastic, (ii) plastic?

(c) A brass wire 2.5 m long of cross-section area 1.0×10^{-3} cm^2 is stretched 1.0 mm by a load of 0.40 kg. Calculate the Young modulus for brass. (Take $g = 10$ m s^{-2}.)

What percentage strain does the wire suffer? Use the value of E to calculate the force required to produce a 4.0% strain in the same wire. Is your answer for the force reliable? If it isn't, would it be greater or less than your answer? Explain.

3. Stress–strain curves for four different materials are shown in Fig. 2.34. Describe what you would feel if a specimen of each was pulled.

4. (a) A 0.50 kg mass is hung from the end of a wire 1.5 m long of diameter 0.30 mm. If the Young modulus for the material of the wire is 1.0×10^{11} Pa, calculate the extension produced. (Take $g = 10$ m s^{-2}.)

(b) Two wires, one of steel and one of phosphor bronze, each 1.5 metres long and of diameter 0.20 cm, are joined end to end to form a composite wire of length 3.0 metres. What

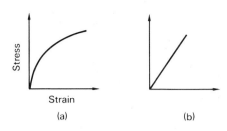

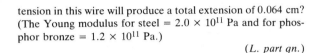

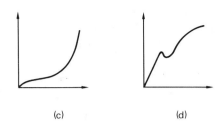

Fig. 2.34   (a)   (b)   (c)   (d)

tension in this wire will produce a total extension of 0.064 cm? (The Young modulus for steel = 2.0×10^{11} Pa and for phosphor bronze = 1.2×10^{11} Pa.)

(*L. part qn.*)

5. Write a short essay on 'strong' materials. Your essay might include discussion of such topics as: the need for strong materials, why some materials are weak, how the weaknesses can be avoided, composite materials such as plywood, reinforced concrete and fibre-glass, and the new technology of composite materials. These are only suggestions: you might write about a few of them, or, if you prefer, about all of them, and you may of course discuss other related issues if you wish.

(*O. and C. Nuffield*)

6. Define *stress, strain, the Young modulus*.

Describe in detail how the Young modulus for a steel wire may be determined by experiment.

A vertical steel wire 350 cm long, diameter 0.100 cm, has a load of 8.50 kg applied at its lower end. Find (*a*) the extension, (*b*) the energy stored in the wire. (Take the Young modulus for steel as 2.00×10^{11} Pa and $g = 9.81$ m s^{-2}.)

(*L.*)

7. Approximate values for the Young modulus, ultimate tensile strength and relative density of some materials are given in the table below.

Material	The Young modulus $E/10^{10}$ Pa	Ultimate tensile strength ($\times 10^8$ Pa)	Relative density
aluminium	7.0	1.0	2.7
carbon fibres	40	17	2.3
glasses	7.0	1.0	2.5
steel	21	10	7.8

(*a*) Explain the meaning and significance of each of these three terms.

(*b*) The tensile strength of glass equals that of aluminium. Why isn't glass used for the same kind of constructional applications as aluminium?

(*c*) In structures like aircraft the designer is concerned with obtaining as much strength and stiffness as possible for a given weight. Using the values in the table derive measures of 'stiffness-to-weight' and 'strength-to-weight' for the four materials listed.

3 Electrical properties

Conduction in solids

Materials exhibit a very wide range of electrical conductivities. The best conductors (silver and copper) are over 10^{23} times better than the worst conductors, i.e. the best insulators (e.g. polythene). Between these extreme cases is the now important group of semiconductors (e.g. germanium and silicon).

The first requirement for conduction is a supply of charge carriers that can wander freely through the material. In most solid conductors, notably metals, we believe that the carriers are loosely held outer electrons. With copper, for example, every atom has one 'free' electron (the one that helps to form the metallic bond, p. 8) which is not attached to any particular atom and so can participate in conduction. On the other hand if all electrons are required to form the bonds (covalent or ionic) that bind the atoms of the material together then that material will be an insulator. In semiconductors only a small proportion of the outer electrons are 'free' to move.

The 'free' electrons in a solid conductor are in a state of rapid motion, moving to and fro within the crystal lattice at speeds calculated to be about 1/1000 of the speed of light. This motion is normally completely haphazard (like that of gas molecules) and as many electrons with a given speed move in one direction as in the opposite direction with the same speed. There is, therefore, no net flow of charge and so no current.

If a battery is connected across the ends of the conductor, an electric field is created in the conductor which causes the electrons to accelerate and gain kinetic energy. Collisions occur between the accelerating electrons and atoms (really positive ions) of the conductor that are vibrating about their mean position

in the crystal lattice but are not free to undergo translational motion. As a result the electrons lose kinetic energy and slow down whilst the ions gain vibrational energy. The net effect is to transfer chemical energy from the battery, via the electrons, to internal energy (see p. 65) of the ions. This shows itself on the macroscopic scale as a temperature rise in the conductor and subsequently energy may pass from the conductor to the surroundings as heat. The electrons are again accelerated and the process is repeated.

The *overall* acceleration of the electrons is zero on account of their collisions. They acquire a *constant average drift velocity* in the direction from negative to positive of the battery and it is this resultant drift of charge that is believed to constitute an electric current. An analogous situation may arise when a ball rolls down a long flight of steps. The acceleration caused by the earth's gravitational field when the ball drops can be cancelled by the force it experiences on 'colliding' with the steps. The ball may roll down the stairs with zero average acceleration, i.e. at constant average speed.

The 'free' electron theory is able to account in a general way for many of the facts of conduction and although in more advanced work it has been extended by the 'band' theory it will be adequate for our present purposes.

Current and charge

In metals, current is the movement of negative charge, i.e. electrons; in gases and electrolytes both positive and negative charges may be involved. Under the action of a battery, charges of opposite sign move in opposite directions and so a convention for current direction has to be chosen. As far as most external effects are concerned, positive charge moving in one direction is the same as negative charge moving in the opposite direction. By agreement all current is assumed to be due to the motion of positive charges and when current arrows are marked on circuits they are directed from the positive to the negative of the supply. If the charge carriers are negative they move in the opposite direction to that of the arrow.

The basic electrical unit is the unit of current—the *ampere* (abbreviated to A); it is defined in terms of the magnetic effect of a current. The unit of electric charge, the *coulomb* (C), is defined in terms of the ampere.

One coulomb is the quantity of electric charge carried past a given point in a circuit when a steady current of 1 ampere flows for one second.

If 2 amperes flow for 1 second, 2×1 coulombs (ampere-seconds) pass; if 2 amperes flow for 3 seconds then 2×3 coulombs pass. In general if a steady current I (in amperes) flows for time t (in seconds) then the quantity Q (in coulombs) of charge that passes is given by

$$Q = It$$

The flow of charge in a conductor is often compared with the flow of water in a pipe. The flow of water in litres per second say, corresponds to the flow of charges in coulombs per second, i.e. amperes.

The charge on an electron, i.e. the electronic charge, is 1.60×10^{-19} C and is much too small as a practical unit. In 1 C there are therefore $1/(1.60 \times 10^{-19})$, i.e. 6.24×10^{18} electronic charges. A current of 1 A is thus equivalent to a drift of 6.24×10^{18} electrons past each point in a conductor every second.

Smaller units of current are the milliampere (10^{-3} A), abbreviated to mA and the microampere (10^{-6} A), abbreviated to μA.

Drift velocity of electrons

On the basis of the 'free' electron theory an expression can be derived for the drift velocity of electrons in a current and an estimate made of its value. The results are surprising.

Consider a conductor of length l and cross-section area A having n 'free' electrons per unit volume each carrying a charge e, Fig. 3.1.

Volume of conductor	$= Al$
Number of 'free' electrons	$= nAl$
Total charge Q of 'free' electrons	$= nAle$

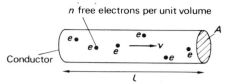

Fig. 3.1

Suppose that a battery across the ends of the conductor causes the charge Q to pass through length l in time t with average drift velocity v. The resulting steady current I is given by

$$I = \frac{Q}{t}$$

$$= \frac{nAle}{t}$$

But $v = l/t$, therefore $t = l/v$.

$$\therefore \quad I = \frac{nAle}{l/v} = nAev$$

$$\therefore \quad v = \frac{I}{nAe}$$

To obtain a value for v consider a current of 1.0 ampere in SWG 28 copper wire of cross-section area 1.1×10^{-7} square metre. If we assume that each copper atom contributes one 'free' electron, it can be shown (see question 2a, p. 60) that $n \simeq 10^{29}$ electrons per cubic metre. Then, since $e = 1.6 \times 10^{-19}$ coulomb (the charge on an electron),

$$v = \frac{I}{nAe}$$

$$= \frac{(1.0 \text{ A})}{(10^{29} \text{ m}^{-3}) \times (1.1 \times 10^{-7} \text{ m}^2) \times (1.6 \times 10^{-19} \text{ C})}$$

$$= \frac{1.0}{10^{29} \times 1.1 \times 10^{-7} \times 1.6 \times 10^{-19}} \cdot \frac{\text{A}}{\text{m}^{-3} \text{ m}^2 \text{ C}}$$

$$\simeq 6 \times 10^{-4} \frac{\text{C s}^{-1}}{\text{m}^{-1} \text{ C}} \qquad (1\text{A} = 1 \text{ C s}^{-1})$$

$$\simeq 6 \times 10^{-4} \text{ m s}^{-1}$$

$$\simeq 0.6 \text{ mm s}^{-1}$$

This is a remarkably small velocity and means that it takes electrons about half an hour to drift 1 m when a current of 1 A flows in this wire. The tiny drift velocity of electrons contrasts with their random speeds due to their vibrational motion (about 1/1000 of the speed of light) and is not to be confused with the speed at which the electric field causing their drift motion travels along a conductor. This is very great and is nearly equal to the speed of light, i.e. 3×10^8 m s^{-1} (See Appendix 7). Current therefore starts to flow almost simultaneously at all points in a circuit.

The same expression for drift velocity holds for charge carriers other than electrons (in fact, it holds for the transport of other things as well as electric charge). In an electrolyte, conduction is due to ions and using the arrangement of Fig. 3.2 information can be obtained about their motion. On applying an electric field (from a 250 volt d.c. supply), the purple stains from the permanganate crystals travel very slowly towards the positive of the supply, and if we make the not unreasonable assumption that the stain travels with the charge carriers then ions too would appear to have tiny drift velocities of a similar value to those calculated for electrons.

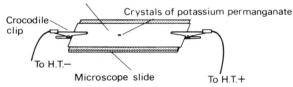

Strip of filter paper soaked in bench ammonium hydroxide solution

Crystals of potassium permanganate

Crocodile clip

To H.T.—

Microscope slide

To H.T.+

Fig. 3.2

Despite the slow movement of the carriers in conductors and their very small charge, large currents are possible. Why?

Potential difference

In an electric circuit electrical energy is converted into other forms of energy. A lamp converts electrical energy into heat and light and an electric motor converts electrical energy into mechanical energy. Such energy conversions, produced by suitable devices, are a useful feature of electrical circuits and form the basis of the definition of the term *potential difference* (p.d.)— an idea that helps us to make sense of circuits.

The potential difference between two points in a circuit is the amount of electrical energy changed to other forms of energy when unit charge passes from one point to the other.

The unit of potential difference is the *volt* and equals the p.d. between two points in a circuit in which 1 joule of electrical energy is converted when 1 coulomb passes from one point to the other. If 2 joules are converted per coulomb then the p.d. is 2 volts. If the passage of 3

coulombs is accompanied by the conversion of 9 joules of energy, the p.d. is 9/3 joules per coulomb, i.e. 3 volts.

Fig. 3.3

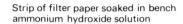

It therefore follows that if the p.d. between A and B in Fig. 3.3 (we more commonly talk about the p.d. *across* AB) is 5 volts then when 4 coulombs pass from A to B, the electrical energy changed will be 5 joules per coulomb, i.e. 5×4 joules. In general if a charge of Q (in coulombs) flows in a part of a circuit across which there is a p.d. of V (in volts) then the energy change W (in joules) is given by

$$W = QV$$

If Q is in the form of a steady current I (in amperes) flowing for time t (in seconds) then $Q = It$ and

$$W = ItV$$

Although it is always the p.d. between two points which is important in electric circuits there are some occasions when it is helpful to consider what is called the *potential at a point*. This involves selecting a convenient point in the circuit and saying it has zero potential. The potentials of all other points are then stated with reference to it, i.e. the potential at any point is then the p.d. between the point and the point of zero potential. In practice one part of a piece of electrical equipment (e.g. a power supply) is often connected to earth; the earth and all points in the circuit joined to it are then taken as having zero potential.

If in Fig. 3.3 positive charge moves (i.e. conventional current flows) from A to B then A is regarded as being at a higher potential than B. Negative charge flow is therefore from a lower to a higher potential, i.e. from B to A. We can look upon p.d. as a kind of electrical 'pressure' that drives conventional current from a point at a higher potential to one at a lower potential.

Resistance

When the same p.d. is applied across different conductors different currents flow. Some conductors offer more opposition or *resistance* to the passage of current than others.

The resistance R of a conductor is defined as the ratio of the potential difference V across it to the current I flowing through it. That is,

$$R = \frac{V}{I}$$

The unit of resistance is the *ohm* (symbol Ω, the Greek letter omega) and is the resistance of a conductor in which the current is 1 ampere when a p.d. of 1 volt is applied across it. Larger units are the *kilohm* (10^3 ohm), symbol kΩ, and the *megohm* (10^6 ohm), symbol MΩ. The ratio V/I is a sensible measure of the resistance of a conductor since the smaller I is for a given V, the greater must be opposition of the conductor, that is, the greater is R.

If the p.d. V across a *metallic* conductor is varied and the corresponding currents I measured, the ratio V/I is found to be constant so long as physical conditions, such as temperature, do not alter. Thus the resistance of such conductors is the same whatever the p.d. applied and we can write

$$\frac{V}{I} = \text{a constant}$$

This means $I \propto V$, i.e. *the current through a metallic conductor is directly proportional to the p.d. between its ends if the temperature and other physical conditions are constant.* This important result is known as Ohm's law and conductors, such as metals and alloys, which obey it are called *ohmic* conductors; a graph of V against I for them is a straight line through the origin. Many conductors do not obey Ohm's law, i.e. they are non-ohmic, and their resistance depends on the p.d. even though their temperature does not change during the measurements. The revolutionary advances in modern electronics are due largely to non-ohmic devices such as transistors.

The resistance of a metal can be regarded as arising from the interaction which occurs between the crystal lattice of the metal and the 'free' electrons as they drift through it under an applied p.d. This interaction is due mainly to collisions between electrons and the vibrating ions of the metal but collisions between defects in the crystal lattice (e.g. impurity atoms and dislocations) also play a part, especially at very low temperatures.

The *conductance* of a specimen is the reciprocal of its resistance and is measured in *siemens* (S).

Types of resistor

Conductors especially constructed to have resistance are called *resistors*, denoted by—⋀⋀⋀—(old symbol) or —▭—(new symbol); they are required for many purposes in electrical circuits. Several types exist.

(*a*) *Carbon composition resistors.* These are made from mixtures of carbon black (a conductor), clay and resin binder (non-conductors) which are pressed and moulded into rods by heating. The resistivity of the

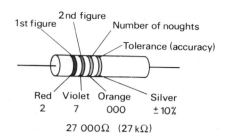

| 1st figure | 2nd figure | Number of noughts | Tolerance (accuracy) |

Red · Violet · Orange · Silver
2 · 7 · 000 · ±10%

27 000Ω (27 kΩ)

Figure	Colour	
0	Black	
1	Brown	
2	Red	
3	Orange	Colours
4	Yellow	of
5	Green	spectrum
6	Blue	
7	Violet	
8	Grey	
9	White	

Fig. 3.4

mixture depends on the proportion of carbon. The stability of such resistors is poor and their values are usually only accurate to within ±10% but they are cheap, small and good enough for many jobs. Three sizes are available with power ratings of $\frac{1}{2}$, 1 and 2 watts.

Values are shown by colour markings as in Fig. 3.4. The tolerance colours are gold ±5%, silver ±10%, no colour ±20%.

This colour code is now being replaced by a code with simpler markings.

Value	0.27 Ω	1 Ω	3.3 Ω	10 Ω	220 Ω
Mark	R27	1R0	3R3	10R	K22

Value	1 kΩ	68 kΩ	100 kΩ	1 MΩ	6.8 MΩ
Mark	1K0	68K	M10	1M0	6M8

Tolerances are indicated by adding a letter; F = ±1%, G = ±2%, J = ±5%, K = ±10%, M = ±20%. Thus 5K6K = 5.6 kΩ ±10%.

(*b*) *Carbon film resistors.* Ceramic rods are heated to about 1000 °C in methane vapour which decomposes and deposits a uniform film of carbon on the rod. The resistance of the film depends on its thickness and can be increased by cutting a spiral groove in it, Fig. 3.5. The film is protected by an epoxy resin coating. The stability and accuracy of this type of resistor is commonly ±2% and the power rating $\frac{1}{8}$ to $\frac{1}{2}$ watt.

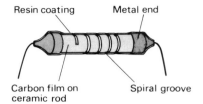

Fig. 3.5

(c) *Wire-wound resistors.* High-accuracy, high-stability resistors are always wire-wound, as are those required to have a large power rating (i.e. over 2 watts). They use the fact that the resistance of a wire increases with its length. Manganin (copper, manganese, nickel) wire is used for high-precision standard resistors because of its low temperature coefficient of resistance (p. 47); constantan or eureka (copper, nickel) wire is used for several purposes and Nichrome. (nickel, chromium) wire for commercial resistors.

Adjustable known resistances, called resistance boxes, are used for electrical measurements in the laboratory. They consist of a number of constantan coils which can be connected in series by switches or plugs to give the required value, Figs. 3.6 and 3.7. It is especially important with resistance boxes not to exceed the maximum safe currents since overheating may change the resistance value or even burn out the coils. (The power limit is about 1 watt per coil and so a 1 ohm coil should not carry more than 1 ampere and a 100 ohm coil more than 0.1 ampere—from power = $I^2 R$, see p. 50).

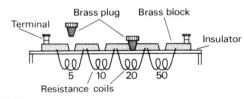

Fig. 3.6

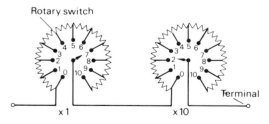

Fig. 3.7

(d) *Variable resistors.* Those used in electronic circuits, often as volume or other controls and sometimes called potentiometers, consist of an incomplete circular track of carbon composition or wire-wound card, with fixed connections to each end and a rotating arm contact which can slide over the track. Figs. 3.8a and b show the outside and inside respectively of a wire-wound potentiometer. If the track is 'linear', the resistance tapped off is proportional to the distance moved by the sliding contact; if it is 'logarithmic' it is proportional to the log of the distance and at the end of the track a small movement of the sliding contact causes a larger increase of resistance than at the start.

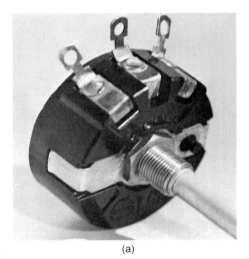

(a)

(b)

Fig. 3.8

Larger current versions as used in many electrical experiments consist of constantan wire wound on a straight ceramic tube with the sliding contact carried on a metal bar above the tube.

There are two ways of using a variable resistor. It may be used as a *rheostat* for controlling the current in a circuit when only one end connection and the sliding contact are required, Fig. 3.9. It can also act as a *potential divider* for controlling the p.d. applied to a device, all three connections then being used. In Fig. 3.10 any fraction of the total p.d. from the battery can be tapped off by varying the sliding contact.

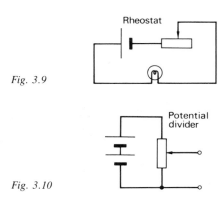

Fig. 3.9

Fig. 3.10

Resistor networks

A network of resistors like that in Fig. 3.11 has a combined or equivalent resistance which can be found experimentally from the ratio of the voltmeter reading to the ammeter reading. Its value may also be calculated.

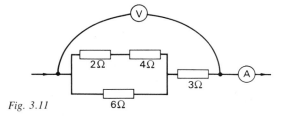

Fig. 3.11

(a) *Resistors in series.* Resistors are in series if the same current passes through each in turn. In Fig. 3.12 if the total p.d. across all three resistors is V and the current is I, the combined resistance R is given by

$$R = \frac{V}{I}$$

The electrical energy changed per coulomb in passing through all the resistors equals the sum of that changed

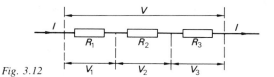

Fig. 3.12

in each resistor. Therefore if V_1, V_2 and V_3 are the p.d.s across R_1, R_2 and R_3 respectively then

$$V = V_1 + V_2 + V_3$$

By the definition of resistance, $R_1 = V_1/I$, $R_2 = V_2/I$ and $R_3 = V_3/I$, therefore $V_1 = IR_1$, $V_2 = IR_2$, $V_3 = IR_3$ and since $V = IR$ we have

$$IR = IR_1 + IR_2 + IR_3$$
$$\therefore \quad R = R_1 + R_2 + R_3$$

(b) *Resistors in parallel.* Here alternative routes are provided to the current which splits and we would expect the combined resistance to be less than the smallest individual resistance. In Fig. 3.13 if I is the total current through the network and I_1, I_2 and I_3 are the currents in the separate branches then since current is not used up in a circuit

$$I = I_1 + I_2 + I_3$$

In a parallel circuit the p.d. across each parallel branch is the same. If this is V then by the definition of resistance $R_1 = V/I_1$, $R_2 = V/I_2$, $R_3 = V/I_3$ and if R is the combined resistance then $R = V/I$. Therefore, $I = V/R$, $I_1 = V/R_1$, $I_2 = V/R_2$ and $I_3 = V/R_3$. Hence

$$\frac{V}{R} = \frac{V}{R_1} + \frac{V}{R_2} + \frac{V}{R_3}$$
$$\therefore \quad \frac{1}{R} = \frac{1}{R_1} + \frac{1}{R_2} + \frac{1}{R_3}$$

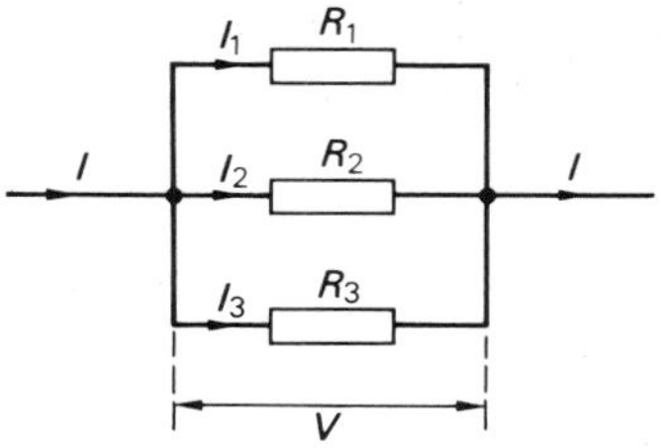

Fig. 3.13

The single resistance R which would have the same resistance as the whole network can be calculated.

For the special case of two equal resistors in parallel we have $R_1 = R_2$ and

$$\frac{1}{R} = \frac{1}{R_1} + \frac{1}{R_1} = \frac{2}{R_1}$$

$$\therefore \quad R = \frac{R_1}{2}$$

In general for n equal resistances R_1 in parallel, the combined resistance is R_1/n.

The combined resistance of the network in Fig. 3.11 is 6 Ω; do you agree?

Using ammeters and voltmeters

Most ammeters and voltmeters are basically galvanometers (i.e. current detectors capable of measuring currents of the order of milliamperes or microamperes) of the moving-coil type which have been modified by connecting suitable resistors in parallel or in series with them as described in the next sections. Moving-coil instruments are accurate, sensitive and reasonably cheap and robust.

Connecting an ammeter or voltmeter should cause the minimum disturbance to the current or p.d. it has to measure. The current in a circuit is measured by breaking the circuit and inserting an ammeter in *series* so that the current passes through the meter. The resistance of an ammeter must therefore be *small* compared with the resistance of the rest of the circuit. Otherwise, inserting the ammeter changes the current to be measured. The perfect ammeter would have zero resistance, the p.d. across it would be zero and no energy would be absorbed by it.

The p.d. between two points A and B in a circuit is most readily found by connecting a voltmeter across the points, i.e. in *parallel* with AB. The resistance of the voltmeter must be *large* compared to the resistance of AB, otherwise the current drawn from the main circuit by the voltmeter (which is required to make it operate) becomes an appreciable fraction of the main current and the p.d. across AB changes. A voltmeter can be treated as a resistor which automatically records the p.d. between its terminals. The perfect voltmeter would have infinite resistance, take no current and absorb no energy.

It is instructive to set up the circuit in Fig. 3.14 for measuring the p.d. tapped off by the potential divider between X and Y, using first a high-resistance voltmeter (e.g. one with a 100 μA movement) for Ⓥ and then a low-resistance voltmeter (e.g. one with a 10 mA movement). The reading in the second case is much lower. What will it be if both voltmeters are connected across XY?

The most straightforward method of measuring resistance uses an ammeter and a voltmeter as in Fig. 3.15a. The voltmeter records the p.d. across R but the ammeter gives the sum of the currents in R and in the voltmeter. If the voltmeter has a much higher resistance than R, the current through it will be small by comparison and the error in calculating R can be neglected. However, if the resistance of the voltmeter is not sufficiently high, perhaps because R is very high, the voltmeter should be connected across both R and the ammeter as in Fig. 3.15b. The ammeter now gives the true current in R. The voltmeter indicates the p.d. across R and the ammeter together, but the resistance of the latter is usually negligible compared with that of R and so the p.d. across it will be so small as to make the error in calculating R negligible.

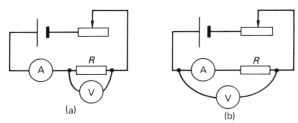

Fig. 3.15

Shunts, multipliers and multimeters

(*a*) *Conversion of a microammeter into an ammeter.* Consider a moving-coil meter which has a resistance (due largely to the coil) of 1000 Ω and which gives a full-scale deflection (f.s.d.) when 100 μA (0.0001 A) passes through it. If we wish to convert it to an ammeter reading 0–1 A this can be done by connecting a resistor (perhaps a misnomer) of very low value in parallel with it. Such a resistor is called a *shunt* and it

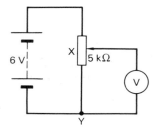

Fig. 3.14

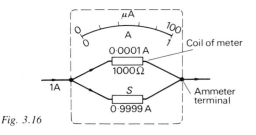

Fig. 3.16

must be chosen so that only 0.0001 A passes through the meter and the rest of the 1 A, namely 0.9999 A, passes through the shunt, Fig. 3.16. A full-scale deflection of the meter will then indicate a current of 1 A.

To obtain the value S of the shunt, we use the fact that the meter and the shunt are in parallel. Therefore,

p.d. across meter = p.d. across shunt

Applying Ohm's law to both meter and shunt

$$0.0001 \times 1000 = 0.9999 \times S \quad \text{(from } V = IR\text{)}$$

$$\therefore \quad S = \frac{0.0001 \times 1000}{0.9999}$$

$$= 0.1 \ \Omega$$

The combined resistance of the meter and the shunt in parallel will now be very small (less than 0.1 Ω) and the current in a circuit will be virtually undisturbed when the ammeter is inserted.

(b) Conversion of a microammeter into a voltmeter. To convert the same moving-coil meter of resistance 1000 Ω and f.s.d. 100 μA to a voltmeter reading 0–1 V, a resistor of high value must be connected in series with the meter. The resistor is called a *multiplier* and it must be chosen so that when a p.d. of 1 V is applied across the meter and resistor in series, only 0.0001 A flows through the meter and a full-scale deflection results, Fig. 3.17.

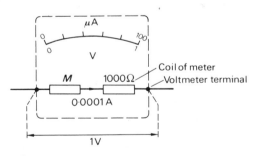

Fig. 3.17

To obtain the value M of the multiplier, we apply Ohm's law when there is an f.s.d. of 0.0001 A. Hence

p.d. across multiplier and meter in series =
$$0.0001 \ (M + 1000)$$

But the meter is to give an f.s.d. when the p.d. across it and the multiplier in series is 1 V. Therefore

$$0.0001(M + 1000) = 1$$

$$\therefore \quad M + 1000 = \frac{1}{0.0001} = 10\,000$$

$$\therefore \quad M = 9000 \ \Omega$$

In Fig. 3.18 a microammeter (20–0–100 μA) with its matching shunts and multipliers is shown.

Voltmeters are often graded according to their 'resistance per volt' at f.s.d. For the above voltmeter, 1 V applied across its terminals produces a full-scale deflection, i.e. a current of 100 μA, and so the resistance of the meter (coil + multiplier) must be 10 000 Ω (since $R = V/I = 1/0.0001 = 10\,000 \ \Omega$). The 'resistance per volt' of the meter is thus 10 000 Ω/V. To be used as a voltmeter with an f.s.d. of 10 V it would need to have a total resistance of 100 000 Ω, i.e. a multiplier of 99 000 Ω to limit the full-scale current to 100 μA—but its resistance for every volt of deflection is still 10 000 Ω. A 100 Ω/V voltmeter has a resistance of 100 Ω for an f.s.d. of 1 V and draws a full-scale current of 10 mA ($I = V/R = \frac{1}{100} = 0.01 \ A = 10 \ mA$). Hence the higher the 'resistance per volt' of a voltmeter the smaller is the current it draws and the less will it disturb the circuit to which it is connected. A good voltmeter should have a resistance of at least 1000 Ω/V.

In electronic circuits resistances of 1 MΩ or higher are encountered and electronic voltmeters which have very high resistances have to be used.

(c) Multimeters. A multi-range instrument or multimeter is a moving-coil galvanometer adapted to measure current, p.d. and resistance. There is a tapped-

Fig. 3.18

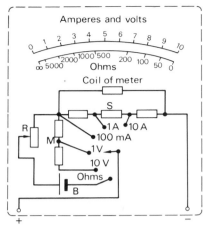

Fig. 3.19

shunt S across the meter and a tapped multiplier M in series with it, Fig. 3.19. A rotary switch allows the various ranges to be chosen.

One other position of the switch is marked 'ohms' and puts a dry cell B (usually 1.5 volts) and a rheostat R in series with the meter. To measure resistance the terminals are short-circuited and R adjusted until the pointer gives a full-scale deflection, i.e. is on the zero of the ohms scale. The unknown resistance then replaces the short circuit across the terminals. The current falls and the pointer indicates the value in ohms. Fig. 3.20 shows a multimeter.

Fig. 3.20

Rough calibration of a voltmeter

The calibration, which is done in two parts, is only approximate but uses the definition of the volt as a joule per coulomb. First, the temperature rise produced in a solid copper drum by a known force acting through a known distance is found and from it the amount of mechanical energy required to cause a 1 °C rise is calculated. Second, the drum is heated electrically for a certain time by a known current and the p.d. to be measured. Assuming that the amounts of electrical energy and mechanical energy needed to cause a 1 °C rise in the drum are the same, the p.d. can be calculated.

(a) *Mechanical experiment.* The apparatus is shown in Fig. 3.21a. The drum can be rotated about a horizontal axis by a handle and has a nylon cord wound round it five or six times. A heavy mass m (e.g. 8 kg) is hung from one end of the cord and a fixed rubber band attached to the other end. The initial temperature of the drum is noted. When the handle is turned the mass should lift slightly from the floor, the rubber band be *just* slack and the drum rotate steadily. The number of turns n to give a temperature rise of θ (about 10 °C) in time t is noted as is the diameter d (in metres) of the drum.

The mass m is supported by a frictional force F between the cord and the drum. If g is the acceleration due to gravity (9.8 m s^{-2}) then $F = mg$. During n revolutions, F acts over a total distance of πdn where πd is the circumference of the drum. The mechanical energy supplied by the person turning the handle is measured by the work and *work = force × distance in direction of force.* Here the work = $F \times \pi dn = mg \times \pi dn$. Therefore the mechanical energy W to produce a temperature rise of 1 °C in the drum is

$$W = \frac{mg\pi dn}{\theta}$$

(b) *Electrical experiment.* The copper drum is allowed to cool to room temperature and is connected into the electrical circuit of Fig. 3.21b with the nylon cord wound round it as before. The voltmeter to be calibrated is connected across the built-in heating coil in the drum and a steady current (0.7 A is suitable for some makes of apparatus) passed so that the temperature rises by about the same as in (a) (i.e. about 10 °C) in roughly the same time t. This ensures the heat losses in each experiment are similar and can be neglected. The electrical energy supplied to produce a temperature rise of 1 °C is W, the same as the mechanical energy required and is given by

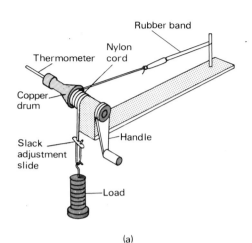

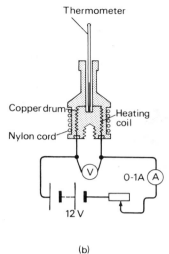

Fig. 3.21   (a)   (b)

$$W = \frac{ItV}{\theta_1}$$

or

$$R = \frac{\rho l}{A}$$

where I is the ammeter reading in amperes, t the time in seconds for a temperature rise of θ_1 and V the unknown p.d. across the coil.

Hence

$$\frac{ItV}{\theta_1} = \frac{mg\pi dn}{\theta}$$

$$\therefore \quad V = \frac{mg\pi dn}{It} \cdot \frac{\theta_1}{\theta}$$

V can thus be calculated in joules per coulomb (volts) and compared with the reading on the voltmeter. In this experiment the same amount of internal energy (not heat as is often stated, see p. 65) is produced in the drum, first from mechanical energy and then from electrical energy.

Resistivity

The resistance of a conductor depends on its size as well as on the material of which it is made. To make fair comparisons of the abilities of different materials to conduct, the resistance of specimens of the same size must be considered.

Experiment shows that the resistance R of a uniform conductor of a given material is directly proportional to its length l and inversely proportional to its cross-section area A. Hence

$$R \propto \frac{l}{A}$$

where ρ is a constant (for fixed temperature and other physical conditions), called the *resistivity* of the material of the conductor.

Hence since $\rho = AR/l$ we can say that the *resistivity of a material is numerically the resistance of a sample of unit length and unit cross-section area, at a certain temperature*. The unit of ρ is ohm metre (Ω m) since those of AR/l are metre^2 × ohm/metre, i.e. ohm metre.

Knowing the resistivity of a material the resistance of *any* specimen of that material may be calculated. For example, if the cross-section area of the live rail of an electric railway is 50 cm^2 and the resistivity of steel is 1.0×10^{-7} Ω m then neglecting the effect of joints, the resistance per kilometre of rail R follows—

$$R = \frac{\rho l}{A}$$

$$= \frac{(1.0 \times 10^{-7} \ \Omega \ \text{m}) \times (10^3 \ \text{m})}{(50 \times 10^{-4} \ \text{m}^2)}$$

$$= \left(\frac{1.0 \times 10^{-7} \times 10^3}{50 \times 10^{-4}}\right) \frac{\Omega \ \text{m} \ \text{m}}{\text{m}^2}$$

$$= 2.0 \times 10^{-2} \ \Omega$$

The resistivities at 20 °C of various materials are given in Table 3.1; their experimental determination is briefly described on p. 52.

The *conductivity* (σ) of a material is the reciprocal of its resistivity (ρ) i.e. $\sigma = 1/\rho$, and has unit ohm^{-1} metre^{-1} (Ω^{-1} m^{-1}).

Table 3.1

Material	Resistivity Ω m	Use
CONDUCTORS		
Metals { Silver	1.6×10^{-8}	Contacts on small switches
Copper	1.7×10^{-8}	Connecting wires
Aluminium	2.7×10^{-8}	Power cables
Tungsten	5.5×10^{-8}	Lamp filaments
Alloys { Manganin	$44 \ \ \times 10^{-8}$	High-precision standard resistors
Constantan or eureka	$49 \ \ \times 10^{-8}$	Resistance boxes, variable resistors
Nichrome (Ni, Cr)	$110 \ \ \times 10^{-8}$	Heating elements
Carbon	$3000 \ \ \times 10^{-8}$	Radio resistors
SEMICONDUCTORS		
Germanium	0.6	Transistors
Silicon	2300	Transistors
INSULATORS		
Glass	$10^{10} - 10^{14}$	
Polystyrene	10^{15}	

Silver is the best conductor, i.e. has the lowest resistivity, and is followed closely by copper which, being much less expensive, is used for electrical connecting wire. Although the resistivity of aluminium is nearly twice that of copper, its density is only about one-third of copper's. The current-carrying-capacity-to-weight ratio of aluminium is therefore greater than that of copper. This accounts for its use in the overhead power cables of the Grid System where aluminium strands are wrapped round a core of steel wires (54 aluminium strands to 7 steel wires for example, Fig. 3.22). The cable then has the strength it requires for suspension in long spans between pylons.

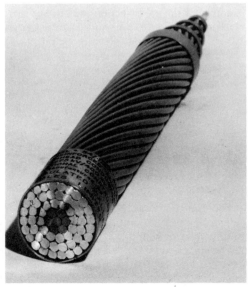

Fig. 3.22

The resistivity of a pure metal is increased by small amounts of 'impurity' and alloys have resistivities appreciably greater than those of any of their constituents. On the other hand, the addition of tiny traces of 'impurities' to pure semiconductors (a process known as 'doping' the semiconductor) reduces their resistivity. 'Impurity' atoms in a crystal lattice act as 'defects' and restrict the movement of charge carriers. When a semiconductor is 'doped' this is more than offset by the production of extra 'free' charges.

Electrical strain gauge

One device which engineers employ to obtain information about the size and distribution of strains in structures such as buildings, bridges and aircraft is the electrical strain gauge. It converts mechanical strain into a resistance change in itself by using the fact that the resistance of a wire depends on its length and cross-section area.

One type of gauge consists of a very fine wire (of an alloy containing mostly nickel, iron and chromium) cemented to a piece of thin paper as in Fig. 3.23. In use it is securely attached with a very strong adhesive to the component under test so that it experiences the same strain as the component. If, for example, an increase of length strain occurs, the gauge wire gets longer and thinner and on both counts its resistance increases. Thick leads connect the gauge to a resistance measuring circuit (e.g. a Wheatstone bridge, p. 51) and previous calibration of the gauge enables the strain to be measured directly. What is the advantage of using a parallel-wire arrangement for the strain gauge?

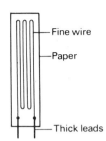

Fine wire

Paper

Thick leads

Fig. 3.23

Fig. 3.24

Strain gauges are used to check theoretical calculations on new aircraft designs. Fig. 3.24 shows a model of a slender wing aircraft with weights and strain gauges attached at various places. The weights simulate the aerodynamic loading when the aircraft is in flight.

Effects of temperature on resistance

(*a*) *Temperature coefficient.* The resistance of a material varies with temperature and the variation can be expressed by its *temperature coefficient of resistance* α. If a material has resistance R_0 at 0°C and its resistance increases by δR due to a temperature rise $\delta \theta$ then α for the material is defined by the equation

$$\alpha = \frac{\delta R}{R_0} \cdot \frac{1}{\delta \theta}$$

In words, α is *the fractional increase in the resistance at 0 °C* (i.e. $\delta R/R_0$) *per unit rise of temperature.* The unit of α is °C^{-1} since δR and R_0 have the same units (ohms) and $\delta R/R_0$ is thus a ratio. For copper $\alpha \simeq 4 \times 10^{-3}$ °C^{-1}, which means that a copper wire having a resistance of 1 ohm at 0 °C increases in resistance by 4×10^{-3} ohm for every 1 °C temperature rise.

Experiment shows that the value of α varies with the temperature at which $\delta \theta$ occurs but, to a good approximation, for metals and alloys we can generally assume it is constant in the range 0 to 100 °C. Thus if a specimen has resistances R_θ and R_0 at temperatures θ and 0 °C respectively then replacing δR by $R_\theta - R_0$ and $\delta \theta$ by θ in the expression for α, we obtain

$$\alpha = \frac{R_\theta - R_0}{R_0 \theta}$$

Rearranging gives $R_\theta - R_0 = R_0 \alpha \theta$

and $\qquad R_\theta = R_0(1 + \alpha \theta)$

When using the equation where accuracy is important

R_0 *should be the resistance at 0 °C.* A calculation shows the procedure when R_0 is not known.

Suppose a copper coil has a resistance of 30 Ω at 20 °C and its resistance at 60 °C is required. Taking α for copper as 4.0×10^{-3} °C^{-1} we have

$$R_{20} = R_0(1 + 20\alpha)$$

and $\qquad R_{60} = R_0(1 + 60\alpha)$

Dividing $\quad \dfrac{R_{60}}{R_{20}} = \dfrac{1 + 60\alpha}{1 + 20\alpha}$

$$\therefore \quad R_{60} = \frac{30(1 + 60 \times 4.0 \times 10^{-3})}{(1 + 20 \times 4.0 \times 10^{-3})}$$

$$= 34.5 \ \Omega$$

If the calculation had *not* been based on the resistance at 0 °C and we had taken the original resistance (i.e. R_{20}) as R_0 then using $R_{60} = R_0(1 + \alpha\theta)$ where $\theta = (60 - 20)$ °C = 40 °C and $R_0 = 30$ Ω, we get

$$R_{60} = 30(1 + 4.0 \times 10^{-3} \times 40)$$

$$= 34.8 \ \Omega$$

This approximate method is quicker but in this example introduces an error of 0.3 in 34.5, i.e. about 1%, which is acceptable for many purposes.

The experimental determination of α is outlined on p. 52. Metals and alloys have *positive temperature* coefficients (they are p.t.c. materials), i.e. their resistance *increases* with temperature rise. The values for pure metals are of the order of 4×10^{-3} per °C or roughly 1/273 per °C, the same as the cubic expansivity of gas. In a tungsten-filament electric lamp the current raises the temperature of the filament to over 2730 °C when lit. The 'hot' resistance of the filament is, therefore, more than ten times the 'cold' resistance. Why doesn't a fuse blow every time lights are switched on? Alloys have much lower temperature coefficients of resistance than pure metals, that for manganin is about

2×10^{-5} per °C and a small temperature change has a small effect on its resistance.

Graphite, semiconductors and most non-metals have *negative temperature coefficients*, i.e. their resistance *decreases* with temperature rise (they are n.t.c. materials).

(*b*) *Superconductors.* When certain metals (e.g. tin, lead) and alloys are cooled to near −273 °C an *abrupt* decrease of resistance occurs. Below a definite temperature, different for each material, the resistance vanishes and a current once started seems to flow for ever. Such materials are called *superconductors* and their use in electrical power engineering and electronics is being explored.

(*c*) *Thermistors* (derived from *therm*al res*istors*). These are devices whose resistance varies quite markedly with temperature. Depending on their composition they can have either n.t.c. or p.t.c. characteristics. The n.t.c. type consists of a mixture of oxides of iron, nickel and cobalt with small amounts of other substances and is used in electronic circuits to compensate for resistance increase in other components when the temperature rises and also as a thermometer for temperature measurement. The p.t.c. type, which is based on barium titanate, can show a resistance increase of 50 to 200 times for a temperature rise of a few degrees. It is useful as a temperature-controlled switch. Why? Fig. 3.25 shows a selection of thermistors.

(*d*) *Electrons, resistance and temperature.* The 'free' electron theory can account qualitatively for the variation of resistance with temperature of different materials. The increased average separation of the ions in a metal which accompanies a temperature rise (see p.74) causes local distortion of the crystal lattice. As a result there is increased interaction between the lattice and the 'free' electrons when they drift under an applied p.d. The average drift speed is reduced and the resistance thus increases. In semiconductors this is more than compensated when greater vibration of the atoms 'frees' more electrons (an insignificant effect in metals) and thereby produces a marked decrease of resistance with temperature rise.

Electromotive force

Batteries and generators are able to maintain one terminal positive (i.e. deficient of electrons) and the other negative (i.e. with an excess of electrons). If we consider the motion of positive charges, then a battery, for example, moves positive charges from a place of low potential (the negative terminal) through the battery to a place of high potential (the positive terminal). The action may be compared with that of a pump causing water to move from a point of low gravitational potential to one of high potential.

A battery or generator therefore *does work on charges* and so energy must be changed within it. (Work is a measure of energy transfer, see p. 66.) In a battery chemical energy is transferred into electrical energy which we consider to be stored in the electric and magnetic fields produced. When current flows in an external circuit this stored electrical energy is changed, for example, to heat, but it is replenished at the same rate at which it is transferred. The electric and magnetic fields thus act as a temporary storage reservoir of electrical energy in the transfer of chemical energy to heat. A battery or dynamo is said to produce an *electromotive force* (e.m.f.), defined in terms of energy transfer.

The electromotive force of a source (a battery, generator, etc.) *is the energy* (chemical, mechanical, etc.) *converted into electrical energy when unit charge passes through it.*

The unit of e.m.f., like the unit of p.d., is the *volt* and equals the e.m.f. of a source which changes 1 joule of chemical, mechanical or other form of energy into electrical energy when 1 coulomb passes through it. A car battery with an e.m.f. of 12 volts supplies 12 joules per coulomb passing through it; a power station generator with an e.m.f. of 25 000 volts is a much greater source of energy and supplies 25 000 joules per coulomb—2 coulombs would receive 50 000 joules and so on. In general, if a charge Q (in coulombs) passes through a source of e.m.f. E (in volts), the electrical energy supplied by the source W (in joules) is

$$W = QE$$

It should be noted that although e.m.f. and p.d. have the same unit, they deal with different aspects of an electric circuit. Whilst e.m.f. applies to a source supplying electrical energy, p.d. refers to the conversion of electrical energy in a circuit. The term e.m.f. is misleading to some extent, since it measures energy per unit charge and not force. It is true, however, that the source of e.m.f. is responsible for moving charges round the circuit.

A voltmeter measures p.d. and one connected across the terminals of an electrical supply such as a battery records what is called the *terminal p.d.* of the battery. If the battery is not connected to an external circuit and the voltmeter has a very high resistance then the current through the battery will be negligible. We

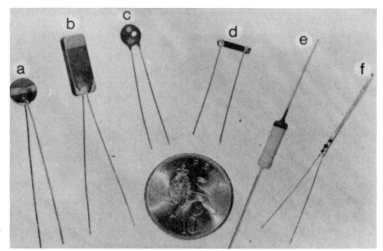

Fig. 3.25
(a) disc n.t.c. type;
(b) plate n.t.c. type;
(c) p.t.c. type; (d) rod
n.t.c. type; (e) rod
voltage dependent re-
sistor; (f) bead-in-glass
type.

can regard the voltmeter as measuring the number of joules of electrical energy the battery supplies per coulomb, i.e. its e.m.f. A working but less basic definition of e.m.f. is to say that it equals the terminal p.d. of a battery or generator *on open circuit*, i.e. when not maintaining current.

Internal resistance

A high-resistance voltmeter connected across á cell on open circuit records its e.m.f. (very nearly), Fig. 3.26a. Let this be E. If the cell is now connected to an external circuit in the form of a resistor R and maintains a steady current I in the circuit, the voltmeter reading falls; let it be V, Fig. 3.26b. V is the terminal p.d. of the cell (but not on open circuit) and it is also the p.d. across R (assuming the connecting leads have zero resistance). Since V is less than E, then not all the energy supplied per coulomb by the cell (i.e. E) is changed in the external circuit to other forms of energy (often heat). What has happened to the 'lost' energy per coulomb?

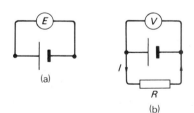

Fig. 3.26

The deficiency is due to the cell itself having some resistance. A certain amount of electrical energy per coulomb is wasted in getting through the cell and so less is available for the external circuit. The resistance of a cell is called its *internal resistance* (r) and taking stock of the energy changes in the complete circuit including the cell, we can say, assuming conservation of energy:

energy *supplied*   energy *changed*   energy *wasted* per
per coulomb by   = per coulomb by   + coulomb on inter-
cell   external circuit   nal resistance of
                                             battery

Or, from the definitions of e.m.f. and p.d.,

$$\text{e.m.f.} = \text{p.d. across } R + \text{p.d. across } r$$

In symbols

$$E \quad = \quad V \quad + \quad v$$

e.m.f   useful   'lost'
           volts   volts

where v is the p.d. across the internal resistance of the cell, a quantity which cannot be measured directly but only by subtracting V from E. From the equation $E = V + v$ we see that the *sum of the p.d.s. across all the resistance in a circuit* (external and internal) *equals the e.m.f.*

Since $V = IR$ and $v = Ir$ we can rewrite the previous equation

$$E = IR + Ir$$

$$\therefore \quad E = I(R + r)$$

Suppose a high-resistance voltmeter reads 1.5 V when connected across a dry battery on open circuit, Fig. 3.27a, and 1.2 V when the same battery is supplying a current of 0.30 A through a lamp of resistance R, Fig. 3.27b. What is (a) the e.m.f. of the battery, (b) the internal resistance of the battery and (c) the value of R?

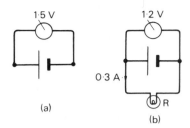

Fig. 3.27

Using symbols with their previous meanings:

(a) Since the terminal p.d. on open circuit equals the e.m.f., we have $E = 1.5$ V.

(b) $E = V + v = V + Ir$ where $V = 1.2$ V and $I = 0.30$ A. Therefore

$$Ir = E - V \quad \text{and} \quad r = \frac{E - V}{I} = \frac{1.5 - 1.2}{0.30} = 1.0 \ \Omega$$

(c) From $V = IR$,

$$R = \frac{V}{I} = \frac{1.2}{0.30} = 4.0 \ \Omega$$

The internal resistance of an electrical supply depends on several factors and is seldom constant as is often assumed in calculations. However, it is sometimes useful to know its rough value and estimates can be made by taking p.d. and current measurements and proceeding as in the above example. Sources such as low-voltage supply units and car batteries from which large currents are required must have very low internal resistances. On the other hand if a 5000 V, E.H.T. power supply does not have an internal resistance of the order of megohms, to limit the current it supplies, it will be dangerous.

The effect of internal resistance can be seen when a bus or car starts with the lights on. Suppose the starter motor requires a current of 100 A from a battery of e.m.f. 12 V and internal resistance 0.04 Ω to start the engine. How many volts are 'lost'? What is the terminal p.d. of the battery with the starter motor working? Why do the lights dim if they are designed to operate on a 12 V supply?

The terminal p.d. of a battery on open circuit as measured by even a very-high-resistance voltmeter is not quite equal to the e.m.f. because the voltmeter must take some current, however small, to give a reading. A small part of the e.m.f. is, therefore, 'lost' in driving current through the internal resistance of the battery. A potentiometer is used to measure e.m.f. to a very high accuracy (p. 54).

Power and heating effect

Current flow is accompanied by the conversion of electrical energy into other forms of energy and it is often necessary to know the *rate* at which a device brings about this conversion.

The power of a device is the rate at which it converts energy from one form into another.

If the p.d. across a device is V and the current through it is I, the electrical energy W converted by it in time t is (from the definition of p.d., p. 38).

$$W = ItV$$

The power P of the device will be

$$P = \frac{W}{t} = \frac{ItV}{t} = IV$$

The unit of power is the *watt* (W) and equals an energy conversion rate of 1 joule per second, i.e. $1 \ \text{W} = 1 \ \text{J s}^{-1}$. In the expression $P = IV$, P will be in watts if I is in amperes and V in volts. A larger unit is the *kilowatt* (kW) which equals 1000 watts.

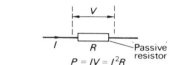

Fig. 3.28

If *all* the electrical energy is converted into heat by the device it is called a 'passive' resistor and the rate of production of heat will also be IV. If its resistance is R, Fig. 3.28, then since $R = V/I$ we have

$$P = IV$$
$$= \frac{V}{R} \cdot V = \frac{V^2}{R}$$
$$= I \cdot IR = I^2R$$

There are thus three alternative expressions for power but the last two are only true when all the electrical energy is changed to heat. The first, $P = IV$, gives the rate of production of all forms of energy. For example if the current in an electric motor is 5 A when the applied p.d. is 10 V then 50 W of electric power is supplied to it. However, it may only produce 40 W of mechanical power, the other 10 W being the rate of production of heat by the motor windings due to their resistance.

(a) *Heating elements and lamp filaments.* The expression $P = V^2/R$ shows that for a *fixed supply p.d.* of

V, the rate of heat production by a resistor increases as R decreases. Now $R = \rho l/A$, therefore $P = V^2 A/\rho l$ and so where a high rate of heat production at constant p.d. is required, as in an electric fire on the mains, the heating element should have a large cross-section area A, a small resistivity ρ and a short length l. It must also be able to withstand high temperatures without oxidizing in air (and becoming brittle). Nichrome is the material which best satisfies all these requirements.

Electric lamp filaments have to operate at even higher temperatures if they are to emit light. In this case, tungsten, which has a very high melting point (3400 °C), is used either in a vacuum or more often in an inert gas (nitrogen or argon). The gas reduces evaporation of the tungsten (why?) and prevents the vapour condensing on the inside of the bulb and blackening it. In modern projector lamps there is a little iodine which forms tungsten iodide with the tungsten vapour and remains as vapour when the lamp is working, thereby preventing blackening.

(*b*) *Fuses.* When current flows in a wire its temperature rises until the rate of loss of heat to the surroundings equals the rate at which heat is produced. If this temperature exceeds the melting point of the material of the wire, the wire melts. A fuse is a short length of wire, often tinned copper, selected to melt when the current through it exceeds a certain value. It thereby protects a circuit from excessive currents.

It can be shown (see question 21, p. 62) that:

(*i*) the temperature reached by a given wire depends only on the current through it and is independent of its length (provided it is not so short for heat loss from the ends where it is supported, to matter); and

(*ii*) the current required to reach the melting point of the wire increases as the radius of the wire increases.

Fuses which melt at progressively higher temperatures can thus be made from the same material by using wires of increasing radius.

(*c*) *The kilowatt-hour* (kWh). For commercial purposes the kilowatt-hour is a more convenient unit of electrical energy than the joule.

The kilowatt-hour is the quantity of energy converted to other forms of energy by a device of power 1 kilowatt in 1 hour.

The energy converted by a device in kilowatt-hours is thus calculated by multiplying the power of the device in kilowatts by the time in hours for which it is used. Hence a 3 kW electric radiator working for 4 hours uses 12 kWh of electrical energy—often called 12 'units'. How many joules are there in 1 kWh?

Wheatstone bridge

(*a*) *Theory.* The Wheatstone bridge circuit enables resistance to be measured more accurately than by the ammeter–voltmeter method (p. 42). It involves making adjustments until a galvanometer is undeflected and so, being a 'null' method, it does not depend on the accuracy of an instrument. Other known resistors are, however, required.

Four resistors P, Q, R, S are joined as in Fig. 3.29*a*. If P is the unknown resistor, Q must be known as must the values of R and S or their ratio. A sensitive galvanometer G and a cell (dry or Leclanché) are connected as shown. One or more of Q, R and S are adjusted until there is no deflection on G. The bridge is then said to be balanced and it can be shown that

$$\frac{P}{Q} = \frac{R}{S}$$

whence P can be found. The proof of this expression follows.

At balance, no current flows through G, therefore the p.d. across BD is zero and so

p.d. across AB = p.d. across AD

Also, current through P = current through $Q = I_1$ and, current through R = current through $S = I_2$. Therefore

$$I_1 \times P = I_2 \times R$$

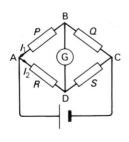

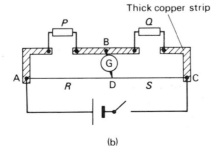

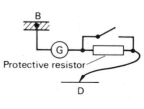

Fig. 3.29   (a)   (b)   (c)

Similarly, p.d. across BC = p.d. across DC. Therefore

$$I_1 \times Q = I_2 \times S$$

Dividing,

$$\frac{P}{Q} = \frac{R}{S}$$

It can be shown that the same condition holds if the cell and G are interchanged.

(*b*) *Metre bridge.* This is the simplest practical form of the Wheatstone bridge, Fig. 3.29*b*. The resistors R and S consist of a wire AC of uniform cross-section and 1 m long, made of an alloy such as constantan, with a resistance of several ohms. The ratio of R to S is altered by changing the position on the wire of the movable contact or 'jockey' D. The other arm of the bridge contains the unknown resistor P and a known resistor Q. Thick copper strips of low resistance connect the various parts. Figs. 3.29*a* and *b* have identical lettering to show their similarity.

The position of D is adjusted until there is no deflection on G, then

$$\frac{P}{Q} = \frac{R}{S} = \frac{\text{resistance of AD}}{\text{resistance of DC}}$$

Since the wire is uniform, resistance will be proportional to length and therefore

$$\frac{P}{Q} = \frac{\text{AD}}{\text{DC}}$$

Four practical points should be noted. Resistor Q should be chosen to give a balance point near to the centre of the wire, say between 30 and 70 cm, for three reasons. First, any errors in reading the balance lengths AD and DC are then small in comparison with their values. Second, where the resistors and metre wire are screwed or soldered to the copper strips there are 'connection' resistances which may, if desired, be determined experimentally and expressed in millimetres of bridge wire. However, when this is not done the 'end corrections', as they are called, will have least effect if neither AD nor DC is small. Third, the bridge is more sensitive near the middle since the unbalance current is larger per mm change of position.

In finding the balance point the cell key should be closed before the jockey makes contact with the wire. This is necessary because due to an effect known as 'self-induction', the currents in the circuit take a short time to grow to their steady values. During this time a momentary deflection of the galvanometer might be obtained even when the bridge is balanced for steady currents.

A high resistor should be joined in series with the galvanometer to protect it from damage whilst the balance point is being found. In the final adjustment it is shorted out and maximum sensitivity of the galvanometer obtained, Fig. 3.29*c*.

Having obtained a balance point, P and Q should be interchanged and a second pair of values for AD and DC obtained and the means taken. This helps to compensate for errors arising from non-uniformity of the wire, from the wrong positioning of the millimetre scale in relation to the wire and from 'end corrections' provided the balance point is near the centre of the wire.

Wheatstone bridge methods are unreliable for finding resistances of less than 1 ohm, due to 'connection' resistance errors becoming appreciable, and are insensitive for resistances greater than 1 megohm unless a highly sensitive galvanometer is used. A modern dial-operated form of Wheatstone bridge for the use of maintenance engineers is shown in Fig. 3.30.

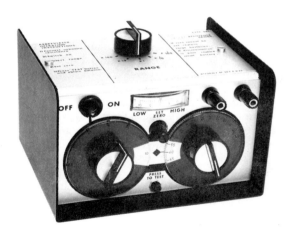

Fig. 3.30

(*c*) *Measurement of resistivity using a metre bridge.* The resistivity ρ of a material can be determined by measuring the resistance R of a known length l of wire and also its average diameter d using a micrometer screw gauge. Then $\rho = AR/l$ where $A = \pi(d/2)^2$.

(*d*) *Measurement of temperature coefficient of resistance.* This may be found for, say, copper by measuring its resistance at different temperatures with the apparatus shown in Fig. 3.31*a*. A graph of resistance against temperature is plotted. Over small temperature ranges it is a straight line and from it the temperature coefficient α is calculated using $\alpha = (R_\theta - R_0)/R_0\theta$, Fig. 3.31*b*.

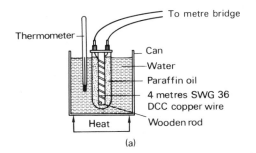

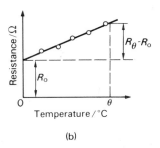

Fig. 3.31   (a)

(b)

Potentiometer

(a) *Theory*. A potentiometer is an arrangement which measures p.d. accurately. It can be adapted to measure current and resistance.

In its simplest form it consists of a length of resistance wire AB of *uniform* cross-section area, lying alongside a millimetre scale, and through which a *steady* current is maintained by a cell, called the driver cell, Fig. 3.32. (This is usually an accumulator because it gives a steady current for a long time.) As a result there is a p.d. between any two points on the wire which is proportional to their distance apart. Part of the p.d. across AB is tapped off and used to counter-balance the p.d. to be measured. If the p.d. across AB is 2 volts and if the wire is uniform and the current steady, what p.d.s. can be tapped off between (*i*) A and P, (*ii*) A and Q, (*iii*) A and R, (*iv*) Q and B?

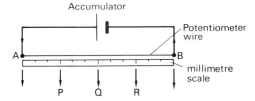

Fig. 3.32

In practice the unknown p.d. V_1 is connected with its positive side to X in Fig. 3.33a if the positive terminal of the driver cell is joined to A, as shown. The negative side of the unknown p.d. goes to a galvanometer G and a jockey C. In AX and YGC there are thus two p.d.s. trying to cause current flow. The one tapped off between A and C from the potentiometer wire acts in an anticlockwise direction whilst the unknown p.d. V_1 tends to drive current in a clockwise direction. The direction of current flow through G therefore depends on whether V_1 is greater or less than the p.d. across AC. When the position of the jockey on the wire is such that there is *no current* through G, these two p.d.s are

equal and the potentiometer is said to be balanced. The balance length l_1 is then measured.

If the resistance of AB per cm is r then the resistance of the balance length l_1 (in cm) is $l_1 r$ and if the steady current through AB from the driver cell is I, we have

unknown p.d. V_1 = p.d. tapped off at balance

$$\therefore \quad V_1 = I \times l_1 r$$

Since I and r are constants

$$V_1 \propto l_1$$

Let the p.d. across the whole potentiometer wire AB be V, then if $l =$ AB

$$V = I \times lr$$

Therefore,

$$\frac{V_1}{V} = \frac{Il_1 r}{Ilr} = \frac{l_1}{l}$$

$$\therefore \quad V_1 = \frac{l_1}{l} \cdot V$$

Knowing V, l and l_1, we can find V_1. When the driver cell is an accumulator of low internal resistance, V may

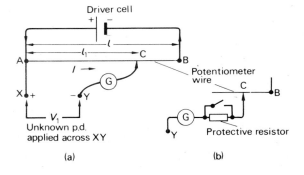

Fig. 3.33

be taken as its e.m.f. If this is 2.0 volts and if $l = 100$ cm and $l_1 = 80$ cm then $V_1 = (80/100) \times 2 = 1.6$ volts. Where higher accuracy is required a slightly different procedure is adopted as we shall presently.

A potentiometer is a kind of voltmeter but is much more accurate than the best dial instrument since its 'scale' (i.e. the wire) may be made as long as we wish and its adjustment being a 'null' method does not depend on the calibration of the galvanometer. It has the further advantage of not altering the p.d. to be measured since at balance no current is drawn by it from the unknown p.d.; it behaves like a voltmeter of infinite resistance. On the other hand the wire form considered here is bulky and slow to use compared with an ordinary voltmeter. Modern potentiometers are dial-operated.

(b) *Practical points.* The following procedure should be noted.

(i) With a large protective resistor in series with the galvanometer, Fig. 3.33b, the circuit is tested by first placing the jockey on one end of the wire and then on the other; the deflections should be in opposite directions. If they are not then either the unknown p.d. is connected the wrong way round or the p.d. across the whole wire is less than the unknown p.d.

(ii) The balance point is found by repeating (i) for pairs of points that get progressively closer together, the protective resistor being shorted out near balance. The jockey should not be drawn along the wire or its uniformity will be lost.

(iii) The balance length is measured from the end A of the wire and should be reasonably long so that the percentage error in measuring it is small.

(c) *Comparison of e.m.f.s of two cells.* The first cell, of e.m.f. E_1, is connected to XY and its balance length l_1 found, Fig. 3.34a. At balance no current is drawn from the cell and so

p.d. at terminals of cell = its e.m.f. = p.d. across AC

Hence

$$E_1 = \frac{l_1}{l} \cdot V$$

where $l = $ AB and $V = $ p.d. across AB due to driver cell.

Replacing the first cell by the second of e.m.f. E_2 and finding its balance length l_2 we have similarly

$$E_2 = \frac{l_2}{l} \cdot V$$

$$\therefore \quad \frac{E_1}{E_2} = \frac{l_1}{l_2}$$

If one of the cells is a Weston standard cell (p. 59) the e.m.f. of the other cell can be calculated accurately and, in effect, the potentiometer is calibrated. The Weston cell maintains a constant e.m.f. E_s of 1.0186 volts (at 20 °C) provided the current taken from it (at out of balance) is less than 10 μA. To ensure this a very high resistance (0.5 MΩ) is connected in series with it, Fig. 3.34b, and shorted out when the balance point is approached. Note that the high resistor does not affect the *position* of balance (since no current flows then) but only the precision with which it can be found.

(d) *Calibrating a voltmeter.* The potentiometer is first calibrated using a standard cell as explained in (c) so that the p.d. per cm of potentiometer wire is known. The circuit of Fig. 3.35 is suitable for calibrating a voltmeter in the range 0.2 to 2.0 volts. In the circuit below XY the variable resistor acts as a potential divider and enables various p.d.s to be applied to both the voltmeter (V) and the potentiometer. The p.d. can be calculated for each balance length.

Design a circuit to calibrate a voltmeter reading up to 10 volts using a 2 volt potentiometer.

In the circuit of Fig. 3.35 the balance lengths for p.d.s of less than 0.2 volt on a potentiometer wire 1–2 metres

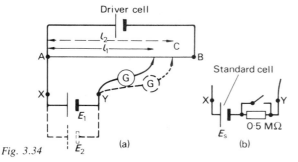

Fig. 3.34

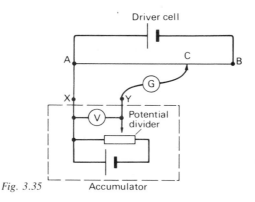

Fig. 3.35

long would be too small for reasonable accuracy. The measurement of small p.d.s (and e.m.f.s) is achieved by reducing the p.d. across the whole potentiometer by a method similar to that now to be considered.

(*e*) *Measuring the e.m.f. of a thermocouple.* The e.m.f.s of thermocouples (p. 59) are of the order of a few millivolts and to ensure that an appreciable balance length is obtained when measuring them a high resistance R is joined in series with the potentiometer wire, Fig. 3.36. The value of R is chosen so that the p.d. across the whole wire AB is just greater than the maximum e.m.f. to be measured.

Let V be the p.d. across R and AB in series and let R_{AB} be the resistance of AB then the current I through AB due to the driver cell is

$$I = \frac{V}{R + R_{AB}}$$

and

$$V_{AB} = IR_{AB} = \frac{V}{R + R_{AB}} R_{AB}$$

If $V = 2$ volts, $R_{AB} = 2$ ohms (as measured by a metre bridge) and $R = 1998$ ohms (a resistance box) then

$$V_{AB} = \frac{2 \times 2}{1998 + 2} = \frac{4}{2000} = \frac{2}{1000} \text{ volts}$$

$$= 2 \text{ mV}$$

If AB is 100 cm long, the p.d. across every cm is $2/100 = 0.02$ mV per cm and the e.m.f. of a thermocouple can then be found from the balance length so long as it does not exceed 2 mV. For example, the e.m.f. of a copper–iron thermocouple can be measured for different hot junction temperatures.

Where greater accuracy is required the potentiometer is calibrated using a standard cell and the circuit of Fig. 3.36 has to be modified as described in many practical books.

(*f*) *Calibrating an ammeter.* The potentiometer is first calibrated using a standard cell so that the p.d. per cm of the wire is known. The circuit is then connected as in Fig. 3.37*a* and the potentiometer used to measure the p.d. V across a suitable standard resistor R in series with the ammeter (A) to be calibrated. The current $I = V/R$; R is chosen to give a balance point near the end of the wire, i.e. V should be nearly 2 volts so that if $I = 0.5$ ampere then $R = 2/0.5 = 4$ ohms.

Unknown connection resistances at the terminals of two-terminal resistors are important in resistances of a few ohms and four-terminal types should be used if high accuracy is required in this type of measurement. The specified resistance exists between L and M in Fig. 3.37*b*; the resistance of the wires connecting L and M to the p.d. terminals does not affect things since at balance the current through them is zero.

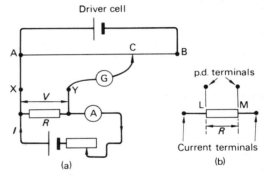

Fig. 3.37

(*g*) *Comparison of resistances.* The ratio of two resistances R_1 and R_2 can be found accurately by using a potentiometer to compare the p.d.s V_1 and V_2 across each when they are carrying the same current I. In the circuit of Fig. 3.38 X and Y are joined across R_1 and R_2 in turn and the corresponding balance lengths l_1 and l_2 measured.

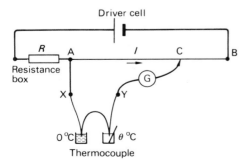

Fig. 3.36

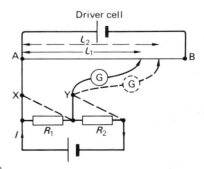

Fig. 3.38

Then

$$\frac{l_1}{l_2} = \frac{V_1}{V_2} = \frac{IR_1}{IR_2} = \frac{R_1}{R_2}$$

To obtain balance lengths near the end of the wire V_1 and V_2 should approach 2 volts. If the value of I needed for this causes overheating of R_1 and R_2 then a smaller current must be used and a suitable resistor connected in series with the wire to make l_1 and l_2 large.

Using four-terminal-type resistors for R_1 and R_2 the potentiometer method is very suitable for resistances of less than 1 ohm since the resistances of connecting wires and terminal connections do not affect the result as they can in bridge methods.

(*h*) *Measuring internal resistance of a cell.* The balance length l is found first with the cell on open circuit, Fig. 3.39 (solid lines). The p.d. across XY therefore equals the p.d. at the terminals of the cell on open circuit, i.e. its e.m.f. E; therefore $E \propto l$. A known resistance R is then connected across the cell (dotted lines) and if l_1 is the new balance length, the p.d. across XY falls and equals the p.d. V across the cell when it maintains current through R; therefore $V \propto l_1$.

Hence

$$\frac{E}{V} = \frac{l}{l_1}$$

If the current through R (at balance) is I and r is the internal resistance of the cell, Ohm's law applied first to the whole circuit and then to R alone gives

$$E = I(R + r) \quad \text{and} \quad V = IR$$

$$\therefore \quad \frac{E}{V} = \frac{R + r}{R} = \frac{l}{l_1}$$

$$\therefore \quad 1 + \frac{r}{R} = \frac{l}{l_1}$$

$$\frac{r}{R} = \frac{l}{l_1} - 1 = \frac{l - l_1}{l_1}$$

Hence r can be calculated.

Electrolysis

(*a*) *Ionic theory.* Liquids which undergo chemical change when a current passes through them are called *electrolytes* and the process is known as *electrolysis*. Solutions in water of acids, bases and salts are electrolytes; liquids that conduct without suffering chemical decomposition are non-electrolytes and molten metals such as mercury are examples.

Conduction in an electrolyte is considered to be due to the movement of positive and negative ions. There is evidence from X-ray crystallography that, in the solid state, compounds such as sodium chloride consist of regular structures of positive and negative ions (p. 10) held together by electrostatic forces. We believe that when such substances are dissolved in water (and some other solvents) the interionic forces are weakened so much by the water that the ions can separate and move about easily in the solution. *Ionization* or *dissociation* is said to have occurred as a result of solution; the ionization of other salts, bases and acids is similarly explained. (*Note.* Whilst ions may exist in a solid, they are not free to move and so we do not consider the solid is ionized.)

In Fig. 3.40 when a p.d. is applied to the plates that dip into the electrolyte, i.e. the electrodes, an electric field is created which causes positive ions to move towards one electrode (the *cathode*), while negative ions are attracted to the other (the *anode*). The two streams of oppositely charged ions, drifting slowly in opposite directions (see p. 38), constitute the current in the electrolyte. At the electrodes, for conduction to continue, either (*i*) the ions must be discharged, i.e. give up their excess electrons if they are negative or accept electrons if they are positive, or (*ii*) fresh ions must be formed from the electrode and pass into solution. In any event the anode must gain electrons and the cathode lose them to maintain electron flow in the external circuit. After being discharged the ions

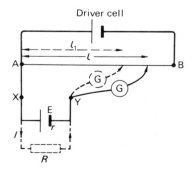

Fig. 3.39

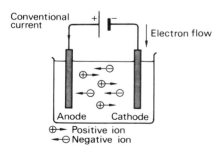

Fig. 3.40

usually come out of solution and are liberated as uncharged matter, being either deposited on the electrodes or released at them as bubbles of gas. (The electrodes must be made of metal or carbon. Why?)

Consider the electrolysis of copper sulphate solution with copper electrodes. The solution contains copper ions with a double positive charge (Cu^{2+}) and sulphate ions with a double negative charge (SO_4^{2-}). Under an applied p.d. the copper ions drift to the cathode where each receives two electrons ($2e$) and forms a copper atom that is deposited on the cathode.

$$Cu^{2+} + (2e \text{ from cathode}) \longrightarrow Cu$$

Sulphate ions collect round the anode and the most likely reaction to occur there, because it involves less energy than any other, is the formation of fresh copper ions by copper atoms of the anode going into solution. The anode is thus able to acquire electrons because every copper atom must lose two electrons to form a copper ion. Also, the fresh copper ions neutralize the negatively charged sulphate ions tending to gather round the anode.

$$Cu \longrightarrow Cu^{2+} + (2e \text{ to anode})$$

The net result is that copper is deposited on the cathode and goes into solution from the anode. In general, metals and hydrogen are liberated at the cathode and non-metals at the anode.

Electrolysis is used in many industrial processes. By allowing chemical reactions to occur at different places in the same solution, it keeps the products separate and makes feasible reactions that are otherwise impossible.

(b) *Specific charge of an ion.* If the same kind of ion always has the same charge and mass then the *mass of a substance liberated or deposited in electrolysis is directly proportional to the total charge passed.* This is confirmed by experiment and is called *Faraday's first law of electrolysis.*

If m is the mass liberated and Q the charge passed for a particular substance, Q/m is a constant for the ions of that substance. It is called the *specific charge of the ion* and is expressed in coulombs per kilogram (C kg^{-1}).

For example, if in the electrolysis of copper sulphate solution a steady current $I = 1.0$ A deposits 1.2 g of copper on the cathode in a time $t = 1$ hour, the specific charge of the copper ion is given by

$$\text{specific charge} = \frac{Q}{m} = \frac{It}{m}$$

$$= \frac{(1.0 \text{ A}) \times (3600 \text{ s})}{(1.2 \times 10^{-3} \text{ kg})}$$

$$= 3.0 \times 10^6 \text{ C kg}^{-1}$$

Knowing that a copper ion carries a double electronic charge, i.e. $2e$, where $e = 1.6 \times 10^{-19}$ C, the mass m of a copper ion is then given by

$$\text{specific charge} = \frac{\text{charge on ion}}{\text{mass of ion}} = \frac{2e}{m}$$

that is, $$3.0 \times 10^6 = \frac{2 \times 1.6 \times 10^{-19}}{m}$$

hence, $$m = \frac{2 \times 1.6 \times 10^{-19}}{3.0 \times 10^6}$$

$$= 1.1 \times 10^{-25} \text{ kg}$$

Large-scale measurements on matter in bulk thus enable us, with the help of theory, to obtain information about atomic-sized particles. The results agree well with those from more direct methods (e.g. mass spectrometer, p. 475).

(c) *The Faraday constant* (F). This is the quantity of electric charge which liberates *one mole* of any single-charged ion. Experiment gives its value as

$$F = 9.65 \times 10^4 \text{ C mol}^{-1}$$

If e is the charge on a hydrogen ion and L is the number of ions in 1 mole of hydrogen ions, i.e. the *Avogadro constant* (p. 5), then

$$F = L\, e$$

since 1 mole of hydrogen ions is liberated by 9.65×10^4 coulombs. X-ray crystallography measurements give $L = 6.02 \times 10^{23}$ per mole and so

$$e = \frac{F}{L} = \frac{9.65 \times 10^4}{6.02 \times 10^{23}} = 1.60 \times 10^{-19} \text{ C}$$

This is the charge on a singly-charged ion and is found to be the same as that on an electron. The above expression gives one of the most accurate ways of obtaining the electronic charge e.

(d) *Ohm's law and electrolytes.* The variation of current with p.d. for an electrolyte may be investigated using the circuit of Fig. 3.41. The p.d. is varied from 0–4 volts by the potential divider and measured with a high-resistance voltmeter.

With copper sulphate solution and copper electrodes, the graph of current against p.d. is a straight line through the origin, Fig. 3.42a, and Ohm's law is fairly well obeyed. The smallest p.d. causes current to flow and supports the ionic theory assumption that electrolytes, as soon as they dissolve, split into ions which are immediately available for conduction.

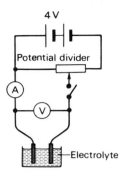

Fig. 3.41

Using water (acidulated) and platinum electrodes there is no appreciable current flow until the p.d. exceeds 1.7 volts. Thereafter increases of p.d. cause proportionate increases of current and hydrogen and oxygen are evolved at the cathode and anode respectively. The current–p.d. graph is shown in Fig. 3.42b. The virtual absence of current for p.d.s below 1.7 volts is attributed to the existence of a *back e.m.f.* of maximum value 1.7 volts which the applied p.d. must exceed before the electrolyte conducts. The back e.m.f. is due to polarization, i.e. the accumulation at the electrodes of products of electrolysis formed when the circuit is first made. In this case hydrogen at the cathode and oxygen at the anode effectively replace the platinum electrodes by gas electrodes and act as a chemical cell with an e.m.f. which opposes the applied p.d. (If the switch in Fig. 3.41 is opened the back e.m.f. is recorded on the voltmeter and falls rapidly.) AB in the graph of Fig. 3.42b is a straight line showing that if allowance is made for the back e.m.f., acidulated water obeys Ohm's law. The equation of AB is

$$V - E = IR$$

where R is the resistance of the electrolyte, I is the current when the applied p.d. is V and E is the back e.m.f.

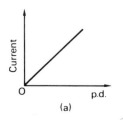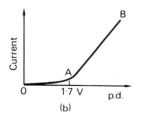

Fig. 3.42

Electric cells

These convert chemical energy into electrical energy and consist of two different metals (or a metal and carbon) separated from each other by an electrolyte. Their e.m.f. depends on the nature and concentration of the chemicals used, their size affects the internal resistance and the amount of electrical energy they can supply, i.e. their *capacity*.

Many different cells have been invented since the first was made by Volta at the end of the eighteenth century. Volta's *simple cell* consisted of plates of copper and zinc in dilute sulphuric acid and had an e.m.f. of about 1 V.

(*a*) *Primary cells.* In general these have to be discarded after use and are popularly called 'dry' batteries though this description is not strictly correct. Some types used today are listed in Table 3.2.

The graphs in Fig. 3.43 show roughly how the e.m.f. of different cells of size AA(U12—Pencell) vary when supplying moderate currents (e.g. 30 mA).

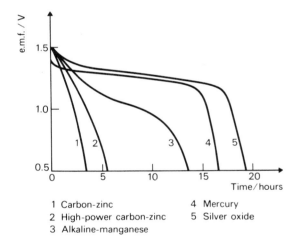

1 Carbon-zinc	4 Mercury
2 High-power carbon-zinc	5 Silver oxide
3 Alkaline-manganese	

Fig. 3.43

(*b*) *Secondary cells.* These can be charged and discharged repeatedly (but not indefinitely) and are generally called accumulators. They supply 'high' continuous currents depending on their *capacity*, which is expressed in ampere-hours (A h) for a particular discharge rate. For example, a cell with a capacity of 30 A h at the 10 hour rate will sustain a current of 3 A for 10 hours, but whilst 1 A would be supplied for more than 30 hours, 6 A would not flow for 5 hours.

The *lead-acid* type is commonest. It has an e.m.f. of 2.0 V which it maintains until it is nearly discharged. A 12 V car battery consists of six in series and may supply

Table 3.2

Cell	e.m.f.	Properties	Use
Carbon–zinc	1.5 V	Relatively cheap; most popular type; e.m.f. falls as current increases; best for low currents or occasional use.	Torch Radio
Alkaline–manganese	1.5 V	Medium price; e.m.f. does not fall so much in use; long 'shelf' life; better for higher continuous currents; lasts about four times longer than same size carbon–zinc cell.	Radio Calculators
Mercury	1.4 V	Large capacity for their size; made as 'buttons'; e.m.f. almost constant till discharged; best for low current use; expensive.	Hearing aids Cameras Watches Calculators
Silver oxide	1.5 V	Similar to mercury cell	As for mercury cell
Weston standard	1.0186 V at 293 K	e.m.f. constant if current does not exceed 10 μA	Laboratory standard of e.m.f.

a current of several hundred ampere for a few seconds to the starter motor. The internal resistance of one cell is of the order of 0.01 Ω, consequently for ordinary currents, the 'lost' volts are negligible.

The *nickel-iron* (*nife*) cell has an e.m.f. of 1.2 V which falls during use but its weight is about half that of a lead-acid cell of the same capacity. Batteries of nife cells are used to drive electrically propelled vehicles such as milk floats.

The *nickel-cadmium* (*nicad*) cell also has an e.m.f. of 1.2 V but it maintains this value in use. 'Button' types are used as rechargeable batteries for calculators and shavers. They are very expensive.

Thermoelectric effect

If two different metals such as copper and iron are joined in a circuit and their junctions are kept at different temperatures, a small e.m.f. is produced and current flows, Fig. 3.44. The effect is known as the thermoelectric or *Seebeck effect* and the pair of junctions is called a *thermocouple*.

The value of the thermo–e.m.f. depends on the metals used and the temperature difference between the junctions; the e.m.f.–temperature difference curve is always approximately a parabola. Fig. 3.45 shows the curves for (*a*) copper–iron and (*b*) iron–constantan which may be obtained using a potentiometer as described previously (p. 55). The iron–constantan curve (although part of a parabola) is almost linear over a large range and produces about 10 times the e.m.f. of a copper–iron couple. The temperature of the hot junction at which the e.m.f. is a maximum is called the *neutral temperature.*

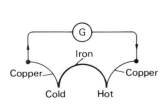

Fig. 3.44

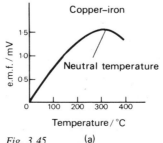

Fig. 3.45 (a)

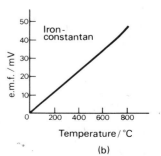

(b)

Thermocouples are used as thermometers particularly for measuring varying temperatures or the temperature at a point (p. 64).

The direct conversion of heat into electricity by metal thermocouples is a very inefficient process but better couples are now available based on semiconductors such as iron disilicide. On account of their reliability, long life and cheapness, these are suitable as small power supply units in space satellites, weather buoys and weather ships. Radiation from a radioactive source (e.g. strontium 90) in the unit falls on the hot junction and produces the necessary temperature rise in it.

QUESTIONS

Current and charge

1. If the heating element of an electric radiator takes a current of 4.0 A, what charge passes each point every minute? If the charge on an electron is 1.6×10^{-19} C how many electrons pass a given point in this time?

2. (a) If the density of copper is 9.0×10^3 kg m^{-3} and 63.5 kg of copper contains 6.0×10^{26} atoms, find the number of 'free' electrons per cubic metre of copper assuming that each copper atom has one 'free' electron.
(b) How many 'free' electrons will there be in a 1.0 m length of copper wire of cross-section area 1.0×10^{-6} m^2 (i.e. 1.0 mm^2)?
(c) Taking the charge on an electron as 1.6×10^{-19} C, what is the total charge of the 'free' electrons per metre of wire?
(d) Assuming that the 'free' electrons are responsible for conduction, how long will the charge in (c) take to travel 1 m when a current of 2.0 A flows?
(e) What is the drift velocity of the 'free' electrons?

3. Explain in terms of the motion of free electrons what happens when an electric current flows through a metallic conductor.
A metal wire contains 5.0×10^{22} electrons per cm^3 and has a cross-sectional area of 1.0 mm^2. If the electrons move along the wire with a mean drift velocity of 1.0 mm s^{-1}, calculate the current in amperes in the wire if the electronic charge is 1.6×10^{-19} C.   (O. and C. part qn.)

Potential difference : resistance : meters

4. (a) What is the p.d. between two points in a circuit if 200 joules of electrical energy are changed to other forms of energy when 25 coulombs of electric charge pass? If the charge flows in 10 seconds, what is the current?
(b) What is the p.d. across an immersion heater which changes 3.6×10^3 joules of electrical energy to heat every second and takes a current of 15 amperes?

5. Three voltmeters V, V_1 and V_2 are connected as in Fig. 3.46.

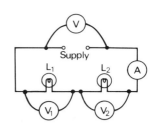

Fig. 3.46

(a) If V reads 12 volts and V_1 reads 8.0 volts, what does V_2 read?
(b) If the ammeter A reads 0.50 ampere, how much electrical energy is changed to heat and light by L_1 in 1 minute?
(c) Copy the diagram and mark with a + the positive terminals of the voltmeters and ammeter for correct connection.

6. A p.d. of V drives current through two resistors of 2 ohms and 3 ohms joined in series, Fig. 3.47.
(a) If voltmeter V_1 reads 4 volts, what is the current in the 2 ohm resistor?
(b) What is the current in the 3 ohm resistor?
(c) What does voltmeter V_2 read?
(d) What is the value of V?
(e) Find the value of the single equivalent resistor which, if it replaced the 2 ohm and 3 ohm resistors in series, would allow the same current to flow when joined to the same p.d. V.

Fig. 3.47

7. Two resistors of 3 ohms and 6 ohms are connected in parallel across a p.d. of 6 volts, Fig. 3.48a.
(a) How are I, I_1 and I_2 related?
(b) What is the p.d. across each resistor?
(c) Find the values of I, I_1 and I_2 in amperes.
(d) Find the value of the single equivalent resistor R which, if it replaced the 3 ohm and 6 ohm resistors in parallel, would allow the value of I found in (c) to flow when the p.d. across it is 6 volts, Fig. 3.48b.

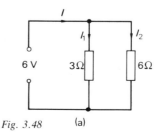
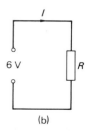

Fig. 3.48   (a)   (b)

8. (*a*) In the circuit of Fig. 3.49*a* what is the p.d. across (*i*) AB, (*ii*) BC?

(*b*) What do these p.d.s become when the circuit is altered as in Fig. 3.49*b*?

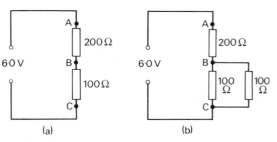

(a) (b)

Fig. 3.49

9. The circuit shown in Fig. 3.50 is used to provide a variable negative voltage, −*V*, with respect to earth, from a −200 V supply to a device drawing negligible current.

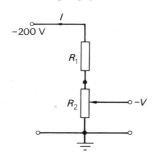

Fig. 3.50

The value of the fixed resistor, R_1, is known only to an accuracy (or tolerance) of ±10% and that of the variable resistor, R_2, to ±20%. Choose values of R_1 and R_2 from the table shown, in order to satisfy the following conditions.

(*a*) The current in R_1 must be as low as possible but greater than 1 mA.

(*b*) It must be possible to vary the voltage (− *V)* over a range of at least 0 to −20 V.

	Resistor values					
R_1 (±10%)/kΩ	100	150	220	330	470	680
R_2 (±20%)/kΩ	10	25	50			

(*J.M.B. Eng. Sc.*)

10. A resistor of 500 ohms and one of 2000 ohms are placed in series with a 60 volt supply. What will be the reading on a voltmeter of internal resistance 2000 ohms when placed across (*a*) the 500 ohms resistor, (*b*) the 2000 ohms resistor?

11. Measurements of p.d. and current on five different 'devices' A, B, C, D, E gave the graphs in Fig. 3.51. Suggest, with reasons, what each might be.

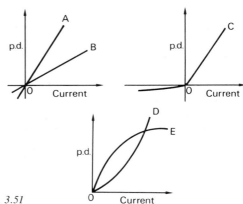

Fig. 3.51

12. If a moving-coil ammeter gives a full-scale deflection for a current of 15 mA and has a resistance of 5.0 ohms, how would you adapt it so that it could be used (*i*) as a voltmeter reading to 1.5 V, (*ii*) as an ammeter reading to 1.5 A? (*W. part qn.*)

Resistivity : temperature coefficient

13. Assuming that the resistivity of copper is half that of aluminium and that the density of copper is three times that of aluminium, find the ratio of the masses of copper and aluminium cables of equal resistance and length.

14. A wire has a resistance of 10.0 ohms at 20.0 °C and 13.1 ohms at 100 °C. Obtain a value for its temperature coefficient of resistance.

15. When the current passing through the Nichrome element of an electric fire is very small its resistance is found to be 50.9 Ω, room temperature being 20.0 °C. In use the current is 4.17 A on a 240 V supply. Calculate (*a*) the rate of energy conversion by the element, (*b*) the steady temperature reached by it. (The temperature coefficient of resistance of Nichrome may be taken to have the constant value 1.70×10^{-4} °C^{-1} over the temperature range involved.)

Electrical energy : e.m.f. : internal resistance

16. How much electrical energy does a battery of e.m.f. 12 volts supply when

(*a*) a charge of 1 coulomb passes through it,

(*b*) a charge of 3 coulombs passes through it,

(*c*) a current of 4 amperes flows through it for 5 seconds?

17. Three accumulators each of e.m.f. 2 volts and internal resistance 0.01 ohm are joined in series and used as the supply for a circuit.

(*a*) What is the total e.m.f. of the supply?

(*b*) How much electrical energy per coulomb is supplied using (*i*) one accumulator, (*ii*) all three accumulators?

(*c*) What is the total internal resistance of the supply?

(*d*) What current would be driven by the supply through a resistance of 1.97 ohms?

18. (a) A flashlamp bulb is marked '2.5 V, 0.30 A' and has to be operated from a dry battery of e.m.f. 3.0 V for the correct p.d. of 2.5 V to be produced across it. Why?

(b) How much heat and light energy is produced by a 100 watt electric lamp in 5 minutes?

(c) What is the resistance of a 240 V, 60 W bulb?

19. The p.d. across the terminals XY of a box is measured by a very-high-resistance voltmeter V. In arrangement A, Fig. 3.52, the voltmeter reads 105 volts. In arrangement B with the same box, the reading of the voltmeter is 100 volts. The inside of the box is not altered between the two arrangements.

Explain what you think may be in the box.

(You may, if you want to, draw in extra features on (copies of) the diagrams above to use in your explanation.)

(O. and C. Nuffield)

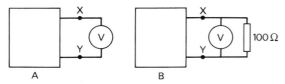

Fig. 3.52

20. What quantitative evidence could you bring forward in favour of the view that a cell may be looked upon as possessing a definite internal resistance? Describe an experiment you would perform to obtain such evidence.

A cell of e.m.f. 2.0 volts and internal resistance 1.0 ohm is connected in series with an ammeter of resistance 1.0 ohm and a variable resistor of R ohm. A voltmeter of resistance 1.0×10^2 ohm is connected across R. Find the value of R and the ammeter reading when the voltmeter reads 1.0 volt. Find also the power delivered to the external circuit. (S.)

21. By considering a wire of radius r, length l and resistivity ρ, through which a current I flows show that

(a) the rate of production of heat by it is $I^2 \rho l/\pi r^2$,

(b) the rate of loss of heat from its surface is $2\pi r l h$ where h is the heat lost per unit area of surface per second,

(c) the steady temperature it reaches is independent of its length and depends only on I.

Wheatstone bridge : potentiometer

22. Two resistance coils, P and Q, are placed in the gaps of a metre bridge. A balance point is found when the movable contact touches the bridge wire at a distance of 35.5 cm from the end joined to P. When the coil Q is shunted with a resistance of 10 ohms, the balance point is moved through a distance of 15.5 cm. Find the values of the resistances P and Q. (W. part qn.)

23. How would you investigate the way in which the current through a metal wire depends on the potential difference between its ends? What conditions should be fulfilled if Ohm's law is to hold?

Explain the theory of the Wheatstone bridge method of comparing resistances.

In an experiment with a simple metre bridge, the unknown X is kept in the left-hand gap and there is a fixed resistance in the right-hand gap. X is heated gradually, and when its temperature is 30 °C, the balance point on the bridge is found to be 51.5 cm from the left-hand end of the slide wire. When its temperature is 100 °C the balance point is 54.6 cm from that end. Find the temperature coefficient of resistance for the material of X, and calculate where the balance point would be if X were cooled to 0 °C. (O.)

24. From the adoption of the fundamental units metre, kilogram, second, trace the steps necessary to define the volt and the ohm in terms of the ampere.

Discuss the suitability of (a) a moving-coil voltmeter, and (b) a slide-wire potentiometer for determining the potential differences in an experiment designed to verify Ohm's law.

Four resistors AB, BD, CD and DA of resistance 4.0 ohms, 8.0 ohms, 4.0 ohms and 8.0 ohms respectively are connected to form a closed loop, and a 6.0 volt battery of negligible resistance is connected between A and C. Calculate (i) the potential difference between B and D and (ii) the value of the additional resistance which must be connected between A and D so that no current flows through a galvanometer connected between B and D. (J.M.B.)

25. Explain in detail how you would measure a small e.m.f. such as that of a thermocouple using a potentiometer method.

A 2 volt cell is connected in series with a resistance R ohms and a uniform wire AB of length 100 cm and resistance 4 ohms. One junction of a thermocouple is connected to A, and the other through a galvanometer to a tapping key. No current flows in the galvanometer when the key makes contact with the mid-point of the wire. If the e.m.f. of the couple was 4 millivolts what was the value of R? If the resistance of R is now increased by 4 ohms, by how much would the balance point change? (S.)

26. A two-metre potentiometer wire is used in an experiment to determine the internal resistance of a voltaic cell. The e.m.f. of the cell is balanced by the fall of potential along 90.6 cm of wire. When a standard resistor of 10.0 ohms is connected across the cell the balance length is found to be 75.5 cm. Draw a labelled circuit diagram and calculate, from first principles, the internal resistance of the cell.

How may the accuracy of this determination be improved? Assume that other electrical components are available if required. (J.M.B.)

4 Thermal properties

Temperature and thermometers

A knowledge of the thermal properties of materials is desirable when deciding, for example, what to use for making an electric storage heater or what to use as lagging in a refrigerator. Before studying some of these properties, certain basic ideas will first be considered.

(*a*) *Defining a temperature scale*. Temperature is sometimes called the degree of hotness and is a quantity which is such that when two bodies are placed in contact, heat flows from the body at the higher temperature to the one at the lower temperature. To measure temperatures, a *temperature scale* has to be established as follows.

(*i*) Some property of matter is selected whose value varies continuously with the degree of hotness. Suitable properties must be accurately measurable over a wide range of temperature with fairly simple apparatus and vary in a similar way to many other physical properties.

(*ii*) Two standard degrees of hotness are chosen— called the *fixed points*—and numbers assigned to them. On the Celsius method of numbering (until 1948 known as the centigrade method) the lower fixed point is the *ice point*, i.e. the temperature of pure ice in equilibrium with air-saturated water at standard atmospheric pressure[1] and is designated as 0 degrees Celsius (0 °C). The upper fixed point is the *steam point*, i.e. the temperature at which steam and pure boiling water are in equilibrium at standard atmospheric pressure and is taken as 100 °C.

(*iii*) The values X_{100} and X_0 of the temperature-measuring property are found at the steam and ice points respectively and $(X_{100} - X_0)$ gives the *fundamental interval* of the scale. If X_θ is the value of the property at some other temperature θ which we wish to know then the value of θ in °C is given by the equation

$$\frac{\theta}{100} = \frac{X_\theta - X_0}{X_{100} - X_0}$$

This equation *defines* temperature θ in °C on the scale based on this particular temperature-measuring property. Note that it has been defined so that equal increases in the value of the property represent equal increases of temperature, i.e. the temperature scale is *defined* so that the property varies uniformly or linearly with temperature measured on its own scale.

Some thermometers using different temperature-measuring properties will now be considered briefly.

(*b*) *Mercury-in-glass thermometer*, Fig. 4.1. The change in length of a column of mercury in a glass capillary tube was one of the first thermometric properties to be chosen. If l_0 and l_{100} are the lengths of a mercury column at 0 °C and 100 °C respectively and if l_θ is the length at some other temperature θ then θ in °C is *defined* on the mercury-in-glass scale by the equation

$$\frac{\theta}{100} = \frac{l_\theta - l_0}{l_{100} - l_0}$$

For example, if a mercury thread has lengths 5.0 cm

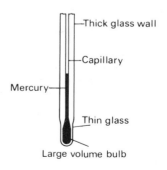

[1] Standard atmospheric pressure is defined to be 1.013×10^5 Pa (1 Pa = 1 pascal = 1 N m^{-2}) and equals the pressure at the foot of a column of mercury 760 mm high of specified density and subject to a particular value of g.

Fig. 4.1

and 20 cm at the ice and steam points and is 8.0 cm long at another temperature θ then

$$\frac{\theta}{100} = \frac{8.0 - 5.0}{20 - 5.0} = \frac{3.0}{15}$$

$$\therefore \quad \theta = \frac{3.0}{15} \times 100 = 20 \ ^\circ\text{C}$$

Inaccuracies arise in mercury thermometers from (i) non-uniformity of the bore of the capillary tube, (ii) the gradual change in the zero due to the bulb shrinking for a number of years after manufacture, and (iii) the mercury in the stem not being at the same temperature as that in the bulb.

Properties of this and other types of thermometer are summarized in Table 4.1, opposite.

(c) *Constant-volume gas thermometer.* If the volume of a fixed mass of gas is kept constant, its pressure changes appreciably when the temperature changes. A temperature θ in °C is *defined* on the constant-volume gas scale by the equation

$$\frac{\theta}{100} = \frac{p_\theta - p_0}{p_{100} - p_0}$$

where p_0, p_θ and p_{100} are the pressures at the ice point, the required temperature θ and the steam point.

A simple constant-volume thermometer is shown in Fig. 4.2. The gas (air in school models, hydrogen, helium or nitrogen in more accurate versions) is in bulb A which is at the temperature to be measured. As the temperature increases the gas expands pushing the mercury down in B and up in tube C. By raising C the mercury level in B is restored to the reference mark R and the volume of gas thus kept constant. The gas pressure is then $h + H$ where H is atmospheric pressure.

In accurate work corrections are made for (i) the gas in the 'dead-space' D not being at A's temperature, (ii)

thermal expansion of A and (iii) capillary effects at the mercury surfaces (see p. 200).

(d) *Platinum resistance thermometer.* The electrical resistance of a pure platinum wire increases with temperature (by about 40% between the ice and steam points) and since resistance can be found very accurately it is a good property on which to base a temperature scale. A temperature θ in °C on the platinum resistance scale is *defined* by the equation

$$\frac{\theta}{100} = \frac{R_\theta - R_0}{R_{100} - R_0}$$

where R_0 and R_{100} are the resistances of the platinum wire at the ice and steam points respectively and R_θ is the resistance at the temperature required.

A platinum resistance thermometer, Fig. 4.3a, consists of a fine platinum wire wound on a strip of mica (an electrical insulator) and connected to thick copper leads. A pair of identical short-circuited dummy leads are enclosed in the same silica tube (which can withstand high temperatures) and compensate exactly for changes in resistance with temperature of the leads to the platinum wire. The thermometer is connected to a Wheatstone bridge circuit (p. 51) as in Fig. 4.3b and if $P = Q$, the resistance of the platinum wire equals that of S.

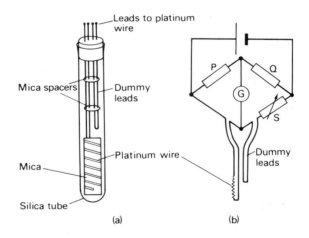

Fig. 4.3

(e) *Thermocouple thermometer.* The thermoelectric effect was considered on p. 59 and is widely used to measure temperature. If great accuracy is not required, especially at high temperatures, the thermocouple can be connected across a galvanometer (rather than to a potentiometer). The meter may be marked to read

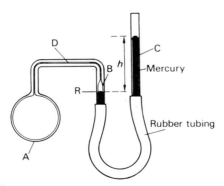

Fig. 4.2

Table 4.1

Thermometer	Range/°C	Comments
Mercury-in-glass	−39 to 500	Simple, cheap, portable, direct reading but not very accurate. Everyday use, clinical work, and weather recording
Constant-volume gas	−270 to 1500	Very wide range, very accurate, very sensitive but bulky, slow to respond and not direct reading. Used as a standard to calibrate other more practical types
Platinum resistance	−200 to 1200	Wide range, very accurate but unsuitable for rapidly changing temperatures because of large heat capacity. Best for small steady temperature differences
Thermocouple	−250 to 1500	Wide range, fairly accurate, robust and compact. Widely used in industry for rapidly changing temperatures and temperatures at a 'point'
Pyrometer	above 1000	The only thermometer for temperatures above 1500 °C

temperatures directly if it is calibrated using the known melting points of metals.

It is not usual to define a thermoelectric scale of temperature but what would be the shape of a graph of thermocouple e.m.f. against temperature measured on such a scale?

(f) Disagreement between scales: thermodynamic scale. Thermometers based on different properties give different values for the same temperature, except at the fixed points where they must agree by definition. All are correct according to their own scales and the discrepancy arises because, not unexpectedly, thermometric properties do not keep in step as the temperature changes. Thus when the length of the mercury column in a mercury-in-glass thermometer is, for example, mid-way between its 0 and 100 °C values (i.e. reading 50 °C) the resistance of a platinum resistance thermometer is not exactly mid-way between its 0 and 100 °C values.

The disagreement between scales, although small in the range 0 to 100 °C, is inconvenient. We could always state the temperature scale involved when giving a temperature, e.g. 50 °C on the mercury-in-glass scale, but a better procedure is to take one scale as a standard in which all temperatures are expressed, however they are measured. The one chosen is called the *absolute thermodynamic scale.* At present it is enough to say that it is the fundamental temperature scale in science and that the SI unit of temperature, the *kelvin* (denoted by K *not* °K) is defined in terms of it. The zero of this scale (0 K) is called *absolute zero* and it is thought that temperatures below this do not exist; certainly so far all attempts to reach it have been unsuccessful, although it has been approached very closely.

On the thermodynamic scale 0 °C = 273.15 K (273 K for most purposes) and 100 °C = 373.15 K, hence a temperature interval of one Celsius degree equals one kelvin.

Heat and internal energy

Temperature is a useful idea when describing some aspects of the behaviour of matter in bulk. It is a quantity which is measurable in the laboratory as we have just seen and is capable of perception by the sense of touch. One of the aims of modern science is to relate macroscopic (i.e. large-scale) properties such as temperature, to the masses, speeds, energies, etc., of the constituent atoms and molecules. That is, to explain the macroscopic in terms of the microscopic.

The kinetic theory regards the atoms of a solid as vibrating to and fro about their equilibrium positions, alternately attracting and repelling one another. Their energy, called *internal energy*, is considered to be partly kinetic and partly potential. The kinetic component is due to the vibratory motion of the atoms and according to the theory depends on the temperature; the potential component is stored in the interatomic bonds that are continuously stretched and compressed as the atoms vibrate and it depends on the forces between the atoms and their separation. In a solid both forms of energy are present in roughly equal amounts and there is continual interchange between them. In a gas, where the intermolecular forces are weak, the internal energy is almost entirely kinetic. The kinetic theory thus links temperature with the kinetic energy of atoms and molecules.

Heat, in science, is defined as *the energy which is transferred from a body at a higher temperature to one at a lower temperature by conduction, convection or radiation.* Like other forms of energy it is measured in joules. When a transfer of heat occurs the internal energy of the body receiving the heat increases and if the kinetic component increases, the temperature of the body rises. Heat was previously regarded as a fluid called 'caloric', which all bodies were supposed to contain. It was measured in calories—a unit now going out of use—one-thousand of which equal the dietician's Calorie.

The internal energy of a body can also be increased by doing work, i.e. by a force undergoing a displacement in its own (or a parallel) direction. Thus the temperature of the air in a bicycle pump rises when it is compressed, i.e. it becomes hotter. Work done by the compressing force has beome internal energy of the air and its temperature rises, as it would by heat transfer. It is therefore impossible to tell whether the temperature rise of a given sample of hot air is due to compression (i.e. work done) or to heat flow from a hotter body.

The expression 'heat in a body', although often used in everyday life, is thus misleading, for it may be that the body has become hot yet no heat flow has occurred. We should talk about the 'internal energy' of the body. It is sometimes said that 'the quantity of heat contained *in* a cup of boiling water is greater than that *in* a spark of white-hot metal'. What is really meant is that the boiling water has more *internal energy* and more heat *can be obtained* from it than from the spark.

The internal energy of a body may be changed in two ways: by doing *work* or by transferring *heat*. Work and heat are both concerned with energy *in the process of transfer* and when the transfer is over, neither term is relevant. Work is energy being transferred by a force moving its point of application, and the force may arise from a mechanical, gravitational, electrical or magnetic source; heat flow arises from a temperature difference.

In a wire carrying a current, electrical energy is changed to internal energy (i.e. more vigorous vibration of the atoms of the wire) and a temperature rise occurs. Subsequently this energy may be given out by the wire to the surroundings as heat. We sum up the whole process by saying that an electric current has a 'heating effect'.

Specific heat capacity

(a) *Definition of c.* Materials differ from one another in the quantity of heat needed to produce a certain rise of temperature in a given mass. The *specific heat capacity c* enables comparisons to be made. Thus if a quantity of heat δQ raises the temperature of a mass m

of a material by $\delta\theta$ then c is defined by the equation

$$c = \frac{\delta Q}{m\,\delta\theta}$$

In words, we can say that c is *the quantity of heat required to produce unit rise of temperature in unit mass.* (In modern terminology the word 'specific' before a quantity means per unit mass.)

The unit of c is joule per kilogram kelvin (J kg^{-1} K^{-1}), since in the above expression δQ is in joule, m in kilogram and $\delta\theta$ in K. Sometimes it is more convenient to consider mass in grams when the unit is J g^{-1} K^{-1}.

In the expression for c, as the temperature rise $\delta\theta$ tends to zero, c approaches the specific heat capacity at a particular temperature and experiment shows that its value for a given material is not constant but varies slightly with temperature. Mean values are thus obtained over a temperature range and to be strictly accurate this range should be stated. For ordinary purposes, however, it is often assumed constant.

The approximate specific heat capacity of water at room temperature is 4.2×10^3 J kg^{-1} K^{-1} (or 4.2 J g^{-1} K^{-1}) and is large compared with the values for most substances. At temperatures approaching absolute zero (0 K) all values of c tend to zero. Values for some other materials at ordinary temperatures are shown in Table 4.2.

Table 4.2

Mean specific heat capacities/J kg^{-1} K^{-1}			
aluminium	9.1×10^2	ice	2.1×10^3
brass	3.8×10^2	rubber	1.7×10^3
copper	3.9×10^2	wood	1.7×10^3
glass (ordinary)	6.7×10^2	alcohol	2.5×10^3
iron	4.7×10^2	glycerine	2.5×10^3
mercury	1.4×10^2	paraffin oil	2.1×10^3
lead	1.3×10^2	turpentine	1.8×10^3

High specific heat capacity is desirable in a material if only a small temperature rise is required for a given heat input. This accounts for the efficiency as a coolant of water in a car radiator and of hydrogen gas in enclosed electric generators (the latter also because of its comparatively good thermal conductivity). Severe demands are often made on the thermal properties of materials used in space travel. In the heat shield of the Apollo series of space craft, Fig. 4.4a, fibre-glass honeycomb was first bonded to the surface of the vehicle and then each cell was packed with silica-reinforced plastic, Fig. 4.4b. On re-entering the earth's atmosphere the shield, which could reach a tempera-

Fig. 4.4

ture of 5000 °C, melts and burns off. Inside the craft the temperature is no more than about 25 °C. The laboratory photograph, Fig. 4.4c, shows the effect of a hot blast at nearly 11 000 °C. What kind of thermal properties should the heat shield have?

(b) *Molar heat capacity*. If heat capacities are referred to 1 mole (p. 5) of the material instead of to unit mass, the quantity obtained by multiplying the specific heat capacity by the atomic or molecular mass (p. 5) is called the *atomic* or *molar heat capacity*. It is very nearly 25 J mol^{-1} K^{-1} for many solids. This fact is known as Dulong and Petit's law. Since 1 mole of any substance contains the same number of atoms or molecules, the heat required per atom or molecule to raise the temperature of many solids by a given amount is about the same. The implication is that the heat capacity of a solid depends on the *number* of atoms or molecules present, not on their mass and is further evidence for the atomic theory of matter.

(c) *Useful equation*. The equation defining specific heat capacity may be written

$$Q = mc(\theta_2 - \theta_1)$$

This expression is useful in heat calculations and gives the quantity of heat Q taken in by a body of mass m and mean specific heat capacity c when its temperature rises from θ_1 to θ_2. It also gives the heat lost by the body when its temperature falls from θ_2 to θ_1. In words, we can say

$$\begin{array}{l} \text{heat given out} \\ \text{(or taken in)} \end{array} = mass \times \begin{array}{l} specific\ heat \\ capacity \end{array} \times \begin{array}{l} temperature \\ change \end{array}$$

Thus if the temperature of a body of mass 0.5 kg and specific heat capacity 400 J kg^{-1} K^{-1} rises from 15 °C to 20 °C (288 K to 293 K) the heat taken in is

$$\begin{aligned} Q &= (0.5\ \text{kg}) \times (400\ \text{J kg}^{-1}\ \text{K}^{-1}) \times (5\ \text{K}) \\ &= (0.5 \times 400 \times 5)\ \text{kg J kg}^{-1}\ \text{K}^{-1}\ \text{K} \\ &= 1000\ \text{J} \end{aligned}$$

(d) *Heat capacity*. The *heat capacity* or *thermal capacity of a body* is a term in common use and is defined as *the quantity of heat needed to produce unit rise of temperature in the body*. It is measured in joules per kelvin (J K^{-1}) and from the definition of specific heat capacity it follows that

$$heat\ capacity = mass \times specific\ heat\ capacity$$

Thus the heat capacity of a copper vessel of mass 0.1 kg and specific heat capacity 390 J kg^{-1} K^{-1} is 39 J K^{-1}.

Measuring specific heat capacities

1. Electrical method

(a) *Solids*. The method is suitable for metals such as copper and aluminium that are good thermal conductors. A cylindrical block of the material is used having holes for an electric heater (12 V, 2–4 A) and a thermometer, Fig. 4.5. The mass m of the block is found and its initial temperature θ_1 recorded. The block is lagged with expanded polystyrene and a suitable steady current switched on as a stop clock is started. The voltmeter and ammeter readings V and I are noted. When the temperature has risen by about 10 K, the current is stopped and the time t taken for which it passed. The highest reading θ_2 on the thermometer is noted.

Assuming that no energy loss occurs we have,

electrical energy supplied by heater
$$= \text{heat received by block}$$
$$ItV = mc(\theta_2 - \theta_1)$$

where c is the specific heat capacity of the metal. Hence

$$c = \frac{ItV}{m(\theta_2 - \theta_1)}$$

Notes. (i) If I is in amperes, t in seconds, V in volts, m in g, θ_1 and θ_2 in K then c is in J g^{-1} K^{-1}.

(ii) The small amount of heat received by the thermometer and heater has been neglected.

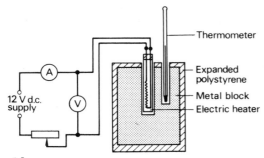

Fig. 4.5

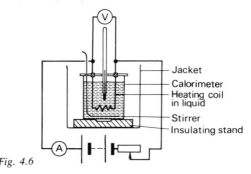

Fig. 4.6

(b) *Liquids.* The apparatus is shown in Fig. 4.6, a calorimeter being a vessel in which heat measurements are made. The procedure is similar to that for solids except that the liquid is stirred continuously during the heating. If m is the mass of liquid, c its specific heat capacity, m_c the mass of the calorimeter and stirrer, c_c the specific heat capacity of the material of the calorimeter and stirrer, and if θ_1, θ_2, I, V and t have their previous meanings then assuming

$$\frac{\text{energy supplied}}{\text{by heater}} = \frac{\text{energy received}}{\text{by liquid}} + \frac{\text{energy received}}{\substack{\text{by calorimeter}\\\text{and stirrer}}}$$

we have

$$ItV = mc(\theta_2 - \theta_1) + m_c c_c(\theta_2 - \theta_1)$$
$$= (mc + m_c c_c)(\theta_2 - \theta_1)$$

whence c can be found if c_c is known.

2. *Method of mixtures*

(a) *Solids.* The solid is weighed to find its mass m, heated in boiling water at temperature θ_3 for 10 minutes, Fig. 4.7a, and then quickly transferred to a calorimeter of mass m_c containing a mass of water m_w at temperature θ_1, Fig. 4.7b. The water is stirred and the highest reading θ_2 on the thermometer noted.

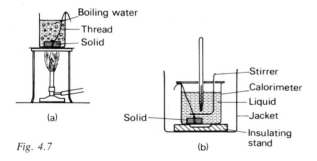

Fig. 4.7

Assuming no heat loss from the calorimeter when the hot solid is dropped into it, we have

$$\substack{\text{heat given out}\\\text{by solid cooling}\\\text{from }\theta_3\text{ to }\theta_2} = \substack{\text{heat received by}\\\text{water warming}\\\text{from }\theta_1\text{ to }\theta_2} + \substack{\text{heat received by}\\\text{calorimeter}\\\text{warming}\\\text{from }\theta_1\text{ to }\theta_2}$$

If c is the specific heat capacity of the solid, c_w that of water and c_c that of the calorimeter, then

$$mc(\theta_3 - \theta_2) = m_w c_w(\theta_2 - \theta_1) + m_c c_c(\theta_2 - \theta_1)$$
$$= (m_w c_w + m_c c_c)(\theta_2 - \theta_1)$$

$$\therefore \quad c = \frac{(m_w c_w + m_c c_c)(\theta_2 - \theta_1)}{m(\theta_3 - \theta_2)}$$

Hence c can be found knowing c_w and c_c.

(b) *Liquids.* In this case a hot solid of known specific heat capacity is dropped into the liquid whose specific heat capacity is required, otherwise the procedure and calculation are the same as for a solid.

3. *Continuous flow method*

The method was devised in 1899 by Callendar and Barnes for measuring the specific heat capacity of water. A simple form of the apparatus is shown in Fig. 4.8. It consists of a wire carrying a steady electric current which heats a liquid flowing at a constant rate through a glass tube from a constant head tank to a collecting vessel. Two thermometers measure the entrance and exit temperatures of the liquid.

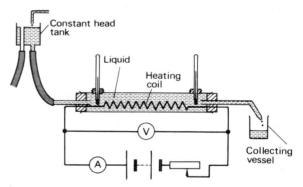

Fig. 4.8

After a time the thermometer readings become steady and none of the electrical energy supplied by the current warms the apparatus (since it is at a constant temperature); it is either used to heat the liquid or is lost to the surroundings.

Let the current in the wire be I_1 and the p.d. across it be V_1, then if θ_1 and θ_2 are the thermometer readings and m_1 is the mass of liquid collected in time t, we have

$$I_1 t V_1 = m_1 c(\theta_2 - \theta_1) + h$$

where c is the specific heat capacity of the liquid and h is the heat lost in time t to the surroundings.

The rate of flow is changed and the p.d. and current adjusted to V_2 and I_2 so that the entrance and exit temperatures are the same as before, i.e. θ_1 and θ_2. This ensures that h is the same as in the first case since the average temperature excess of the apparatus over the surroundings is unaltered. If m_2 is now the mass collected in the same time t then

$$I_2 t V_2 = m_2 c(\theta_2 - \theta_1) + h$$

Subtracting the two equations

$$(I_1 V_1 - I_2 V_2)t = c(m_1 - m_2)(\theta_2 - \theta_1)$$

$$\therefore \quad c = \frac{(I_1 V_1 - I_2 V_2)t}{(m_1 - m_2)(\theta_2 - \theta_1)}$$

The advantages of this method are (*i*) the heat capacity of the apparatus does not need to be known and (*ii*) consideration of heat loss is unnecessary. The chief difficulty is ensuring the liquid is mixed sufficiently to keep θ_2 constant.

In their more elaborate form of the apparatus Callendar and Barnes surrounded the glass tube by a vacuum jacket to make h very small, the currents and p.d.s were measured accurately by potentiometer methods and the temperatures were taken with platinum resistance thermometers (p. 64).

4. Mechanical method

The first part of the experiment described on p. 44 to calibrate a voltmeter can also be used to find the specific heat capacity of a solid in the form of the material of the drum. If M is the mass of the drum, c its specific heat capacity and θ the temperature *rise* produced by n revolutions with a mass m attached we can say,

$$\text{mechanical energy supplied} = mc\pi dn$$

and

$$\text{heat needed for a temperature rise } \theta = Mc\theta$$

Assuming all the mechanical energy appears as heat,

therefore

$$Mc\theta = mg\pi dn$$

$$\therefore \quad c = \frac{mg\pi dn}{M\theta}$$

Heat loss and cooling corrections

In experiments with calorimeters certain precautions can be taken to minimize heat losses. These include (*i*) polishing the calorimeter to reduce radiation loss, (*ii*) surrounding it by an outer container or a jacket of a poor heat conductor to reduce convection and conduction loss and (*iii*) supporting it on an insulating stand or supports to minimize conduction.

When the losses, despite all precautions, are not small, or where great accuracy is required, an estimate can be made of the temperature that would have been reached, i.e. a 'cooling correction' is made which when added to the observed maximum temperature gives the estimated maximum temperature had no heat been lost. Alternatively the need to make a cooling correction can be eliminated, as in the continuous flow method described above, or in other ways, one of which is explained under the second procedure.

1. Graphical method. As well as being suitable for electrical heating experiments, this method is convenient when finding the specific heat capacity of a bad thermal conductor (e.g. glass or rubber) by the method of mixtures. In the latter case the hot solid is slow to transfer heat to the calorimeter and water and some time elapses before the mixture reaches its maximum temperature. During this time appreciable cooling occurs even if the calorimeter is lagged.

To make the cooling correction, the temperature is taken at half-minute intervals starting *just before* the hot solid is added to the calorimeter and ending when the temperature has fallen by at least 1 °C from its observed maximum value. A graph of temperature against time is plotted. In that shown in Fig. 4.9, θ_1 is the initial temperature of the calorimeter and contents (i.e. room temperature) and θ_2 is the observed maximum temperature. The dotted line shows how the temperature might have risen if no heat were lost.

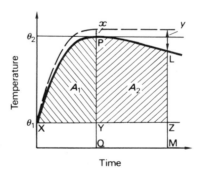

Fig. 4.9   Time

The cooling correction required is x. To obtain it, PQ is drawn through the top of the curve parallel to the temperature axis and similarly LM further along the curve so that y is 1 °C. XYZ is then drawn through θ_1, parallel to the time axis. The areas A_1 and A_2 are found by counting the squares on the graph paper and it can be shown that the cooling correction is given by

$$x = \frac{A_1}{A_2} \times y$$

where $y = 1$ °C for convenience. The estimated maximum temperature is then $\theta_2 + x$.

This method is based on the assumption that the rate of loss of heat is directly proportional to the temperature difference between the body (e.g. calorimeter) and its surroundings. This is true for heat loss by (a) conduction (see p. 77), (b) convection so long as it is forced (i.e. a draught) or if natural, provided the temperature difference is small (see below) and (c) radiation if the temperature excess is small.

In electrical heating experiments, temperature–time readings are taken during and immediately after the heating, and the cooling correction obtained from a graph as explained above.

2. Initial cooling method. If the calorimeter and its contents are cooled to about 5 °C below room temperature and then heated steadily during the experiment to about 5 °C above, the heat gained from the surroundings during the first half of the time will be nearly equal to that lost to the surroundings during the second half. No cooling correction is then necessary. The method is suitable when finding the specific heat capacity of a liquid by electrical heating.

Cooling laws and temperature fall

(a) *Five-fourths power law.* For cooling in still air by natural convection the five-fourths power law holds. It states

$$rate\ of\ loss\ of\ heat \propto (\theta - \theta_0)^{5/4}$$

where θ is the temperature of the body in surroundings at temperature θ_0. If the temperature excess $(\theta - \theta_0)$ is small, the relation becomes approximately linear.

(b) *Newton's law of cooling.* Under conditions of forced convection, i.e. in a steady draught, Newton's law applies. It states

$$rate\ of\ loss\ of\ heat \propto (\theta - \theta_0)$$

and is true for quite large temperature excesses.

(c) *Rate of fall of temperature.* As well as the temperature excess, the rate of loss of heat from a body depends on the area and nature of its surface (i.e. whether it is dull or shiny). Hence for a body having a uniform temperature θ and a surface area A we can say, if Newton's law holds

$$rate\ of\ loss\ of\ heat = kA(\theta - \theta_0)$$

where k is a constant depending on the nature of the surface.

If the temperature θ of the body falls we can also write

$$rate\ of\ loss\ of\ heat = mc \times rate\ of\ fall\ of\ temperature$$

where m is the mass of the body and c is its specific heat capacity. Hence

$$rate\ of\ fall\ of\ temperature = kA(\theta - \theta_0)/(mc)$$

The mass of a body is proportional to its volume and so the rate of fall of temperature of a body is proportional to the ratio of its surface area to its volume, i.e. is inversely proportional to a linear dimension. A small body, therefore, cools faster than a large one (its temperature falls faster), as everyday experience confirms. In calorimeter experiments the use of large apparatus, etc., minimizes the effect of errors due to heat loss.

Latent heat

The heat which a body absorbs in melting, evaporating or sublimating and gives out in freezing or condensing is called *latent* (hidden) *heat* because it does not produce a change of temperature in the body—it causes a change of state or phase. Thus when water is boiling its temperature remains steady at 100 °C (at s.t.p.) even although heat, called latent heat of vaporization, is being supplied to it. Similarly the temperature of liquid naphthalene stays at 80 °C whilst it is freezing; there is no fall of temperature until all the liquid has solidified but heat, called latent heat of fusion, is still being given out by the liquid.

The kinetic theory sees the supply of latent heat to a melting solid as enabling the molecules to overcome sufficiently the forces between them for the regular crystalline structure of the solid to be broken down. The molecules then have the greater degree of freedom and disorder that characterize the liquid state. Thus, whilst heat which increases the kinetic energy component of molecular internal energy causes a temperature rise, the supply of latent heat is regarded as increasing the potential energy component since it allows the molecules to move both closer together and farther apart.

When vaporization of a liquid occurs a large amount of energy is needed to separate the molecules and allow them to move around independently as gas molecules. In addition some energy is required to enable the vapour to expand against the atmospheric pressure. The energy for both these operations is supplied by the latent heat of vaporization and like latent heat of fusion, we regard it as increasing the potential energy of the molecules.

(a) *Specific latent heat of fusion.* This is defined as *the quantity of heat required to change unit mass of a substance from solid to liquid without change of temperature.* It is denoted by the symbol l_m and is measured in

J kg⁻¹ or J g⁻¹.

The specific latent heat of fusion of ice can be determined by the method of mixtures. A calorimeter of mass m_c is two-thirds filled with a mass m_w of water warmed to about 5 °C above room temperature. The temperature θ_1 of the water is noted, then a sufficient number of small pieces of ice, carefully dried in blotting paper, are added one at a time and the mixture stirred, until the temperature is about 5 °C below room temperature. The lowest temperature θ_2 is noted. The calorimeter and contents are then weighed to find the mass of ice m added.

The heat given out by the calorimeter and warm water in cooling from θ_1 to θ_2 does two things. First it supplies the latent heat needed to melt the ice at 0 °C to water at 0 °C and second it provides the heat to raise the now melted ice from 0 °C to the final temperature of the mixture θ_2. Hence

heat given out by calorimeter and water cooling	=	heat used to melt ice at 0 °C	+	heat used to warm melted ice from 0 °C to θ_2

If c_c and c_w are the specific heat capacities of the calorimeter and water respectively and l_m is the specific latent heat of fusion of ice then

$$m_c c_c(\theta_1 - \theta_2) + m_w c_w(\theta_1 - \theta_2) = m l_m + m c_w(\theta_2 - 0)$$

Hence

$$(m_c c_c + m_w c_w)(\theta_1 - \theta_2) = m(l_m + c_w\theta_2)$$

$$\therefore \quad l_m + c_w\theta_2 = \frac{(m_c c_c + m_w c_w)(\theta_1 - \theta_2)}{m}$$

and

$$l_m = \frac{(m_c c_c + m_w c_w)(\theta_1 - \theta_2)}{m} - c_w\theta_2$$

For ice the accepted value of l is 334 J g⁻¹. No cooling correction is necessary (see 'Initial cooling method' on p. 71) but the temperature of the mixture must not be taken more than 5 °C below room temperature otherwise water vapour in the air may condense to form dew on the calorimeter and give up latent heat to it.

(b) *Specific latent heat of vaporization.* This is *the quantity of heat required to change unit mass of a substance from liquid to vapour without change of temperature.* It is denoted by l_v and measured in J kg⁻¹ or J g⁻¹.

A value can be found for l_v by a continuous flow-type method using the apparatus of Fig. 4.10. The liquid is heated electrically by a coil carrying a steady current I and having a p.d. V across it. Vapour passes down the inner tube of a condenser where it is changed back to

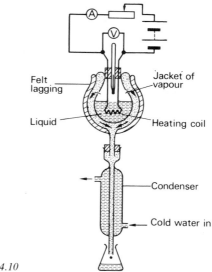

Felt lagging

Jacket of vapour

Liquid

Heating coil

Condenser

Cold water in

Fig. 4.10

liquid by cold water flowing through the outer tube.

After the liquid has been boiling for some time it becomes surrounded by a 'jacket' of vapour at its boiling point and a steady state is reached when the rate of vaporization equals the rate of condensation. All the electrical energy supplied is then used to supply latent heat to the liquid (and none to raise its temperature) and to make good any heat loss from the 'jacket'. If a mass m of liquid is now collected in time t from the condenser, we have

$$ItV = m l_v + h$$

where l_v is the specific latent heat of vaporization of the liquid and h is the heat lost from the 'jacket' in time t. The 'jacket' of vapour makes h small and if it is neglected l_v can be found. Alternatively it may be eliminated as in the Callendar and Barnes experiment (p. 69) by a second determination with a different power input.

The specific latent heat of vaporization of water is 2.3×10^3 J g⁻¹.

(c) *Bonding energy and latent heat.* The bonding energy for two atoms or molecules is the amount of energy that has to be supplied to pull them apart and make their potential energy zero. In Fig. 1.6b (p. 9) it equals E_0, the minimum value of the p.e. at the equilibrium separation r_0.

If a molecule has n neighbours, the total bonding energy for each molecule with its neighbours is nE_0. For a mole, containing L molecules (the Avogadro constant) the energy needed to break all the bonds would be $\frac{1}{2}nLE_0$ (the $\frac{1}{2}$ is necessary because, for any pair of molecules A and B, A is regarded as a neighbour of B and then B as a neighbour of A—otherwise each bond would be considered twice).

Taking the conversion of water to steam as an example. 2.3×10^6 J are needed to vaporize 1 kg of water (at 373 K), so, since 1 mole of water has mass 0.018 kg, the energy to be supplied is given by

$$\tfrac{1}{2} n L E_0 = (2.3 \times 10^6 \text{ J kg}^{-1})(0.018 \text{ kg mol}^{-1})$$
$$= 4.14 \times 10^4 \text{ J mol}^{-1}$$

But $L = 6.02 \times 10^{23}$ mol^{-1} and for a liquid $n \simeq 10$, therefore

$$E_0 = \frac{2 \times 4.14 \times 10^4 \text{ J mol}^{-1}}{10 \times 6.02 \times 10^{23} \text{ mol}^{-1}}$$
$$= 1.4 \times 10^{-20} \text{ J}$$

This is a rough estimate of the bonding energy of water.

Heat calculations

1. A piece of copper of mass 100 g is heated to 100 °C and then transferred to a well-lagged copper can of mass 50.0 g containing 200 g of water at 10.0 °C. Neglecting heat loss, calculate the final steady temperature of the water after it has been well stirred. Take the specific heat capacities of copper and water as 4.00×10^2 J kg^{-1} K^{-1} and 4.20×10^3 J kg^{-1} K^{-1} respectively.

Let the final steady temperature $= \theta$
Fall in temperature of piece of copper $= (100 - \theta)$ °C
Rise in temperature of can and water $= (\theta - 10)$ °C

Expressing masses in kg,

heat given out by copper $= 0.1 \times 400 \times (100 - \theta)$ J
heat received by copper can $= 0.05 \times 400 \times (\theta - 10)$ J
heat received by water $= 0.2 \times 4200 \times (\theta - 10)$ J
heat given out $=$ heat received

Therefore

$$40(100 - \theta) = 20(\theta - 10) + 840(\theta - 10)$$
$$= (20 + 840)(\theta - 10)$$
$$4000 - 40\theta = 860\theta - 8600$$
$$900\theta = 12\,600$$
Hence $\theta = 14.0$ °C

2. The rate of flow of liquid through a continuous flow calorimeter is 15×10^{-3} kg s^{-1} and the electric heating element dissipates 200 W, a steady difference of temperature of 3 °C being maintained. To maintain the same temperatures, 80 W is necessary when the flow is reduced to 5.0×10^{-3} kg s^{-1}. *Assuming the temperature of the surroundings to be the same in the two cases, calculate the specific heat capacity of the liquid and the rate at which heat is lost to the surroundings.*

In the steady state the apparatus absorbs no heat, hence

electrical energy supplied per second	=	heat received by water per second	+	heat lost to surroundings per second

Since 1 W $= 1$ J s^{-1}, in the first case the electrical energy supplied per second $= 200$ J s^{-1}. Also, heat received by water per second $= mc(\theta_1 - \theta_2)$ where $m = 15 \times 10^{-3}$ kg s^{-1}, $c =$ specific heat capacity of liquid in J kg^{-1} K^{-1} and $(\theta_1 - \theta_2) = 3$ °C $= 3$ K. If h is the heat lost to the surroundings per second, then

$$200 \text{ J s}^{-1} = (15 \times 10^{-3} \text{ kg s}^{-1})\, c\, (3 \text{ K}) + h \quad (1)$$

In the second case

$$80 \text{ J s}^{-1} = (5.0 \times 10^{-3} \text{ kg s}^{-1})\, c\, (3 \text{ K}) + h \quad (2)$$

Subtracting (2) from (1)

$$(200 - 80) \text{ J s}^{-1} = (15 \times 10^{-3} - 5.0 \times 10^{-3}) \text{ kg s}^{-1} \times c \times (3 \text{ K})$$

$$\therefore \quad 120 \text{ J s}^{-1} = (10 \times 10^{-3} \text{ kg s}^{-1})\, c\, (3 \text{ K})$$

$$\therefore \quad c = \frac{120 \text{ J s}^{-1}}{(10 \times 10^{-3} \text{ kg s}^{-1})(3 \text{ K})}$$

$$= \frac{120}{30 \times 10^{-3}} \frac{\text{J s}^{-1}}{\text{kg s}^{-1} \text{ K}}$$

$$= 4.0 \times 10^3 \text{ J kg}^{-1} \text{ K}^{-1}$$

Substituting for c in (1)

$$200 \text{ J s}^{-1} = (15 \times 10^{-3} \text{ kg s}^{-1})(4.0 \times 10^3 \text{ J kg}^{-1} \text{ K}^{-1})(3 \text{ K}) + h$$

$$= (15 \times 10^{-3} \times 4.0 \times 10^3 \times 3) \text{ kg s}^{-1} \text{ J kg}^{-1} \text{ K}^{-1} \text{ K} + h$$

$$= 180 \text{ J s}^{-1} + h$$

$$\therefore \quad h = (200 - 180) \text{ J s}^{-1} = 20 \text{ J s}^{-1}$$

3. When a current of 2.0 A is passed through a coil of constant resistance 15 Ω immersed in 0.5 kg of water at 0 °C in a vacuum flask, the temperature of the water rises to 8 °C in 5 minutes. If instead the flask originally contained 0.25 kg of ice and 0.25 kg of water, what current must be passed through the coil if this mixture is to be heated to the same temperature in the same time? (Specific heat capacity of water $= 4.2 \times 10^3$ J kg^{-1} K^{-1}; specific latent heat of fusion of ice $= 3.3 \times 10^5$ J kg^{-1}.)

Assuming no heat is lost from the vacuum flask then in time t,

$$\frac{\text{electrical energy supplied}}{} = \frac{\text{heat received by water}}{} + \frac{\text{heat received by vacuum flask}}{} \quad (1)$$

But, electrical energy supplied $= I^2Rt$ joules where $I = 2$ A, $R = 15$ Ω and $t = 5 \times 60 = 300$ s. Also, heat received by water $= mc(\theta_1 - \theta_2)$ where $m = 0.5$ kg, $c = 4.2 \times 10^3$ J kg^{-1} K^{-1} and $\theta_1 - \theta_2 = 8$ °C $= 8$ K. If h is the heat received by the vacuum flask in time t, then from (1)

$$(2 \text{ A})^2(15 \text{ V})(300 \text{ s}) = (0.5 \text{ kg})(4.2 \times 10^3 \text{ J kg}^{-1}\text{K}^{-1})$$
$$(8 \text{ K}) + h$$
$$(4 \times 15 \times 300) \text{ J} = (0.5 \times 4.2 \times 10^3 \times 8)$$
$$\text{kg J kg}^{-1}\text{K}^{-1} \text{ K} + h$$
$$\therefore \quad h = (18\ 000 - 16\ 800) \text{ J}$$
$$= 1\ 200 \text{ J}$$

In the second part, in the same time t,

$$\frac{\text{electrical energy supplied}}{} = \frac{\text{heat to melt 0.25 kg ice at 0 °C}}{} + \frac{\text{heat to warm 0.5 kg water from 0 °C to 8 °C}}{} + \frac{\text{heat received by vacuum flask (i.e. } h)}{}$$

$$\therefore \quad I^2(15 \ \Omega)(300 \text{ s}) = (0.25 \text{ kg})(3.3 \times 10^5 \text{ J kg}^{-1}) +$$
$$(0.5 \text{ kg})(4.2 \times 10^3 \text{ J kg}^{-1} \text{ K}^{-1})$$
$$(8 \text{ K}) + 1200 \text{ J}$$

where I is the current required to warm the ice and water to 8 °C in 300 s.

$$\therefore \quad I^2(15 \times 300) \ \Omega \text{ s} = (0.25 \times 3.3 \times 10^5)\text{J} +$$
$$(0.5 \times 4.2 \times 10^3 \times 8) \text{ J}$$
$$+ 1200 \text{ J}$$
$$\therefore \quad I^2(4500) \ \Omega \text{ s} = 82\ 500 \text{ J} + 16\ 800 \text{ J} + 1200 \text{ J}$$
$$= 100\ 500 \text{ J}$$
$$I^2 = \frac{100\ 500}{4\ 500} \ \frac{\text{J}}{\Omega \text{ s}}$$
$$= 22.3 \text{ V C } \Omega^{-1} \text{ s}^{-1}$$
$$I = 4.7 \text{ A}$$
$$(\text{since A} = \text{C s}^{-1} = \text{V } \Omega^{-1})$$

Expansion of solids

(a) *Linear expansion.* The change of length which occurs with temperature change in a solid has to be allowed for in the design of many devices. The variation is described by the *linear expansivity* α. Thus if a solid of length l increases in length by δl due to a temperature rise $\delta\theta$, α for the material is defined by the equation

$$\alpha = \frac{\delta l}{l} \cdot \frac{1}{\delta\theta}$$

In words, α is *the fractional increase of length* (i.e. $\delta l / l$) *per unit rise of temperature.* The unit of α is K^{-1} since δl and l have the same units (metres) and so $\delta l / l$ is a ratio.

As the temperature rise $\delta\theta$ tends to zero, α approaches the linear expansivity at a particular temperature and experiment shows that its value for a given material is not constant but varies slightly with temperature. Mean values are therefore obtained over a temperature range and in accurate work this range is stated. For ordinary purposes α can be assumed constant in the range 0 to 100 °C. The mean value for copper (at room temperatures) is 1.7×10^{-5} K^{-1}; a copper rod 1 metre long therefore increases in length by 1.7×10^{-5} metre for every 1 °C (1 K) temperature rise.

A useful expression is obtained if we consider a solid of original length l_0 which increases to l_θ for a temperature *rise* of θ. Replacing δl by $l_\theta - l_0$ and $\delta\theta$ by θ in the expression for α we get

$$\alpha = \frac{l_\theta - l_0}{l_0\theta}$$

Rearranging $\qquad l_\theta - l_0 = l_0\alpha\theta$

$$\therefore \quad l_\theta = l_0(1 + \alpha\theta)$$

Note that l_0 is the *original* length and, since values of α are very small, it need not be the length at 0 °C. (Unlike R_0 in the temperature coefficient of resistance formula $R_\theta = R_0(1 + \alpha\theta)$ on p. 47, which is generally taken to be the resistance at 0 °C.) For example, if the temperature of a 2 metre long copper rod rises from 15 °C to 25 °C then $l_0 = 2$ m, $\theta = (25 - 15) = 10$ °C $= 10$ K, $\alpha = 1.7 \times 10^{-5}$ K^{-1} and

$$l_\theta - l_0 = l_0\alpha\theta$$
$$= (2 \text{ m})(1.7 \times 10^{-5} \text{ K}^{-1})(10 \text{ K})$$
$$= 2 \times 1.7 \times 10^{-5} \times 10 \quad \text{m K}^{-1} \text{ K}$$
$$= 3.4 \times 10^{-4} \text{ m}$$
$$= 0.34 \text{ mm}$$

Thermal expansion of a solid can be explained on the atomic scale with the help of Fig. 4.11a which shows that the repelling forces between atoms increases more rapidly than the attractive forces as their separation varies. A rise of temperature increases the amplitude of vibration of the atoms about their equilibrium position

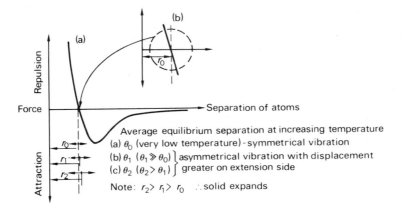

Average equilibrium separation at increasing temperature
(a) θ_0 (very low temperature) - symmetrical vibration
(b) θ_1 $(\theta_1 \gg \theta_0)$ ⎱ asymmetrical vibration with displacement
(c) θ_2 $(\theta_2 > \theta_1)$ ⎰ greater on extension side

Note: $r_2 > r_1 > r_0$ ∴ solid expands

Fig. 4.11

but this will be greater on the extension side of that position, for the reason just given, and the average separation of the atoms increases. At low temperatures the amplitudes of oscillation are small and symmetrical, Fig. 4.11b, about the equilibrium position and so we would not expect expansion with increase of temperature—a fact confirmed by experiment. A further conclusion from this argument is that linear expansivities should increase with rising temperature, as they do.

(b) *Area expansion.* The change of area of a surface with temperature change is described by the *area* or *superficial expansivity* β. Thus if an area A increases by δA due to a temperature rise $\delta\theta$ then β is given by

$$\beta = \frac{\delta A}{A} \cdot \frac{1}{\delta\theta}$$

In words, β is *the fractional increase of area* (i.e. $\delta A/A$)*per unit rise of temperature.*

The variation of area with temperature is given by an equation similar to that for linear expansion, thus

$$A_\theta = A_0(1 + \beta\theta)$$

where A_0 and A_θ are the original and new areas respectively and θ is the temperature rise.

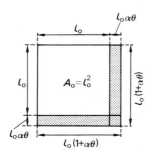

Fig. 4.12

It can be shown that for a given material $\beta \simeq 2\alpha$. Consider a square plate of side l_0, Fig. 4.12. We have

$$A_0 = l_0^2$$

A temperature rise of θ causes the length of each side to become $l_0(1 + \alpha\theta)$ if the material is isotropic, i.e. has the same properties in all directions. Hence the new area A_θ is

$$A_\theta = l_0^2(1 + \alpha\theta)^2$$
$$= l_0^2(1 + 2\alpha\theta + \alpha^2\theta^2)$$

Now $\alpha^2\theta^2$ is very small compared with $2\alpha\theta$ and since $A_0 = l_0^2$ we have

$$A_\theta \simeq A_0(1 + 2\alpha\theta)$$

Comparing this with $A_\theta = A_0(1 + \beta\theta)$ it follows that

$$\beta \simeq 2\alpha$$

(c) *Volume expansion.* Changes of volume of a material with temperature are expressed by the *cubic expansivity* γ. Thus if a volume V increases by δV for a temperature rise $\delta\theta$ then γ is given by

$$\gamma = \frac{\delta V}{V} \cdot \frac{1}{\delta\theta}$$

In words, γ is *the fractional increase of volume* (i.e. $\delta V/V$) *per unit rise of temperature.*

The equation $V_\theta = V_0(1 + \gamma\theta)$ is also useful and it can be shown that for a given material $\gamma \simeq 3\alpha$. The proof involves calculating the volume change of a cube in terms of the linear expansion of each side—in a similar manner to that adopted for areas. Cubic and area expansivities for solids are not given in tables of physical constants since they are readily calculated from linear expansivities. The comment on the constancy of α (p. 74) also applies to β and γ.

A hollow body such as a bottle expands as if it were solid throughout, otherwise it would not retain the same shape when heated.

Thermal stress

Forces are created in a structure when thermal expansion or contraction is resisted. An idea of the size of such forces can be obtained by considering a metal rod of initial length l_0, cross-section area A and linear expansivity α, supported between two fixed end plates, Fig. 4.13. If the temperature of the rod is raised by θ °C, it tries to expand but is prevented by the plates and a compressive stress arises in it.

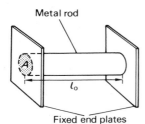

Metal rod

Fixed end plates

Fig. 4.13

Removing one of the plates would allow the rod to expand freely and its new length l_θ would be $l_0 + l_0\alpha\theta$. We can therefore look upon the plate, when fixed in position, as exerting a force F on the rod and reducing its length from l_θ to l_0. Then (from p. 20),

$$\text{compressive strain} = \frac{\text{change in length}}{\text{original length}}$$

$$= \frac{l_\theta - l_0}{l_\theta} = \frac{l_0\alpha\theta}{l_\theta}$$

Also,

$$\text{compressive stress} = \frac{F}{A}$$

If the Young modulus for the material of the rod is E (the compressive modulus is the same as the tensile one for small compressions) then

$$E = \frac{\text{stress}}{\text{strain}} = \frac{Fl_\theta}{Al_0\alpha\theta}$$

The difference between l_θ and l_0 would be small compared with either and so to a good approximation $l_\theta = l_0$. Therefore

$$E = \frac{F}{A\alpha\theta} \quad \text{and} \quad F = EA\alpha\theta$$

For a steel girder with $E = 2.0 \times 10^{11}$ Pa (N m^{-2}), $A = 100$ cm$^2 = 10^{-2}$ m^2, $\alpha = 1.2 \times 10^{-5}$ K^{-1} and a temperature rise $\theta = 20$ °C $= 20$ K,

$$F = (2 \times 10^{11} \text{ N m}^{-2}) \times (10^{-2} \text{ m}^2) \times (1.2 \times 10^{-5} \text{ K}^{-1}) \times (20 \text{ K})$$

$$= 2 \times 10^{11} \times 10^{-2} \times 1.2 \times 10^{-5} \times 20$$
$$\text{N m}^{-2} \text{ m}^2 \text{ K}^{-1} \text{ K}$$

$$= 4.8 \times 10^5 \text{ N}$$

A sizeable force. The original length of the rod does not affect the force but long rods tend to buckle at lower compressive stresses than short ones.

Thermal stress is put to good use in the technique of shrink fitting in which, for example, a large gear wheel is fitted on to a shaft of the same material. The diameter of the central hole in the wheel is smaller at room temperature than the outside diameter of the shaft. If the shaft is cooled with solid carbon dioxide ('dry ice') at −78 °C it can be fitted into the wheel. At room temperature the shaft is under compression and the wheel under tension and a tight-fitting joint results.

Expansion of liquids

(a) *Real and apparent expansion.* Only the cubic expansivity has meaning in the case of a liquid since its shape depends on the containing vessel. The cubic expansivities of liquids are generally greater than those for solids and like the latter they vary with temperature. Mean values are therefore obtained for the temperature range considered.

The expansion of a liquid is complicated by the fact that the vessel expands as well and makes the expansion of the liquid appear less than it really is. Consequently two expansivities are defined.

The apparent cubic expansivity γ_{app} of a liquid in a vessel of a particular material is the apparent fractional increase of volume per unit rise of temperature.

The real or absolute cubic expansivity γ_{real} of a liquid is the actual fractional increase of volume per unit rise of temperature and is always greater than γ_{app}.

It can be shown that

$$\gamma_{real} \approxeq \gamma_{app} + \gamma_{material\ of\ vessel}$$

This relationship enables the apparent expansion of a liquid relative to its container to be calculated. For example, if the temperature of 100 cm^3 of mercury in a glass vessel is raised from 10 °C to 100 °C and γ_{real} for mercury $= 1.82 \times 10^{-4}$ K^{-1} and $\alpha_{glass} = 8.00 \times 10^{-6}$ K^{-1} then since

$$\gamma_{glass} = 3\alpha_{glass} = 2.40 \times 10^{-5} \text{ K}^{-1}$$

we have $\gamma_{app} \simeq \gamma_{real} - \gamma_{glass}$

$$\simeq 1.82 \times 10^{-4} - 2.40 \times 10^{-5}$$

$$\simeq 1.58 \times 10^{-4} \text{ K}^{-1}$$

Hence

apparent expansion of mercury $= V_0 \gamma_{app} \theta$

$$= (100 \text{ cm}^3)(1.58 \times 10^4 \text{ K}^{-1})(90 \text{ K})$$

$$= 1.42 \text{ cm}^3$$

(*b*) *Variation of density with temperature.* It is sometimes more useful to know how the density rather than the volume of a liquid changes with temperature. Consider a fixed mass m of liquid of real cubic expansivity γ which occupies volume V_0 and has density ρ_0 at a certain temperature. Let the temperature *rise* be θ causing the volume to increase to V_θ and the density to decrease to ρ_θ then as for a solid,

$$V_\theta = V_0(1 + \gamma\theta)$$

$$\therefore \quad \frac{V_\theta}{V_0} = 1 + \gamma\theta$$

But $\quad m = V_0\rho_0 = V_\theta\rho_\theta$

$$\therefore \quad \frac{V_\theta}{V_0} = \frac{\rho_0}{\rho_\theta}$$

Thus $\quad \dfrac{\rho_0}{\rho_\theta} = 1 + \gamma\theta$

or $\quad \rho_0 = \rho_\theta(1 + \gamma\theta)$

(*c*) *Anomalous expansion of water.* At 4 °C the density of water is a maximum; from 4 °C to 0 °C it expands and the expansion is said to be anomalous, i.e. abnormal. It is this abnormal behaviour of water which results in convection currents ceasing when all the water in a pond has reached 4 °C, assuming, of course, a surface air temperature of that value or below. The expansion between 4 °C and 0 °C is explained on the assumption that at 4 °C the expansion due to the dissociation of complex molecules such as H_4O_2 and H_6O_3, already present in the water, more than cancels out the contraction due to the fall of temperature.

Water is also unusual because it expands on freezing, every 100 cm^3 of water becoming 109 cm^3 of ice. This accounts for the bursting of water pipes in very cold weather.

Thermal conductivity

Heat transfer, i.e. the passage of energy from a body at a higher temperature to one at a lower temperature, occurs by the three processes of conduction, convection and radiation although evaporation and condensation may often play an important part. In some cases the aim of the heat engineer is to encourage heat flow (as in a boiler) while in others it is to minimize it (e.g. lagging a house). Here we shall discuss *conduction*, i.e. the transfer of energy due to the temperature difference between neighbouring parts of the same body.

(*a*) *Definition of k.* Consider a *thin* slab of material of thickness δx and uniform cross-section area A between whose faces a *small* temperature difference $\delta\theta$ is maintained, Fig. 4.14. If a quantity of heat δQ passes through the slab by conduction in time δt, the *thermal conductivity k* of the material is defined by the equation

$$\frac{\delta Q}{\delta t} = -kA\frac{\delta\theta}{\delta x}$$

The negative sign indicates that heat flows towards the lower temperature, i.e. as x increases θ decreases thus making $\delta\theta/\delta x$—called the *temperature gradient*—negative. Inserting the negative sign ensures that $\delta Q/\delta t$ and k will be positive. In words, we may define k as *the rate of flow of heat through a material per unit area, per unit temperature gradient.*

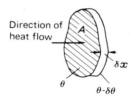

Direction of heat flow

Fig. 4.14

The unit of k is J s^{-1} m^{-1} K^{-1} or W m^{-1} K^{-1} as can be seen by inserting units for the various quantities in the previous expression and remembering that 1 watt is 1 joule per second. Its value for copper is 390 W m^{-1} K^{-1} and for asbestos 0.08 W m^{-1} K^{-1} (both at room temperature).

When $\delta\theta \to 0$, k approaches the thermal conductivity at a particular temperature and experiment shows that its value for a given material varies slightly with temperature. (In the limiting case when $\delta x \to 0$, a cross-section is then being considered and the equation defining k can be more precisely written in calculus notation as $dQ/dt = = - kA(d\theta/dx)$.) If measurements are not made over too great a temperature range, a constant mean value for k is usually assumed.

(b) *Temperature gradients.* When heat has been passing along a conductor for some time from a source of fixed high temperature, a steady state may be reached with the temperature at each point of the conductor becoming constant.

In the *unlagged* bar of Fig. 4.15a the quantity of heat passing in a given time through successive cross-sections decreases due to heat loss from the sides. The lines of heat flow are divergent and the temperature falls faster near the hotter end. For steady state conditions a graph of temperature θ against distance x from the hot end is as shown. The temperature gradient at any point is given by the slope of the tangent at that point (in calculus notation by $d\theta/dx$).

In a *lagged* bar whose sides are well wrapped with a good insulator, Fig. 4.15b, heat loss from the sides is negligible and the rate of flow of heat is the same all along the bar. The lines of heat flow are parallel and, in the steady state, the temperature falls at a constant rate as shown. The temperature gradient in this case is the slope of the graph, i.e. $\delta\theta/\delta x$.

There is a useful expression applicable to many simple problems in which the *lines of heat flow are parallel*; they are in a lagged bar and in a plate of large cross-section area. Why the latter? Consider a conductor of length x, cross-section area A and thermal conductivity k whose opposite ends are maintained at temperatures θ_2 and θ_1 ($\theta_2 > \theta_1$). From what has been said in the previous paragraph it follows that the quantity of heat Q passing any point in time t when the *steady state* has been reached is given by

$$\frac{Q}{t} = kA\left(\frac{\theta_2 - \theta_1}{x}\right)$$

We will now use this expression, which is sometimes called *Fourier's law.*

(c) *Composite slab problem.* Suppose we wish to find the rate of flow of heat through a plaster ceiling which measures 5 m $\times$ 3 m $\times$ 15 mm (i) without and (ii) with a 45 mm thick layer of insulating fibre-glass if the inside and outside surfaces are at the surrounding air temperatures of 15 °C and 5 °C respectively. ($k_{plaster}$ = 0.60 W m^{-1} K^{-1} and $k_{fibre-glass}$ = 0.040 W m^{-1} K^{-1}.)

Assuming steady states are reached and lines of heat flow are parallel we can use rate of flow of heat = $Q/t = kA(\theta_2 - \theta_1)/x$. We have $A = 5 \times 3 = 15$ m^2.

(i) Without fibre-glass, Fig. 4.16a, $x = 15$ mm = 0.015 m,

$$\frac{Q}{t} = \frac{(0.60 \text{ W m}^{-1} \text{ K}^{-1})(15 \text{ m}^2) \times (10 \text{ °C})}{(0.015 \text{ m})}$$

$$= \left(\frac{0.60 \times 15 \times 10}{0.015}\right) \frac{\text{W m}^{-1} \text{ K}^{-1} \text{ m}^2 \text{ °C}}{\text{m}}$$

$$= 6.0 \times 10^3 \text{ W}$$

(In practice it will be very much less than this, see later.)

(ii) With fibre-glass, Fig. 4.16b. Let the temperature of the plaster–fibre-glass boundary be θ. *The rate of flow of heat is the same through both materials.*

$$\therefore \quad \frac{Q}{t} = \frac{(0.60 \text{ W m}^{-1} \text{ K}^{-1})(15 \text{ m}^2)(15 \text{ °C} - \theta)}{(0.015 \text{ m})}$$

$$= \frac{(0.04 \text{ W m}^{-1} \text{ K}^{-1})(15 \text{ m}^2)(\theta - 5 \text{ °C})}{(0.045 \text{ m})}$$

Solving for θ we get

$$\theta = 14.8 \text{ °C}$$

UNLAGGED BAR

LAGGED BAR

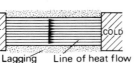

Line of heat flow

Lagging   Line of heat flow

Slope of tangent gives temperature gradient at A

$\delta\theta$ δx

Distance x

Distance x

Fig. 4.15

(a)

(b)

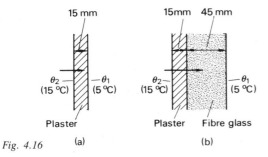

Fig. 4.16   (a)   (b)

Substituting θ in the equation above

$$\frac{Q}{t} = \frac{0.60 \times 15 \times (15 - 14.8)}{0.015}$$

$$= 1.2 \times 10^2 \text{ W}$$

The previous example shows that when heat flows through a composite slab the temperature fall per mm is greater across the poorer conductor. This is of practical importance where a good conductor is in contact with a bad conductor such as air (or a liquid); the latter in fact controls the rate of conduction of heat through the good conductor. If it is assumed that the surface of a good conductor is at the same temperature as that of the surrounding air, heat flows will be obtained that are of the order of one hundred times too large. In Fig. 4.17 most of the temperature drop occurs in the layer of gas between the flame and the boiler plate and in any scale deposited on the plate by the water.

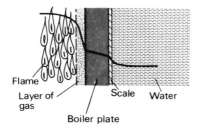

Fig. 4.17

(d) *Mechanisms of thermal conduction.* In solids (and liquids) two processes seem to be involved. The first concerns atoms and the second 'free' electrons.

Atoms at a higher temperature vibrate more vigorously about their equilibrium positions in the lattice than their colder neighbours. But because they are coupled to them by interatomic bonds they pass on some of their vibratory energy and cause them to vibrate more energetically as well. These in turn affect other atoms and thermal conduction occurs. However,

the process is generally *slow* because atoms, compared with electrons, are massive and the increases in vibratory motion are therefore fairly small. Consequently materials such as electrical insulators, in which this is the main conduction mechanism, are usually poor thermal conductors.

This effect is often regarded as resembling the passage of elastic waves through the material and because light waves are considered to have a dual nature, sometimes behaving as particles called photons, so too are these elastic waves considered to have particle-like forms called *phonons*. On this view, thermal conduction is said to be due to phonons having collisions with, and transferring energy to, atoms in the lattice.

The second process concerns materials with a supply of 'free' electrons. In these the electrons share in any gain of energy due to temperature rise of the material and their velocities increase much more than those of the atoms in the lattice since they are considerably lighter. They are thus able to move over larger distances and pass on energy *quickly* to cooler parts. Materials such as electrical conductors in which this mechanism predominates are therefore good thermal conductors; 'free' electrons are largely responsible for both properties. The analogy between them is developed further after the motion of fluids has been considered (p. 215).

Lest it be thought that only metals are good thermal conductors it should be stated that phonons can be a very effective means of heat transfer, especially at low temperatures. Thus at about $-180\,^\circ\text{C}$ synthetic sapphire (Al_2O_3) is a better conductor than copper.

Methods of measuring k

Since we use the expression $Q/t = kA(\theta_2 - \theta_1)/x$ we must ensure that (i) the specimen is in a steady state and (ii) the lines of heat flow are parallel. By measuring the rate of flow of heat Q/t through the specimen, the temperature gradient $(\theta_2 - \theta_1)/x$ and the cross-section area A, k can be calculated.

(a) *Good conductors.* The problem here is to obtain a measurable temperature gradient and is solved by using a *bar* of the material which is long compared with its diameter. Large x and small A then make $(\theta_2 - \theta_1)$ sufficiently big whilst still giving a satisfactory heat flow rate.

The apparatus due to Searle is shown in Fig. 4.18. The bar under study is heavily lagged and heated at one end by steam (or by an electrical heating coil in some arrangements). A spiral copper tube soldered to the bar carries a steady flow of water which is warmed at

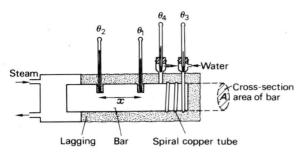

Fig. 4.18

the cold end by the heat conducted down the bar. Two thermometers record the entrance and exit temperatures θ_3 and θ_4 of the water. Which will be the greater? The thermometers giving the temperatures θ_2 and θ_1 at a known separation x on the bar are inserted in holes containing mercury. Why?

When all four thermometer readings are constant they are noted and the mass m of water flowing through the copper tube in time t is found with a measuring cylinder and stop-watch. If Q is the quantity of heat flowing down the bar also in time t and A is the cross-section area of the bar then

$$\frac{Q}{t} = kA\left(\frac{\theta_2 - \theta_1}{x}\right)$$

This heat is taken from the bar by the mass m of cooling water of specific heat capacity c and so

$$Q = mc(\theta_4 - \theta_3)$$

Hence
$$kAt\left(\frac{\theta_2 - \theta_1}{x}\right) = mc(\theta_4 - \theta_3)$$

from which k can be found.

In this method once a steady state has been reached the rate of flow of heat and the temperature gradient are the same for *any* section of the bar, since it is lagged. It is then possible to measure each at different parts of the bar.

For accurate work thermocouples bound to the bar replace the mercury thermometers.

(b) *Poor conductors.* In this case even a thin specimen gives a measurable temperature gradient; the

difficulty is getting an adequate rate of heat flow. It is overcome by using a thin disc of the material (i.e. x small and A large); this shape also helps to reduce heat loss from the sides of the specimen thereby giving parallel lines of heat flow.

A simple form of the apparatus adapted from one due to Lees, is shown in Fig. 4.19a. A disc D of the material under test rests on a thick brass slab B containing a thermometer and is heated from above by a steam-chest C whose thick base also carries a thermometer.

In the first part of the experiment steam is passed until temperatures θ_2 and θ_1 are steady. We can then say that the heat passing per second through D, to B from C, equals the heat lost per second by B at temperature θ_1 to the surroundings. If this is Q then

$$Q = kA\left(\frac{\theta_2 - \theta_1}{x}\right)$$

A, x, θ_2 and θ_1 can all be measured.

In the second part D is removed and B heated directly from C until its temperature is about 5 °C above what it was in the first part, i.e. about 5 °C above θ_1. C is now removed and a thick felt pad placed on B, Fig. 4.19b. Temperature–time readings are taken and a cooling curve plotted. The rate of fall of temperature of B (in °C per second) at θ_1 equals the slope (b/a) of the tangent to the curve at θ_1, Fig. 4.19c. If we assume that the heat lost per second by B at θ_1, cooling under more or less the same conditions as in the first part (since the felt minimizes heat loss from the top surface) is Q then

$$Q = mc\left(\frac{b}{a}\right)$$

where m is the mass of B and c its specific heat capacity. Hence k can be found from

$$kA\left(\frac{\theta_2 - \theta_1}{x}\right) = mc\left(\frac{b}{a}\right)$$

Note that whilst B's temperature is steady in the first part when it is supplied with heat from C, in the second part it is falling since it is drawing on its own internal energy.

(a)

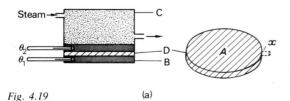

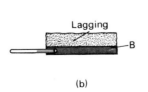

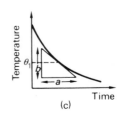

Fig. 4.19   (a)   (b)   (c)

QUESTIONS

Temperature: thermometers

1. Explain what is meant by a *scale of temperature* and how a temperature is defined in terms of a specified property.

When a particular temperature is measured on scales based on different properties it has a different numerical value on each scale except at certain points. Explain why this is so and state (*a*) at what points the values agree and (*b*) what scale of temperature is used as a standard.

Explain the principles of two different types of thermometer one of which is suitable for measuring a rapidly varying temperature and the other for measuring a steady temperature whose value is required to a high degree of accuracy. Give reasons for your choice of thermometer in each case. Experimental details are not required. (*L.*)

2. (*a*) What is the value of the temperature θ in °C on the scale of a platinum resistance thermometer if $R_0 = 2.000$ ohms, $R_{100} = 2.760$ ohms and $R_\theta = 2.480$ ohms?

(*b*) The resistance of a wire at a temperature θ °C measured on a standard scale is given by

$$R_\theta = R_0(1 + A\theta + 10^{-3} A\theta^2)$$

where A is a constant. When the thermometer is at a temperature of 50.0 °C on the standard scale, what will be the temperature indicated on the resistance scale?

(*J.M.B. part qn.*)

Specific heat capacity: latent heat

3. A current of 2.50 A passing through a heating coil immersed in 180 g of paraffin (specific heat capacity 2.00 J g^{-1} K^{-1}) contained in a 100 g calorimeter (specific heat capacity 0.400 J g^{-1} K^{-1}) raises the temperature from 5 °C below room temperature to 5 °C above room temperature in 100 s. What should be the reading of a voltmeter connected across the heating coil? (*S.*)

4. When water was passed through a continuous flow calorimeter the rise in temperature was from 16.0 to 20.0 °C, the mass of water flowing was 100 g in 1 minute, the potential difference across the heating coil was 20.0 V and the current was 1.50 A. Another liquid at 16.0 °C was then passed through the calorimeter and to get the same change in temperature the potential difference was changed to 13.0 V, the current to 1.20 A and the rate of flow to 120 g in 1 minute. Calculate the specific heat capacity of the liquid if the specific heat capacity of water is assumed to be 4.20 J g^{-1} K^{-1}.

State two advantages of the continuous flow method of calorimetry. (*J.M.B.*)

5. Give a labelled diagram of a continuous-flow apparatus which could be used to determine the specific heat capacity of water. The diagram should include the electrical circuit, but a description of the apparatus is *not* required.

In such an experiment, the following readings were taken:

Current in heating coil	2.0 A	1.5 A
Potential difference across coil	6.0 V	4.5 V
Mass of water collected	42.3 g	70.2 g
Time of flow	60 s	180 s
Inlet temperature	38.0 °C	38.0 °C
Outlet temperature	42.0 °C	42.0 °C

Explain how each reading would be taken, and use the figures to obtain a value for the specific heat capacity of water in J g^{-1} K^{-1}. If each temperature reading was subject to an uncertainty of ± 0.1 °C, find the resulting percentage uncertainty in the specific heat capacity due to this cause alone.

(*C. part qn.*)

6. Describe how you would determine experimentally the specific heat capacity of *either* copper *or* water by a direct mechanical method.

(You may assume, as necessary, that the values of other specific heats are known.)

A metal disc of radius 0.050 m and thickness 0.10 m is turned at 10 revolutions per second against a friction band, the tensions on the two sides of the band being 45 N and 5.0 N respectively. The density of the metal is 8.0×10^3 kg m^{-3} and its specific heat capacity 4.0×10^2 J kg^{-1} K^{-1}. Find the rate at which the temperature of the disc rises initially.

For what speed of revolution would a disc of the same dimensions, but consisting of metal of density 2.7×10^3 kg m^{-3} and specific heat capacity 8.0×10^2 J kg^{-1} K^{-1}, give the same initial rate of temperature rise when rotated against the friction band under the same tensions as before?

(*O.*)

7. Define the terms *specific heat capacity, specific latent heat of vaporization*.

Describe how the specific latent heat of vaporization of a liquid such as alcohol may be determined by an electrical method explaining how a correction for heat losses may be made in the experiment.

A well-lagged copper calorimeter of mass 100 g contains 200 g of water and 50.0 g of ice all at 0 °C. Steam at 100 °C containing condensed water at the same temperature is passed into the mixture until the temperature of the calorimeter and its contents is 30.0 °C. If the increase in mass of the calorimeter and its contents is 25.0 g calculate the percentage of condensed water in the wet steam. (Assume that specific latent heat of vaporization of water at 100 °C = 2.26×10^3 J g^{-1}; specific latent heat of fusion of ice at 0 °C = 3.34×10^2 J g^{-1}; specific heat capacity of copper = 0.400 J g^{-1} K^{-1}; mean specific heat capacity of water = 4.18 J g^{-1} K^{-1}. (*L.*)

Expansion

8. The steel cylinder of a car engine has an aluminium alloy piston. At 15 °C the internal diameter of the cylinder is exactly 8.0 cm and there is an all-round clearance between the piston and the cylinder wall of 0.050 mm. At what temperature will they fit perfectly? (Linear expansivities of steel and the

aluminium alloy are 1.2×10^{-5} and 1.6×10^{-5} K^{-1} respectively.)

9. The height of the mercury column in a barometer is 76.46 cm as read at 15 °C by a brass scale which was calibrated at 0 °C. Calculate the error caused by the expansion of the scale and hence find the true height of the column. (Linear expansivity of brass is 1.900×10^{-5} K^{-1}.)

10. Calculate the minimum tension with which platinum wire of diameter 0.10 mm must be mounted between two points in a stout invar frame if the wire is to remain taut when the temperature rises 100 °C. Platinum has linear expansivity 9.0×10^{-6} K^{-1} and the Young modulus 1.7×10^{11} Pa. The thermal expansion of invar may be neglected.

(*O. and C. part qn.*)

11. (*a*) Distinguish between the *real* and *apparent* cubic expansivities.

A glass vessel contains some tungsten and is then filled with mercury to a certain mark. It is found that the mercury level remains at this mark despite changes of temperature. What is the ratio of volumes of the mercury and tungsten? (Linear expansivities of tungsten and glass are 4.4×10^{-6} and 8.0×10^{-6} K^{-1} respectively, and real cubic expansivity of mercury is 1.8×10^{-4} K^{-1}.)　　(*L. part qn.*)

(*b*) The density of a certain oil at 15 °C is 1.03 g cm^{-3} and its cubic expansivity is 8.50×10^{-4} K^{-1}; the density of water at 4 °C is 1.00 g cm^{-3} and its mean cubic expansivity over the range concerned is 2.10×10^{-4} K^{-1}. Find the temperature at which drops of the oil will just float in water.

(*O. part qn.*)

Thermal conductivity

12. A cubical container full of hot water at a temperature of 90 °C is completely lagged with an insulating material of thermal conductivity 6.4×10^{-4} W cm^{-1} K^{-1} (6.4×10^{-2} W m^{-1} K^{-1}). The edges of the container are 1.0 m long and the thickness of the lagging is 1.0 cm. Estimate the rate of flow of heat through the lagging if the external temperature of the lagging is 40 °C. Mention any assumptions you make in deriving your result.

Discuss qualitatively how your result will be affected if the thickness of the lagging is increased considerably assuming that the temperature of the surrounding air is 18 °C.

(*J.M.B.*)

13. Explain what is meant by *temperature gradient*.

An ideally lagged compound bar 25 cm long consists of a copper bar 15 cm long joined to an aluminium bar 10 cm long and of equal cross-sectional area. The free end of the copper is maintained at 100 °C and the free end of the aluminium at 0 °C. Calculate the temperature gradient in each bar when steady state conditions have been reached.

Thermal conductivity of copper = 3.9 W cm^{-1} K^{-1}.
Thermal conductivity of aluminium = 2.1 W cm^{-1} K^{-1}.

(*J.M.B.*)

14. Define thermal conductivity and explain how you would measure its value for a good conductor. Explain why the method which you describe is not suitable for a poor conductor.

The walls of a container used for keeping objects cool consist of two thicknesses of wood 0.50 cm thick separated by a space 1.0 cm wide packed with a poorly conducting material. Calculate the rate of flow of heat per unit area into the container if the temperature difference between the internal and external surfaces is 20 °C. (Thermal conductivity of wood = 2.4×10^{-3} W cm^{-1} K^{-1}, of the poorly conducting material = 2.4×10^{-4} W cm^{-1} K^{-1}.)　　(*A.E.B.*)

15. The ends of a straight uniform metal rod are maintained at temperatures of 100 °C and 20 °C, the room temperature being below 20 °C. Draw sketch-graphs of the variation of the temperature of the rod along its length when the surface of the rod is (*a*) lagged, (*b*) coated with soot, (*c*) polished. Give a qualitative explanation of the form of the graphs.

A liquid in a glass vessel of wall area 595 cm^2 and thickness 2.0 mm is agitated by a stirrer driven at a uniform rate by an electric motor rated at 100 W. The efficiency of conversion of electrical to mechanical energy in the motor is 75%. The temperature of the outer surface of the glass is maintained at 15.0 °C. Estimate the equilibrium temperature of the liquid, stating any assumptions you make.

Thermal conductivity of glass = 0.840 W m^{-1} K^{-1}.　　(*C.*)

5 Optical properties

Introduction

The scientific study of light and optical materials is involved in the making of spectacles, cameras, projectors, binoculars, microscopes and telescopes. Whilst the most important of all optical materials are the various kinds of glass, others such as plastics, Polaroid, synthetic and natural crystals have useful applications.

In this chapter we shall consider the behaviour of certain optical components and instruments. Light will be treated as a form of energy which travels in straight lines called *rays*, a collection of rays being termed a *beam*. The ray treatment of light is known as geometrical optics and is developed from

    (i) rectilinear propagation, i.e. straight-line travel
    (ii) the laws of reflection
    (iii) the laws of refraction.

When light comes to be regarded as waves it will be seen that shadows cast by objects are not as sharp as rectilinear propagation suggests. However, for the present it will be sufficiently accurate to assume that light does travel in straight lines so long as we exclude very small objects and apertures (those with diameters less than about 10^{-2} mm).

Reflection at plane surfaces

(*a*) *Laws of reflection*. When light falls on a surface it is partly reflected, partly transmitted and partly absorbed. Considering the part reflected, experiments

with rays of light and mirrors show that two laws hold.

    1. *The angle of reflection equals the angle of incidence*, i.e. $i_1 = i_2$ in Fig. 5.1.

    2. *The reflected ray is in the same plane as the incident ray and the normal to the mirror at the point of incidence*, i.e. the reflected ray is not turned to either side of the normal as seen from the incident ray.

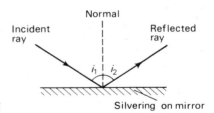

Fig. 5.1

Note that the angles of incidence and reflection are measured to the normal to the surface and not with the surface itself.

(*b*) *Regular and diffuse reflection*. A mirror in the form of a highly polished metal surface or a piece of glass with a deposit of silver on its back surface reflects a high percentage of the light falling on it. If a parallel beam of light falls on a plane (i.e. flat) mirror in the direction IO, it is reflected as a parallel beam in the direction OR and *regular reflection* is said to occur, Fig. 5.2a. Most objects, however, reflect light diffusely and the rays in an incident parallel beam are reflected in many directions as in Fig. 5.2b. Diffuse reflection is due

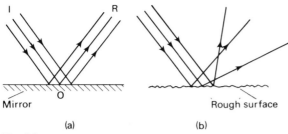

Mirror   Rough surface

(a)   (b)

Fig. 5.2

to the surface of the object not being perfectly smooth like a mirror and although at each point on the surface the laws of reflection are observed, the angle of incidence and therefore the angle of reflection varies from point to point.

(c) *Rotation of a mirror.* When a mirror is rotated through a certain angle, the reflected ray turns through *twice* that angle. This is a useful fact which can be proved by considering the plane mirror of Fig. 5.3a. When it is in position MM, the ray IO, incident at angle i, is reflected along OR so that $\angle RON = i$.

If the mirror is now rotated through an angle θ to position M'M' and the *direction of the incident ray kept constant*, Fig. 5.3b, the angle of incidence $\angle ION'$ becomes $(i + \theta)$ since the angle between the first and second positions of the normals, i.e. $\angle NON'$, is also θ. Let OR' be the new direction of the reflected ray then $\angle R'ON' = (i + \theta)$. The reflected ray is thus turned through $\angle ROR'$ and

$$\begin{aligned}\angle ROR' &= \angle R'ON - \angle RON \\ &= \angle R'ON' + \angle N'ON - \angle RON \\ &= (\theta + i) + \theta - i \\ &= 2\theta \\ &= \text{twice the angle of rotation of the mirror}\end{aligned}$$

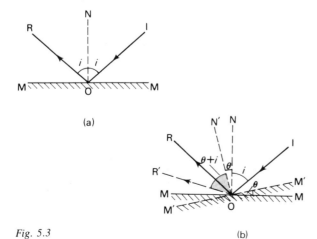

Fig. 5.3

(a)

(b)

(d) *Optical lever and light-beam galvanometers.* Some sensitive galvanometers use a beam of light in conjunction with a small mirror as a pointer. The arrangement is called an *optical lever*, it uses the 'rotation of a mirror' principle and increases the ability of the meter to detect small currents, i.e. makes it more sensitive.

A tiny mirror M is attached to the part (e.g. a coil of wire) of the meter which rotates when a current flows in it, Fig. 5.4. A beam of light from a fixed lamp falls on M and is reflected onto a scale S. For a given current, the longer the pointer (i.e. the reflected beam) the greater will be the deflection observed on S. Besides being almost weightless the arrangement has the additional advantage of doubling the rotation of the moving part since the angle the reflected beam turns through is twice the angle of rotation of the mirror, the direction of the incident ray remaining fixed.

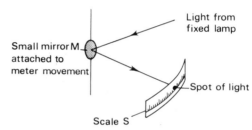

Small mirror M attached to meter movement

Light from fixed lamp

Spot of light

Scale S

Fig. 5.4

Images in plane mirrors

(a) *Point object.* The way in which the image of a point object is seen in a plane mirror is shown in Fig. 5.5. Rays from the object at O are reflected according to the laws of reflection so that they *appear* to come from a point I behind the mirror and this is where the observer imagines the image to be. The image at I is called an unreal or *virtual* image because the rays of light do not actually pass through it, they only seem to come from it. It would not be obtained on a screen placed at I as would a *real* image which is one where rays really do meet. (The image produced on a cinema screen by a projector is a real image.) Rays OA and AE are real rays, IA is a virtual ray that appears to have travelled a certain path but in fact has not; it gives rise to a virtual image.

Everyday observation suggests and experiment shows that *the image in a plane mirror is as far behind the mirror as the object is in front and that the line joining the object to the image is perpendicular to the mirror*, i.e. in Fig. 5.5 ON = NI and OI is at right

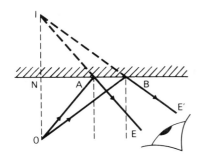

Fig. 5.5

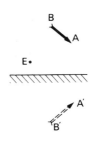

Fig. 5.7

angles to the mirror. However, you should be able to show, using the first law of reflection and congruent triangles, that this is so and also that *all* rays from O, after reflection, appear to intersect at I. A perfect image is thus obtained, i.e. all rays from the point object pass through *one* point on the image; a plane mirror is one of the few optical devices achieving this.

It is possible for a plane mirror to give a real image. Thus in Fig. 5.6*a* a convergent beam is reflected so that the reflected rays actually pass through a point I in front of the mirror. There is a real image at I which can be picked up on a screen. At the point O, towards which the incident beam was converging before it was intercepted by the mirror, there is considered to be a *virtual object*. Later we will find it useful on occasion to treat a convergent beam in this way. Comparing Figs. 5.6*a* and *b*, we see that in the first, a *convergent beam* regarded as a virtual point object gives a real point image, whilst in the second, a *divergent beam* from a real point object gives a virtual point image. In both cases object and image are equidistant from the mirror.

is at A′, the two points being equidistant from the mirror. The image of point B is at B′. If an eye at E views the object directly it sees A on the right of B, but if it observes the image in the mirror A′ is on the left. The right-hand side of the object thus becomes the left-hand side of image and vice versa. The image is said to be *laterally inverted*, i.e. the wrong way round, as you can check by looking at yourself in a mirror.

(*c*) *No parallax method of locating images*. Suppose the object is a small pin O placed in front of a plane mirror M. To find the position of its virtual image I, a large locating pin P is placed behind M, Fig. 5.8*a*, and moved towards or away from M until P and the image of O always appear to move together when the observer moves his head from side to side. P and I are then coincident and P is at the position of the image of O. When P and I do not coincide there is relative movement, called *parallax*, between them when the observer's head is moved sideways. The location of I can be achieved more quickly by remembering that if P is farther from M than I, then P appears to move in the *same* direction as the observer, Fig. 5.8*b*.

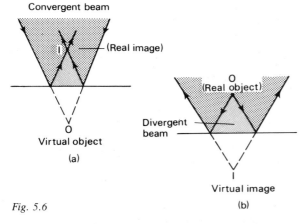

Fig. 5.6

(*b*) *Extended object*. Each point on an extended (finite-sized) object produces a corresponding point image. In Fig. 5.7 the image of a point A on the object

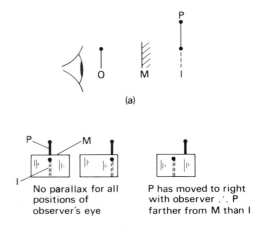

Fig. 5.8

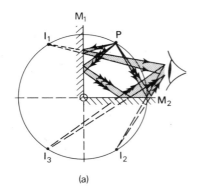

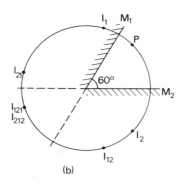

Fig. 5.9   (a)   (b)

The no parallax method is used to find real as well as virtual images as we shall see later when curved mirrors and lenses are considered.

(*d*) *Inclined mirrors.* Two mirrors M_1 and M_2 at 90° form *three* images of an object P placed between them, Fig. 5.9*a*. I_1 is formed by a single reflection at M_1, I_2 by a single reflection at M_2 and I_3 by reflections at M_1 and M_2. The line joining each image and its object is perpendicularly bisected by the mirror involved (we can think of I_3 as being either the image of I_1 acting as an object for mirror M_2 extended to the left or the image of I_2 as an object for mirror M_1 extended downwards) and so it follows that $OP = OI_1 = OI_2 = OI_3$. Justify this. Hence P, I_1, I_2 and I_3 lie on a circle centre O.

Two mirrors at 60° to each other form *five* images, Fig. 5.9*b*. As the angle between the mirrors decreases, the number of images increases and in general for an angle θ (which is such that $360/\theta$ is an integer) it can be shown that $[(360/\theta) - 1]$ images are formed. When the mirrors are parallel $\theta = 0°$ and in theory an infinite number of images should be obtained, all lying on a straight line passing through the object and perpendicular to the mirrors, Fig. 5.10. In practice some light is lost at each reflection and a limited number only is seen. If the distances of P from M_1 and M_2 are a and b respectively, prove that the separation of the images is successively $2a$, $2b$, $2a$, $2b$, etc.

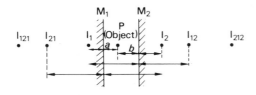

Fig. 5.10

Curved mirrors

Curved mirrors are used as car driving-mirrors and as reflectors in car headlamps, searchlights and flash-lamps. They are an essential component of the largest telescopes. We shall consider mainly spherical mirrors, i.e. those which are part of a spherical surface.

(*a*) *Terms and definitions.* There are two types of spherical mirror, concave and convex, Figs. 5.11*a* and *b*. In a concave mirror the centre C of the sphere of which the mirror is a part is in front of the reflecting surface, in a convex mirror it is behind. C is the *centre of curvature* of the mirror and P, the centre of the mirror surface, is called the *pole*. The line CP produced is the *principal axis*. AB is the *aperture* of the mirror.

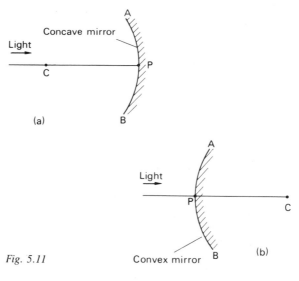

Fig. 5.11

Observation shows that a *narrow* beam of rays, parallel and near to the principal axis, is reflected from a concave mirror so that all rays converge to a point F on the principal axis, Fig. 5.12. F is called the *principal*

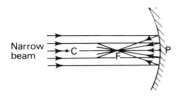

Fig. 5.12

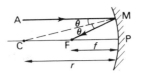

Fig. 5.14

focus of the mirror and it is a *real* focus since light actually passes through it. Concave mirrors are also known as converging mirrors because of their action on a parallel beam of light.

A narrow beam of rays, parallel and near to the principal axis, falling on a convex mirror is reflected to form a divergent beam which *appears* to come from a point F behind the mirror, Fig. 5.13. A convex mirror thus has a *virtual* principal focus; it is also called a diverging mirror.

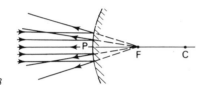

Fig. 5.13

Rays which are close to the principal axis and make small angles with it, i.e. are nearly parallel to the axis, are called *paraxial* rays. Our treatment of spherical mirrors will be restricted at present to such rays, which, in effect, means we shall consider only mirrors of small aperture. In diagrams, however, they will be made larger for clarity.

Spherical mirrors form a point image of *all* paraxial rays from a point object (as we will see shortly, p. 89), as well as bringing paraxial rays that are parallel to the principal axis to a point focus F.

(*b*) *Relation between f and r.* The distance PC from the pole to the centre of curvature of a spherical mirror is called its *radius of curvature* (*r*); the distance PF from the pole to the principal focus is its *focal length* (*f*). A simple relation exists between *f* and *r*.

In Fig. 5.14 a ray AM, parallel to the principal axis of a concave mirror of small aperture, is reflected through the principal focus F. If C is the centre of curvature, CM is the normal to the mirror at M because the radius of a spherical surface is perpendicular to the surface. Hence by the first law of reflection

$$\angle AMC = \angle CMF = \theta$$

But $\qquad \angle AMC = \angle MCF \quad$ (alternate angles)

$$\therefore \quad \angle CMF = \angle MCF$$

$\triangle$FCM is thus isosceles and FC = FM. The rays are paraxial and so M is very close to P; therefore to a good approximation FM = FP.

$$\therefore \quad FC = FP \quad or \quad FP = \tfrac{1}{2}CP$$

That is $\qquad\qquad f = r/2$

Thus the *focal length of a spherical mirror is approximately half the radius of curvature.* Check this relation for a convex mirror using Fig. 5.15.

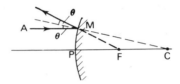

Fig. 5.15

Images in spherical mirrors

In general the *position* of the image formed by a spherical mirror and its *nature* (i.e. whether it is real or virtual, inverted or upright, magnified or diminished) depend on the distance of the object from the mirror. Information about the image in any case can be obtained either by drawing a ray diagram or by calculation using formulae.

(*a*) *Ray diagrams.* We shall assume that small objects on the principal axes of mirrors of small aperture are being considered so that all rays are paraxial. Point images will therefore be formed of points on the object.

To construct the image, *two* of the following three rays are drawn from the *top of the object*.

(*i*) A ray parallel to the principal axis which after reflection actually passes through the principal focus or appears to diverge from it.

(*ii*) A ray through the centre of curvature which strikes the mirror normally and is reflected back along the same path.

(*iii*) A ray through the principal focus which is reflected parallel to the principal axis, i.e. a ray travelling the reverse path to that in (*i*).

Since we are considering paraxial rays the mirror must be represented by a *straight line* in accurate

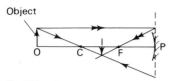

(i) *Object* beyond C.
Image between C and F, real,
inverted, diminished.

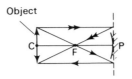

(ii) *Object* at C.
Image at C, real, inverted,
same size.

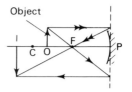

(iii) *Object* between C and F.
Image beyond C, real,
inverted, magnified.

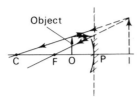

(iv) *Object* between F and P.
Image behind mirror virtual,
upright, magnified.

Fig. 5.16

Notes: 1. In (i) and (iii) O and I are interchangeable; such positions of object and image are called *conjugate points*.

2. C is a *self-conjugate point*—as (ii) shows, object and image are coincident at C.

3. If the object is at infinity (i.e. a long way off), a real image is formed at F. Conversely an object at F gives a real image at infinity.

4. In all cases the foot of the object is on the principal axis and its image also lies on this line.

diagrams. It should also be appreciated that the rays drawn are *constructional* rays and are not necessarily those by which the image is seen.

The diagrams for a concave mirror are shown in Fig. 5.16 and for a convex mirror in Fig. 5.17. In the latter case no matter where the object is, the image is always virtual, upright and diminished. Fig. 5.18 shows that a convex mirror gives a wider *field of view* than a plane mirror which explains its use as a driving mirror and on the stairs of double-decker buses. It does make the estimation of distances more difficult, however, because there is only small movement of the image for large movement of the object.

(b) The mirror formula. In Figs. 5.19*a* and *b* a ray OM from a point object O on the principal axis is reflected at M so that the angles θ, made by the incident and reflected rays with the normal CM, are equal. A ray OP strikes the mirror normally and is reflected back along PO. The intersection I of the reflected rays MI and PO in (*a*) gives a *real* point image of O, and of MI and PO both produced backwards in (*b*) gives a *virtual* point image of O.

Let angles α, β and γ be as shown. In $\triangle$CMO, since the exterior angle of a triangle equals the sum of the interior opposite angles,

Concave

$$\beta = \alpha + \theta, \quad \therefore \theta = \beta - \alpha$$

In $\triangle$CMI
$$\gamma = \beta + \theta$$
$$\therefore \quad \theta = \gamma - \beta$$
$$\therefore \quad \beta - \alpha = \gamma - \beta$$
$$\therefore \quad 2\beta = \gamma + \alpha \tag{1}$$

Convex

$$\theta = \alpha + \beta$$

In $\triangle$CMI
$$\gamma = \theta + \beta$$
$$\therefore \quad \theta = \gamma - \beta$$
$$\therefore \quad \alpha + \beta = \gamma - \beta$$
$$\therefore \quad 2\beta = \gamma - \alpha \tag{1$'$}$$

If the mirror is of small aperture, the rays are *paraxial*, M will be close to P and α, β and γ are small. Then α (in radians) $\simeq \tan \alpha$ (1 radian = 57.3°)

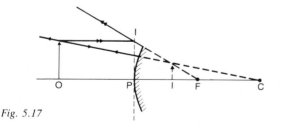

Fig. 5.17

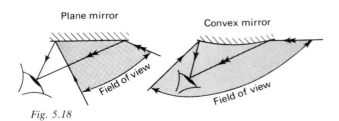

Plane mirror

Convex mirror

Fig. 5.18

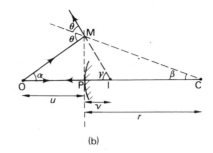

Fig. 5.19   (a)   (b)

$$\therefore \quad \alpha \simeq \frac{MP}{OP} \quad$$ where OP is the object distance

Similarly $\quad \beta \simeq \frac{MP}{CP} \quad$ where CP is the radius of curvature of the mirror

$$\gamma \simeq \frac{MP}{IP} \quad$$ where IP is the image distance

Substituting in (1)

$$\frac{2MP}{CP} = \frac{MP}{IP} + \frac{MP}{OP}$$

$$\therefore \quad \frac{2}{CP} = \frac{1}{IP} + \frac{1}{OP} \qquad (2)$$

Substituting in (1)'

$$\frac{2MP}{CP} = \frac{MP}{IP} - \frac{MP}{OP}$$

$$\therefore \quad \frac{2}{CP} = \frac{1}{IP} - \frac{1}{OP} \qquad (2)'$$

If we now introduce a *sign convention* so that distances are given a positive or a negative sign, the same equation is obtained for both concave and convex mirrors irrespective of whether objects and images are real or virtual. We shall adopt the 'real is positive' rule which states:

A real object or image distance is positive.
A virtual object or image distance is negative.

The focal length of a concave mirror is thus positive (since its principal focus is real) and of a convex mirror negative. The radius of curvature takes the same sign as the focal length.

If we now let *u, v* and *r* stand for the *numerical values* and *signs* of the object and image distances and radius of curvature respectively, then for both cases we get the same algebraic relationship

$$\frac{1}{v} + \frac{1}{u} = \frac{2}{r}$$

Also, since $r = 2f$, we have

$$\frac{1}{v} + \frac{1}{u} = \frac{1}{f}$$

Notes. (*i*) The formula is independent of the angle the incident ray makes with the axis, therefore *all* paraxial rays from point object O must, after reflection, pass through I to give a point image.

(*ii*) When numerical values for *u, v, r* or *f* are substituted in the formula, the appropriate sign *must* also be included; the sign (as well as the value) of the distance to be found comes out in the answer and so even if the sign is known from other information it must *not* be inserted in the equation.

(*iii*) Only two cases have been considered but it can be shown that the same formula holds for others, e.g. a concave mirror forming a virtual image of a real object, a convex mirror giving a real image of a virtual object (i.e. of converging light).

(*c*) *Magnification.* The lateral, transverse or linear magnification *m* (abbreviated to magnification) produced by a mirror is defined by

$$m = \frac{\text{height of image}}{\text{height of object}}$$

In Fig. 5.20, II' is the real image formed by a concave mirror of a finite object OO'. A paraxial ray from the top O' of the object, after reflection at say P, passes through the top I' of the image. Since the principal axis is the normal to the mirror at P, $\angle O'PO = \angle I'PI$, by the first law of reflection. Triangles O'PO and I'PI are

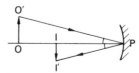

Fig. 5.20

therefore similar and so

$$\frac{\text{height of image}}{\text{height of object}} = \frac{I'I}{O'O} = \frac{IP}{OP}$$

$$\therefore \quad m = \frac{v}{u}$$

For example, if the image distance is twice the object distance, the image is twice the height of the object.

Notes. (*i*) No signs need be inserted in this formula for m, i.e. it is a numerical and not an algebraic formula.

(*ii*) The same result can be derived for other cases.

Mirror calculations

1. An object is placed 15 cm from (a) a concave mirror, (b) a convex mirror, of radius of curvature 20 cm. Calculate the image position and magnification in each case.

(*a*) Concave mirror

The object is real, therefore $u = + 15$ cm.
Since the mirror is concave $r = + 20$ cm, therefore $f = + 10$ cm.
Substituting values and signs in $1/v + 1/u = 1/f$

$$\frac{1}{v} + \frac{1}{(+15)} = \frac{1}{(+10)}$$

$$\therefore \quad \frac{1}{v} = \frac{1}{10} - \frac{1}{15} = \frac{1}{30}$$

$$\therefore \quad v = +30 \text{ cm}$$

The image is real since v is positive and it is 30 cm in front of the mirror. Also,

$$\text{magnification } m = \frac{v}{u} \quad \text{(numerically)}$$

$$= \frac{30}{15} = 2.0$$

The image is twice as high as the object (see Fig. 5.16*iii*).

(*b*) Convex mirror

We have $u = + 15$ cm but $r = - 20$ cm and $f = - 10$ cm since the mirror is convex.
Substituting as before in $1/v + 1/u = 1/f$

$$\frac{1}{v} + \frac{1}{(+15)} = \frac{1}{(-10)}$$

$$\therefore \quad \frac{1}{v} = -\frac{1}{10} - \frac{1}{15} = -\frac{5}{30}$$

$$\therefore \quad v = -\frac{30}{5} = -6.0 \text{ cm}$$

The image is virtual since v is negative and it is 6.0 cm behind the mirror. Also

$$m = \frac{v}{u} \quad \text{(numerically)}$$

$$= \frac{6}{15} = \frac{2}{5}$$

The image is two-fifths as high as the object (see Fig. 5.17).

2. When an object is placed 20 cm from a concave mirror, a real image magnified three times is formed. Find (a) the focal length of the mirror, (b) where the object must be placed to give a virtual image three times the height of the object.

(*a*) The object is real, therefore $u = + 20$ cm.
Also,

$$m = 3 = \frac{v}{u} = \frac{v}{20} \quad \text{(numerically)}$$

The image is real

$$\therefore \quad v = + 3 \times 20 = + 60 \text{ cm}$$

Substituting in $1/v + 1/u = 1/f$

$$\frac{1}{(+60)} + \frac{1}{(+20)} = \frac{1}{f}$$

$$\therefore \quad \frac{1}{f} = \frac{4}{60}$$

$$\therefore \quad f = +15 \text{ cm}$$

(*b*) Let the *numerical* value of the object distance $= x$.
Therefore, $u = + x$ since the object is real and $v = - 3x$ since the image is virtual; $f = + 15$ cm.
Using the mirror formula

$$\frac{1}{(+x)} + \frac{1}{(-3x)} = \frac{1}{(+15)}$$

$$\therefore \quad \frac{3}{3x} - \frac{1}{3x} = \frac{1}{15}$$

$$\therefore \quad \frac{2}{3x} = \frac{1}{15}$$

$$\therefore \quad x = 10 \text{ cm}$$

The object should be 10 cm in front of the mirror.

Note. By letting x be the numerical value of u we are able to substitute for u since we know its sign—a useful dodge.

Methods of measuring f for spherical mirrors

(a) Concave mirror

Rough method. The image formed by the mirror of a distant (several metres away) window is focused sharply on a screen, Fig. 5.21a. The distance between the mirror and the screen is f since rays of light from a point on such an object are approximately parallel, Fig. 5.21b.

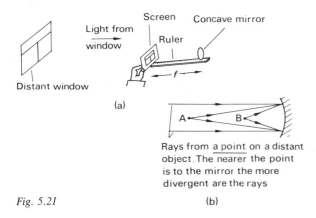

Fig. 5.21

Rays from a point on a distant object. The nearer the point is to the mirror the more divergent are the rays

(a)　　　　　　　(b)

Self-conjugate point method. The position of an object is adjusted until it coincides in position with its own image. In this position it is at the centre of curvature and distant r, i.e. $2f$, from the mirror. A point at which an object and its image coincide is said to be 'self-conjugate'.

The object can be a pin moved up and down above the mirror until there is no parallax between it and its

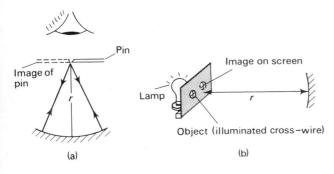

Fig. 5.22

real, inverted image, Fig. 5.22a. If an illuminated object is used, it is moved to and from the mirror until a clear image is obtained on a screen beside the object, Fig. 5.22b.

The pin/no parallax method generally gives more accurate results.

Mirror formula method. Several values of the image distance v corresponding to different values of the object distance u are found using either the pin/no parallax method or an illuminated object and screen. For each pair of values, f is calculated from $1/f = 1/v + 1/u$ and the average taken.

A better plan is to plot a graph of $1/v$ against $1/u$ and draw the best straight line AB through the points, Fig. 5.23. Each pair of values of u and v gives two points on the graph because u and v are interchangeable; they are called 'conjugate points'. The intercepts OA and OB on the axes are both equal to $1/f$. Prove this.

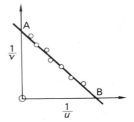

Fig. 5.23

(b) Convex mirror

Auxiliary converging lens method. A convex mirror normally forms a virtual image of a real object. Such an image cannot be located by a screen and is not easy to find by a pin/no parallax method. With the help of a converging lens, however, a real image can be obtained.

In Fig. 5.24 the lens L forms a real image at C of an object O when the convex mirror is absent. This image is located and the distance LC noted. The mirror is then placed between L and C and moved until O coincides in position with its own image.

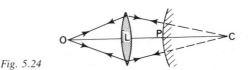

Fig. 5.24

The light from L is then falling normally on the mirror and is retracing its path to form a real inverted image at O. If produced, the rays from L must pass through the centre of curvature of the mirror since they are normal to the mirror. C, the position of the image of O formed by L alone, must therefore also be the centre of curvature of the mirror and so PC = r. Distance LP is measured and then $r = 2f = $ PC = LC − LP.

Refraction at plane surfaces

(a) *Laws of refraction*. When light passes from one medium, say air, to another, say glass, Fig. 5.25, part is reflected back into the first medium and the rest passes into the second medium with its direction of travel changed. The light is said to be bent or *refracted* on entering the second medium and the angle of refraction is the angle made by the refracted ray OB with the normal ON. There are two laws of refraction.

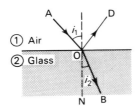

Fig. 5.25

1. *For two particular media, the ratio of the sine of the angle of incidence to the sine of the angle of refraction is constant*, i.e. $\sin i_1 / \sin i_2 = $ a constant in Fig. 5.25. (This is known as Snell's law after its discoverer.)

2. *The refracted ray is in the same plane as the incident ray and the normal to the mirror at the point of incidence but on the opposite side of the normal from the incident ray.*

The constant ratio $\sin i_1 / \sin i_2$ is called the *refractive index* for light passing from the first to the second medium. If the media containing the incident and refracted rays are denoted by ① and ② respectively, the refractive index is written as $_1n_2$. That is,

$$_1n_2 = \frac{\sin i_1}{\sin i_2}$$

The ratio depends on the colour of the light and is usually stated for yellow light. If medium ① is a vacuum (or in practice, air) we refer to the *absolute* refractive index of medium ② and denote it by $_vn_2$ or $_an_2$ or simply by n_2. The absolute refractive index of water is 1.33, of crown glass about 1.5 and of air at normal pressure about 1.0003—which is 1 near enough, the same as for a vacuum.

The greater the refractive index (absolute) of a medium the greater is the change in direction suffered by a ray of light when it passes from air to the medium. Refraction is therefore greater from air to crown glass than from air to water. In both cases the refracted ray is bent *towards* the normal, i.e. towards ON in Fig. 5.26a and the light is travelling into an 'optically denser' medium. A ray travelling from glass or water to air is bent *away from* the normal, Fig. 5.26b.

Refraction can be attributed to the fact that light has different speeds in different media.

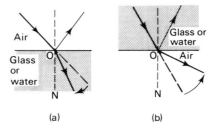

Fig. 5.26

(b) *Refractive index relationships*

(i) $_1n_2 = 1/_2n_1$. Consider a ray AO travelling from air (medium ①) to glass (medium ②) and refracted along OB as in Fig. 5.27. We have

$$_1n_2 = \frac{\sin i_1}{\sin i_2}$$

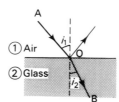

Fig. 5.27

Since light rays are reversible, a ray travelling along BO in glass (medium ②) will be refracted along OA in air (medium ①), hence

$$_2n_1 = \frac{\sin i_2}{\sin i_1}$$

$$\therefore \quad _1n_2 = \frac{1}{_2n_1}$$

For example, if the refractive index from air to water $(_an_w)$ is 4/3 then that from water to air $(_wn_a)$ is 3/4.

(ii) $_1n_3 = {_1n_2} \times {_2n_3}$. Suppose a ray AB travels from air (medium ①), to glass (medium ②), to water (medium ③), to air (medium ①), as in Fig. 5.28. Experiment shows that *if the boundaries of the media are parallel,* the emergent ray DE, although laterally displaced, is parallel to the incident ray AB. The incident and emergent angles are thus equal and are denoted by i_1. We then have

$$_1n_2 = \frac{\sin i_1}{\sin i_2} \qquad _2n_3 = \frac{\sin i_2}{\sin i_3} \qquad _3n_1 = \frac{\sin i_3}{\sin i_1}$$

$$\therefore \quad _1n_2 \times {_2n_3} \times {_3n_1} = \frac{\sin i_1}{\sin i_2} \times \frac{\sin i_2}{\sin i_3} \times \frac{\sin i_3}{\sin i_1} = 1$$

$$\therefore \quad _1n_2 \times {_2n_3} = \frac{1}{{_3n_1}}$$

From (i), $\qquad _1n_3 = \dfrac{1}{{_3n_1}}$

$$\therefore \quad _1n_3 = {_1n_2} \times {_2n_3}$$

The order of the subscripts aids memorization of the relation. For example, if we wish to know the refractive index for water to glass ($_wn_g$) and we know air to water ($_an_w = \frac{4}{3}$) and air to glass ($_an_g = \frac{3}{2}$) then

$$_wn_g = {_wn_a} \times {_an_g} = \frac{1}{{_an_w}} \times {_an_g}$$

$$= \tfrac{3}{4} \times \tfrac{3}{2} = \tfrac{9}{8}$$

Will a ray of light be bent towards or away from the normal on travelling from water to glass?

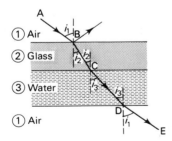

Fig. 5.28

(iii) $n_1 \sin i_1 = n_2 \sin i_2$. This is a more symmetrical form of Snell's law, useful in calculations, which will now be derived.

From Fig. 5.27 we can say

$$_1n_2 = \frac{\sin i_1}{\sin i_2}$$

From (ii) above,

$$_1n_2 = {_1n_a} \times {_an_2} \quad \text{(a = air or a vacuum)}$$

$$= \frac{_an_2}{_an_1} = \frac{n_2}{n_1}$$

where n_1 and n_2 are absolute refractive indices of media ① and ② respectively.

$$\therefore \quad \frac{n_2}{n_1} = \frac{\sin i_1}{\sin i_2}$$

$$\therefore \quad n_1 \sin i_1 = n_2 \sin i_2$$

For example, if a ray of light is incident on a water-glass boundary at 30° then $i_1 = i_w = 30°$ and if $n_1 = n_w = \frac{4}{3}$ and $n_2 = n_g = \frac{3}{2}$, the angle of refraction $i_2 = i_g$ is given by

$$n_w \sin i_w = n_g \sin i_g$$

i.e. $\qquad \tfrac{4}{3} \sin 30 = \tfrac{3}{2} \sin i_g$

$$\therefore \quad \sin i_g = \tfrac{4}{3} \times \tfrac{1}{2} \times \tfrac{2}{3} \quad (\sin 30 = \tfrac{1}{2})$$

$$= \tfrac{4}{9}$$

$$\therefore \quad i_g = 26°$$

(c) *Real and apparent depth.* Because of refraction the apparent depth of a pool of clear water, when viewed from above the surface, is less than its real depth; also, an object under water is not where it seems to be to an outside observer.

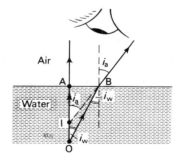

Fig. 5.29

In Fig. 5.29 rays from a point O under water are bent away from the normal at the water–air boundary and appear to come from I, the image of O. For refraction at B from water to air,

$$n_w \sin i_w = n_a \sin i_a = \sin i_a \qquad (n_a = 1)$$

$$\therefore \quad n_w = \frac{\sin i_a}{\sin i_w}$$

$$\therefore \quad n_w = \frac{AB/IB}{AB/OB} = \frac{OB}{IB}$$

If the observer is directly above O, i_w and i_a are small, rays OB and IB are close to OA, thus making OB ≃ OA and IB ≃ IA. Hence

$$n_w = \frac{OA}{IA} \quad \text{(approximately)}$$

$$= \frac{\text{real depth}}{\text{apparent depth}}$$

Taking $n_w = \frac{4}{3}$, what will be the apparent depth of a pond actually 2 metres deep?

The distance OI is called the *displacement d* of the object and if t is the real depth then

$$d = OA - IA = t - \frac{t}{n_w}$$

$$= t\left(1 - \frac{1}{n_w}\right)$$

The same expression gives the displacement of an object which is some distance in air below a parallel-sided block of material, as can be seen from Fig. 5.30. When viewed through several media whose boundaries are parallel, the total displacement is the sum of the displacements that would be produced by each medium if the others were absent.

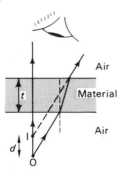

Fig. 5.30

A pool of water appears even shallower when viewed obliquely rather than from vertically above. As the observer moves, the image of a point O traces out a curve, called a *caustic*, whose apex is at I_1, Fig. 5.31.

(*d*) *Multiple images in mirrors.* Several images are seen when an object is viewed obliquely in a thick glass mirror with silvering on the back surface. In Fig. 5.32, I_1 is a faint image of object O formed by the weak

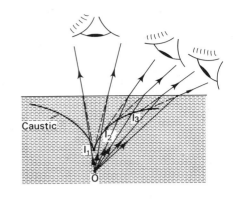

Fig. 5.31

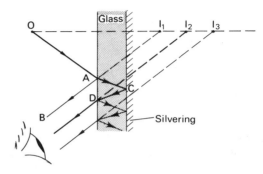

Fig. 5.32

reflected ray AB from the front surface of the mirror. I_2, the main reflection, is bright and is due to the refracted ray AC being reflected at the back (silvered) surface and again refracted at the front surface. I_3 and other weaker images are formed as shown. The net effect of these multiple reflections and refractions is to reduce the sharpness of the primary image I_2. Front-silvered mirrors eliminate the secondary images but are liable to be scratched and to tarnish.

(*e*) *Mirages.* These are often seen as small, distant pools of water on a hot tar-macadam road, particularly when vision is very oblique as it is for anyone in a car. They are caused by refraction in the atmosphere.

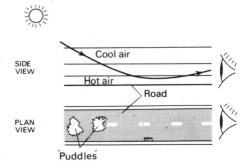

Fig. 5.33

The air near a road heated by the sun is hot; higher up the air is cool and its density greater. Consequently rays from the sky travelling towards the road are gradually refracted away from the normal as they pass from denser to less dense air. Upward bending of the light occurs, Fig. 5.33, and the blue light from the sky then seems to an observer to have been reflected from the road and gives the appearance of puddles.

Mirages are sometimes seen in the desert as distant, shimmering lakes.

Total internal reflection

(a) *Critical angle.* For small angles of incidence a ray of light travelling from one medium to another of smaller refractive index, say from glass to air, is refracted away from the normal, Fig. 5.34a; a weak internally reflected ray is also formed. Increasing the angle of incidence increases the angle of refraction and at a certain angle of incidence c, called the *critical angle*, the refracted ray just emerges along the surface of the glass and the angle of refraction is 90°, Fig. 5.34b. At this stage the internally reflected ray is still weak but just as c is exceeded it suddenly becomes bright and the refracted ray disappears, Fig. 5.34c. *Total internal reflection* is now said to be occurring since all the incident light is reflected inside the optically denser medium.

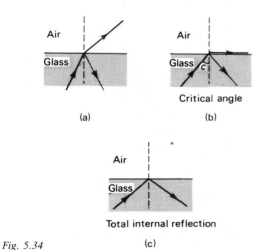

Fig. 5.34

Applying Snell's law in the form $n_1 \sin i_1 = n_2 \sin i_2$ to the critical ray at a glass–air boundary, we have

n_1 = refractive index of glass = n_g

i_1 = critical angle for glass = c

n_2 = refractive index of air = 1

i_2 = angle of refraction in air = 90°

$\therefore \; n_g \sin c = 1 \sin 90 = 1 \quad (\sin 90 = 1)$

$\therefore \; n_g = \dfrac{1}{\sin c}$

Taking $n_g = \frac{3}{2}$ (crown glass), $\sin c = \frac{2}{3}$ and so $c \simeq 42°$. Thus if the incident angle in the crown glass exceeds 42°, total internal reflection occurs. Can it occur when a ray of light in glass ($n_g = \frac{3}{2}$ say) is incident on a boundary with water ($n_w = \frac{4}{3}$)?

(b) *Totally reflecting prisms.* The disadvantages of plane mirrors (p. 94), silvered on either the back or front surface, can be overcome by using right-angled isosceles prisms (angles 90°, 45°, 45°) as reflectors.

The critical angle of crown glass is about 42° and a ray OA incident normally on face PQ of such a prism, Fig. 5.35a, suffers total internal reflection at face PR since the angle of incidence in the optically denser medium is 45°. A bright ray AB emerges at right angles to face QR since the angle of reflection at QR is also 45°. The prism thus reflects the ray through 90°.

Light can be reflected through 180° and an erect image obtained of an inverted one (as in prism binoculars, p. 119) if the prism is arranged as in Fig. 5.35b.

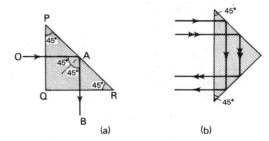

Fig. 5.35

(c) *Fibre optics.* Light can be confined within a bent glass rod by total internal reflection and so 'piped' along a twisted path, as in Fig. 5.36. The beam is reflected from side-to-side practically without loss (except for that due to absorption in the glass) and emerges only at the end of the rod where it strikes the surface almost normally, i.e. at an angle less than the critical angle. A single, very thin, solid glass-fibre behaves in the same way and if several thousand are taped together a flexible light pipe is obtained that can be used, as it has been in medicine and engineering, to illuminate some otherwise inaccessible spot. One difficulty which arises in a bundle of fibres is leakage of

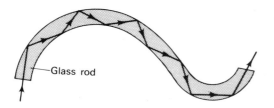

Fig. 5.36

light at places of contact between the fibres. This can be reduced by coating each fibre with glass of lower refractive index than its own, thereby encouraging total internal reflection.

If it is desired to transport an *image* and not simply to transport *light*, the fibres must occupy the same positions in the bundle relative to each other. Such bundles are more difficult to make and cost more. Fig. 5.37*a* shows part of the end-section of a bundle of fibres and Fig. 5.37*b* is a fibre-optics viewing instrument with bent light pipe and an inset of the word 'optics' as seen by a camera above the eyepiece at the top. Figure 5.37*c* shows a motorway sign illuminated by light pipes. Fibre optics could become an important research technique in medicine and engineering.

Refraction through prisms

A prism has two plane surfaces inclined to each other as are LMQP and LNRP in Fig. 5.38. Angle MLN is called the *refracting angle* of the prism, LP is the *refracting edge* and any plane such as XYZ which is perpendicular to LP is a *principal plane*.

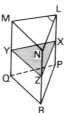

Fig. 5.38

The importance of the prism really depends on the fact that the angle of deviation suffered by light at the first refracting surface, say LMQP, is not cancelled out by the deviation at the second surface LNRP (as it is in a parallel-sided block), but is added to it. This is why it can be used in a spectrometer, an instrument for analysing light into its component colours. In what

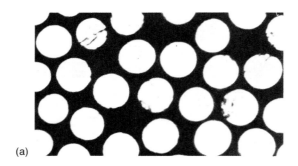

(a)

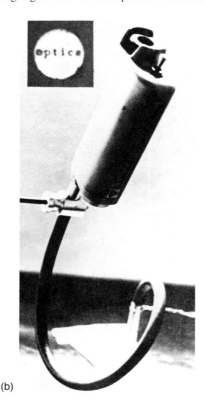

(c)

(b)

Fig. 5.37

follows, expressions for the angle of deviation will be obtained and subsequently used.

(*a*) *General formulae.* In Fig. 5.39, EFGH is a ray lying in a principal plane XYZ of a prism of refracting angle A and passing from air, through the prism and back to air again. KF and KG are normals at the points of incidence and emergence of the ray.

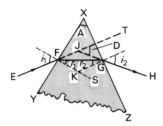

Fig. 5.39

For refraction at XY

angle of deviation = angle JFG = $i_1 - r_1$

For refraction at XZ

angle of deviation = angle JGF = $i_2 - r_2$

Since both deviations are in the same direction, the total deviation D is given by

angle TJH = angle JFG + angle JGF

i.e. $D = (i_1 - r_1) + (i_2 - r_2)$ (1)

Another expression arises from the geometry of Fig. 5.39.

In quadrilateral XFKG

angle XFK + angle XGK = 180°

∴ A + angle FKG = 180°

But since FKS is a straight line

angle GKS + angle FKG = 180°

∴   angle GKS = A

In triangle KFG, angle GKS is an exterior angle

∴   angle GKS = $r_1 + r_2$

∴ $A = r_1 + r_2$ (2)

Equations (1) and (2) are true for any prism. (The position and shape of the third side of the prism does not affect the refraction under consideration and so is shown as an irregular line in Fig. 5.39.)

(*b*) *Minimum deviation.* It is found that the angle of deviation D varies with the angle of incidence i_1 of the ray incident on the first refracting face of the prism. The variation is shown in Fig. 5.40*a* and for one angle of incidence it has a minimum value D_{min}. At this value the *ray passes symmetrically through the prism* (a fact that can be proved theoretically as well as be shown experimentally), i.e. the angle of emergence of the ray from the second face equals the angle of incidence of the ray on the first face: $i_2 = i_1 = i$, Fig. 5.40*b*. It therefore follows that $r_1 = r_2 = r$. Hence from equation (1) of the previous section the angle of minimum deviation D_{min} is given by

$$D_{min} = (i - r) + (i - r) = 2(i - r)   (3)$$

Also, from equation (2)

$$A = r + r = 2r$$

$$\therefore \quad r = \frac{A}{2}$$

Substituting for r in (3)

$$D_{min} = 2i - A$$

$$\therefore \quad i = \frac{A + D_{min}}{2}$$

If n is the refractive index of the material of the prism then

$$n = \frac{\sin i}{\sin r}$$

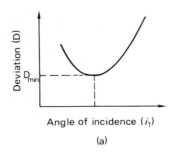

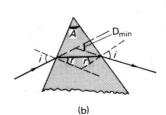

Fig. 5.40   (a)   (b)

$$n = \frac{\sin\left[(A + D_{min})/2\right]}{\sin\left(A/2\right)}$$

Thus if $A = 60°$ and $D_{min} = 40°$, then $(A + D_{min})/2 = 50°$ and so $n = \sin 50/\sin 30 = 1.5$.

Two points for you to consider. First, no values of D are shown on the graph of Fig. 5.40a for small values of i (less than about 30° for a crown glass prism of refracting angle 60° and $n = 1.5$). Why? Second, the above formula for minimum deviation only holds for a prism of angle A less than twice the critical angle. Why?

(c) *Small-angle prism.* The expression for the deviation in this case will be used later for developing lens theory.

Consider a ray falling almost normally in air on a prism of small angle A (less than about 6° or 0.1 radian) so that angle i_1 in Fig. 5.41 is small. Now $n = \sin i_1/\sin r_1$ where n is the refractive index of the material of the prism, therefore r_1 will also be small. Hence, since the sine of a small angle (like the tangent) is nearly equal to the angle in radians, we have

$$i_1 = nr_1$$

Also, $A = r_1 + r_2$ (see p. 97), and so if A and r_1 are small, r_2 and i_2 will also be small. From $n = \sin i_2/\sin r_2$ we can say

$$i_2 = nr_2$$

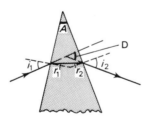

Fig. 5.41   (angles exaggerated for clarity)

The deviation D of a ray passing through any prism is given by (p. 97)

$$D = (i_1 - r_1) + (i_2 - r_2)$$

Substituting for i_1 and i_2

$$D = nr_1 - r_1 + nr_2 - r_2$$
$$= n(r_1 + r_2) - (r_1 + r_2)$$
$$= (n - 1)(r_1 + r_2)$$

But $A = r_1 + r_2$

$$\therefore D = (n - 1)A$$

This expression shows that for a given angle A *all* rays entering a *small-angle* prism at *small angles of incidence* suffer the *same* deviation.

(d) *Dispersion.* Newton found that when a beam of white light (e.g. sunlight) passes through a prism it is spread out by the prism into a band of all the colours of the rainbow from red to violet. The band of colours is called a *spectrum* and the separation of the colours by the prism is known as *dispersion*. He concluded that white light is a mixture of light of various colours and identified red, orange, yellow, green, blue, indigo, violet.

Red is deviated least by the prism and violet most as shown by the exaggerated diagram of Fig. 5.42a. The refractive index of the material of the prism for violet light is thus greater than for red light since the angle of incidence in the air is the same for red and violet rays.

A method of producing a *pure* spectrum, i.e. one in which the different colours do not overlap (as they do when a prism is used on its own), is shown in Fig. 5.42b. A diverging beam of white light, emerging from a very narrow slit, is made parallel by lens L_1 and then dispersed by the prism into a number of different coloured parallel beams, each travelling in a slightly different direction. Lens L_2 brings each colour to a separate focus on a screen. The spectrum is a series of monochromatic images of the slit and the narrower this is the purer the spectrum. (L_1 and L_2 are achromatic doublets, see p. 109.)

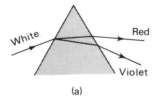

(a)

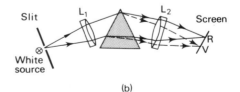

Fig. 5.42   (b)

Methods of measuring n

(a) *Real and apparent depth method* (solids and liquids). A travelling microscope is focused on a pencil dot O on a sheet of white paper lying on the bench and the reading on the microscope scale noted, Fig. 5.43. Let it be x. If the refractive index of glass is required, a block of the material is placed over the dot and the microscope refocused on the image I of O as seen through the block. Let the reading be y. Finally the microscope is focused on the top T of the block, made visible by a sprinkling of lycopodium powder.

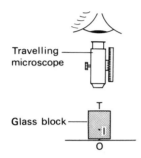

Fig. 5.43

Suppose the reading is now z, then

$$\text{real depth of O} = OT = z - x$$

$$\text{apparent depth of O} = IT = z - y$$

$$\therefore \quad n = \frac{\text{real depth}}{\text{apparent depth}} = \frac{z - x}{z - y}$$

The refractive index of a liquid in a beaker can be found by a similar procedure.

The method satisfies the assumption made in deducing the expression for n (p. 94) because the microscope collects only rays very close to the normal OT; accuracy of $\pm 1\%$ is possible if the microscope has a small depth of focus.

(b) *Minimum deviation method* (solids and liquids). A solid prism of the material is placed on the table of a spectrometer, A and D_{min} measured as described on p. 120 and n calculated from $n = \sin [(A + D_{min})/2]/\sin (A/2)$. The method is suitable for liquids if a hollow prism with perfectly parallel, thin walls is used. Accuracy of $\pm 0.1\%$ is possible.

(c) *Concave mirror method* (liquids). The centre of curvature C of the mirror is first found by moving an object pin up and down above the mirror until it coincides in position with its image (Method 2, p. 91). Some liquid is then poured into the mirror and the object pin moved until point O is found where it again coincides with its image. In Fig. 5.44 ray ONB must be retracing its own path after striking the mirror normally at B and if BN is produced it will pass through C.

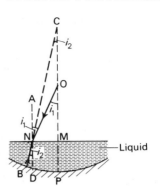

Fig. 5.44

For the refraction at N

i_1 = angle of incidence = $\angle$ONA = $\angle$NOM
(alt. angles)

i_2 = angle of refraction = $\angle$BND = $\angle$NCM
(corr. angles)

The refractive index n of the liquid is given by

$$n = \frac{\sin i_1}{\sin i_2} = \frac{\sin NOM}{\sin NCM} = \frac{NM/NO}{NM/NC} = \frac{NC}{NO}$$

If ray ONB is close to the principal axis CP of the mirror then to a good approximation NC = MC and NO = MO

$$\therefore \quad n = \frac{MC}{MO}$$

Both distances can be measured and n found. The method is useful when only a small quantity of liquid is available.

Thin lenses

Lenses are of two basic types, *convex* which are thicker in the middle than at the edges and *concave* for which the reverse holds. Fig. 5.45 shows examples of both types, bounded by spherical or plane surfaces.

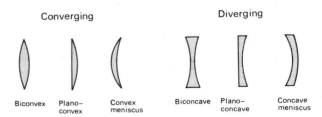

Fig. 5.45

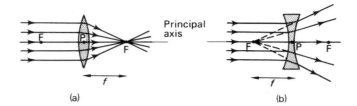

Fig. 5.46　(a)　(b)

The *principal axis* of a spherical lens is the line joining the centres of curvature of its two surfaces, Fig. 5.46. For the present our treatment will be confined to *paraxial* rays, i.e. rays close to the axis and making very small angles with it. In effect this means we shall only consider lenses of small aperture but in diagrams both angles and lenses will be made larger for clarity. The case of wide angle beams will be considered briefly later (p. 108).

The *principal focus* F of a thin lens is the point on the principal axis towards which all paraxial rays, parallel to the principal axis, converge in the case of a convex lens or from which they appear to diverge in the case of a concave lens, after refraction, Fig. 5.46*a* and *b*. Since light can fall on either surface, a lens has two principal foci, one on each side, and these are equidistant from its centre P (if the lens is thin and has the same medium on both sides, e.g. air). The distance FP is the *focal length f* of the lens. A convex lens is a *converging* lens[1] and has real foci. A concave lens is a *diverging* lens and has virtual foci.

A parallel beam at a small angle to the axis of a lens is refracted to converge to, or to appear to diverge from, a point in the plane containing F, perpendicular to the axis and known as the *focal plane*, Fig. 5.47*a* and *b*.

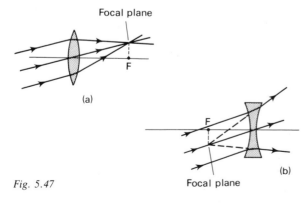

Focal plane

(a)

F

F

(b)

Focal plane

Fig. 5.47

[1] This is only true if the convex lens has a greater refractive index than the surrounding medium. In water a biconvex air lens diverges light.

As we shall see shortly the important property of a lens is that it focuses *all* paraxial rays from a point object (and not just parallel, paraxial rays) to form a point image.

Images formed by thin lenses

Information as to the position and nature of the image in any case can be obtained either from a ray diagram or by calculation.

(*a*) *Ray diagrams.* To construct the image of a small object perpendicular to the axis of a lens, *two* of the following three rays are drawn from the top of the object.

(*i*) A ray parallel to the principal axis which after refraction passes through the principal focus or appears to diverge from it.

(*ii*) A ray through the centre of the lens (called the *optical centre*) which continues straight on undeviated (it is only slightly displaced laterally because the middle of the lens acts like a thin parallel-sided block), Fig. 5.48.

Fig. 5.48

(*iii*) A ray through the principal focus which is refracted parallel to the principal axis, i.e. a ray travelling the reverse path to that in (*i*).

The diagrams for a converging lens are shown in Fig. 5.49*a* to *e* and for a diverging lens in Fig. 5.49*f*. The latter, like a convex mirror, always forms a virtual, upright and diminished image whatever the object position. Note that a thin lens is represented by a straight line at which all the refraction is considered to occur; in practice it is usually refracted both on entering and leaving the lens.

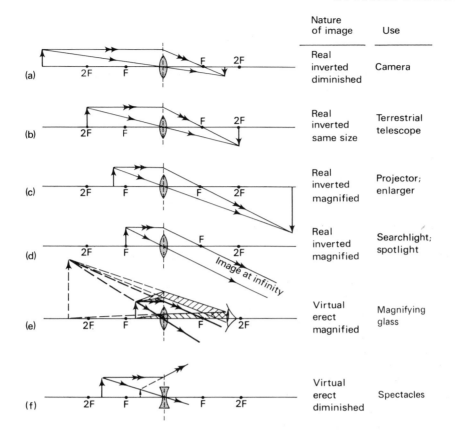

	Nature of image	Use
(a)	Real inverted diminished	Camera
(b)	Real inverted same size	Terrestrial telescope
(c)	Real inverted magnified	Projector; enlarger
(d)	Real inverted magnified	Searchlight; spotlight
(e)	Virtual erect magnified	Magnifying glass
(f)	Virtual erect diminished	Spectacles

Fig. 5.49

It must also be emphasized that the lines drawn are *constructional* ones; two narrow cones of rays that actually enter the eye of an observer from the top and bottom of an object are shown shaded in Fig. 5.49*e*. They are obtained by working back from the eye, from right to left here.

(*b*) *Simple formula for a thin lens.* We can regard a thin lens as made up of a large number of small-angle prisms whose angles increase from zero at the middle of the lens to a small value at its edge. Consider one such prism at distance *h* from the optical centre P of a lens, Fig. 5.50. If a paraxial ray parallel to the axis is incident on this prism it suffers small deviation *D* (since the prism is small-angled) and is refracted through the principal focus F. Hence, since the tangent of a small angle equals the small angle in radians,

$$D = \frac{h}{FP} \tag{1}$$

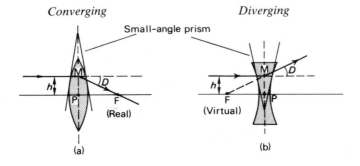

Converging   *Diverging*

Small-angle prism

Fig. 5.50

Converging

Diverging

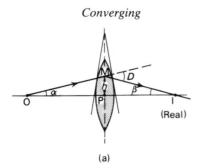

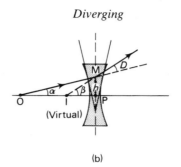

(Real)

(Virtual)

Fig. 5.51

(a)

(b)

Now consider a point object O on the axis which gives rise to a point image I, Fig. 5.51. If a paraxial ray from O is incident on the small-angle prism at distance h from the axis, it must also suffer deviation D (since all rays entering a small-angle prism at small angles of incidence suffer the same deviation (p. 98).

Let angles α and β be as shown. In triangle IOM, since the exterior angle of a triangle equals the sum of the interior opposite angles,

Converging	*Diverging*
$D = \alpha + \beta$	$D = \beta - \alpha$
$= \dfrac{h}{OP} + \dfrac{h}{IP}$	$= \dfrac{h}{IP} - \dfrac{h}{OP}$

Therefore from (1)

$$\frac{h}{FP} = \frac{h}{OP} + \frac{h}{IP} \quad (2) \qquad \bigg| \qquad \frac{h}{FP} = \frac{h}{IP} - \frac{h}{OP} \quad (2)'$$

If we introduce the 'real is positive' sign convention given on p. 89, the *focal length of a converging lens is positive* and of *a diverging lens negative*. If u, v and f stand for the *numerical values* and *signs* of the object and image distances and focal length respectively then for *both* cases we get the *algebraic* relationship

$$\frac{1}{v} + \frac{1}{u} = \frac{1}{f}$$

Notes. (*i*) The formula is independent of the angle the incident ray makes with the axis, therefore *all* paraxial rays from point object O must, after refraction, pass through I to give a point image, i.e. the small angle prisms to which the lens is equivalent deviate the various rays from O by varying amounts depending on the angle of the prism but always so that they all pass through I.

(*ii*) When numerical values for u, v and f are inserted in the formula, the appropriate sign *must* also be included.

(*c*) *Magnification.* The lateral, transverse or linear magnification m (abbreviated to magnification) produced by a lens is defined by

$$m = \frac{\text{height of image}}{\text{height of object}}$$

In Fig. 5.52 II′ is the real image formed by a converging lens of a finite object OO′. Triangles O′PO and I′PI are similar, therefore

$$\frac{\text{height of image}}{\text{height of object}} = \frac{I'I}{O'O} = \frac{IP}{OP}$$

$$\therefore \quad m = \frac{v}{u}$$

This is a numerical formula and no signs need be inserted.

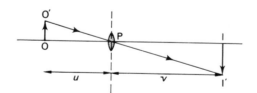

Fig. 5.52

Lens calculations

1. An object is placed 20 cm *from (a) a converging lens, (b) a diverging lens, of focal length 15* cm. *Calculate the image position and magnification in each case.*

(*a*) *Converging lens*
The object is real, therefore $u = +20$ cm.
Since the lens converges, $f = +15$ cm.
Substituting values and signs in $1/v + 1/u = 1/f$

$$\frac{1}{v} + \frac{1}{(+20)} = \frac{1}{(+15)}$$

$$\therefore \quad \frac{1}{v} = \frac{1}{15} - \frac{1}{20} = \frac{4}{60} - \frac{3}{60} = \frac{1}{60}$$

$$\therefore \quad v = +60 \text{ cm}$$

The image is real since v is positive and it is 60 cm from the lens.
Also,

$$\text{magnification } m = \frac{v}{u} \quad (\text{numerically})$$

$$= \frac{60}{20} = 3.0$$

The image is three times as high as the object (see Fig. 5.49c).

(b) *Diverging lens*
We have $u = +20$ cm and $f = -15$ cm.
Substituting as before in $1/v + 1/u = 1/f$

$$\frac{1}{v} + \frac{1}{(+20)} = \frac{1}{(-15)}$$

$$\therefore \quad \frac{1}{v} = -\frac{1}{15} - \frac{1}{20} = -\frac{7}{60}$$

$$\therefore \quad v = -\frac{60}{7} = -8.6 \text{ cm}$$

The image is virtual since v is negative and it is 8.6 cm from the lens.
Also,

$$m = \frac{v}{u} \quad (\text{numerically})$$

$$= \frac{60/7}{20} = \frac{3}{7}$$

The image is three-sevenths as high as the object (see Fig. 5.49f).

2. *An object is placed 6.0 cm from a thin converging lens A of focal length 5.0 cm. Another thin converging lens B of focal length 15 cm is placed coaxially with A and 20 cm from it on the side away from the object. Find the position, nature and magnification of the final image.*

For lens A,

$$u = +6.0 \text{ cm}, \qquad f = +5.0 \text{ cm}$$

Substituting in $1/v + 1/u = 1/f$

$$\frac{1}{v} + \frac{1}{(+6)} = \frac{1}{(+5)}$$

$$\therefore \quad \frac{1}{v} = \frac{1}{5} - \frac{1}{6} = \frac{6}{30} - \frac{5}{30} = \frac{1}{30}$$

$$\therefore \quad v = +30 \text{ cm}$$

Image I_1 in Fig. 5.53 is real, therefore *converging* light falls on lens B and I_1 acts as a virtual object for B. Applying $1/v + 1/u = 1/f$ to B we have,

$$u = -(AI_1 - AB) = -(30 - 20) = -10 \text{ cm},$$
$$f = +15 \text{ cm}$$

$$\therefore \quad \frac{1}{v} + \frac{1}{(-10)} = \frac{1}{(+15)}$$

$$\therefore \quad \frac{1}{v} = \frac{1}{15} + \frac{1}{10} = \frac{2}{30} + \frac{3}{30} = \frac{5}{30}$$

$$\therefore \quad v = +6.0 \text{ cm}$$

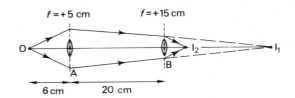

Fig. 5.53

Image I_2 is real and is formed 6.0 cm beyond B.

$$\text{Magnification by A} = m_1 = \frac{v}{u} = \frac{30}{6} = 5.0$$

$$\text{Magnification by B} = m_2 = \frac{v}{u} = \frac{6}{10} = \frac{3}{5}$$

$$\text{Total magnification } m = m_1 \times m_2 = 5 \times \frac{3}{5}$$

$$= 3.0$$

The final image is three times the size of the object. (*Note:* m_1 and m_2 are multiplied and *not* added. Why?)

Other thin lens formulae

(a) *Full formula for a thin lens.* We require to find a relationship, sometimes called the 'lens-maker's formula', between the focal length of a thin lens, the radii of curvature of its surfaces and the refractive index of the lens material. It will be assumed that (i) the lens can be replaced by a system of small-angle prisms and (ii) all rays falling on the lens are paraxial, i.e. the lens has a small aperture and all objects are near the axis.

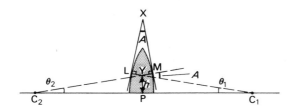

Fig. 5.54

Consider the prism of small angle A which is formed by the tangents XL and XM to the lens surfaces at L and M, Fig. 5.54. XL and XM are perpendicular to the radii of curvature C_1L and C_2M respectively, C_1 and C_2 being the centres of curvature of the surfaces. Therefore angle LXM (i.e. A) between the tangents equals angle MYC_1 between the radii

$$\therefore \quad \angle MYC_1 = A = \theta_1 + \theta_2$$

(ext. angle of triangle
$\qquad YC_2C_1 = $ sum of int. opp. angles)

But since θ_1 and θ_2 are small we can say $\theta_1 = \tan \theta_1$ and $\theta_2 = \tan \theta_2$

$$\therefore \quad A = \frac{h}{C_1P} + \frac{h}{C_2P} \qquad (h = YP)$$

The deviation D produced in *any* ray incident at a small angle on a prism of small angle A and refractive index n is (p. 98)

$$D = (n - 1)A$$
$$= (n - 1)\left(\frac{h}{C_1P} + \frac{h}{C_2P}\right)$$

If we now consider a ray parallel to the axis and at height h above it, it suffers the same deviation D as any other paraxial ray and since it is refracted through the principal focus F, $D = h/FP$ (from equation (1), p. 101). Hence

$$\frac{h}{FP} = (n - 1)\left(\frac{h}{C_1P} + \frac{h}{C_2P}\right)$$

$$\therefore \quad \frac{1}{FP} = (n - 1)\left(\frac{1}{C_1P} + \frac{1}{C_2P}\right)$$

Introducing a sign convention for distances converts this numerical relationship to an algebraic one applicable to all lenses and cases. Thus if f, r_1 and r_2 stand for the numerical values and *signs* of the focal length and radii of curvature respectively of the lens then we have

$$\frac{1}{f} = (n - 1)\left(\frac{1}{r_1} + \frac{1}{r_2}\right)$$

In the 'real is positive' convention the rule for the sign of a radius of curvature is—*a surface convex to the less dense medium has a positive radius while a surface concave to the less dense medium has a negative radius.* A positive surface thus converges light, a negative one diverges it.

A numerical example may help. For the convex meniscus lens of Fig. 5.55a we have $n = 1.5$, $r_1 = + 10$ cm (since it is convex to the air on its left), $r_2 = - 15$ cm (since it is concave to the air on its right).

$$\therefore \quad \frac{1}{f} = (1.5 - 1)\left(\frac{1}{(+10)} + \frac{1}{(-15)}\right) = \frac{1}{2}\left(\frac{3}{30} - \frac{2}{30}\right)$$
$$= \frac{1}{60}$$
$$\therefore \quad f = +60 \text{ cm}$$

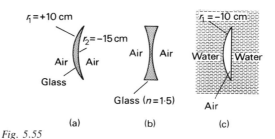

Fig. 5.55

Now calculate f for the biconcave lens in Fig. 5.55b whose radii of curvature are each 20 cm. (*Ans.* -20 cm.)

A more general form of the formula is

$$\frac{1}{f} = \left(\frac{n_2}{n_1} \sim 1\right)\left(\frac{1}{r_1} + \frac{1}{r_2}\right)$$

where n_2 is the refractive index of the lens material and n_1 that of the surrounding medium. $(n_2/n_1 \sim 1)$ *is always taken to be positive* since refractive indices do not have signs. ($\sim$ means the 'difference between'.)

For a plano-convex *air* lens of radius 10 cm, in *water* of refractive index $\frac{4}{3}$ we have $n_2 = 1$, $n_1 = \frac{4}{3}$, $r_1 = - 10$ cm (since it is concave to the air), $r_2 = \infty$, Fig. 5.55c.

$$\therefore \quad \frac{1}{f} = \left(\frac{1}{4/3} \sim 1\right)\left(\frac{1}{(-10)} + \frac{1}{\infty}\right) = \frac{1}{4} \times \left(\frac{-1}{10} + 0\right)$$

$$\therefore \quad f = -40 \text{ cm} \quad \text{(a diverging lens)}$$

(b) *Focal length of two thin lenses in contact.* Combinations of lenses in contact are used in many optical

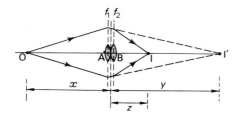

Fig. 5.56

instruments to improve their performance. In Fig. 5.56, A and B are two thin lenses in contact, of focal lengths f_1 and f_2. Paraxial rays from point object O on the principal axis are refracted through A and would, in the absence of B, give a real image of O at I'. Hence for A, $u = +x$ and $v = +y$. From the simple formula for a thin lens

$$\frac{1}{(+x)} + \frac{1}{(+y)} = \frac{1}{f_1}$$

For B, I' acts as a *virtual object* (i.e. converging light falls on B from A) giving a real image of O at I and so for B, $u = -y$ and $v = +z$

$$\therefore \quad \frac{1}{(-y)} + \frac{1}{(+z)} = \frac{1}{f_2}$$

Adding

$$\frac{1}{(+x)} + \frac{1}{(+z)} = \frac{1}{f_1} + \frac{1}{f_2}$$

Considering the combination, I is the real image formed of O by both lenses, therefore $u = +x$ and $v = +z$

$$\therefore \quad \frac{1}{(+x)} + \frac{1}{(+z)} = \frac{1}{f}$$

where f is the combined focal length, i.e. the focal length of the single lens that would be exactly equivalent to the two in contact

$$\therefore \quad \frac{1}{f} = \frac{1}{f_1} + \frac{1}{f_2}$$

For example if a converging lens of 5.0 cm focal length is in contact with a diverging lens of 10 cm focal length, then $f_1 = +5.0$ cm, $f_2 = -10$ cm and the combined focal length f is given by

$$\frac{1}{f} = \frac{1}{(+5)} + \frac{1}{(-10)} = \frac{1}{5} - \frac{1}{10} = +\frac{1}{10}$$

$$\therefore \quad f = +10 \text{ cm} \quad \text{(a converging combination)}$$

(c) *Power of a lens*. The shorter the focal length of a lens, the more does it converge or diverge light. The *power F* of a lens is defined as the reciprocal of its focal length f in metres.

$$F = \frac{1}{f}$$

The unit of power is now the *radian per metre* (rad m⁻¹) since $f = h/D$ (p. 101), where the distance h is in metres and the angle of deviation D is in radians. (The former unit was the dioptre, 1 dioptre = 1 radian per metre.) The power of a lens of focal length (*i*) 1 metre is 1 rad m⁻¹, (*ii*) 25 cm (0.25 m) is 4.0 rad m⁻¹. The sign of F is the same as f, i.e. positive for a converging lens and negative for a diverging one.

Opticians obtain the power of a lens using a 'lens measurer', Fig. 5.57. This has three legs, the centre one being spring-loaded and connected to a pointer moving over a scale. By measuring the surface curvature, the power may be obtained quickly and accurately of any lens made of material of a certain refractive index. Lenses of materials of other refractive indices are catered for by using a scale of refractive indices along with the instrument.

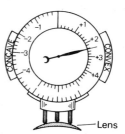

Fig. 5.57

Whilst f is useful for constructing ray diagrams, it is more convenient to use F when calculating the combined effect of several optical parts. Thus the combined power F of three thin lenses of power F_1, F_2 and F_3 in contact is

$$F = F_1 + F_2 + F_3$$

Methods of measuring f for lenses

(a) Converging lens
Rough method. The image formed by the lens of a *distant* window is focused sharply on a screen. The distance between the lens and the screen is f. Why?

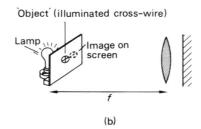

Fig. 5.58   (a)   (b)

Plane mirror method. Using the arrangement shown in Fig. 5.58, a pin or illuminated object is adjusted until it coincides in position with its image, located by no parallax or by a screen. The rays from the object must emerge from the lens and fall on the plane mirror normally to retrace their path. The object is therefore at the principal focus.

Lens formula method. Several values of the image distance v, corresponding to different values of the object distance u are found using either the pin/no parallax method or an illuminated object and screen. For each pair of values, f is calculated from $1/f = 1/v + 1/u$ and the average taken.

A better plan is to plot a graph of $1/v$ against $1/u$, draw the best straight line AB through the points, Fig. 5.59. The intercepts OA and OB on the axes are both equal to $1/f$ since when $1/u = 0$, $1/v = OA = 1/f$, from the simple lens equation.

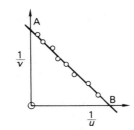

Fig. 5.59

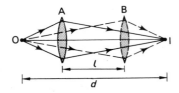

Fig. 5.60

Two-position method. The image I of an illuminated object O is obtained on a screen, Fig. 5.60. With the object and screen in the same position, the lens is moved from A to B where another sharp image is obtained. (The image is magnified for position A of the lens and diminished for B. Why?) The distances between O and I, d, and between A and B, l, are measured then f calculated from

$$f = \frac{d^2 - l^2}{4d}$$

By changing d, a set of values of d and l can be found to give an average value of f.

To derive the above expression for f we use the fact that O and I are interchangeable, i.e. are conjugate points, therefore OA = BI and OB = AI. For position A of the lens

$$u = OA = OI - AB - BI = d - l - u$$

$$\therefore \quad 2u = d - l \qquad \therefore \quad u = (d - l)/2$$

$$v = AI = OI + AB - AI = d + l - v$$

$$\therefore \quad 2v = d + l \qquad \therefore \quad v = (d + l)/2$$

Substituting for u and v in $1/v + 1/u = 1/f$ gives

$$\frac{1}{(d + l)/2} + \frac{1}{(d - l)/2} = \frac{1}{f}$$

Hence f follows. The method is suitable when u and v cannot be measured because the faces of the lens are inaccessible due, for example, to the lens being in a tube. *N.B.* A converging lens cannot form a real image on a screen if (*i*) the object is inside the principal focus or (*ii*) the distance between the object and the screen is less than $4f$ (this can be proved theoretically and experimentally). When the separation is $4f$, the object and image are then each distant $2f$ from the lens on opposite sides. In the 'two-position' method for f, the separation of the object and screen must exceed $4f$.

Magnification method. Using an object of known size (e.g. an illuminated transparent scale) direct measurement is made of the size of the image produced on a screen by the lens. The magnification *m* can thus be obtained directly.

Multiplying both sides of $1/v + 1/u = 1/f$ by v we get

$$1 + \frac{v}{u} = \frac{v}{f}$$

$$\therefore \quad 1 + m = \frac{v}{f}$$

$$\therefore \quad m = \frac{v}{f} - 1$$

A set of values of *m* and *v* are obtained and a graph of *m* against *v* plotted. It should be a straight line whose slope is $1/f$ and intercept on the *v*-axis (when $m = 0$) is f, Fig. 5.61.

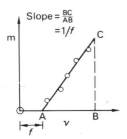

Fig. 5.61

(b) Diverging lens

Auxiliary converging lens method. A diverging lens normally forms a virtual image of a real object. Such an image cannot be located by a screen and is not easily found by a pin/no parallax method. However with the help of a converging lens, a real image can be obtained.

In Fig. 5.62 the converging lens forms a real image at I′ of an object O when the diverging lens is absent. This image is located and its position noted. The diverging lens is then placed between C and I′ and the converging beam of light falling on it behaves as a virtual object at I′. The diverging lens forms a real image of I′ at I,

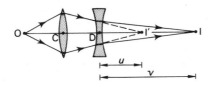

Fig. 5.62

which is located. For the diverging lens $u = -\text{I}′\text{D}$ and $v = +\text{ID}$ whence *f* can be calculated.

Methods of measuring *r* for a lens surface

(a) Converging lens (Boys' method). The lens is floated on mercury and an object O moved until its image is also formed at O. Rays from O must be refracted at the top surface to be incident normally on the bottom surface and thence be reflected to retrace their own path to O, Fig. 5.63*a*.

If refraction was allowed to occur at *both* surfaces of the lens, the ray OAB, since it falls normally on the bottom surface, would pass straight through the lens along BD and form a *virtual* image of O at C, the centre of curvature of the bottom surface, Fig. 5.63*b*.

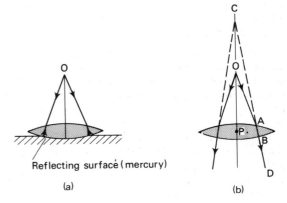

Fig. 5.63

Knowing the focal length *f* of the lens and measuring $u (= + \text{OP})$, $v (= - \text{CP})$, which is the radius of curvature of the bottom surface, can be calculated from the simple *lens* equation $1/v + 1/u = 1/f$. The radius of the other surface is found in the same way by turning the lens over.

Notes. (*i*) Although use is made of *reflection* at the bottom surface of the lens, the calculation is based on what would occur if refraction through the lens took place.

(*ii*) The refractive index *n* of the lens material can be found if r_1, r_2 and *f* are known, using $1/f = (n - 1)(1/r_1 + 1/r_2)$.

(b) Diverging lens. The position is found in which an object coincides with its image formed by the weak *reflection* at the surface of the lens acting as a concave mirror. The object is then at the centre of curvature of the surface. The method is simply the 'self-conjugate point' method for a concave mirror (p. 91).

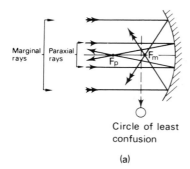

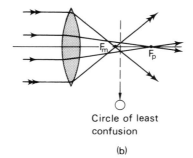

Fig. 5.64   (a)   (b)

Defects in images

So far our discussion of the formation of images by spherical mirrors and lenses has been confined to paraxial rays; we have assumed that the mirror or lens had a small aperture and that object points were on or near the principal axis. In such cases it is more or less true to say that point images are formed of point objects. However, when rays are non-paraxial and objects are extended and mirrors and lenses are of large aperture, the image can differ in shape, sharpness and colour from the object. Two of the most important image defects or *aberrations* will be considered.

(*a*) *Spherical aberration*. This arises with mirrors and lenses of large aperture and results in the image of an object point not being a point. The defect is due to the fact that the focal length of the mirror or lens for marginal rays is *less* than for paraxial rays—a property of a spherical surface.

Consider a point object at infinity (i.e. a long distance off) on the principal axis of a mirror or lens whose aperture is not small. The incident rays are parallel to the axis and are reflected or refracted so that the marginal rays farthest from the axis come to a focus at F_m whilst the paraxial rays give a point focus at F_p, Fig. 5.64a and b. All the reflected or refracted rays are tangents to a surface, called a *caustic surface*, which has an apex at F_p. (A caustic curve may be seen on the surface of a cup of tea in bright light, the inside of the cup acting as the mirror.) The nearest approach to a sharp image is the *circle of least confusion*, i.e. the smallest circle through which passes all the reflected or refracted rays.

In general, the image of any object point, on or off the axis, is a circular 'blur' and not a point. The distance F_mF_p in Fig. 5.64 is the longitudinal spherical aberration of the mirror or lens for the particular object distance.

Whilst it is not possible to construct a mirror which

always forms a point image of a point object on the axis, an ellipsoidal mirror achieves this for one definite point on the axis for both paraxial and marginal rays. In Fig. 5.65a, ABC represents an ellipsoidal mirror with foci F_1 and F_2; *all* rays from F_1 are reflected through F_2. A parabola is an ellipse with one focus at infinity and so a paraboloidal mirror brings all rays from an object point *on* the axis at infinity to a point focus, thus accounting for its use as the objective in an astronomical telescope, Fig. 5.65b. (It should be noted, however, that it does not form a point image of a point object *off* the axis.) Searchlights and car headlamps have paraboloidal reflectors which produce a roughly parallel beam from a small light source at the focus. A perfectly parallel beam does not spread out as the distance from the reflector increases and its intensity does not therefore decrease on this account.

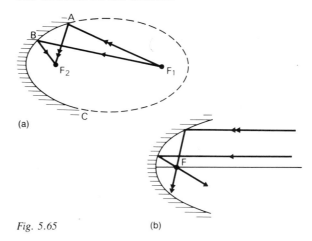

Fig. 5.65   (b)

In a lens spherical aberration can be *minimized* if the angles of incidence at each refracting surface are kept small, thus, in effect, making all rays paraxial. This is achieved by sharing the deviation of the light as equally as possible between the surfaces. Fig. 5.66 shows

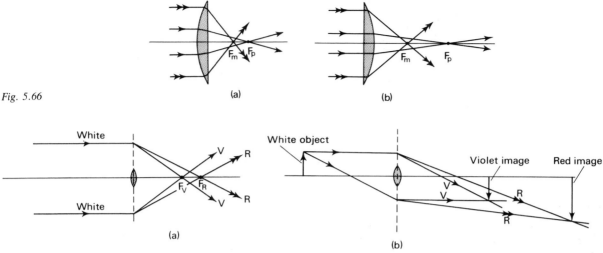

Fig. 5.66

(a)

(b)

Fig. 5.67

parallel light falling on a plano-convex lens; spherical aberration is smaller in (a) than in (b). Why? Why would it be better to have the convex side towards the object if the lens was used as a telescope objective but the other way round for a microscope objective?

Spherical aberration can also be reduced by placing a stop in front of the lens or mirror to cut off marginal rays but this has the disadvantage of making the image less bright.

(b) *Chromatic aberration.* This defect occurs only with lenses and causes the image of a white object to be blurred with coloured edges. A lens has a greater focal length for red light than for violet light, as shown in the exaggerated diagram of Fig. 5.67a. (This can be seen from $1/f = (n - 1)(1/r_1 + 1/r_2)$ bearing in mind that $n_{violet} > n_{red}$.) Thus a converging lens produces a series of coloured images of an extended white object, of slightly different sizes and at different distances from the lens, Fig. 5.67b. The eye, being most sensitive to yellow–green light, would focus the image of this colour on a screen but superimposed on it would be the other images, all out of focus.

Chromatic aberration can be eliminated for *two* colours (and reduced for all) by an *achromatic doublet.* This consists of a converging lens of crown glass combined with a diverging lens of flint glass. One surface of each lens has the same radius of curvature to allow them to be cemented together with Canada balsam and thereby reduce light loss by reflection, Fig. 5.68. The flint glass of the diverging lens produces the same dispersion as the crown glass of the converging

lens but in the opposite direction and with less deviation of the light, so that overall the combination is converging. In Fig. 5.69, the dispersions (exaggerated) θ_1 and θ_2 are equal and opposite; for the deviations, $D_1 > D_2$.

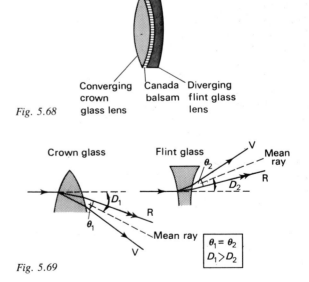

Fig. 5.68

Converging crown glass lens

Canada balsam

Diverging flint glass lens

Fig. 5.69

Crown glass

Flint glass

Mean ray

V

R

D_2

θ_2

D_1

θ_1

Mean ray

V

R

$\theta_1 = \theta_2$
$D_1 > D_2$

The eye and its defects

The construction of the human eye is shown in Fig. 5.70. The image on the retina is formed by successive

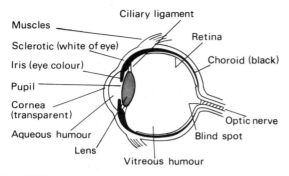

Ciliary ligament

Muscles

Retina

Sclerotic (white of eye)

Iris (eye colour)

Choroid (black)

Pupil

Cornea (transparent)

Aqueous humour

Optic nerve

Blind spot

Lens

Vitreous humour

Fig. 5.70

refraction at the surfaces between the air, the cornea, the aqueous humour, the lens and the vitreous humour. The brain interprets the information transmitted to it as electrical impulses from the retinal image and appreciates by experience that an inverted image means an upright object, Fig. 5.71. In good light the eye automatically focuses the image of an object on a very small region towards the centre of the retina called the fovea. The fovea permits the best observation of detail—to about two minutes of arc, i.e. to $\frac{1}{10}$ mm at about 20 cm. The periphery of the retina can only detect much coarser detail but it is more sensitive to dim light. Objects at different distances are focused by the ciliary ligaments changing the shape of the lens—a process known as *accommodation*. It becomes more convex to view nearer objects.

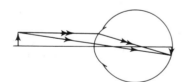

Fig. 5.71

The farthest point which can be seen distinctly by the unaided eye is called the *far point*—infinity for the normal eye; the nearest point that can be focused distinctly by the unaided eye is called the *near point*—25 cm for a normal adult eye but less for younger people. The distance of 25 cm is known as the *distance of most distinct vision*. The range of accommodation of the normal eye is thus from 25 cm to infinity and when relaxed it is focused on the latter point.

(*a*) *Short sight* (Myopia). The short-sighted person sees near objects clearly but his far point is closer than infinity. The image of a distant object is focused in front of the retina because the focal length of the eye is too short for the length of the eyeball, Fig. 5.72a. The defect is corrected by a *diverging* spectacle lens whose focal length *f* is such that it produces a virtual image at the far point of the eye of an object at infinity, Fig. 5.72b. Thus if the far point is 200 cm, $v = -200$ cm, $u = \infty$ and from $1/f = 1/v + 1/u$ we get $1/f = -1/200 + 1/\infty$, therefore $f = -200$ cm.

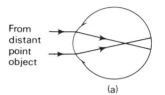

From distant point object

(a)

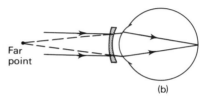

Far point

Fig. 5.72

(b)

(*b*) *Long sight* (Hypermetropia). The long-sighted person sees distant objects clearly but his near point is more than 25 cm from the eye. The image of a near object is focused behind the retina because the focal length of the eye is too long for the length of the eyeball, Fig. 5.73a. The defect is corrected by a *converging* spectacle lens of focal length *f* which gives a virtual image at the near point of the eye for an object at 25 cm, Fig. 5.73b. For example, if the near point is 50 cm, $u = +25$ cm, $v = -50$ cm and $1/f = -1/50 + 1/25 = +1/50$. Therefore $f = +50$ cm.

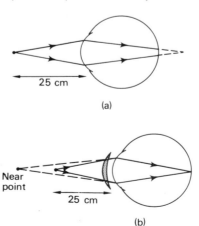

25 cm

(a)

Near point

25 cm

Fig. 5.73

(b)

(*c*) *Presbyopia*. In this defect, which often develops with age, the eye loses its power of accommodation and two pairs of spectacles may be needed, one for distant objects and the other for reading. Sometimes 'bifocals' are used which have a diverging top part to correct for distant vision and a converging lower part for reading, Fig. 5.74.

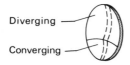

Diverging
Converging

Fig. 5.74 Bifocal lens

(*d*) *Astigmatism*. If the curvature of the cornea varies in different directions, rays in different planes from an object are focused in different positions by the eye and the image is distorted. The defect is called astigmatism and anyone suffering from it will see one set of lines in Fig. 5.75 more sharply than the others. It may be possible to correct it with a non-spherical spectacle lens whose curvature increases the effect of that of the cornea in its direction of minimum curvature or decreases it in the maximum curvature direction.

Fig. 5.75

(*e*) *Contact lenses*. These consist of tiny, unbreakable plastic lenses held to the cornea by the surface tension (p. 196) of eye fluid and in recent years they have increased in popularity. Fig. 5.76*a* shows one balanced on a finger tip. As well as being safer for sportsmen, they may help certain eye defects which spectacles cannot. Thus if the cornea is conical-shaped, vision is very distorted but if a contact lens is fitted and the space between the lens and the cornea filled with a saline solution of the same refractive index as the cornea, normal vision results, Fig. 5.76*b*.

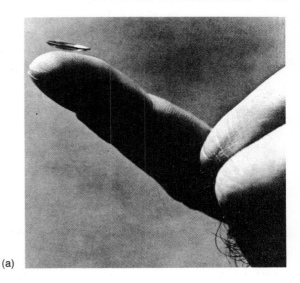

(a)

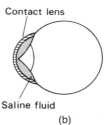

Contact lens

Saline fluid

Fig. 5.76 (b)

Magnifying power of optical instruments

Previously, when considering the magnification produced by mirrors and lenses, we used the idea of *linear magnification m* and showed that it was given by $m = v/u$. However, if the image is formed at infinity, as it can be with some optical instruments, then *m* should be infinite! The difficulty is that we cannot get to the image to view it and in such cases *m* is therefore not a very helpful indication of the improvement produced by the instrument. A more satisfactory term is clearly necessary to measure this.

The *apparent* size of an object depends on the size of its image on the retina and, as Fig. 5.77 shows, this depends not so much on the actual size of the object as on the angle it subtends at the eye, i.e. on the *visual angle*. Thus, AB is larger than CD but because it subtends the same visual angle as CD, it *appears* to be of equal size. The *angular magnification* or *magnifying power M* of an optical instrument is defined by the equation

$$M = \frac{\beta}{\alpha}$$

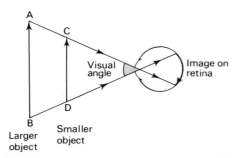

Fig. 5.77

where β = angle subtended at the eye by the *image* formed when using the instrument, and α = angle subtended at the unaided eye (i.e. without the instrument) by the *object* at some 'stated distance'.

In the case of a telescope the 'stated distance' has got to be where the object (e.g. the moon) is; for a microscope it is usually taken to be the 'distance of most distinct vision' (i.e. 25 cm away) since it is at that distance the object is seen most distinctly by the normal, unaided eye.

The difference between the magnifying power M and the magnification m should be noted. M is the ratio of the *apparent* sizes of image and object and involves a comparison of visual angles; m is the ratio of the *actual* sizes of image and object. They do not necessarily have the same value but in some cases they do.

Magnifying glass

This is also called the simple microscope and consists of a converging lens forming a virtual, upright, magnified image of an object placed inside its principal focus, Fig. 5.78a. The image appears largest and clearest when it is at the near point.

Assuming rays are paraxial and that the eye is close to the lens, we can say $\beta = h_1/D$ where h_1 is the height of the image and D is the *magnitude* of the distance of most distinct vision (usually 25 cm). If the object is viewed at the near point by the unaided eye, Fig. 5.78b,

we have $\alpha = h/D$ where h is the height of the object. Hence, the magnifying power M is given by

$$M = \frac{\beta}{\alpha} = \frac{h_1/D}{h/D} = \frac{h_1}{h}$$

In this case, $M = m$ where the linear magnification $m = v/u$; v and u being the image and object distances respectively.

$$\therefore \quad M = \frac{v}{u}$$

If $1/v + 1/u = 1/f$ is multiplied throughout by v, we get $v/u = v/f - 1$

$$\therefore \quad M = \frac{v}{f} - 1$$

It follows that a lens of short focal length has a large magnifying power. For example if $f = + 5.0$ cm and $v = - D = - 25$ cm (since image is virtual) then

$$M = \frac{v}{f} - 1 = \frac{-D}{f} - 1 = \frac{-25}{+5} - 1 = -6.0$$

The magnifying power is 6.0, i.e. $(D/f + 1)$; since M is a number the negative sign can be omitted.

You should draw a ray diagram for a magnifying glass forming a virtual image at *infinity* (where must the object be placed?) and use it to show that in this case $M = D/f$ numerically, i.e. M is one less than when the image is at the near point.

What is the effect, if any, on M if the eye is moved back from the lens when the image is at (i) the near point, (ii) infinity?

Compound microscope

The focal length of a lens can be decreased and its magnifying power thereby increased by making its surfaces more curved. However, serious distortion of the image results from excessive curvature and to obtain greater magnifying power a compound micro-

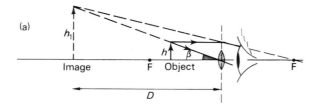

Fig. 5.78

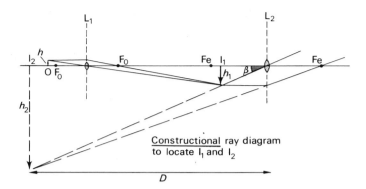

Fig. 5.79

scope is used consisting of two separated, converging lenses of short focal lengths.

The lens L_1 nearer to the object, called the 'objective', forms a real, magnified, inverted image I_1 of an object O placed just outside its principal focus F_o. I_1 is just inside the principal focus F_e of the second lens L_2, called the 'eyepiece', which acts as a magnifying glass and produces a magnified, virtual image I_2 of I_1. The microscope is said to be in 'normal adjustment' when I_2 is at the near point. Fig. 5.79 shows the usual constructional rays (p. 100), drawn to locate I_1 and I_2; note that the object is seen inverted.

(a) *Magnifying power.* We shall assume that (i) all rays are paraxial, (ii) the eye is close to the eyepiece and (iii) the microscope is in normal adjustment. $M = \beta/\alpha$ where, in this case,

β = angle subtended at the eye by I_2 *at the near point*
  = h_2/D (h_2 being the height of I_2 and D the distance of most distinct vision), and

α = angle subtended at the eye by O at the near point, without the microscope
  = h/D (h being the height of O).

Hence

$$M = \frac{\beta}{\alpha} = \frac{h_2/D}{h/D} = \frac{h_2}{h}$$

$$= \frac{h_2}{h_1} \times \frac{h_1}{h}$$

(where h_1 is the height of I_1). Now h_2/h_1 is the linear magnification m_e produced by the eyepiece and h_1/h is the linear magnification m_o due to the objective. Thus, $M = m_e \times m_o$, i.e. when the microscope is in normal adjustment with the final image at the near point, the magnifying power equals the linear magnification (as it does for a magnifying glass with the image at the near point). It follows that M will be large if f_o and f_e are small.

Many school-type microscopes have a low-power objective ($f_o \simeq 16$ mm) magnifying 10 times and a high-power objective ($f_o \simeq 4$ mm) magnifying 40 times. When used with a ×10 eyepiece, the overall magnifying power is therefore 100 on low power and 400 on high power.

If prolonged observation is to be made it is more restful for the eye to view the final image I_2 at infinity instead of at the near point. The intermediate image I_1 must then be at the principal focus F_e of the eyepiece so that the emergent rays from the eyepiece are parallel. It can be shown that the magnifying power is then slightly less than for normal adjustment.

(b) *Resolving power.* This is the ability of an optical instrument to reveal detail, i.e. to form separate images of objects that are very close together. To increase the magnifying power unduly without also increasing the resolving power has been likened to stretching an elastic sheet on which a picture has been painted—the picture gets bigger but no more detail is seen.

It can be shown that the resolving power of a microscope is greater (i) the greater the angle θ subtended at the objective by a point in the object, Fig. 5.80, and (ii) the shorter the wavelength of the light used. Both these factors impose a definite limit on the resolving power and, having regard to this limit, the maximum useful magnifying power is about 600. In practice, in the interests of eye comfort, this value is often exceeded (over ×2000 is attainable in the best instruments).

Fig. 5.80

(c) *The eye ring.* The best position for an observer to place his eye when using a microscope is where it

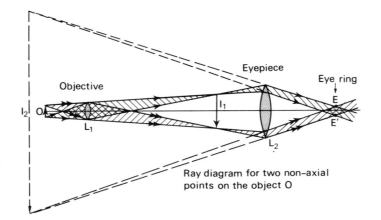

Fig. 5.81

gathers most light from that passing through the objective—the image is then brightest and the field of view greatest.

In Fig. 5.81 the paths of two cones of rays are shown coming from the top and bottom of an object, filling the whole aperture of the objective and passing through the microscope. (Constructional rays, not shown here, are first drawn as in Fig. 5.79 to locate I_1 and I_2.) The only position at which the eye would receive both these cones (and also those falling on the objective from all other object points) is at EE' where they cross. All light from the objective refracted by the eyepiece will pass through a small circle of diameter EE' which must therefore be the image of the objective formed by the eyepiece. This image at EE' is called the *eye ring* (or exit-pupil) and it is the best position for an observer's eye.

Ideally EE' should equal the diameter of the average eye pupil and in a microscope a circular opening of this size is often fixed just beyond the eyepiece to indicate the eye ring position. If, for example, the objective is 15 cm from the eyepiece of focal length 1 cm then the distance v of the eye ring from the eyepiece is given by $1/v + 1/(+ 15) = 1/(+ 1)$, i.e. $v = 1.1$ cm.

(*d*) *A calculation.* You are strongly advised to work out numerical problems from first principles and not to quote formulae. It is also helpful to draw a diagram.

The objective and the eyepiece of a microscope may be treated as thin lenses with focal lengths of 2.0 cm and 5.0 cm respectively. If the distance between them is 15 cm and the final image is formed 25 cm from the eyepiece, calculate (i) the position of the object and (ii) the magnifying power of the microscope.

Let the positions of the object O, the first image I_1 and the final image I_2 be as in Fig. 5.82*a*. A ray from

the top of O through the optical centre P_1 of the objective passes through the top of I_1 and a ray from the top of I_1 through the optical centre P_2 of the eyepiece passes through the top of I_2 when produced backwards. Let h and h_1 be the heights of O and I_1 respectively.

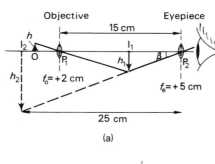

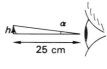

Fig. 5.82   (b)

(*i*) Consider the eyepiece. I_1 acts as the object and the final image I_2 is *virtual*. We have $v = -25$ cm and $f_e = +5.0$ cm

$$\therefore \frac{1}{(-25)} + \frac{1}{u} = \frac{1}{(+5)}$$

$$\therefore \frac{1}{u} = \frac{1}{5} + \frac{1}{25} = \frac{6}{25}$$

$$\therefore u = I_1P_2 = +4\tfrac{1}{6} \text{ cm}$$

Consider the objective. Image I_1 is *real* and at a distance $P_1 I_1$ from the objective where $P_1 I_1 = P_1 P_2 - I_1 P_2 = 15 - 4\frac{1}{6} = 10\frac{5}{6}$ cm. Hence $v = P_1 I_1 = +10\frac{5}{6} = +65/6$ cm. Also $f_o = +2.0$ cm

$$\therefore \quad \frac{1}{(+65/6)} + \frac{1}{u} = \frac{1}{(+2)}$$

$$\therefore \quad \frac{1}{u} = \frac{1}{2} - \frac{6}{65} = \frac{53}{130}$$

$$\therefore \quad u = OP_1 = +\frac{130}{53} \text{ cm}$$

$$= +2\frac{24}{53} \text{ cm} \quad (\simeq 2.5 \text{ cm})$$

The object O is about 2.5 cm from the objective.

(*ii*) Assuming the observer is close to the eyepiece the angle subtended at his eye is given by

$$\beta = \frac{h_1}{I_1 P_2} = \frac{h_1}{(25/6)} = \frac{6h_1}{25}$$

The angle α subtended at the observer's eye when he views the object at his near point (assumed to be 25 cm away) without the microscope is given by $\alpha = h/25$, Fig. 5.82*b*. Therefore

$$\text{magnifying power } M = \frac{\beta}{\alpha}$$

$$= \frac{6h_1/25}{h/25} = \frac{6h_1}{h}$$

But $\quad \dfrac{h_1}{h} = \dfrac{P_1 I_1}{P_1 O} = \dfrac{10\frac{5}{6}}{2\frac{24}{53}} = \dfrac{65/6}{130/53} = \dfrac{53}{12}$

$$\therefore \quad M = 6 \times \frac{53}{12} = 27$$

Refracting astronomical telescope

A lens-type astronomical telescope consists of two converging lenses; one is an objective of long focal length and the other an eyepiece of short focal length. The objective L_1 forms a real, diminished, inverted image I_1 of a distant object at its principal focus F_o since the rays incident on L_1 from a *point* on such an object can be assumed parallel. The eyepiece L_2 acts as a magnifying glass and forms a magnified virtual image of I_1 and, when the telescope is in normal adjustment, this image is at infinity. I_1 must therefore be at the principal focus F_e of L_2, hence F_o and F_e coincide.

In Fig. 5.83 three *actual* rays are shown coming from the top of a distant object and passing through the top of I_1 in the focal plane of L_1. They must emerge parallel from L_2 to appear to come from the top of the final image at infinity. They must also be parallel to the line joining the top of I_1 to the optical centre of L_2.

(*a*) *Magnifying power.* It will be assumed that (*i*) all rays are paraxial, (*ii*) the eye is close to the eyepiece and (*iii*) the telescope is in normal adjustment. Now $M = \beta/\alpha$ and in this case

β = angle subtended at the eye by the final image at infinity
   = angle subtended at the eye by I_1
   = h_1/f_e

(h_1 being the height of I_1), and

α = angle subtended at the eye by the object without the telescope
   = angle subtended at the objective by the object

(since the distance between L_1 and L_2 is very small compared with the distance of the object from L_1)

$$\alpha = h_1/f_o$$

Hence $$M = \frac{\beta}{\alpha} = \frac{h_1/f_e}{h_1/f_o}$$

$$\therefore \quad M = \frac{f_o}{f_e}$$

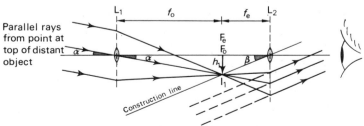

Parallel rays from point at top of distant object

Construction line

To top of final virtual image at infinity

Fig. 5.83

Notes. (*i*) The above expression for M is true only for normal adjustment; the separation of the objective and eyepiece is then $f_o + f_e$.

(*ii*) A telescope is in normal adjustment when the final image is formed at infinity; a microscope is in normal adjustment with the final image at the near point.

For high magnifying power the objective should have a large focal length and the eyepiece a small one. The largest lens telescope in the world is at the Yerkes Observatory, U.S.A.; the objective has a focal length of about 20 metres and the most powerful eyepiece has a focal length of about 6.5 mm. The maximum value of M is therefore $20 \times 10^3/6.5 \simeq 3000$.

If it is desired to form the final image at the near point, i.e. telescope not in normal adjustment, the eyepiece must be moved so that I_1 is closer to it than F_e. The magnifying power is then slightly greater than f_o/f_e.

(*b*) *Resolving power.* It can be shown that the ability of a telescope to reveal detail increases as the diameter of the objective increases. However, large lenses are not only difficult to make but they tend to sag under their own weight. The objective of the Yerkes telescope has a diameter of 1 metre which is about the maximum possible. There is no point in increasing the magnifying power of a telescope unduly if the resolving power cannot also be increased.

(*c*) *The eye ring.* As in the case of the microscope, the eye ring is in the best position for the eye and is the circular image of the objective formed by the eyepiece. All rays incident on the objective which leave the telescope pass through it. In Fig. 5.84*a* two cones of rays are shown coming from the top and bottom of a distant object and crossing at the eye ring EE'.

If the telescope is in normal adjustment, the separation of the lenses is $f_o + f_e$ and from similar triangles in Fig. 5.84*b*

$$\frac{AB}{EE'} = \frac{f_o}{f_e}$$

But the magnifying power M for a telescope in normal adjustment is f_o/f_e, hence

$$M = \frac{\text{diameter of objective}}{\text{diameter of eye ring}}$$

This expression enables M to be found simply by illuminating the objective with a sheet of frosted glass and a lamp, locating the image of the objective formed by the eyepiece (i.e. the eye ring) on a screen and measuring the diameter.

(*d*) *Brightness of image.* A telescope increases the light-gathering power of the eye and in the case of a point object, such as a star, forms a brighter image. Thus, when the diameter of the objective is doubled, the telescope collects four times more light from a given star (why?) and since a point image is formed of a point object, whatever the magnifying power, the star appears brighter. Many more stars are therefore visible than would otherwise be seen and in fact the range of a telescope is proportional to the diameter of its objective.

The brightness of the background is not similarly increased because it acts as an extended object and, as we will now see, a telescope does not increase the brightness of such an object.

When the diameter of the eye ring equals the diameter of the pupil of the eye, almost all the light entering the telescope enters the eye. If M is the magnifying power of the telescope, the diameter of the

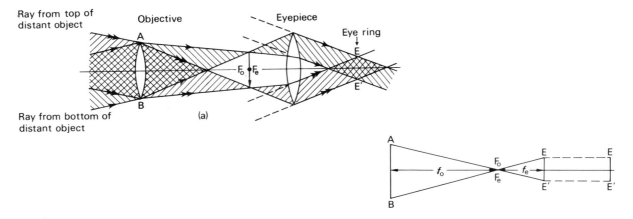

Fig. 5.84

objective is M times the diameter of the eye ring and the area of the objective is M^2 times greater. M^2 times more light enters the eye via the telescope than would enter it from the object directly. However the image has an area M^2 times that of the object since the telescope makes the object appear M times as high and M times as wide as it does to the unaided eye at the same distance. The brightness of the image cannot therefore exceed that of the object and is less, due to loss of light in the instrument. The contrast between a star and its background is thus increased by a telescope.

(e) A calculation.

An astronomical telescope has an objective of focal length 100 cm and an eyepiece of focal length 5.0 cm. Calculate the power when the final image of a distant object is formed (i) at infinity, (ii) 25.0 cm from the eyepiece.

(*i*) When the final image is at infinity, the telescope is in normal adjustment and the magnifying power is given by

$$M = \frac{f_o}{f_e} = \frac{100}{5.0} = 20$$

(*ii*) Let the position of I_1 be as shown in Fig. 5.85. A ray from the top of the distant object through the optical centre P_1 of the objective passes through the top of I_1. Also a ray from the top of I_1 through the optical centre P_2 of the eyepiece passes through the top of the final image when produced backwards.

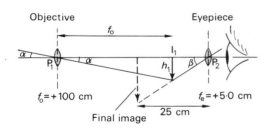

Fig. 5.85

For the eyepiece we can say $v = -25.0$ cm (final image *virtual*) and $f_e = +5.0$ cm.

$$\therefore \quad \frac{1}{(-25)} + \frac{1}{u} = \frac{1}{(+5)}$$

$$\therefore \quad \frac{1}{u} = \frac{1}{5} + \frac{1}{25} = \frac{6}{25}$$

$$\therefore \quad u = I_1P_2 = 4\tfrac{1}{6} \text{ cm}$$

If the eye is close to the eyepiece the angle β subtended at the eye is given by $\beta = h_1/I_1P_2 = h_1/(25/6) = 6h_1/25$. The angle α subtended at the unaided eye by the object = angle subtended at the objective by the object (see p. 115) = $h_1/f_o = h_1/100$.

$$\text{Hence} \qquad M = \frac{\beta}{\alpha} = \frac{6h_1/25}{h_1/100} = \frac{6 \times 100}{25}$$

$$= 24$$

Reflecting astronomical telescope

The largest modern astronomical telescopes use a concave mirror of long focal length as the objective instead of a converging lens but the principle is the same as the refracting telescope. One arrangement, called the Newtonian form after the inventor of the reflecting telescope, is shown in Fig. 5.86. Parallel rays from a distant point object on the axis are reflected first at the objective and then at a small plane mirror to form a real image I_1 which can be magnified by an eyepiece or photographed by having a film at I_1. The plane mirror, whose area is negligible compared with that of the concave mirror, deflects the light sideways without altering the effective focal length f_o of the objective. In normal adjustment the magnifying power is f_o/f_e where f_e is the focal length of the eyepiece.

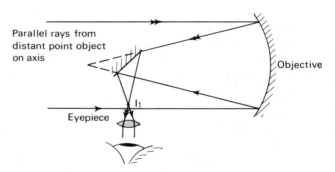

Fig. 5.86

The advantages of reflecting telescopes are
(*i*) no chromatic aberration since no refraction occurs at the objective,
(*ii*) no spherical aberration for a point object on the axis at infinity if a paraboloidal mirror is used (see p. 108),
(*iii*) a mirror can have a much larger diameter than a lens (since it can be supported at the back) thereby giving greater resolving power and a brighter image of a point object,

Fig. 5.87

(*iv*) only one surface requires to be ground (compared with two for a lens), thus reducing costs.

The largest reflecting optical telescope in the world on Mount Palomar, California, Fig. 5.87, has a concave paraboloidal mirror of diameter 5 metres. It is made of low expansion glass which took six years to grind and the reflecting surface is coated with aluminium. Photographs of nebulae up to a distance of 10^{10} light-years away (1 light-year is the distance travelled by light in 1 year) can be taken. It is used in conjunction with spectrometers, cameras and other instruments in temperature-controlled, air-conditioned surroundings.

For general astronomical work lens telescopes are more easily handled than large mirror telescopes; the latter are used only where high resolving power is required.

Other telescopes

(*a*) *Terrestrial telescope.* The final image in an astronomical telescope is inverted and whilst this is not a

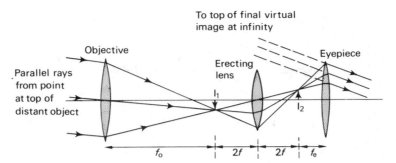

Fig. 5.88

handicap for looking at a star, it is when viewing objects on earth.

A terrestrial telescope is a refracting astronomical telescope with an intermediate 'erecting' lens arranged as in Fig. 5.88 to be at a distance of $2f$ (where f is the focal length of the erecting lens) from the inverted image I_1 formed by the objective. An erect image I_2 of the same size as I_1 is formed at $2f$ beyond the erecting lens and acts as an 'object' for the eyepiece in the usual way. A disadvantage of this arrangement is the increase in length of the telescope by $4f$.

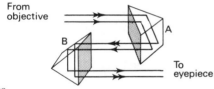

Fig. 5.89

(b) Prism binoculars. These consist of a pair of refracting astronomical telescopes with two totally re-flecting prisms (angles 90°, 45° and 45°) between each objective and eyepiece as in Fig. 5.89. Prism A causes lateral inversion and prism B inverts vertically so that the final image is the same way round and the same way up as the object. Each prism reflects the light through

180° making the effective length of each telescope three times the distance between the objective and the eyepiece. Good magnifying power is thus obtained with compactness.

Prism binoculars marked '7 × 50' have a magnifying power of 7 and objectives of diameter 50 mm.

(c) Galilean telescope. A final erect image is obtained in the Galilean telescope using only two lenses—a converging objective of large focal length f_o and a diverging eyepiece of small focal length f_e.

The image of a distant object would, in the absence of the eyepiece, be formed by the objective at I_1, where $P_1I_1 = f_o$, Fig. 5.90. With the eyepiece in position at a distance f_e from I_1, the separation of the lenses is $f_o - f_e$ (numerically) and rays falling on the eyepiece emerge parallel, so that to the eye the top of the final image is *above* the axis of the telescope. An upright image at infinity is thus obtained. The converging light falling on the eyepiece behaves like a virtual object at I_1 and a virtual image of it is formed.

If the telescope is in normal adjustment, i.e final image at infinity, the magnifying power is f_o/f_e, as for an astronomical telescope. In Fig. 5.90 the ray from the top of I_1 passing through the centre P_2 of the eyepiece goes to the top of the final image at infinity. It must therefore be parallel to the three parallel rays emerging from the eyepiece. The angle β subtended at the eye

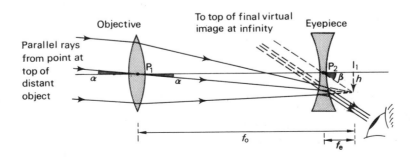

Fig. 5.90

(close to the telescope) is thus given by $\beta = h/I_1P_2 = h/f_e$ where h is the height of I_1. Also, the angle α subtended at the unaided eye by the object is very nearly equal to the angle subtended at the objective by the object, hence $\alpha = h/I_1P_1 = h/f_o$. Thus, the magnifying power M is

$$M = \frac{\beta}{\alpha} = \frac{h/f_e}{h/f_o} = \frac{f_o}{f_e}$$

The Galilean telescope is shorter than the terrestrial telescope but the field of view is very limited because the eye ring is between the lenses (why?) and so inaccessible to the eye. Opera glasses consist of two telescopes of this type.

Spectrometer

The spectrometer is designed primarily to produce and make measurements on the spectra of light sources and is generally used with a diffraction grating but a prism can be employed. It also provides a very accurate method of measuring refractive index.

The instrument consists of (*i*) a fixed collimator with a movable slit of adjustable width (to produce a parallel beam of light from the source illuminating the slit), (*ii*) a turntable (having a circular scale) on which the grating or prism is placed and (*iii*) a telescope (with a vernier scale) rotatable about the same vertical axis as the turntable, Fig. 5.91. The converging lenses in the collimator and telescope are achromatic.

Four preliminary adjustments must first be made.

(*a*) *Adjustments*

(*i*) *Eyepiece.* This is moved in the tube containing the cross-wires until the cross-wires are clearly seen. An image formed on the wires will then be distinct.

(*ii*) *Telescope.* A distant object (e.g. a vertical line of mortar between bricks in a building outside) is viewed through the telescope and the distance of the objective from the cross-wire eyepiece adjusted by a thumb-screw until there is no parallax between the image of the distant object and the cross-wires. Parallel rays entering the telescope are now brought to a focus at the cross-wires.

(*iii*) *Collimator.* The telescope is turned into line with the collimator and the slit, illuminated with sodium light, is moved in or out of the collimator tube until there is no parallax between the image of the slit and the cross-wires. The slit is then at the principal focus of the collimator lens which is producing a parallel beam.

(*iv*) *Levelling the table.* The method adopted depends on whether a grating or a prism is to be used; we will consider the latter at present. The prism is placed on the turntable with one face (AB in Fig. 5.92*a*) perpendicular to the lines on the table which join levelling screws S_1 and S_2. With the telescope at right angles to the collimator the table is rotated until a *reflected* image of the slit from AB enters the telescope. S_1 is then adjusted so that this image is in the centre of the field of view, as in Fig. 5.92*b*. Keeping the telescope in the same position, the turntable is rotated to give a reflected image of the slit which is obtained in the telescope from face AC of the prism. S_3 *only* is adjusted to centralize the image of the slit in the field of view. The turntable is then level, i.e. the refracting edge of the prism is now parallel to the axis of rotation of the telescope.

(*b*) *Measurement of refractive index*

By finding the refracting angle A of a prism and its angle of minimum deviation D_{min}, the refractive index n of the material of the prism can be found for light of one colour (i.e. monochromatic light) from

$$n = \frac{\sin\left[(A + D_{min})/2\right]}{\sin(A/2)} \qquad \text{(p. 98)}$$

To measure A, the slit is made as narrow as possible and the prism set on the turntable as in Fig. 5.93*a* so that the incident light is reflected from both faces AB and AC. The telescope is rotated in turn into positions T_1 and T_2 so that the images of the slit, *reflected* from AB and AC respectively, coincide with the intersection of the cross-wires. From the diagram we see that if three dotted lines are drawn parallel to the incident beam of light then $A = \alpha + \beta$ and the angle between T_1 and T_2 is $2(\alpha + \beta) = 2A$. Hence A is half the angle read on the telescope scale between positions T_1 and T_2.

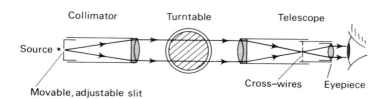

Collimator   Turntable   Telescope

Source

Cross-wires   Eyepiece

Fig. 5.91   Movable, adjustable slit

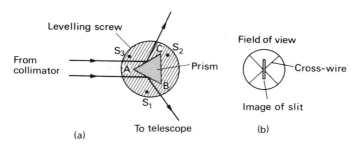

Fig. 5.92 (a) (b)

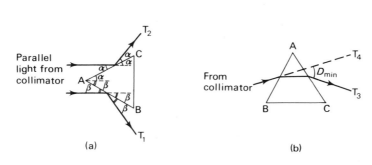

Fig. 5.93 (a) (b)

To measure D_{min} a monochromatic source, e.g. a sodium lamp or flame, must be used and the prism set on the turntable as in Fig. 5.93b. The minimum deviation position of the prism is found by rotating the table so that the telescope is as near the straight-through position as possible whilst still receiving a *refracted* image of the slit at the intersection of the cross-wires. The scale reading in this position T_3 is noted, the prism removed and the straight-through reading taken with the telescope and collimator in line, position T_4. D_{min} is the angle between T_3 and T_4. (Alternatively the minimum deviation position on each side can be found and the difference halved.)

Camera

A typical arrangement is shown in Fig. 5.94. The lens system has to have a field of view of about 50° (compared with 1° or so for an average microscope objective) and so the reduction of aberrations is a major consideration. Very large apertures give blurred images because of aberrations; so do very small apertures but due to the phenomenon called diffraction. The best images are therefore generally obtained with intermediate apertures. For some types of optical systems (e.g. eyes, cameras, enlargers) aberrations are more significant and the aperture has to be reduced to obtain clear images. For others (e.g. telescopes) diffraction is usually more significant and apertures have to be made as large as is practicable.

Cheap cameras use a meniscus lens, which is usually an achromatic doublet, and a stop to restrict the aperture. More expensive cameras have a lens system of several components designed to minimize the various aberrations. Focusing of objects at different distances is achieved by slightly altering the separation of the lens from the film.

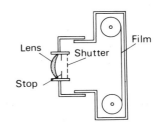

Fig. 5.94

In many cameras the amount of light passing through the lens can be altered by an aperture control or stop of variable width. This has a scale of f-numbers with all or some of the following settings—1.4, 2, 2.8, 4, 5.6, 8, 11, 16, 22, 32. These are such that reducing the f-number by one setting, say from 8 to 5.6, *doubles* the area of the aperture, i.e. the *smaller* the f-number the *larger* the aperture. An f-number of 4 means the diameter d of the aperture is $\frac{1}{4}$ the focal length f of the lens, i.e. $d = f/4$.

The aperture affects (*i*) the exposure time and (*ii*) the depth of field. Consider (*i*). Using the next lower f-number halves the exposure time needed to produce the same illumination on the film (since the area of the aperture has been doubled). The exposure required depends on the lighting conditions and must be brief if the object is moving. In better cameras exposure time can be varied.

The *depth of field* (often called the depth of focus) is the range of distances in which the camera can more or less focus objects simultaneously. A landscape photograph needs a large depth of field whilst in a family group it may be desirable to have the background out of focus. The depth of field is increased by reducing the lens aperture as can be seen from Fig. 5.95, in which the images formed by a lens of point objects O_1 and O_2 are at I_1 and I_2 respectively. The diameter of the circular patch of light on a film in focus for I_1 is, for the out-of-focus I_2, AB if the whole aperture is used but only A'B' if the lens is stopped down.

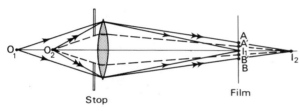

Fig. 5.95

Projector

A projector is designed to throw on a screen a magnified image of a film or transparency. It consists of two main parts—an illumination system and a projection lens, Fig. 5.96.

The image on the screen is usually so highly magnified that very strong but uniform illumination of the film with white light is necessary if the image is also to be bright. This is achieved by directing light from a high power, intense, tungsten iodine filament lamp on to the film by means of a curved reflector and a condenser lens system arranged as shown. Since the screen is generally a considerable distance away, the film, in-

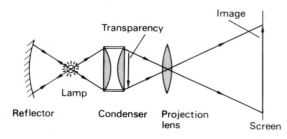

Fig. 5.96

verted, must be just outside the principal focus of the projection lens. Any chance of the image of the lamp appearing on the screen is also removed. To keep the projector as short as possible, the condenser has a short focal length f and the lamp is placed at a distance of $2f$ from it. What will be approximately (i) the separation of the condenser and the projection lens and (ii) the focal length of the projection lens if the film is close to the condenser?

Radio telescopes

Certain 'objects' in space, among them the sun, emit radio signals which can be picked up by huge aerials called radio telescopes. These do not give visual pictures as do optical telescopes but produce electrical signals which are often recorded graphically. There are various types of radio telescopes: the famous one at Jodrell Bank, Cheshire, Fig. 5.97, consists of a steerable metal reflector or 'dish', 75 metres in diameter; others have isolated aerials distributed over a large area.

The resolving power of any telescope depends on the diameter of the objective (lens, mirror or aerial) and the wavelength of the radiation from the object. The larger the diameter and the shorter the wavelength, the closer together can two distant points be and still be separated by the telescope. The Mount Palomar optical telescope has an objective (concave mirror) of diameter 5 metres but its resolving power is more than 1000 times that of the Jodrell Bank radio telescope with an aerial of diameter 75 metres. This is because the shortest wavelength of the radio signals from outer space that can penetrate the ionized layers in the upper atmosphere is about 1 cm, whereas the mean wavelength of light is 6×10^{-5} cm. (The longest wavelength that can pass through the earth's radio 'window' is about 30 metres.)

One of the strongest radio sources in our own-star-system (the Galaxy) is the Crab nebula, a mass of luminous gas in the constellation of the Bull. This nebula is believed to be the remains of a star which underwent a tremendous explosion, becoming a *supernova* and shining so brightly that it was observed in 1054, according to Chinese records, in broad daylight for several months. The radio signals arise from the highly excited gas which is still expanding outwards from the explosion centre. Some other radio 'stars' may also be due to old supernovae but most lie outside our galaxy and cannot usually be identified with ordinary stars.

Whilst interstellar dust and gas stop light reaching us from distant stars, they do not block radio waves which

Fig. 5.97

can bring information about regions we cannot see. Very cold, rarefied hydrogen gas emits 21 cm long waves and radio astronomers have established that invisible clouds of hydrogen in this state are very widely distributed in all space. By studying its distribution in our own galaxy, we now know that the latter is spiral-shaped and rotating like a huge Catherine-wheel.

Cosmology is concerned with how the universe began and there are two main theories. The 'evolutionary' (or 'big bang') theory suggests that millions of years ago all matter in the universe was concentrated into a very small volume of space, referred to as the 'primeval atom', which exploded, throwing out matter in all directions. From the debris the galaxies of stars gradually formed and the expanding universe is a result of this explosion. The 'steady-state' theory on the other hand proposes that new matter is continually being created out of nothing to fill up the empty space arising from the expansion of the universe. In this case the universe would always 'look' the same.

It is not easy to decide between the two theories but the problem may be resolved by radio astronomy. Some galaxies are known to be radio sources and the signals we receive give information about them, not as they are now, but as they were millions of years ago when the radio signals left them. If the evolutionary theory is correct we would expect the distant galaxies to be much closer together than those nearer to us; the more distant galaxies let us 'look back' through a longer time. If the steady-state theory applies there should be no difference in the average 'density' of galaxies near and far. Work done so far tends to support the evolutionary theory but the issue is still very open.

Two recent discoveries of radio astronomy are *quasars* and *pulsars*. Quasars (quasi-stellar) are very distant 'objects' that are small compared with ordinary galaxies but are very powerful radio sources. Pulsars also emit strong radio signals but in sharp regular pulses at rates varying from 30 pulses per second to 1 in 4 seconds. They are thought to be a very long way off, planet-sized and their source of energy is a mystery.

Electron microscopes

The electron microscope is analogous in principle to the optical microscope but its performance is far superior. The maximum magnifying power attainable with the best optical microscope is about 2000, for an electron microscope up to 100 000 is typical. The former can resolve detail about 10^{-6} m across, in the electron instrument it is very much smaller, about 10^{-9} m. Since atomic dimensions are of the order of 10^{-9} m (1 nm), this means that in some cases an electron microscope can reveal separate molecules.

The similarity between the paths of light in an optical microscope and the paths of electrons in an electron microscope can be seen from Fig. 5.98. In the latter, electrons are produced by an electron gun and the 'lenses' are electromagnets designed so that their fields focus the electron beam to give an image on a fluorescent screen or photographic plate. The focal lengths of the 'lenses' are variable and are determined by the current through the 'lens' coils.

When the object is struck by electrons, more penetrate in some parts than in others depending on the thickness and density of the part. The image is brightest where most electrons have been transmitted. The object must be very thin, otherwise too much electron scattering occurs and no image forms. Also, the whole arrangement is highly evacuated. Why? An air-lock device permits objects to be inserted and removed without loss of vacuum.

The high resolving power of an electron microscope arises from the fact that just as light can be considered to have both wave-like and particle-like properties, so *moving* electrons seem to have the characteristics of both particles and waves. (Electrons as well as light can give interference and diffraction effects.) The wavelength associated with a moving electron depends only on its speed which in turn depends on the p.d. accelerating it in the electron gun. It can be shown that for a p.d. of 100 kV, common in many electron microscopes, the wavelength is about 3.5×10^{-12} m. The resolving power of a microscope increases as the wavelength of the radiation falling on the object decreases (p. 113) and therefore, if we compare the above very small wavelength with that of light (say 6×10^{-5} m), we see why the electron microscope shows much greater detail.

Two recent developments in electron microscopy are

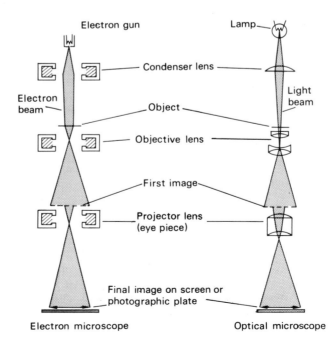

Fig. 5.98   Electron microscope   Optical microscope

the high voltage electron microscope and the scanning electron microscope. The first operates at 1 million volts and enables thicker specimens to be studied. In the second, the surface of a relatively large specimen is scanned by the electron beam; photographs taken with this instrument are shown on p. 4.

QUESTIONS

Spherical mirrors

1. If a concave mirror has a focal length of 10 cm, find the two positions where an object can be placed to give, in each case, an image twice the height of the object.

2. A convex mirror of radius of curvature 40 cm forms an image which is half the height of the object. Find the object and image positions.

3. A concave mirror of radius of curvature 25 cm faces a convex mirror of radius of curvature 20 cm and is 30 cm from it. If an object is placed midway between the mirrors find the nature and position of the image formed by reflection first at the concave mirror and then at the convex mirror.

4. Describe an experiment to determine the radius of curvature of a convex mirror by an optical method. Illustrate your answer with a ray diagram and explain how the result is derived from the observations.
   A small convex mirror is placed 0.60 m from the pole and on the axis of a large concave mirror, radius of curvature 2.0 m. The position of the convex mirror is such that a real image of a distant object is formed in the plane of a hole drilled through the concave mirror at its pole. Calculate (a) the radius of curvature of the convex mirror, (b) the height of the real image if the distant object subtends an angle of 0.50° at the pole of the concave mirror. Draw a ray diagram to illustrate the action of the convex mirror in producing the image of a non-axial point of the object and suggest a practical application of this arrangement of mirrors. (*J.M.B.*)

Refraction at plane surfaces

5. A ray of light in air passes successively through parallel-sided layers of water and glass. If the angle of incidence in air is 60° and the refractive indices of water and glass are 4/3 and 3/2 respectively, calculate (a) the angle of refraction in the water, (b) the angle of incidence at the water–glass boundary and (c) the angle of refraction in the glass.

6. Find by calculation what happens to a ray of light which falls at an angle of (i) 30°, (ii) 60°, on a glass–water surface if the refractive index of the glass is 3/2 and of water 4/3.

7. The refractive indices of crown glass and of a certain liquid are 1.51 and 1.63 respectively. Determine the conditions under which total internal reflection can occur at a surface separating them.

8. (a) A ray of light is incident at 45° on one face of a 60° prism of refractive index 1.5. Calculate the total deviation of the ray.
   (b) A ray of light just undergoes total internal reflection at the second face of a prism of refracting angle 60° and refractive index 1.5. What is its angle of incidence on the first face?

9. (a) A 60.0° prism is made of glass whose refractive index for a certain light is 1.65. At what angle of incidence will minimum deviation occur? Between what limits must the angle of incidence lie, if light is to pass through the prism by refraction at adjacent faces? (*W. part qn.*)
   (b) A ray of monochromatic light is incident at an angle of 30.0° on a prism of which the refractive index for the given light is 1.52. What is the maximum refracting angle for the prism if the light is just to emerge from the opposite face?
   (*A.E.B. part qn.*)

10. If the refractive index of diamond for sodium light is 2.42, calculate the refracting angle of a diamond prism which will cause the greatest possible deviation of a beam of sodium light after two refractions (with no total internal reflection). Explain your reasoning. (*C. part qn.*)

Lenses: the eye

11. (a) The filament of a lamp is 80 cm from a screen and a converging lens forms an image of it on the screen, magnified three times. Find the distance of the lens from the filament and the focal length of the lens.
    (b) An erect image 2.0 cm high is formed 12 cm from a lens, the object being 0.5 cm high. Find the focal length of the lens.

12. Explain what is meant by (a) a virtual image, (b) a virtual object, in geometrical optics. Illustrate your answer by describing the formation of (i) a virtual image of a real object by a thin converging lens, (ii) a real image of a virtual object by a thin diverging lens. In each instance draw a ray diagram showing the passage of two rays through the lens for a non-axial object point. (*J.M.B.*)

13. A lens forms the image of a distant object on a screen 30 cm away. Where should a second lens, of focal length 30 cm, be placed so that the screen has to be moved 8.0 cm towards the first lens for the new image to be in focus?

14. The radii of curvature of the faces of a thin converging meniscus lens of glass of refractive index 3/2 are 15 cm and 30 cm. What is the focal length of the lens (a) in air, (b) when completely surrounded by water of refractive index 4/3?

15. Explain why a sign convention is adopted in geometrical optics. Describe a convention and explain its use in solving the following problem.
    An equi-convex lens *A* is made of glass of refractive index 1.5 and has a power of 5.0 rad m^{-1}. It is combined in contact with a lens *B* to produce a combination whose power is 1.0 rad m^{-1}. The surfaces in contact fit exactly. The refractive index of the glass in lens *B* is 1.6. What are the radii of the four surfaces? Draw a diagram to illustrate your answer. (*W.*)

16. Draw a diagram to explain what is meant by (a) the *principal axis*, (b) the *focal length* of a thin converging lens.

A small luminous object is placed on the axis of a thin plano-convex lens (made of glass of refractive index 1.6) on the side of the lens nearer to the plane face. When at a distance of 30 cm from the lens it coincides with the real inverted image formed by light which has undergone two refractions at the plane face and one reflection at the curved face. Find the position and nature of the image of this object formed by light transmitted directly by the lens. (*J.M.B.*)

17. A person can focus objects between 60.0 cm and 500 cm from his eyes. What spectacles are needed to make his far point infinity? What is now his range of vision?

18. What spectacles are required by a person whose near and far points are 40.0 cm and 200 cm away respectively to bring his near point to a distance of 25.0 cm? Find his new range of vision.

Optical instruments

19. Explain the difference between the terms *magnifying power* and *magnification*, as used about optical systems. Illustrate this, by calculating both, in the case of an object placed 5.0 cm from a simple magnifying glass of focal length 6.0 cm, assuming that the minimum distance of distinct vision for the observer is 25 cm. (*S.*)

20. (a) Explain the terms *magnifying power* and *resolving power* in connection with a microscope.

(b) A compound microscope is formed from two lenses of focal lengths 1.0 and 5.0 cm. A small object is placed 1.1 cm from the objective and the microscope adjusted so that the final image is formed 30 cm from the eyepiece. Calculate the angular magnification of the instrument. (Assume that the nearest distance of distinct vision is 25 cm.) (*A.E.B. part qn.*)

21. Describe, with the help of diagrams, how (a) a single biconvex lens can be used as a magnifying glass, (b) two biconvex lenses can be arranged to form a microscope. State (i) one advantage, (ii) one disadvantage, of setting the microscope so that the final image is at infinity rather than at the near point of the eye.

A centimetre scale is set up 5.0 cm in front of a biconvex lens whose focal length is 4.0 cm. A second biconvex lens is placed behind the first, on the same axis, at such a distance that the final image formed by the system coincides with the scale itself and that 1.0 mm in the image covers 2.4 cm in the scale. Calculate the position and focal length of the second lens. (*O. and C.*)

22. State what is meant by *normal adjustment* in the case of an astronomical telescope.

Trace the paths of three rays from a distant non-axial point source through an astronomical telescope in normal adjustment.

Define the *magnifying power* of the instrument, and, by reference to your diagram, derive an expression for its magnitude.

A telescope consists of two thin converging lenses of focal lengths 100 cm and 10.0 cm respectively. It is used to view an object 2.00×10^3 cm from the objective. What is the separation of the lenses if the final image is 25.0 cm from the eye-lens? Determine the magnifying power for an observer whose eye is close to the eye-lens. (*J.M.B.*)

23. What should be the focal length of the objective of an astronomical telescope if the eyepiece is of focal length 5.0 cm and the lenses are to be fixed 85 cm apart in normal adjustment (final image at infinity)? What will then be the magnifying power obtained? (*No proofs required.*)

With such a telescope all the light received by the objective which also passes through the eyepiece eventually passes through a small circular region, called the exit pupil (eye ring), a short distance beyond the eyepiece. The exit pupil coincides in position and size with the image which would be formed by the eyepiece of the objective lens if the latter were a self-luminous object. Give arguments to justify these statements. How far behind the eyepiece will the exit pupil be in the case given above?

It is generally reckoned best to have an objective of such a size that the exit pupil can coincide in size and position with the pupil of the observer's eye. If his pupil (at night) has a diameter of 8.0 mm, what should be the diameter of the objective for this telescope?

What are the advantages of using an objective of as large a diameter as possible? What are the disadvantages of a large objective? (*S.*)

24. What is meant by the *f-number* of a camera lens? The stop of a camera lens is reduced from $f/8$ to $f/22$. State and explain in what ratio the illumination of the image on the film is changed. Explain also, with the aid of a diagram, why the 'depth of focus' is increased.

An image of a distant object is formed on a screen by an optical system consisting of a converging lens of focal length 15 cm placed co-axially 9.0 cm from a diverging lens of focal length 7.5 cm, the light being incident on the converging lens. Compare the size of this image with that produced of the same object by the converging lens alone. (*L.*)

Objective-type revision questions for Part 1

The first figure of a question number gives the relevant chapter, e.g. **2.3** is the third question for chapter 2.

Multiple choice

Select the response which you think is correct.

1.1. The density of aluminium is 2.7 g cm^{-3}, its atomic mass is 27 and the Avogadro constant is 6.0×10^{23} atoms per mole. If aluminium atoms are assumed to be spheres, packed so that they occupy three-quarters of the total volume, the volume of an aluminium atom in cm^3 is

A $4 \times 27/(3 \times 2.7 \times 6.0 \times 10^{23})$
B $3 \times 27/(4 \times 2.7 \times 6.0 \times 10^{23})$
C $3 \times 2.7/(4 \times 27 \times 6.0 \times 10^{23})$
D $4 \times 27 \times 6.0 \times 10^{23}/(3 \times 2.7)$
E $3 \times 2.7 \times 6.0 \times 10^{23}/(4 \times 2.7)$

1.2. The weakest form of bonding in materials is

A van der Waals B ionic C covalent D metallic

2.1. Which one of the following is the Young modulus (in Pa) for the wire having the stress-strain curve of Fig. 1?

A 36×10^{11} B 8.0×10^{11} C 2.0×10^{11}
D 0.50×10^{11} E 16×10^{11}

2.2. Which one of the following statements is a correct statement about the evidence provided by Fig. 1?

A The wire only obeys Hooke's law between O and A and after A it becomes much more difficult to stretch it.
B The wire does not obey Hooke's law between O and A and after A it becomes much more difficult to stretch it.
C The wire only obeys Hooke's law between O and A and after A it becomes much easier to stretch it.
D The wire does not obey Hooke's law between O and A and after A it becomes much easier to stretch it.

E The wire does not obey Hooke's law at all and is no harder or easier to stretch before A than after A.

(J.M.B.)

2.3. Wires X and Y are made from the same material. X has twice the diameter and three times the length of Y. If the elastic limits are not reached when each is stretched by the same tension, the ratio of energy stored in X to that in Y is

A 2:3 B 3:4 C 3:2 D 6:1 E 12:1

2.4. Reasons for the good stiffness and strength of three different materials are given below. Select from the list of five materials the *one* to which *each* statement best applies (three answers).

(*i*) It has 'foreign' atoms in the lattice which oppose dislocation movement.
(*ii*) It has high covalent bond density.
(*iii*) It has long-chain molecules lying more or less parallel along their length.

A steel B rubber C copper D polythene E glass

3.1. The drift velocity of the free electrons in a conductor is independent of one of the following. Which is it?

A The length of the conductor.
B The number of free electrons per unit volume.
C The cross-sectional area of the conductor.
D The electronic charge.
E The current.

3.2. The value of X in ohms which gives zero deflection on the galvanometer in Fig. 2 is

A 3 B 6 C 15 D 18 E 27

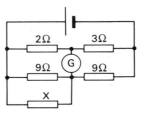

Fig. 2

3.3. If X, Y and Z in Fig. 3 are identical lamps, which of the following changes to the brightnesses of the lamps occur when switch S is closed?

A X stays the same   Y decreases

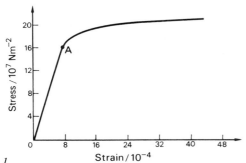

Fig. 1

127

B X increases   Y stays the same
C X increases   Y decreases
D X decreases   Y increases
E X decreases   Y decreases

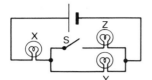

Fig. 3

3.4. A moving-coil galvanometer has a resistance of 10 Ω and gives a full scale deflection for a current of 0.01 A. It could be converted into a voltmeter reading up to 10 V by connecting a resistor of value

A 0.10 Ω   in parallel with it   **B** 90 Ω   in series with it
C 0.10 Ω   in series with it   **D** 990 Ω   in parallel with it
E 990 Ω   in series with it   (*J.M.B.*)

3.5. Which of the graphs in Fig. 4 best shows the variation of current with time in a tungsten filament lamp, from the moment current flows?

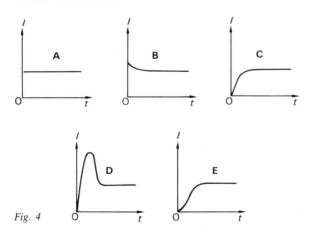

Fig. 4

3.6. Two resistors are connected in parallel as shown in Fig. 5. A current passes through the parallel combination. The power dissipated in the 5.0 Ω resistor is 40 W. Which one of the following is the power dissipated in watts in the 10 Ω resistor?

A 10   **B** 20   **C** 40   **D** 80

(*J.M.B. Eng. Sc.*)

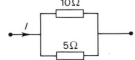

Fig. 5

3.7. If each resistor in Fig. 6 is 2 Ω, the effective resistance in ohms between X and Y is

A 2/5   **B** 1   **C** 2   **D** $2\frac{2}{3}$ **E** $3\frac{1}{2}$

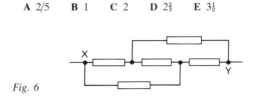

Fig. 6

3.8. A cylindrical copper rod is re-formed to twice its original length. Which one of the following statements describes the way in which the resistance is changed?

A The resistance remains constant.
B The resistance increases by a factor of two.
C The resistance increases by a factor of four.
D The resistance increases by a factor of eight.

(*J.M.B. Eng. Sc.*)

3.9. A thermocouple thermometer is to be designed using the circuit of Fig. 7. AB is a potentiometer wire of resistance 5.0 Ω and ED is a thermocouple whose e.m.f. is 20 mV at 400 °C and zero at 0 °C. For a temperature measurement range from 0 °C to 400 °C, the required value for resistor R in ohms is

A 195   **B** 295   **C** 395   **D** 495

(*J.M.B. Eng. Sc.*)

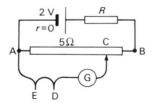

Fig. 7

4.1. Choose from the following statements one which does *not* apply to the platinum resistance thermometer.

A It can give high accuracy.
B It is suitable for measuring the temperature in a small object.
C It has a high heat capacity.
D It can cover a wide range of temperature.
E It can only be used for steady temperatures.

4.2. Spheres P and Q are uniformly constructed from the same material which is a good conductor of heat and the radius of Q is twice the radius of P. The rate of fall of temperature of P is x times that of Q when both are at the same surface temperature. The value of x is

A $\frac{1}{4}$ **B** $\frac{1}{2}$ **C** 2   **D** 4   **E** 8

4.3. Heat flows through the bar XYZ in Fig. 8a, the ends X and Z being maintained at fixed temperatures (temperature at X > temperature at Z). If only the part YZ is lagged, which graph in Fig. 8b shows the variation of temperature (θ) with distance along XZ for steady state conditions?

(a)

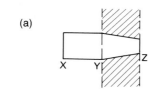

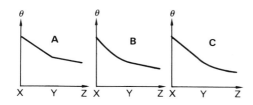

(b)

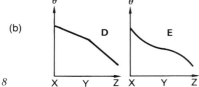

Fig. 8

4.4. The dimensions of specific heat capacity are

A $ML^2T\theta^{-1}$ **B** $L^2T^{-2}\theta^{-1}$ **C** $L^2T^2\theta^{-1}$
D $M^{-1}LT^{-1}\theta^{-1}$ **E** $ML^{-1}T^{-2}\theta^{-1}$

5.1. A lens of focal length 12 cm forms an upright image three times the size of a real object. The distance in cm between the object and image is

A 8.0   **B** 16   **C** 24   **D** 32   **E** 48

5.2. When a lens is inserted between an object and a screen which are a fixed distance apart the size of the image is either 6 cm or $\frac{2}{3}$ cm. The size of the object in cm is

A 2   **B** 3   **C** 4   **D** $4\frac{1}{3}$ **E** 9

5.3. Which *one* of the following combinations of lenses is used as a compound microscope? (The objective is listed first.)

A long focus converging and shorter focus converging
B long focus converging and shorter focus diverging
C long focus converging and long focus converging
D short focus converging and longer focus converging
E short focus converging and longer focus diverging

(*J.M.B.*)

Multiple selection

In each question one or more of the responses may be correct. Choose one letter from the answer code given.

Answer **A** if (*i*), (*ii*) and (*iii*) are correct
Answer **B** if only (*i*) and (*ii*) are correct
Answer **C** if only (*ii*) and (*iii*) are correct
Answer **D** if (*i*) only is correct
Answer **E** if (*iii*) only is correct

3.10. In the potentiometer circuit of Fig. 9 the galvanometer reveals a current in the direction shown wherever the sliding contact touches the wire. This could be caused by

(*i*) E_1 being too low
(*ii*) 3.0 Ω being too high
(*iii*) A break in PQ

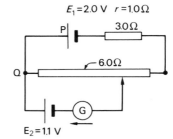

Fig. 9 $E_2 = 1.1$ V

5.4. A microscope with a short focus objective

(*i*) allows more light to be collected
(*ii*) keeps the distance between objective and eyepiece small
(*iii*) gives high magnifying power

Part 2 MECHANICS

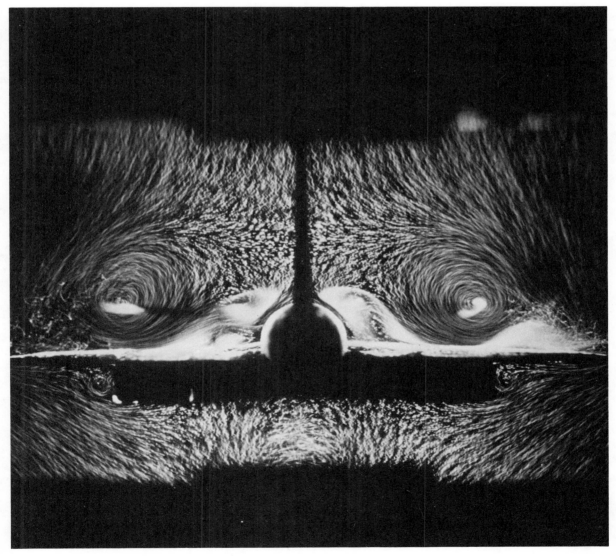

6 Statics and dynamics

Mechanics

Mechanics is concerned with the action of forces on a body. If the forces balance they are said to be in *equilibrium* and the branch of mechanics which deals with such cases is called *statics*—the subject reviewed in the first part of this chapter. In the second part of the chapter we will consider the effects of forces which are not in equilibrium—a study known as *dynamics*.

Many engineering and technological problems such as designing buildings, bridges (like the Forth road and rail bridges shown in Fig. 6.1), roads, reservoirs, jet engines and aircraft require the application of the principles of mechanics so that structures with the necessary strength are obtained using the minimum of material. Not only are these principles useful for dealing with the world of ordinary experience but, suitably

Fig. 6.1

supplemented, they enable us to deal with the physics of the atom on the one hand, and astronomy and space travel on the other.

Composition and resolution of forces

(*a*) *Scalar and vector quantities.* A scalar quantity has magnitude only and is completely described by a certain number of appropriate units. A vector quantity has both magnitude and direction; it can be represented by a straight line whose length represents the magnitude of the quantity on a particular scale and whose direction (shown by an arrow) indicates the direction of the quantity.

For example, if the points X and Y in Fig. 6.2*a* are 2 metres apart, the statement that XY = 2 m fully describes the *distance* between them; distance is a scalar. However the *displacement* **XY** between the points is 2 m in a direction 30° east of north, Fig. 6.2*b*; displacement is a vector (like other vectors it is often printed in bold type). Other examples of scalars are mass, time, density, speed, energy. Force, velocity (displacement per unit time) and momentum are vectors. What kind of quantity is (*i*) temperature, (*ii*) acceleration?

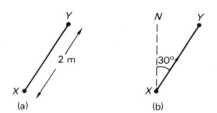

(*a*)   (*b*)

Fig. 6.2

(*b*) *Parallelogram of forces.* Scalars and vectors require different mathematical treatment. Thus scalars are added arithmetically but vectors are added geometrically by the *parallelogram law* which ensures their directions as well as their magnitudes are taken into account. The law will be illustrated by the addition of two displacements.

Suppose we walk from A to B and then from B to C as in Fig. 6.3*a* so that we suffer successive displacements **AB** and **BC**. The resultant displacement is given by **AC** in magnitude and direction. The same resultant displacement (i.e. **AC**) would be obtained if we started from the same point A and drew AD equal to BC in magnitude and direction and then drew DC equal to AB, Fig. 6.3*b*. The sum of two vectors therefore equals the diagonal of the parallelogram of which the vectors

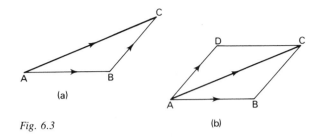

(a)   (b)

Fig. 6.3

are adjacent sides. How would you subtract two vectors?

The parallelogram law for the addition (composition) of forces is stated as follows.

If two forces acting at a point are represented in magnitude and direction by the sides of a parallelogram drawn from the point, their resultant is represented by the diagonal of the parallelogram drawn from the point.

(*c*) *Resolution of forces.* The reverse process to the addition of two vectors by the parallelogram law, is the splitting or *resolving* of one vector into two components. It is particularly useful in the case of forces when the components are taken at right angles to each other.

Suppose the force *F* is represented by OA in Fig. 6.4*a* and that we wish to find its components along OX and OY ($\angle$ XOY = 90°). A perpendicular AB is dropped from A on to OX and another AC from A on to OY, to give rectangle (parallelogram) OCAB. OB and OC are the required components or resolved parts. If $\angle$ AOB = θ then

$$\cos \theta = OB/OA = OB/F$$
$$\therefore \quad OB = F \cos \theta$$

and
$$\sin \theta = AB/OA = OC/F$$
$$\therefore \quad OC = F \sin \theta$$

The two mutually perpendicular forces $F \cos \theta$ and F

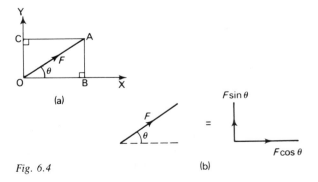

(a)   (b)

Fig. 6.4

sin θ are thus equivalent to F, Fig. 6.4b. The total effect of F along OX is represented by $F \cos \theta$. Also note that if $\theta = 0$, $F \cos \theta = F$ and $F \sin \theta = 0$, hence a force has no effect in a perpendicular direction. Resolving a force (or any vector) gives two quite independent forces and is a process we shall use frequently.

The component of a force of 10 N in a direction making an angle of 60° with it, is 10 cos 60 N (i.e. 5 N); in a direction perpendicular to this component the effective value of the force is 10 sin 60 N (i.e. $5\sqrt{3}$ N).

Moments and couples

A force applied to a hinged or pivoted body changes its rotation about the hinge or pivot. Experience shows that the turning effect or *moment* or *torque* of the force is greater the greater the magnitude of the force and the greater the distance of its point of application from the pivot.

The moment or torque of a force about a point is measured by the product of the force and the perpendicular distance from the line of action of the force to the point.

Thus in Fig. 6.5a if OAB is a trap-door hinged at O and acted on by forces P and Q as shown then

$$\text{moment of } P \text{ about } O = P \times \text{OA}$$

and, $\qquad \text{moment of } Q \text{ about } O = Q \times \text{OC}$

Note that the perpendicular distance must be taken. Alternatively, we can resolve Q into components $Q \cos \theta$ perpendicular to OB and $Q \sin \theta$ along OB, Fig. 6.5b. The moment of the latter about O is zero since its line of action passes through O. For the former we have

$$\text{moment of } Q \cos \theta \text{ about } O = Q \cos \theta \times \text{OB}$$
$$= Q \times \text{OC}$$
$$(\text{since } \cos \theta = \text{OC/OB})$$

which is the same as before. Moments are measured in

newton metres (N m) and are given a positive sign if they tend to produce clockwise rotation.

A *couple* consists of two equal and opposite parallel forces whose lines of action do not coincide; it always tends to change rotation. A couple is applied to a water tap to open it. From Fig. 6.6 we can say that the moment or torque of the couple $P - P$ about O

$$= P \times \text{OA} + P \times \text{OB} \quad (\text{both are clockwise})$$
$$= P \times \text{AB}$$

Hence,

moment of couple = one force × perpendicular distance between forces

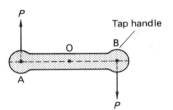

Fig. 6.6

Equilibrium of coplanar forces

(a) *General conditions for equilibrium.* If a body is acted on by a number of coplanar forces (i.e. forces in the same plane) and is in equilibrium (i.e. there is rest or unaccelerated motion) then

(i) the components of the forces in both of any two directions (usually taken at right angles) must balance, and
(ii) the sum of the clockwise moments about any point equals the sum of the anticlockwise moments about the same point.

The first statement is a consequence of there being no translational motion in any direction and the second follows since there is no rotation of the body. In brief, if a body is in equilibrium the forces and the moments must *both* balance.

The following worked example (and also that on p. 137) shows how the conditions for equilibrium are used to solve problems.

(b) *Worked example. A sign of mass 5.0 kg is hung from the end B of a uniform bar AB of mass 2.0 kg. The bar is hinged to a wall at A and held horizontal by a wire joining B to a point C which is on the wall vertically above A. If angle ABC = 30°, find the force in the wire and that exerted by the hinge. (g = 10 m s⁻².)*

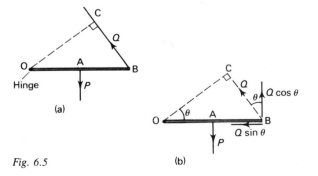

Fig. 6.5

The weight of the sign will be 50 N and of the bar 20 N (since $W = mg$). The arrangement is shown in Fig. 6.7a. Let P be the force in the wire and suppose Q, the force exerted by the hinge, makes angle θ with the bar. The bar is uniform and so its weight acts vertically downwards at its centre G. Let the length of the bar be $2l$.

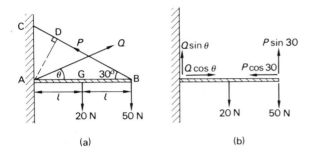

(a)   (b)

Fig. 6.7

(*i*) *There is no rotational acceleration*, therefore taking moments about A we have

clockwise moments = anticlockwise moment

i.e.

$$20 \times l + 50 \times 2l = P \times AD \text{ (AD is perpendicular to BC)}$$
$$\therefore \quad 120l = P \times AB \sin 30 \ (\sin 30 = AD/AB)$$
$$= P \times 2l \times 0.5$$
$$\therefore \quad P = 1.2 \times 10^2 \text{ N}$$

Note: by taking moments about A there is no need to consider Q since it passes through A and so has zero moment.

(*ii*) *There is no translational acceleration*, therefore the vertical components (and forces) must balance, likewise the horizontal components. Hence resolving Q and P into vertical and horizontal components (which now replace them, Fig. 6.7b) we have:

Vertically

$$Q \sin \theta + P \sin 30 = 20 + 50$$
$$\therefore \quad Q \sin \theta = 70 - 120(1/2)$$
$$Q \sin \theta = 10 \qquad (1)$$

Horizontally

$$Q \cos \theta = P \cos 30 = 120(\sqrt{3}/2)$$
$$\therefore \quad Q \cos \theta = 60\sqrt{3} \qquad (2)$$

Dividing (1) by (2)

$$\tan \theta = 10/(60\sqrt{3})$$
$$\therefore \quad \theta = 5.5°$$

Squaring (1) and (2) and adding

$$Q^2(\sin^2 \theta + \cos^2 \theta) = 100 + 10\,800$$
$$\therefore \quad Q^2 = 10\,900 \qquad (\sin^2 \theta + \cos^2 \theta = 1)$$
and
$$Q = 1.0(4) \times 10^2 \text{ N}$$

(*c*) *Structures.* Forces act at a joint in many structures and if these are in equilibrium then so too are the joints. The joint O in the bridge structure of Fig. 6.8 is in equilibrium under the action of forces P and Q exerted by the girders and the normal force S exerted by the bridge support at O. The components of the forces in two perpendicular directions at the joint must balance. Hence

$$S = Q \sin \theta \quad \text{and} \quad P = Q \cos \theta$$

If θ and S are known (the latter from the weight and loading of the bridge) then P and Q (which the bridge designer may wish to know) can be found. Other joints may be treated similarly (see question 4, p. 153).

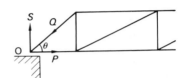

Fig. 6.8

Laws of friction

Frictional forces act along the surface between two bodies whenever one moves or tries to move over the other and in a direction so as to oppose relative motion of the surfaces. Sometimes it is desirable to reduce friction to a minimum but in other cases its presence is essential. For example, it is the frictional push of the ground on the soles of our shoes that enables us to walk. Otherwise our feet would slip backwards as they do when we try to walk on an icy road. The study of friction, wear and lubrication, now called *tribology*, is a matter of great importance to industry and is the subject of much research.

(*a*) *Coefficients of friction.* Friction between two solid surfaces can be studied using the apparatus of Fig. 6.9 in which the plank tends to move or does move, depending on the force applied to the crank, whilst the block remains at rest. The frictional force between the block and the plank is measured by the spring balance.

As the crank is gently wound the spring balance reading increases and reaches a maximum value when the plank is about to move. This maximum force between the surfaces is called the *limiting* frictional

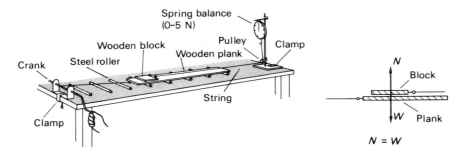

Fig. 6.9

force. When motion does start, the spring balance reading usually decreases slightly showing that the *sliding, kinetic* or *dynamic* frictional force (all terms are used) is rather smaller than the limiting value. The block can be set on edge to see if friction depends on the area of contact between the surfaces. The normal force N exerted by the plank on the block equals the weight W of the block. The effect on the frictional force of varying N can be found by placing weights on the block.

The results of such experiments are summarized in the following laws of friction, which hold approximately.

1. *The frictional force between two surfaces opposes their relative motion.*

2. *The frictional force does not depend on the area of contact of the surfaces if the normal reaction is constant.*

3. (a) *When the surfaces are at rest the limiting frictional force F is directly proportional to the normal force N, i.e. F ∝ N (or F/N = constant).*

(b) *When motion occurs the dynamic frictional force F' is directly proportional to the normal force N i.e. F' ∝ N (or F'/N = constant) and is reasonably independent of the relative velocity of the surfaces.*

The coefficients of limiting and dynamic friction are denoted by μ and μ' respectively and are defined by the equations

$$\mu = F/N \quad \text{and} \quad \mu' = F'/N$$

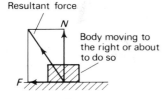

Resultant force

Body moving to the right or about to do so

Fig. 6.10

For two given surfaces μ' is usually less than μ but they are often assumed equal. For wood on wood μ is about 0.2 to 0.5. In general a surface exerts a frictional force and the resultant force on a body on the surface has two components—a normal force N perpendicular to the surface and a frictional force F along the surface, Fig. 6.10. If the surface is smooth, as is sometimes assumed in mechanics calculations, $\mu = 0$ and so $F = 0$. Therefore a smooth surface only exerts a force at right angles to itself, i.e. a normal force N.

The coefficient of limiting friction, μ, can also be found by placing the block on the surface and tilting the latter to the angle θ at which the block is just about to slip, Fig. 6.11a. The three forces acting on the block are its weight mg, the normal force N of the surface and the limiting frictional force $F(=\mu N)$. They are in equilibrium and if mg is resolved into components $mg \sin \theta$ along the surface and $mg \cos \theta$ perpendicular to the surface, Fig. 6.11b, then

$$F = \mu N = mg \sin \theta$$
and
$$N = mg \cos \theta$$
Dividing
$$\mu = \tan \theta$$

Hence μ can be found by measuring θ, called the *angle of friction.*

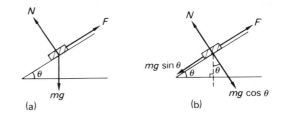

(a)   (b)

Fig. 6.11

(b) *Worked example. A uniform ladder 4.0 m long, of mass 25 kg, rests with its upper end against a smooth vertical wall and with its lower end on rough ground. What must be the least coefficient of friction between the ground and the ladder for it to be inclined at 60° with the horizontal without slipping? (g = 10 m s⁻².)*

The weight (mg) of the ladder is 250 N and the forces acting on it are shown in Fig. 6.12. The wall is smooth and so the force S of the wall on the ladder is normal to the wall. Since the ladder is uniform its weight W can be taken to act at its mid-point G. If it is about to slip there will be a force exerted on it by the ground which can be resolved into a normal force N and a limiting frictional force $F = \mu N$, where μ is the required coefficient of friction.

The forces are in equilibrium.

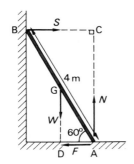

Fig. 6.12

Resolving vertically

$$N = W = 250 \text{ newtons}$$

Resolving horizontally

$$F = \mu N = S$$

Taking moments about A

$$S \times AC = W \times AD$$

$$S \times 4.0 \cos 30 = 250 \times 2.0 \sin 30 = 250$$

$$\therefore \quad S = 125/\sqrt{3} \text{ newtons}$$

Hence

$$\mu = \frac{S}{N} = \frac{125}{250\sqrt{3}}$$

$$= 0.29$$

Nature of friction

Close examination of the flattest and most highly polished surfaces shows that they have hollows and humps more than one hundred atoms high. When one solid is placed on another, contact therefore only occurs at a few places of small area, Fig. 6.13. From

Fig. 6.13

electrical resistance measurements of two metals in contact it is estimated that in the case of steel, the actual area that is touching may be no more than 1/10 000th of the apparent area.

The pressures at the points of contact are extremely high and cause the humps to flatten out (being plasticly deformed) until the increased area of contact enables the upper solid to be supported. It is thought that at the points of contact small, cold welded 'joints' are formed by the strong adhesive forces between molecules which are very close together. These have to be broken before one surface can move over the other, thus accounting for law 1 (p. 137). Measurements show that changing the apparent area of contact of the bodies has little effect on the actual area for the same normal force, so explaining law 2. It is also found that the actual area is proportional to the normal force and since this theory suggests that the frictional force depends on the actual area we might expect the frictional force to be proportional to the normal force—as law 3 states.

Velocity and acceleration

(*a*) *Speed and velocity.* If a car travels from A to B along the route shown in Fig. 6.14*a*, its *average speed* is defined as *the actual distance travelled*, i.e. AXYZB, *divided by the time taken*. The speed at any instant is found by considering a very short time interval. Speed has magnitude only and is a scalar quantity.

Velocity is defined as *the distance travelled in a particular direction divided by the time taken*. The average velocity of the car in the direction AB is the distance between A and B, i.e. the length of the straight line AB, Fig. 6.14*b*, divided by the time actually taken for the journey from A to B. Thus AB is the displacement of the car; in this case it is not the

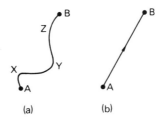

Fig. 6.14

actual path followed from A to B, although of course it could be in other cases. Velocity can therefore also be defined as *the change of displacement in unit time*; it has both magnitude and direction and is a vector quantity.

The SI unit of velocity is 1 metre per second (1 m s^{-1}); $10 \text{ m s}^{-1} = 36 \text{ km h}^{-1}$. In Fig. 6.14 if the route AXYZB is 40 kilometres and the car takes 1 hour for the journey, its average *speed* is 40 km h^{-1}. If B is 25 km north-east of A as the crow flies, its average *velocity* is only 25 km h^{-1} towards the north-east.

The velocity v of a body which undergoes a very small displacement δs in the very small time δt is given by the equation

$$v = \frac{\delta s}{\delta t}$$

Or more strictly, in calculus notation, the velocity v at an instant is defined by

$$v = \lim_{\delta t \to 0} \left(\frac{\delta s}{\delta t} \right) = \frac{ds}{dt}$$

Velocity is therefore the rate of change of displacement.

A body which covers equal distances in the same straight line in equal time intervals, no matter how short these are, is said to be moving with constant or *uniform velocity*. Only a body moving in a straight line can have uniform velocity. The direction of motion of a body travelling in a curved path is continually changing and so it cannot have uniform velocity even though its speed may be constant.

(*b*) *Acceleration.* A body is said to accelerate when its velocity changes. Thus if a very small velocity change δv occurs in a very small time interval δt, the acceleration a of the body is

$$a = \frac{\text{change in velocity}}{\text{time taken for change}} = \frac{\delta v}{\delta t}$$

More correctly, in calculus notation, the instantaneous acceleration a is defined by

$$a = \lim_{\delta t \to 0} \left(\frac{\delta v}{\delta t} \right) = \frac{dv}{dt}$$

In words, *acceleration is the rate of change of velocity*. For a car accelerating towards the north from 10 m s^{-1} to 20 m s^{-1} in 5.0 seconds we can say

$$\text{average acceleration} = \frac{(20 - 10) \text{ m s}^{-1}}{5.0 \text{ s}}$$

$$= 2.0 \text{ m s}^{-2} \text{ towards the north}$$

That is, on average, the velocity of the car increases by 2.0 m s^{-1} every second. Since $10 \text{ m s}^{-1} = 36 \text{ km h}^{-1}$ and $20 \text{ m s}^{-1} = 72 \text{ km h}^{-1}$, we could also say the average acceleration is $(72 - 36) \text{ km h}^{-1}/5 \text{ s} = 7.2 \text{ km h}^{-1}$ per second $= 7.2 \text{ km h}^{-1} \text{ s}^{-1}$.

Equations for uniform acceleration

The acceleration of a body is uniform if its velocity changes by equal amounts in equal times. We will now derive three useful equations for a body moving in a straight line with uniform acceleration.

Suppose the velocity of the body increases steadily from u to v in time t then the uniform acceleration a is given by

$$a = \frac{\text{change of velocity}}{\text{time taken}}$$

$$= \frac{v - u}{t}$$

$$\therefore \quad v = u + at \qquad (1)$$

Since the velocity is increasing steadily, the average velocity is the mean of the initial and final velocities, i.e.

$$\text{average velocity} = \frac{u + v}{2}$$

If s is the displacement of the body in time t, then since average velocity = displacement/time = s/t, we can say

$$\frac{s}{t} = \frac{u + v}{2}$$

$$\therefore \quad s = \tfrac{1}{2}(u + v)t$$

But

$$v = u + at$$

$$\therefore \quad s = \tfrac{1}{2}(u + u + at)t$$

or

$$s = ut + \tfrac{1}{2}at^2 \qquad (2)$$

If we eliminate t from (2) by substituting $t = (v - u)/a$ from (1), we get on simplifying

$$v^2 = u^2 + 2as \qquad (3)$$

Knowing any three of u, v, a, s and t the others can be found.

Velocity–time graphs

Acceleration is rate of change of velocity (in calculus notation dv/dt) and equals at any instant the slope of the velocity–time graph. In Fig. 6.15, curve ① has zero slope and represents uniform velocity, curve ② is a

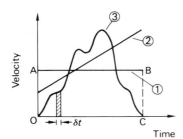

Fig. 6.15

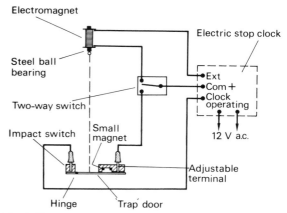

Fig. 6.16

straight line of constant slope and represents uniform acceleration, while curve ③ is for variable acceleration since its slope varies.

The distance travelled by a body during any interval of time can also be found from a velocity–time graph, a fact which is especially useful in cases of non-uniform acceleration since the three equations of motion do not then apply. For the constant velocity case, curve ①, the distance travelled in time OC = velocity × time = OA × OC = area OABC. In curve ③, if we consider a small enough time interval δt, the velocity is almost constant and the distance travelled in δt will be the area of the very thin shaded strip. By dividing up the whole area under curve ③ into such strips it follows that the *total distance travelled in time OC equals the area between the velocity–time graph and the time-axis*.

Motion under gravity

(a) *Free fall*. Experiments show that at a particular place all bodies falling freely under gravity, in a vacuum or when air resistance is negligible, have the *same* constant acceleration irrespective of their masses. This acceleration towards the surface of the earth, known as the *acceleration due to gravity*, is denoted by g. Its magnitude varies slightly from place to place on the earth's surface and is approximately 9.8 m s^{-2}. The velocity of a freely falling body therefore increases by 9.8 m s^{-1} every second; in the equations of motion g replaces a.

A direct determination of g may be made using the apparatus of Fig. 6.16 in which the time for a steel ball-bearing to fall a known distance from rest is measured (to about 0.005 second) by an electric stop clock. When the two-way switch is changed to the 'down' position, the electromagnet releases the ball and simultaneously the clock starts. At the end of its fall the ball opens the 'trap-door' on the impact switch and the clock stops. Air resistance is negligible for a dense object such as a ball-bearing. The result is found from $s = ut + \frac{1}{2}at^2$ since $u = 0$ and $a = g$.

The measurement of g using a simple pendulum is described on page 182.

(b) *Vertical projection*. The velocity of a body projected upwards from the ground decreases by 9.8 m s^{-1} (near enough to 10 m s^{-1}) every second—neglecting the effect of air resistance. Hence if a ball is thrown straight upward with an initial velocity of 30 m s^{-1} then in just over 3 seconds it will have zero velocity and be at its highest point.

(c) *Sign convention for displacement, velocity and acceleration*. The three equations for uniform acceleration can be used to solve problems on falling and rising bodies so long as a sign convention is adopted.

Vector quantities have magnitudes and directions. By their nature magnitudes are positive and direction is stated in terms of angles. However when we are dealing with one-dimensional effects, such as linear motion, the only way to indicate direction is by signs. If we take *downward as being positive*, a downward acceleration, velocity or displacement is positive. Hence $g = +10$ m s^{-2} and this is true whether the body is rising and slowing down or falling and speeding up.

Consider an example. Suppose a ball is thrown straight up with an initial velocity of 40 m s^{-1} and we wish to find its velocity and height after 2 seconds. We have $u = -40$ m s^{-1}, $t = 2$ s, $a = g = +10$ m s^{-2}.

Since
$$v = u + at$$
$$v = -40 + 10 \times 2$$
$$= -20 \text{ m s}^{-1}$$

The ball has a velocity of 20 m s^{-1} upward. Also

$$s = ut + \tfrac{1}{2}at^2$$
$$= -40 \times 2 + \tfrac{1}{2} \times 10 \times 4$$
$$= -60 \text{ m}$$

It rises 60 m in 2 seconds.

Projectiles

In Fig. 6.17 a multiflash photograph is shown of the motion of two balls, one released from rest and the other projected simultaneously with a horizontal velocity. It is clear that the vertical motion of the projected ball (a constant acceleration = g) is unaffected by its horizontal motion (a constant velocity). The two motions are quite independent of each other.

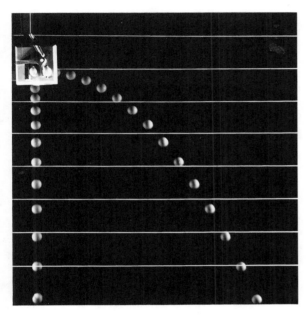

Fig. 6.17

Consider a body projected obliquely from O with velocity u at an angle θ to the horizontal, Fig. 6.18. Suppose we wish to know the height attained by the body and its horizontal range. If we resolve u into horizontal and vertical components $u \cos \theta$ and $u \sin \theta$ respectively, each component can be considered independently of the other.

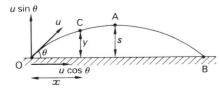

Fig. 6.18

Vertical motion. Whilst rising, the body is subject to a constant acceleration $a = -g$. (Here it is convenient to take downward directed vectors as negative, which explains why g has a negative sign.) If s is the height attained then, since the initial velocity is $u \sin \theta$ and the final velocity zero, we have from the third equation of motion ($v^2 = u^2 + 2as$)

$$0 = u^2 \sin^2 \theta - 2gs$$

$$\therefore \quad s = \frac{u^2 \sin^2 \theta}{2g}$$

Also, if t is the time to reach the highest point A, it follows from the first equation of motion ($v = u + at$) that

$$0 = u \sin \theta - gt$$

$$\therefore \quad t = \frac{u \sin \theta}{g}$$

The time taken by the body to fall to the horizontal level of O is also t. Therefore

$$\text{time of flight} = 2t = 2\frac{u \sin \theta}{g}$$

Horizontal motion. Neglecting air resistance, the horizontal component $u \cos \theta$ remains constant during the flight since g has no effect in a horizontal direction. The horizontal distance travelled, OB,

$$= \text{horizontal velocity} \times \text{time of flight}$$

$$= \frac{u \cos \theta \times 2u \sin \theta}{g}$$

$$= \frac{2u^2 \sin \theta \cos \theta}{g}$$

$$= \frac{u^2 \sin 2\theta}{g} \qquad (\sin 2\theta = 2 \sin \theta \cos \theta)$$

For a given velocity of projection the range is a maximum when $\sin 2\theta = 1$, i.e. when $\theta = 45°$ and has the value u^2/g.

Trajectory. Let the body be at point C (co-ordinate x, y) at time t after projection from O. Therefore

$$x = tu \cos \theta$$

and

$$y = tu \sin \theta - \tfrac{1}{2}gt^2$$

Substituting for t in the second equation we get

$$y = \frac{x}{u \cos \theta} u \sin \theta - \frac{gx^2}{2u^2 \cos^2 \theta}$$

$$\therefore \quad y = x \tan \theta - \frac{gx^2}{2u^2 \cos^2 \theta}$$

This is of the form $y = ax + bx^2$ which is the equation of a parabola (a and b are constants for a given velocity and angle of projection). In practice air resistance causes slight deviation from a parabolic path.

Newton's laws of motion

Newton (1642–1727) studied and developed Galileo's (1564–1642) ideas about motion and subsequently stated the three laws which now bear his name. He established the subject of dynamics. His laws are a set of statements and definitions that we believe to be true because the results they predict are found to be in very exact agreement with experiment over a wide range of conditions. We do not regard them as absolutely true and more exact laws are required for certain extreme cases.

(*a*) *First law. If a body is at rest it remains at rest or if it is in motion it moves with uniform velocity* (i.e. constant speed in a straight line) *until it is acted on by a resultant force.*

The second part of the law appears to disagree with certain everyday experiences which suggest that a steady effort has to be exerted on a body, e.g. a bicycle, even to keep it moving with constant velocity (let alone to accelerate it), otherwise it comes to rest. The law on the other hand states that a moving body retains its motion naturally and if any change occurs (i.e. if it is accelerated) some outside agent—a force—must be responsible. It seems that the question to be asked about a moving body is not 'what keeps it moving' but 'what changes or stops its motion'. Frequently it is friction and if a body does move under near-frictionless conditions, its velocity is in fact almost uniform. This is shown in Fig. 6.19 which is a photograph of a puck, illuminated at equal time intervals by a flashing xenon lamp, moving on a cushion of carbon dioxide gas across a clean, level glass plate.

This law really defines a force as something which changes the state of rest or uniform motion of a body. Contact may be necessary, as when we push a body with our hands, or it may not be, as in the case of gravitational, electric and magnetic forces.

(*b*) *Mass.* The first law implies that matter has a built-in reluctance to change its state of rest or motion. This property, possessed by all bodies, is called *inertia*. Its effects are evident when a vehicle suddenly stops, causing the passengers to lurch forward (tending to keep moving), or starts, jerking the passengers backwards (since they tend to remain at rest).

The *mass* of a body is a measure of its inertia; a large mass requires a large force to produce a certain acceleration. The unit of mass is the *kilogram* (kg) and is the mass of a piece of platinum-iridium carefully preserved at Sèvres, near Paris. *In principle* the mass m of a body can be measured by comparing the accelerations a and a_0 produced by the same force in the body and the standard kilogram (m_0) respectively. The ratio of the two masses is then defined by

$$\frac{m}{m_0} = \frac{a_0}{a}$$

whence m can be calculated. *In practice* this is neither quick nor accurate and mass is most readily found using a beam balance to compare the weight of the body with that of a standard. It can be shown (p. 143) that the mass of a body is proportional to its weight and so a beam balance also compares masses.

The second law indicates how forces can be measured.

(*c*) *Second law. The rate of change of momentum of a body is proportional to the resultant force and occurs in the direction of the force.*

The momentum of a body of constant mass m moving with velocity u is, by definition, mu. That is

$$momentum = mass \times velocity$$

Suppose a force F acts on the body for time t and changes its velocity from u to v, then

$$change\ of\ momentum = mv - mu$$

$$\therefore \quad rate\ of\ change\ of\ momentum = \frac{m(v - u)}{t}$$

Hence, by the second law

$$F \propto \frac{m(v - u)}{t}$$

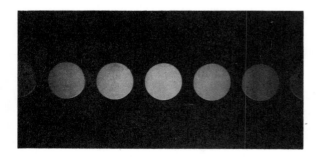

Fig. 6.19

If a is the acceleration of the body then

$$a = \frac{v - u}{t}$$

$$\therefore \quad F \propto ma$$

or

$$F = kma$$

where k is a constant. Now, *one newton is defined as the force which gives a mass of 1 kilogram an acceleration of 1 metre per second per second.* Hence if $m = 1$ kg and $a = 1$ m s^{-2} then $F = 1$ N and substituting these values in $F = kma$ we obtain $k = 1$. Thus *with these units*

$$k = 1 \quad \text{and} \quad F = ma$$

This expression is one form of Newton's second law and it indicates that a force can be measured by finding the acceleration it produces in a known mass. It can be verified experimentally using, for example, a tickertape timer and trolleys on a runway. Two points should be noted when using $F = ma$ to solve numerical problems. First, F is the *resultant* (or unbalanced) force causing acceleration a in a certain direction and second, F must be in newtons, m in kilograms and a in metres per second squared.

(*d*) *Weight.* The weight W of a body is the force of gravity acting on it towards the centre of the earth. Weight is thus a *force*, not to be confused with mass which is independent of the presence or absence of the earth. If g is the acceleration of the body towards the centre of the earth then we can substitute F (force accelerating the body) $= W$ and $a = g$ in $F = ma$, hence

$$W = mg$$

Thus if $g = 9.8$ m s^{-2}, a body of mass 1 kg has a weight of 9.8 N (roughly 10 N). The mass m of a body is constant but its weight mg varies with position on the earth's surface since g varies from place to place. Weight can be measured by a calibrated spring balance.

Note. To be strictly accurate the value of mg recorded by a spring balance is not quite equal to W if W is the gravitational attraction on the body directed towards the *centre of the earth*. We will see later that due to the rotation of the earth the observed direction of g is not exactly towards the earth's centre and its observed value is slightly different from the true value in that direction (p. 169). Hence mg does not equal W in magnitude and direction but the differences are extremely small.

If two bodies of masses m_1 and m_2 have weights W_1 and W_2 at the same place then

$$W_1 = m_1 g \quad \text{and} \quad W_2 = m_2 g$$

$$\therefore \quad \frac{W_1}{W_2} = \frac{m_1}{m_2}$$

That is, the *weight of a body is proportional to its mass*, a fact we use when finding the mass of a body by comparing its weight with that of standard masses on a beam balance.

(*e*) *Third law. If body A exerts a force on body B, then body B exerts an equal but opposite force on body A.*

The law is stating that forces never occur singly but always in pairs as a result of the interaction between two bodies. For example, when you step forward from rest your foot pushes backwards on the earth and the earth exerts an equal and opposite force forward on you. Two bodies and two forces are involved. The comparatively small force you exert on the large mass of the earth produces no noticeable acceleration of the earth, but the equal force it exerts on your very much smaller mass causes you to accelerate. It is important to note that the equal and opposite forces *do not act on the same body*; if they did, there could never be any resultant forces and all acceleration would be impossible.

If you pull a string attached to a block with a force P to the right, Fig. 6.20, the string pulls you with an equal force P to the left. Generally we can assume the string transmits the force unchanged and so there is another pair of equal and opposite forces at the block. The string exerts a pull P to the right *on the block* and the block exerts an equal pull to the left *on the string*—one force acts on the block and the other on the string. The string is pulled outwards at both ends and is in a state of tension.

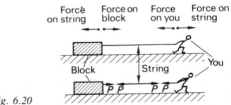

Fig. 6.20

Two pairs of forces exist when a book lies at rest on a table; draw diagrams to show what they are.

$F = ma$ calculations

1. A Saturn V rocket develops an initial thrust of 3.3×10^7 N and has a lift-off mass of 2.8×10^6 kg.

Find the initial acceleration of the rocket at lift-off.
(Take $g = 10$ m s^{-2}.)

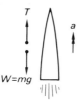

Fig. 6.21

Let T be the initial thrust on the rocket, m its mass and W its weight, Fig. 6.21. Then

$$W = mg = 2.8 \times 10^6 \text{ kg} \times 10 \text{ m s}^{-2}$$
$$= 2.8 \times 10^7 \text{ N}$$

The *resultant* upwards force *on the rocket* is $(T - W)$ and if a is the initial vertical acceleration, then from $F = ma$ we have

$$(T - W) = ma$$
$$\therefore \quad a = \frac{T - W}{m}$$
$$= \frac{3.3 \times 10^7 - 2.8 \times 10^7}{2.8 \times 10^6} \quad \frac{\text{N}}{\text{kg}}$$
$$= 1.8 \text{ m s}^{-2} \quad (\text{N kg}^{-1})$$

Note. We can apply $F = ma$ here since the rocket is instantaneously at rest. In general this is not possible because the mass of the rocket changes.

2. Two blocks A and B are connected as in Fig. 6.22 on a horizontal frictionless floor and pulled to the right with an acceleration of 2.0 m s^{-2} by a force P. If $m_1 = 50$ kg and $m_2 = 10$ kg, what are the values of T and P?

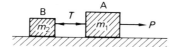

Fig. 6.22

The forces acting *on the blocks* are shown. Apply $F = ma$ to each.

For B, $\qquad T = m_2 a = 10 \times 2 = 20$ N

For A, $\quad P - T = m_1 a = 50 \times 2 = 100$ N

$$\therefore \quad P = 120 \text{ N}$$

3. A helicopter of mass M and weight W rises with vertical acceleration, a, due to the upward thrust U generated by its rotor. The crew and passengers of total mass m and total weight w exert a combined force R on the floor of the helicopter. Write an equation for the motion of (a) the helicopter, (b) the crew and passengers.

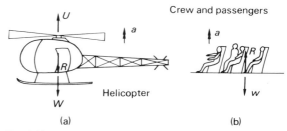

Fig. 6.23

The forces acting *on the helicopter* are the upwards force U due to the action of the rotor on the surrounding air, its weight W downwards due to the earth and the force R downwards exerted on the floor by the crew and passengers, Fig. 6.23a.

$$\therefore \quad \textit{Resultant} \text{ upwards force on helicopter}$$
$$= U - W - R$$

Hence, by the second law

$$U - W - R = Ma \qquad (1)$$

The forces acting *on the crew and passengers* are the upwards push of the floor of the helicopter (which by the third law must equal the downwards push R of the crew and passengers on the floor) and their weight w downwards, Fig. 6.23b.

$$\therefore \quad \textit{Resultant} \text{ upwards force on crew and passengers}$$
$$= R - w$$

Hence, by second law

$$R - w = ma \qquad (2)$$

The required equations are (1) and (2).

Momentum

The *momentum* of a body was previously defined as the mass of the body multiplied by its velocity. If SI units are used, Newton's second law may be written

force = rate of change of momentum

In symbols

$$F = \frac{mv - mu}{t}$$

where F is the force acting on a body of mass m which increases its velocity from u to v in time t.
Hence

$$Ft = mv - mu$$

The quantity Ft is called the *impulse* of the force on the body. It is a vector and, like linear momenta, impulses in opposite directions must be given positive and negative signs. In words, the *impulse–momentum equation* is

$$\text{impulse} = \text{change of momentum}$$

The equation shows that impulse and momentum have the same units, i.e. N s or kg m s^{-1}.

These ideas are important in games. The good cricketer or tennis player 'follows through' with the bat or racquet when striking the ball. The force applied then acts for a longer time, the impulse is greater and so also is the change of momentum (and velocity) of the bat. On the other hand when a cricket ball is caught its momentum is reduced to zero. This is achieved by an impulse in the form of an opposing force acting for a certain time and whilst any number of combinations of force and time will give a particular impulse, the 'sting' can be removed from the catch by drawing back the hands as the ball is caught. A smaller force is thus applied for a longer time.

In collisions of this and other types, the force is not constant but builds up to a maximum value as the deformation of the colliding bodies increases. It does, however, have an average value.

Conservation of momentum

(a) *Principle.* Suppose a body A of mass m_1 and velocity u_1 collides with another body B of mass m_2 and velocity u_2 moving in the same direction, Fig. 6.24a. If A exerts a force F to the *right* on B for time t then by Newton's third law, B will exert an equal but opposite force F on A, also for time t (since the time of contact is the same for each) but to the *left*.

The bodies thus receive equal but opposite impulses Ft and so it follows from the impulse–momentum equation that the changes of momentum must be equal and opposite. The total momentum change of A and B is therefore zero, or in other words the *total momentum of A and B together remains* constant despite the collision. Thus if A has a reduced velocity v_1 after the collision and B has an increased velocity v_2, both in the same direction as before, Fig. 6.24b, then

$$m_1u_1 + m_2u_2 = m_1v_1 + m_2v_2$$

This important result, known as the *principle of conservation of momentum*, has been deduced from Newton's second and third laws and is a universal rule of the physical world which still holds even in certain extreme (relativistic) conditions where Newton's laws fail. It applies not only to collisions but to any interaction between two or more bodies. Thus in an explosion such as occurs when a gun is fired, the backward momentum component of the gun in a horizontal direction equals the component of the forward momentum of the shell and propellant gases so that the *total* momentum of the gun–shell system remains zero even though the momentum of each part changes.

No *external* agent must act on the interacting bodies otherwise momentum may be added to the system. Sometimes momentum does appear to be gained (or lost). For example a body falling towards the earth increases its downwards momentum but the body is interacting with the earth (the external agent) which gains an equal amount of upward momentum from the attraction of the body on the earth. The complete system consists of the body *and* the earth and their total momentum remains constant. Similarly when a car comes to rest we believe that all the momentum it loses is transferred by the action of friction to the earth, although we cannot easily prove this.

The general statement of the principle is as follows.

When bodies in a system interact the total momentum remains constant provided no external force acts on the system.

(b) *Experimental test.* The principle can be investigated experimentally using a linear air track (which enables Perspex vehicles to move with negligible friction) and multiflash photography or electric stop clocks to measure velocities. Fig. 6.25a shows an air track (supplied with air by a domestic vacuum cleaner) and two vehicles with drinking straws attached so that a multiflash photograph can be taken using a xenon stroboscope which flashes at regular intervals.

In Fig. 6.25b a collision is shown between a vehicle of mass 'two' moving in from the left and one of mass 'three' from the right; the top markers give the velocities before the collision and the bottom ones after when

Before collision   After collision

(a)   (b)

Fig. 6.24

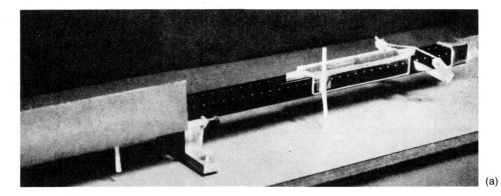

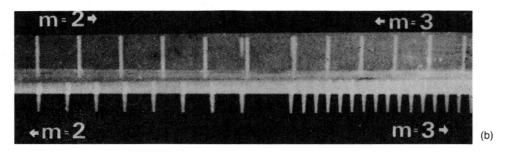

Fig. 6.25

the vehicles have 'rebounded' and are moving in opposite directions. (In Fig. 6.25*a* the hinged shutter is shown in the up position revealing to the camera the bottom half of one straw. At the exact instant of collision the shutter is rotated from one position to the other so that only one half of each straw is visible at any time. The velocities before and after the collision are thus obtained.)

Make measurements on Fig. 6.25*b* to see if momentum is conserved in this collision. In any attempt to verify the principle of conservation of momentum friction must be negligibly small. Why?

(*c*) *Speed of an air-rifle pellet.* An estimate is made by firing the pellet of mass *m* and speed *v* into

'Plasticine' on a model railway truck of total mass *M* on a friction-compensated runway, Fig. 6.26*a*, and measuring the speed *V* of the truck. Assuming conservation of momentum we have

momentum of pellet to the right before collision
= momentum of pellet and truck to the right after collision

$$\therefore \quad mv = (M + m)V$$

V is found from the time taken by a card (say 10 cm long) attached to the truck to pass through a beam of light which switches on a millisecond scaler (or electric stop clock) when it interrupts the beam.

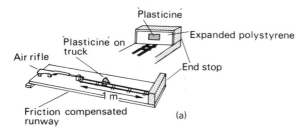

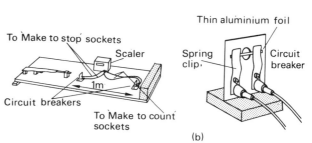

Fig. 6.26

A check may be made on v and so also on the principle of conservation of momentum, by timing the pellet directly using the scaler and two aluminium foil 'circuit-breakers' 1 metre apart, Fig. 6.26b.

Rocket and jet propulsion

The principle of both is illustrated by the behaviour of an inflated balloon when released with its neck open. If the neck is closed there is a state of balance inside the balloon with equal pressure at all points, Fig. 6.27a. When the neck is opened the pressure on the surface opposite the neck is now unbalanced and the balloon is forced to move in the opposite direction to that of the escaping air, Fig. 6.27b. According to the principle of conservation of momentum the air and the balloon have equal but opposite amounts of momentum, that is

$$m_{air} \times v_{air} = m_{balloon} \times v_{balloon}$$

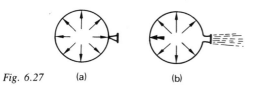

Fig. 6.27   (a)   (b)

In a rocket and a jet engine a stream of gas is produced at very high temperature and pressure and then escapes at high velocity through an exhaust nozzle. The thrust arises from the large increases in momentum of the exhaust gases. A rocket carries its own supplies of oxygen (liquid) and fuel (e.g. kerosene or liquid hydrogen), Fig. 6.28a. The mass of a rocket is not constant but decreases appreciably as it uses fuel (often at a rate of over 3000 kg s^{-1}). The acceleration consequently increases. A jet engine uses the surrounding air for its oxygen supply and so is unsuitable for space travel. Fig. 6.28b is a simplified drawing of one type of jet engine (gas turbine). The compressor draws in air at the front, compresses it, fuel (often paraffin) is injected and the mixture burns to produce hot exhaust gases which escape at high speed from the rear of the engine. These cause forward propulsion and drive the turbine which in turn rotates the compressor.

Momentum calculations

The concepts of impulse and momentum are useful when considering collisions and explosions, i.e. situations in which forces (called impulsive forces) act for a short time.

1. A jet of water emerges from a hose pipe of cross-section area 5.0×10^{-3} m^2 with a velocity of 3.0 m s^{-1} and strikes a wall at right angles. Calculate the force on the wall assuming the water is brought to rest and does not rebound. (Density of water = 1.0×10^3 kg m^{-3}.)

If the water arrives with a velocity of 3.0 m s^{-1}, 3.0 m^3 hits every square metre of the wall per second.

Hence volume of water striking wall per second

$$= (3.0 \text{ m s}^{-1})(5.0 \times 10^{-3} \text{ m}^2)$$

$$= 1.5 \times 10^{-2} \text{ m}^3 \text{ s}^{-1}$$

Therefore mass of water striking wall per second

$$= 1.5 \times 10^{-2} \times 1.0 \times 10^3 \text{ kg s}^{-1}$$

$$= 15 \text{ kg s}^{-1}$$

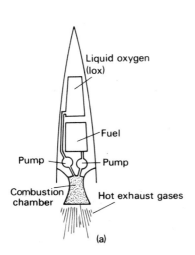

Liquid oxygen (lox)

Fuel

Pump   Pump

Combustion chamber   Hot exhaust gases

Fig. 6.28   (a)

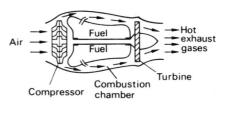

Air

Fuel

Fuel

Hot exhaust gases

Turbine

Compressor   Combustion chamber

(b)

Velocity change of water on striking wall

$$= 3.0 - 0 = 3.0 \text{ m s}^{-1}$$

Therefore momentum change per second of water on striking wall

$$= (15 \text{ kg s}^{-1})(3.0 \text{ m s}^{-1})$$

$$= 45 \text{ kg m s}^{-2}$$

But   force = momentum change per second

$$= 45 \text{ N}$$

(In practice the horizontal momentum of the water is seldom completely destroyed and so the answer is only approximate.)

2. A railway truck A of mass 2×10^4 kg travelling at 0.5 m s^{-1} collides with another truck B of half its mass moving in the opposite direction with a velocity of 0.4 m s^{-1}. If the trucks couple automatically on collision, find the common velocity with which they move, Fig. 6.29.

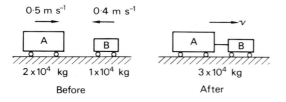

Fig. 6.29

Total momentum *to the right* of A and B before collision

$$= 2 \times 10^4 \times 0.5 - 1 \times 10^4 \times 0.4 \text{ kg m s}^{-1}$$

$$= 0.6 \times 10^4 \text{ kg m s}^{-1}$$

(If the momentum of A is taken as positive, that of B must be negative.)
Total momentum *to the right* of A and B after collision

$$= 3 \times 10^4 \times v \text{ kg m s}^{-1}$$

By the principle of a conservation of momentum

$$3 \times 10^4 \times v = 0.6 \times 10^4$$

$$\therefore \quad v = 0.2 \text{ m s}^{-1}$$

3. A jet engine on a test bed takes in 20.0 kg of air per second at a velocity of 100 m s^{-1} and burns 0.80 kg of fuel per second. After compression and heating the exhaust gases are ejected at 500 m s^{-1} relative to the aircraft. Calculate the thrust of the engine.

Velocity change of 20 kg of air

$$= (500 - 100) = 400 \text{ m s}^{-1}$$

Therefore momentum change per second of 20 kg of air

$$= 20 \text{ kg s}^{-1} \times 400 \text{ m s}^{-1}$$

The initial velocity of the fuel is zero and so its velocity change is 500 m s^{-1}.

$\therefore$ Momentum change per second of 0.80 kg of fuel

$$= 0.80 \text{ kg s}^{-1} \times 500 \text{ m s}^{-1}$$

$\therefore$ Total momentum change per second of air and fuel

$$= (20 \times 400 + 0.80 \times 500) \text{ kg m s}^{-2}$$

$$= 8.4 \times 10^3 \text{ kg m s}^{-2}$$

But
force (thrust) = total change of momentum per second

$\therefore$ thrust of engine $= 8.40 \times 10^3$ N

$$(1 \text{ kg m s}^{-2} = 1 \text{ N})$$

Note. If the engine is in an aircraft flying at 100 m s^{-1}, taking in air at this speed, the thrust would be about the same.

Work, energy and power

(*a*) *Work.* In science the term work has a definite meaning which differs from its everyday one. For example someone holding a heavy weight at rest may say and feel he is doing hard work but in fact none is being done on the weight in the scientific sense.

Work is done in science when a force moves its point of application along the direction of its line of action.

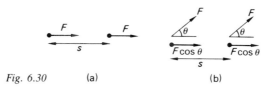

Fig. 6.30   (a)   (b)

In the simple case of Fig. 6.30a, the constant force F and the displacement s are in the same direction and we define the work W done by the force on the body by

$$W = Fs$$

If the force does not act in the direction in which motion occurs but at an angle θ to it as in Fig. 6.30b, then the work done is defined as the product of the component of the force in the direction of motion and the displacement in that direction. That is,

$$W = (F \cos \theta)s$$

When $\theta = 0$, $\cos \theta = 1$ and so $W = Fs$, in agreement with the first equation. When $\theta = 90°$, $\cos \theta = 0$ and F has no component in the direction of motion and so no work is done. Thus the work done by the force of gravity when a body is moved horizontally is zero.

If the force varies, the work done can be obtained from a force–displacement graph in which the component of the force in the direction of the displacement is plotted, Fig. 6.31. Suppose the force is F when the displacement is x, then the work done during a further, very small displacement δx (which is so small that F can be considered constant during it) is $F\delta x$, i.e. the shaded area. By dividing up the whole area under the curve into narrow strips we see that the total work done during displacement s is represented by area OABC.

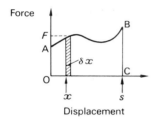

Fig. 6.31

Work can be either positive or negative. It is positive if the force (or a component of it) acts in the same direction as the displacement (Figs. 6.30*a* and *b*), but negative if it is oppositely directed (θ is then >90° and $\cos \theta$ is negative). The work done by friction when it opposes one body sliding over another is negative.

The unit of work is *1 joule* and is *the work done by a force of 1 newton when its point of application moves through a distance of 1 metre in the direction of the force.* Thus

1 joule (J) = 1 newton metre (N m)

Work is a scalar although force and displacement are both vectors.

(*b*) *Energy.* When a body A does work by exerting a force on another body B, the body A is said to lose energy, equal in amount to the work it performs. Energy is therefore often defined as *that which enables a body to do work*; it is measured in joules, like work. When an interchange of energy occurs between two

bodies we can look upon *the work done as measuring the quantity of energy transferred between them.* Thus if body A does 5 joules of work on body B then the energy transfer from A to B is 5 joules.

(*c*) *Power.* The power of a machine is *the rate at which it does work*, i.e. the rate at which it converts energy from one form to another. The unit of power is the *watt* (W) and equals a rate of working of 1 joule per second, i.e. 1 W = 1 J s^{-1}.

The two basic reasons for bodies having mechanical energy will now be considered.

Kinetic and potential energy

(*a*) *Kinetic energy.* This is the energy a body has because of its motion. For example a moving hammer does work against the resistance of the wood into which a nail is being driven. An expression for kinetic energy can be obtained by calculating the amount of work the body will do while it is being brought to rest.

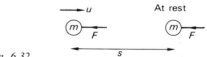

Fig. 6.32

Consider a body of constant mass m moving with velocity u. Let a constant force F act on it and bring it to rest in a distance s, Fig. 6.32. Since the final velocity v is zero, from $v^2 = u^2 + 2as$ we have

$$0 = u^2 + 2as$$

$$\therefore \quad a = -\frac{u^2}{2s}$$

The negative sign shows that the acceleration a is opposite in direction to u (as we would expect). The acceleration in the direction of F is thus $+u^2/2s$. The original kinetic energy of the body equals the work W it does against F, hence

kinetic energy of body $= W = Fs$

$$= mas \quad \text{(since } F = ma\text{)}$$

$$= ms\frac{u^2}{2s} \quad \left(\text{since } a = \frac{u^2}{2s}\right)$$

$$= \tfrac{1}{2}mu^2$$

Thence the kinetic energy of a body of mass m moving with speed u is $\tfrac{1}{2}mu^2$. Conversely if work is done on a body the gain of kinetic energy when its velocity increases from zero to u can be shown to be $\frac{1}{2}mu^2$.

In general if the velocity of a body of mass m increases from u to v when work is done on it by a force F acting over a distance s, then

$$Fs = \tfrac{1}{2}mv^2 - \tfrac{1}{2}mu^2$$

This is called the *work–energy equation* and may be stated

work done by the forces acting on the body = change in kinetic energy of the body

(b) *Potential energy.* This is the energy a system of bodies has because of the relative positions of its parts, i.e. due to its configuration. It arises when a body experiences a force in a field such as the earth's gravitational field. In that case the body occupies a position with respect to the earth and the potential energy is regarded as a joint property of the body–earth system and not of either body separately. The relative positions of the parts of the system, i.e. of the body and earth, determine its potential energy; the greater the separation the greater the potential energy.

Normally we are only concerned with differences of potential energy. In the gravitational case it is convenient to consider that the potential energy is zero when the body is at the surface of the earth. The potential energy when a body of mass m is at height h above ground level equals the work which must be done against the downward pull of gravity to raise the body to this height. A force, equal and opposite to mg, has to be exerted on the body over displacement h (assuming g is constant near the earth's surface). Therefore

work done by external force against gravity
$$= \text{force} \times \text{displacement}$$
$$= mgh$$
$$\therefore \quad \text{potential energy} = mgh$$

On returning to ground level an amount of potential energy equal to mgh would be lost. A good example of this occurs when the water in a mountain reservoir falls to a lower level and does work by driving a power station turbine.

A stretched or compressed spring is also considered to have potential energy.

Conservation of energy

If a body of mass m is thrown vertically upwards with velocity u at A, it has to do work against the constant force of gravity, Fig. 6.33. When it has risen to B let its

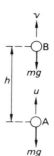

Fig. 6.33

reduced velocity be v. By the definition of kinetic energy (k.e.)

loss of k.e. between A and B
$$= \text{work done by body against } mg$$

By the definition of potential energy (p.e.)

gain of p.e. between A and B
$$= \text{work done by body against } mg$$

$$\therefore \quad \text{loss of k.e.} = \text{gain of p.e.}$$

$$\therefore \quad \tfrac{1}{2}mu^2 - \tfrac{1}{2}mv^2 = mgh$$

This is called the *principle of conservation of mechanical energy* and may be stated as follows.

The total amount of mechanical energy (k.e. + p.e.) which the bodies in an isolated system possess is constant.

It applies only to frictionless motion, i.e. to *conservative systems*. Otherwise in the case of a rising body, work has to be done against friction as well as against gravity and the body gains less p.e. than when friction is absent. Furthermore, the gain of p.e. would depend on the path taken; it does not in a conservative system.

Work done against frictional forces is generally accompanied by a temperature rise. This suggests that we might include in our energy accountancy what we have called *internal energy*. This would then extend the energy conservation principle to non-conservative systems and we can then say, for example,

loss of k.e. = gain of p.e. + gain of internal energy

The mechanics of a body seen to be in motion has thus been related to a phenomenon which is apparently not mechanical and in which motion is not directly detected. (However, we *believe* that internal energy is random molecular kinetic and potential energy, see p. 65.) In a similar way, the idea of energy has been extended to other areas of physics and is now a unifying theme. In fact, physics is sometimes said to be the study of energy transformations, measured in terms of the work done by the forces created in the transformation.

The principle of conservation of mechanical energy is a special case of the more general *principle of conservation of energy*—one of the fundamental laws of science.

Energy may be transformed from one form to another, but it cannot be created or destroyed, i.e. the total energy of a system is constant.

Energy calculations

The work–energy equation $Fs = \frac{1}{2}mv^2 - \frac{1}{2}mu^2$ is useful for solving problems when the distance over which a force acts is known.

1. A car of mass 1.0×10^3 kg travelling at 72 km h^{-1} on a horizontal road is brought to rest in a distance of 40 m by the action of the brakes and frictional forces. Find (a) the average stopping force, (b) the time taken to stop the car.

A speed of 72 km h^{-1} = 72×10^3 m/3600 s = 20 m s^{-1}

(a) If the car has mass m and initial speed u, then

$$\text{kinetic energy lost by car} = \tfrac{1}{2}mu^2$$

If F is the average stopping force and s the distance over which it acts, then

$$\text{work done by car against } F = Fs$$

But $\qquad Fs = \frac{1}{2}mu^2$

$\therefore \quad F \times 40 \text{ m} = \frac{1}{2} \times (1.0 \times 10^3 \text{ kg}) \times (20 \text{ m s}^{-1})^2$

$\therefore \quad F = \dfrac{1.0 \times 10^3 \times 400}{2 \times 40} \ \dfrac{\text{kg m}^2\text{ s}^{-2}}{\text{m}}$

$\qquad = 5.0 \times 10^3 \text{ N}$

(b) Assuming constant acceleration and substituting $v = 0$, $u = 20$ m s^{-1} and $s = 40$ m in $v^2 = u^2 + 2as$ we have

$$0 = 20^2 + 2a \times 40$$

$\therefore \quad a = -5.0 \text{ m s}^{-1}$

(the negative sign indicates the acceleration is in the opposite direction to the displacement).

Using $v = u + at$

$$0 = 20 - 5.0t$$

$\therefore \quad t = 4.0 \text{ s}$

2. A bullet of mass 10 g travelling horizontally at a speed of 1.0×10^2 m s^{-1} embeds itself in a block of wood of mass 9.9×10^2 g suspended by strings so that it can swing freely. Find (a) the vertical height through which the block rises, (b) how much of the bullet's energy becomes internal energy. ($g = 10$ m s^{-2}.)

(a) The bullet is brought to rest very quickly due to the resistance offered by the block and we shall assume that the block (with the bullet embedded) hardly moves until the bullet is at rest. Momentum is conserved in the collision and so

$$mu = (M + m)v$$

where m and M are the masses of the bullet and block respectively, u is the velocity of the bullet before impact and v is the velocity of the block + bullet as they move off.

$\therefore \quad 10 \times 10^{-3} \text{ kg} \times 1.0 \times 10^2 \text{ m s}^{-1}$
$\qquad\qquad\qquad = (990 + 10) \times 10^{-3} \text{ kg} \times v$

$\therefore \quad v = 1.0 \text{ m s}^{-1}$

When the block has swung to its maximum height h, all its kinetic energy has become potential energy—if frictional forces are neglected. Conservation of energy therefore holds and we can say

$$\tfrac{1}{2}(M + m)v^2 = (M + m)gh$$

$\therefore \quad h = \dfrac{v^2}{2g}$

$\qquad = \dfrac{(1 \text{ m s}^{-1})^2}{2 \times 10 \text{ m s}^{-2}} = \dfrac{1}{2 \times 10} \ \dfrac{\text{m}^2\text{ s}^{-2}}{\text{m s}^{-2}}$

$\therefore \quad h = 5.0 \times 10^{-2} \text{ m}$

(b) Original kinetic energy of bullet $= \frac{1}{2}mu^2$

$\qquad = \frac{1}{2} \times (10 \times 10^{-3} \text{ kg}) \times (1.0 \times 10^2 \text{ m s}^{-1})^2$

$\qquad = \frac{1}{2} \times 10 \times 10^{-3} \times 1.0 \times 10^4 \ \text{kg m}^2\text{ s}^{-2}$

$\qquad = 50 \text{ J}$

Kinetic energy of block + bullet after impact

$\qquad = \tfrac{1}{2}(M + m)v^2$

$\qquad = \tfrac{1}{2}(1000 \times 10^{-3} \text{ kg})(1 \text{ m s}^{-1})^2$

$\qquad = 0.50 \text{ J}$

Therefore

$\text{internal energy produced} = \text{loss of kinetic energy}$

$\qquad\qquad\qquad = 50 - 0.50$

$\qquad\qquad\qquad = 49.5 \text{ J}$

Elastic and inelastic collisions

Whilst momentum is always conserved in a collision, there is generally a change of some kinetic energy, usually to internal energy or to a very small extent to sound energy. Collisions in which a loss of kinetic energy occurs (see previous worked example) are said to be *inelastic*. In a *perfectly elastic* collision, kinetic energy is conserved.

The linear air track (or trolleys and ticker tape timers) can be used to investigate what happens to kinetic energy, as well as to momentum, in different collisions. Nearly perfect elastic collisions are obtained

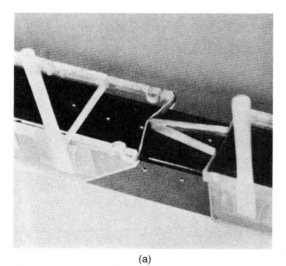

(a)

(b)

Fig. 6.34

if a rubber band is fitted to the front of one vehicle and the other vehicle has a pointed end, Fig. 6.34a. By contrast, completely inelastic (i.e. no-bounce) collisions occur if a needle is fitted to one vehicle and 'Plasticine' inserted in a hole (in line with the needle) in the other, Fig. 6.34b. By making measurements on Fig. 6.25b (p. 146) find the total kinetic energy before and after the collision and say what kind of collision has occurred.

The head-on collisions of air–track vehicles are one dimensional. An oblique, two-dimensional collision between a moving, magnetic, 'dry-ice' puck and a stationary one of *equal mass* is shown in Fig. 6.35a. Make measurements to see if momentum is conserved (remember that momentum is a vector quantity). Compare the kinetic energy before and after the collision and comment on the result. What is the angle between the directions of motion of the pucks after the collision?

Collisions between atoms and other atomic particles were first studied in a cloud chamber; many such

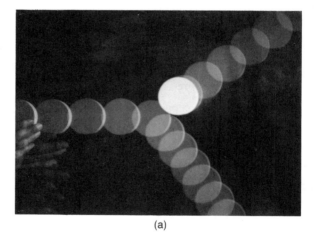

(a)

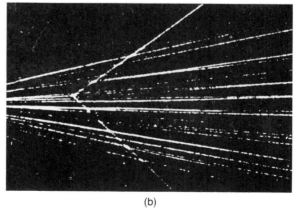

(b)

Fig. 6.35

collisions are elastic. Fig. 6.35*b* shows a cloud chamber photograph of an alpha particle colliding with a helium nucleus. Compare this with Fig. 6.35*a* and assuming atomic particles and magnetic pucks behave similarly, comment on (*i*) the mass of an alpha particle compared with that of a helium nucleus, (*ii*) the type of collision which has occurred.

Electrons can have elastic or inelastic collisions with the atoms of a gas and the latter give information about the electronic structure of atoms.

QUESTIONS

Assume $g = 10$ m s^{-2}

Statics

1. Distinguish between scalar and vector quantities, giving two examples of each. How may the resultant of a number of vector quantities be obtained? Why are vector quantities frequently resolved into rectangular components?

It is possible for the product of two vector quantities to be a scalar. Give an example of two such quantities.

Show how, by considering the sail to be a flat plane and by resolving the thrust on the sail due to the wind into components, it is possible to explain how a yacht may make progress upwind. Why is it an advantage for a yacht to have a large and heavy keel? (*Hint:* see p. 213.) (*A.E.B.*)

2. State the conditions of equilibrium of a body acted on by a system of coplanar forces.

An aerial attached to the top of a radio mast 20 m high exerts a horizontal force on it of 3.0×10^2 N. A stay-wire from the mid-point of the mast to the ground is inclined at 60° to the horizontal. Assuming the action of the ground on the mast can be regarded as a single force, find (*a*) the force exerted on the mast by the stay-wire, (*b*) the magnitude and direction of the action of the ground.

3. A uniform ladder 5.0 m long and having mass 40 kg rests with its upper end against a smooth vertical wall and with its lower end 3.0 m from the wall on rough ground. Find the magnitude and direction of the force exerted at the bottom of the ladder.

4. Find the forces in the members of a pin-jointed structure shown in Fig. 6.36. State whether each force is tensile or compressive. Given that the maximum safe stress for the material used in each member is 8×10^7 N m^{-2}, for the member with the highest load calculate the minimum cross-sectional area. (*J.M.B. Eng. Sc.*)

Dynamics

5. A dart player stands 3.00 m from the wall on which the board hangs and throws a dart which leaves his hand with a horizontal velocity at a point 1.80 m above the ground. The dart strikes the board at a point 1.50 m from the ground. Assuming air resistance to be negligible, calculate (i) the time of the flight of the dart, (*ii*) the initial speed of the dart and (*iii*) the speed of the dart when it hits the board.
(*A.E.B. part qn.*)

6. A projectile is fired from ground level with a velocity of 500 m s^{-1} at 30° to the horizontal. Calculate its horizontal range, the greatest height it reaches and the time taken to rise to that height. (Neglect air resistance.)

7. A body slides, with constant velocity, down a plane inclined at 30° with the horizontal. Show in a diagram the forces acting on the body, and find the coefficient of kinetic friction between the body and the plane.

If the plane were now tilted so as to make an angle of 60° with the horizontal, with what acceleration would the body slide down the plane? What force, applied parallel to this plane, would be required to cause the body to move up the plane with a constant velocity? (*W.*)

8. An object of mass m rests on the floor of a lift which is ascending with acceleration a. Draw a diagram to show the external forces acting on the object, and write down its equation of motion. How do these forces arise? Show graphically how their magnitudes vary with the acceleration of the lift. What force constitutes the second member of the action–reaction pair in the case of each of these external forces?

9. Five identical cubes, each of mass m, lie in a straight line, with their adjacent faces in contact, on a horizontal surface, as shown in Fig. 6.37.

Suppose the surface is frictionless and that a constant force P is applied from left to right to the end face of A.

What is the acceleration of the system and what is the resultant force acting on each cube? What force does cube C exert on cube D?

If friction is present between the cubes and the surface, draw a graph to illustrate how the total frictional force varies as P increases uniformly from zero. (*W.*)

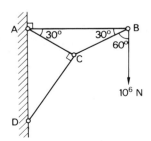

Fig. 6.36

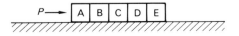

Fig. 6.37

10. State Newton's second law of motion.

A stream of water travelling horizontally at 30 m s^{-1} is ejected from a hole of cross-sectional area 40 cm^2 and is directed against a vertical wall. Calculate the force exerted on the wall assuming that the water does not rebound.

What is the power of the pump needed to give the ejected water the necessary kinetic energy?

Density of water = 1.0 g cm^{-3} = 1.0 × 10^3 kg m^{-3}

(*J.M.B.*)

11. Sketch a graph of the relationship between the kinetic energy E (plotted on the vertical axis) and the distance travelled x (plotted on the horizontal axis) for a body of mass m sliding from rest with negligible friction down a uniform slope which makes an angle of 30° with the horizontal. What is the gradient of the graph equal to? (*S.*)

12. Define linear momentum and state the principle of conservation of linear momentum. Explain briefly how you would attempt to verify this principle by experiment.

Sand is deposited at a uniform rate of 20 kilograms per second and with negligible kinetic energy on to an empty conveyor belt moving horizontally at a constant speed of 10 metres per minute. Find (*a*) the force required to maintain constant velocity, (*b*) the power required to maintain constant velocity and (*c*) the rate of change of kinetic energy of the moving sand. Why are the latter two quantities unequal?

(*O. and C.*)

13. Write down an expression for the kinetic energy of a body.

(*a*) A car of mass 1.00 × 10^3 kg travelling at 20 m s^{-1} on a horizontal road is brought to rest by the action of its brakes in a distance of 25 m. Find the average retarding force.

(*b*) If the same car travels up an incline of 1 in 20 at a constant speed of 20 m s^{-1}, what power does the engine develop if the frictional resistance is 100 N?

14. State the principle of conservation of linear momentum and show how it follows from Newton's laws of motion.

A stationary radioactive nucleus of mass 210 units disintegrates into an alpha particle of mass 4 units and a residual nucleus of mass 206 units. If the kinetic energy of the alpha particle is E, calculate the kinetic energy of the residual nucleus. (*J.M.B.*)

7 Circular motion and gravitation

Motion in a circle

In everyday life, in atomic physics and in astronomy and space travel there are many examples of bodies moving in paths which if not exactly circular are nearly so. In this chapter we will see how ideas developed for dealing with straight-line motion enable us to tackle circular motion.

A body which travels equal distances in equal times along a circular path has constant speed but *not* constant velocity. This is due to the way we have defined speed and velocity; speed is a scalar quantity, velocity is a vector quantity. Fig. 7.1 shows a ball attached to a string being whirled round in a horizontal circle. The velocity of the ball at P is directed along the tangent at P; when it reaches Q its velocity is directed along the tangent at Q. If the speed is constant the *magnitudes* of the velocities at P and Q are the same but their *directions* are different and so the velocity of the ball has changed. A change of velocity is an acceleration and a body moving uniformly in a circular path or arc is therefore accelerating.

since the velocity changes not only when the speed changes, but also when the direction of motion changes, then, for example, a car rounding a bend (even at constant speed) is accelerating.

Two useful expressions

We will use these from time to time when dealing with circular motion.

(a) *Angles in radians* : $s = r\theta$. Angles can be measured in radians as well as degrees. In Fig. 7.2 the angle θ, in radians, is defined by the equation

$$\theta = \frac{s}{r}$$

If $s = r$ then $\theta = 1$ radian (rad). Therefore 1 radian is the angle subtended at the centre of a circle by an arc equal in length to the radius. When $s = 2\pi r$ (the circumference of a circle of radius r) then $\theta = 2\pi$ radians $= 360°$.

$$\therefore \quad 1 \text{ radian } = 360°/2\pi \simeq 57°$$

Fig. 7.1

Fig. 7.2

In everyday language acceleration usually means going faster and faster, i.e. involves a change of speed. However, in physics it means a change of velocity and

From the definition of a radian it follows that the length s of an arc which subtends an angle θ at the

155

centre of a circle of radius r, is given by

$$s = r\theta$$

where θ is in radians.

(b) *Angular velocity* : $v = r\omega$. The speed of a body moving in a circle can be specified either by its speed along the tangent at any instant, i.e. by its linear speed, or, by its angular velocity. This is the angle swept out in unit time by the radius joining the body to the centre of the circle. It is measured in radians per second (rad s^{-1}).

We can derive an expression connecting angular velocity and linear speed. Consider a body moving uniformly from A to B in time t so that radius OA rotates through an angle θ, Fig. 7.2. The angular velocity ω of the body about O is

$$\omega = \frac{\theta}{t}$$

If arc AB has length s and if v is the constant speed of the body then

$$v = \frac{s}{t}$$

But from (a), $s = r\theta$ where r is the radius of the circle

$$\therefore \quad v = \frac{r\theta}{t}$$

But $\omega = \theta/t$

$$\therefore \quad v = r\omega$$

If $r = 3$ m and $\omega = 1$ revolution per second $= 2\pi$ rad s^{-1} then the linear speed $v = 6\pi$ m s^{-1}.

Deriving $a = v^2/r$

To obtain an expression for the acceleration of a small body (i.e. a particle) describing circular motion, consider such a body moving with *constant speed* v in a circle of radius r, Fig. 7.3a. If it travels from A to B in a

short interval of time δt then, since distance = speed × time,

$$\text{arc AB} = v\,\delta t$$

Also, by the definition of an angle in radians

$$\text{arc AB} = r\,\delta\theta \qquad (\delta\theta = \angle AOB)$$

$$\therefore \quad r\,\delta\theta = v\,\delta t$$

$$\therefore \quad \delta\theta = \frac{v\,\delta t}{r} \qquad (1)$$

The vectors $\mathbf{v}_A$ and $\mathbf{v}_B$ drawn tangentially at A and B represent the velocities at these points. The *change* of velocity between A and B is obtained by subtracting $\mathbf{v}_A$ from $\mathbf{v}_B$. That is

$$\text{change of velocity} = \mathbf{v}_B - \mathbf{v}_A$$

But $$\mathbf{v}_B - \mathbf{v}_A = \mathbf{v}_B + (-\mathbf{v}_A)$$

Hence, to subtract vector $\mathbf{v}_A$ from vector $\mathbf{v}_B$ we *add* vectors $\mathbf{v}_B$ and $(-\mathbf{v}_A)$ by the parallelogram law.

In Fig. 7.3b, XY represents $\mathbf{v}_B$ in magnitude (v) and direction (BD); YZ represents $(-\mathbf{v}_A)$ in magnitude (v) and direction (CA). The resultant, which gives the change of velocity, is then seen from the figure to be, in effect, vector XZ.

Since one vector $(-\mathbf{v}_A)$ is perpendicular to OA and the other $\mathbf{v}_B$ is perpendicular to OB, $\angle XYZ = \angle AOB = \delta\theta$. If δt is very small, $\delta\theta$ will also be small and XZ in Fig. 7.3b will have almost the same length as arc XZ in Fig. 7.3c which subtends angle $\delta\theta$ at the centre of a circle of radius v. Arc XZ $= v\,\delta\theta$ (from definition of radian) and so

$$XZ = v\,\delta\theta$$

But from (1) $$\delta\theta = \frac{v\,\delta t}{r}$$

$$\therefore \quad XZ = \frac{v^2}{r}\,\delta t$$

The *magnitude* of the acceleration a between A and B is

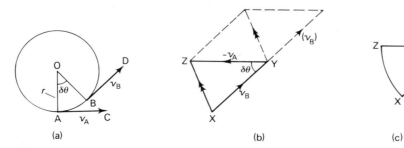

Fig. 7.3   (a)   (b)   (c)

$$a = \frac{\text{change of velocity}}{\text{time interval}} = \frac{XZ}{\delta t}$$

$$\therefore \quad a = \frac{v^2}{r}$$

If ω is the angular velocity of the body, $v = r\omega$ and we can also write

$$a = \omega^2 r$$

The *direction* of the acceleration is *towards the centre* O of the circle as can be seen if δt is made so small that A and B all but coincide; vector XZ is then perpendicular to v_A (or v_B), i.e. along AO (or BO). We say the body has a *centripetal acceleration* (i.e. centre-seeking).

Does a body moving uniformly in a circle have *constant* acceleration? (Remember that acceleration is a vector.)

Centripetal force

Since a body moving in a circle (or a circular arc) is accelerating, it follows from Newton's first law of motion that there must be a force acting on it to cause the acceleration. This force, like the acceleration, will also be directed towards the centre and is called the *centripetal force*. It causes the body to deviate from the straight-line motion which it would naturally follow if the force were absent. The value F of the centripetal force is given by Newton's second law, that is

$$F = ma = \frac{mv^2}{r}$$

where m is the mass of the body and v is its speed in the circular path of radius r. If the angular velocity of the body is ω we can also say, since $v = r\omega$,

$$F = m\omega^2 r$$

When a ball attached to a string is swung round in a horizontal circle, the centripetal force which keeps it in a circular orbit arises from the tension in the string. We can think of the tension as tugging continually on the body and 'turning it in' so that it remains at a fixed distance from the centre. If the ball is swung round faster, a larger force is needed and if it is greater than the tension the string can bear, it breaks and the ball continues to travel along a tangent to the circle at the point of breaking, Fig. 7.4.

Other examples of circular motion will be discussed presently but in all cases it is important to appreciate that the forces acting on the body must provide a resultant force of magnitude mv^2/r towards the centre. What is the nature of the centripetal force for (*a*) a car rounding a bend, (*b*) a space capsule circling the earth?

One arrangement for testing $F = mv^2/r$ experimentally is shown in Fig. 7.5. The turntable, driven by the electric motor, is *gradually* speeded up and the spring extends until the truck just reaches the stop at the end of the track. The speed v of the *truck* in orbit is found by measuring the time for one revolution of the turntable and then, with the turntable at rest, the radius

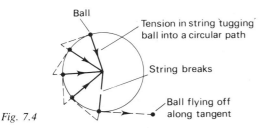

Fig. 7.4

Ball

Tension in string tugging ball into a circular path

String breaks

Ball flying off along tangent

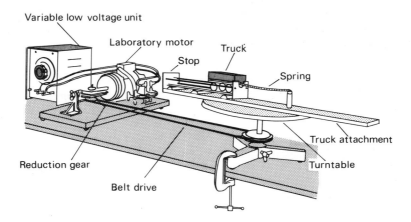

Variable low voltage unit

Laboratory motor

Stop

Truck

Spring

Reduction gear

Belt drive

Truck attachment

Turntable

Fig. 7.5

r of the circle described by the truck (i.e. the distance from the centre of the turntable to the centre of the truck). Knowing the mass m of the truck, mv^2/r can be calculated.

The tension in the stretched spring is the centripetal force and this can be found by measuring with a spring balance, the tension required to extend the spring *by the same amount* as it is when the truck is at the end stop, Fig. 7.6. The value obtained should agree with the value of mv^2/r to within a few per cent.

The mass of the truck can be altered by loading it with lead plates and the experiment repeated for each mass.

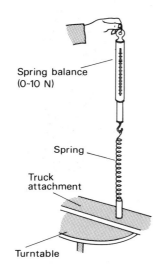

Fig. 7.6

Rounding a bend

If a car is travelling round a circular bend with uniform speed on a horizontal road, the resultant force acting on it must be directed to the centre of its circular path, i.e. it must be the centripetal force. This force arises from the interaction of the car with the air and the ground. The direction of the force exerted by the air on the car will be more or less opposite to the instantaneous direction of motion. The other and more

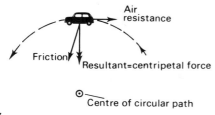

Fig. 7.7

important horizontal force is the frictional force exerted inwards by the ground on the tyres of the car, Fig. 7.7. The resultant of these two forces is the centripetal force.

The successful negotiation of a bend on a flat road therefore depends on the tyres and the road surface being in a condition that enables them to provide a sufficiently high frictional force—otherwise skidding occurs. Safe cornering that does not rely on friction is achieved by 'banking' the road.

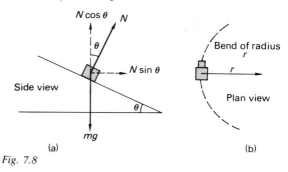

Fig. 7.8

The problem is to find the angle θ at which a bend should be banked so that the centripetal force acting on the car arises entirely from a component of the normal force N of the road, Fig. 7.8a. Treating the car as a particle and resolving N vertically and horizontally we have, since $N \sin \theta$ *is* the centripetal force,

$$N \sin \theta = \frac{mv^2}{r}$$

where m and v are the mass and speed respectively of the car and r is the radius of the bend, Fig. 7.8b. Also, the car is assumed to remain in the same horizontal plane and so has no vertical acceleration, thus

$$N \cos \theta = mg$$

Hence, by division

$$\tan \theta = \frac{v^2}{gr}$$

The equation shows that for a given radius of bend, the angle of banking is only correct for one speed. In a race track, the banking becomes steeper towards the outside and the driver can select a position according to his speed, Fig. 7.9.

A bend in a railway track is also banked, in this case so that at a certain speed no lateral thrust has to be exerted by the outer rail on the flanges of the wheels of the train, otherwise the rails are strained. The horizontal component of the normal force of the rails on the train then provides the centripetal force.

Fig. 7.9

An aircraft in straight, level flight experiences a lifting force at right angles to the surface of its wings which balances its weight. To turn, the ailerons are operated so that the aircraft banks and the horizontal component of the lift supplies the necessary centripetal force, Fig. 7.10. The aircraft's weight is now opposed only by the vertical component of the lift and height will be lost unless the lift is increased by, for example, increasing the speed.

their backs against the wall. The drum is spun at increasing speed about its central vertical axis and at a certain speed the floor is pulled downwards. The occupants do not fall but remain 'pinned' against the wall of the rotor.

The forces acting on a passenger of mass m are shown in Fig. 7.11. N is the normal force of the wall on the passenger and is the centripetal force needed to keep him moving in a circle. Hence if r is the radius of

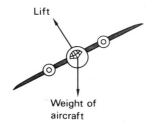

Fig. 7.10

Other examples of circular motion

(a) *The rotor*. This device is sometimes present in amusement parks. It consists of an upright drum of diameter about 4 metres inside which people stand with

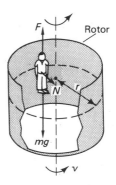

Fig. 7.11

the rotor and v the speed of the passenger then

$$N = \frac{mv^2}{r}$$

F is the frictional force acting upwards between the passenger and rotor wall and since there is no vertical motion of the passenger

$$F = mg$$

If μ is the coefficient of limiting friction between passenger and wall, we have $F = \mu N$

$$\therefore \quad \mu N = mg$$

$$\therefore \quad \mu = \frac{mg}{N} = \frac{mg}{mv^2/r}$$

$$\therefore \quad \mu = \frac{gr}{v^2}$$

This equation gives the minimum coefficient of friction required to prevent the passenger slipping; it does not depend on the passenger's weight. A typical value of μ between clothing and a rotor wall (of canvas) is about 0.40 and so if $r = 2$ m, v must be about 7 m s^{-1} (or more). What will be the angular velocity of the drum? How many revolutions will it make per minute?

(b) *Looping the loop.* A pilot who is not strapped into his aircraft can loop the loop without falling out at the top of the loop. A bucket of water can be swung round in a vertical circle without spilling. A ball-bearing can loop the loop on a length of curtain rail in a vertical plane. All three effects have similar explanations.

Consider the bucket of water when it is at the top of the loop, A in Fig. 7.12. If the weight mg of the water is *less than* mv^2/r, the normal force N of the bottom of the bucket on the water provides the rest of the force required to maintain the water in its circular path. However, if the bucket is swung more slowly then mg will be greater than mv^2/r and the 'unused' part of the weight causes the water to leave the bucket. What

provides the centripetal force for the *water* when the bucket is at (*i*) B, (*ii*) C and (*iii*) D?

(c) *Centrifuges.* These separate solids suspended in liquids or liquids of different densities. The mixture is in a tube, Fig. 7.13a, and when it is rotated at high speed in a horizontal circle the less dense matter moves towards the centre of rotation. On stopping the rotation, the tube returns to the vertical position with the less dense matter at the top. Cream is separated from milk in this way.

The action uses the fact that if a horizontal tube of liquid is rotated, the force exerted by the closed end must be greater than when the tube was at rest so that it can provide the necessary centripetal force acting radially inwards. In Fig. 7.13b the liquid pressure at B is greater than at A and a pressure gradient exists along the tube. For any part of the liquid the force due to the pressure difference supplies exactly the centripetal force required. If this part of the liquid is replaced by matter of smaller density (and thus of smaller mass), the force is too large and the matter moves inwards.

During the launching and re-entry of space vehicles accelerations of about 8g occur and the resulting large forces which act on the surface of the astronaut's body cause blood to drain from some parts and congest others. If the brain is deprived, loss of vision and unconsciousness may follow. Tests with large man-carrying centrifuges in which passengers are subjected to high centripetal accelerations show that a person will tolerate 15g for a few minutes when his body is perpendicular to the direction of the acceleration but only 6g when in the direction of acceleration. What will be the best position for an astronaut to adopt at lift-off and re-entry?

Moment of inertia

In most of the cases of circular motion considered so far we have treated the body as a 'particle' so that all of it, in effect, revolves in a circle of the same radius. When this cannot be done we have to regard the rotating body

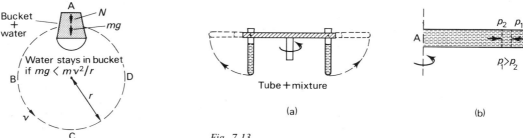

Fig. 7.12

Fig. 7.13

as a system of connected 'particles' moving in circles of different radii. The way in which the mass of the body is distributed then affects its behaviour.

This may be shown by someone who is sitting on a freely rotating stool with a heavy weight in each hand. When he extends his arms, Fig. 7.14, the speed of rotation decreases but increases again when he brings them in. The angular velocity of the system clearly depends on how the mass is distributed about the axis of rotation. A concept is needed to express this property.

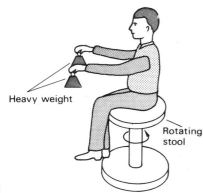

Fig. 7.14

The mass of a body is a measure of its in-built opposition to any change of linear motion, i.e. mass measures inertia. The corresponding property for rotational motion is called the *moment of inertia*. The more difficult it is to change the angular velocity of a body rotating about a particular axis the greater is its moment of inertia about that axis. Experiment shows that a wheel with most of its mass in the rim is more difficult to start and stop than a uniform disc of equal mass rotating about the same axis; the former has a greater moment of inertia. Similarly the moment of inertia of the person on the rotating stool is greater when his arms are extended. It should be noted that moment of inertia is a property of a body rotating about a particular axis; if the axis changes so does the moment of inertia.

We now require a measure of moment of inertia which takes into account the distribution of mass about the axis of rotation and which plays a role in rotational motion, similar to that played by mass in linear, i.e. straight-line, motion.

Kinetic energy of a rotating body

Suppose the body of Fig. 7.15 is rotating about an axis through O with constant angular velocity ω. A particle

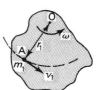

Fig. 7.15

A, of mass m_1, at a distance r_1 from O, describes its own circular path and if v_1 is its linear velocity along the tangent to the path at the instant shown, then $v_1 = r_1\omega$ and

$$\text{kinetic energy of A} = \tfrac{1}{2}m_1v_1^2$$
$$= \tfrac{1}{2}m_1r_1^2\omega^2$$

The kinetic energy of the whole body is the sum of the kinetic energies of its component particles. If these have masses m_1, m_2, m_3, etc., and are at distances r_1, r_2, r_3, etc., from O then, since all particles have the same angular velocity ω (the body being rigid), we have

kinetic energy of whole body

$$= \tfrac{1}{2}m_1r_1^2\omega^2 + \tfrac{1}{2}m_2r_2^2\omega^2 + \tfrac{1}{2}m_3r_3^2\omega^2 + \ldots$$
$$= \tfrac{1}{2}\omega^2(\textstyle\sum mr^2)$$

where $\sum mr^2$ represents the sum of the mr^2 values for all the particles of the body. The quantity $\sum mr^2$ depends on the mass and its distribution and is taken as a measure of the moment of inertia of the body about the axis in question. It is denoted by the symbol I and so

$$I = \textstyle\sum mr^2$$

Therefore,

$$\text{kinetic energy of body} = \tfrac{1}{2}I\omega^2$$

Comparing this with the expression $\tfrac{1}{2}mv^2$ for linear kinetic energy we see that the mass m is replaced by the moment of inertia I and the velocity v is replaced by the angular velocity ω. The unit of I is kg m^2.

Values of I for regular bodies can be calculated (using calculus); that for a uniform rod of mass m and length l about an axis through its centre is $ml^2/12$. About an axis through its end it is $ml^2/3$.

It must be emphasized that rotational kinetic energy ($\tfrac{1}{2}I\omega^2$) is not a new type of energy but is simply the sum of the linear kinetic energies of all the particles of the body. It is a convenient way of stating the kinetic energy of a rotating rigid body.

The mass of a flywheel is concentrated in the rim, thereby giving it a large moment of inertia. When rotating, its kinetic energy is therefore large and explains why it is able to keep an engine (e.g. in a car) running at a fairly steady speed even though energy is supplied intermittently to it. Some toy cars have a small lead flywheel which is set into rapid rotation by a brief push across a solid surface. The kinetic energy of the flywheel will then keep the car in motion for some distance.

Work done by a couple

Rotation is changed by a couple, that is, by two equal and opposite parallel forces whose lines of action do not coincide. It is often necessary to find the work done by a couple so that the energy transfer occurring as a result of its action on a body is known.

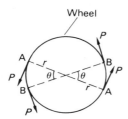

Fig. 7.16

Consider the wheel in Fig. 7.16 of radius r on which the two equal and opposite forces P act tangentially and rotation through angle θ (in radians) occurs.

Work done by each force = force × distance

$$= P \times \text{arc AB} = P \times r\theta$$

∴ total work done by couple $= Pr\theta + Pr\theta = 2Pr\theta$

But, torque (or moment) of couple $= P \times 2r = 2Pr$ (p. 135). Therefore

work done by couple = torque × angle of rotation

$$= T\theta$$

For example, if $P = 2.0$ N, $r = 0.50$ m and the wheel makes 10 revolutions then $\theta = 10 \times 2\pi$ rad and $T = P \times 2r = 2.0$ N $\times 2 \times 0.50$ m $= 2$ N m. Hence, work done by couple $= T\theta = 2 \times 20\pi = 1.3 \times 10^2$ J.

In general, if a couple of torque T about a certain axis acts on a body of moment of inertia I through an angle θ about the same axis and its angular velocity increases from 0 to ω, then

work done by couple = kinetic energy of rotation

$$T\theta = \tfrac{1}{2}I\omega^2$$

Angular momentum

(*a*) *Definition.* In linear motion it is often useful to consider the (linear) momentum of a body. In rotational motion, *angular momentum* is important.

Consider a rigid body rotating about an axis O and having angular velocity ω at some instant, Fig. 7.17. Let A be a particle of this body, distant r_1 from O and having linear velocity v_1 as shown, then the linear momentum of A $= m_1v_1 = m_1\omega r_1$ (since $v_1 = \omega r_1$).

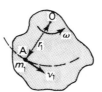

Fig. 7.17

The angular momentum of A about O is defined as the *moment of its momentum* about O. Hence

angular momentum of A $= r_1 \times m_1\omega r_1 = \omega m_1 r_1^2$

∴ total angular momentum of rigid body $= \sum \omega mr^2$

$$= \omega \sum mr^2$$

$$= I\omega$$

where I is the moment of inertia of the body about O. Angular momentum is thus the analogue of linear momentum (mv), with I replacing m and ω replacing v.

(*b*) *Newton's second law.* A body rotates when it is acted on by a couple. The rotational form of Newton's second law of motion may be written (by analogy with $F = ma$),

$$T = I\alpha$$

where T is the torque or moment of the couple causing rotational acceleration α. In terms of momentum, the second law can be stated, for linear motion,

force = rate of change of linear momentum

i.e.
$$F = \frac{\mathrm{d}(mv)}{\mathrm{d}t}$$

and for rotational motion,

torque = rate of change of angular momentum

i.e.
$$T = \frac{\mathrm{d}(I\omega)}{\mathrm{d}t}$$

(c) *Conservation.* A similar argument to that used to deduce the principle of conservation of linear momentum from Newton's third law can be employed to derive the principle of conservation of angular momentum. It may be stated as follows.

The total angular momentum of a system remains constant provided no external torque acts on the system.

Ice skaters, ballet dancers, acrobats and divers use the principle. The diver in Fig. 7.18 leaves the high-diving board with outstretched arms and legs and some initial angular velocity about his centre of gravity. His angular momentum ($I\omega$) remains constant since no external torques act on him (gravity exerts no torque about his centre of gravity). To make a somersault he must increase his angular velocity. He does this by pulling in his legs and arms so that I decreases and ω therefore increases. By extending his arms and legs again, his angular velocity falls to its original value. Similarly a skater can whirl faster on ice by folding his arms.

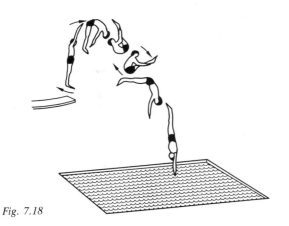

Fig. 7.18

The principle of conservation of angular momentum is useful for dealing with large rotating bodies such as the earth, as well as tiny, spinning particles such as electrons.

(d) *Worked example. A shaft rotating at 3.0 × 10³ revolutions per minute is transmitting a power of 10 kilowatts. Find the magnitude of the driving couple.*

Work done per second by driving couple
= power transmitted by shaft

Hence, since 1 W = 1 J s⁻¹,

$$T\theta = 10 \times 10^3 \text{ J s}^{-1} \quad \text{(see p. 162)}$$

where T is the moment of the couple and θ is the angle

through which the shaft rotates in 1 second. Now

3.0 × 10³ revs per minute
$$= 3.0 \times 10^3/60 \text{ revs per second}$$
$$= 50 \text{ revs per second}$$

$\therefore \ \theta = 50 \times 2\pi \text{ rad s}^{-1}$ (since 2π rad = 360° = 1 rev)

$$\therefore \ T = \frac{10 \times 10^3}{50 \times 2\pi} \text{ N m}$$
$$= 32 \text{ N m}$$

Kepler's laws

About 1542 the Polish monk Copernicus proposed that the earth, rather than being the centre of the universe as was generally thought, revolved round the sun, as did the other planets. This heliocentric (sun-centred) model was greatly developed by Kepler who, following on prolonged study of observations made by the Danish astronomer Tycho Brahé over a period of twenty years, arrived at a very complete description of planetary motion. He announced his first two laws in 1609 and the third in 1619.

1. *Each planet moves in an ellipse which has the sun at one focus.*
2. *The line joining the sun to the moving planet sweeps out equal areas in equal times.*
3. *The squares of the times of revolution of the planets (i.e their periodic times T) about the sun are proportional to the cubes of their mean distance (r) from it (i.e. r³/T² is a constant).*

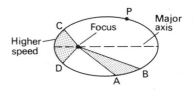

Fig. 7.19

In Fig 7.19, if planet P takes the same time to travel from A to B as from C to D then the shaded areas are equal. Strictly speaking the distances in law 3 should be the semi-major axes of the ellipses but all the orbits are sufficiently circular for the mean radius to be taken. The third column of Table 7.1 shows the constancy of r^3/T^2.

Table 7.1

Planet	Mean radius of orbit r (metres)	Period of revolution T (seconds)	r^3/T^2
Mercury	5.79×10^{10}	7.60×10^6	3.36×10^{18}
Venus	1.08×10^{11}	1.94×10^7	3.35
Earth	1.49×10^{11}	3.16×10^7 (1 year)	3.31
Mars	2.28×10^{11}	5.94×10^7 (1.9 years)	3.36
Jupiter	7.78×10^{11}	3.74×10^8 (11.9 years)	3.36
Saturn	1.43×10^{12}	9.30×10^8 (29.5 years)	3.37
Uranus	2.87×10^{12}	2.66×10^9 (84.0 years)	3.34
Neptune	4.50×10^{12}	5.20×10^9 (165 years)	3.37
Pluto	5.90×10^{12}	7.82×10^9 (248 years)	3.36

Kepler's three laws enabled planetary positions, both past and future, to be determined accurately without the complex array of geometrical constructions used previously which were due to the Greeks. His work was also important because by stating his empirical laws (i.e. laws based on observation, not on theory) in mathematical terms he helped to establish the equation as a form of scientific shorthand.

Gravity and the moon

Kepler's laws summed up neatly *how* the planets of the solar system behaved without indicating *why* they did so. One of the problems was to find the centripetal force which kept a planet in its orbit round the sun, or the moon round the earth, in a way which agreed with Kepler's laws.

Newton reflected (perhaps in his garden when the apple fell) that the earth exerts an inward pull on nearby objects causing them to fall. He then speculated whether this same force of gravity might not extend out farther to pull on the moon and keep it in orbit. If it did, might not the sun also pull on the planets in the same way with the same kind of force? He decided to test the idea first on the moon's motion—as we will do now.

If r is the radius of the moon's orbit round the earth and T is the time it takes to complete one orbit, i.e. its period, Fig. 7.20, then using accepted values we have

$$r = 3.84 \times 10^8 \text{ m}$$

$$T = 27.3 \text{ days}$$

$$= 27.3 \times 24 \times 3600 \text{ s}$$

(The time between full moons is 29.5 days but this is due to the earth also moving round the sun. The moon has therefore to travel a little farther to reach the same position relative to the sun. Judged against the background of the stars, the moon takes 27.3 days to make one complete orbit of the earth, which is its true period T.)

The speed v of the moon along its orbit (assumed circular) is

$$v = \frac{\text{circumference of orbit}}{\text{period}} = \frac{2\pi r}{T}$$

$$= \frac{2\pi \times 3.84 \times 10^8}{27.3 \times 24 \times 3600} \text{ m s}^{-1}$$

$$= 1.02 \times 10^3 \text{ m s}^{-1}$$

The moon's centripetal acceleration a will be

$$a = \frac{v^2}{r} = \frac{(1.02 \times 10^3 \text{ m s}^{-1})^2}{3.84 \times 10^8 \text{ m}}$$

$$= 2.72 \times 10^{-3} \text{ m s}^{-2}$$

The acceleration due to gravity at the earth's surface is 9.81 m s^{-2} and so if gravity is the centripetal force for the moon it must weaken between the earth and the moon. The simplest assumption would be that gravity halves when the distance doubles and at the moon it would be $1/60$ of 9.81 m s^{-2} since the moon is 60 earth-radii from the centre of the earth and an object at the earth's surface is 1 earth-radius from the centre. But $9.81/60 = 1.64 \times 10^{-1}$ m s^{-2}, which is still too large.

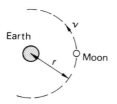

Fig. 7.20

The next relation to try would be an inverse square law in which gravity is one-quarter when the distance doubles, one-ninth when it trebles and so on. At the moon it would be $1/60^2$ of 9.81 m s^{-2}, i.e. $9.81/3600 = 2.72 \times 10^{-3}$ m s^{-2}—the value of the moon's centripetal acceleration.

Law of universal gravitation

Having successfully tested the idea of inverse square law gravity for the motion of the moon round the earth, Newton turned his attention to the solar system.

His proposal, first published in 1687 in his great work the *Principia* (Mathematical principles of natural knowledge), was that the centripetal force which keeps the planets in orbit round the sun is provided by the gravitational attraction of the sun for the planets. This, according to Newton, was the same kind of attraction as that of the earth for an apple. Gravity—the attraction of the earth for an object—was thus a particular case of gravitation. In fact, Newton asserted that every object in the universe attracted every other object with a gravitational force and that this force was responsible for the orbital motion of celestial (heavenly) bodies.

Newton's hypothesis, now established as a theory and known as the *law of universal gravitation*, may be stated quantitatively as follows.

Every particle of matter in the universe attracts every other particle with a force which is directly proportional to the product of their masses and inversely proportional to the square of their distances apart.

The gravitational attraction F between two particles of masses m_1 and m_2, distance r apart is thus given by

$$F \propto \frac{m_1 m_2}{r^2} \quad \text{or} \quad F = G\frac{m_1 m_2}{r^2}$$

where G is a constant, called the *universal gravitational constant*, and assumed to have the same value everywhere for all matter.

Newton believed the force was directly proportional to the mass of each particle because the force on a falling body is proportional to its mass ($F = ma = mg = m \times$ constant, therefore $F \propto m$), i.e. to the mass of the *attracted* body. Hence, from the third law of motion, he argued that since the falling body also attracts the earth with an equal and opposite force that is proportional to the mass of the earth, then the gravitational force between the bodies must also be proportional to the mass of the *attracting* body. The moon test justified the use of an inverse square law relation between force and distance.

The law applies to *particles* (i.e. bodies whose dimensions are very small compared with other distances involved), but Newton showed that the attraction exerted at an external point by a sphere of uniform density (or a sphere composed of uniform concentric shells) was the same as if its whole mass were concentrated at its centre. We tacitly assumed this for the earth in the previous section and will use it in future.

The gravitational force between two ordinary objects (say two 1 kg masses 1 metre apart) is extremely small and therefore difficult to detect. What does this indicate about the value of G in SI units? What will be the units of G in the SI system?

Testing gravitation

To test $F = Gm_1m_2/r^2$ for the sun and planets the numerical values of all quantities on both sides of the equation need to be known. Newton neither had reliable information about the masses of the sun and planets nor did he know the value of G and so he could not adopt this procedure. There are alternatives however.

(*a*) *Deriving Kepler's laws.* The behaviour of the solar system is summarized by Kepler's laws and any theory which predicts these would, for a start, be in agreement with the facts.

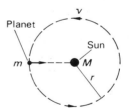

Fig. 7.21

Suppose a planet of mass m moves with speed v in a circle of radius r round the sun of mass M, Fig. 7.21. Hence

$$\text{gravitational attraction of sun for planet} = G\frac{Mm}{r^2}$$

If this is the centripetal force keeping the planet in orbit then

$$G\frac{Mm}{r^2} = \frac{mv^2}{r}$$

$$\therefore \quad \frac{GM}{r} = v^2$$

If T is the time for the planet to make one orbit

$$v = \frac{2\pi r}{T}$$

$$\therefore \quad \frac{GM}{r} = \frac{4\pi^2 r^2}{T^2}$$

$$\therefore \quad GM = \frac{4\pi^2 r^3}{T^2}$$

Hence $\quad \dfrac{r^3}{T^2} = \dfrac{GM}{4\pi^2}$

Since GM is constant for any planet, r^3/T^2 is constant, which is Kepler's third law. We have considered a circular orbit but more advanced mathematics gives the same result for an elliptical one.

The first law can be derived by showing that if inverse square law gravitation holds, a planet moves in an orbit which is a conic section (i.e. a circle, ellipse, parabola or hyperbola) with the sun at one focus. Also, it may be shown that when a planet is acted on by *any* force, not just an inverse square law one, directed from the planet towards the sun, the radius covers equal areas in equal times—which is the second law.

(*b*) *Discovery of other planets.* Theories can never be proved correct, they are only disproved by making predictions which conflict with observations. A good theory should lead to new discoveries. Newton's theory of gravitation has not only enabled us to work out problems connected with space travel, leading to new knowledge about the solar system, but it has also resulted in the discovery of planets not known in his day.

The planets must exert gravitational pulls on one another, but except in the case of the larger planets like Jupiter and Saturn, the effect is only slight. The French scientist Laplace showed after Newton's time how to predict the effect of these disturbances (called perturbations) on Kepler's simple elliptical orbits.

The planet Uranus, discovered in 1781, showed small deviations from its expected orbit even after allowance had been made for the effects of known neighbouring planets. Two astronomers, Adams in England and Leverrier in France, working quite independently, predicted from the law of gravitation, the position, size and orbit of an unknown planet that could cause the observed perturbations. A search was made and the new planet located in 1846 in the predicted position by the Berlin Observatory. Thus Neptune was discovered.

In 1930, history was repeated when American astronomers discovered Pluto from perturbations of the orbit of Neptune.

Masses of the sun and planets

The theory of gravitation enables us to obtain information about the mass of any celestial body having a satellite. If the value of the gravitational constant G is known, the actual mass can be calculated. Otherwise only a comparison is possible. A determination of G was not made until after Newton's death.

The principle is simply to measure all the quantities in $F = Gm_1m_2/r^2$ except G which can then be calculated. The earliest determinations used a measured mountain as the 'attracting' mass and a pendulum as the 'attracted' one. The first laboratory experiment was performed by Cavendish in 1798. He measured the very small gravitational forces exerted on two small lead balls (m_1 and m_2) by two larger ones (M_1 and M_2) using a torsion balance, Fig. 7.22. In this, the force twists a calibrated wire. Modern measurements give the value

$$G = 6.7 \times 10^{-11} \text{ N m}^2 \text{ kg}^{-2} \text{ (or m}^3 \text{ s}^{-2} \text{ kg}^{-1})$$

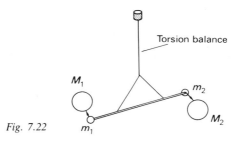

Torsion balance

Fig. 7.22

(*a*) *Mass of the sun.* Consider the earth of mass m_e moving with speed v_e round the sun of mass m_s in a circular orbit of radius r_e, Fig. 7.23*a*. The gravitational attraction of the sun for the earth is the centripetal force. Hence

$$G\frac{m_s m_e}{r_e^2} = \frac{m_e v_e^2}{r_e}$$

$$\therefore \quad m_s = \frac{v_e^2 r_e}{G}$$

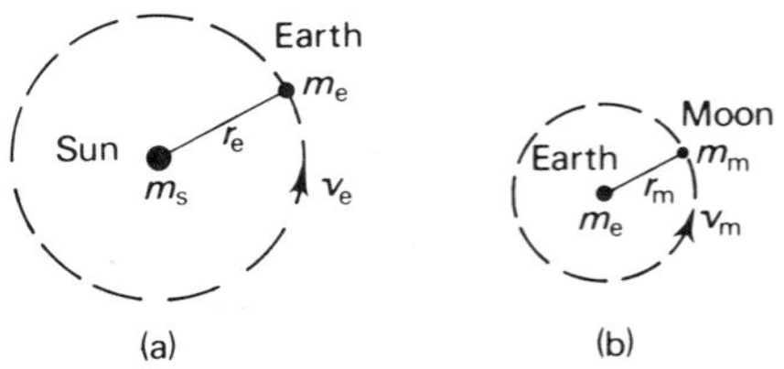

Sun m_s Earth m_e r_e v_e

(a)

Earth m_e Moon m_m r_m v_m

(b)

Fig. 7.23

If T_e is the time for the earth to make one orbit, then

$$v_e = \frac{2\pi r_e}{T_e} \quad \text{and} \quad m_s = \frac{4\pi^2}{G} \cdot \frac{r_e^3}{T_e^2}$$

Substituting for G, $r_e = 1.5 \times 10^{11}$ m and $T_e = 3.0 \times 10^7$ s (1 year), we get

$$m_s = 2.0 \times 10^{30} \text{ kg}$$

(b) *Mass of the earth*. Considering the moon of mass m_m moving with speed v_m round the earth of mass m_e in a circular orbit of radius r_m, Fig. 7.23b, we similarly obtain

$$G\frac{m_e m_m}{r_m^2} = \frac{m_m v_m^2}{r_m}$$

$$\therefore \quad m_e = \frac{v_m^2 r_m}{G}$$

Also if T_m is the period of the moon then $v_m = 2\pi r_m / T_m$ and so

$$m_e = \frac{4\pi^2}{G} \cdot \frac{r_m^3}{T_m^2}$$

Substituting for G, $r_m = 4.0 \times 10^8$ m and $T_m = 2.4 \times 10^6$ s (1 month) we find that

$$m_e = 6.0 \times 10^{24} \text{ kg}$$

The ratio of the mass of the sun to that of the earth is $2.0 \times 10^{30} : 6.0 \times 10^{24}$, i.e. 330 000:1. Table 7.2 gives the relative masses and densities of various bodies.

Table 7.2

	Mass (earth = 1)	Density (water = 1)
Sun	330 000	1.4
Moon	0.012	3.3
Mercury	0.056	6.1
Venus	0.82	5.1
Earth	1.0	5.5
Mars	0.11	4.1
Jupiter	320	1.4
Saturn	95	0.7
Uranus	15	1.6
Neptune	17	2.3
Pluto	0.8(?)	?

Newton's work

(a) *Scientific explanation*. The charge is sometimes made that science does not get down to underlying causes and give the 'true' reasons. In many cases this is so. Newton's work raises the question of what is meant by scientific explanation.

Consider gravitation. Newton did not really explain why a body falls or why the planets move round the sun. He attributed these effects to something called 'gravitation' and this, like other basic scientific ideas, seems by its very nature to defy explanation in any simpler terms. It appears that we must accept it as a fundamental concept of science which is very useful because it enables us to regard apparently different phenomena—the falling of an apple and the motion of the planets—as having the same 'cause'.

A scientific explanation is very often an idea or concept that provides a connecting link between effects and so simplifies our knowledge. Explanations in terms of such concepts as energy, momentum, molecules, atoms, electrons, fall into this category. Concepts which do not cast their net wide are of little value in science.

(b) *Influence of Newton's work*. Starting from the laws of motion and gravitation Newton created a model of the universe which explained known facts, led to new discoveries and produced a unified body of knowledge. He united the physics of 'heaven and earth' by the same set of laws and so brought to a grand climax the work begun by Copernicus, Kepler and Galileo.

The success of Newtonian mechanics had a profound influence on both scientific and philosophical thought for 200 years. There arose a widespread belief that using scientific laws the future of the whole universe could be predicted if the positions, velocities and accelerations of all the particles in it were known at a certain time. This 'mechanistic' outlook regarded the universe as a giant piece of clockwork, wound up initially by the 'divine power' and now ticking over according to strict mathematical laws.

Today scientists are humbler and probability has replaced certainty. Also, whilst Newtonian mechanics is still perfectly satisfactory for the world of ordinary experience, it has been supplemented by two other theories. The *theory of relativity* has joined it for situations in which bodies are moving at very high speeds and *quantum mechanics* enables us to deal with the physics of the atom.

Earth's gravitational field

An action-at-a-distance effect in which one body A exerts a force on another body B not in contact with it, can be regarded as due to a 'field of force' in the region around A. The *field* may be considered to be the *interpretation* and the *force* on B the *observation*. (Body

A will of course experience a force by being in the field due to B.)

We can think of the sun and all other celestial and terrestrial bodies as each having a gravitational field which exerts a force on any other body in the field. The *strength of a gravitational field is defined as the force acting on unit mass placed in the field.* Thus if a body of mass m experiences a force F when in the earth's field, the strength of the earth's field is F/m (in newtons per kilogram). Measurement shows that if $m = 1$ kg, then $F = 9.8$ N (at the earth's surface); the strength of the earth's field is therefore 9.8 N kg^{-1}. However if a mass m falls freely under gravity its acceleration g would be $F/m = 9.8$ m s^{-2} (since $F = ma = mg$).

We thus have two ways of looking at g. When considering bodies falling freely we can think of it as an acceleration (of 9.8 m s^{-2}), but when a body of known mass is *at rest* or is *unaccelerated* in the earth's field and we wish to know the gravitational force (in newtons) acting on it we regard g as the earth's gravitational field strength (of 9.8 N kg^{-1}).

At our level of study the *field* concept tends to be more useful when dealing with electric and magnetic effects whilst the *force* concept is generally employed for gravitational effects.

Acceleration due to gravity

(*a*) *Relation between g and G.* A body of mass m at a place on the earth's surface where the acceleration due to gravity is g, experiences a force $F = mg$ (i.e. its weight) due to its attraction by the earth, Fig. 7.24. Assuming the earth behaves as if its whole mass M were concentrated at its centre O, then, by the law of gravitation, we can also say that F is the gravitational pull of the earth on the body. Hence

$$F = G\frac{Mm}{r^2}$$

where r is the radius of the earth.

$$\therefore \quad mg = G\frac{Mm}{r^2}$$

$$\therefore \quad g = \frac{GM}{r^2}$$

Fig. 7.24

It is worth noting that the mass m in $F = ma = mg$ is called the *inertial mass* of the body; it measures the opposition of the body to change of motion, i.e. its inertia. The mass of the same body when considering the law of gravitation is known as the *gravitational mass.* Experiments show that to a high degree of accuracy these two masses are equal for a given body and so we can, as we have done here, represent each by m.

(*b*) *Variation of g with height.* If g' is the acceleration due to gravity at a distance a from the centre of the earth where $a > r$, r being the earth's radius, then from (*a*),

$$g' = \frac{GM}{a^2} \quad \text{and} \quad g = \frac{GM}{r^2}$$

Dividing
$$\frac{g'}{g} = \frac{r^2}{a^2}$$

or
$$g' = \frac{r^2}{a^2}g$$

Above the earth's surface, the acceleration due to gravity g' thus varies inversely as the square of the distance a from the centre of the earth (since r and g are constant), i.e. it decreases with height as shown in Fig. 7.25.

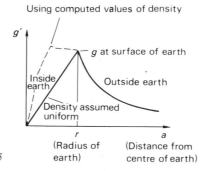

Fig. 7.25

At height h above the surface, $a = r + h$

$$\therefore \quad g' = \frac{r^2}{(r + h)^2}g = \frac{1}{(1 + h/r)^2}g$$

$$= \left(1 + \frac{h}{r}\right)^{-2}g$$

If h is very small compared with r (6400 km) we can neglect powers of (h/r) higher than the first. Hence

$$g' = \left(1 - \frac{2h}{r}\right)g$$

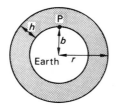

Fig. 7.26

(c) *Variation of g with depth.* At a point such as P below the surface of the earth, it can be shown that if the shaded spherical shell in Fig. 7.26 has uniform density, it produces no gravitational field inside itself. The gravitational acceleration g_1 at P is then due entirely to the sphere of radius b and if this is assumed to be of uniform density, then from (a)

$$g_1 = \frac{GM_1}{b^2} \quad \text{and} \quad g = \frac{GM}{r^2}$$

where M_1 is the mass of the sphere of radius b. The mass of a uniform sphere is proportional to its radius cubed, hence

$$\frac{M_1}{M} = \frac{b^3}{r^3}$$

But
$$\frac{g_1}{g} = \frac{M_1}{M} \cdot \frac{r^2}{b^2}$$

$$\therefore \quad \frac{g_1}{g} = \frac{b}{r} \quad \text{or} \quad g_1 = \frac{b}{r} g$$

Thus, assuming the earth has uniform density, the acceleration due to gravity g_1 is directly proportional to the distance b from the centre, i.e. it decreases linearly with depth, Fig. 7.25. At depth h below the earth's surface, $b = r - h$

$$\therefore \quad g_1 = \left(\frac{r - h}{r}\right)g = \left(1 - \frac{h}{r}\right)g$$

In fact, because the earth's density is not constant, g_1 actually *increases* for all depths now obtainable as shown by part of the dotted curve in Fig. 7.25.

(d) *Variation of g with latitude.* The observed variation of g over the earth's surface is largely due to (i) the equatorial radius of the earth exceeding its polar radius by about 21 km and thereby making g greater at the poles than at the equator where a body is farther from the centre of the earth, and (ii) the effect of the earth's rotation which we will now consider.

A body of mass m at any point of the earth's surface (except at the poles) must have a centripetal force acting on it. This force is supplied by part of the earth's gravitational attraction for it. On a stationary earth the gravitational pull of the earth on m would be mg where g is the acceleration due to gravity under such conditions. However, because of the earth's rotation, the observed gravitational pull is less than this and equals mg_0 where g_0 is the *observed* acceleration due to gravity. Hence

$$\text{centripetal force on body} = mg - mg_0$$

At the equator, the body is moving in a circle of radius r where r is the earth's radius and it has the same angular velocity ω as the earth. The centripetal force is then $m\omega^2 r$ and so

$$mg - mg_0 = m\omega^2 r$$

$$\therefore \quad g - g_0 = \omega^2 r$$

Substituting $r = 6.4 \times 10^6$ m and $\omega = 1$ revolution in 24 hours $= 2\pi/(24 \times 3600)$ rad s^{-1}, we get $g - g_0 = 3.4 \times 10^{-2}$ m s^{-2}. Assuming the earth is perfectly spherical this is also the difference between the polar and equatorial values of the acceleration due to gravity. (At the poles $\omega = 0$ and so $g = g_0$.) The observed difference is 5.2×10^{-2} m s^{-2}, of which 1.8×10^{-2} m s^{-2} arises from the non-sphericity of the earth.

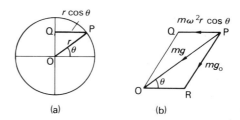

Fig. 7.27

At latitude θ on an assumed spherical earth, the body describes a circle of radius $r \cos \theta$, Fig. 7.27a. The magnitude of the centripetal force required is thus $m\omega^2 r \cos \theta$ and is smaller than at the equator since ω has the same value. However its direction is along PQ whereas mg acts along PO towards the centre of the earth. The observed gravitational pull mg_0 is therefore less than mg by a force $m\omega^2 r \cos \theta$ along PQ and will be in a different direction from mg. The value and direction of mg_0 must be such that when it is compounded by the parallelogram law with $m\omega^2 r \cos \theta$ along PQ, it gives mg along PO, Fig. 7.27b. The direction of g_0 as shown by a falling body or a plumb line is not exactly towards the centre of the earth except at the poles and the equator.

Artificial satellites

(a) *Satellite orbits.* The centripetal force which keeps an artificial satellite in orbit round the earth is the gravitational attraction of the earth for it. For a satellite of mass m travelling with speed v in a circular orbit of radius R (measured from the centre of the earth), we have

$$\frac{mv^2}{R} = \frac{GMm}{R^2}$$

where M is the mass of the earth.

$$\therefore \quad v^2 = \frac{GM}{R}$$

But

$$g = \frac{GM}{r^2} \quad \text{(see p. 168)}$$

where r is the radius of the earth and g is the acceleration due to gravity at the earth's surface.

$$\therefore \quad v^2 = \frac{gr^2}{R}$$

If the satellite is close to the earth, say at a height of 100–200 km, then $R \simeq r$ and

$$v^2 = gr$$

Substituting $r = 6.4 \times 10^6$ m (6400 km) and $g = 9.8$ m s^{-2}

$$v = \sqrt{gr} = \sqrt{9.8 \text{ m s}^{-2} \times 6.4 \times 10^6 \text{ m}}$$

$$= 7.9 \times 10^3 \text{ m s}^{-1}$$

$$\therefore \quad v \simeq 8 \text{ km s}^{-1}$$

The time for the satellite to make one complete orbit of the earth, i.e. its period T, is

$$T = \frac{\text{circumference of earth}}{\text{speed}} = \frac{2\pi r}{v}$$

$$= \frac{2\pi \times 6.4 \times 10^6 \text{ m}}{7.9 \times 10^3 \text{ m s}^{-1}}$$

$$\therefore \quad T \simeq 5000 \text{ s} \simeq 83 \text{ minutes}$$

We can regard a satellite in orbit as being continually pulled in by gravity from a straight-line tangent path to a circular path, Fig. 7.28. It 'falls' again and again from the tangents instead of continuing along them; its horizontal speed is such that it 'falls' by the correct distance to keep it in a circle. Although the satellite has an acceleration towards the centre of the earth, i.e. in a vertical direction, it has no vertical velocity because it 'falls' at the same rate as the earth's surface falls away

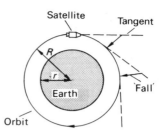

Fig. 7.28

underneath it. With respect to the earth's surface its velocity in a vertical direction is zero since the distance between the satellite and the earth's surface remains constant. In practice it is very difficult to achieve an exactly circular orbit.

(b) *Launching a satellite.* To be placed in orbit a satellite must be raised to the desired height and given the correct speed and direction by the launching rocket. A typical launching sequence using a two-stage rocket might be as follows.

At lift-off, the rocket, with a manned or unmanned space capsule on top, is held down by clamps on the launching pad for a few seconds until the exhaust gases have built up an upward thrust which exceeds the rocket's weight. The clamps are then removed by remote control and the rocket accelerates upwards. Fig. 7.29 shows the lift-off of a Saturn V rocket and a manned Apollo space capsule on their way to earth orbit before a trip to land astronauts on the moon. To penetrate the dense lower part of the atmosphere by the shortest possible route, the rocket rises vertically initially and after this is gradually tilted by the guidance system. The first-stage rocket, which may burn for about 2 minutes producing a speed of 3 km s^{-1} or so, lifts the vehicle to a height of around 60 km, then separates and falls back to earth, landing many kilometres from the launching site.

The vehicle now coasts in free flight (unpowered) to its orbital height, say 160 km, where it is momentarily moving horizontally (i.e. parallel to earth's surface immediately below). The second-stage rocket then fires and increases the speed to that required for a circular orbit at this height (about 8 km s^{-1}). By firing small rockets, the capsule is separated from the second stage which follows behind, also in orbit.

The equation $v^2 = gr^2/R$ for circular orbits shows that each orbit requires a certain speed and the greater the orbit radius R the smaller the speed v.

Some notable space flights are given in Table 7.3. Synchronous satellites have a period of 24 hours,

Fig. 7.29

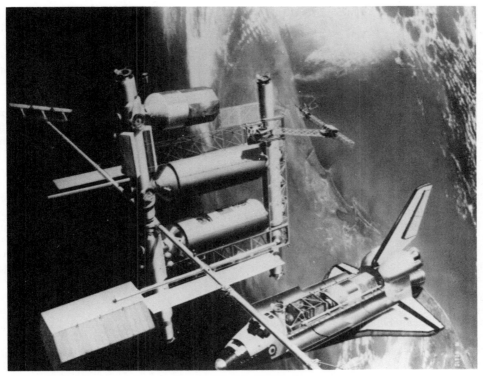

Fig. 7.30

Table 7.3

Name	Launch date	Descent date or lifetime	Period (min)	Height (km)	Notes
Sputnik 1	4 Oct 1957	4 Jan 1968 (?)	96.2	215–939	Mass 83.6 kg First artificial earth satellite
Explorer 1	31 Jan 1958	11 years	114.8	356–2548	Mass 14.0 kg Discovered inner Van Allen belt
Vostok 1	12 Apr 1961	12 Apr 1961	89.3	169–315	First manned spaceflight. One orbit by Yuri Gagarin
Mercury 6	20 Feb 1962	20 Feb 1962	88.6	159–265	First U.S. manned spaceflight. Three orbits by John Glenn
Early Bird	6 Apr 1965	10^6 years	1437	35 000–36 000	First commercial synchronous communications satellite: 'stationary' between Africa and South America
Pageos A	24 June 1966	50 years	181.4	4207–4271	Inflated sphere 30 metres in diameter. Visible to the naked eye at times given in national daily newspapers
Apollo 11	16 July 1969	24 July 1969			First men to land on the moon and return to earth with 'moon samples'
Columbia	12 April 1981	14 April 1981			First re-usable U.S. space shuttle to orbit and return to the earth.

exactly the same as that of the earth, and so remain in the same position above the earth, apparently stationary. By acting as relay stations, they make continuous, world-wide communications (e.g. of television programmes) possible. Fig. 7.30 shows an artist's impression of the U.S. space shuttle Columbia servicing an orbiting space station.

Weightlessness

An astronaut orbiting the earth in a space vehicle with its rocket motors off is said to be 'weightless'. If weight means the pull of the earth on a body, then the statement, although commonly used, is misleading. A body is not truly weightless unless it is outside the earth's (or any other) gravitational field, i.e. at a place where $g = 0$. In fact it is gravity which keeps an astronaut and his vehicle in orbit. To appreciate what 'experiencing the sensation of weightlessness' means we will consider similar situations on earth.

We are made aware of our weight because the ground (or whatever supports us) exerts an *upward* push on us as a result of the *downward* push our feet exert on the ground. It is this upward push which makes us 'feel' the force of gravity. When a lift suddenly starts upwards the push of the floor on our feet increases and we feel heavier. On the other hand if the support is reduced we seem to be lighter. In fact *we judge our weight from the upward push exerted on us by the floor*. If our feet are completely unsupported we experience weightlessness. Passengers in a lift which

has a continuous downward acceleration equal to g would get no support from the floor since they, too, would be falling with the same acceleration as the lift. There is no upward push on them and so no sensation of weight is felt. The condition is experienced briefly when we jump off a wall or dive into a swimming pool.

An astronaut in an orbiting space vehicle is not unlike a passenger in a freely falling lift. The astronaut is moving with *constant speed* along the orbit, but since he is travelling in a circle he has a centripetal acceleration—of the same value as that of his space vehicle and equal to g at that height. The walls of the vehicle exert no force on him, he is unsupported, the physiological sensation of weight disappears and he floats about 'weightless'. Similarly any object released in the vehicle does not 'fall'; anything not in use must be firmly fixed and liquids will not pour. Summing up, to be strictly correct we should not use the term 'weightless' unless by weight we mean the force exerted on (or by) a body by (or on) its support and generally we do not.

It is important to appreciate that although 'weightless' a body still has mass and it would be just as difficult to push it in space as on earth. An astronaut floating in his vehicle could still be injured by hitting a hard but weightless object.

Speed of escape

The faster a ball is thrown upwards the higher does it rise before it is stopped and pulled back by gravity. To escape from the earth into outer space we will show shortly that an object must have a speed of just over 11 km s^{-1}—called the *escape speed*.

The multi-stage rockets in use at present burn their fuel in a comparatively short time to obtain the best performance. They behave rather like objects thrown upwards, i.e. like projectiles. For a journey to, say, the moon, they therefore have to attain the escape speed. In many cases this is done by first putting the final vehicle into a 'parking' orbit round the earth with a speed of 8 km s^{-1} and then firing the final rocket again to reach escape speed in the appropriate direction.

The attainment of the escape speed is not a necessary condition however. The essential thing is that a certain amount of *energy* is required to escape from the earth and if rockets were available which could develop large power over a long time, escape would still be possible without ever achieving escape speed. In fact if we had a long enough ladder and the necessary time and energy we could walk to the moon!

The escape speed is obtained from the fact that the potential energy gained by the body equals its loss of kinetic energy, if air resistance is neglected. The work done measures the energy change. Let m be the mass of the escaping body and M the mass of the earth. The force F exerted on the object by the earth when it is distance x from the centre of the earth is

$$F = G\frac{Mm}{x^2}$$

Therefore work done δW by gravity when the body moves a further short distance δx upwards is

$$\delta W = -F\,\delta x = -G\frac{Mm}{x^2}\,\delta x$$

(the negative sign shows the force acts in the opposite direction to the displacement, see p. 149). Therefore

$$\text{total work done while body escapes} = \int_r^\infty -G\frac{Mm}{x^2}\,\mathrm{d}x$$

$$(r = \text{radius of earth})$$

$$= -GMm\left[-\frac{1}{x}\right]_r^\infty = GMm\left[\frac{1}{x}\right]_r^\infty$$

$$= -\frac{GMm}{r}$$

If the body leaves the earth with speed v and just escapes from its gravitational field

$$\tfrac{1}{2}mv^2 = \frac{GMm}{r}$$

$$\therefore \quad v = \sqrt{\frac{2GM}{r}}$$

But

$$g = \frac{GM}{r^2} \qquad\qquad (p.\ 168)$$

$$\therefore \quad v = \sqrt{2gr}$$

Substituting $r = 6.4 \times 10^6$ m and $g = 9.8$ m s^{-2}, we get $v \simeq 11$ km s^{-1}.

Possible paths for a body projected at different speeds from the earth are shown in Fig. 7.31.

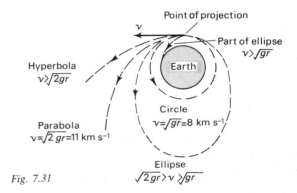

Fig. 7.31

Air molecules at s.t.p. have an average speed of about 0.5 km s^{-1} which, being much less than the escape speed, ensures that the earth's gravitational field is able to maintain an atmosphere of air round the earth. The average speed of hydrogen molecules at s.t.p. is more than three times that of air molecules and explains their rarity in the earth's atmosphere. The moon has no atmosphere. Can you suggest a possible reason?

QUESTIONS

Assume $g = 10$ m s^{-2} unless stated otherwise.

Circular motion

1. A particle moves in a semicircular path AB of radius 5.0 m with constant speed 11 m s^{-1}, Fig. 7.32. Calculate (*a*) the time taken to travel from A to B (take $\pi = 22/7$), (*b*) the average velocity, (*c*) the average acceleration.

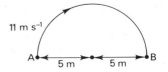

Fig. 7.32

2. The turntable of a record player makes 45 revolutions per minute. Calculate (a) its angular velocity in rad s^{-1}, (b) the linear velocity of a point 0.12 m from the centre.

3. What is meant by a *centripetal force*? Why does such a force do no work in a circular orbit?

(a) An object of mass 0.50 kg on the end of a string is whirled round in a horizontal circle of radius 2.0 m with a constant speed of 10 m s^{-1}. Find its angular velocity and the tension in the string.

(b) If the same object is now whirled in a vertical circle of the same radius with the same speed, what are the maximum and minimum tensions in the string?

4. Explain exactly how the centripetal acceleration is caused in the following cases:

(a) a train on a circular track;

(b) a conical pendulum.

A small bob of mass 0.1 kg is suspended by an inextensible string of length 0.5 m and is caused to rotate in a horizontal circle of radius 0.4 m which has its centre vertically below the point of suspension. Show on a sketch the two forces acting on the bob as seen by an outside observer. Find (i) the resultant of these forces, (ii) the period of rotation of the bob.

(*C. part qn.*)

5. Explain why a particle moving with uniform speed in a circle must be acted upon by a centripetal force. Derive from first principles an expression for the magnitude of this force.

A pilot of mass 84.0 kg loops the loop (i.e. executes a vertical circle) in his aircraft at a steady speed of 300 km hr^{-1}. Account for the centripetal force acting on him at (a) the highest, (b) the lowest points of the circle and calculate its value. Also calculate the magnitude of the force with which the pilot is pressed into the seat at the highest and lowest points of the loop if the radius of the circle is 0.580 km.

(*J.M.B.*)

Moment of inertia

6. A bicycle wheel has a diameter of 0.50 m, a mass of 0.80 kg and a moment of inertia about its axle of 4.0×10^{-2} kg m^2.

Assuming $\pi = 22/7$ find the values of the following quantities when the wheel rolls, at 7 rotations per second without slipping, over a horizontal surface:

(a) the angular velocity in radians per second,

(b) the linear velocity of the centre of gravity,

(c) the instantaneous linear velocity of the topmost point on the wheel,

(d) the total kinetic energy of the wheel. (*Hint:* the wheel has both rotational and translational kinetic energy.)

(*J.M.B.*)

7. Explain, in non-mathematical terms, the physical significance of *moment of inertia*. Illustrate your answer by reference to two examples in which moments of inertia are involved.

A flywheel of moment of inertia 6.0×10^{-3} kg m^2 is rotating with an angular velocity of 20 rad s^{-1}. Calculate the steady couple required to bring it to rest in 10 revolutions.

(*L. part qn.*)

8. Derive an expression for the kinetic energy of a rotating rigid body.

An electric motor supplies a power of 5×10^2 W to drive an unloaded flywheel of moment of inertia 2 kg m^2 at a steady speed of 6×10^2 revolutions per minute. How long will it be before the flywheel comes to rest after the power is switched off assuming the frictional couple remains constant?

9. (a) Define *angular momentum*. State the principle of conservation of angular momentum.

A figure skater is spinning about a vertical axis with his arms extended vertically upwards. Will he spin faster or slower when he allows his arms to fall until they are horizontal? Has his kinetic energy been increased or decreased? How do you account for the change?

(b) A horizontal disc rotating freely about a vertical axis makes 90 revolutions per minute. A small piece of putty of mass 2.0×10^{-2} kg falls vertically on to the disc and sticks to it at a distance of 5.0×10^{-2} m from the axis. If the number of revolutions per minute is thereby reduced to 80, calculate the moment of inertia of the disc.

Gravitation

10. The mass of the earth is 5.98×10^{24} kg and the gravitational constant is 6.67×10^{-11} m^3 kg^{-1} s^{-2}. Assuming the earth is a uniform sphere of radius 6.37×10^6 m, find the gravitational force on a mass of 1.00 kg at the earth's surface.

11. Define moment of inertia and angular momentum.

A small planet, mass m, moves in an elliptical orbit round a large sun, mass M, which is at the focus F_1 of the ellipse, Fig. 7.33.

Write down an expression for the force acting on the planet when it is at a position A, at a distance r from the sun. Indicate the direction of this force on a copy of Fig. 7.33, and also mark on it the directions of the planet's velocity and acceleration when at A. What is the magnitude of this acceleration?

According to Kepler's second law, the planet is moving faster at B than at C. Account for this with reference to the principle of conservation of energy.

What is the moment of inertia of the planet, when at B, about F_1? The velocities at B and C are v_B and v_C. Use the principle of conservation of angular momentum to deduce the ratio v_B/v_C.

(*S.*)

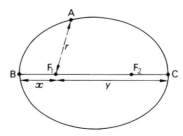

Fig. 7.33

12. It is proposed to place a communications satellite in a circular orbit round the equator at a height of 3.59×10^7 m above the earth's surface. Find the period of revolution of the satellite in hours and comment on the result. (Use the values given in *Question 10* for the radius and mass of the earth and the constant of gravitation.)

13. State Newton's Law of Gravitation. If the acceleration due to gravity, g_m, at the moon's surface is 1.70 m s^{-2} and its radius is 1.74×10^6 m, calculate the mass of the moon.

To what height would a signal rocket rise on the moon, if an identical one fired on earth could reach 200 m? (Ignore atmospheric resistance.) Explain your reasoning.

Explain, using algebraic symbols and stating which quantity each represents, how you could calculate the distance D of the moon from the earth (Mass M_e) if the moon takes t seconds to move once round the earth.

What is meant by 'weightlessness', experienced by an astronaut orbiting the earth, and how is it caused? Explain also whether he would have the same experience when falling freely back to earth in his capsule just prior to re-entry in the earth's atmosphere. (Gravitational constant = 6.67×10^{-11} m^3 kg^{-1} s^{-2}; acceleration due to gravity = 9.81 m s^{-2}.)
(S.)

14. Explaining each step in your calculation and pointing out the assumptions you make, use the information below to estimate the mean distance of the moon from the earth.

Period of rotation of the moon around the earth = 27.3 days
Radius of earth = 6.37×10^3 km
Acceleration due to gravity at earth's surface, g = 9.81 m s^{-2}
(*J.M.B. part qn.*)

15. The graph (Fig. 7.34) shows how the force of attraction on a 1 kg mass towards the earth varies with its distance from the centre of the earth. It shows that the force at a distance equal to the radius R of the earth is 10 newtons.

(*a*) Calculate the force on the mass at distances of (*i*) 9R (*ii*) 10R from the earth's centre.

(*b*) Shade in on (a copy of) the graph an area which gives the energy change in moving the mass from 3R to 2R.

(*c*) Make a rough estimate of the increase in kinetic energy when the mass falls from 10R to 9R.   (*O. and C. Nuffield*)

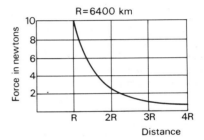

Fig. 7.34

8 Mechanical oscillations

Introduction

In previous chapters linear and circular motion were considered. Another common type of motion is the to-and-fro repeating movement called a *vibration* or *oscillation*.

Examples of oscillatory motion are provided by a swinging pendulum, the balance wheel of a watch, a mass on the end of a vibrating spring, the strings and air columns of musical instruments when producing a note. Sound waves are transmitted by the oscillation of the particles of the medium in which the sound is travelling.

We also believe that the atoms in a solid vibrate about fixed positions in their lattice.

Engineers as well as scientists need to know about vibrations. They can occur in turbines, aircraft, cars, tall buildings and chimneys and were responsible for the collapse of the Tacoma Narrows suspension bridge in America in 1940 when a moderate gale set the bridge oscillating until the main span broke up, Fig. 8.1a and b. In metal structures they can cause fatigue failure (p. 27).

In a mechanical oscillation there is a continual interchange of potential and kinetic energy due to the

Fig. 8.1(a)

system having (*i*) *elasticity* (or springiness) which allows it to store p.e. and (*ii*) *mass* (or inertia) which enables it to have k.e. Thus, when a body on the lower end of a spiral spring, Fig. 8.2, is pulled down and released, the elastic restoring force pulls the body up and it accelerates towards its equilibrium position O with increasing velocity. The accelerating force decreases as the body approaches O (since the spring is stretched less) and so the *rate* of change of velocity (i.e. the acceleration) decreases.

At O the restoring force is zero but because the body has inertia it overshoots the equilibrium position and continues to move upwards. The spring is now compressed and the elastic restoring force acts again but downwards towards O this time. The body therefore slows down and at an increasing rate due to the restoring force increasing at greater distances from O. The body eventually comes to rest above O and repeats its motion in the opposite direction, p.e. stored as elastic energy of the spring being continually changed to k.e. of the moving body and vice versa. The motion would continue indefinitely if no energy loss occurred, but energy is lost. Why?

The time for a complete oscillation from A to B and back to A, or from O to A to O to B and back to O

again is the *period T* of the motion. The *frequency f* is the number of complete oscillations per unit time and a little thought (perhaps using numbers) will indicate that

$$f = \frac{1}{T}$$

An oscillation (or cycle) per second is a *hertz*. The maximum displacement OA or OB is called the *amplitude* of the oscillation.

Some other simple oscillatory systems are shown in Fig. 8.3. It is worth trying to discover experimentally (*i*) which have a constant period (compared with the oscillating balance wheel of a watch), (*ii*) what factors determine the period (or frequency) of the oscillation and (*iii*) whether 'time-traces' of their motions can be obtained and what they look like.

Simple harmonic motion

In Fig. 8.4, N is a body oscillating in a straight line about O, between A and B; N could be a mass hanging from a spiral spring. Previously, in linear motion we considered accelerations that were constant in magnitude and direction and in circular motion the accelera-

Fig. 8.1(b)

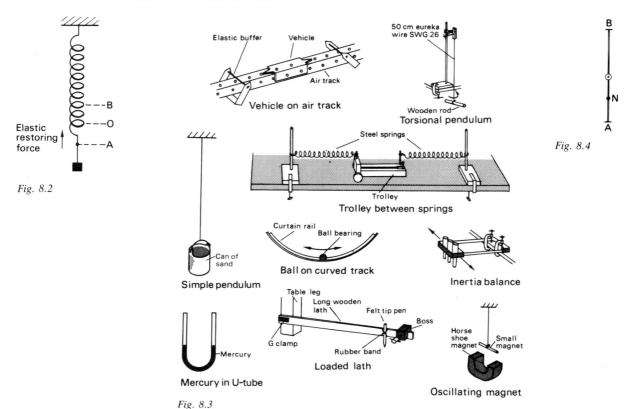

Fig. 8.2

Elastic restoring force

Vehicle on air track

Elastic buffer   Vehicle   Air track

Torsional pendulum

50 cm eureka wire SWG 26

Wooden rod

Trolley between springs

Steel springs   Trolley

Fig. 8.4

Simple pendulum

Can of sand

Mercury in U-tube

Mercury

Ball on curved track

Curtain rail   Ball bearing

Inertia balance

Loaded lath

Table leg   Long wooden lath   Felt tip pen   Boss   G clamp   Rubber band

Oscillating magnet

Horse shoe magnet   Small magnet

Fig. 8.3

tions (centripetal) were constant in magnitude if not in direction. In oscillatory motion, the accelerations, like the displacements and velocities, change periodically in both magnitude and direction.

Consider first displacements and velocities. When N is below O, the displacement (measured from O) is downwards; the velocity is directed downwards when N is moving away from O but upwards when it moves towards O and is zero at A and B. When N is above O, the displacement is upwards and the velocity upwards or downwards according to whether N is moving away from or towards O.

The variation of acceleration can be seen by considering a body oscillating on a spiral spring. The magnitude of the elastic restoring force increases with displacement but always acts towards the equilibrium position (i.e. O); the resulting acceleration must therefore behave likewise, increasing with displacement but being directed to O whatever the displacement. Thus if N is below O, the displacement is downwards and the acceleration upwards, but if the displacement is upwards the acceleration is downwards. If we adopt the sign convention that quantities acting downwards are positive and those acting upwards are negative then

acceleration and displacement always have opposite signs in an oscillation. Fig. 8.5 summarizes these facts and should be studied carefully.

The simplest relationship between the magnitudes of the acceleration and the displacement would be one in

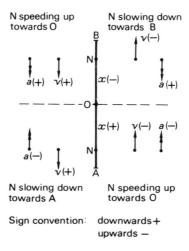

N speeding up towards O

N slowing down towards B

$v(-)$

$a(+)$ $v(+)$ $x(-)$ $a(+)$

$x(+)$ $v(-)$ $a(-)$

$a(-)$ $v(+)$

N slowing down towards A

N speeding up towards O

Sign convention: downwards +
upwards −

Fig. 8.5

which the the acceleration a of the body is directly proportional to its displacement x. Such an oscillation is said to be a *simple harmonic motion* (s.h.m.) and is defined as follows.

If the acceleration of a body is directly proportional to its distance from a fixed point and is always directed towards that point, the motion is simple harmonic.

The equation relating acceleration and displacement can be written

$$a \propto -x$$

or

$$a = -\text{constant} \cdot x$$

The negative sign indicates that although the acceleration is larger at larger displacements it is always in the opposite direction to the displacement, i.e. towards O. What kind of motion would be represented by the above equation if a positive sign replaced the negative sign? (It would not be an oscillation.)

In practice many mechanical oscillations are nearly simple harmonic, especially at small amplitudes, or are combinations of such oscillations. In fact any system which obeys Hooke's law will exhibit this type of motion when vibrating. The equation for s.h.m. turns up in many problems in sound, optics, electrical circuits and even in atomic physics. In calculus notation it is written

$$\frac{\mathrm{d}^2x}{\mathrm{d}t^2} = -\text{constant} \cdot x$$

where $a = \mathrm{d}v/\mathrm{d}t = \mathrm{d}^2x/\mathrm{d}t^2$. Using calculus this second order differential equation can be solved to give expressions for displacement and velocity. However, in the next section we shall use a simple geometrical method which links circular motion and simple harmonic motion.

Equations of s.h.m.

Suppose a point P moves round a circle of radius r and centre O with uniform angular velocity ω, its speed v round the circumference will be constant and equal to ωr, Fig. 8.6a. As P revolves, N, the foot of the perpendicular from P on the diameter AOB, moves from A to O to B and returns through O to A as P completes each revolution. Let P and N be in the positions shown at time t after leaving A, with radius OP making angle θ with OA and distance ON being x. We will now show that N describes s.h.m. about O.

(*a*) *Acceleration.* The motion of N is due to that of P, therefore the acceleration of N is the *component* of the

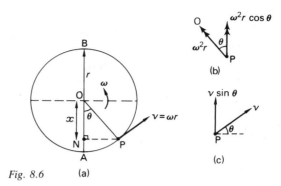

Fig. 8.6   (a)

acceleration of P parallel to AB. The acceleration of P is $\omega^2 r$ (or v^2/r) along PO and so the component of this parallel to AB is $\omega^2 r \cos \theta$, Fig. 8.6b. Hence the acceleration a of N is

$$a = -\omega^2 r \cos \theta$$

The negative sign, as explained before, indicates mathematically that a is always directed towards O. Now $x = r \cos \theta$

$$\therefore a = -\omega^2 x$$

Since ω^2 is a positive constant, this equation states that the acceleration of N towards O is directly proportional to its distance from O. N thus describes s.h.m. about O as P moves round the circle—called the *auxiliary circle*—with constant speed.

The table below gives values of a for different values of x and we see that a is zero at O and a maximum at the limits A and B of the oscillation where the direction of motion changes.

x	0	$+r$	$-r$
a	0	$-\omega^2 r$	$+\omega^2 r$

Using the arrangement of Fig. 8.7 the shadow of a

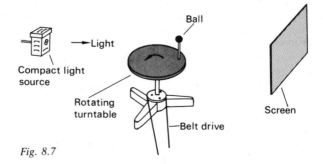

Fig. 8.7

ball moving steadily in a circle can be viewed on a screen. The shadow moves with s.h.m. and represents the *projection* of the ball on the screen.

(b) *Period.* The period T of N is the time for N to make one complete oscillation from A to B and back again. In the same time P will travel once round the auxiliary circle and therefore

$$T = \frac{\text{circumference of auxiliary circle}}{\text{speed of P}}$$

$$= \frac{2\pi r}{v} = \frac{2\pi}{\omega} \quad \text{(since } v = \omega r\text{)}$$

For a particular s.h.m. ω is constant and so T is constant and independent of the amplitude r of the oscillation. If the amplitude increases, the body travels faster and so T remains unchanged. A motion which has a constant period whatever the amplitude is said to be *isochronous* and this property is an important characteristic of s.h.m.

(c) *Velocity.* The velocity of N is the *component* of P's velocity parallel to AB which

$$= -v \sin \theta \quad \text{(see Fig. 8.6c)}$$

$$= -\omega r \sin \theta \quad \text{(since } v = \omega r\text{)}$$

Since $\sin \theta$ is positive when $0° < \theta < 180°$, i.e. N moving upwards, and negative when $180° < \theta < 360°$, i.e. N moving downwards, the negative sign ensures that the velocity is negative when acting upwards and positive when acting downwards (see Fig. 8.5). The variation of the velocity of N with time t (assuming P, and so N, start from A at zero time)

$$= -\omega r \sin \omega t \quad \text{(since } \theta = \omega t\text{)}$$

The variation of the velocity of N with displacement x

$$= -\omega r \sin \theta$$

$$= \pm \omega r \sqrt{1 - \cos^2 \theta} \quad \text{(since } \sin^2 \theta + \cos^2 \theta = 1\text{)}$$

$$= \pm \omega r \sqrt{1 - (x/r)^2}$$

$$= \pm \omega \sqrt{r^2 - x^2}$$

Hence the velocity of N is

$$\pm \omega r \text{ (a maximum) when } x = O$$

$$\text{zero when } x = \pm r$$

(d) *Displacement.* This is given by

$$x = r \cos \theta$$

$$= r \cos \omega t$$

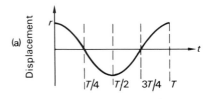

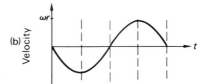

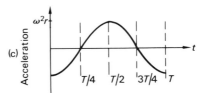

Fig. 8.8

The graph of the variation of the displacement of N with time (i.e. its 'time-trace') is shown in Fig. 8.8a and like those for velocity and acceleration in Figs. 8.8b and c it is sinusoidal. Note that when the velocity is zero the acceleration is a maximum and vice versa. We say there is a *phase difference* of a quarter of a period (i.e. $T/4$) between the velocity and the acceleration. What is the phase difference between the displacement and the acceleration?

(e) *Summary*

$$\text{acceleration} = -\omega^2 x = -\omega^2 r \cos \omega t$$

$$\text{velocity} = \pm \omega \sqrt{r^2 - x^2} = -\omega r \sin \omega t$$

$$\text{displacement} = r \cos \omega t$$

$$\text{period} = 2\pi/\omega$$

These equations are true for *any* s.h.m.

Expression for ω

Consider the equation $a = -\omega^2 x$. We can write (ignoring signs)

$$\omega^2 = \frac{a}{x} = \frac{ma}{mx} = \frac{ma/x}{m}$$

where m is the mass of the system. The force causing the acceleration a at displacement x is ma, therefore

ma/x is the force per unit displacement. Hence

$$\omega = \sqrt{\frac{force\ per\ unit\ displacement}{mass\ of\ oscillating\ system}}$$

The period T of the s.h.m. is given by

$$T = \frac{2\pi}{\omega}$$

$$= 2\pi\sqrt{\frac{mass\ of\ oscillating\ system}{force\ per\ unit\ displacement}}$$

This expression shows that T increases if (*i*) the *mass* of the oscillating system *increases* and (*ii*) the *force per unit displacement decreases*, i.e. if the elasticity factor decreases.

An oscillation is simple harmonic if its equation of motion can be written in the form

$$a = - (\text{positive constant}) . x$$

For convenience the 'positive constant' is usually represented by ω^2 since $T = 2\pi/\omega$. Hence ω is the square root of the 'positive constant' in the acceleration–displacement equation.

Mass on a spring

(*a*) *Period of oscillations*. The extension of a spiral spring which obeys Hooke's law is directly proportional to the extending tension. A mass m attached to the end of a spring exerts a downward tension mg on it and if it stretches it by an amount l as in Fig. 8.9*a*, then if k is the tension required to produce unit extension (called the *spring constant* and measured in N m^{-1}) the stretching tension is also kl and so

$$mg = kl$$

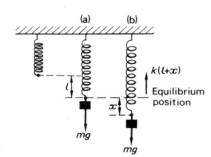

Fig. 8.9

Suppose the mass is now pulled down a further distance x below its equilibrium position, the stretching tension acting downwards is $k(l + x)$ which is also the tension in the spring acting upwards, Fig. 8.9*b*. Hence resultant restoring force *upwards* on mass

$$= k(l + x) - mg$$

$$= kl + kx - kl \quad (\text{since } mg = kl)$$

$$= kx$$

When the mass is released it oscillates up and down. If it has an acceleration a at extension x then by Newton's second law

$$-kx = ma$$

The negative sign indicates that at the instant shown a is upwards (negative on our sign convention) while the displacement x is downwards (i.e. positive).

$$\therefore \quad a = -\frac{k}{m}x = -\omega^2 x$$

where $\omega^2 = k/m = $ a positive constant since k and m are fixed. The motion is therefore simple harmonic about the equilibrium position so long as Hooke's law is obeyed. The period T is given by

$$T = \frac{2\pi}{\omega} = 2\pi\sqrt{\frac{m}{k}}$$

It follows that $T^2 = 4\pi^2 m/k$. If the mass m is varied and the corresponding periods T found, a graph of T^2 against m is a straight line but it does not pass through the origin as we might expect from the above equation. This is due to the mass of the spring itself being neglected in the above derivation. Its effective mass and a value of g can be found experimentally.

(*b*) *Measurement of g and effective mass of spring*. Let m_s be the effective mass of the spring then

$$T = 2\pi\sqrt{\frac{m + m_s}{k}}$$

But $\qquad\qquad mg = kl$

Substituting for m in the first equation and squaring, we get

$$T^2 = \frac{4\pi^2}{k}\left(\frac{kl}{g} + m_s\right)$$

$$\therefore \quad l = \frac{g}{4\pi^2} . T^2 - \frac{g\ m_s}{k}$$

By measuring (*i*) the static extension l and (*ii*) the corresponding period T, using several different masses in turn, a graph of l against T^2 can be drawn. It is a straight line of slope $g/4\pi^2$ and intercept gm_s/k on the l axis, Fig. 8.10. Thus g and m_s can be found. Theory

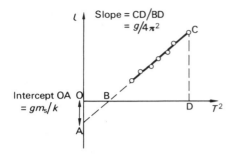

Fig. 8.10

suggests that the effective mass of a spring is about one-third of its actual mass.

Simple pendulum

(a) *Period of oscillations.* The simple pendulum consists of a small bob (in theory a 'particle') of mass m suspended by a light inextensible thread of length l from a fixed point B, Fig. 8.11. If the bob is drawn aside slightly and released, it oscillates to-and-fro in a vertical plane along the arc of a circle. We shall show that it describes s.h.m. about its equilibrium position O.

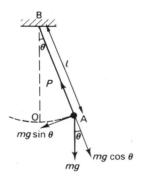

Fig. 8.11

Suppose at some instant the bob is at A where arc $OA = x$ and $\angle OBA = \theta$. The forces on the bob are P and the weight mg of the bob acting vertically downwards. Resolving mg radially and tangentially at A we see that the tangential component $mg \sin \theta$ is the unbalanced restoring force acting towards O. (The component $mg \cos \theta$ balances P.) If a is the acceleration of the bob along the arc at A due to $mg \sin \theta$ then the equation of motion of the bob is

$$- mg \sin \theta = ma$$

The negative sign indicates that the force is towards O while the displacement x is measured along the arc from O in the opposite direction.

When θ is small, $\sin \theta = \theta$ in radians (e.g. if $\theta = 5°$, $\sin \theta = 0.0872$ and $\theta = 0.0873$ rad) and $x = l\theta$ (see p. 155). Hence

$$-mg\, \theta = -mg\frac{x}{l} = ma$$

$$\therefore \quad a = -\frac{g}{l}x = -\omega^2 x \quad \text{(where } \omega^2 = g/l)$$

The motion of the bob is thus simple harmonic *if the oscillations are of small amplitude*, i.e. θ does not exceed 10°. The period T is given by

$$T = \frac{2\pi}{\omega} = \frac{2\pi}{\sqrt{g/l}}$$

$$= 2\pi\sqrt{\frac{l}{g}}$$

T is therefore independent of the amplitude of the oscillations and at a given place on the earth's surface where g is constant, it depends only on the length l of the pendulum.

A multiflash photograph of a single swing of a simple pendulum is shown in Fig. 8.12.

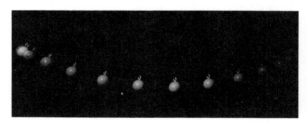

Fig. 8.12

(b) *Measurement of g.* A fairly accurate determination of g can be made by measuring T for different values of l and plotting a graph of l against T^2. A straight line AB is then drawn so that the points are evenly distributed about it, Fig. 8.13. It should pass through the origin and its slope BC/CA gives an average value of l/T^2 from which g can be calculated since

$$T = 2\pi\sqrt{\frac{l}{g}}$$

$$\therefore \quad T^2 = 4\pi^2\frac{l}{g}$$

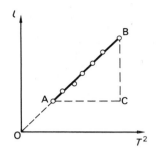

Fig. 8.13

$$\therefore \quad g = 4\pi^2 \frac{l}{T^2} = 4\pi^2 \frac{BC}{CA}$$

The experiment requires (*i*) 100 oscillations to be timed, (*ii*) an angle of swing less than 10°, (*iii*) the length *l* to be measured to the centre of the bob, (*iv*) the oscillations to be counted as the bob passes the equilibrium position O. Why?

S.H.M. calculations

1. A particle moving with s.h.m. has velocities of 4 cm s^{-1} and 3 cm s^{-1} at distances of 3 cm and 4 cm respectively from its equilibrium position. Find (a) the amplitude of the oscillation, (b) the period, (c) the velocity of the particle as it passes through the equilibrium position.

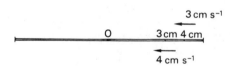

Fig. 8.14

(*a*) Using the previous notation and taking the case shown in Fig. 8.14, the equation for the velocity is

$$\text{velocity} = -\omega \sqrt{r^2 - x^2} \qquad \text{(p. 180)}$$

if we take velocities and displacements to the left as being negative and those to the right positive.

When $x = +3$ cm, velocity $= -4$ cm s^{-1}; therefore

$$-4 = -\omega \sqrt{r^2 - 9}$$

When $x = +4$ cm, velocity $= -3$ cm s^{-1}; therefore

$$-3 = -\omega \sqrt{r^2 - 16}$$

Squaring and dividing these equations we get

$$\frac{16}{9} = \frac{r^2 - 9}{r^2 - 16}$$

Hence $\qquad r = \pm 5$ cm

(*b*) Substituting for *r* in one of the velocity equations we find

$$\omega = 1 \text{ s}^{-1}$$

$$\therefore \quad T = \frac{2\pi}{\omega} = 2\pi \text{ s}$$

(*c*) At the equilibrium position $x = 0$

$$\therefore \quad \text{velocity} = \pm \omega \sqrt{r^2 - x^2}$$

$$= \pm \omega r$$

$$= \pm 5 \text{ cm s}^{-1}$$

2. A light spiral spring is loaded with a mass of 50 g and it extends by 10 cm. Calculate the period of small vertical oscillations. ($g = 10$ m s^{-2})

The period *T* of the oscillations is given by

$$T = 2\pi \sqrt{\frac{m}{k}}$$

where $\qquad m = 50 \times 10^{-3}$ kg

and $\qquad k = $ force per unit displacement

$$= \frac{50 \times 10^{-3} \times 10 \text{ N}}{10 \times 10^{-2} \text{ m}} = 5.0 \text{ N m}^{-1}$$

$$\therefore \quad T = 2\pi \sqrt{\frac{50 \times 10^{-3}}{5}} = 2\pi \sqrt{10^{-2}} \text{ s}$$

$$= 2\pi \times 10^{-1} \text{ s}$$

$$= 0.63 \text{ s}$$

3. A simple pendulum has a period of 2.0 s and an amplitude of swing 5.0 cm. Calculate the maximum magnitudes of (a) the velocity of the bob, (b) the acceleration of the bob.

(*a*) $T = 2\pi/\omega$, therefore $\omega = 2\pi/T = 2\pi/2 = \pi$ s^{-1}. Velocity is a maximum at the equilibrium position where $x = 0$ and

$$= \pm \omega \sqrt{r^2 - x^2}$$

$$= \pm \pi \sqrt{25} \qquad \text{(since } r = \pm 5 \text{ cm)}$$

$$= \pm 5\pi \text{ cm s}^{-1}$$

$$= \pm 16 \text{ cm s}^{-1}$$

(*b*) Acceleration is a maximum at the limits of the

swing where $x = r = \pm 5.0$ cm and

$$= -\omega^2 r$$

$$= -\pi^2 \times 5 \text{ cm s}^{-2}$$

$$= -50 \text{ cm s}^{-2}$$

Energy of s.h.m.

In an oscillation there is a constant interchange of energy between the kinetic and potential forms and if the system does no work against resistive forces (i.e. is undamped) its total energy is constant, as we shall now show.

(a) *Kinetic energy.* The velocity of a particle N of mass m at a distance x from its centre of oscillation O, Fig. 8.15,

$$= +\omega \sqrt{r^2 - x^2}$$

$$\therefore \quad \text{k.e. at displacement } x = \tfrac{1}{2} m \omega^2 (r^2 - x^2)$$

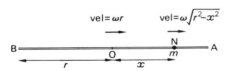

Fig. 8.15

(b) *Potential energy.* As N moves out from O towards A (or B) work is done against the force (e.g. the tension in a stretched spring) trying to restore it to O. Thus N loses k.e. and gains p.e. When $x = 0$, the restoring force is zero; at displacement x, the force is $m\omega^2 x$ (since the acceleration has magnitude $\omega^2 x$). Hence

average force on N while moving to displacement x
$$= \tfrac{1}{2} m \omega^2 x$$

$\therefore$ work done = average force ×
displacement in direction of force

$$= \tfrac{1}{2} m \omega^2 x \times x = \tfrac{1}{2} m \omega^2 x^2$$

$\therefore$ p.e. at displacement $x = \tfrac{1}{2} m \omega^2 x^2$

(c) *Total energy.* At displacement x we have

total energy = k.e. + p.e.

$$= \tfrac{1}{2} m \omega^2 (r^2 - x^2) + \tfrac{1}{2} m \omega^2 x^2$$

$$= \tfrac{1}{2} m \omega^2 r^2$$

This is constant, does not depend on x and is directly proportional to the product of (*i*) the mass, (*ii*) the

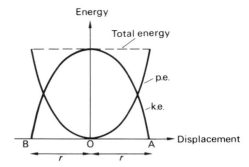

Fig. 8.16

square of the frequency and (*iii*) the square of the amplitude. Fig. 8.16 shows the variation of k.e., p.e. and total energy with displacement.

In a simple pendulum all the energy is kinetic as the bob passes through the centre of oscillation and at the top of the swing it is all potential.

Damped oscillations

The amplitude of the oscillations of, for example, a pendulum gradually decreases to zero due to the resistive force that arises from the air. The motion is therefore not a perfect s.h.m. and is said to be *damped* by air resistance; its energy becomes internal energy of the surrounding air.

The behaviour of a mechanical system depends on the extent of the damping. The damping of the mass on the spring in Fig. 8.17 is greater than when it is in air. Undamped oscillations are said to be *free*, Fig. 8.18a. If a system is slightly damped oscillations of decreasing amplitude occur, Fig. 8.18b. When heavily damped no oscillations occur and the system returns very slowly to its equilibrium position, Fig. 8.18c. When the time taken for the displacement to become zero is a minimum, the system is said to be *critically damped*, Fig. 8.18d.

The motion of many devices is critically damped on purpose. Thus the shock absorbers on a car critically

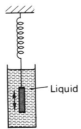

Fig. 8.17

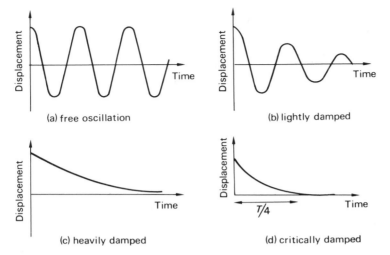

Fig. 8.18

(a) free oscillation

(b) lightly damped

(c) heavily damped

(d) critically damped

damp the suspension of the vehicle and so resist the setting up of vibrations which could make control difficult or cause damage. In the shock absorber of Fig. 8.19 the motion of the suspension up or down is opposed by viscous forces when the liquid passes through the transfer tube from one side of the piston to the other. The damping of a car can be tested by applying your weight to the suspension momentarily; the car should rapidly return to its original position without vibrating.

Instruments such as balances and electrical meters are critically damped (i.e. dead-beat) so that the pointer moves quickly to the correct position without oscillating. The damping is often produced by electromagnetic forces.

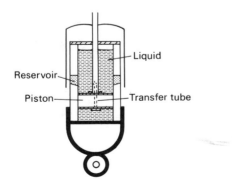

Fig. 8.19

Forced oscillation and resonance

(*a*) *Barton's pendulums*. The assembly consists of a number of paper cone pendulums (made by folding paper circles of about 5 cm diameter) of lengths varying from $\frac{1}{4}$ m to $\frac{3}{4}$ m, each loaded with a plastic curtain ring. All are suspended from the same string as a 'driver' pendulum which has a heavy bob and a length of $\frac{1}{2}$ m, Fig. 8.20.

The driver pendulum is pulled well aside and released so that it oscillates in a plane perpendicular to that of the diagram. The motion settles down after a short time and all the pendulums oscillate with very nearly the *same* frequency as the driver but with *different* amplitudes. This is a case of *forced oscillation*.

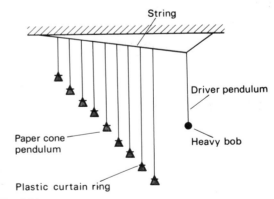

Fig. 8.20

The pendulum whose length equals that of the driver has the greatest amplitude; its natural frequency of oscillation is the same as the frequency of the driving pendulum. This is an example of *resonance* and the driving oscillator then transfers its energy most easily to the other system, i.e. the paper cone pendulum of the same length.

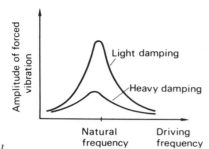

Fig. 8.21

The amplitudes of oscillations also depend on the extent to which the system is damped. Thus removing the rings from the paper cone pendulums reduces their mass and so increases the damping. All amplitudes are then found to be reduced and that of the resonant frequency is less pronounced. These results are summarized by the graphs in Fig. 8.21 which indicate that the sharpest resonance is given by a lightly damped system. Fig. 8.22*a* and *b* are time-exposure photographs taken with the camera looking along the line of swinging pendulums towards and at the same level as the bob of the driver. Is (*a*) more or less damped than (*b*)?

Careful observation shows that the resonant pendulum is always a quarter of an oscillation behind the driver pendulum, i.e. there is a phase difference of a quarter of a period. The shorter pendulums are nearly in phase with the driver, while those that are longer

than the driver are almost half a period behind it. This is evident from the instantaneous photograph of Fig. 8.22*c*, taken when the driver is at maximum displacement to the left.

(*b*) *Hacksaw blade oscillator.* The arrangement, shown in Fig. 8.23, provides another way of finding out what happens when one oscillator (a loaded hacksaw blade) is driven by another (a heavy pendulum), as often occurs in practice. The positions of the mass on the blade and the pendulum bob can both be adjusted to alter the natural frequencies. By using different rubber bands the degree of coupling may be varied, as can the damping, by turning the postcard. The motion of the driver is maintained by gentle, timely taps just below its support.

There is scope for investigating (*i*) the transient oscillations that occur as the motion starts and before the onset of steady conditions, (*ii*) resonance, (*iii*) phase relationships, (*iv*) damping and (*v*) coupling.

(*c*) *Examples of resonance.* These are common throughout science and are generally useful. Thus in the production of musical sounds from air columns in wind instruments resonance occurs, in many cases, between the vibrations of air columns and of small vibrating reeds. Electrical resonance occurs when a radio circuit is tuned by making its natural frequency for electrical oscillations equal to that of the incoming radio signal.

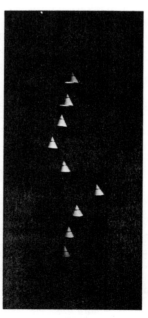

Fig. 8.22   (a)   (b)   (c)

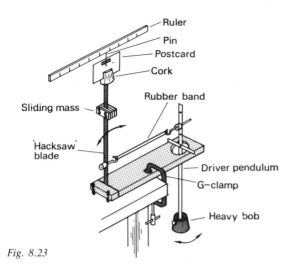

Ruler
Pin
Postcard
Cork
Sliding mass
Rubber band
Hacksaw blade
Driver pendulum
G-clamp
Heavy bob

Fig. 8.23

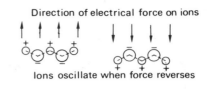

Direction of electrical force on ions

Row of ions in a crystal

Ions oscillate when force reverses

Fig. 8.24

Information about the strength of chemical bonds between ions can be obtained by a resonance effect. If we regard electromagnetic radiation (e.g. light, infrared, etc.) as a kind of oscillating electrical disturbance which, when incident on a crystal, subjects the ions to an oscillating electrical force, then, with radiation of the correct frequency, the ions could be set into oscillation by resonance, Fig. 8.24. Energy would be absorbed from the radiation and the absorbed frequency could be found using a suitable spectrometer. With sodium chloride absorption of infrared radiation occurs.

Resonance in mechanical systems is a source of trouble to engineers. The Tacoma Narrows Bridge disaster (p. 176) was really caused by the wind producing an oscillating resultant force in resonance with a natural frequency of the bridge. An oscillation of large amplitude built up and destroyed the structure. Fig. 8.25 shows the tail of a wind tunnel model of an aircraft being tested for resonance, i.e. shaken at different frequencies. It is important that the natural frequencies of vibration of an aircraft do not equal any that may be produced by the forces experienced in flight. Otherwise resonance might occur and undue stress result.

Fig. 8.25

S.H.M.—a mathematical model

Real oscillators such as a motor cycle on its suspension, a tall chimney swaying in the wind, atoms (or ions) vibrating in a crystal, only approximate to the ideal type of motion we call s.h.m.

S.H.M. describes in mathematical terms a kind of motion which is not fully realized in practice. It is a mathematical model and is useful because it represents well enough many real oscillations. This is due to its simplicity; complications such as damping, variable mass and variable stiffness (elastic modulus) are omitted and the only conditions imposed on the system are that the restoring force should be directed towards the centre of the motion and be proportional to the displacement.

A more complex model might, for example, take damping into account and as a result be a better description of a particular oscillator, but it would probably not be so widely applicable. On the other hand if a model is too simple it may be of little use for dealing with real systems. A model must have just the correct degree of complexity. The mathematical s.h.m. model has this and so is useful in practice.

Film loops (8 mm)

1. *Tacoma Narrows Bridge collapse*—Ealing Scientific Ltd, Bushey Mill Lane, Watford, Herts.
2. *Wind-induced oscillations*—Penguin Education, Harmondsworth, Middlesex.

QUESTIONS

1. Write a short account of *simple harmonic motion* explaining the terms *amplitude, time period* and *frequency*.

A particle of mass m moves such that its displacement from the equilibrium position is given by $y = a \sin \omega t$ where a and ω are constants. Derive an expression for the kinetic energy of the particle at a time t and show that its value is a maximum as the particle passes through the equilibrium position.

A steel strip clamped at one end vibrates with a frequency of 30 Hz, and an amplitude of 4.0 mm at the free end. Find (a) the velocity of the free end as it passes through the equilibrium position and (b) the acceleration at the maximum displacement.   (*A.E.B.*)

2. A simple pendulum of length 80 cm is oscillating with an amplitude of 4.0 cm. Calculate the velocity of the bob as it passes through the mid-point of its oscillation. Explain why the tension in the string as it passes through this point is different from that in the string when the pendulum hangs vertically at rest and state which tension has the greater value.
(*L. part qn.*)

3. Define *simple harmonic motion*.

Describe an experiment to measure the acceleration due to gravity at the earth's surface, using a simple pendulum. Derive the equation used.

A pendulum of length 130 cm has a periodic time of T_1. The bob is now pulled aside and made to move as a conical pendulum in a horizontal circle of radius 50.0 cm. The period of rotation is T_2. Find the ratio $T_1 : T_2$.   (*L.*)

4. Define simple harmonic motion and explain the meaning of the terms amplitude, period, and frequency.

A body of mass 0.10 kg hangs from a long spiral spring. When pulled down 10 cm below its equilibrium position A and released, it vibrates with simple harmonic motion with a period of 2.0 s.

(a) What is its velocity as it passes through A?

(b) What is its acceleration when it is 5.0 cm above A?

(c) When it is moving upwards, how long a time is taken for it to move from a point 5.0 cm below A to a point 5.0 cm above?

(d) What are the maximum and minimum values of its kinetic energy, and at what points of the motion do they occur?

(e) What is the value of the total energy of the system and does it vary with time?   (*W.*)

5. Describe an experiment to demonstrate the effects of damping on the oscillatory motion of a vibrating system undergoing (a) free and (b) forced harmonic oscillations. Draw labelled diagrams to illustrate the results you would expect to obtain.

What is the physical origin of the damping mechanisms in the case of (a) the oscillations of a simple pendulum in air, (b) the vibrations of a bell sounding in air and (c) oscillatory currents in an electrical circuit?

A spring is supported at its upper end. When a mass of 1.0 kg is hung on the lower end the new equilibrium position is 5.0 cm lower. The mass is then raised 5.0 cm to its original position and released. Discuss as fully as you can the subsequent motion of the system.   (*O. and C.*)

6. A mass of 2.0 kg is hung from the lower end of a spiral spring and extends it by 0.40 m. When the mass is displaced a further short distance x and released, it oscillates with acceleration a towards the rest position. If $a = -kx$ and if the tension in the spring is always directly proportional to its extension, what is the value of the constant k? (Earth's gravitational field strength is 9.8 N kg^{-1}.)

7. Define *simple harmonic motion* and state where the magnitude of the acceleration is (a) greatest, (b) least.

Some sand is sprinkled on a horizontal membrane which can be made to vibrate vertically with simple harmonic motion. When the amplitude is 0.10 cm the sand just fails to make continuous contact with the membrane. Explain why this phenomenon occurs and calculate the frequency of vibration. ($g = 10$ m s^{-2})   (*J.M.B.*)

8. If a mass hanging on a vertical spring which obeys Hooke's Law is given a small vertical displacement it will oscillate vertically, above and below its equilibrium position. Why is the motion of the mass simple harmonic motion?

Explain why the oscillations of a simple pendulum, consisting of a mass m hanging from a thread of length l, are almost perfectly simple harmonic, and why the deviations from perfect simple harmonic motion become greater if the amplitude of the swing is increased.

A metre ruler is clamped to the top of a table so that most of its length overhangs and is free to vibrate with vertical simple harmonic motion. Calculate the maximum velocity of the tip of the ruler if the amplitude of the vibration is 5.0 cm and the frequency is 4.0 Hz. What is the maximum possible amplitude if a small object placed on the ruler at the vibrating end is to maintain contact with it throughout the vibration? If this amplitude were very slightly exceeded, at what stage in the vibration would contact between the object and the ruler be broken? Explain clearly why it would happen at this point. ($g = 10$ m s^{-2}) (S.)

9. By reference to a particular system explain what is meant by (a) forced vibrations, (b) resonance. What are the effects of damping? (J.M.B.)

10. Because of the effect on the comfort of the ride and the noise experienced by drivers, a manufacturer wishes to investigate the various vibrations (of the body, wheels, springs, door panels and so on) which can arise in motor cars. The manufacturer has two cars, one of which is reported to be much more uncomfortable and noisy than the other.

Draft an outline plan of a programme of tests to:

(a) identify and compare vibrations in the cars,
(b) investigate the reasons for discomfort and noise,
(c) provide guidance for designers of new models.

You may assume that any test equipment needed can be made available but you have to say for what the equipment is needed. Your plan should include an explanation of why the various tests are proposed.

You are asked to plan the tests and *not* to predict their results nor to explain how the car design might be improved.
(O. and C. Nuffield)

9 Fluids at rest

Introduction

Liquids and gases can flow and are called *fluids*. Before considering their behaviour at rest, certain basic terms will be defined, by way of revision.

The *density* ρ of a sample of a substance of mass m and volume V is defined by the equation

$$\rho = \frac{m}{V}$$

In words, density is the *mass per unit volume*. The density of water (at 4 °C) is 1.00 g cm^{-3} or, in SI units, 1.00×10^3 kg m^{-3}; the density of mercury (at room temperature) is about 13.6 g cm^{-3} or 13.6×10^3 kg m^{-3}.

The term *relative density* is sometimes used and is given by

$$relative\ density = \frac{density\ of\ material}{density\ of\ water}$$

It is a ratio and has no unit. The relative density of mercury is thus 13.6.

Fig. 9.1

If a force acts on a surface (like the weight of a brick on the ground) it is often more useful to consider the *pressure* exerted rather than the force. The pressure p caused by a force F acting normally on a surface of area A is defined by

$$p = \frac{F}{A}$$

Pressure is therefore *force per unit area*; the SI unit is the *pascal* (Pa) which equals a pressure of 1 newton per square metre.

Pressure in a liquid

In designing a dam like that in Fig. 9.1 the engineer has to know, among other things, about the size and point of action of the resultant force exerted on the dam by the water behind it. This involves making calculations based on the expression for liquid pressure.

(*a*) *Expression for liquid pressure.* The pressure exerted by a liquid is experienced by any surface in contact with it. It increases with depth because the liquid has weight and we define the *pressure at a point in a liquid as the force per unit area on a very small area round the point*. Thus if the force is δF and the small area δA then the pressure p at the point is given by

$$p = \frac{\text{force}}{\text{area}} = \frac{\delta F}{\delta A}$$

More exactly, in calculus terms, the defining equation is

$$p = \underset{\delta A \to 0}{\text{limit}} \frac{\delta F}{\delta A}$$

Fig. 9.2

An expression for the pressure p at a depth h in a liquid of density ρ can be found by considering an extremely small horizontal area δA, Fig. 9.2. The force δF acting vertically downwards on δA equals the weight of the liquid column of height h and uniform cross-section area δA above it. We can say

volume of liquid column = $h\,\delta A$

mass of liquid column $\quad = h\,\delta A\,\rho$

weight of liquid column $= h\,\delta A\,\rho\,g$

where g is the acceleration due to gravity (or the strength of the earth's gravitational field). Hence

$$\delta F = h\,\delta A\,\rho\,g$$

and $\qquad p = \dfrac{\delta F}{\delta A} = \dfrac{h\,\delta A\,\rho\,g}{\delta A}$

$$\therefore \quad p = h\rho g$$

Thus the pressure at a point in a liquid depends only on the depth and the density of the liquid. If h is in m, ρ in kg m^{-3} and g in m s^{-2} (or N kg^{-1}) then p is in Pa.

Notes. (*i*) In the above derivation, δA was considered to be horizontal but it can be shown that the same result is obtained for any other orientation of δA, i.e. *the pressure at a point in a liquid acts equally in all directions*—as experiment confirms.

(*ii*) The force exerted on a surface in contact with a liquid at rest is *perpendicular* to the surface at all points. Otherwise, the equal and opposite force exerted by the containing surface on the liquid would have a component parallel to the surface which would cause the liquid to flow.

(*b*) *Transmission of pressure.* A liquid can transmit any external pressure applied to it to all its parts. Use is made of this property in the hydraulic press to produce a large force from a small one. In its simplest form it consists of a narrow cylinder connected to a wide cylinder, both containing liquid (usually oil) and fitted with pistons A and B as in Fig. 9.3a. If, for example, A has a cross-sectional area of 1×10^{-4} m^2 (i.e. 1 cm^2) and a downwards force of 1 N is applied to it, a pressure of 1×10^{-4} Pa (i.e. 1 N cm^{-2}) in excess of atmospheric pressure is transmitted through the liquid. If B has a cross-sectional area of 100×10^{-4} m^2 (i.e. 100 cm^2), it experiences an upwards force of 100 N. Piston B acts against a fixed plate and in a large press like that in Fig. 9.3b, it is used to forge motor car parts from steel.

The same principle operates in hydraulic jacks for lifting cars in a garage and also in the hydraulic braking system of a car. In the latter, the force applied to the brake pedal causes a piston to produce an increase of pressure in an oil-filled cylinder and this is transmitted through oil-filled pipes to four other pistons which apply the brake-shoes or discs to the car wheels. This results in the same pressure being applied to all wheels and minimizes the risk of the car pulling to one side or skidding.

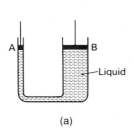

(a)

(b)

Fig. 9.3   (c)

(c) *High pressure water jet cutting*. This is a new cutting technique which is entirely dust-free. It can be used with a wide range of materials such as slate, stone, brake-lining material, Formica, rubber, foams and is especially advantageous for materials like asbestos, the dust from which can cause respiratory disease. In the equipment shown, Fig. 9.3c, the cutting jet is guarded by a Perspex cover, the jet pressure is about 3.5×10^8 Pa (3500 times normal atmospheric pressure), the water flow is 3.0 litres per minute and the width of cut can be varied from about 0.3 mm to 0.9 mm.

Liquid columns

(*a*) U-*tube manometer.* An open U-tube containing a suitable liquid can be used to measure pressures, for example, the pressure of the gas supply, and is called a U-tube manometer. It uses the fact that the pressure in a column of liquid is directly proportional to the height of the column. In the U-tube manometer of Fig. 9.4*a* the pressure *p* to be measured acts on the surface of the liquid at A and balances the pressure of the liquid column BC of height *h*, *plus* atmospheric pressure *P*, acting on B. Hence

$$p = P + h\rho g$$

where ρ is the density of the liquid in the manometer. For small pressures, water or a light oil is used, for medium pressures mercury is suitable. The amount by which *p* exceeds atmospheric pressure, i.e. $(p - P)$, equals the pressure due to the column of liquid BC, i.e. $h\rho g$. Consequently it is often convenient to state a pressure as a number of mm of water or mercury rather than in Pa.

When the absolute pressure is required and not the excess over atmospheric, the limb of the tube open to the atmosphere is replaced by a closed, evacuated one, Fig. 9.4*b*. The height *h* of the liquid column then gives the absolute pressure directly. The principle is used in the measurement of atmospheric pressure by a mercury barometer, the short limb being replaced by a reservoir of mercury, Fig. 9.4*c*. The column of mercury is supported by the pressure of the air on the surface of the mercury in the reservoir and any change in this causes the length of the column to vary.

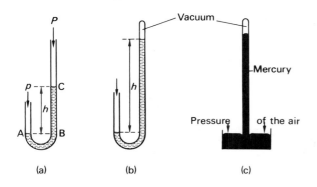

Fig. 9.4

(*b*) *Balancing liquid columns.* If a U-tube contains two immiscible liquids of different densities the surfaces of the liquids are not level. In Fig. 9.5 the column of water AB is balanced by the column of paraffin CD. Hence

pressure at B = pressure at D
$$\therefore \quad h_1 \rho_1 g = h_2 \rho_2 g$$

where h_1, ρ_1 and h_2, ρ_2 are the heights and densities of the water and paraffin respectively. It follows that

$$\frac{\rho_2}{\rho_1} = \frac{h_1}{h_2}$$

Measurement of h_1 and h_2 thus gives a simple way of finding the relative density of paraffin. The limbs of the U-tube need not have the same diameter (if not too small). Why not?

Miscible liquids can be separated by mercury as in Fig. 9.6. In this case the heights of the columns are adjusted by adding liquid until the mercury surfaces are exactly level. Measurements of h_1 and h_2 are then made from this level.

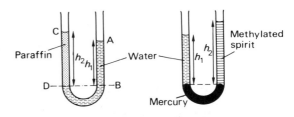

Fig. 9.5　　　　　　　　*Fig. 9.6*

Archimedes' principle

When a body is immersed in a liquid it is buoyed up and appears to lose weight. The upward force is called the *upthrust* of the liquid on the body and is due to the pressure exerted by the liquid on the lower surface of the body being greater than that on the top surface since pressure increases with depth. The law summarizing such effects was discovered over 2000 years ago by Archimedes. It also applies to bodies in gases and is stated as follows.

When a body is completely or partly immersed in a fluid it experiences an upthrust, or apparent loss in weight, which is equal to the weight of fluid displaced.

A numerical case is illustrated in Fig. 9.7*a*; more briefly we can say: *upthrust = weight of fluid displaced.*

The principle can be verified experimentally, or deduced theoretically by considering the pressures exerted by a liquid on the top and bottom surfaces of a

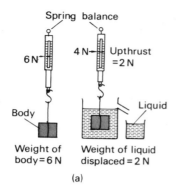

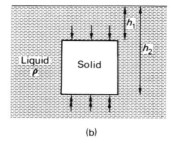

Fig. 9.7   (a)   (b)

rectangular-shaped solid, Fig. 9.7b—as you can confirm for yourself.

(*a*) *Floating bodies.* If a body floats partly immersed in a liquid (e.g. a ship), completely immersed in a liquid (e.g. a submarine) or a gas (e.g. a balloon), it appears to have zero weight and we can say

upthrust on body = weight of floating body

By Archimedes' principle

upthrust on body = weight of fluid displaced

Hence

weight of floating body = weight of fluid displaced

This result, sometimes called the 'principle of flotation', is a special case of Archimedes' principle and can be stated thus.

A floating body displaces its own weight of fluid.

If the body cannot do this, even when completely immersed, it sinks.

The stability of a floating body such as a ship when it heels over depends on the relative positions of the ship's centre of gravity G, through which its weight W acts, and the centre of gravity of the displaced water,

called the *centre of buoyancy* B, through which the upthrust U acts. In Fig. 9.8*a* the ship is on an even keel and B and G are on the same vertical line. If it heels over the shape of the displaced water changes, causing B to move and thereby setting up a couple which tends either to return the ship to its original position or to make it heel over more. The point of intersection of the vertical line from B with the central line of the ship is called the *metacentre* M. If M is above G as in Fig. 9.8*b*, the couple has an anticlockwise moment which acts to decrease the ship's heel and the equilibrium is stable. If M is below G as in Fig. 9.8*c*, equilibrium is unstable since the couple has a clockwise moment which causes further listing. For maximum stability G should be low and M high.

(*b*) *Hydrometer.* This is an instrument which uses the principle of flotation to give a rapid measurement of the relative density of a liquid, e.g. the acid in a lead accumulator. It consists of a narrow glass stem, a large buoyancy bulb and a smaller bulb loaded with lead shot to keep it upright when floating, Fig. 9.9. The relative density is found by floating the hydrometer in the liquid and taking the reading on the scale inside the stem at the level of the liquid surface. The instrument shown is for use in the range 1.30 to 1.00, the numbers increase downwards (why?) and the scale is uneven.

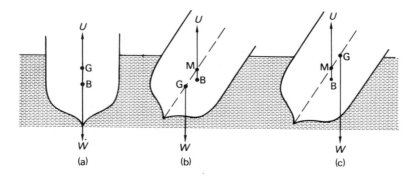

Fig. 9.8   (a)   (b)   (c)

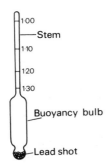

1·00

Stem

1·10

1·20

1·30

Buoyancy bulb

Fig. 9.9

Lead shot

Atmospheric pressure

The pressure due to a gas arises from the bombardment of the walls of the containing vessel by its molecules. In a small volume of gas the pressure is uniform throughout but in a large volume such as the atmosphere, gravity causes the density of the gas and therefore its pressure to be greater in the lower regions than in the upper regions. In fact atmospheric pressure at a height of about 6 km is half its sea-level value even though the atmosphere extends to a height of 150 km or so.

(a) *Value.* The statement that atmospheric pressure is '760 mm of mercury' means that it equals the pressure at the bottom of a column of mercury 760 mm high. Sometimes pressure is expressed in torrs (after Torricelli who made the first mercury barometer), 1 torr = 1 mmHg. To find the value of atmospheric pressure in SI units we use $p = h\rho g$ where $h = 0.760$ m, $\rho = 13.6 \times 10^3$ kg m^{-3}, $g = 9.81$ N kg^{-1}. Thus

$$p = 0.760 \times 13.6 \times 10^3 \times 9.81 \ \text{m} \times \text{kg m}^{-3} \times \text{N kg}^{-1}$$

$$= 1.01 \times 10^5 \ \text{Pa}$$

Standard pressure or 1 atmosphere is defined as the pressure at the foot of a column of mercury 760 mm high of specified density and subject to a particular value of g; it equals 1.01325×10^5 Pa.

(b) *Temperature corrections to the barometric height.* The temperatures at weather stations are generally different and before their pressure readings are compared the barometric height at each is often 'reduced' to the height it would have been at 0 °C. This may involve correcting for

(i) the change in density of mercury with temperature and
(ii) the metal scale on the barometer being used at a temperature other than its calibration temperature.

Consider (i). Suppose H_θ is the height recorded at the station at temperature θ and H_0 is the height of the mercury column to give the *same* pressure when the mercury is at 0 °C. We have

$$H_\theta \rho_\theta g = H_0 \rho_0 g$$

where ρ_θ and ρ_0 are the densities of mercury at θ and 0 °C respectively. Hence

$$H_0 = H_\theta \times \frac{\rho_\theta}{\rho_0}$$

But $\qquad\qquad \rho_0 = \rho_\theta(1 + \gamma\theta) \qquad$ (see p. 77)

where γ is the real cubic expansivity of mercury

$$\therefore \quad H_0 = \frac{H_\theta}{1 + \gamma\theta}$$

Consider (ii). If the scale is correct at 0 °C, the true reading at θ will be $H_\theta(1 + \alpha\theta)$ and not H_θ (see p. 82, question 9) where α is the linear expansivity of the metal scale. The final corrected height H_0 is given by

$$H_0 = \frac{H_\theta(1 + \alpha\theta)}{1 + \gamma\theta}$$

A numerical example will illustrate the use of this expression. Suppose the height of a mercury barometer read with a steel scale which is correct at 0 °C is 754 mm at 20 °C. What is the height reduced to 0 °C? Taking $\alpha_{\text{steel}} = 1.20 \times 10^{-5}$ K^{-1} and $\gamma_{\text{mercury}} = 1.80 \times 10^{-4}$ K^{-1} and substituting in the expression we get

$$H_0 = \frac{754(1 + 1.20 \times 10^{-5} \times 20)}{1 + 1.80 \times 10^{-4} \times 20}$$

$$= \frac{754(1 + 2.40 \times 10^{-4})}{1 + 3.60 \times 10^{-3}}$$

$$\simeq 754(1 + 2.40 \times 10^{-4})(1 - 3.60 \times 10^{-3})$$

since by the Binomial theorem $1/(1 + x) \simeq (1 - x)$ if x is small; therefore

$$H_0 \simeq 754(1 + 2.40 \times 10^{-4} - 3.60 \times 10^{-3})$$

neglecting $(2.40 \times 10^{-4}) \times (3.60 \times 10^{-3})$. Therefore

$$H_0 \simeq 754 \times 0.997$$

$$\simeq 752 \ \text{mm}$$

Pressure gauges and vacuum pumps

The U-tube manometer can be used to measure moderate pressures, i.e. those in the region of atmospheric. High and low pressures require special gauges.

(a) *Bourdon gauge.* This measures pressures up to about 2000 atmospheres. It consists of a curved metal tube, sealed at one end, to which the pressure to be measured is applied, Fig. 9.10. As the pressure increases the tube uncurls and causes a rack and pinion to move a pointer over a scale.

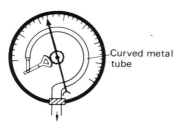

Fig. 9.10   To unknown pressure

(b) *McLeod gauge.* It is used to measure the very low pressures produced by vacuum pumps. The principle is to compress a sample of the gas whose pressure is required until its pressure is measurable on a mercury manometer and then, knowing the initial and final volumes of the gas, to apply Boyle's law.

A measurement is made by lowering the mercury reservoir on the gauge thus allowing gas from the evacuated vessel at unknown pressure p to enter bulb B and capillary tube C, Fig. 9.11. It is next raised until the mercury reaches a fixed mark M on C, Fig. 9.12. The pressure of the compressed gas is then $p + h$. If V is the volume of B and all C and v is the volume of C above mark M, applying Boyle's law we have

$$pV = (p + h)v$$

Hence p can be found since the values of V and v are determined during manufacture of the gauge and h is read off directly. The capillary tube C has the same internal diameter (bore) as D thereby eliminating errors due to surface tension (see p. 199) when h is measured.

If V/v is of the order of 10^4, what will be the value of h when the pressure is 10^{-3} mmHg (torr)? The McLeod gauge can be used down to about 10^{-4} mmHg; for lower pressures (down to about 10^{-12} mmHg) an instrument called an *ion gauge* is used.

(c) *Rotary vacuum pump.* One rotary pump can produce pressures of about 10^{-2} mmHg; with two in series 10^{-4} mmHg is attainable. The pump comprises an eccentrically mounted cylindrical rotor inside and in contact at one point with a cylindrical stator, Fig. 9.13. Two spring-loaded vanes attached to the rotor press against the wall of the stator. The whole is immersed in a special low vapour pressure oil which both seals and lubricates the pump. As the rotor revolves, driven by an electric motor, each vane in turn draws gas into the increasing volume of space A on the intake side and then compresses it in space B where it is ejected from the outlet valve.

Lower pressures, down to 10^{-12} mmHg, require the use of a *diffusion pump*.

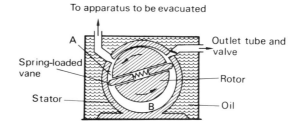

Fig. 9.13

Surface tension

(a) *Some effects.* Various effects suggest that the surface of a liquid behaves like a stretched elastic skin, i.e. it is in a state of tension. For example, a steel needle will float if it is placed *gently* on the surface of a bowl of water, despite its greater density. (What else helps to support its weight?) The effect, called *surface tension*, enables certain insects to run over the surface of a pond without getting wet, Fig. 9.14a.

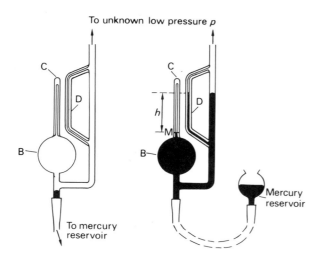

Fig. 9.11   *Fig. 9.12*

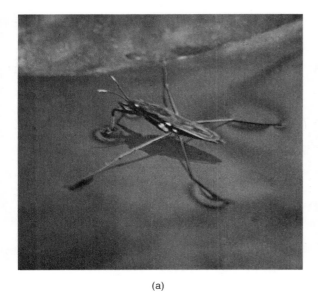

(a)

Fig. 9.14   (b)

Small liquid drops are nearly spherical, as can be seen when water drips from a tap, Fig. 9.14*b*; a sphere has the minimum surface area for a given volume. (What distorts larger drops?) This tendency of a liquid surface to shrink and have a minimum area can also be shown by the arrangement of Fig. 9.15*a*. When the soap film *inside* the loop of thread is punctured, the thread is pulled into the shape of a circle, Fig. 9.15*b*. Since a circle has the maximum area for a given perimeter, the area of the film outside the thread is a minimum.

(*b*) *Definition and unit.* We can conclude from the circular shape of the thread in the previous demonstration that the liquid (soap solution) is pulling on the thread at right angles all along its circumference, Fig. 9.16*a*. This suggests that we might define the surface tension of a liquid in the following way.

Imagine a straight line of length *l* in the surface of a liquid. If the force acting at right angles to this line and

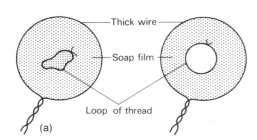

Fig. 9.15

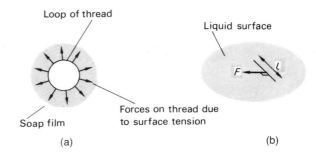

Fig. 9.16

in the surface is F, Fig. 9.16*b*, then the surface tension γ of the liquid is defined by

$$\gamma = \frac{F}{l}$$

In words, γ *is the force per unit length acting in the surface perpendicular to one side of a line in the surface.* The unit of γ is newton metre^{-1} (N m^{-1}). Its value depends on, among other things, the temperature of the liquid. At 20 °C, for water $\gamma = 72.6 \times 10^{-3}$ N m^{-1} and for mercury $\gamma = 465 \times 10^{-3}$ N m^{-1}.

It must be emphasized that normally surface tension acts equally on both sides of *any* line in the surface of a liquid; it creates a state of tension in the surface. The *effects* of surface tension are evident only when liquid is absent from one side of the line. For example in Fig. 9.17, to keep wire AB at rest an external force F has to be applied to the right to counteract the unbalanced surface tension forces acting to the left. A film has two surfaces and so for a frame of width l, the surface tension force is $2\gamma l$ where γ is the surface tension of the liquid.

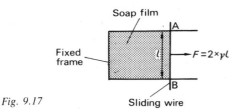

Fig. 9.17

Or again, when a drop of methylated spirit or soap solution is dropped into the centre of a dish of water whose surface has been sprinkled with lycopodium powder, the powder rushes out to the sides leaving a clear patch. The effect is due to the surface tension of water being greater than that of meths or soap solution so that there is unbalance between the surface tension forces round the boundary of the two liquids. The powder is thus carried away from the centre by the water.

(*c*) *Molecular explanation.* Molecules in the surface of a liquid are farther apart than those in the body of the liquid, i.e. the surface layer has a lower density than the liquid in bulk. This follows because the increased separation of molecules which accompanies a change from liquid to vapour is not a sudden transition. The density of the liquid must therefore decrease *through* the surface.

The intermolecular forces in a liquid, like those in a solid, are both attractive and repelling and these bal-

ance when the spacing between molecules has its equilibrium value. However, from the intermolecular force–separation curve (Fig. 4.11*a*, p. 75) we see that when the separation is greater than the equilibrium value (r_0), the attractive force between molecules exceeds the repelling force. This is the situation with the more widely spaced surface layer molecules of a liquid. They experience attractive forces on either side due to their neighbours which puts them in a state of tension, Fig. 9.18. The liquid surface thus behaves like a stretched elastic skin. If the tension-creating bonds between molecules are severed on one side by parting the liquid surface, then there is a resultant attractive force on the molecules due to the molecules on their other side. The effect of surface tension is then apparent.

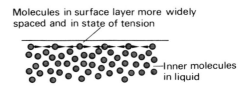

Fig. 9.18

The value of γ for a liquid does not increase when its surface area increases because more molecules enter the surface layer thereby keeping the molecular separation constant. Otherwise, any increase of separation would increase the attractive force between molecules and so also the surface tension.

Liquid surfaces

(*a*) *Shape of liquid surfaces.* The surface of a liquid must be at right angles to the resultant force acting on it, otherwise there would be a component of this force parallel to the surface which would cause motion. Normally a liquid surface is horizontal, i.e. at right angles to the force of gravity, but where it is in contact with a solid it is usually curved.

To explain the shape of the surface in Fig. 9.19 consider the liquid at B adjoining a vertical solid wall.

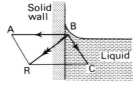

Fig. 9.19

It experiences an attractive force BC due to neighbouring liquid molecules; this is the *cohesive* force studied previously which binds liquid molecules together and makes them behave as a liquid. An attractive force BA is also exerted by neighbouring molecules of the solid; this is called the *adhesive* force and if it is greater than the cohesive force then the resultant force BR on the liquid at B will act to the left of the wall in the direction shown. The liquid surface at B has to be at right angles to this direction and so curves upwards. Since there is then equilibrium the resultant force must be balanced by appropriate intermolecular repulsive forces. At points on the liquid surface farther from the wall, the adhesive forces are smaller, the resultant force more nearly vertical and so the surface more nearly horizontal.

By contrast, when the cohesive force between molecules of the liquid is greater than the adhesive force between molecules of the liquid and molecules of the solid, the resultant force BR acts as in Fig. 9.20 and the surface curves downwards at the wall. This is the case with mercury against glass.

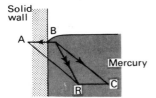

Fig. 9.20

The *angle of contact* θ is defined as the angle between the solid surface and the tangent plane to the liquid surface, measured *through the liquid*. The liquid in Fig. 9.21a has an acute angle of contact with this particular solid ($\theta < 90°$), while that in Fig. 9.21b has an obtuse angle of contact ($\theta > 90°$). Water, like many organic liquids, has zero angle of contact with a *clean* glass surface, i.e. the adhesive force is so much greater than the cohesive force that the water surface is parallel to the glass where it meets it, Fig. 9.21c. On a horizontal clean glass surface water tends to spread indefinitely and form a very thin film. Contamination of a surface

affects the angle of contact appreciably; the value for water on greasy glass may be about 10° and causes it to form drops rather than spread. Mercury in contact with clean glass has an angle of contact of about 140° and tends to form drops instead of spreading over glass.

Liquids with acute angles of contact are said to 'wet' the surface, those with obtuse angles of contact do not 'wet' it. Fig. 9.22a shows a drop of a liquid which 'wets' the surface and b shows a drop on a surface which it does not 'wet'.

Fig. 9.22

(*b*) *Practical applications of spreading*. The behaviour of liquids in contact with solids is important practically. In soldering a good joint is formed only if the molten solder (a tin-lead alloy) 'wets' and spreads over the metal involved. Spreading occurs most readily if the liquid solder has a small surface tension. The use of a flux (e.g. resin) with the solder cleans the metal surface and acts as a 'wetting agent' which assists spreading. Metals like aluminium have an almost permanent oxide skin that resists the action of a flux and makes good soldered joints by normal methods very difficult.

'Wetting agents' play a key role in painting and spraying where the paint must not form drops but remain in a layer once spread out. The use of spreading agents (e.g. stearic acid) also assists lubricating oils to adhere to axles, bearings, etc.

If detergents are to remove the dirt particles that are held to fabrics usually by grease, they must be able to spread over the fabric before they can dislodge the grease. Detergent solutions should therefore, on this account, have low surface tensions and small angles of contact. By contrast, fabrics are weatherproofed by treatment with a silicone preparation which causes water to collect in drops and not to spread.

Capillarity

Surface tension causes a liquid with an angle of contact less than 90° to rise in a fine bore (capillary) tube above the level outside. The narrower the tube the greater the elevation, Fig. 9.23. The effect is called *capillarity* and is of practical importance.

Why does the rise occur? In Fig. 9.24a, round the boundary where the liquid surface meets the tube,

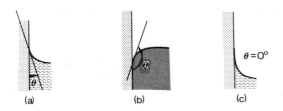

Fig. 9.21

Fig. 9.23

surface tension forces exert a *downwards* pull on the tube since they are not balanced by any other surface tension forces. The tube therefore exerts an equal but *upwards* force on the liquid, Fig. 9.24b, and causes it to rise (Newton's third law of motion). The liquid stops rising when the weight of the raised column acting vertically downwards equals the *vertical component* of the upwards forces exerted by the tube on the liquid, Fig. 9.24c.

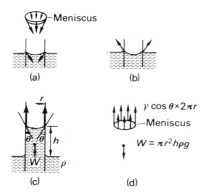

Fig. 9.24

If the liquid has density ρ, surface tension γ and angle of contact θ and if the rise is h in a tube of radius r then neglecting the small amount of liquid in the meniscus,

weight W of liquid column $= \pi r^2 h \rho g$

vertical component of supporting forces
$$= \gamma \cos \theta \times 2\pi r$$

since these forces act round a circumference of $2\pi r$, Fig. 9.24d. Hence

$$\pi r^2 h \rho g = 2\pi r\, \gamma \cos \theta$$

$$\therefore \quad h = \frac{2\gamma \cos \theta}{r\rho g}$$

A more rigorous treatment shows that this equation only holds for very fine bore tubes in which the curvature of the meniscus is everywhere spherical; it

also brings out the fact that r is the radius of the tube at the meniscus (p. 202).

An estimate of the rise h can be obtained by substituting values for γ, r, ρ and g in this expression. For example, for water in a very clean glass tube $\theta = 0°$, i.e. the water surface meets the tube vertically, and so $\cos \theta = 1$. Also $\rho = 1.0$ g cm^{-3} $= 1.0 \times 10^3$ kg m^{-3} and $\gamma = 7.3 \times 10^{-2}$ Nm^{-1}. If the capillary tube has radius 0.50×10^{-3} m then

$$h = \frac{2 \times 7.3 \times 10^{-2} \times 1}{0.50 \times 10^{-3} \times 1.0 \times 10^3 \times 9.8} \quad \frac{\text{N m}^{-1}}{\text{m kg m}^{-3}\,\text{N kg}^{-1}}$$

$$= 3.0 \times 10^{-2}\ \text{m}$$

$$= 30\ \text{mm}$$

If θ is greater than 90°, the meniscus is convex upwards, $\cos \theta$ is negative and the expression shows that h will also be negative. This means the liquid falls in the capillary tube below the level of the surrounding liquid. Mercury in a glass capillary tube usually behaves in this way, Fig. 9.25. What happens when $\theta = 0°$?

Fig. 9.25

The drying action of blotting paper is due to the ink rising up the pores of the paper by capillarity. It also helps in soldering by causing the molten solder to penetrate any cracks. However, for this to happen the above expression for h indicates that the solder should have a high surface tension, a property which does not encourage spreading (see p. 199). Compromise is clearly necessary here as in the dyeing of fabrics where success depends largely on the dye penetrating into the fabric by capillarity.

Bubbles and drops

A study of bubbles and drops not only helps with the determination of the surface tension of liquids, as we shall see in a later section, but it also has practical relevance. Thus, the formation of gas bubbles plays an important part in the manufacture of expanded plastics such as polystyrene. In oil-fired boilers pressure burners depend on droplet formation for fast and efficient burning of the vapour. In steam heating systems the efficiency of heat transfer from the steam would be higher if instead of condensing as a film, which it does, it condensed in drops, and attempts are at present being made to achieve drop condensation.

A soap bubble blown on the end of a tube and then left open to the atmosphere gradually gets smaller, showing that the air is being forced out. Surface tension tries to make the film contract and thereby causes the pressure inside the bubble to exceed that outside. An expression can be obtained for the *excess pressure* inside a spherical soap bubble.

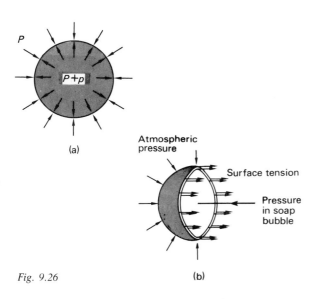

Fig. 9.26

Consider a bubble of radius r, blown from a soap solution of surface tension γ. Let atmospheric pressure be P and suppose the pressure inside the bubble exceeds P by p, i.e. is $(P + p)$, Fig. 9.26a. Consider the equilibrium of *one half* of the bubble; there are two sets of opposing forces.

(*i*) Atmospheric pressure acts in different directions over the surface of the hemisphere but the resultant force in Fig. 9.26b acts *horizontally to the right* since the vertical components cancel (if any vertical variation of atmospheric pressure is neglected). It can be shown (see later *Note*) that the force exerted by a fluid in a certain direction on a curved surface equals the force on the projection of the surface on to a plane whose direction is perpendicular to the required direction. Here the projection of the hemisphere in a horizontal direction is a circle and so the horizontal force due to atmospheric pressure equals the product of the pressure and the area of projection i.e. $P \times \pi r^2$. Also, surface tension forces are exerted by the right-hand

hemisphere (not shown in Fig. 9.26b) on the circular rim of the left-hand hemisphere along *both* its inside and outside surfaces. (Similar surface tension forces are exerted on the right-hand hemisphere by the left-hand one.) This force equals $2\gamma \times 2\pi r$ and so the total horizontal force to the right is $P\pi r^2 + 4\gamma\pi r$.

(*ii*) The pressure $(P + p)$ acts on the curved inside surface of the left-hand hemisphere and produces a *horizontal force to the left* equal to $(P + p)\pi r^2$.

Hence, if the horizontal forces balance, we have

$$P\pi r^2 + 4\gamma\pi r = (P + p)\pi r^2$$

$$\therefore \quad 4\gamma\pi r = p\pi r^2$$

The excess pressure p inside the bubble is then

$$p = \frac{4\gamma}{r}$$

Taking γ for a soap solution as 2.5×10^{-2} N m^{-1}, the excess pressure inside a bubble of radius 1.0 cm or 1.0×10^{-2} m is

$$p = \frac{4 \times 2.5 \times 10^{-2}}{1.0 \times 10^{-2}} \frac{\text{N m}^{-1}}{\text{m}}$$

$$= 10 \text{ Pa}$$

If two soap bubbles of different radii are blown separately using the apparatus of Fig. 9.27, and then connected by opening taps T_1 and T_2 (T_3 being closed), the smaller bubble A gradually collapses while the larger bubble B expands. Why? Equilibrium is attained when A has become a small curved film of radius equal to that of bubble B.

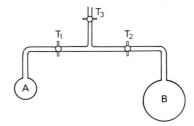

Fig. 9.27

A spherical drop of liquid in air or a bubble of gas in a liquid has only one surface and the excess pressure inside it is $2\gamma/r$, the proof being similar to that given above for a soap bubble.

Note. Force on a curved surface in a fluid exerting a uniform pressure. Consider a volume of fluid with end A hemispherical and end B plane, as in Fig. 9.28. The

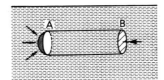

Fig. 9.28

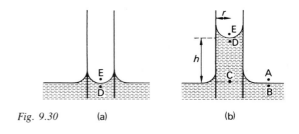

Fig. 9.30   (a)   (b)

force on surface B is horizontal and must equal the horizontal component of the resultant force on surface A (the vertical components at end A cancel) if the volume of fluid is in equilibrium. But surface B is the projection of surface A on a vertical plane, i.e. in a direction at right angles to the horizontal. Hence the force in, say, a horizontal direction on a curved surface equals the force on its projection on to a vertical plane.

Pressure difference across a spherical surface

It can be shown (in more advanced books) that due to surface tension, the pressure on the concave side of any spherical liquid surface of radius r exceeds that on the convex side by $2\gamma/r$ where γ is the surface tension of the liquid, Fig. 9.29. Bubbles and drops are special cases of this more general result which is useful when considering certain effects such as capillarity. Earlier we treated capillarity in terms of forces; the excess pressure method is more informative.

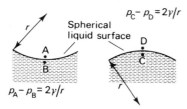

Fig. 9.29

(a) *Capillary rise formula.* Consider a liquid of surface tension γ in a capillary tube. If the meniscus is everywhere spherical (as it will be in a very narrow tube) and if the angle of contact is zero, then the radius of the meniscus will equal the radius of the tube.

When the tube is first placed in the liquid, the curvature of the surface makes the pressure at D just below the meniscus less than that at E, which is atmospheric, Fig. 9.30a. Liquid will flow into the tube because of this pressure difference and capillary rise occurs.

Let p_A, p_B, p_C, p_D and p_E be the pressures at A, B, C,

D and E respectively, Fig. 9.30b. We have, once equilibrium is established,

$$p_A = p_B$$

(no pressure difference across a flat surface)

$$= p_C$$

$$= p_D + h\rho g$$

where ρ is the density of the liquid, r the radius of the tube and h the capillary rise. Also,

$$p_E = p_D + \frac{2\gamma}{r}$$

But $p_E = p_A$ (both are atmospheric pressure which is constant if we ignore the pressure due to the column AE of air)

$$\therefore \quad h\rho g = \frac{2\gamma}{r}$$

$$\therefore \quad h = \frac{2\gamma}{r\rho g}$$

If the angle of contact θ is zero this expression is the same as that obtained previously by the force method (p. 200). However, this derivation shows that the expression only holds strictly for tubes in which the meniscus is everywhere spherical, i.e. for fine bore tubes. The fact that r is the radius of the tube at the meniscus is also made clear.

(b) *Worked example.* A U-tube with limbs of diameters 5.0 mm and 2.0 mm *contains water of surface tension* 7.0×10^{-2} N m^{-1}, *angle of contact zero and density* 1.0×10^{3} kg m^{-3}. *Find the difference in levels, Fig. 9.31* ($g = 10$ m s^{-2}).

If the menisci are spherical they will be hemispheres since the angle of contact is zero; their radii will then equal the radii of the limbs. The pressure on the concave side of each surface exceeds that on the convex side by $2\gamma/r$ where γ is the surface tension and r is the radius of the limb concerned.

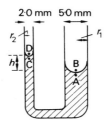

Fig. 9.31

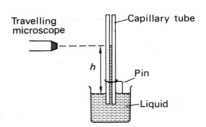

Fig. 9.32

Now $r_1 = 2.5$ mm $= 2.5 \times 10^{-3}$ m and $r_2 = 1.0$ mm $= 1.0 \times 10^{-3}$ m. Hence

$$p_B - p_A = \frac{2 \times 7.0 \times 10^{-2}}{2.5 \times 10^{-3}} = 56 \text{ Pa}$$

$$\therefore \quad p_A = P - 56$$

where $p_B = P$ = atmospheric pressure. Also

$$p_D - p_C = \frac{2 \times 7.0 \times 10^{-2}}{1.0 \times 10^{-3}} = 140 \text{ Pa}$$

$$\therefore \quad p_C = P - 140 \quad \text{(since } p_D = P\text{)}$$

$$\therefore \quad p_A - p_C = (P - 56) - (P - 140)$$

$$= 84 \text{ Pa}$$

But

$$p_A = p_C + h\rho g$$

$$\therefore \quad h\rho g = 84 \text{ Pa}$$

$$\therefore \quad h = \frac{84}{10^3 \times 10} \text{ m}$$

$$= 8.4 \text{ mm}$$

Methods of measuring γ

(a) *Capillary rise method.* The expression $\gamma = hr\rho g/(2 \cos \theta)$ is used and so θ must be known. In the case of a liquid for which $\theta = 0°$ the expression becomes

$$\gamma = \frac{hr\rho g}{2}$$

Knowing the density ρ of the liquid and g, only h and r remain to be determined.

The apparatus is shown in Fig. 9.32, the capillary tube is previously cleaned thoroughly by immersing in caustic soda, dilute nitric acid and distilled water in turn. A travelling microscope is focused first on the bottom of the meniscus in the tube and then, with the beaker removed, on the tip of the pin which previously just touched the surface of the liquid in the beaker. Hence h is obtained.

To find r the tube is broken at the meniscus level and the average reading of two diameters at right angles taken with the travelling microscope.

Surface tension decreases rapidly with temperature and so the temperature of the liquid should be stated.

(b) *Jaeger's method.* This method measures the excess pressure required to blow an air bubble in the liquid under investigation and then γ is calculated using $p = 2\gamma/r$.

The pressure inside the apparatus, Fig. 9.33a, is gradually increased by allowing water to enter the flask from the dropping funnel and the increase is recorded on the manometer (containing a low density liquid such as xylol). An air bubble grows at the end of the capillary tube in the beaker of test liquid and as it does so the pressure rises to a maximum and then falls as the bubble breaks away. The maximum pressure will occur when the radius of the bubble is a minimum. Assuming that the bubble is then hemispherical with radius equal to that of the bore of the tube, Fig. 9.33b,

$$\text{pressure inside bubble} = P + h\rho g$$

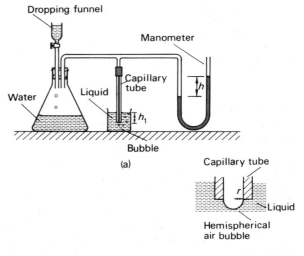

Fig. 9.33

(a)

(b)

where P is the atmospheric pressure, h the *maximum* manometer reading and ρ is the density of the liquid in the manometer. Also

$$\text{pressure in liquid outside bubble} = P + h_1\rho_1 g$$

where h_1 is the depth of the end of the capillary tube in the test liquid of density ρ_1. Hence

$$\text{excess pressure in bubble} = (P + h\rho g) - (P + h_1\rho_1 g)$$
$$= h\rho g - h_1\rho_1 g$$

But, excess pressure $= \dfrac{2\gamma}{r}$

where γ is the surface tension of the liquid and r is the radius of the tube at the end.

$$\therefore \quad \frac{2\gamma}{r} = (h\rho - h_1\rho_1)\,g$$

$$\therefore \quad \gamma = \frac{gr}{2}(h\rho - h_1\rho_1)$$

It is essential to measure h, h_1 and r carefully. One way of obtaining h is to arrange two pins so that their points mark the liquid levels in the manometer at the instant of maximum pressure and then to measure the distance between them afterwards with a travelling microscope. The same instrument should be used to find h_1 and r. Best results are achieved when a bubble is formed every few seconds.

Every bubble has a fresh surface and so the risk of contamination is small if the tube is clean. Also, measurements at different temperatures are easily made by changing and maintaining at any required value the temperature of the liquid in the beaker. The method is most suitable for accurate comparisons between different liquids and for one liquid at different temperatures. Absolute measurements are not reliable because the assumption that the minimum bubble radius equals the radius of the tube is not quite true.

Surface energy

(*a*) *Definition.* Molecules in the surface of a liquid are farther apart than those inside the liquid (p. 198). The p.e.-separation curve for two molecules (Fig. 1.6*b*, p. 9) shows that if this is so, then the mutual intermolecular p.e. of surface molecules is greater, i.e. less negative, than that of molecules in the interior.

When a new surface is formed, energy must therefore be supplied by an external agent to increase the separation of the new surface molecules. This energy becomes the molecular p.e. of these molecules and is called the (free) *surface energy* of the liquid. It is denoted by σ and defined by the equation

$$\sigma = \frac{W}{A}$$

where W is the energy required to create a new area A of surface, i.e. it is the *energy needed to create unit area of new surface*. Its unit is J m^{-2}. Included in W is any energy taken in from the surroundings to keep the surface temperature constant, i.e. the new surface is formed under isothermal conditions.

(*b*) *Bonding energy and surface energy.* An expression relating these two quantities can be derived from theoretical considerations.

We have seen (p. 9) that the bonding energy E_o for two molecules is the energy needed to break the bonds between them. If each molecule in a liquid has n near neighbours, nE_o is the energy required to break the bonds between one molecule and its neighbours.

When a molecule is pulled from the interior of a liquid to form a new surface, bonds are broken and remade during the process. At the surface only about *half* its bonds have been remade and so the energy supplied to pull away the other half is roughly $\frac{1}{2}nE_o$. If the new surface contains N molecules per unit area, there are $N/2$ pairs of molecules (see p. 72) and so the total energy needed to create unit area of new surface. i.e. σ is given by

$$\sigma = \tfrac{1}{2}N \times \tfrac{1}{2}nE_o = \tfrac{1}{4}nNE_o$$

Substitution of numerical values for a particular liquid gives an answer for σ of the right rough order of magnitude.

(*c*) *Surface tension and surface energy.* We will show that these two quantities are equivalent.

In Fig. 9.34 a film of liquid, of surface tension γ, is shown stretched across a horizontal frame PQRS. The force on the sliding wire PQ of length l is $\gamma \times 2l$, since the film has two surfaces. If PQ is moved by an external

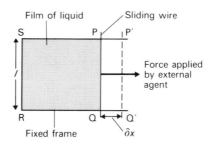

Fig. 9.34

force to P'Q' through a distance δx against the surface tension, the new area of surface $A = 2l \times \delta x$ (the film has two sides), and

work done W to enlarge surface $= 2\gamma l \times \delta x$

This equals the increase in the surface energy σ, and we then have

$$\sigma = \frac{W}{A} = \frac{2\gamma l \times \delta x}{2l \times \delta x} = \gamma$$

Thus, σ and γ are numerically equal; the latter is the more useful idea in practice, as we have seen.

QUESTIONS

Pressure: Archimedes

1. Define *pressure at a point* in a fluid. In what unit is it measured?

State an expression for the pressure at a point at depth h in a liquid of density ρ. Does it also hold for a gas?

What force is exerted on the bottom of a tank of uniform cross-section area 2.0 m² by water which fills it to a depth of 0.50 m? (Density of water = 1.0×10^3 kg m⁻³; $g =$ 10 N kg⁻¹.)

Find the extra force on the bottom of the tank when a block of wood of volume 1.0×10^{-1} m³ and relative density 0.50 floats on the surface.

2. (*a*) State Archimedes' principle.

(*b*) A string supports a solid copper block of mass 1 kg (density 9×10^3 kg m⁻³) which is completely immersed in water (density 1×10^3 kg m⁻³). Calculate the tension in the string.

3. A specimen of an alloy of silver and gold, whose densities are 10.50 and 18.90 g cm⁻³ respectively, weighs 35.20 g in air and 33.13 g in water. Find the composition, by mass, of the alloy, assuming that there has been no volume change in the process of producing the alloy. (*W. part qn.*)

4. A simple hydrometer, consisting of a loaded glass bulb fixed at the bottom of a glass stem of uniform cross section, sinks in water of density 1.0 g cm⁻³ so that a certain mark X on the stem is 4.0 cm below the surface. It sinks in a liquid of density 0.90 g cm⁻³ until X is 6.0 cm below the surface. It is then placed in a liquid of density 1.1 g cm⁻³. How far below the surface will X be? (Neglect surface tension effects.) (*S.*)

5. A barometer is exactly 760.0 mm high with the temperature at 20 °C. What would be its height for the same pressure if the temperature were 0 °C?

Calculate the expansion of a brass scale 760.0 mm long for a temperature rise of 20 °C.

What height would have been measured for the barometer when it was exactly 760.0 mm high at 20 °C if a brass scale had been used which was correct at 0 °C?

(Linear expansivity of brass 1.800×10^{-5} K⁻¹; cubic expansivity of mercury 1.800×10^{-4} K⁻¹.) (*S.*)

6. A simple reciprocating exhaust pump has a piston area 0.002 m² and a stroke length 0.2 m. It is directly connected to a vessel of volume 0.01 m³, containing air at atmospheric pressure.

Calculate the minimum number of strokes needed to reduce the pressure in the vessel to 0.01 atmosphere, assuming isothermal conditions.

Discuss briefly the validity of this assumption and state any further assumptions made. (*J.M.B. Eng. Sc.*)

Surface tension

7. Explain (*a*) in terms of molecular forces why the water is drawn up above the horizontal liquid level round a steel needle which is held vertically and partly immersed in water, (*b*) why, in certain circumstances, a steel needle will rest on a water surface. In each case show the relevant forces on a diagram. (*J.M.B.*)

8. Explain, using a simple molecular theory, why the surface of a liquid behaves in a different manner from the bulk of the liquid.

Giving the necessary theory, explain how the rise of water in a capillary tube may be used to determine the surface tension of water.

A microscope slide measures 6.0 cm × 1.5 cm × 0.20 cm. It is suspended with its face vertical and with its longest side horizontal and is lowered into water until it is half immersed. Its apparent weight is then found to be the same as its weight in air. Calculate the surface tension of water assuming the angle of contact to be zero. (*A.E.B.*)

9. A clean glass capillary tube of internal diameter 0.60 mm is held vertically with its lower end in water and with 80 mm of the tube above the surface. How high does the water rise in the tube?

If the tube is now lowered until only 30 mm of its length is above the surface, what happens? Surface tension of water is 7.2×10^{-2} N m⁻¹.

10. Describe and explain *two* experiments of a different nature to illustrate the phenomenon of surface tension.

Give a quantitative definition of *surface tension* and explain what is meant by *angle of contact*.

The internal diameter of the tube of a mercury barometer is 3.00 mm. Find the corrected reading of the barometer after allowing for the error due to surface tension, if the observed reading is 76.56 cm. (Surface tension of mercury = 4.80×10^{-1} N m⁻¹; angle of contact of mercury with glass = 140°; density of mercury = 13.6×10^3 kg m⁻³.) (*L.*)

10 Fluids in motion

Introduction

The study of moving fluids is important in engineering. A large quantity of liquid may have to flow rapidly through a pipe from one location to another or air entering the inlet of a machine, e.g. a jet engine, may have to be transported to the outlet, undergoing changes of pressure, temperature and speed as it passes. In all such cases of *mass transport* a knowledge of the conditions existing at various points in the system is essential for efficient design.

Fluid dynamics is a complex subject which we shall only touch upon in this chapter.

Viscosity

If adjacent layers of a material are displaced laterally over each other as in Fig. 10.1*a*, the deformation of the material is called a *shear*. Basically the simplest type of fluid flow involves shear as we shall now see.

All liquids and gases (except *very* low density gases) stick to a solid surface so that when they flow the velocity must gradually decrease to zero as the wall of the pipe or containing vessel is approached. (The existence of a stationary layer may be inferred from the fact that whilst large particles of dust can be blown off a shelf, small particles remain which can be wiped off subsequently with the finger.) A fluid is therefore sheared when it flows past a solid surface and the *opposition set up by the fluid to shear* is called its

viscosity. Liquids such as syrup and engine oil which pour slowly are more viscous than water.

Viscosity is a kind of internal friction exhibited to some degree by all fluids. It arises in liquids because the forced movement of a molecule relative to its neighbours is opposed by the intermolecular forces between them.

When the fluid particles passing successively through a given point in a fluid always follow the same path afterwards, the flow is said to be *steady*. 'Streamlines' can be drawn to show the direction of motion of the particles and are shown in Fig. 10.1*b* for steady flow of the water at various depths near the centre of a wide river. The layer of water in contact with the bottom of the river must be at rest (or the river bed would be rapidly eroded) and the velocities of higher layers increase the nearer the layer is to the surface. The length of the streamlines represents the magnitude of the velocities. The water suffers shear, a cube becoming a rhombus, Fig. 10.1*c*, as if acted on by tangential forces at its upper and lower faces. Steady flow thus involves parallel layers of fluid sliding over each other with different velocities thereby creating viscous forces acting tangentially (as shear forces do) between the layers and impeding their motion.

Coefficient of viscosity

To obtain a definition of viscosity we consider two plane parallel layers of liquid separated by a very

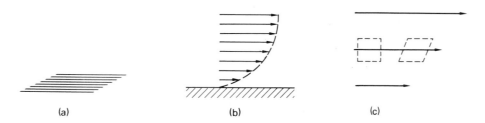

Fig. 10.1   (a)   (b)   (c)

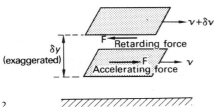

Fig. 10.2

small distance δy and having velocities $v + \delta v$ and v, Fig. 10.2. The *velocity gradient* (i.e. change of velocity/distance) in a direction perpendicular to the velocities is $\delta v/\delta y$. The slower, lower layer exerts a tangential retarding force F on the faster upper layer and experiences itself an equal and opposite tangential force F due to the upper layer (Newton's third law). The *tangential stress* between the layers is therefore F/A where A is their area of contact. The *coefficient of viscosity* η is defined by the equation

$$\eta = \frac{tangential\ stress}{velocity\ gradient} = \frac{F/A}{\delta v/\delta y}$$

In words η is the *tangential force per unit area of fluid which resists the motion of one layer over another when the velocity gradient between the layers is unity.* If $\delta v/\delta y$ is small and F/A large, η is large and the fluid very viscous. For many pure liquids (e.g. water) and gases η is independent of the velocity gradient at a particular temperature, i.e. η is constant and so the *tangential stress is directly proportional to the velocity gradient.* Fluids for which this is true are called *Newtonian fluids* since Newton first suggested this relationship might hold. For some liquids such as paints, glues and liquid cements, η decreases as the tangential stress increases and these are said to be *thixotropic*.

The equation defining η shows that it can be measured in newton second metre^{-2} (N s m^{-2}) i.e. in pascal second (Pa s). Check this. At 20 °C, η for water is 1.0×10^{-3} Pa s and for glycerine 8.3×10^{-1} Pa s. Experiment shows that the coefficient of viscosity of a liquid usually decreases rapidly with temperature rise.

Viscosity is an essential property of a lubricating oil if it is to keep apart two solid surfaces in relative motion. Too high viscosity on the other hand causes unnecessary resistance to motion. 'Viscostatic' oils have about the same value of η whether cold or hot.

It should be noted that viscous forces are called into play as soon as fluid flow starts. If the external forces causing the flow are constant, the rate of flow becomes constant and a steady state is attained with the resisting viscous forces equal to the applied force. The viscous forces stop the flow when the applied force is removed.

Poiseuille's formula: steady and turbulent flow

The streamlines for *steady* flow in a circular pipe are shown in Fig. 10.3. Everywhere they are parallel to the axis of the pipe and represent velocities varying from zero at the wall of the pipe to a maximum at its axis. The surfaces of equal velocity are the surfaces of concentric cylinders.

Fig. 10.3

An expression for the volume of liquid passing per second, V, through a pipe when the flow is *steady*, can be obtained by the method of dimensions (Appendix 1). It is reasonable to assume that V depends on (*i*) the coefficient of viscosity η of the liquid, (*ii*) the radius r of the pipe and (*iii*) the pressure gradient p/l causing the flow, where p is the pressure difference between the ends of the pipe and l is its length, Fig. 10.4. We have

$$V = k\eta^x r^y (p/l)^z$$

where x, y and z are the indices to be found and k is a dimensionless constant. The dimensions of V are $[L^3T^{-1}]$, of η $[ML^{-1}T^{-1}]$, of r $[L]$, of p $[MLT^{-2}/L^2]$, i.e. $[ML^{-1}T^{-2}]$ (since pressure = force/area), and of l $[L]$. Hence the dimensions of p/l are $[ML^{-2}T^{-2}]$.

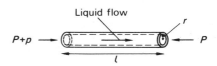

Fig. 10.4 P = atmospheric pressure

Equating dimensions,

$$[L^3T^{-1}] = [ML^{-1}T^{-1}]^x[L]^y[ML^{-2}T^{-2}]^z$$

Equating indices of M, L and T on both sides,

$$0 = x + z$$
$$3 = -x + y - 2z$$
$$-1 = -x - 2z$$

Solving, we get $x = -1$, $y = 4$ and $z = 1$. Hence

$$V = \frac{kpr^4}{\eta l}$$

The value of k cannot be obtained by the method of dimensions but a fuller analysis shows that it equals $\pi/8$ and so the complete expression is

$$V = \frac{\pi p r^4}{8\eta l}$$

This is known as *Poiseuille's formula* since he made the first thorough experimental investigation of the *steady* flow of liquid through a pipe in 1844.

So far we have considered only steady flow. When the velocity of flow exceeds a certain critical value the motion becomes *turbulent*, the liquid is churned up and the streamlines are no longer parallel and straight. The change from steady to turbulent flow can be studied with the apparatus of Fig. 10.5. The flow of water along the tube T is controlled by the clip C. Potassium permanganate solution from the reservoir R is fed into the water flowing through T by a fine jet J. At low flow velocities a fine coloured stream is observed along the centre of T, but as the rate of flow increases it starts to break up and the colour rapidly spreads throughout T indicating the onset of turbulence.

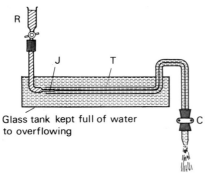

Fig. 10.5

R
J T
Glass tank kept full of water to overflowing
C

Reynolds's number (*Re*) is useful in the study of the stability of fluid flow.

It is defined by the equation

$$Re = \frac{v l \rho}{\eta}$$

where η and ρ are the viscosity and density respectively of the fluid, v is the speed of the bulk of the fluid and l is a characteristic dimension of the solid body concerned. For a cylindrical pipe l is usually the diameter ($2r$) of the pipe. Experiment shows that for cylindrical pipes, when

$Re < 2200$,   flow is steady

$Re \simeq 2200$,   flow is unstable (critical velocity v_c)

$Re > 2200$,   flow is usually turbulent

Hence large η and small v, r and ρ promote steady flow.

Poiseuille's formula holds for velocities of flow below v_c.

Motion in a fluid: Stokes' law

The streamlines for a fluid flowing *slowly* past a stationary solid sphere are shown in Fig. 10.6. When the sphere moves slowly rather than the fluid, the pattern is similar but the streamlines then show the apparent motion of the fluid particles as seen by someone on the moving sphere. In this latter case it is known that the layer of fluid in contact with the sphere moves with it, thus creating a velocity gradient between this layer and other layers of the fluid. Viscous forces are thereby brought into play and constitute the resistance experienced by the moving sphere.

Viscous fluid

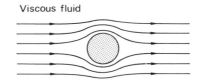

Fig. 10.6

If we make the plausible assumption that the viscous retarding force F depends on (*i*) the viscosity η of the fluid, (*ii*) the velocity v and radius r of the sphere, then an expression can be derived for F by the method of dimensions. Thus

$$F = k\eta^x v^y r^z$$

where x, y and z are the indices to be found and k is a dimensionless constant. The dimensional equation is

$$[MLT^{-2}] = [ML^{-1}T^{-1}]^x [LT^{-1}]^y [L]^z$$

Equating indices of M, L and T on both sides,

$$1 = x$$
$$1 = -x + y + z$$
$$-2 = -x - y$$

Solving, we get $x = 1$, $y = 1$ and $z = 1$. Hence

$$F = k\eta v r$$

A detailed treatment, first done by Stokes, gives $k = 6\pi$ and so

$$F = 6\pi\eta v r$$

This expression, called *Stokes' law*, only holds for *steady motion* in a fluid of *infinite extent* (otherwise the walls and bottom of the vessel affect the resisting force).

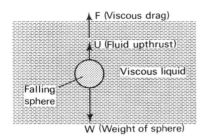

Fig. 10.7

Now consider the sphere falling vertically under gravity in a viscous fluid. Three forces act on it, Fig. 10.7,

(*i*) its weight *W*, acting downwards,

(*ii*) the upthrust *U* due to the weight of fluid displaced, acting upwards, and

(*iii*) the viscous drag *F*, acting upwards.

The resultant downward force is $(W - U - F)$ and causes the sphere to accelerate until its velocity, and so the viscous drag, reach values such that

$$W - U - F = 0$$

The sphere then continues to fall with a constant velocity, known as its *terminal velocity*, of say v_t. Now

$W = \frac{4}{3}\pi r^3 \rho g$ where ρ is the density of the sphere

$U = \frac{4}{3}\pi r^3 \sigma g$ where σ is the density of the fluid

Also, *if steady conditions still hold* when velocity v_t is reached then by Stokes' law

$$F = 6\pi \eta r v_t$$

Hence

$$\frac{4}{3}\pi r^3 \rho g - \frac{4}{3}\pi r^3 \sigma g - 6\pi \eta r v_t = 0$$

$$\therefore \quad v_t = \frac{2r^2(\rho - \sigma)g}{9\eta}$$

All that has been said so far applies to steady flow. As the velocity of the sphere increases, a critical velocity v_c is reached when the flow breaks up, eddies are formed as in Fig. 10.8*a* and the motion becomes *turbulent*. At velocities greater than v_c the resistance to motion, called the *drag*, increases sharply and is roughly proportional to the square of the velocity. (Below v_c Stokes' law indicates that the resistance is proportional to the velocity.) For highly turbulent flow resistance is dependent on density, not viscosity; this is the ordinary case of air resistance to vehicles.

By modifying the shape of a body the critical velocity can be raised and the drag thereby reduced at a particular speed if steady flow replaces turbulent flow. This is called *streamlining* the body and Fig. 10.8*b* shows how it is done for a sphere. The pointed tail can be regarded as filling the region where eddies occur in turbulent motion, thus ensuring that the streamlines merge again behind the sphere. Streamlining is particularly important in the design of high-speed aircraft.

The air flow past a model 'Mini' car in a wind tunnel is shown in Fig. 10.9. The flow is visualized by streams of finely condensed paraffin vapour giving the appearance of smoke trails. Features illustrated are flow separation part-way along the bonnet, vortex ('whirlwind') flow across the foot of the windscreen, flow separation just off the top of the windscreen roof joint and unsteady flow behind the car (as shown by the dispersion of the smoke trails).

Methods of measuring η

(*a*) *Using Poiseuille's formula.* The method is suitable for a liquid obtainable in large quantities and which flows easily, e.g. water. The liquid passes *slowly* from a constant head apparatus through a capillary tube and the volume collected in a certain time is found, Fig. 10.10*a*.

By altering the position of tube T, the rates of flow for different pressure differences can be measured. If a graph of volume delivered per second *V* against pressure difference *p* is plotted, the onset of turbulence, to which Poiseuille's formula does not apply, will be shown by non-linearity, Fig. 10.10*b*. The slope of the

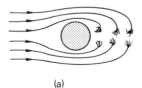

Fig. 10.8

(a)

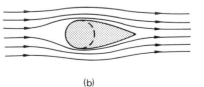

(b)

Fig. 10.9

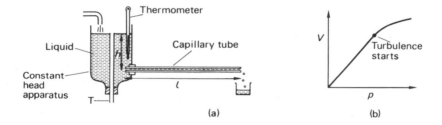

Fig. 10.10   (a)   (b)

linear part of the graph gives an average value for V/p where $p = h\rho g$, ρ being the density of the liquid and h the pressure head. Knowing the length l of the capillary tube and its radius r, the viscosity η of the liquid at the particular temperature is calculated from

$$\eta = \frac{\pi p r^4}{8 V l}$$

To measure r with the care required (since it is small and appears to the fourth power) a long thread of mercury is introduced into the tube and its length and mass found. A narrow bore capillary tube is used so that steady flow is obtained with pressure differences that are not so small as to be difficult to measure accurately.

(b) *Ostwald's viscometer.* Viscosities can be easily and rapidly *compared* using this instrument, Fig. 10.11. A certain volume of liquid is introduced via E and sucked up into bulb B until its upper level is above

mark A. It is then allowed to flow under its own weight through capillary CD and the time t_1 found for the upper level to fall between marks A and C. This is also the time for a volume of liquid equal to the volume V of

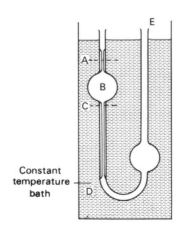

Fig. 10.11

the viscometer between A and C to flow through CD. The experiment is repeated and t_2 found with the *same volume* of another liquid (or with the same liquid at a different temperature if the variation of η with temperature of a given liquid is being studied).

The pressure difference causing the flow decreases during the flow but since the viscometer always contains the same volume of liquid, the average difference of level, say h, is always the same. Hence for liquids of densities ρ_1 and ρ_2 the average pressure differences are $h\rho_1 g$ and $h\rho_2 g$ respectively. Thus for steady flow of the first liquid of viscosity η_1

$$\frac{V}{t_1} = \frac{\pi h \rho_1 g r^4}{8\eta_1 l}$$

and for the second of viscosity η_2

$$\frac{V}{t_2} = \frac{\pi h \rho_2 g r^4}{8\eta_2 l}$$

where r and l are the radius and length of the capillary respectively. Therefore

$$\frac{\eta_1}{\eta_2} = \frac{\rho_1 t_1}{\rho_2 t_2}$$

The method is widely used in practice because of its simplicity and accuracy.

(c) *Using Stokes' law.* The viscosity of a liquid such as glycerine or a heavy oil, whose high viscosity makes the previous methods unsuitable, may be found by timing a small ball-bearing falling with its terminal velocity through the liquid. So long as the terminal velocity does not exceed the critical velocity, i.e. the flow is steady, Stokes' law applies and we can therefore say

$$v_t = \frac{2r^2(\rho - \sigma)g}{9\eta} \qquad \text{(p. 209)}$$

where v_t is the terminal velocity, r and ρ the radius and density of the ball-bearing and σ the density of the liquid. The viscosity η can then be calculated.

To satisfy as far as possible the assumption in Stokes' law that the liquid is infinite in extent, the vessel of

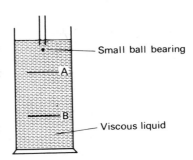

Fig. 10.12

liquid must be wide compared with the diameter of the ball-bearing (less than 2 mm for a wide measuring cylinder); it should also be deep, Fig. 10.12. The terminal velocity v_t is obtained by finding the average time t taken by balls of the same size to fall from mark A (which is far enough below the surface for the ball to have its terminal velocity at A) to mark B (which is not too near the bottom of the vessel). Then $v_t = \text{AB}/t$.

To reduce the chance of air bubbles adhering to the falling ball it should be dipped in some of the liquid and thereby coated, before dropping. The temperature of the liquid should also be kept constant.

Bernoulli's equation

The pressure is the same at all points on the same horizontal level in a fluid at rest; this is not so when the fluid is in motion. The pressure at different points in a liquid flowing through (a) a uniform tube and (b) a tube with a narrow part, is shown by the height of liquid in the vertical manometers in Figs. 10.13a and b. In (a) the pressure drop along the tube is steady and maintains the flow against the viscosity of the liquid. In (b) the pressure falls in the narrow part B but rises again in the wider part C. If the liquid can be assumed to be incompressible, the same volume of liquid passes through B in a given time as enters A and so the velocity of the liquid must be greater in B than in A or C. Therefore a decrease of pressure accompanies an increase of velocity. This may be shown by blowing into

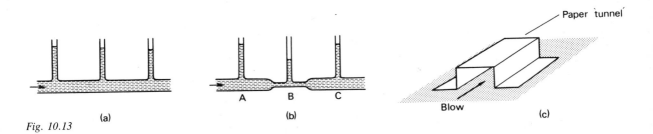

Fig. 10.13 (a) (b) (c)

a 'tunnel' made from a sheet of paper, Fig. 10.13c. The faster one blows the more does the tunnel collapse.

A useful relation can be obtained between the pressure and the velocity at different parts of a fluid in motion.

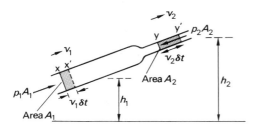

Fig. 10.14

Suppose a fluid flows through a non-uniform tube from X to Y, Fig. 10.14, and its velocity changes from v_1 at X where the cross-section area is A_1 to v_2 at Y where the cross-section is A_2. The flow of fluid between X and Y is caused by the forces acting on its ends which arise from the pressure exerted on it by the fluid on either side of it. At X, if the fluid pressure is p_1, there is a force $p_1 A_1$ acting in the direction of flow and at Y, if the fluid pressure is p_2, a force $p_2 A_2$ opposes the flow. Consider a small time interval δt in which the fluid at X has moved to X' and that at Y to Y'.

At X, work done during δt *on* the fluid XY by $p_1 A_1$ pushing it into the tube

= force × distance moved = force × velocity × time

= $p_1 A_1 \times v_1 \times \delta t$

At Y, work done during δt *by* the fluid XY emerging from the tube against $p_2 A_2 = p_2 A_2 \times v_2 \times \delta t$. Therefore

net work W done *on* the fluid = $(p_1 A_1 v_1 - p_2 A_2 v_2)\delta t$

If the fluid is incompressible, volume between X and X' equals volume between Y and Y', i.e.

$$A_1 \times v_1 \, \delta t = A_2 \times v_2 \, \delta t$$

$$\therefore \quad W = (p_1 - p_2)A_1 v_1 \, \delta t$$

As a result of work done on it, the fluid gains p.e. and k.e. when XY moves to X'Y'.

Gain of p.e. = p.e. of X'Y' − p.e. of XY

= p.e. of X'Y + p.e. of YY' −
 p.e. of XX' − p.e. of X'Y

= p.e. of YY' − p.e. of XX'

Gain of p.e. = $(A_2 v_2 \, \delta t \rho)gh_2 - (A_1 v_1 \, \delta t \rho)gh_1$
(since p.e. = mgh)

= $A_1 v_1 \, \delta t \rho g(h_2 - h_1)$
(since $A_1 v_1 \, \delta t = A_2 v_2 \, \delta t$)

where h_1 and h_2 are the heights of XX' and YY' above an arbitrary horizontal reference level and ρ is the density of the fluid. Similarly,

gain of k.e. = k.e. of YY' − k.e. of XX'

= $\frac{1}{2}(A_2 v_2 \, \delta t \rho)v_2^2 - \frac{1}{2}(A_1 v_1 \, \delta t \rho)v_1^2$
(since k.e. = $\frac{1}{2}mv^2$)

= $\frac{1}{2}A_1 v_1 \, \delta t \rho(v_2^2 - v_1^2)$

If the fluid is non-viscous (i.e. inviscid) no work is done against viscous forces to maintain the flow, no change of internal energy of the fluid occurs and by the principle of conservation of energy we have

net work done *on* fluid = gain of p.e. + gain of k.e.

$$\therefore \quad (p_1 - p_2)A_1 v_1 \, \delta t = A_1 v_1 \, \delta t \rho g(h_2 - h_1) \\ + \frac{1}{2}A_1 v_1 \, \delta t \rho(v_2^2 - v_1^2)$$

$$p_1 - p_2 = \rho g(h_2 - h_1) + \frac{1}{2}\rho(v_2^2 - v_1^2)$$

or $\quad p_1 + h_1\rho g + \frac{1}{2}\rho v_1^2 = p_2 + h_2\rho g + \frac{1}{2}\rho v_2^2$

This is *Bernoulli's equation* and it is usually stated by saying that *along a streamline in an incompressible, inviscid fluid*

$$p + h\rho g + \tfrac{1}{2}\rho v^2 = constant$$

In deriving the equation we have in effect assumed that the pressure and velocity are uniform over any cross-section of the tube. This is not so for a real (viscous) fluid and so it only applies strictly to a single streamline in the fluid. In addition, actual fluids, especially gases, are compressible. The equation has therefore to be applied with care or the results will be misleading.

Applications of Bernoulli

(a) Jets and nozzles. Bernoulli's equation suggests that for fluid flow where the potential energy change $h\rho g$ is very small or zero, as in a horizontal pipe, *the pressure falls when the velocity rises.* The velocity increases at a constriction—a slow stream of water from a tap can be converted into a fast jet by narrowing the exit with a finger—and the greater the change in cross-sectional area, the greater is the increase of velocity and so the greater is the pressure drop. Several devices with jets and nozzles use this effect; Fig. 10.15

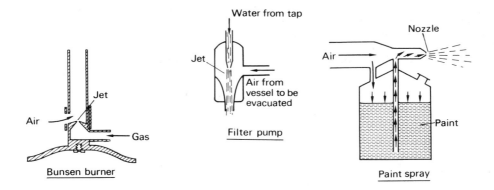

Fig. 10.15

shows the action of a Bunsen burner, a filter pump and a paint spray.

(*b*) *Spinning ball.* If a tennis ball is 'cut' or a golf ball 'sliced' it spins as it travels through the air and experiences a sideways force which causes it to curve in flight. This is due to air being dragged round by the spinning ball, thereby increasing the air flow on one side and decreasing it on the other. A pressure difference is thus created, Fig. 10.16. The swing of a spinning cricket ball is complicated by its raised seam.

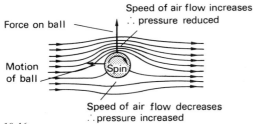

Fig. 10.16

(*c*) *Aerofoil.* This is a device which is shaped so that the relative motion between it and a fluid produces a force perpendicular to the flow. Examples of aerofoils are aircraft wings, turbine blades and propellors.

The shape of the aerofoil section in Fig. 10.17 is such that fluid flows faster over the top surface than over the bottom, i.e. the streamlines are closer above than below the aerofoil. By Bernoulli, it follows that the pressure underneath is increased and that above reduced. A resultant upwards force is thus created, normal to the flow and it is this force which provides most of the 'lift' for an aircraft. Its value increases with the angle between the wing and the air flow (called the 'angle of attack') until at a certain angle the flow separates from the upper surface, lift is lost almost completely, drag increases sharply, the flow downstream becomes very turbulent and the aircraft stalls.

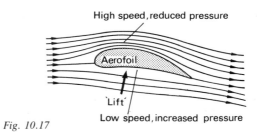

Fig. 10.17

The sail of a yacht 'tacking' into the wind is another example of an aerofoil. The air flow over the sail produces a pressure increase on the windward side and a decrease on the leeward side. The resultant force is roughly normal to the sail and can be resolved into a component F producing forward motion and a greater component S acting sideways, Fig. 10.18. The keel produces a lateral force to balance S.

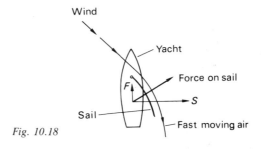

Fig. 10.18

Flowmeters

These measure the rate of flow of a fluid through a pipe. Two types will be considered.

(*a*) *Venturi meter.* This consists of a horizontal tube

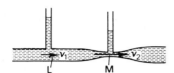

Fig. 10.19

with a constriction and replaces part of the piping of a system, Fig. 10.19. The two vertical tubes record the pressures (above atmospheric) in the fluid flowing in the normal part of the tube and in the constriction.

If p_1 and p_2 are the pressures and v_1 and v_2 the velocities of the fluid (density ρ) at L and M on the same horizontal level, then assuming Bernoulli's equation holds

$$p_1 + \tfrac{1}{2}\rho v_1^2 = p_2 + \tfrac{1}{2}\rho v_2^2 \qquad \text{(since } h_1 = h_2\text{)}$$

$$\therefore \quad p_1 - p_2 = \tfrac{1}{2}\rho(v_2^2 - v_1^2)$$

If A_1 and A_2 are the cross-sectional areas at L and M and if the fluid is incompressible, the same volume passes each section of the tube per second

$$\therefore \quad A_1 v_1 = A_2 v_2$$

Hence $\quad p_1 - p_2 = \tfrac{1}{2}\rho v_1^2\left(\dfrac{A_1^2}{A_2^2} - 1\right)$

Knowing A_1, A_2, ρ and $(p_1 - p_2)$, v_1 can be found and so also the rate of flow $A_1 v_1$. Why is the above equation not valid for (*i*) a gas, (*ii*) a heavy oil, (*iii*) very rapid flow?

(*b*) *Pitot tube.* The pressure exerted by a moving fluid, called the *total pressure*, can be regarded as having two components: the *static pressure* which it would have if it were at rest and the *dynamic pressure* which is the pressure equivalent of its velocity. The Pitot tube measures total pressure and in essence is a manometer with one limb parallel to the flow and open to the oncoming fluid, Fig. 10.20. The fluid at the open end is at rest and a 'stagnant' region exists there. The total pressure is also called the *stagnation pressure*. The static pressure is measured by a manometer connected at right angles to the pipe or surface over which the fluid is passing.

In Bernoulli's equation

$$p + h\rho g + \tfrac{1}{2}\rho v^2 = constant$$

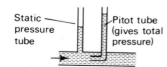

Static pressure tube

Pitot tube (gives total pressure)

Fig. 10.20

the static pressure is given by $p + h\rho g$ or by p if the flow is horizontal, the dynamic pressure by $\tfrac{1}{2}\rho v^2$ and the total pressure by $p + \tfrac{1}{2}\rho v^2$. Hence

$$total\ pressure - static\ pressure = p + \tfrac{1}{2}\rho v^2 - p = \tfrac{1}{2}\rho v^2$$

$$\therefore \quad v = \sqrt{\frac{2}{\rho}(total\ pressure - static\ pressure)}$$

This expression enables a value for the velocity of flow v of an incompressible, inviscid fluid to be calculated from the readings of Pitot-static tubes. In real cases v varies across the diameter of the pipe carrying the fluid (because of its viscosity) but it can be shown that if the open end of the Pitot tube is offset from the axis of the pipe by $0.7 \times$ radius of the pipe, then v is the *average* flow velocity.

Fluid flow calculations

1. A garden sprinkler has 150 small holes each 2.0 mm² in area, Fig. 10.21. If water is supplied at the rate of 3.0×10^{-3} m³ s⁻¹, what is the average velocity of the spray?

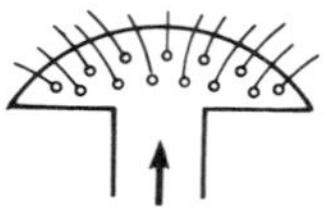

Fig. 10.21

Volume of water per second from sprinkler

= volume supplied per second

= 3×10^{-3} m³ s⁻¹

= total area of sprinkler holes × average velocity of spray

= 300×10^{-6} m² × average velocity of spray

Therefore average velocity of spray

$$= \frac{3 \times 10^{-3}}{300 \times 10^{-6}} \frac{m^3\ s^{-1}}{m^2}$$

$$= 10\ m\ s^{-1}$$

2. Obtain an estimate for the velocity of emergence of a liquid from a hole in the side of a wide vessel 10 cm below the liquid surface.

Consider the general case in which the hole is at depth h below the surface of the liquid of density ρ, Fig. 10.22. If the liquid is incompressible and inviscid and

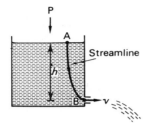

Fig. 10.22

the motion is steady we can apply Bernoulli's equation to points A and B on the streamline AB.

At A $\quad p_1$ = atmospheric pressure = P

$\quad\quad\quad h_1 = h$

$\quad\quad\quad v_1 = 0$

(assuming the rate of fall of the surface can be neglected compared with the speed of emergence since the vessel is wide).

At B

$\quad p_2 =$ pressure of air into which

$\quad\quad\quad\quad\quad\quad$ the liquid emerges = P

$\quad h_2 = 0$

$\quad v_2 = v$

Substituting in Bernoulli's equation

$$P + h\rho g + 0 = P + 0 + \tfrac{1}{2}\rho v^2$$

$$\therefore\ h\rho g = \tfrac{1}{2}\rho v^2$$

From which we see that the potential energy lost by unit volume of liquid (mass ρ) in falling from the surface to depth h is changed to kinetic energy. The velocity of emergence is given by

$$v^2 = 2gh$$

and is the same as the vertical velocity which would be acquired in free fall—a statement known as *Torricelli's theorem*. In fact, v is always less than $\sqrt{2gh}$ due to the viscosity of the liquid.

If $h = 10$ cm = 0.1 m and $g = 9.8$ m s^{-2} then

$$v = \sqrt{2 \times 9.8 \times 0.1}\ \text{m s}^{-1}$$

$$= 1.4\ \text{m s}^{-1}$$

Flow of mass, energy and charge

The transfer of mass, energy or electrical charge from one place to another is an important engineering problem. For example, the transportation of all three occurs in the generation and distribution of electricity.

The flow of each quantity is expressed by the same general expression

$$flow\ rate \propto \frac{'pressure'\ causing\ flow}{resistance\ to\ flow}$$

(*a*) *Mass.* Mass transport in the form of *steady* fluid flow along a pipe is given by Poiseuille's formula (p. 207)

$$V = \frac{\pi p r^4}{8\eta l}$$

where V is the volume of fluid passing per second, p the pressure difference between the ends of the pipe of length l and radius r and η is the coefficient of viscosity of the fluid.

If ρ_m is the density of the fluid then the mass passing per second is $\rho_m V$ and we can say

$$mass\ flow\ rate = p\,\frac{\rho_m \pi r^4}{8\eta l} = \frac{p}{R_m}$$

where R_m is a constant which incorporates the resistance to flow, i.e. η, and equals $8\eta l/(\rho_m \pi r^4)$. Hence

$$fluid\ mass\ flow\ rate \propto \frac{fluid\ pressure\ difference}{resistance\ to\ flow}$$

(*b*) *Energy (heat).* The transport of heat by conduction is expressed by Fourier's law (p. 78)

$$\frac{Q}{t} = \frac{kA(\theta_2 - \theta_1)}{x}$$

where Q is the quantity of heat passing in time t down a lagged bar of cross-sectional area A, length x and thermal conductivity k when its opposite ends are at steady temperatures θ_2 and θ_1. Hence

$$heat\ flow\ rate = (\theta_2 - \theta_1)\frac{kA}{x} = \frac{\theta_2 - \theta_1}{R_e}$$

where R_e is a constant incorporating the resistance to flow, i.e. $1/k$, and equals $x/(kA)$. Hence

$$heat\ flow\ rate \propto \frac{temperature\ difference}{resistance\ to\ flow}$$

(*c*) *Charge.* The rate of electric charge flow, i.e. current, in metallic conductors obeys Ohm's law (p. 38)

$$I = \frac{Q}{t} = \frac{V}{R}$$

where I is the current and V the potential difference across the ends of a conductor of resistance R. Since $R = \rho l/A$ where ρ is the resistivity of the material of the conductor of length l and cross-sectional area A we can also write

$$I = V\frac{A}{\rho l}$$

Hence

$$\text{electric charge flow rate} \propto \frac{potential\ difference}{resistance\ to\ flow}$$

Further comparison of these transport effect expressions shows that for all three we can also say

$$\text{flow rate} = \frac{driving}{factor} \times \frac{physical}{factor} \times \frac{conductivity}{factor}$$

In particular,

$$\text{mass flow rate} = p \times \frac{\pi r^2}{l} \times \frac{\rho_m r^2}{8\eta}$$

$$\text{heat flow rate} = (\theta_2 - \theta_1) \times \frac{A}{x} \times k$$

$$\text{electric charge flow rate} = V \times \frac{A}{l} \times \frac{1}{\rho}$$

In each case the physical factor equals (cross-sectional area of flow path)/(length of flow path).

Film (16 mm): *Fluid flow*—Unilever Film Library, Unilever House, London, E.C.4.

QUESTIONS

1. What do you understand by the *dimensions* of a physical quantity? Derive the dimensions of coefficient of viscosity.

Explain the value, and the limitations, of the method of dimensions as a means of checking, and sometimes deriving, the form of equations involving physical quantities.

Confirm the dimensional consistency of the following statements in which η represents the coefficient of viscosity of a liquid, a and l are lengths, ρ a density, p a pressure-difference and v a speed:

(*a*) The product $lv\rho/\eta$, known as Reynolds's number, is dimensionless.

(*b*) According to Poiseuille, the volume of liquid flowing per second steadily through a capillary tube is $\pi pa^4/8l\eta$.

(*c*) For a sphere of density ρ falling steadily under gravity through an expanse of liquid of density ρ'

$$6\pi\eta av = \tfrac{4}{3}\pi a^3(\rho - \rho')g \qquad (O.)$$

2. An aluminium sphere is suspended by a thread below the surface of a liquid. Show on a sketch the forces acting on the sphere, and explain its equilibrium. (No formal proof is required.)

The thread is now cut. Show on a second sketch, or explain in words, the forces which act on the sphere when it is in motion.

The following figures for x, the total distance travelled by the sphere in the liquid at time t, were obtained:

t(s)	1.0	2.0	3.0	4.0	5.0
x(cm)	3.60	10.3	18.6	27.9	37.4

Draw a graph to display the relation between x and t, and explain its form. Find the terminal velocity of the sphere. Give a qualitative account of how you would expect the graph to be modified if the temperature of the liquid were increased. (C.)

3. Define *coefficient of viscosity η* and obtain its dimensions in terms of M, L and T.

Stokes's law for the viscous force F acting on a sphere of radius a falling with velocity v through a large expanse of fluid of coefficient of viscosity η is expressed by the equation

$$F = 6\pi a\eta v$$

Show that this equation is correct dimensionally and state why it is true only for sufficiently low velocities.

Explain why a sphere released in a fluid will fall with diminishing acceleration until it attains a constant terminal velocity.

Calculate this velocity for an oil drop of radius 3.0×10^{-6} m falling through air of coefficient of viscosity 1.8×10^{-5} Pa s, given that the density of the oil is 8.0×10^2 kg m^{-3} and that the density of air may be neglected. (L.)

4. In the simplified petrol engine carburettor shown in Fig. 10.23, air is drawn into the carburettor by the action of the engine piston, and the petrol enters at the point of minimum cross-sectional area. If the throat of the venturi section has an area of 78 mm^2, calculate the area of the fuel jet required to produce an air–fuel mass ratio of 12:1.

The density of air is 1.2 kg m^{-3} and that of petrol is 7.8×10^2 kg m^{-3}. (*J.M.B. Eng. Sc.*)

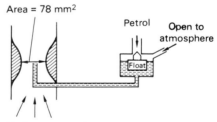

Fig. 10.23   Air from atmosphere

5. A toy designer has submitted a design for a water pistol with a barrel area 75 mm^2 and jet area 1.0 mm^2. The manufacturer required that when the pistol was fired horizontally, the jet should be able to hit a target 3.5 m away not more than 1.0 m below the firing line. Given that the average child is able to exert a force of 10 N on the plunger, has the designer satisfied the requirements? You may neglect barrel friction and energy loss at the exit jet.

Atmospheric pressure = 1.0×10^5 Pa, density of water = 1.0×10^3 kg m^{-3}, $g = 9.8$ m s^{-2}. (*J.M.B. Eng. Sc.*)

Objective-type revision questions for Part 2

The first figure of a question number gives the relevant chapter, e.g. **7.3** is the third question for chapter 7.

Multiple choice

Select the response which you think is correct.

6.1. A pendulum bob suspended by a string from the point P, Fig. 10, is in equilibrium under the action of three forces: W, the weight of the bob; T, the tension in the string; and F, a horizontally applied force. Which one of the following statements is untrue?

A $F^2 + W^2 = T^2$ **B** F and W are the components of T
C $W = T \cos \theta$ **D** $F = W \tan \theta$

(J.M.B. Eng. Sc.)

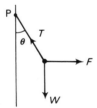

Fig. 10

6.2. Forces of 3 N, 4 N and 12 N act at a point in mutually perpendicular directions. The magnitude of the resultant force in newtons is

A 5   **B** 11   **C** 13   **D** 19
E indeterminate from information given

6.3. Which graph in Fig. 11 best represents the variation of velocity with time of a ball which bounces vertically on a hard surface, from the moment when it rebounds from the surface?

6.4. A ball is projected horizontally at 15 m s^{-1} from a point 20 m above a horizontal surface ($g = 10$ m s^{-2}). The magnitude of its velocity in m s^{-1} when it hits the surface is

A 10   **B** 15   **C** 20   **D** 25   **E** 35

6.5. A trolley of mass 60 kg moves on a frictionless horizontal surface and has kinetic energy 120 J. A mass of 40 kg is lowered vertically on to the trolley. The total kinetic energy of the system is now

A 60 J   **B** 72 J   **C** 120 J   **D** 144 J
E another answer

7.1. A mass of 2.0 kg describes a circle of radius 1.0 m on a smooth horizontal table at a uniform speed. It is joined to the centre of the circle by a string which can just withstand 32 N. The greatest number of revolutions per minute the mass can make is

A 38   **B** 4   **C** 76   **D** 240   **E** 16

7.2. In order to turn in a horizontal circle an aircraft banks so that

A there is a resultant force on the wings from the centre of the circle
B the weight of the aircraft has a component towards the centre of the circle
C the drag on the plane is reduced
D the lifting force on the wings has a component towards the centre of the circle

(J.M.B. Eng. Sc.)

7.3. If a small body of mass m is moving with angular velocity ω in a circle of radius r, what is its kinetic energy?

A $m\omega r/2$ **B** $m\omega^2 r/2$ **C** $m\omega r^2/2$ **D** $m\omega^2 r^2/2$
(J.M.B. Eng. Sc.)

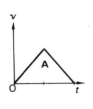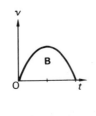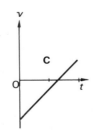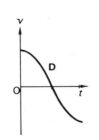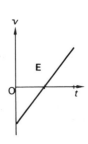

Fig. 11

7.4. Planet X is twice the radius of planet Y and is of material of the same density. The ratio of the acceleration due to gravity at the surface of X to that at the surface of Y is

A 1:4   **B** 1:2   **C** 2:1   **D** 4:1   **E** 8:1

8.1. The frequency of oscillation of a mass m suspended at the end of a vertical spring having a spring constant k is directly proportional to

A mk **B** m/k **C** m^2k **D** $1/(mk)^{1/2}$ **E** $(k/m)^{1/2}$

8.2. The graph of Fig. 12 shows how the displacement of a particle describing s.h.m. varies with time. Which one of the following statements is, from the graph, false?

A The restoring force is zero at time $T/4$.
B The velocity is a maximum at time $T/2$.
C The acceleration is a maximum at time T.
D The displacement is a maximum at time T.
E The kinetic energy is zero at time $T/2$.

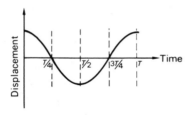

Fig. 12

9.1. When a capillary tube of uniform bore is dipped in water the water level in the tube rises 10 cm higher than in the vessel. If the tube is lowered until its open end is 5.0 cm above the level in the vessel, the water in the tube appears as in (Fig. 13)

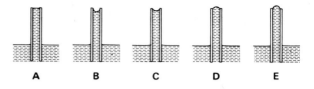

Fig. 13

10.1 Spheres X and Y of the same material fall at their terminal velocities through a liquid without causing turbulence. If Y has twice the radius of X, the ratio of the terminal velocity of Y to that of X is

A 1:4   **B** 1:2   **C** 1:1   **D** 2:1   **E** 4:1

Multiple selection

In each question one or more of the responses may be correct. Choose one letter from the answer code given.

> *Answer **A** if (i), (ii) and (iii) are correct*
> *Answer **B** if only (i) and (ii) are correct*
> *Answer **C** if only (ii) and (iii) are correct*
> *Answer **D** if (i) only is correct*
> *Answer **E** if (iii) only is correct*

6.6. In the equation $Ft = mv - mu$

(i) the dimensions of F are MLT^2
(ii) the dimensions of mv are MLT^{-1}
(iii) the dimensions of all three terms are the same

8.3. The period of a simple pendulum oscillating in a vacuum depends on

(i) the mass of the pendulum bob
(ii) the length of the pendulum
(iii) the acceleration due to gravity

Part 3 FIELDS

11 Electric fields

Simple electrostatics

(*a*) *Electric charges.* In general, when any two different materials are rubbed together they exert forces on each other and each is said to have acquired an 'electric charge'. Electrostatics is the study of electric charges at rest. Experiments show that there are two kinds of charge and that *like charges repel, unlike charges attract.* The two kinds cancel one another out and in this respect are opposite. One type is taken to be positive and the other negative.

The allocation of signs to charges was made quite arbitrarily many years ago and, with the materials used today, the choice makes *polythene* rubbed with wool *negatively* charged and *cellulose acetate* (and also Perspex) rubbed with wool *positively* charged. Previously ebonite (rubbed with fur) and glass (rubbed with silk) were used to obtain negative and positive charges respectively. The forces between charged strips can be investigated as in Fig. 11.1.

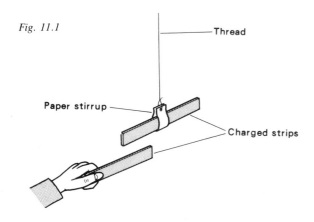

Fig. 11.1

Thread

Paper stirrup

Charged strips

The sign convention adopted leads to the electron having a negative charge and the proton a positive one.

According to modern theory, an atom normally contains equal numbers of electrons and protons, making it electrically neutral. Electrification by rubbing may be explained by supposing that electrons are transferred from one material to the other. For example, when cellulose acetate is rubbed with wool, electrons go from the surface of the acetate to the wool, thus leaving the acetate deficient of electrons, i.e. positively charged, and making the wool negatively charged. Equal amounts of opposite charges should therefore be produced (see p. 255).

(*b*) *Insulators and conductors.* On the electron theory, all the electrons in the atoms of electrical insulators) such as polythene, cellulose acetate, Perspex, ebonite and glass) are considered to be firmly bound to their nuclei and the removal or addition of electrons at one place does not cause the flow of electrons elsewhere. That is, the charge is confined to the region where it was produced (e.g. by rubbing) or placed. Electrical conductors (e.g. metals) have electrons that are quite free from individual atoms (although fairly strongly bound to the material as a whole) and if such materials gain electrons, these can move about in them. Loss of electrons by a conductor causes a redistribution of those left. A charge on a conductor therefore spreads over the entire surface. The 'free' electron theory is adequate for our present purposes but later we will see that it has been extended by the more advanced 'band' theory which explains the behaviour of insulators, semiconductors and conductors in terms of energy levels.

The human body and the earth are comparatively good conductors and if we try to charge a metal rod by rubbing, it must be well insulated and not held in the hand. Otherwise any charge produced is conducted away through the body to earth. Water also conducts and its presence on the surface of many materials (e.g.

glass) that are otherwise insulators accounts for the charge leakage which often occurs. Many modern plastics (e.g. polythene, Perspex, cellulose acetate) are water-repellent.

(*c*) *Electrostatic induction.* A negatively charged polythene strip held close to an insulated, uncharged conductor such as a small aluminized expanded-polystyrene ball attracts it. This may be explained by saying that electrons are repelled to the far side of the ball leaving the near side positively charged, Fig. 11.2*a*. The attraction between the negative charge on the polythene strip and the induced positive charge is greater than the repulsion between the strip and the more distant negative charge. The effect is called *electrostatic induction.* It accounts for the attraction of scraps of paper by a plastic comb, charged by being drawn through the hair.

(*d*) *Electrophorus.* This is a device for producing charges by electrostatic induction, Fig. 11.2*b*. It consists of a circular metal plate with an insulating handle, placed on an insulating sheet (e.g. of polythene) previously charged by rubbing. When the plate is earthed by touching it with the finger and then removed, it has a large charge of opposite sign to that on the insulating sheet. This charge can be transferred to another conductor and the electrophorus plate recharged as before, again and again.

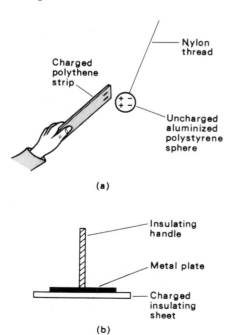

(a)

(b)

Fig. 11.2

It may seem strange that the metal plate does not become charged by contact with the insulating sheet and have the same sign of charge as it. However, it appears that contact between even plane surfaces occurs at only a few points so that, in fact, the metal plate is charged by induction. This would account for it becoming oppositely charged.

(*e*) *Static and current electricity.* Static charges produced by rubbing insulators (or insulated conductors) give the same effects when they move as do electric currents due to a battery. However, in electrostatics we are usually dealing with quite a small charge (a few microcoulombs) but a large p.d. (thousands of volts); in current electricity the opposite is usually true.

Electrostatics today

Electrostatics was the first branch of electricity to be investigated and for a long time was regarded as a subject of no practical value. In recent years this has changed and it now has important industrial applications.

The electrostatic precipitation of flue-ash that would otherwise be discharged into the atmosphere from modern coal-fired power stations is a vital factor in the reduction of pollution. An average power station produces about 30 000 kg (30 tonnes) of flue-ash per hour. Power station precipitators are shown in Fig. 11.3 between the chimney and the main building which contains the coal bunkers, boiler house and turbine hall. The precipitators are built to remove 99 per cent of the ash from the flue gases before they reach the power station chimney. A precipitator is made up of a number of wires and plates. The wires are negatively charged and give a similar charge to the particles of ash which are then attracted to the positive plates. These are mechanically shaken to remove the ash which is collected and used as a by-product.

Electrostatic precipitation is also important in the steel, cement and chemical industries where flue gas outputs are high. Electrostatic spraying of paints, plastics and powders is also possible and lends itself to automation.

In nuclear physics research, electrostatic generators of, for example, the van de Graaff type (p. 256) are employed to produce p.d.s of up to 14 million volts for accelerating atomic particles. Their use in this field has done much to renew interest in electrostatics.

A knowledge of electrostatics is important in the design of cathode-ray tubes for radar and television, in electrical prospecting for minerals and in surveying sites for large structures. Electrostatic loudspeakers

Fig. 11.3

and microphones are in common use as are electrostatic office copying machines.

Electric charges can build up due to friction on aircraft in flight and on plastic sheeting in industry, creating a potential explosion hazard unless preventive steps are taken. In the case of aircraft the rubber tyres are made slightly conducting so that the charge leaks away harmlessly at touch-down. The crackling which occurs when a nylon garment is removed from the body or when someone steps from a car with plastic seat covers is also due to static charges causing the insulation of the surrounding air to break down. A flash of lightning is nature's most spectacular electrostatic event.

Coulomb's law

(a) *Statement.* A knowledge of the forces that exist between charged particles is necessary for an understanding of the structure of the atom and of matter. The magnitude of the forces between charged spheres was first investigated quantitatively in 1785 by Coulomb, a French scientist. The law he discovered may be stated as follows:

The force between two point charges is directly proportional to the product of the charges divided by the square of their distance apart.

The law applies to point charges. Sub-atomic particles such as electrons and protons may be regarded as approximating to point charges. Later we shall see that a *uniformly* charged conducting sphere behaves—so far as external effects are concerned—as if the charges were concentrated at its centre. It is therefore sometimes considered to be a point charge but there must not be any charges nearby to disturb the uniform distribution of charge on it, i.e. it must be an *isolated* charged spherical conductor. In practice two small spheres will only approximate to point charges when they are far apart. A point charge, like a point mass, is a convenient theoretical simplification.

Coulomb's law may be stated in mathematical terms as

$$F \propto \frac{Q_1 Q_2}{r^2}$$

where F is the electric (or Coulomb) force between two point charges Q_1 and Q_2, distance r apart.

(b) *Experimental test.* Two small metallized spheres X and Y are used. X is glued to the bottom of a 'V' of a metre length of fine nylon thread so that it can only swing at right angles to the plane of suspension, Fig. 11.4a. The position of the centre (or edge) of the shadow of X on the scale on the screen is noted. Sphere Y is glued to the end of an insulating rod.

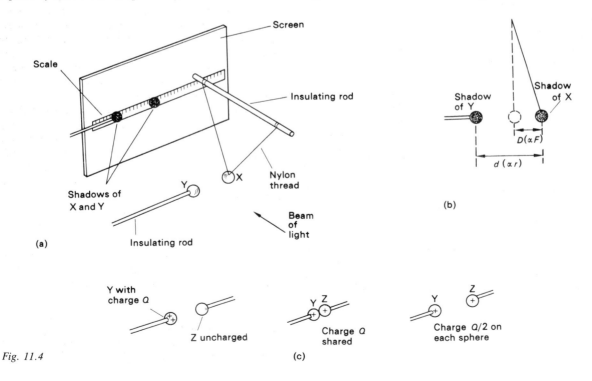

Fig. 11.4

Both spheres are given the same charge by touching each in turn with, say, the plate of an electrophorus. When Y is brought up to X, repulsion occurs, Fig. 11.4*b*. The positions of the centres (or edges) of the shadows of X and Y are noted. Distance *d* is proportional to the separation (*r*) of the spheres and it can be shown that the deflection *D* (the distance between the centres—or edges—of the first and second positions of the shadow of X) is proportional to the force between X and Y. If a few readings of *d* and *D* are taken and $D \times d^2$ found to be constant, then $F \propto 1/r^2$.

To test if $F \propto Q_1Q_2$, Y is touched by another exactly similar but uncharged sphere Z, Fig. 11.4*c*, so that its charge is halved (Z taking the other half). Using one of the previous separations (*d*) between X and Y, the force should therefore become half what it was before. It will be one quarter of its original value if the charge on X is also halved by sharing its charge with the uncharged Z.

For reasonable success this experiment requires the insulators to be thoroughly dry (if necessary by using a hair dryer) to prevent loss of charge by leakage. The readings should therefore be taken in quick succession. The most convincing evidence for Coulomb's law, however, is provided not directly, but indirectly by experimental verification of deductions from the law.

(*c*) *Permittivity*. The force between two charges also depends on what separates them; its value is always reduced when an insulating material replaces a vacuum. To take this into account a medium is said to have *permittivity*, denoted by ε (epsilon) and ε is included in the denominator of the expression for Coulomb's law. A material with high permittivity is one which reduces appreciably the force between two charges compared with the vacuum value.

SI units, which are used in this book, are 'rationalized'. This means that the values of certain constants are adjusted so that π does not occur in formulae in frequent use, but it does occur in others. The formulae thus conveniently simplified usually refer to situations in which there is plane symmetry (later we shall see that uniform electric and magnetic fields are in this category) and where we would not logically expect π to occur: in an unrationalized system it does. In a rationalized system 4π or π appear if there is spherical or cylindrical symmetry respectively (certain non-uniform fields are examples).

Spherical symmetry is associated with point charges (p. 229) and the equation for Coulomb's law is rationalized by including 4π in the denominator. Hence

$$F = \frac{1}{4\pi\varepsilon} \cdot \frac{Q_1Q_2}{r^2}$$

If F is in newtons, r in metres, Q_1 and Q_2 in coulombs (1 coulomb being the charge flowing per second through a conductor in which there is a steady current of 1 ampere) then the unit of ε is $C^2 N^{-1} m^{-2}$ since $\varepsilon = Q_1Q_2/(4\pi F r^2)$.

The permittivity of a vacuum is denoted by ε_0 (epsilon nought) and is called the *permittivity of free space*. The numerical value of ε_0 is found experimentally by an indirect method (p. 242) which does not involve the difficult task of measuring the force between known 'point' charges at a given separation in a vacuum. The result is

$$\varepsilon_0 = 8.85 \times 10^{-12} \ C^2 \ N^{-1} \ m^{-2}$$

We can also write

$$1/(4\pi\varepsilon_0) = 8.98 \times 10^9 \ N \ m^2 \ C^{-2}$$

and so
$$F \simeq 9 \times 10^9 \ Q_1Q_2/r^2$$

The permittivity of air at s.t.p. is $1.0005 \ \varepsilon_0$ and we can usually take ε_0 as the value for air. A more widely used unit for permittivity is the *farad per metre* (F m^{-1}), as explained later (p. 238).

Electric field strength

(*a*) *Definition*. A resultant force changes motion. Many everyday forces are pushes or pulls between bodies in contact. In other cases forces arise between bodies that are separated from one another. Electric, magnetic and gravitational effects involve such action-at-a-distance forces and to deal with them physicists find the idea of a *field of force* (or simply a *field*) useful. Fields of these three types have common features as well as important differences.

An electric field is a region where an electric charge experiences a force—just as a hayfield is a region in which hay is found. If a very small, positive point charge Q is placed at any point in an electric field and it experiences a force F, then the *field strength E* (also called the *E-field*) at that point is defined by the equation

$$E = \frac{F}{Q}$$

In words, the magnitude of E is the force per unit charge and its direction is that of F (i.e. of the force which acts on a positive charge). Field strength E is thus a vector. Note that we refer to E as the force *per* unit charge and not as the force *on* unit charge. A finite charge (such as a unit charge) might affect the field by inducing charges on neighbouring bodies and so we must *imagine* a very small test charge $+Q$ to be placed at the point since we require to know E before $+Q$ was introduced into the field.

If F is in newtons (N) and Q is in coulombs (C) then the unit of E is the newton per coulomb (N C^{-1}). A commoner but equivalent unit is the *volt per metre* (V m^{-1}), as we shall see later.

(*b*) *E due to a point charge.* The magnitude of E due to an isolated positive point charge $+Q$, at a point P distance r away, in a medium of permittivity ε, can be calculated by imagining a very small charge $+Q_0$ to be placed at P, Fig. 11.5. By Coulomb's law, the force F on Q_0 is

$$F = \frac{1}{4\pi\varepsilon} \cdot \frac{QQ_0}{r^2}$$

But E is the force per unit charge, that is

$$E = \frac{F}{Q_0}$$

$$= \frac{1}{4\pi\varepsilon} \cdot \frac{Q}{r^2}$$

E is directed away from $+Q$, as shown. If a point charge $-Q$ replaced $+Q$, E would be directed towards $-Q$ since unlike charges attract.

Medium of permittivity ε

Fig. 11.5

The above expression shows that E decreases with distance from the point charge according to an inverse square law. The field due to an isolated point charge is thus non-uniform but it has the same values at equal distances from the charge and so has spherical symmetry. In Fig. 11.6, if the magnitude of the field strength due to point charge $+Q$ is E at A, what is it at (*i*) B, (*ii*) C?

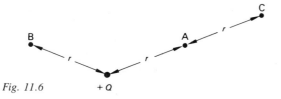

Fig. 11.6

(*c*) *Field strength and charge density.* So far as external effects are concerned an isolated spherical conductor having a charge Q uniformly distributed over its surface behaves like a point charge Q at its centre (p. 229). If r is the radius of the sphere, the field

strength E at its surface is thus given by

$$E = \frac{1}{4\pi\varepsilon} \cdot \frac{Q}{r^2}$$

The charge per unit area of the surface of the conductor is called the *charge density* σ (sigma) and since a sphere has surface area $4\pi r^2$ we have $\sigma = Q/(4\pi r^2)$. Therefore $Q = 4\pi r^2\sigma$ and so

$$E = \frac{\sigma}{\varepsilon}$$

This expression will be used later (p. 241). It has been derived by considering a sphere but it gives E at the surface of any charged conductor. Thus it can be seen to apply to a plane surface if the radius of the sphere, which does not appear in the expression, is allowed to tend to infinity.

Field lines

An electric field can be represented and so visualized by electric field lines. These are drawn so that (*i*) the field line at a point (or the tangent to it if it is curved) gives the direction of E at that point, i.e. the direction in which a positive charge would accelerate, and (*ii*) the number of lines per unit cross-section area is proportional to E. The field line is imaginary but the field it represents is real.

Electric field patterns similar to the magnetic field patterns given by iron filings can be obtained using tiny 'needles' of an insulating substance such as semolina powder or grass seeds, suspended by previous stirring in fresh castor oil in a glass dish. An electric field is created by applying a high p.d. from a van de Graaff

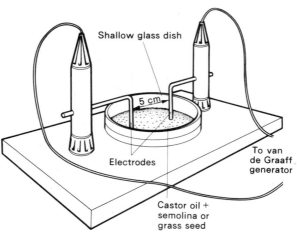

Fig. 11.7

generator to metal electrodes dipping in the oil, Fig. 11.7. The powder or seed orientates itself to form different patterns according to the shape of the electrodes.

Some electric field patterns are shown in Fig. 11.8. A uniform field is one in which E has the same magnitude and direction at all points, there is plane symmetry and the field lines are parallel and evenly spaced. In Fig. 11.8a the field is uniform between the plates away from the edges. When E varies in magnitude and direction with position, the field is non-uniform. In Fig. 11.8b the field is non-uniform but radial and there is spherical symmetry.

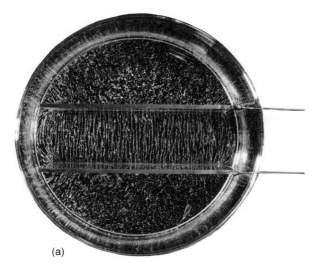

(a)

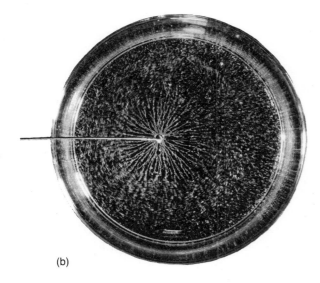

(b)

Fig. 11.8

Electric field patterns are useful in designing electronic devices such as cathode-ray tubes. The engineer is often able to sketch intuitively the pattern for a given electrode arrangement and so predict the probable behaviour of the device.

Field lines are also referred to as 'lines of force' and the term is appropriate when considering electric and gravitational fields because the field lines do indicate the direction in which a charge or mass experiences a force. This is not so in the magnetic case, as will be seen later, and therefore in general it is preferable to talk about field lines rather than lines of force.

Electric potential

Information about the field may be given by stating the *field strength* at any point; alternatively the *potential* can be quoted. Before discussing this idea some basic mechanics will be revised briefly.

(*a*) *Work and energy.* In science, work is done when the point of application of a force (or a component of it) undergoes a displacement in its own direction. The product of the force (or its component) F and the displacement s is taken as a measure of the work done W, i.e. $W = F \times s$. When F is in newtons and s in metres, W is in newton-metres or joules.

If a body A exerts a force on body B and work is done, a transfer of energy occurs *which is measured by the work done*. Thus, if we raise a mass m through a vertical height h, the work done W by the force we apply (i.e. by mg) is $W = mgh$ (assuming the earth's gravitational field strength g is constant). The energy transfer is mgh and we consider that the system gains and stores that amount of gravitational potential energy in its gravitational field. This energy is obtained from the conversion of chemical energy by our muscular activity. When the mass falls the system loses gravitational potential energy and, neglecting air resistance, there is a transfer of kinetic energy to the mass equal to the work done by gravity.

(*b*) *Meaning of potential.* A charge in an electric field experiences a force and if it moves work will, in general, be done. If a positive charge is moved from A to B in a direction opposite to that of the field E, Fig. 11.9a, an external agent has to do work against the forces of the field and energy has to be supplied. As a result, the system (of the charge in the field) gains an amount of electrical potential energy equal to the work done. This is analogous to a mass being raised in the earth's gravitational field g, Fig. 11.9b. When the charge is allowed to return from B to A, work is done by the forces of the field and the electrical potential energy previously gained by the system is lost. If, for example, the motion is in a vacuum, an equivalent amount of kinetic energy is transferred to the charge.

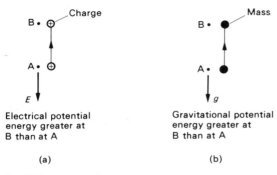

Electrical potential energy greater at B than at A

(a)

Gravitational potential energy greater at B than at A

(b)

Fig. 11.9

In general, the potential energy associated with a charge at a point in an electric field depends on the location of the point and the magnitude of the charge (since the force acting depends on the latter, i.e. $F = QE$). Therefore if we state the magnitude of the charge we can describe an electric field in terms of the potential energies of that charge at different points. A unit positive charge is chosen and the change of potential energy which occurs when such a charge is moved from one point to another is called the change of *potential* of the field itself.

Hence the potential at B in Fig. 11.9a exceeds that at A by the work which must be done against the electric force to take unit positive charge from A to B. To be strictly accurate, however, as we were when we defined E (p. 225), we should refer to the work done *per* unit charge when a *very small* charge moves from one point to the other since the introduction of a unit charge would in general modify the field.

If for theoretical purposes we select as the zero of potential the potential at an infinite distance from any electric charges, then the *potential at a point in a field can be defined as the work done per unit positive charge moving from infinity to the point*, always assuming the charge does not affect the field. The choice of the zero of potential is purely arbitrary and whilst infinity may be a few hundred metres in some cases, in atomic physics where distances of 10^{-10} m are involved, it need only be a very small distance away from the charge responsible for the field.

Potential is a property of a *point* in a field and is a scalar since it deals with a quantity of work done or potential energy per unit charge. The symbol for potential is V and the unit a *joule per coulomb* (J C^{-1}) or *volt* (V).

Just as a mass moves from a point of higher gravitational potential to one of lower potential (i.e. to fall towards the earth's surface), so a positive charge is urged by an electric field to move from a point of higher electrical potential to one of lower potential. Negative charges move in the opposite direction if free to do so.

(*c*) *Potential and field strength compared.* When describing a field, potential is usually a more useful quantity than field strength because, being a scalar, it can be added directly when more than one field is concerned. Field strength is a vector and addition by the parallelogram law is more complex. Also, it is often more important to know what energy changes occur (rather than what forces act) when charges move in a field and these are readily calculated if potentials are known (see p. 232).

Equipotentials

All points in a field which have the same potential can be imagined as lying on a surface—called an *equipotential* surface. When a charge moves on such a surface no energy change occurs and no work is done. The force due to the field must therefore act at right angles to the

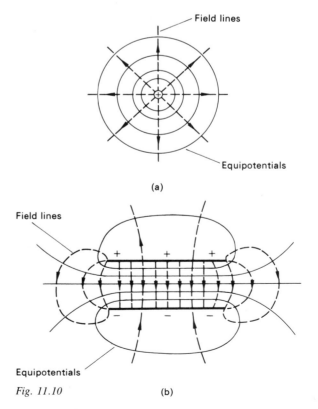

(a)

(b)

Fig. 11.10

equipotential surface at any point and so equipotential surfaces and field lines always intersect at right angles.

A field can therefore be represented pictorially by field lines and by equipotential surfaces (or lines in two dimensional diagrams). Equipotential surfaces for a point charge are concentric spheres (circles in two dimensions), Fig. 11.10a; there is spherical symmetry. The plane symmetry of a uniform field is seen in Fig. 11.10b. If equipotentials are drawn so that the change of potential from one to the next is constant, then the spacing will be closer where the field is stronger. To perform a certain amount of work in such regions a shorter distance need be travelled.

The surface of a conductor in electrostatics (i.e. one in which no current is flowing) must be an equipotential surface since any difference of potential would cause a redistribution of charge in the conductor until no field existed in it.

Potential due to a point charge

We wish to find the potential at A in the field of—and distant r from—an isolated point charge $+Q$ situated at O in a medium of permittivity ε, Fig. 11.11. Imagine a very small point charge $+Q_0$ is moved by an external agent from C, distance x from A, through a very small distance δx to B without affecting the field due to $+Q$.

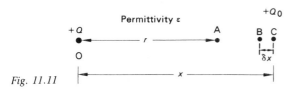

Fig. 11.11

Assuming the force F on Q_0 due to the field remains constant over δx, the work done δW by the external agent over δx against the force of the field is

$$\delta W = F(-\delta x)$$

The negative sign is inserted to show that the displacement δx is in the opposite direction to that in which F acts. By Coulomb's law

$$F = \frac{QQ_0}{4\pi\varepsilon} \cdot \frac{1}{x^2}$$

$$\therefore \quad \delta W = \frac{QQ_0}{4\pi\varepsilon} \cdot \frac{(-\delta x)}{x^2}$$

The total work done W in bringing Q_0 from infinity to A is

$$W = \frac{-QQ_0}{4\pi\varepsilon} \int_\infty^r \frac{dx}{x^2} = \frac{-QQ_0}{4\pi\varepsilon} \left[\frac{-1}{x} \right]_\infty^r$$

$$= \frac{QQ_0}{4\pi\varepsilon} \cdot \frac{1}{r}$$

The potential V at A is the work done per unit positive charge brought from infinity to A. Hence

$$V = \frac{W}{Q_0} = \frac{1}{4\pi\varepsilon} \cdot \frac{Q}{r}$$

What would be the analogous expression for the gravitational potential V at a distance r from a point mass M?

Potential due to a conducting sphere

(a) An expression. A charge $+Q$ on an isolated conducting sphere is uniformly distributed over its surface (due to the repulsion of like charges) and has a radial electric field pattern, Fig. 11.12a. The field at any point outside the sphere is exactly the same as if the whole charge were concentrated at a point charge $+Q$ at the centre of the sphere, Fig. 11.12b. (This can be shown to follow from Coulomb's law; a spherical mass similarly behaves as if its whole mass were concentrated at its centre.)

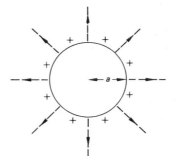

(a) Charged sphere with charge $+Q$ on its surface

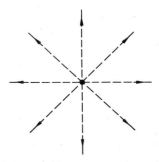

Fig. 11.12

(b) Point charge $(+Q)$

From the expression already obtained for a point charge, we can say that the potential V at a point P distance r from the centre of the sphere, is

$$V = \frac{1}{4\pi\varepsilon} \cdot \frac{Q}{r}$$

If the radius of the sphere is a, the potential V at its surface is

$$V = \frac{1}{4\pi\varepsilon} \cdot \frac{Q}{a}$$

At all points inside the sphere the field strength is zero, otherwise field lines would link charges of opposite sign in the sphere and such a state of affairs is impossible under static conditions in a conductor. (This may also be shown to be a result of Coulomb's law.) It follows that no work is done when a charge is moved between any two points inside the sphere. The potential is thus the same at all points throughout the sphere and equal to that at the surface. As well as talking about the potential at a point in a field we also consider an insulated conductor to have a potential.

The variation of E and V at points outside and inside a positively charged conducting sphere should therefore be as shown in the graphs of Fig. 11.13. Outside the sphere E varies as $1/r^2$ and, as we have just seen, theory predicts that V varies as $1/r$.

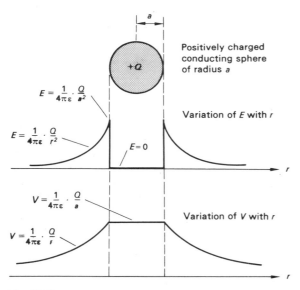

Fig. 11.13

(b) *Flame probe investigation.* The results for V outside may be investigated experimentally using the apparatus of Fig. 11.14. A probe,[1] in the form of a small gas flame at the point of a hypodermic needle, is connected to an electroscope calibrated to measure potentials (see p. 251). The potential measured is that at the probe. The conducting sphere is charged to 1500 V from an e.h.t. power supply and the potentials (V) noted from the electroscope when the probe is at different distances (r) from the centre of the sphere. (In particular, at a distance of twice the radius from the centre of the sphere the potential should be 750 V.) The sphere should be 'isolated' (as we shall presently) by being well away from walls, bench tops, the experimenter etc., and the lead from the probe must also be clear of the bench. If the results confirm that $V \propto 1/r$ then this may be taken as indirect evidence of Coulomb's law since the $1/r$ law for potential is a consequence of a $1/r^2$ law for E.

(c) *Effect of neighbouring bodies.* The electric potential at a point in a field due to a charged body is not determined solely by the charge unless the body is 'isolated' as we assume in theory. In practice it is affected by the presence of other bodies, charged or uncharged, and the surrounding material. Thus the potential at a point near a positively charged body increases when another positively charged body approaches and decreases when a negatively charged body is brought up.

In Fig. 11.15 the variation of potential with distance from a positively charged sphere A is shown before and after an uncharged conductor BC is brought near. Initially each point on BC is at the potential which previously existed at that point. Momentarily, therefore, B is at a higher potential than C and so, since BC is a conductor, electrons flow from C to B (i.e. from a lower to a higher potential) until the potential is the same all over BC—called the *potential of the conductor*. Electrostatic induction has occurred and the induced negative charge at B lowers the potential there, as well as at all points between A and B (including that of A). The induced positive charge at C raises the potential at C and at points beyond. The constant potential of BC lies between the original potentials at B and C.

Potential difference

The idea of potential applies not only to the electric fields produced in air or a vacuum by static charges (on an insulator or a conductor) but also to those in a wire

---

[1] The construction and action of the probe are described in Appendix 8.

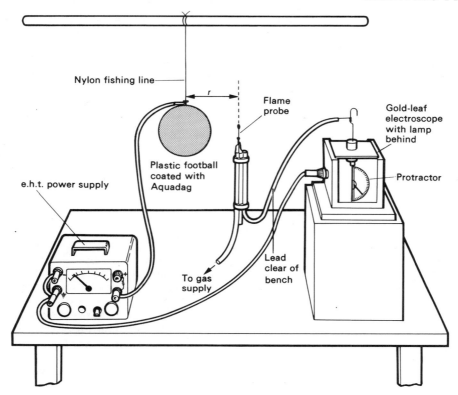

Fig. 11.14

having a battery or power supply across its ends and which cause charges to move as an electric current in the wire. The term *potential* is useful in both electrostatics and in current electricity.

However, there is an important difference between the two cases. In electrostatics when a charge moves in the direction of the field, the potential energy lost by

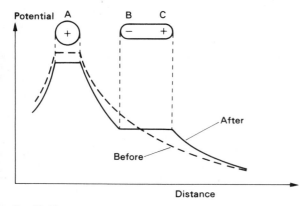

Fig. 11.15

the field-charge system can be regained if the charge is moved by an external agent in the opposite direction. In current electricity, energy lost by the electric field inside a conductor is irrecoverable since the heat produced cannot be converted back into other forms of energy by reversing the current.

In practice, especially in current electricity, we are usually concerned with the difference of potential or p.d. between two points in an electric field so that the zero of potential does not matter. The p.d. between two points is the *work done per unit charge* (or the energy change per unit charge) passing from one point to the other. The same symbol is used for p.d. as for potential, i.e. V, and the same unit, i.e. joule per coulomb (J C^{-1}) or volt (V).

If the p.d. between two points in an electric field is 10 volts then 10 joules of work are done per coulomb of charge moving from one point to the other and an energy change of 10 joules per coulomb occurs. In general if V is the p.d. (in V) between two points in an electric field, the energy change W occurring when a charge Q (in C) moves through the p.d. is given by

$$W = QV$$

Two examples follow to show how, using this expression, energy changes in electric fields can be calculated.

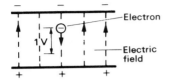

Fig. 11.16

First, suppose we wish to know the energy change when an electron 'falls' through a p.d. of 1 V in an electric field in a vacuum, i.e. it travels between two points whose p.d.s. differ by 1 V, Fig. 11.16. The electron is accelerated by the force acting on it due to the field and work is done. The energy change W is found from $W = QV$ where $Q = 1.6 \times 10^{-19}$ C (the electronic charge) and $V = 1.0$ V $= 1.0$ J C^{-1}

$$\therefore \quad W = (1.6 \times 10^{-19} \text{ C}) (1.0 \text{ J C}^{-1})$$

$$= 1.6 \times 10^{-19} \text{ J}$$

This tiny amount of energy is called an *electron-volt* (eV) and is a unit of energy (not SI) much used in atomic physics.

Second, an example from current electricity. If a p.d. of 12 V maintains a current of 3.0 A through a resistor, the electrical energy W (from the electric field in the resistor) changed to heat per second is obtained from

$$W = QV = ItV \text{ (since } Q = It)$$

where $I = 3.0$ A $= 3.0$ C s^{-1}, $t = 1.0$ s and $V = 12$ V $= 12$ J C^{-1}.

$$\therefore \quad W = (3.0 \text{ C s}^{-1}) (1.0 \text{ s}) (12 \text{ J C}^{-1})$$

$$= 3.0 \times 1.0 \times 12 \quad \text{C s}^{-1} \times \text{s} \times \text{J C}^{-1}$$

$$= 36 \text{ J}$$

Relation between E and V

Consider a charge $+Q$ at a point A in an electric field where the field strength is E, Fig. 11.17. The force F on Q is given by

$$F = EQ$$

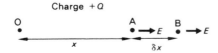

Fig. 11.17

If Q moves a very short distance δx from A to B in the direction of E, then (assuming E is constant over AB) the work done δW by the electric force on Q is

$$\delta W = \text{force} \times \text{distance}$$

$$= F \, \delta x$$

$$= EQ \, \delta x$$

If the p.d. between B and A is δV, we have by the definition of p.d.

$$\delta V = \text{work done per unit charge}$$

$$= -\frac{\delta W}{Q} = -\frac{EQ \, \delta x}{Q}$$

That is,

$$\delta V = - E \, \delta x$$

The negative sign is inserted to show that if displacements in the direction of E are taken to be positive, then when δx is positive, δV is negative, i.e. the potential decreases. On the other hand, if the charge is moved in a direction opposite to that of E, δx is negative and δV is positive, indicating an increase of potential as occurs in practice.

In the limit, as $\delta x \to 0$, E becomes the field strength at a point (A) and in calculus notation

$$E = - \frac{dV}{dx}$$

dV/dx is called the *potential gradient* in the x-direction and so the field strength at a point equals the negative of the potential gradient there. Potential gradient is a vector and is measured in volts per metre (V m^{-1}). It follows that this is also a unit of E (as well as N C^{-1}—an equivalent but less-used unit).

In a uniform field E is constant in magnitude and direction at all points, hence dV/dx is constant, i.e. the potential changes steadily with distance. The field near the centre of two parallel metal plates is uniform and if this is created by a p.d. V between plates of separation d, then

$$E = - \frac{V}{d}$$

where E is the field strength at any point in the *uniform* region (not at the edges). For example, if $V = 2.0 \times 10^3$ V and $d = 1.0$ cm $= 1.0 \times 10^{-2}$ m,

$$E = V/d \text{ (numerically)}$$

$$= (2.0 \times 10^3 \text{ V})/(1.0 \times 10^{-2} \text{ m})$$

$$= 2.0 \times 10^5 \text{ V m}^{-1} \text{ (or N C}^{-1})$$

Gravitational analogy

Analogies exist between electric and gravitational fields.

(a) *Inverse square law of force.* Coulomb's law is similar in form to Newton's law of universal gravitation. Both are inverse square laws with $1/(4\pi\varepsilon)$ in the electric case corresponding to the gravitational constant G. The main difference is that whilst electric forces can be attractive or repulsive, gravitational forces are always attractive. Two types of charge are known but there is only one type of matter. By comparison with electric forces, gravitational forces are extremely weak (p. 236).

Coulomb's law

$$F = \frac{1}{4\pi\varepsilon} \cdot \frac{Q_1 Q_2}{r^2}$$

Newton's law

$$F = G \cdot \frac{m_1 m_2}{r^2}$$

(b) *Field strength.* The field strength at a point in a gravitational field is defined as the force acting per unit mass placed at the point. Thus if a mass m in kilograms experiences a force F in newtons at a certain point in the earth's field, the strength of the field at that point will be F/m in newtons per kilogram. This is also the acceleration a the mass would have in metres per second squared if it fell freely under gravity at this point (since $F = ma$). The gravitational field strength and the acceleration due to gravity at a point thus have the same value (i.e. F/m) and the same symbol, g, is used for both. At the earth's surface $g = 9.8$ N kg^{-1} = 9.8 m s^{-2} (vertically downwards).

Electric field strength
(at distance r from point charge Q)

$$E = \frac{1}{4\pi\varepsilon} \cdot \frac{Q}{r^2} \text{ (in N C}^{-1})$$

Gravitational field strength
(at distance r from point mass m)

$$g = G \cdot \frac{m}{r^2} \text{ (in N kg}^{-1})$$

Expressing forces in terms of field strengths we also have

Electric force
(on a charge Q)

$$F = QE$$

Gravitational force
(on a mass m)

$$F = mg$$

(c) *Field lines and equipotentials.* These can also be drawn to represent gravitational fields but such fields are so weak, even near massive bodies, that there is no method of plotting field lines similar to those used for electric (and magnetic) fields. Field lines for the earth are directed towards its centre and the field is spherically symmetrical. Over a small part of the earth's surface the field can be considered uniform, the lines being vertical, parallel and evenly spaced. Figs. 11.18a and b represent uniform electric and gravitational fields.

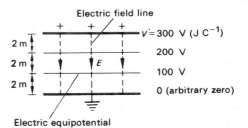

$$E = \frac{V}{d} = \frac{300 \text{ V}}{6 \text{ m}} = 50 \text{ V m}^{-1}$$

(a)

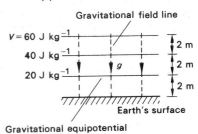

$$g = \frac{V}{d} = \frac{60 \text{ J kg}^{-1}}{6 \text{ m}} = 10 \text{ N kg}^{-1}$$

(b)

Fig. 11.18

(d) *Potential and p.d.* Electric potentials and p.d.s are measured in joules per coulomb (J C^{-1}) or volts; gravitational potentials and p.d.s are measured in joules per kilogram (J kg^{-1}). If the p.d. between two points in the earth's gravitational field is 20 J kg^{-1}, the work done and the change of potential energy will be 20 J when 1 kg moves from one point to the other. In general if V is the p.d. between two points, the energy change W which occurs when a mass m moves from one point to the other is given by $W = mV$. From this expression the energy required to send a rocket from one point in space to another can be calculated (see p. 235).

Electrical p.d. and energy change

$$W = QV$$

Gravitational p.d. and energy change

$$W = mV$$

As a mass moves away from the earth the potential energy of the earth-mass system increases, transfer of energy from some other source being necessary. If infinity is taken as the zero of gravitational potential (i.e. a point well out in space where no more energy is needed for the mass to move further away from the earth) then the potential energy of the system will have a negative value except when the mass is at infinity. At every point in the earth's field the potential is therefore negative (see expression below), a fact which is characteristic of fields that exert attractive forces. Fields giving rise to repulsive forces cause positive potentials.

Electric potential
(at distance r from point charge Q)

$$V = \frac{1}{4\pi\varepsilon} \cdot \frac{Q}{r}$$

Gravitational potential
(at distance r from point mass m)

$$V = -G \cdot \frac{m}{r}$$

Fig. 11.19a shows the variation of gravitational potential V with distance r from the centre of a spherical mass or the variation of electrical potential V with distance r from the centre of a negatively charged sphere, i.e. it applies to an attractive force field. Fig. 11.19b shows the variation of electrical potential V with distance r from the centre of a positively charged sphere, i.e. it refers to a repulsive force field.

Field treatment of space travel

The motion of spacecraft and satellites in the earth's gravitational field is often considered in terms of 'forces'. As an alternative, a 'field' treatment may be given which involves working out energy changes from potential differences.

(*a*) *Satellite orbits.* The gravitational potential V at a distance r from the centre of the earth is $-GM/r$ where M is the earth's mass. Potential equals the potential energy per unit mass (or charge, in the electrical case, see p. 228) and so a satellite of mass m at a distance r from the centre of the earth has potential energy $E_p = -GMm/r$. If the satellite is moving with speed v in a circular orbit of radius r, Fig. 11.20, its kinetic energy $E_k = \frac{1}{2}mv^2$. Hence the total energy E of the satellite in orbit is given by

$$E = E_p + E_k = -GMm/r + \tfrac{1}{2}mv^2$$

From the expression for centripetal force and the law of gravitation we have

$$\frac{mv^2}{r} = \frac{GMm}{r^2}$$

$$\therefore \quad mv^2 = \frac{GMm}{r}$$

Substituting for mv^2 in the expression for E, we can say

$$E = -\frac{GMm}{r} + \tfrac{1}{2} \cdot \frac{GMm}{r} = -\tfrac{1}{2} \cdot \frac{GMm}{r}$$

The potential energy E' of a mass m on the earth's surface is

$$E' = -GMm/R$$

where R is the radius of the earth. Hence to put a satellite into circular orbit of radius r it must be given the extra amount of energy $(E - E')$ where

$$E - E' = -\tfrac{1}{2} \cdot \frac{GMm}{r} - \frac{(-GMm)}{R}$$

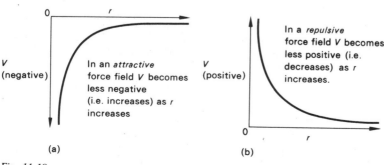

(a)

Fig. 11.19

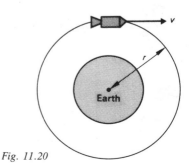

Fig. 11.20

$$E - E' = GMm \left(\frac{1}{R} - \frac{1}{2r} \right)$$

If the orbit is close to the earth's surface (e.g. at a height of 100–200 km), $r \simeq R$ and

$$E - E' = \tfrac{1}{2} \cdot \frac{GMm}{R}$$

If this extra energy is all given in the form of kinetic energy ($\tfrac{1}{2}mu^2$) at lift-off, then

$$\tfrac{1}{2}mu^2 = \tfrac{1}{2}\frac{GMm}{R}$$

$$\therefore \quad u^2 = \frac{GM}{R}$$

Taking $G = 6.7 \times 10^{-11}\,\text{N m}^2\,\text{kg}^{-2}$, $M = 6.0 \times 10^{24}\,\text{kg}$ and $R = 6.4 \times 10^6\,\text{m}$, and substituting in the expression for u^2, we get $u = 8.0\,\text{km s}^{-1}$. The speed the satellite must be given at launching to go into a circular orbit close to the earth's surface will be greater than $8.0\,\text{km s}^{-1}$ since air resistance has been neglected.

In practice circular orbits are seldom attained: most are elliptical.

(b) *Speed of escape.* Some values are given in Table 11.1 of the gravitational potential ($-GM/r$) at different distances r from the centre of the earth.

As r increases, the potential becomes less negative, i.e. it increases, and the potential difference between any two values of r equals the change of potential energy when a mass of 1 kg moves from one point to the other. For example, if a spacecraft travels from the earth's surface ($r = 6.4 \times 10^6\,\text{m}$) to the moon ($r = 400 \times 10^6\,\text{m}$), the gain of potential energy per kg is $(63 - 1) \times 10^6 = 62 \times 10^6\,\text{J kg}^{-1}$. The spacecraft loses this amount of kinetic energy and so the energy needed per kg to get the craft from the earth's surface to the distance of the moon is $62 \times 10^6\,\text{J kg}^{-1}$.

To escape to 'infinity' $63 \times 10^6\,\text{J kg}^{-1}$ is required, and if this is given to a spacecraft of mass m at lift-off as kinetic energy, we can say

$$\tfrac{1}{2}mu^2 = 63 \times 10^6\,m$$

where u is the 'escape' speed from the surface of the earth. Hence

$$u \simeq 11\,\text{km s}^{-1}$$

Potential and field strength calculations

1. Find (a) the potential and (b) the field strength at points A and B, Fig. 11.21, due to two small spheres X and Y, 1.0 m apart in air and carrying charges of $+2.0 \times 10^{-8}\,\text{C}$ and $-2.0 \times 10^{-8}\,\text{C}$ respectively. Assume the permittivity of air $= \varepsilon_0$ and $1/(4\pi\varepsilon_0) = 9.0 \times 10^9\,\text{N m}^2\,\text{C}^{-2}$.

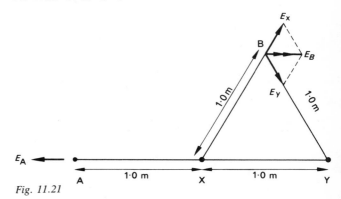

Fig. 11.21

(a) Potential at A due to X

$$= \frac{1}{4\pi\varepsilon_0} \cdot \frac{Q}{r}$$

$$= \frac{(9.0 \times 10^9\,\text{N m}^2\,\text{C}^{-2}) \times (+2.0 \times 10^{-8}\,\text{C})}{1.0\,\text{m}}$$

$$= \frac{9.0 \times 10^9 \times 2.0 \times 10^{-8}}{1.0} \qquad \frac{\text{N m}^2\,\text{C}^{-2}\,\text{C}}{\text{m}}$$

$$= +1.8 \times 10^2\,\text{N m C}^{-1} \qquad (1\,\text{N m} = 1\,\text{J})$$

$$= +1.8 \times 10^2\,\text{V} \quad (1\,\text{J C}^{-1} = 1\,\text{V})$$

Similarly the potential at A due to Y

$$= \frac{9.0 \times 10^9 \times (-2.0 \times 10^{-8})}{2.0}\,\text{V}$$

$$= -0.90 \times 10^2\,\text{V}$$

Potential is a scalar quantity and is added algebraically, therefore the potential V_A at A due to X and Y is

$$V_A = (1.8 - 0.90) \times 10^2\,\text{V}$$

$$= 0.90 \times 10^2\,\text{V} = 90\,\text{V}$$

Table 11.1

$r/10^6$ m	6.4	6.6	20	40	400	∞
$(-GM/r)/10^6$ J kg^{-1}	-63	-61	-20	-10	-1.0	0.00

Since B is equidistant from equal and opposite charges, the potential V_B at B due to X and Y is

$$V_B = (1.8 - 1.8) \times 10^2 \text{ V} = 0$$

(b) Field strength at A due to X

$$= \frac{1}{4\pi\varepsilon_0} \cdot \frac{Q}{r^2}$$

$$= \frac{9.0 \times 10^9 \times 2.0 \times 10^{-8}}{1.0^2} \quad \frac{\text{N m}^2 \text{ C}^{-2} \text{ C}}{\text{m}^2}$$

$$= 1.8 \times 10^2 \text{ N C}^{-1} \text{ (or V m}^{-1}) \quad \text{towards the left}$$

Similarly the field strength at A due to Y

$$= 0.45 \times 10^2 \text{ N C}^{-1} \text{ (or V m}^{-1}) \text{ towards the right}$$

The resultant field strength E_A at A due to X and Y is

$$E_A = (1.8 - 0.45) \times 10^2 \text{ N C}^{-1} \text{ (or V m}^{-1})$$

$$= 1.3(5) \times 10^2 \text{ N C}^{-1} \text{ (or V m}^{-1}) \text{ towards the left}$$

At B, the field strengths due to X and Y have the directions shown and the resultant field strength E_B is their vector sum. Hence

$$E_B = E_X \cos 60 + E_Y \cos 60$$

$$= E_X \text{ (since } E_X = E_Y \text{ and } \cos 60 = \tfrac{1}{2})$$

$$= \frac{9.0 \times 10^9 \times 2.0 \times 10^{-8}}{1.0^2}$$

$$= 1.8 \times 10^2 \text{ N C}^{-1} \text{ (or V m}^{-1}) \quad \text{towards the right}$$

2. *Two large horizontal, parallel metal plates are 2.0 cm apart in vacuo and the upper is maintained at a positive potential relative to the lower so that the field strength between them is 2.5×10^5 V m^{-1}. (a) What is the p.d. between the plates? (b) If an electron of charge 1.6×10^{-19} C and mass 9.1×10^{-31} kg is liberated from rest at the lower plate, what is its speed on reaching the upper plate?*

(a) If E is the field strength (assumed uniform) and V is the p.d. between two plates distance d apart, we have from $E = V/d$ (p. 232) that

$$V = Ed$$

$$= (2.5 \times 10^5 \text{ V m}^{-1}) \times (2.0 \times 10^{-2} \text{ m})$$

$$= 5.0 \times 10^3 \text{ V}$$

(b) The energy change (i.e. work done) W which occurs when a charge Q moves through a p.d. of V in an electric field is given by $W = QV$. There is a transfer of electrical potential energy from the field to kinetic energy of the electron. We have

$$QV = \tfrac{1}{2}mv^2$$

where v is the required speed and m is the mass of the electron.

Therefore

$$\tfrac{1}{2}mv^2 = QV$$

$$\therefore \quad v = \sqrt{\frac{2QV}{m}}$$

$$= \sqrt{\frac{(2 \times 1.6 \times 10^{-19} \text{ C}) \times (5.0 \times 10^3 \text{ V})}{(9.1 \times 10^{-31} \text{ kg})}}$$

$$= \sqrt{\frac{2 \times 1.6 \times 10^{-19} \times 5.0 \times 10^3}{9.1 \times 10^{-31}}} \quad \frac{\text{C J C}^{-1}}{\text{kg}}$$

$$(1 \text{ V} = 1 \text{ J C}^{-1})$$

$$= \sqrt{\frac{16}{9.1} \times 10^{15}} \text{ m}^2 \text{ s}^{-2}$$

$$(1 \text{ J} = 1 \text{ N m} = 1 \text{ kg m s}^{-2} \text{ m})$$

$$= 4.2 \times 10^7 \text{ m s}^{-1}$$

3. *Compare the electrical and gravitational forces between a proton of charge $+e$ and mass M at a distance r from an electron of charge $-e$ and mass m (e.g. as in a hydrogen atom), given that*

$$e = 1.6 \times 10^{-19} \text{ C}$$
$$m = 9.1 \times 10^{-31} \text{ kg}$$
$$M = 1.7 \times 10^{-27} \text{ kg}$$
$$1/(4\pi\varepsilon_0) = 9.0 \times 10^9 \text{ N m}^2 \text{ C}^{-2}$$
$$G = 6.7 \times 10^{-11} \text{ N m}^2 \text{ kg}^{-2}$$

Electrical attraction $= \dfrac{1}{4\pi\varepsilon_0} \cdot \dfrac{Q_1 Q_2}{r^2} = \dfrac{1}{4\pi\varepsilon_0} \cdot \dfrac{e^2}{r^2}$

Gravitational attraction $= \dfrac{GmM}{r^2}$

Hence,

$$\frac{\text{electrical attraction}}{\text{gravitational attraction}} = \frac{1}{4\pi\varepsilon_0} \cdot \frac{e^2}{GmM}$$

$$= \frac{(9.0 \times 10^9 \text{ N m}^2 \text{ C}^{-2})}{(6.7 \times 10^{-11} \text{ N m}^2 \text{ kg}^{-2})} \times$$

$$\frac{(1.6 \times 10^{-19} \text{ C})^2}{(9.1 \times 10^{-31} \text{ kg}) \times (1.7 \times 10^{-27} \text{ kg})}$$

$$\simeq 10^{39} : 1$$

The gravitational force is insignificant compared with the electrical force.

QUESTIONS

1. An isolated conducting spherical shell of radius 10 cm in vacuo carries a positive charge of 1.0×10^{-7} C. Calculate (a) the electric field intensity, and (b) the potential, at a point on the surface of the conductor. Sketch a graph to show how one of these quantities varies with distance along a radius from the centre to a point well outside the spherical shell. Point out the main features of the graph. (Permittivity of free space $= 8.9 \times 10^{-12}$ F m^{-1}.) (J.M.B.)

2. Considering a hydrogen atom to consist of a proton and an electron at an average separation of 0.50×10^{-10} m, find (a) the electrical potential at 0.50×10^{-10} m from the proton, (b) the potential energy of the electron, and (c) the energy required to remove the electron from the atom assuming the electron is at rest. (Charge on proton $= +1.6 \times 10^{-19}$C; charge on electron $= -1.6 \times 10^{-19}$ C; $1/(4\pi\varepsilon_0) = 9.0 \times 10^9$ N m^2 C^{-2}.) *Note.* In practice other factors as well as the potential energy have to be considered when calculating the energy needed to remove an electron from an atom. Suggest one.

3. Four infinite conducting plates A, B, C and D, of negligible thickness, are arranged parallel to one another so that the distance between adjacent plates is 2.0 cm. The outer plates A and D are earthed and the inner ones B and C are maintained at potentials of +20 V and +60 V respectively. Draw a graph showing how the potential between the plates varies as a function of position along a line perpendicular to the plates. What is the magnitude and direction of the electric field between each pair of adjacent plates? (O. and C. part qn.)

4. Many electrical insulators are ionic compounds consisting of arrays of dipoles, i.e. pairs of equal but oppositely charged ions a small distance apart. The properties of the dipoles account for the fact that such compounds have very strong internal electric fields but almost zero external fields.

In Fig. 11.22a, A and B form an ionic dipole with charges of $+1.6 \times 10^{-19}$ C and -1.6×10^{-19} C respectively and at a separation of 2.0×10^{-10} m. What is the field strength due to the dipole at (a) X, mid-way between A and B, and (b) Y, 50×10^{-10} m to the right of A? Find the ratio of these two field strengths. $(1/(4\pi\varepsilon_0) = 9 \times 10^9$ N m^2 C$^{-2})$.

In Fig. 11.22b A is a single charge of $+1.6 \times 10^{-19}$ C. What is the field at (c) X, and (d) Y? Again, find the ratio.

Compare the two ratios. Does the external field due to a dipole decrease more rapidly with distance than that of a single charge?

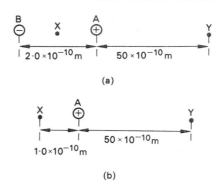

(a)

(b)

Fig. 11.22

5. The diagram, Fig. 11.23, shows a pair of flat, wide conducting plates. They are parallel and are connected to a steady potential difference of 2000 V. An oil drop between the plates moves from D_1 to D_2 along a straight line at right angles to the plates and between their centres.

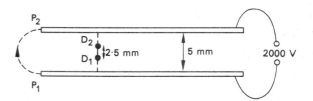

Fig. 11.23

(i) If the drop carries five electron charges, each 1.6×10^{-19} C, and moves 2.5 mm (half the distance between the plates), how much electrical energy is transformed?

Give the unit of energy appropriate to your answer.

(ii) State what you can about the amount of electrical energy transformed if the drop were to move along the curved path P_1P_2 from one plate to the other outside the plates.

Explain your statement. (O. and C. Nuffield)

6. An electric field is maintained between two parallel metal plates by a p.d. of 1000 V. A charge of 10^{-15} C is moved by the forces of the field from one plate to the other. (i) How much electrical energy is transferred from the field charge system? (ii) How much work is done by the electrical force? (iii) If the plates are 20 mm apart what is the value of the electrical force assuming it is constant, i.e. the field between the plates is uniform? (iv) What is the field strength in N C^{-1}? (v) What is the field strength in V m^{-1}?

12 Capacitors

Capacitance

(a) *Definition.* Charge given to an isolated conductor may be thought of as being 'stored' on it. The amount it will take depends on the electric field thereby created at the surface of the conductor. If this is too great there is breakdown in the insulation of the surrounding medium, resulting in sparking and discharge of the conductor.

In terms of potential we can say that the smaller the change of potential of a conductor when a certain charge is transferred to it, the more charge it can 'store' before breakdown occurs. The change in potential due to a given charge depends on the size of the conductor, the material surrounding it and the proximity of other conductors.

The idea that an insulated conductor in a particular situation has a certain *capacitance*, or charge-storing ability, is useful and is defined as follows: if the potential of an insulated conductor changes by V when given a charge Q, the *capacitance* C of the conductor is

$$C = \frac{Q}{V}$$

In words, *capacitance equals the charge required to cause unit change in the potential of a conductor*. If Q is in coulombs (C) and V in volts (V) then the unit of C is a coulomb per volt ($C\,V^{-1}$) or a *farad* (F). This is a very large unit and the microfarad (1 μF = 10^{-6} F), the nanofarad (1 nF = 10^{-9} F) or the picofarad (1 pF = 10^{-12} F) are generally used.

Experiment (see p. 239) and theory both show that for an insulated conductor in a given situation, $V \propto Q$, i.e. C is a constant.

(b) *Capacitance of an isolated conducting sphere.* The potential V of such a sphere of radius a, in a medium of permittivity ε and having a charge Q, is given by

$$V = \frac{1}{4\pi\varepsilon} \cdot \frac{Q}{a} \qquad \text{(p. 230)}$$

But

$$C = \frac{Q}{V}$$

$$\therefore \quad C = 4\pi\varepsilon \cdot a$$

The capacitance of a sphere is therefore proportional to its radius. We also see from this expression that if C is in farads and a in metres then ε can be expressed in farads per metre (F m^{-1}), since $\varepsilon = C/(4\pi a)$.

(c) *Practical zero of potential.* The theoretical zero of potential is taken, as we have seen, as the potential of points at infinity. However, when making actual measurements this is an impracticable zero and the potential of the earth (itself a conductor) is adopted as the practical zero.

When a charged conductor X of small capacitance and an uncharged conductor Y of large capacitance are connected, they end up with the same potential which is less than X's was originally and X loses most of its charge to Y. There is charge flow until the potentials of X and Y are the same.

If a is the radius of the earth, its capacitance C is

$$C = 4\pi\varepsilon \cdot a$$

$$= (1/9 \times 10^{-9} \text{ F m}^{-1}) \times (6 \times 10^6 \text{ m})$$

$$= 7 \times 10^{-4} \text{ F}$$

$$= 700 \ \mu\text{F}$$

This is a large capacitance compared with that of other conductors used in electrostatics. Therefore when a charged conductor is 'earthed'—say by touching it (the human body conducts)—it loses most of its charge to the earth, i.e. is discharged and acquires earth potential, i.e. zero potential. The earth has such a large capacitance that any change in its potential due to the loss or gain of charge, because of connection to another conductor, is negligible. It thus provides a satisfactory practical zero of potential.

Capacitors

A capacitor is designed to store electric charge and basically consists of two conductors, such as a pair of parallel metal plates, separated by an insulator. The symbol for a capacitor is ⊣⊢ and the conductors are usually referred to as 'plates' whatever their form.

(*a*) *Action of a capacitor.* A capacitor is readily 'charged' by applying a p.d. across the plates from a battery or other power supply. Insight into the process can be obtained from the circuit of Fig. 12.1 using a 500 μF electrolytic capacitor (see p. 245).

When the circuit is completed, equal-sized flicks occur on the meters indicating momentary but identical current flow through each. The current directions round the circuit are the same, which suggests that as much charge, in the form of a current pulse, leaves one plate of the capacitor as enters the other. Thus if charge $+Q$ has flowed on to X, an equal charge $+Q$ has flowed off Y.

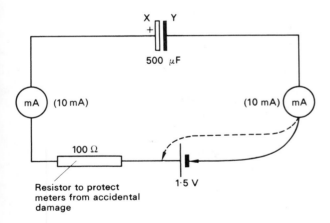

Fig. 12.1

In terms of electron flow we can say that when the supply is connected to the capacitor, electrons flow from the negative of the supply on to plate Y and from plate X towards the positive of the supply at the same rate. The positive and negative charges that appear on X and Y respectively oppose further electron flow and as these charges build up, the p.d. between X and Y

increases until it equals the p.d. of the supply (1.5 V). Electron flow then stops. There is never a steady current and no permanent meter deflections.

We say the capacitor has charge Q, meaning that one plate has charge $+Q$ and the other charge $-Q$; as much charge flows off one plate as flows on to the other, Fig. 12.2.

Making the connection shown by the dotted line causes the meters to again give equal, momentary deflections but in the opposite direction to that indicated previously. Electrons flow back from Y to X until the positive charge on X is neutralized. This momentary pulse of current thus discharges the capacitor, leaving zero charge on its plates.

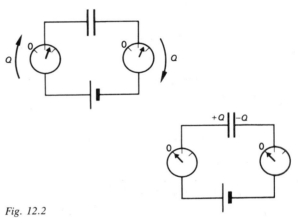

Fig. 12.2

(*b*) *Factors affecting the charge stored.* These may be investigated by the circuit of Fig. 12.3*a* using a parallel-plate capacitor in the form of two square metal plates each of side 25 cm, kept about 1.5 mm apart by four small polythene spacers (5 mm × 5 mm) at the corners, Fig. 12.3*b*. The capacitor is alternately charged, usually from a 12 V smooth d.c. supply (e.g. dry batteries), and discharged through a sensitive light-beam galvanometer 400 times a second by a *reed switch*.

A reed switch is shown in Fig. 12.3*c* and when *rectified* a.c. passes through the coil surrounding it, the reed and magnetic contact become oppositely magnetized and attract on the conducting half-cycle. On the non-conducting half-cycle the reed, no longer magnetized, springs back to the non-magnetic contact. The number of charge-discharge actions per second equals the frequency of the a.c. supply to the coil. If this is high enough current pulses follow one another so rapidly that the galvanometer deflection is steady and represents the average current I through it. When Q is the

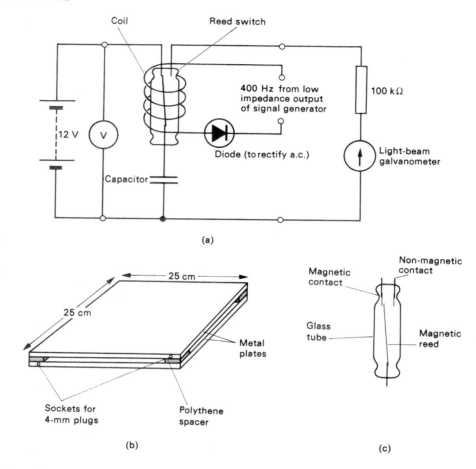

Fig. 12.3   (b)   (c)

charge stored on the capacitor and released on discharge through the galvanometer and when f is the switching frequency (i.e. that of the a.c. supply) then

$$I = \text{charge passing per second}$$

$$= Qf$$

If f is constant then $I \propto Q$. The 100 kΩ protective resistor prevents excessive current pulses when the reed switch contacts close; the capacitor will not discharge completely if the resistor is too large. How could you check that discharge is complete?

If the charging p.d. V is varied, we find that the charge Q on the capacitor, i.e. the galvanometer reading I, is directly proportional to V.

When V is kept constant and the separation d of the plates varied (using small insulating spacers at the corners of the plates), it can be inferred that Q is inversely proportional to d if d is not too large. (A graph of Q against $1/d$ does not pass through the origin because of the capacitance between the upper plate and

the bench or other nearby objects.)

If V and d are fixed and the area of overlap A of the plates is varied, we can conclude that Q is directly proportional to A. Finally, inserting a solid slab of insulator (e.g. Perspex) in the space between the plates causes Q to increase (if V, d and A are fixed).

This investigation may also be performed using a d.c. amplifier electrometer (Appendix 9).

Summarizing the results, we have

$$Q \propto V, \ Q \propto 1/d, \ Q \propto A$$

Hence

$$Q \propto \frac{VA}{d}$$

The capacitance C of a capacitor is defined (as for an insulated conductor) as the charge stored per unit p.d. between its plates. That is,

$$C = \frac{Q}{V}$$

Hence

$$C \propto \frac{A}{d} = \text{constant} \cdot \frac{A}{d}$$

It will be seen shortly the capacitance of a parallel-plate capacitor is given by

$$C = \frac{\varepsilon A}{d}$$

The constant of proportionality thus being ε, the permittivity of the insulating medium (also called the *dielectric*) between the plates.

Measurement of capacitance

A reed switch is used as in the circuit of Fig. 12.3*a* (p. 240) but for capacitances of a few microfarads the light-beam galvanometer is replaced by a milliammeter (1 mA) and a protective resistor of 220 Ω is suitable. For smaller capacitances a microammeter (100 μA) with a 2 kΩ resistor is required. In both cases the reed switch can be operated from a 50 Hz supply.

The capacitance C is found from

$$C = \frac{Q}{V} = \frac{I}{fV} \quad \text{(since } I = Qf\text{)}$$

where I and V are the two meter readings and f is the frequency of the supply.

The effect of connecting capacitors in series and in parallel may be investigated.

Parallel-plate capacitor

An expression for the capacitance of a parallel-plate capacitor is required.

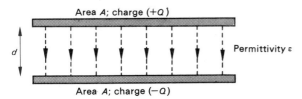

Area A; charge $(+Q)$

d Permittivity ε

Area A; charge $(-Q)$

Fig. 12.4

Consider a capacitor with plates of common area A, separated by a medium of thickness d and permittivity ε, Fig. 12.4. If one plate has charge $+Q$ and the other $-Q$, the charge density σ is Q/A. Assuming the field between the plates is uniform, the field strength E is the same at all points and is given by

$$E = \frac{\sigma}{\varepsilon} = \frac{Q}{A\varepsilon} \qquad \text{(see p. 226)}$$

Further if V is the p.d. between the plates then

$$E = \frac{V}{d} \qquad \text{(see p. 232)}$$

Therefore,

$$\frac{Q}{A\varepsilon} = \frac{V}{d}$$

Hence

$$\frac{Q}{V} = \frac{A\varepsilon}{d}$$

$$\therefore \quad C = \frac{A\varepsilon}{d} \quad \text{(since } Q = VC\text{)}$$

C will be in farads if A is in m^2, d in m and ε in F m^{-1}. It is worth noting that since the field between the plates is uniform, there is plane symmetry and π does not appear in the formula for C. It is an example of the simplification of a common formula by the use of rationalized units (p. 225). In practice, the expression is not strictly true due to non-uniformity of the field at the edges of the plates.

Large capacitance values are obtained by having 'plates' with a large overlapping area that are very close together and are separated by a dielectric with high permittivity.

Permittivity

Earlier we saw that the force between two charges depended on the intervening medium and the idea of permittivity ε was introduced to take this into account. We will now consider this idea further, especially in relation to capacitors.

(a) Relative permittivity or dielectric constant ε_r. Experiment shows that inserting an insulator or dielectric between the plates of a capacitor increases its capacitance (p. 240). If C_0 is the capacitance of a capacitor when a vacuum separates its plates and C is the capacitance of the same capacitor with a dielectric filling the space between the plates, the relative permittivity ε_r of the dielectric is defined by

$$\varepsilon_r = \frac{C}{C_0}$$

Taking a parallel-plate capacitor as an example we have

$$\varepsilon_r = \frac{C}{C_0} = \frac{\varepsilon A/d}{\varepsilon_0 A/d} = \frac{\varepsilon}{\varepsilon_0}$$

where ε is the permittivity of the dielectric and ε_0 is that of a vacuum (i.e. of free space). The expression for the capacitance of a parallel-plate capacitor with a dielectric of relative permittivity ε_r can therefore be written as

$$C = \frac{A\varepsilon_r\varepsilon_0}{d}$$

Relative permittivity has no units, unlike ε and ε_0 which have; it is a pure number without dimensions. For air at atmospheric pressure $\varepsilon_r = 1.0005$, which is near enough 1 and so for most purposes $\varepsilon_{air} = \varepsilon_0$. Table 12.1 gives some values of ε_r.

Table 12.1

Dielectric	Relative permittivity ε_r
Vacuum	1.0000
Air at s.t.p.	1.0005
Polythene	2.3
Perspex	2.6
Paper (waxed)	2.7
Mica	7
Water (pure)	80
Barium titanate	1200

The near impossibility of removing all the impurities dissolved in water makes it unsuitable in practice as a dielectric.

(b) *Action of a dielectric.* The molecules of a dielectric which is between the plates of a charged capacitor are in an electric field. The positive nuclei are urged in the direction of the field and the negative electrons in the opposite direction. As a result the molecules become distorted or *polarized* by the field with one end having an excess of positive charge and the other of negative charge. Electric dipoles are thus formed, Fig. 12.5a.

Inside the dielectric the positive and negative ends of adjacent dipoles cancel each other's effects but at the surfaces of the dielectric unneutralized charges appear which have opposite signs to those on the plates, Fig. 12.5b. Thus the positive potential of the positive plate is reduced by the negative charge at one surface of the dielectric and the negative potential of the negative plate is made less negative by the positive charge at the other surface of the dielectric. The p.d. between the plates is thereby less than it would otherwise be and so more charge is required before the p.d. between the plates equals the applied (i.e. charging) p.d. The capacitance is thus increased.

Some molecules, called *polar* molecules, are polarized even in the absence of an electric field and consequently they increase the capacitance even more than does a dielectric with *non-polar* molecules. The high relative permittivity of water arises from the H_2O molecule being polar.

(c) *Measurement of ε_0.* The circuit is the same as that for investigating the factors affecting the charge stored by a capacitor (p. 239) and is shown again in Fig. 12.6. As before, a capacitor in the form of a large pair of parallel metal plates is alternately charged from a smooth d.c. supply (0–12 V) and discharged through a light-beam galvanometer (on its most sensitive range, e.g. × 1) about 400 times a second by a vibrating reed switch.

The capacitance C of a parallel-plate air capacitor having a plate overlap area A and plate separation d is given by

$$C = \frac{\varepsilon_{air}A}{d}$$

$$\therefore \quad \varepsilon_{air} = \frac{Cd}{A}$$

where ε_{air} is the permittivity of air which, to a good approximation, is ε_0. If Q is the charge stored on the capacitor when the applied p.d. is V (measured on the voltmeter) then since $Q = VC$ we have

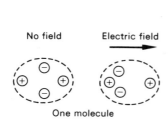

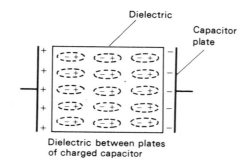

Fig. 12.5

(a)

No field   Electric field

One molecule

Dielectric

Capacitor plate

Dielectric between plates of charged capacitor

(b)

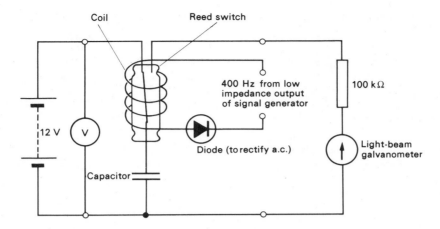

Coil   Reed switch

400 Hz from low
impedance output
of signal generator

100 kΩ

12 V

Diode (to rectify a.c.)

Light-beam
galvanometer

Capacitor

Fig. 12.6

$$\varepsilon_0 = \frac{Qd}{AV}$$

The steady current I recorded by the galvanometer is Qf, where f is the switching frequency, provided the capacitor is fully charged and discharged during each contact of the reed switch. Hence from $I = Qf$ we get

$$\varepsilon_0 = \frac{Id}{fVA}$$

Knowing the current sensitivity of the galvanometer (marked on the instrument or supplied by the manufacturer, and typically 20–25 mm μA^{-1}), I can be found in amperes. The switching frequency f is obtained from the dial reading on the signal generator and may be checked using a CRO and a low voltage 50 Hz mains supply (see p. 490). The area A of one of the plates is readily measured as is the separation d if the top capacitor plate rests on four small insulating spacers (each about 1.5 mm thick) at the corners of the bottom plate.

Very small charges are involved and, to prevent leakage, the connection from the top plate to the reed switch should not touch anything. The accepted value of ε_0 is 8.85×10^{-12} F m^{-1}.

The permittivity of different materials can also be found by this method if the experiment is done with the material completely filling the space between the plates. The permittivity ε of the material is then $\varepsilon = I_1 d/(fVA)$ where I_1 is the galvanometer current. The ratio of the galvanometer currents with and without the material between the plates gives the relative permittivity ε_r. That is,

$$\varepsilon_r = \frac{\varepsilon}{\varepsilon_0} = \frac{I_1 d/(fVA)}{Id/(fVA)} = \frac{I_1}{I}$$

Both ε_0 and ε_r may be determined in a similar manner using a d.c. amplifier electrometer (Appendix 9).

Types of capacitor

Capacitors are used in electric circuits for various purposes, as we will see later. Different types have different dielectrics. The choice of type depends on the value of capacitance and stability (i.e. ability to keep the same value with age, temperature change, etc.) needed and on the frequency of any alternating current (a.c.) that will flow in the capacitor (this affects the power loss).

For every dielectric material there is a certain potential gradient at which it breaks down and a spark passes. The working p.d. of a capacitor is thus determined by the thickness of the dielectric. Liquid and gaseous dielectrics recover when the applied p.d. is reduced below the breakdown value; only some solid dielectrics do.

(*a*) *Paper, plastic, ceramic and mica capacitors.* Waxed paper, plastics (e.g. polystyrene), ceramics (e.g. talc with barium titanate added) and mica (which occurs naturally and splits into very thin sheets of uniform thickness) are all used as dielectrics. Typical constructions are shown in Fig. 12.7 and actual capacitors in Fig. 12.8.

The losses in paper capacitors limit their use to frequencies less than 1 MHz (10^6 Hz); also, their stability is poor—up to 10 per cent changes occurring with age. Plastic, ceramic and mica types have better stability (1 per cent) and can be used at much higher frequencies.

Capacitance values for these four types seldom exceed a few microfarads and in the case of mica the limit is about 0.01 μF (10 000 pF).

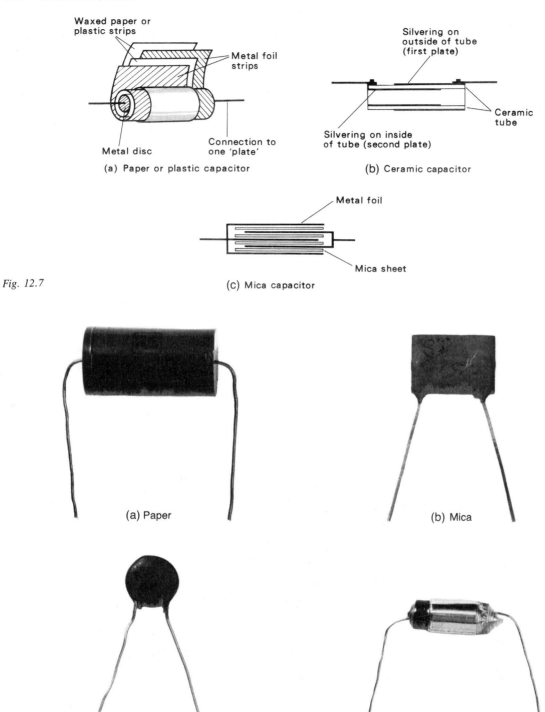

Waxed paper or plastic strips

Metal foil strips

Metal disc

Connection to one 'plate'

(a) Paper or plastic capacitor

Silvering on outside of tube (first plate)

Ceramic tube

Silvering on inside of tube (second plate)

(b) Ceramic capacitor

Metal foil

Mica sheet

(c) Mica capacitor

Fig. 12.7

(a) Paper

(b) Mica

(c) Ceramic

(d) Plastic

Fig. 12.8

(*b*) *Electrolytic capacitors*. These have capacitances up to 100 000 μF and are quite compact because the dielectric can have a thickness as small as 10^{-4} mm and not suffer breakdown even for applied p.d.s of a few hundred volts.

The dielectric is a film of aluminium oxide formed by passing a current through a strip of paper soaked with aluminium borate solution, separating two aluminium foil electrodes, Fig. 12.9*a*. The oxide forms on the anode, which acts as one plate of the capacitor. The borate solution, being an electrolyte, is the other; connection to it is made via the cathode (the other piece of aluminium foil).

In use, a 'leakage' current of the order of 1 mA must always pass through the capacitor, in the correct direction, to maintain the dielectric. The anode is marked with a + to show that it must be at a higher potential than the cathode when the capacitor is in circuit. Care must be taken to ensure that there is direct current (d.c.) in the circuit and that the capacitor is correctly connected.

Some electrolytic capacitors are shown in Fig. 12.9*b*; the case is often of aluminium and acts as the negative terminal. They are not used in a.c. circuits where the frequency exceeds about 10 kHz. Their stability is poor (10–20 per cent) but in many cases this does not matter.

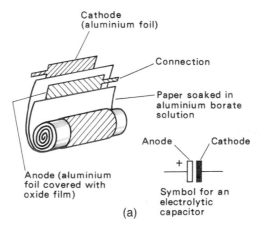

(a)

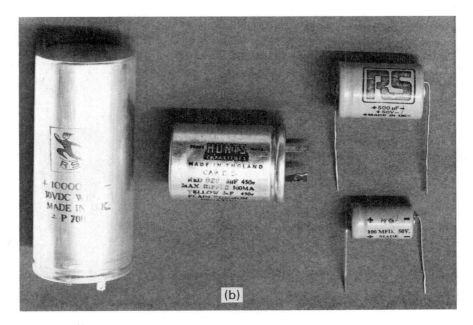

(b)

Fig. 12.9

(c) *Air capacitors.* Air is used as the dielectric in variable capacitors. These consist of two sets of parallel metal plates, one set is fixed and the other moves on a spindle within—but not touching—the fixed set, Fig. 12.10. The interleaved area varies, thus changing the capacitance.

Losses in an air dielectric are very small at all frequencies. However, relatively large thicknesses are needed because breakdown occurs at a potential gradient which is small compared with those of other dielectrics. Variable capacitors are used to tune radio receivers.

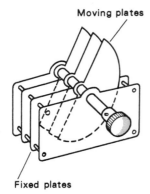

Moving plates

Fig. 12.10   Fixed plates

Capacitor networks

A network of capacitors has a combined or equivalent capacitance which can be calculated.

(a) *Capacitors in parallel.* In Fig. 12.11a the three capacitors of capacitance C_1, C_2 and C_3 are in parallel. *The applied p.d. V is the same across each but the charges are different* and are given by

$$Q_1 = VC_1 \qquad Q_2 = VC_2 \qquad Q_3 = VC_3$$

The total charge Q on the three capacitors is

$$Q = Q_1 + Q_2 + Q_3$$
$$Q = V(C_1 + C_2 + C_3)$$

If C is the capacitance of the single equivalent capacitor, it would have charge Q when the p.d. across it is V,

Fig. 12.11b. Hence

$$Q = VC$$
$$\therefore C = C_1 + C_2 + C_3$$

The combined capacitance of capacitors of 1 μF, 2 μF and 3 μF in parallel is 6 μF.

It should be noted that the charges on capacitors in parallel are in the ratio of their capacitances, i.e.

$$Q_1 : Q_2 : Q_3 = C_1 : C_2 : C_3$$

The expression for capacitors in parallel is similar to that for resistors in series.

(b) *Capacitors in series.* The capacitors in Fig. 12.12a are in series and have capacitances C_1, C_2 and C_3. Suppose a p.d. V applied across the combination causes the motion of charge from plate Y to plate A so that a charge $+Q$ appears on A and an equal but opposite charge $-Q$ appears on Y. This charge $-Q$ will induce a charge $+Q$ on plate X if the plates are large and close together. The plates X and M and the connection between them form an insulated conductor whose net charge must be zero and so $+Q$ on X induces a charge $-Q$ on M. In turn this charge induces $+Q$ on L and so on.

Capacitors in series thus *all have the same charge* and the p.d.s across each are given by

$$V_1 = \frac{Q}{C_1} \qquad V_2 = \frac{Q}{C_2} \qquad V_3 = \frac{Q}{C_3}$$

The total p.d. V across the network is

$$V = V_1 + V_2 + V_3$$
$$\therefore \quad V = \frac{Q}{C_1} + \frac{Q}{C_2} + \frac{Q}{C_3}$$
$$= Q\left(\frac{1}{C_1} + \frac{1}{C_2} + \frac{1}{C_3}\right)$$

If C is the capacitance of the single equivalent capacitor, it would have charge Q when the p.d. across it is V, Fig. 12.12b.

$$V = \frac{Q}{C}$$

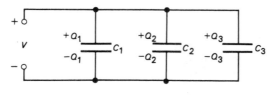

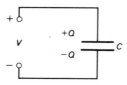

Fig. 12.11   (a)   (b)

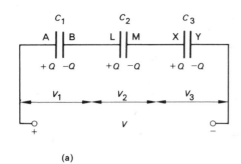

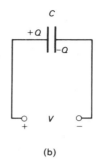

Fig. 12.12   (a)   (b)

Therefore

$$\frac{Q}{C} = Q\left(\frac{1}{C_1} + \frac{1}{C_2} + \frac{1}{C_3}\right)$$

$$\therefore \quad \frac{1}{C} = \frac{1}{C_1} + \frac{1}{C_2} + \frac{1}{C_3}$$

The combined capacitance of capacitors of $1\ \mu F$, $2\ \mu F$ and $3\ \mu F$ in series is $6/11\ \mu F$, i.e. the equivalent capacitance is less than the smallest capacitance.

The expression for capacitors in series is similar to that for resistors in parallel.

Notes. 1. For capacitors in parallel the p.d. across each is the same.

2. For capacitors in series each has the same charge.

Energy of a charged capacitor

Some basic facts can be established from experiments with a large electrolytic capacitor before the theory is developed.

(a) Discharge through a motor. If a 10 000 μF capacitor is charged from a 10 V supply and then discharged through a small electric motor, a light load can be raised, Fig. 12.13. Energy stored in the capacitor is changed into mechanical energy. Are any other forms of energy produced?

(b) Discharge through lamps. Using the circuit of Fig. 12.14a a 10 000 μF capacitor is charged from a 3 V supply then discharged through a 2.5 V, 0.3 A lamp, and the brightness is noted. If the procedure is repeated with two lamps in series (or in parallel) the brightness of each is much less than before. With a 6 V charging supply, two lamps in series flash brighter and longer than one did at 3 V. However, *four* lamps arranged as in Fig. 12.14b light up to about the same brightness and for about the same time as one lamp did on 3 V. How many times more energy is stored in the capacitor at 6 V than at 3 V? What value of charging p.d. is needed to light *nine* lamps similarly to one lamp at 3 V?

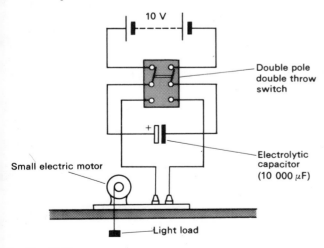

Fig. 12.13

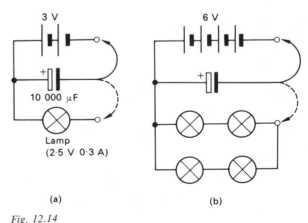

Fig. 12.14

(c) Discharge through a heating coil. The capacitor in Fig. 12.15 is charged and then discharged through the resistance coil made from 2 m of constantan wire

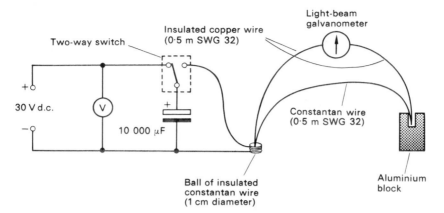

Fig. 12.15

(SWG 32). The resulting temperature rise in the coil is detected by a copper-constantan thermocouple and indicated by a light-beam galvanometer on its most sensitive range (× 1). One discharge at 20 V gives roughly the same deflection as four successive discharges at 10 V.

From this experiment and the previous one we can say that doubling the p.d. to which a capacitor is charged quadruples the energy stored, i.e. if the p.d. is V then the energy is proportional to V^2. Can you say why?

(*d*) *Graphical treatment.* The charge on a capacitor is directly proportional to the p.d. across it ($Q = VC$). A graph of p.d. against charge is therefore a straight line through the origin, Fig. 12.16.

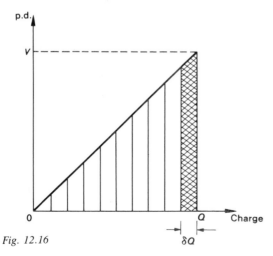

Fig. 12.16

Suppose a capacitor has capacitance C and that when the p.d. is V the charge is Q. If the capacitor starts to discharge and initially a very small charge δQ passes from the negative to the positive plate, then by the definition of p.d. the resulting energy loss (or work done) is V. δQ—assuming δQ is so small that the decrease in V is negligible. Hence from the graph,

energy loss = area of shaded strip

If the capacitance discharges completely so that Q and V fall to zero, we have

total energy loss = area of all strips

$\qquad\qquad$ = area of triangle below graph

$\qquad\qquad$ = $\frac{1}{2}QV$

This is the energy stored in the capacitor. We can also write, since $Q = VC$, that

total energy = $\frac{1}{2}V^2C = \frac{1}{2}Q^2/C$

The energy stored is thus proportional to V^2, as suggested by the previous experiments. If Q is in coulombs, V in volts and C in farads, the energy is in joules. How much energy is stored in a 10 000 μF capacitor charged to 30 V? The energy of a charged capacitor is considered to be stored as electrical energy in the field in the medium between the plates.

The expression $\frac{1}{2}QV$ is analogous to $\frac{1}{2}Fe$ for the energy stored in a wire that obeys Hooke's law, when a tension F causes an extension e.

Capacitor calculations

1. In the circuit of Fig. 12.17a, $C_1 = 2\,\mu F$, $C_2 = C_3 = 0.5\,\mu F$ and E is a 6 V battery. For each capacitor calculate (a) the charge on it, and (b) the p.d. across it.

The combined capacitance C_4, of C_2 and C_3 in parallel is given by

$C_4 = C_2 + C_3 = 0.5\,\mu F + 0.5\,\mu F = 1\,\mu F$

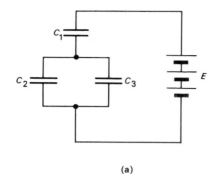

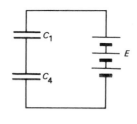

Fig. 12.17   **(a)**   **(b)**

The circuit may now be redrawn as in Fig. 12.17b in which C_1 and C_4 are in series. Their charges Q_1 and Q_4 will be equal, hence

$$Q_1 = Q_4 = V_1C_1 = V_4C_4$$

where V_1 and V_4 are the p.d.s across C_1 and C_4 respectively. Therefore

$$\frac{V_1}{V_4} = \frac{C_4}{C_1} = \frac{1}{2}$$

But $V_1 + V_4 = 6$

∴ $V_1 = 2$ V and $V_4 = 4$ V

Also $Q_1 = V_1C_1 = (2 \text{ V}) \times (2 \times 10^{-6} \text{ F})$
               $= 4 \times 10^{-6}$ C

∴ $Q_4 = 4 \times 10^{-6}$ C $= 4 \,\mu$C

The p.d. across the combined capacitance C_4 equals that across each of C_2 and C_3. Hence

$$V_2 = V_3 = V_4 = 4 \text{ V}$$

Now $Q_2 = V_2C_2$ and $Q_3 = V_3C_3$

∴ $\frac{Q_2}{Q_3} = \frac{C_2}{C_3} = \frac{0.5}{0.5}$ (since $V_2 = V_3$)

         $= 1$

But $Q_2 + Q_3 = Q_4 = 4 \times 10^{-6}$ C

∴ $Q_2 = Q_3 = 2 \times 10^{-6}$ C $= 2 \,\mu$C

Therefore
$$Q_1 = 4 \,\mu\text{C} \quad\quad Q_2 = Q_3 = 2 \,\mu\text{C}$$
$$V_1 = 2 \text{ V} \quad\quad V_2 = V_3 = 4 \text{ V}$$

2. A 10 μF capacitor is charged from a 30 V supply and then connected across an uncharged 50 μF capacitor. Calculate (a) the final p.d. across the combination, and (b) the initial and final energies, Fig. 12.18a and b.

(a) Initial charge Q on C_1

     $= V_1C_1 = (30 \text{ V}) \times (10 \times 10^{-6} \text{ F})$

     $= 3.0 \times 10^{-4}$ C $= 300 \,\mu$C

When C_1 and C_2 are connected (in parallel) the charge Q is shared in the ratio of their capacitances. That is,

$$\frac{Q_1}{Q_2} = \frac{C_1}{C_2} = \frac{10}{50} = \frac{1}{5}$$

But $Q = Q_1 + Q_2 = 300 \,\mu$C

∴ $Q_1 = 50 \,\mu$C and $Q_2 = 250 \,\mu$C

The final, common p.d. V is given by

$$V = \frac{Q_1}{C_1} \left(\text{or } \frac{Q_2}{C_2} \right) = \frac{50 \times 10^{-6} \text{ C}}{10 \times 10^{-6} \text{ F}} = 5.0 \text{ V}$$

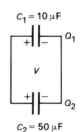

Fig. 12.18   **(a)**   **(b)**

(b) Initial energy $= \frac{1}{2}QV_1 = \frac{1}{2} \times 3.0 \times 10^{-4} \times 30$ J

                     $= 4.5 \times 10^{-3}$ J

Final energy $= \frac{1}{2}Q_1V + \frac{1}{2}Q_2V$

                 $= \frac{1}{2}V(Q_1 + Q_2) = \frac{1}{2}QV$

                 $= \frac{1}{2} \times 3.0 \times 10^{-4} \times 5.0$ J

                 $= 0.75 \times 10^{-3}$ J

Note. The apparent loss of energy is due to the production of heat when charge flows in the wires connecting the capacitors.

Discharge of a capacitor

Many electronic circuits involve capacitors charging and discharging through resistors. The way in which these processes occur is typical of other growth and decay effects in physics (e.g. radioactive decay), chemistry, biology and economics. All may be described either graphically or mathematically (using calculus) by the same equation. Here we shall consider mainly the decay of charge on a capacitor.

(*a*) *Exponential decay curve.* First, using the circuit of Fig. 12.19*a* it can be shown that the larger the values of *C* and *R*, the slower the discharge.

To obtain a decay curve a convenient value for *C* is 500 μF and for *R* 100 kΩ. The capacitor is charged to about 10 V and then readings taken on the microammeter (100 μA) at 10 second intervals during discharge.

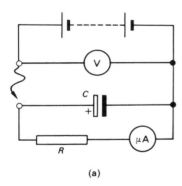

(a)

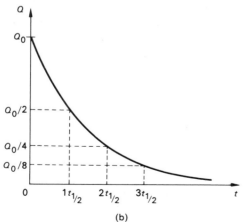

(b)

Fig. 12.19

The decay curve of charge *Q* against *t*, Fig. 12.19*b*, will have the same shape as the decay curve of *I* against *t* since $Q = VC$ (where *Q* is the charge on the capacitor and *V* the p.d. across it at time *t*) and $V = IR$ (*V* causes *I* through *R*), therefore $Q = IRC$. But *R* and *C* are constant and so $Q \propto I$.

Examination of the decay curve shows that it decreases by the same fraction in successive equal time intervals. Thus if it falls from Q_0 to $Q_0/2$ in time $t_{\frac{1}{2}}$, it will also fall from $Q_0/2$ to $Q_0/4$ in the next time interval $t_{\frac{1}{2}}$ and so on. Time $t_{\frac{1}{2}}$ is therefore the time for the charge to fall by half and is known as the *half-life* of the decay process. For the values of *R* and *C* used, $t_{\frac{1}{2}} \simeq 35$ s. In general the time for the charge to fall by *any* fraction (not just a half) is constant and this is a characteristic of any quantity which decays by an *exponential* law. The graph in Fig. 12.19*b* is an exponential decay curve.

An alternative arrangement for obtaining a decay curve using a d.c. amplifier is given in Appendix 9.

(*b*) *Calculus treatment.* If at a certain time *t* during the discharge of a capacitor of capacitance *C* through a resistance *R*, Fig. 12.20, the p.d. across the capacitor is *V* and the charge on it is *Q*, then

$$Q = VC$$

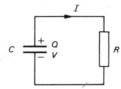

Fig. 12.20

The discharge current *I* at time *t* equals the rate of loss of charge and in calculus notation

$$I = -\frac{dQ}{dt}$$

The negative sign shows that *Q* decreases as *t* increases. We also have that

$$V = IR$$

From these three equations

$$Q = -CR\frac{dQ}{dt}$$

Rearranging and integrating

$$\int_0^t \frac{dt}{CR} = -\int_{Q_0}^Q \frac{dQ}{Q} = \int_Q^{Q_0} \frac{dQ}{Q}$$

where Q_0 is the initial charge. Therefore

$$\frac{t}{CR} = \log_e \left(\frac{Q_0}{Q} \right)$$

$$\therefore \quad Q \quad = Q_0 \, e^{-t/CR}$$

This is the equation of the exponential decay curve.
When $Q = Q_0/2$, $t = t_{\frac{1}{2}}$ and

$$\frac{Q_0}{2} = Q_0 e^{-t_{\frac{1}{2}}/CR}$$

$$\therefore \quad \tfrac{1}{2} = e^{-t_{\frac{1}{2}}/CR}$$

$$\therefore \quad e^{t_{\frac{1}{2}}/CR} = 2$$

$$\therefore \quad t_{\frac{1}{2}} = CR \log_e 2 = 0.693 \, CR$$

In the circuit of Fig. 12.19a, if $C = 500 \; \mu\text{F} = 500 \times 10^{-6} \; \text{F} = 5 \times 10^{-4} \; \text{F}$ and $R = 100 \; \text{k}\Omega = 10^5 \; \Omega$, then $CR = 50 \; \text{s}$

$$\therefore \quad t_{\frac{1}{2}} = 34.7 \; \text{s}$$

It can be seen that CR has units of time by expressing C and R in terms of the units from which they are derived. Thus

$$C(\text{farads}) = \frac{Q(\text{coulombs})}{V(\text{volts})} = \frac{It(\text{amperes} \times \text{seconds})}{V(\text{volts})}$$

$$R(\text{ohms}) = \frac{V(\text{volts})}{I(\text{amperes})}$$

$$\therefore \quad \text{F} \times \Omega = \left(\frac{\text{A} \times \text{s}}{\text{V}} \right) \times \left(\frac{\text{V}}{\text{A}} \right) = \text{s}$$

The product CR is called the *time constant* of the circuit; it is an important and useful quantity to know. When $t = CR$ we have

$$Q = Q_0 \, e^{-t/CR} = \frac{Q_0}{e}$$

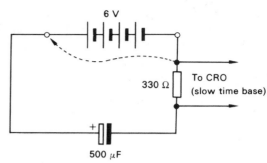

Fig. 12.21

Now e, the base of natural logs, equals 2.7 (approximately) and so

$$Q = 0.37 \, Q_0$$

That is, CR is the time for Q to fall to 37 per cent of its initial value.

When a capacitor charges up through a resistor, the charge (and p.d.) grows exponentially. A circuit for viewing the charge and discharge curves on a CRO is given in Fig. 12.21.

Electroscopes

The *gold-leaf* electroscope, Fig. 12.22a, was used in most of the early work on electrostatics. The *Braun* electroscope, Fig. 12.22b, is a more robust instrument.

Basically an electroscope is a capacitor, the leaf (or pointer) and rod to which it is attached forming one plate and the case the other. When a p.d. is applied to the 'plates', a small charge flows on to the leaf and an equal but opposite charge is induced on the inside of the case. The forces of the resulting electric field between the 'plates' deflect the leaf, and the deflection is a measure of the p.d. between leaf etc. and case.

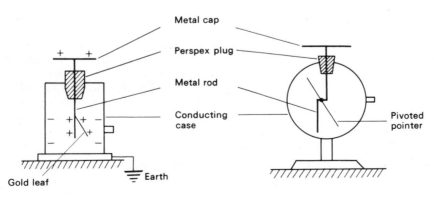

Fig. 12.22   (a)   (b)

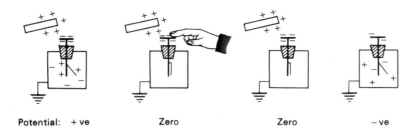

Fig. 12.23   Potential: + ve   Zero   Zero   − ve

If the case is earthed, as it will be if it is made of wood and stands on a wooden bench, then the *electroscope records potential*.

The last point can be shown by charging by induction an electroscope with an earthed case. Thus when a positively charged body is brought near the electroscope cap, the electric field due to the charge raises the potential of the cap and leaf and deflection of the latter occurs. If the cap is now earthed by touching it with the finger, the leaf falls. Although the cap has a negative (induced) charge, its effect is cancelled by that of the positively charged body so making the potential of the cap and leaf zero. It remains at zero when the finger is removed. When the positively charged body is removed, the negative charge on the cap gives the cap and leaf a negative potential (nearly equal in magnitude but opposite in sign to that produced previously by the positively charged body) and there is again a deflection. Fig. 12.23 shows the various stages.

If the case stands on an insulator and the cap is given a charge, what happens if the cap and case are (*i*) connected, Fig. 12.24*a*, (*ii*) not connected, Fig. 12.24*b*?

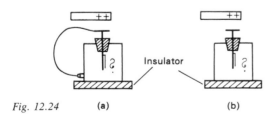

Fig. 12.24   (a)   (b)

The d.c. amplifier as an electrometer

An electrometer measures p.d.s, including those produced electrostatically. An electroscope with a scale for reading deflections is an example of such an instrument.

The d.c. amplifier, one make of which is shown in Fig. 12.25, can also be used as an electrometer. It consists of an amplifier whose stages are *directly coupled* (hence d.c.) and are not joined by capacitors as in a normal amplifier. Basically a d.c. amplifier acts as a very high resistance voltmeter (about 10^{13} Ω) but it can be adapted to measure current and charge. In all cases, however, a p.d. of up to 1 V (for the instrument shown) has to be applied to the input terminals and this controls the output current which is recorded on a meter (1 mA or 100 μA). The meter has to be calibrated to read the input p.d. directly—as described below.

To measure current, a resistor R is connected across the input as in Fig. 12.26*a* and the calibrated output meter then gives the p.d. V across R. The current I in R will be V/R. If $R = 10^{11}$ Ω, a meter reading corresponding to a 1 V input indicates that $I = V/R = 1/10^{11} = 10^{-11}$ A. (The much higher input resistance of the amplifier—about 10^{13} Ω—in parallel with R can be ignored.)

The measurement of charge requires a known capacitor C across the input as in Fig. 12.26*b*. The p.d. V across C, i.e. the input p.d., is again shown by the calibrated output meter. The unknown charge Q on the capacitor (which is responsible for the p.d.) is obtained from $Q = VC$. For example, if $C = 10^{-8}$ F (0.01 μF) then when the output meter gives a reading of 1 mA, the input p.d. $V = 1$ V and so $Q = 10^{-8}$ C, the input capacitance of the amplifier being neglected. Almost complete transfer of charge to the amplifier capacitor C will occur if the capacitance of the object bringing the charge to the input terminal is small compared with that of the amplifier capacitor. Also, the input resistance of the amplifier is so high that leakage of charge from the amplifier capacitor through the instrument is negligible. There is, therefore, ample time to make a reading since the p.d. across the capacitor remains constant for a long time.

Electrostatics demonstrations with a d.c. amplifier

The instructions refer to the instrument shown in Fig. 12.25.

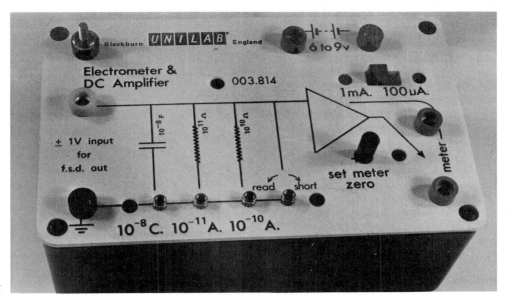

Fig. 12.25

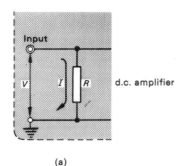

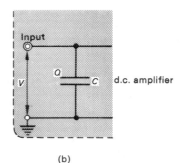

Fig. 12.26   (a)   (b)

(*a*) *Calibration.* Connect a 1 mA or 100 μA meter to the *output sockets* and push the *meter slide switch* to the appropriate setting. Connect a 9 V (or 6 V) battery, observing the correct polarity. The instrument is now on.

Turn the *short-read contact screw* fully clockwise. Adjust the *meter zero control* so that the meter reads zero. Unscrew the short-read contact screw anticlockwise one turn.

Connect the potential divider circuit of Fig. 12.27 so that 1 V is applied to the *input sockets* (as shown by the high resistance voltmeter $\textcircled{V}$). The output meter should give a full-scale deflection from a left-hand zero position. Remove the potential divider circuit. The instrument is ready for use.

Once calibrated the electrical zero on the output meter can be moved anywhere on the scale by adjusting the *meter zero control.*

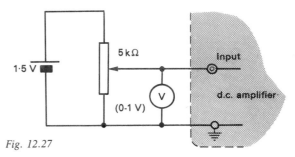

Fig. 12.27

(*b*) *Positive and negative charges.* Alter the *meter zero control* to give a centre-zero on the meter. Insert a brass rod (e.g. 25 mm long) in the input socket and set the electrometer on the 10^{-8} C range by screwing down the *contact screw* which connects the 10^{-8} F capacitor across the input.

Draw a cellulose acetate (or Perspex) strip, charged

positively by rubbing, across the brass rod so that it charges the input capacitor, the meter deflects to the right.

Discharge the capacitor by turning the *short-read screw* fully clockwise to 'short'. The meter returns to centre zero. Turn back the *short-read screw* one turn to 'read'. Repeat the procedure with a negatively charged polythene strip. The meter should deflect to the left.

(*c*) *Spooning charge*. Show that charge is a *quantity* of something that can be measured and passed from place to place. Transfer a 'spoonful' of positive charge from the positive terminal of a high voltage supply (e.g. 1 kV) to the electrometer input using, say, a 30 mm diameter metal sphere on an insulating handle as the

'spoon', Fig. 12.28. Each additional 'spoonful' produces the same increase of deflection.

The effect of different sized spoons and different supply p.d.s can be investigated.

(*d*) *Electrostatic induction*. Arrange two metal spheres (30 mm in diameter) on insulated stands, initially touching each other, Fig. 12.29. Bring a charged polythene strip near to one sphere, separate the spheres and remove the strip. Show that the charges induced on the spheres are equal in magnitude but opposite in sign by touching each in turn on the brass rod in the *input socket* of the electrometer (meter still on centre zero).

(*e*) *Distribution of charge on conductors*. The charge

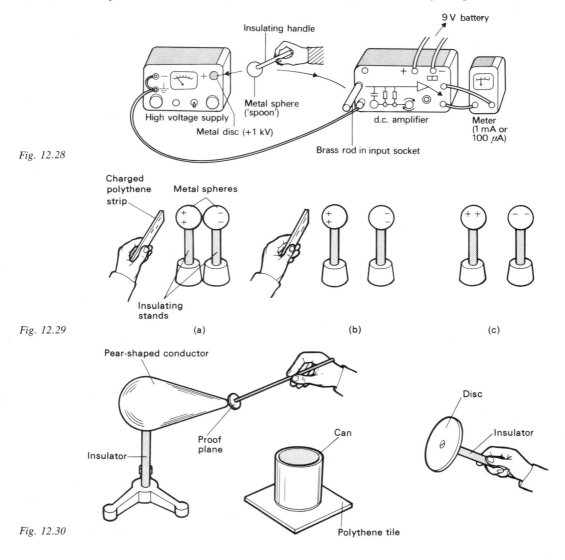

Fig. 12.28

Fig. 12.29   (a)   (b)   (c)

Fig. 12.30

per unit area of the surface of a conductor, i.e. the *charge density*, is greatest on the most highly convex parts. Show this by transferring charge with a proof-plane (a small metal disc on an insulating handle) from different parts of various conductors on insulating stands, Fig. 12.30. The pear-shaped conductor and the can may be charged four or five times by an electrophorus.

The charge acquired by the proof-plane is proportional to the charge density on the surface touched since the proof-plane in effect becomes part of the charged surface. Although the charge density on the surface of a conductor may vary from place to place, the potential is the same at all points.

(*f*) *Faraday's ice-pail.* Faraday investigated electrostatic induction using, by chance, a pail he had for storing ice. To repeat his experiment, plug a hollow metal sphere (75 mm diameter) into the *input socket* of the electrometer. Discharge a Perspex rod by quickly passing it through a flame. Plug a 30 mm sphere into the rod, charge the sphere from a polythene strip and lower it into the hollow sphere without touching it, Fig. 12.31. The meter deflects and does not change with the position of the charged sphere so long as it is well inside; nor does it alter when the charged body touches the inside of the hollow sphere and is then removed from it. Remove the sphere and show that it is completely discharged.

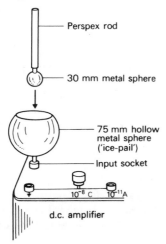

Fig. 12.31

The conclusions to be drawn are:

(*i*) a charged body enclosed in a hollow conductor induces an equal charge of opposite sign on the inside of the conductor and an equal charge of the same sign as its own on the outside, Fig. 12.32, and

(*ii*) inside a hollow conductor the net charge is zero; any excess charge resides on the outside surface.

The experiment also shows that all the charge on a conductor can be transferred to a hollow can if it touches the can *well down inside*; only part of the charge would be transferred if the outside were touched.

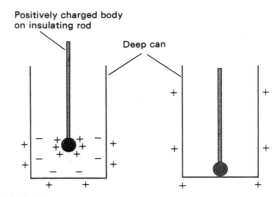

Fig. 12.32

(*g*) *Conservation of charge.* Use the 'ice-pail' as in (*f*). First discharge strips of cellulose acetate and polythene in a flame, then rub them together inside the 'pail', Fig. 12.33. There should be little deflection until one or other of the strips is removed, when roughly equal and opposite deflections are obtained.

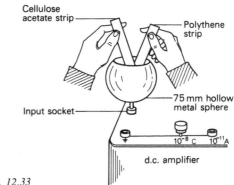

Fig. 12.33

(*h*) *Charge sharing.* Plug the 25 mm long brass rod into the input socket again and set the *meter zero control* for an end-zero on the meter. Charge the internal 10^{-8} F capacitor by connecting 1 V to the *input sockets*; there should be a full-scale deflection on the meter. Remove the charging connections and connect externally across the *input sockets* a 10^{-8} F (0.01 μF) capacitor. The meter should fall to half-scale, showing the charge has been shared equally between capacitors of the same value. Try larger and smaller capacitors.

Action at points

The point of a pin is highly curved and if the charge density is sufficiently great, an intense electric field arises near the point. Background ionizing radiation produces electrons in the air which are accelerated by the intense field and cause ionization of the air by collision (p. 473). Ions having the same sign as the charge on the pin are strongly repelled from it to create an electric 'wind'. Ions with charges of opposite sign to that on the pin are attracted to the point and neutralize its charge. The net result of the 'action at points' is the apparent loss of charge from the pin to the surrounding air by what is termed a *point* or *corona discharge*.

The arrangement of Fig. 12.34 may be used to show a point discharge from a pin and collection of some of the charge by an electroscope in the path of the 'wind'. The electric 'windmill' of Fig. 12.35 provides another demonstration of the electric 'wind'. The 'windmill' revolves rapidly when connected to a high potential due to the

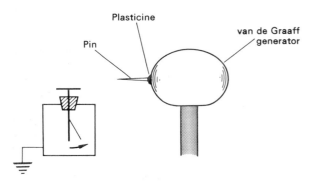

Fig. 12.34

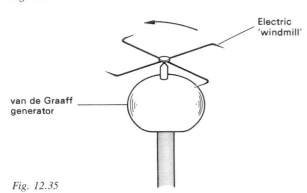

Fig. 12.35

'wind' streaming away from each point. Why should apparatus working at high p.d.s not have sharply curved parts?

As well as losing charge, a sharp point can also collect it. Thus, when a negatively charged polythene

strip is held close to the point of a pin taped to the cap of an electroscope, Fig. 12.36, a deflection is obtained which remains on removing the polythene. Electrostatic induction occurs as shown, then due to 'action at points' a positive ion 'wind' is created, neutralizing part of the charge on the strip. At the same time negative ions in the air are attracted to the point, neutralizing its induced positive charge. Consequently the electroscope gains negative charge, apparently collected by the point of the pin.

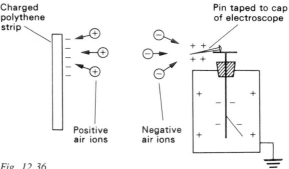

Fig. 12.36

van de Graaff generator

(*a*) *Action.* Large generators of this type are used to develop p.d.s of up to 14 million volts for accelerating atomic particles in nuclear physics research. In the generator of Fig. 12.37 beams of protons (or deuterons) from a source in the dome at the top are accelerated down the column and cause nuclear reactions when they hit 'targets' of different materials at the bottom.

A simplified school version is shown in Fig. 12.38. Initially a positive charge is produced on the motor-driven Perspex roller by friction between it and the rubber belt. This charge induces a negative charge on the earthed comb of metal points P (by drawing electrons from earth) which is then sprayed off by 'action at points' on to the outside of the belt and carried upwards. The comb of metal points Q is connected to the inside of an insulated metal sphere and when the negative charge on the upward moving belt reaches Q, positive charge is induced in Q due to the repulsion of negative charge to the surface of the sphere. By 'action at points' Q has thus apparently drawn off negative charge from the belt on to the sphere. The belt becomes positively charged as it moves down, due to Q spraying charge on to it and because of friction between the belt and the polythene roller.

A large force is set up as a result of repulsion between the negative charge on the collecting sphere

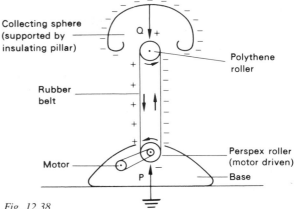

Fig. 12.38

Fig. 12.37

and the negative charge on the upward moving belt. The motor driving the roller therefore has to do work against this repulsion.

(*b*) *Estimating the p.d.* School-type generators can produce p.d.s of up to a few hundred thousand volts, the insulation of the surrounding air being an important limiting factor.

A rough idea of the p.d. can be obtained by finding the greatest separation of the collecting and discharging

spheres for a spark to pass between them. Under dry, dust-free conditions at s.t.p., spark discharges 1 cm, 7 cm and 12 cm long occur between two *rounded* electrodes when the p.d.s. are roughly 30 kV, 250 kV and 750 kV respectively.

Another method, not requiring the assumption of the above information, involves connecting a micro-ammeter (100 μA) in series with the collecting sphere and base of the generator and measuring the steady current I which passes with the generator working at a certain speed, Fig. 12.39*a*. The meter is then removed and the discharging sphere brought up until regular sparking *just* occurs between it and the collecting sphere, Fig. 12.39*b*, with the generator running at the same speed as before. The number of sparks passing per second is then found with a stop watch.

If Q is the charge which builds up on the collecting sphere and passes per spark and t is the interval between sparks, then the discharge may be regarded as equivalent to a steady current of value Q/t. Assuming

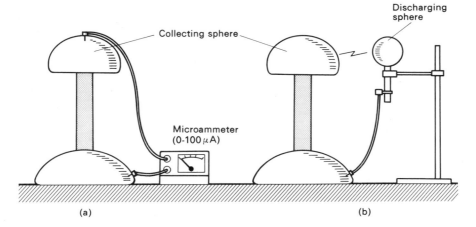

Fig. 12.39   (a)   (b)

that this current is the same as the current I recorded by the microammeter, we can say $I = Q/t$. Suppose there were two sparks per second, then $t = 0.5$ s. Also taking $I = 10 \, \mu\text{A} = 1.0 \times 10^{-5}$ A, then

$$Q = It = (1.0 \times 10^{-5} \text{ A}) \times (0.5 \text{ s}) = 5.0 \times 10^{-6} \text{ C}$$

But $Q = VC$ where V is the potential of the collecting sphere and C is its capacitance. For an isolated sphere (which the collecting 'sphere' is not), $C = 4\pi\varepsilon_0 a$ where a is the radius of the sphere (p. 238). If $a = 13$ cm $= 13 \times 10^{-2}$ m then $C = (4\pi \times 9.0 \times 10^{-12} \text{ F m}^{-1}) \times (13 \times 10^{-2} \text{ m})$ since $\varepsilon_0 \simeq 9.0 \times 10^{-12}$ F m^{-1}. Therefore $C \simeq 1.5 \times 10^{-11}$ F

$$\therefore \quad V = \frac{Q}{C} = \frac{5.0 \times 10^{-6} \text{ C}}{1.5 \times 10^{-11} \text{ F}}$$

$$\simeq 3.3 \times 10^5 \text{ V} \quad (330 \text{ kV})$$

QUESTIONS

1. (a) In the circuit of Fig. 12.40a what is the p.d. across each capacitor? What is the total charge stored? What is the capacitance of the single capacitor which would store the same charge as the two capacitors together?

(b) In Fig. 12.40b if a charge of 4 μC flows from the 6 V battery to plate P of the 1 μF capacitor, what charge flows from (i) Q to R, and (ii) S to the battery? What is the p.d. across (iii) C_1, and (iv) C_2? What is the capacitance of (v) C_2, and (vi) the single capacitor which is equivalent to C_1 and C_2 in series, and what charge would it store?

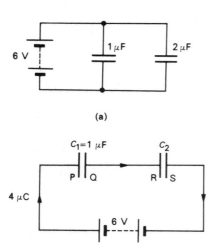

(a)

(b)

Fig. 12.40

2. (a) What is the capacitance of a small metal sphere of radius 1 cm (take $4\pi\varepsilon_0 = 10^{-10}$ F m^{-1})? (b) What charge is stored on it when it briefly touches the 1 kV terminal of an

e.h.t. power supply? (c) If the sphere now shares its charge with a 0.001 μF capacitor, what is the ratio of the charge on the capacitor to that on the sphere? (d) What, in effect, will be the charge on (i) the capacitor, and (ii) the sphere? (e) If the capacitor had had a capacitance of 1 pF (10^{-12} F) how would the charge have been shared?

3. Describe an experiment to compare the capacitances of two capacitors.

A capacitor is charged through a large series resistance. Sketch a curve of the increase in potential difference across the capacitor with time.

Under what circumstances is it necessary to know the capacitance of an electroscope or electrostatic voltmeter?

A steady potential of 100 V is maintained across the combination of a capacitor of capacitance 5.0 μF in series with one of 2.0 μF. Calculate (a) the potential difference across each, (b) the charge on each, and (c) the energy stored in each capacitor. (A.E.B.)

4. (a) A parallel-plate capacitor consists of two square plates each of side 25 cm, 3.0 mm apart. If a p.d. of 200 V is applied, calculate the charge on the plates with (i) air, and (ii) paper of relative permittivity 2.5, filling the space between them. ($\varepsilon_0 = 8.9 \times 10^{-12}$ F m^{-1}.)

(b) A tubular 0.10 μF capacitor is to be made from a sheet of plastic 2.0 cm wide and 1.0×10^{-3} cm thick rolled between metal foil of the same width. What length of plastic is required if its relative permittivity is 2.4 and $\varepsilon_0 = 8.9 \times 10^{-12}$ F m^{-1}?

5. Define electric field strength.

How would you demonstrate experimentally that (a) a charge given to a hollow conductor is confined to the outside surface, and (b) the potential is constant over the surfaces of the charged conductor?

A fine layer of silver is deposited on each side of a sheet of mica 2.4 cm^2 in area to form a capacitor of capacitance 1.2×10^{-4} μF. If the dielectric constant (relative permittivity) of the mica is 6.0 find the thickness of the sheet of mica. If the insulation of the mica breaks down when subjected to an electric field strength of 4.0×10^5 V cm^{-1} and it is desirable in practice never to exceed one-half of this field strength, what is the maximum working voltage of the capacitor? ($\varepsilon_0 = 8.9 \times 10^{-12}$ F m^{-1}). (L.)

6. A 100 μF capacitor is charged from a supply of 1000 V, disconnected from the supply and then connected across an uncharged 50 μF capacitor. Calculate the energy stored initially and finally in the two capacitors. What conclusion do you draw from a comparison of these results? (J.M.B.)

7. (a) Explain how field strength is related to potential difference in an electrostatic field.

The field strength E close to a plane conductor in free space, carrying charge density σ, is σ/ε_0. Use this information to derive an expression for the capacitance of a parallel-plate capacitor in free space where ε_0 is the permittivity of free space.

(b) A 2 μF capacitor is required which will work at 1000 V d.c. but the only capacitors available are each 2 μF with a working voltage of 400 V d.c. Describe how such capacitors could be combined to give an equivalent capacitor of the required rating. (*J.M.B.Eng. Sc.*)

8. Define *electric potential* and *capacitance of an insulated conductor*. Explain carefully, with reference to your definitions, why the presence of a similar earthed conductor near and parallel to an isolated sheet of metal considerably increases the capacitance of the sheet.

Fig. 12.41 shows a capacitor in a circuit in which a vibrating switch first charges it up to 200 V and then discharges it through a galvanometer, repeating the sequence 50 times a second. The mean current registered by the galvanometer is 11.0 μA. Use this information to find the capacitance of the capacitor in farads.

The capacitor in fact consists of two flat square metal sheets, 50 cm × 50 cm placed parallel one over the other and 2.0 mm apart. Calculate the magnitude of ε_0, the permittivity of free space (or air), and state the units in which it is measured.

If a capacitor is to be accurately measured with such a circuit it is essential that the part of the circuit between B and the galvanometer shall be well insulated from the part between A and the battery. What must be the minimum resistance of any possible leakage path between these parts if the error introduced is not to exceed 1 per cent in measuring this capacitor? (*S.*)

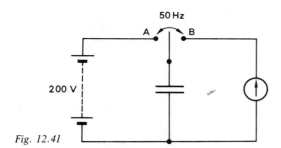

50 Hz

A   B

200 V

Fig. 12.41

9. Discuss the essential differences between current and static electricity.

Describe and explain the working of a device which can supply a current of the order of 10^{-6} A and maintain a potential difference of the order of 10^6 V.

A high voltage generator continuously supplies charge to a spherical dome of radius 15 cm, which gives 10 sparks each minute to a nearby earthed sphere. While the generator is operating at the same rate, the dome is connected to earth through a galvanometer, which indicates a steady current of 2.0×10^{-6} A. Calculate

(a) the charge passed in each spark,

(b) the maximum potential difference between the dome and earth,

(c) the distance from the surface of the dome to the surface of the nearby earthed sphere.

Make the simplifying assumptions that an isolated spherical dome of 1 cm radius has a capacitance of 10^{-12} F, and that in this system the mean potential gradient needed for sparks to pass is 4×10^4 V cm^{-1}. (*C.*)

10. What is meant by *the electrostatic potential at a point*?

Illustrate, by a suitably-labelled sketch in each case, how it is possible to set up electrostatic systems in which (a) a conductor at earth potential carries a net positive charge, (b) a conductor at earth potential has regions of both negative and positive charge, and (c) a conductor with no net charge is at a positive potential with respect to earth.

How would you verify experimentally that both negative and positive charges are present in case (b)?

A radioactive source emits beta particles (electrons) at a substantially constant rate of 3.7×10^4 per second. If the source, which is a metal sphere 1 mm in diameter, is electrically insulated, how long will it take for its potential to rise by 1 V, assuming 90 per cent of the beta particles emitted escape from the source? The capacitance in SI units of an isolated spherical conductor of radius r is $4\pi\varepsilon_0 r$.

Charge on the electron = 1.6×10^{-19} C; $\varepsilon_0 = 8.9 \times 10^{-12}$ F m^{-1}.) (*O. and C.*)

13 Magnetic fields

Importance of electromagnetism

The discovery in 1819 by Professor Oersted at the University of Copenhagen that an electric current is accompanied by magnetic effects saw the birth of electromagnetism. Since then, the subject has been an extremely fruitful field of study for the imagination of scientists and the creative genius of engineers.

In science, Faraday announced the discovery of electromagnetic induction in 1831 and developed the idea of a 'field' for dealing with such action-at-a-distance effects. Clerk Maxwell put Faraday's field ideas into mathematical form and predicted electromagnetic waves. Einstein pondered over the mysteries of electromagnetism, especially the need for relative motion, and was led to the theory of relativity.

In electrical engineering the dynamo, the motor and the transformer were the outcome of Faraday's work. Edison was responsible for opening the first power station in 1882, designed to supply electricity to domestic consumers in New York. Shortly afterwards Tesla invented the induction motor which does not require brushes (to supply current to its rotor) and which is now the indispensable servant of industry. In 1887 Hertz produced radio-type electromagnetic waves and then Marconi, contrary to all theoretical predictions, transmitted them across the Atlantic. And so the remarkable story continues with, for example, the development today of the linear motor which, unlike conventional rotary motors, is flat and eliminates the energy losses occurring when rotary motion is converted to the often required linear motion.

Modern civilization is heavily indebted to electromagnetism, the topic of this chapter and the next.

Fields due to magnets

The magnetic properties of a magnet appear to originate at certain regions in the magnet which we call the *poles*: in a bar magnet these are near the ends. Experiments show that (*i*) magnetic poles are of two kinds, (*ii*) like poles repel each other and unlike poles attract, (*iii*) poles always seem to occur in equal and opposite pairs,

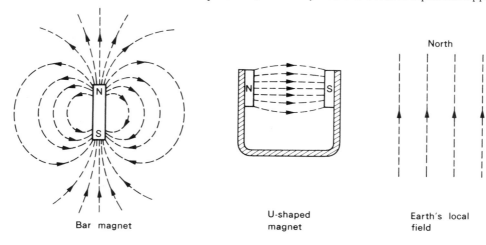

Fig. 13.1

Bar magnet

U-shaped magnet

Earth's local field

North

and (*iv*) when no other magnet is near, a freely-suspended magnet sets so that the line joining its poles (i.e. its magnetic axis) is approximately parallel to the earth's north-south axis.

The last fact suggests that the earth itself behaves like a large permanent magnet and it makes it appropriate to call the pole of a magnet which points (more or less) towards the earth's geographical North Pole, the *north pole* of the magnet and the other the *south pole*. What kind of magnetic pole must be near the earth's geographical North Pole?

The space surrounding a magnet where a magnetic force is experienced is called a *magnetic field*. The direction of a magnetic field at a point is taken as the direction of the force that acts on a north magnetic pole there. A magnetic field can be represented by magnetic *field lines* drawn so that (*i*) the line (or the tangent to it if it is curved) gives the direction of the field at that point and (*ii*) the number of lines per unit cross-section area is an indication of the 'strength' of the field. Arrows on the lines show the direction of the field and since a north pole is repelled by the north pole of a magnet and attracted by the south, the arrows point away from north poles and towards south poles.

Field lines can be obtained quickly with iron filings or accurately plotted using a small compass (i.e. a pivoted magnet). Some typical field patterns are shown in Fig. 13.1. The field round a bar magnet varies in strength and direction from point to point, i.e. is non-uniform. Locally the earth's magnetic field is uniform; the lines are parallel, equally spaced and point north.

Earth's magnetic field

The magnitude and direction of the earth's field varies with position over the earth's surface and it also seems to be changing gradually with time. The pattern of field lines is similar to that which would be given if there was a strong bar magnet at the centre of the earth, Fig. 13.2*a*. At present there is no generally accepted theory of the earth's magnetism but it may be caused by electric currents circulating in its core due to convection currents arising from radioactive heating inside the earth.

In Britain the earth's field is inclined downwards at an angle of about 70° to the horizontal—called the *angle of dip* α. At the magnetic poles (which are near the geographical poles but whose positions change slowly) it is vertical and α = 90°. At the magnetic equator it is parallel to the earth's surface, i.e. horizontal and α = 0°.

It is convenient to resolve the earth's 'field strength' B_R into horizontal and vertical components, B_H and B_V respectively. We then have from Fig. 13.2*b* that

$$B_H = B_R \cos \alpha$$

and

$$B_V = B_R \sin \alpha$$

Also

$$\tan \alpha = \frac{B_V}{B_H}$$

Instruments such as compass needles whose motion is confined to a horizontal plane are affected by B_H only.

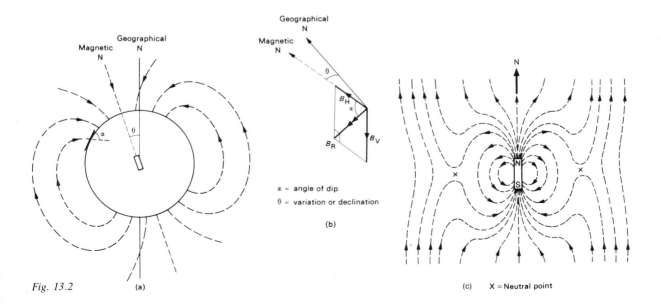

α = angle of dip
θ = variation or declination

(b)

Fig. 13.2   (a)   (c)   X = Neutral point

A *neutral point* is a place where two magnetic fields are equal and opposite and the resultant force is zero. Two such points are shown in Fig. 13.2*c* in the combined field due to the earth and a bar magnet with its N pole pointing N. Where will they be when the N pole of the magnet points S?

Fields due to currents

A conductor carrying an electric current is surrounded by a magnetic field. The field patterns due to conductors of different shapes can be obtained as for a magnet using iron filings or a plotting compass.

The lines due to a straight wire are circles, concentric with the wire, Fig. 13.3*a*. The *right-hand screw rule* is a useful aid for predicting the direction of the field knowing the direction of the current. It states that *if a right-handed screw moves forward in the direction of the current (conventional), then the direction of rotation of the screw gives the direction of the lines*. Figs. 13.3*b* and *c* illustrate the rule; in *b* the current is flowing out of the paper and the dot in the centre of the wire is the point of an approaching arrow; in *c* the current is flowing into the paper and the cross is the tail of a receding arrow.

The field pattern due to a current in a plane circular coil is shown in Fig. 13.4*a*, and one due to a current in a long cylindrical coil—called a *solenoid*—in Fig. 13.4*b*. In both cases the field direction is again given by the right-hand screw rule. A solenoid produces a field similar to that of a bar magnet; in Fig. 13.4*b* the left-hand end behaves like the north pole of a bar magnet and the right-hand end like the south pole. In general, the field produced by any magnet can also be produced by a current in a suitably-shaped conductor. It is this fact which leads us to believe that *all* magnetic effects are due to electric currents, which in the case of permanent magnets arise from the motion of electrons within the atoms themselves.

In brief, we can say that a static electric charge gives rise to an electric field, whilst a moving electric charge creates a magnetic field as well as an electric field.

Force on a current in a magnetic field

A magnet in a uniform magnetic field experiences two equal but opposite parallel forces, i.e. a couple, which gives it an angular acceleration and—provided there is damping—it ultimately comes to rest with its axis parallel to the field, Fig. 13.5. In a non-uniform field, the poles of the magnet are acted on by unequal forces and these may cause translational as well as rotational acceleration. A current-carrying conductor behaves like a magnet and so it is not surprising to find that it too experiences a force in a magnetic field. In the next section we will see how this force can be used to define and measure the strength of a magnetic field. But first some basic experimental facts.

(*a*) *Fleming's left-hand rule*. Using the apparatus of Fig. 13.6 it can be observed that when current passes through the 'bridge' wire, it shoots along the 'rails' if the magnetic field of the U-shaped magnet is perpendicular to it (as shown). There is no motion if the field

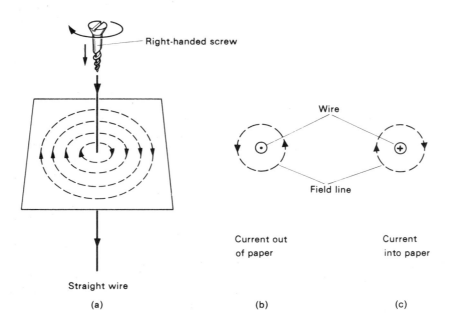

Right-handed screw

Straight wire

Wire

Field line

Current out of paper

Current into paper

Fig. 13.3 (a) (b) (c)

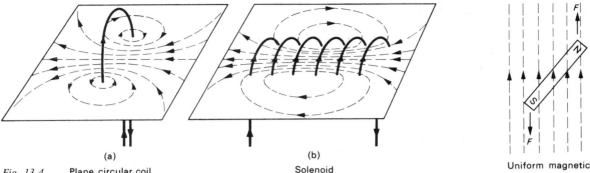

Fig. 13.4   Plane circular coil   (a)

(b)   Solenoid

Fig. 13.5   Uniform magnetic field

is parallel to it. From such experiments we conclude that the force on a current-carrying conductor (*i*) is always perpendicular to the plane containing the conductor and the direction of the field in which it is placed, and (*ii*) is greatest when the conductor is at right angles to the field.

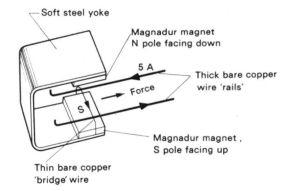

Fig. 13.6

The facts about the relative directions of current, field and force are summarized by Fleming's left-hand (or motor) rule. It states that *if the thumb and first two fingers of the left hand are held each at right angles to the other, with the First finger pointing in the direction of the Field and the seCond finger in the direction of the Current, then the Thumb predicts the direction of the Thrust or force*, Fig. 13.7.

(*b*) *Factors affecting the force.* The effect of the current, the length of the conductor and the strength and orientation of the magnetic field may be investigated by passing current through one arm PQ of a copper wire frame balanced on two razor blades via which the current enters and leaves, Fig. 13.8*a*. The ends

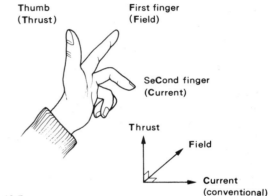

Fig. 13.7

of the wire in the arm RS of the frame are held together by an insulator and no current flows in RS. If the magnet is arranged so that PQ rises when current passes, the balance can be restored and the force measured by placing a suitable rider (a short length of wire) on PQ.

To see how the force depends on the *current* (measured by an ammeter which strictly speaking should be calibrated using a current balance, see p. 272, it is convenient to find the currents required to balance one, two and three identical riders on PQ. The effect on the force of doubling and trebling the *length* of conductor carrying the same current in the same magnetic field can be found by placing a second and a third, equally strong magnet beside the first, Fig. 13.8*b*. If the magnetic field is now produced by passing a current through a coil of many turns (e.g. 1100), Fig. 13.8*c*, increasing the coil current can reasonably be expected to increase the *field strength*. The response of the force on PQ to this may then be observed and also the effect of the *angle* between the direction of the field and PQ.

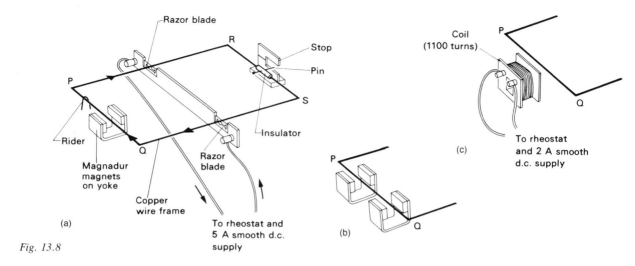

Fig. 13.8

Magnetic flux density

(*a*) *Definition of B.* The experiments described in the previous section show that the force *F* on a wire lying at right angles to a magnetic field is directly proportional to the current *I* in the wire and to the length *l* of the wire in the field. It also depends on the magnetic field, a fact we can use to define the strength of the field.

Electric field strength *E* is defined as the force per unit charge; gravitational field strength *g* is force per unit mass. An analogous quantity for magnetic fields is the *flux density* or *magnetic induction, B* (also called the *B-field*), defined as *the force acting per unit current length*, i.e. the force acting per unit length on a conductor which carries unit current and is at right angles to the direction of the magnetic field. In symbols, *B* is defined by the equation

$$B = \frac{F}{Il}$$

Thus if $F = 1$ newton when $I = 1$ ampere and $l = 1$ metre then $B = 1$ newton per ampere metre, i.e. the SI unit of *B* is the newton per ampere metre (N A^{-1} m^{-1}), which is given the special name of 1 *tesla* (T).

B is a vector whose direction at any point is that of the field line at the point. Its magnitude may be represented pictorially by the number of field lines passing through unit area; the greater this is, the greater the value of *B*.

(*b*) *Expression for force on a current.* Rearranging the equation defining *B* we see that the force *F* on a conductor of length *l*, carrying a current *I* and lying at right angles to a magnetic field of flux density *B*, is

given by

$$F = BIl$$

If the conductor and field are not at right angles, but make an angle θ with one another, Fig. 13.9, the expression becomes $BIl \sin \theta$. When $\theta = 90°$, $\sin \theta = 1$ and $F = BIl$ as before. If $\theta = 0$, the conductor and field are parallel and, as experiment confirms, $F = 0$.

Fig. 13.9

(*c*) *Measuring B using a current balance.* The copper wire frame used previously to investigate the factors affecting the force on a current in a magnetic field (p. 263) is a very simple 'current balance' (see p. 272) and enables us to measure the flux density of a field. For example, suppose a rider of mass 0.084 g is required to counterpoise the frame when arm PQ, of length 25 cm and carrying a current of 1.2 A, is inside and in series with a flat, wide solenoid, Fig. 13.10. We then have

$$F = \text{force on arm PQ}$$
$$= \text{weight of rider} = \text{mass of rider} \times g$$
$$= (0.084 \times 10^{-3} \text{ kg}) \times (10 \text{ N kg}^{-1})$$
$$= 0.084 \times 10^{-2} \text{ N}$$

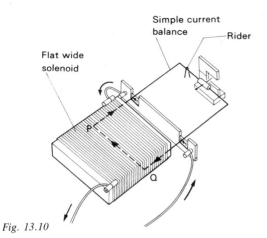

Fig. 13.10

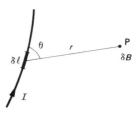

Fig. 13.11

$$I = 1.2 \text{ A}$$

$$l = 0.25 \text{ m}$$

The flux density B (assumed uniform) inside the solenoid is given by

$$B = \frac{F}{Il} = \frac{8.4 \times 10^{-4} \text{ N}}{1.2 \text{ A} \times 0.25 \text{ m}}$$

$$= 2.8 \times 10^{-3} \text{ T}$$

The field in the gap between two Magnadur magnets on a mild steel yoke (see Fig. 13.8a) is about ten times stronger than this, whilst in the earth's magnetic field in Britain, $B_R \simeq 5.3 \times 10^{-5}$ T and $B_H \simeq 1.8 \times 10^{-5}$ T. A flux density of 1 T would be produced by a fairly strong magnet.

Suggest (i) a source of error in the above measurement of B and (ii) how the result could be made more reliable.

The Biot-Savart law

In the previous section we saw how B could be found experimentally; it can also be calculated using the Biot-Savart law.

The calculation involves considering a conductor as consisting of a number of very short lengths, each of which contributes to the total field at any point. Biot and Savart *stated* that for a very short length δl of conductor, carrying a steady current I, the magnitude of the flux density δB at a point P distant r from δl is

$$\delta B \propto \frac{I \, \delta l \sin \theta}{r^2}$$

where θ is the angle between δl and the line joining it to

P, Fig. 13.11. The direction of δB, given by the right-hand rule, is at right angles to the plane containing δl and P and, for the current direction shown, acts into the paper. The product $I \, \delta l$ is called a 'current element'. It is difficult to prove Biot and Savart's law directly for an infinitesimally small conductor, i.e. a current element, but experimentally verifiable deductions can be made from it for 'life-sized' conductors as we shall see shortly.

Before the Biot-Savart law can be used in calculations it has to be expressed as an equation and a constant of proportionality introduced. Now δB depends on one other factor besides I, δl, θ and r, and that is the medium through which the magnetic force acts. This factor can be incorporated in the Biot-Savart law if we regard the constant of proportionality as a property of the medium. The constant is called the *permeability* of the medium and is denoted by μ (mu).

The permeability of a vacuum is denoted by μ_0 (mu nought), its value is *defined* (from the definition of the ampere, p. 272) to be $4\pi \times 10^{-7}$ and its unit is the *henry per metre* (H m^{-1}), as we shall see later (p. 303). Air and most other materials except ferromagnetics have nearly the same permeability as a vacuum. (Note that whilst the value of ε_0—the permittivity of free space—is found by experiment, that for μ_0 is defined.)

Rationalization (see p. 225) is achieved by introducing 4π in the denominator. The Biot-Savart equation for a current element becomes

$$\delta B = \frac{\mu_0 I \, \delta l \sin \theta}{4\pi r^2}$$

It is an inverse square relation.

Calculation of flux density

In most cases the calculation requires the use of calculus but for a circular coil the value of B at the centre can be found fairly simply.

(i) *Circular coil.* Suppose the coil is in air, has radius r, carries a steady current I and is considered to consist of current elements of length δl. Each element is at

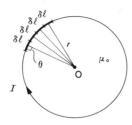

Fig. 13.12

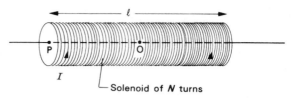

Fig. 13.14

distance r from the centre O of the coil and is at right angles to the line joining it to O, i.e. $\theta = 90°$, Fig. 13.12. At O the total flux density B is the sum of the flux densities δB due to all the elements. That is

$$B = \sum \frac{\mu_0 I \, \delta l \sin \theta}{4\pi r^2} = \frac{\mu_0 I \sin \theta}{4\pi r^2} \sum \delta l$$

But $\Sigma \, \delta l = $ total length of the coil $= 2\pi r$ and $\sin \theta = \sin 90 = 1$. Hence

$$B = \frac{\mu_0 I \, 2\pi r}{4\pi r^2} = \frac{\mu_0 I}{2r}$$

If the coil has N turns each of radius r

$$B = \frac{\mu_0 N I}{2r}$$

(ii) *Very long straight wire*. It can be shown (see Appendix 10) that the value of B at a perpendicular distance a from a very long straight wire carrying a current I, Fig. 13.13, is

$$B = \frac{\mu_0 I}{2\pi a}$$

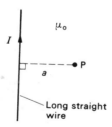

Fig. 13.13

In this case the field is non-uniform, there is cylindrical symmetry (Fig. 13.3a) and 2π appears quite appropriately. If $I = 10$ A and $a = 1.0$ cm $= 0.010$ m, then $B = 2.0 \times 10^{-4}$ T.

(iii) *Very long solenoid*. If the solenoid has N turns, length l and carries a current I, the flux density B at a point O on the axis near the centre of the solenoid, Fig. 13.14, is found to be given by

$$B = \frac{\mu_0 N I}{l}$$

$$= \mu_0 n I$$

where $n = N/l = $ number of turns per unit length. B thus equals μ_0 multiplied by the *ampere-turns per metre*. The field is fairly uniform over the cross-section of the solenoid for most of its length (Fig. 13.4b) and the flux density is given by the above expression. The absence of π from it may be noted and is what we expect from the plane symmetry of a uniform field using a rationalized system.

For a solenoid of 400 turns, 25 cm long, carrying a current of 5.0 A we have $n = 400/0.25 = 1600$ turns per metre, $I = 5.0$ A and so $B \simeq 1.0 \times 10^{-2}$ T.

At P, a point at the end of a long solenoid,

$$B = \frac{\mu_0 \, n \, I}{2}$$

That is, B at a point at the end of the solenoid's axis is half the value at the centre.

These expressions are strictly true only for an infinitely long solenoid: in practice they are accurate enough for most purposes if the length of the solenoid is at least ten times its diameter.

(iv) *Helmholtz coils*. Two flat coaxial coils having the same radius r, the same number of turns N and separated by a distance r, Fig. 13.15, produce an almost

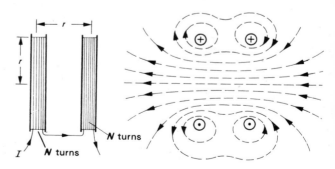

Fig. 13.15

uniform field over some distance near their common axis midway between them when the same current I flows round each in the same direction. It is given by

$$B = \frac{8\mu_0 \, NI}{5\sqrt{5} \, r}$$

$$\simeq 0.72 \frac{\mu_0 NI}{r}$$

Helmholtz coils are used in a school laboratory measurement of e/m for electrons.

Magnetic field measurements

Measurement of B for different current-carrying conductors can be made and compared with the values and variations predicted by the expressions derived from the Biot and Savart law. Whilst a simple current balance of the type used previously (Fig. 13.10, p. 265) is satisfactory for defining B, it is not sensitive enough for showing how B varies from one point to another in a field. However, two other devices are convenient. One is the search coil, used with alternating fields produced by alternating currents and the other is the Hall probe for steady fields produced by direct currents.

(a) Search coil and a.c. A typical search coil has 5000 turns of wire and an external diameter of no more than 1.5 cm so that it samples the field over a small area. If it is connected to the Y-plates of a CRO (time base off), the length of the vertical trace is proportional to the maximum value of the alternating flux density, or any such component, *acting along the axis of the search coil at right angles to its face.* (The technique uses an effect called *mutual induction* which will be considered in the next chapter, p. 282. The CRO is in fact being used as a voltmeter to measure an e.m.f. induced in the search coil by the alternating magnetic field. The sensitivity of the method increases with the frequency of the a.c. in the conductor producing the field and 10 kHz is a convenient frequency to use.) Fig. 13.16 shows two types of mounting for a search coil.

(b) Hall probe and d.c. The Hall probe consists of a slice of semiconducting material on the end of a rod, Fig. 13.17, connected to a light-beam galvanometer via a circuit box, Fig. 13.19b. When a battery in the box is switched on, current is supplied to the slice. The 'balance' control is adjusted for zero deflection on the galvanometer and the probe is then inserted in the field to be measured. The change in the galvanometer

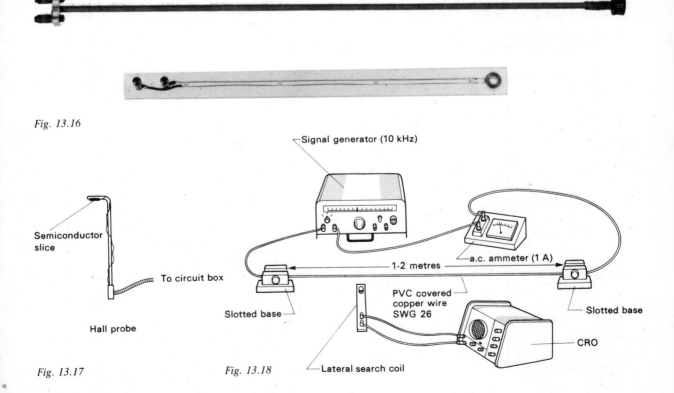

Fig. 13.16

Fig. 13.17

Fig. 13.18

reading is proportional to the flux density (or its component) *at right angles to the slice.* The action of the probe depends on the Hall effect to be discussed shortly (p. 270).

(c) *Investigations.* If actual values of B are required, the search coil or Hall probe would have to be calibrated (How?). For studying comparisons and relationships this is not necessary.

A long straight wire may be investigated using the arrangement in Fig. 13.18. What is the direction of B near the wire? Is B the same all along the wire at a constant distance from it? Is B directly proportional to the current in the wire? Does B vary as $1/a$ (where a is the perpendicular distance from the wire), as theory suggests?

Using a Slinky spring as a solenoid, Fig. 13.19a, find if the field is uniform across the width of the solenoid but zero outside. Is B constant along most of the length of the solenoid? Can we accept the theoretical prediction that for a long solenoid B at the end is half that at the centre? How does B at the centre alter if the solenoid is stretched, the current remaining constant? Answers to some of these questions may also be found with the apparatus of Fig. 13.19b.

A 'magnetic field board' like that in Fig. 13.20 enables circular coils to be studied. Do the direction and magnitude of B vary over the plane of the coil? When the current and radius are constant how does B at a given point vary with the number of turns? With a certain number of turns and current, is B at the centre inversely proportional to the radius of the coil? The

field along the axis should vary with distance x along the axis from the centre as $r^2/(x^2 + r^2)^{3/2}$ where r is the coil radius. Does it?

For a compact coil of, say, 240 turns how do the values of B compare at P and Q, Fig. 13.21? Is there a relationship between B and x along the axis of the coil?

Force on a charge in a magnetic field

An electric current is regarded as a drift of charges and so it is reasonable to assume that the force experienced by a current-carrying conductor in a magnetic field is the resultant of the forces acting on the charges constituting the current. Accepting this, an expression for the force on a single charged particle moving in a magnetic field can be derived.

Consider a length l of conductor containing n charged particles each of charge Q and average drift velocity v, Fig. 13.22. Each particle will take the same time t to travel distance l, and so

$$v = l/t$$

The total charge passing through any cross-section of the conductor in time t is nQ. Therefore the current I is given by

$$I = \text{total charge passing a section/time}$$
$$= nQ/t$$
$$= nQv/l \quad (\text{since } t = l/v)$$

If the conductor makes an angle θ with a uniform

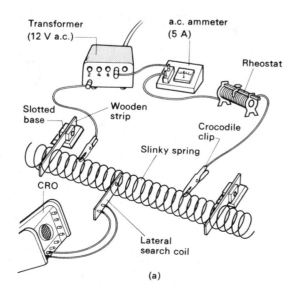

(a)

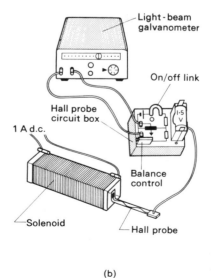

(b)

Fig. 13.19

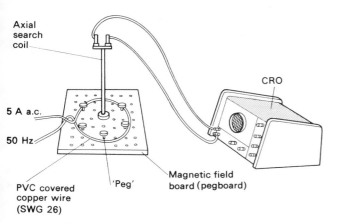

Axial search coil

CRO

5 A a.c.

50 Hz

PVC covered copper wire (SWG 26)

'Peg'

Magnetic field board (pegboard)

Fig. 13.20

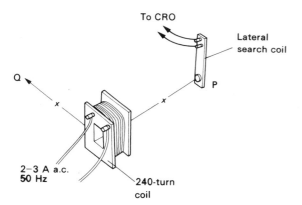

To CRO

Lateral search coil

Q

P

2–3 A a.c. 50 Hz

240-turn coil

Fig. 13.21

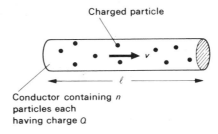

Charged particle

v

l

Conductor containing n particles each having charge Q

Fig. 13.22

(a)

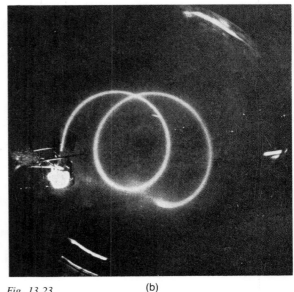

(b)

Fig. 13.23

magnetic field of flux density B, then

$$\text{force on conductor} = BIl \sin \theta \qquad \text{(p. 264)}$$

$$= B\left(\frac{nQv}{l}\right) l \sin \theta$$

$$= BnQv \sin \theta$$

Hence the force on one charged particle is $BQv \sin \theta$. If the charge moves at right angles to the field, $\theta = 90°$, $\sin \theta = 1$ and the force F is given by

$$F = BQv$$

If B is in teslas (T), Q in coulombs (C) and v in metres per second (m s^{-1}), then F is in newtons (N). The direction of the force is given by Fleming's left-hand rule (remembering that a negative charge moving one way is equivalent to conventional current flowing in the opposite direction) and is at right angles both to the field and to the direction of motion.

This force is also responsible for the deflection of a

beam of charged particles travelling through a magnetic field in a vacuum, e.g. electrons in a cathode-ray tube. Since it is always at right angles to the path of the beam it only changes the direction of motion but not the speed. When the particles enter a uniform field at right angles, they are deflected into a circular path; for other angles (except 0°) they describe a helix. The circular and helical paths of electrons in a fine beam tube (p. 462) are shown in Figs. 13.23a and b.

Charged particles thus tend to become 'trapped' in magnetic fields and their existence in certain regions round the earth, known as the van Allen radiation belts, is due to the earth's magnetic field. The behaviour of charged particles in magnetic fields is of considerable importance in atomic physics.

Hall effect

A current-carrying conductor in a magnetic field has a small p.d. across its sides, in a direction at right angles to the field. In the slab of conductor shown in Fig. 13.24a, the p.d., called the *Hall* p.d., appears across XY.

(a) *Explanation.* The effect was discovered in 1879 by Hall and can be attributed to the forces experienced by the charge carriers in the conductor. These act at right angles to the directions of the magnetic field and the current (and are given by Fleming's left-hand rule) and cause the charge carriers to be pushed sideways, thus increasing their concentration towards one side of the conductor. As a result, a p.d. (and an electric field) is produced across the conductor.

With the magnetic field and conventional current having the directions shown in Fig. 13.24b and c, the forces act upwards and side X of the conductor develops the same sign as the charge carriers, i.e. positive in (b) and negative in (c). The Hall p.d. therefore reveals the sign of the charge carriers (more exactly, the majority charge carriers) in the conductor. In metals such as copper and aluminium they are negative; in semiconductors they may be positive or negative.

(b) *Demonstration of Hall effect.* Germanium, a semiconductor, is used because the effect is very much greater with it than with metals (for reasons given later); p.d.s of about 0.1 V are obtainable. Pure germanium has very few 'free' electrons for current carrying but if it is 'doped' with a small proportion of atoms of another element (e.g. antimony) which provides extra electrons, n-type germanium is formed with enhanced conduction properties. p-type germanium is 'doped' with atoms (e.g. those of indium) which cause the germanium to behave as if it had positive charge carriers.

In the circuit of Fig. 13.25 a current of about 50 mA is passed through the Hall slice (a piece of doped germanium) via the current terminals P and Q. If the Hall p.d. connections X and Y on the slice are not directly opposite each other, a p.d. will exist between them even in the absence of a magnetic field and there will be a current through the microammeter. In this case the 'balance' control should be altered to zero the meter and bring X and Y to the same potential. If a Magnadur magnet is then placed face downwards over the slice, the microammeter should indicate a Hall p.d., the sign of which for p-type germanium will be opposite to that for n-type. Application of the left-hand rule enables the sign of the charge carriers to be determined.

(c) *Expression for Hall p.d.* The Hall p.d. and electric field build up until the repulsive electrical force they cause on a charge carrier drifting through the conductor is equal and opposite to the deflecting magnetic force on it and no further charge displacement occurs.

If V_H is the Hall p.d. across the width d of the conductor and E is the final value of the electric field strength, then since electric field strength = potential gradient, we have

$$E = V_H/d$$

If the charge carriers each have charge Q and drift velocity v then

electric force on each drifting charge = $EQ = V_H Q/d$

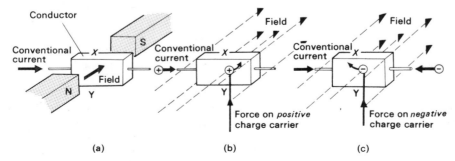

Fig. 13.24   **(a)**   **(b)**   **(c)**

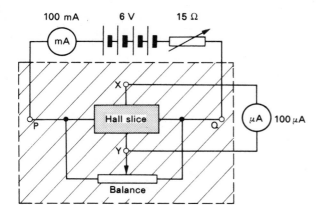

Fig. 13.25

magnetic force on each drifting charge $= BQv$

At equilibrium,

$$V_H Q/d = BQv$$

$$\therefore \ V_H = Bvd$$

The case of positive charge carriers is illustrated in Fig. 13.26. Draw the corresponding diagrams for a negative charge carrier. It can be shown (page 37) for a conductor of cross-sectional area A carrying current I and having n charge carriers per unit volume, that

$$I = nAQv$$

Substituting for v we get $\quad V_H = \dfrac{BId}{nAQ}$

But $A = d \times t$ where t is the thickness of the conductor

$$\therefore \ V_H = \dfrac{BI}{nQt}$$

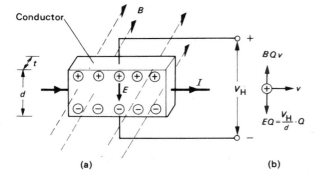

Fig. 13.26

The expression shows that V_H is greatest in materials for which n is small, i.e. semiconductors, since those have a smaller number of charge carriers per unit volume than metals.

(d) *Number of charge carriers per atom.* If V_H is measured for, say, a thin strip of aluminium foil carrying a large current in a strong magnetic field, n may be found. Thus if $B = 1.4$ T, $I = 10$ A, $t = 0.060$ mm $= 6.0 \times 10^{-5}$ m and $V_H = 10 \ \mu$V $= 10 \times 10^{-6}$ V, then taking $Q =$ electronic charge $= 1.6 \times 10^{-19}$ C we get

$$n = \frac{BI}{V_H \, Qt}$$

$$= \frac{1.4 \times 10}{10 \times 10^{-6} \times 1.6 \times 10^{-19} \times 6.0 \times 10^{-5}}$$

$$= 1.4 \times 10^{29} \text{ electrons per metre}^3$$

A calculation using the Avogadro constant (6.0×10^{23} atoms mol^{-1}), the relative atomic mass and the density of aluminium (27 and 2.7 g cm^{-3} respectively) shows that there are 6.0×10^{28} aluminium atoms per cubic metre. Check this. The number of 'free' (conduction) electrons per atom must therefore be between 2 and 3 ($[1.4 \times 10^{29}]/[6.0 \times 10^{28}] = 2.3$). Chemical evidence suggests that aluminium is trivalent, i.e. has three 'outer' electrons (which participate in chemical reactions). Hall effect measurements do not conflict with this conclusion.

The ampere and current balances

(a) *Force between currents.* The ampere is the basic electrical unit of the SI system and has to be defined, like any other unit, so that it is accurately reproducible. This is achieved by basing the definition on the force between two long, straight, parallel current-carrying conductors, i.e. by using the magnetic effect, and making measurements with a current balance.

The forces between currents may be demonstrated with two long narrow strips of aluminium foil arranged as in Fig. 13.27. When the currents flow in opposite directions the strips repel and when in the same direction they attract. In brief, *unlike currents repel, like currents attract*.

To derive an expression for the force, consider two long, straight, parallel conductors, distance a apart in air, carrying currents I_1 and I_2 respectively, Fig. 13.28. The magnetic field at the right-hand conductor due to the current I_1 in the left-hand one is directed into the

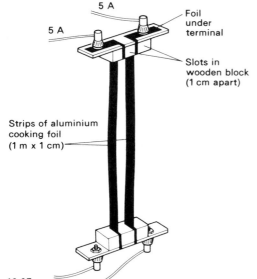

5 A

5 A

Foil under terminal

Slots in wooden block (1 cm apart)

Strips of aluminium cooking foil (1 m x 1 cm)

Fig. 13.27

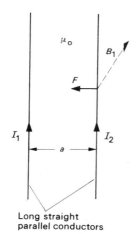

μ_0

B_1

F

I_1

I_2

a

Fig. 13.28

Long straight parallel conductors

paper and its flux density B_1 is given by

$$B_1 = \frac{\mu_0 I_1}{2\pi a}$$ (p. 266)

The force F acting on length l of the right-hand conductor (carrying current I_2) is therefore

$$F = B_1 I_2 l$$
$$= \frac{\mu_0 I_1 I_2 l}{2\pi a}$$

The left-hand conductor experiences an equal and opposite force due to being in the field of the right-hand conductor.

(*b*) *The ampere and* μ_0. The definition of the ampere is based on the previous expression and may be stated as follows:

The ampere is the constant current which, flowing in two infinitely long, straight, parallel conductors of neg-ligible circular cross-section, placed in a vacuum 1 metre apart, produces between them a force of 2×10^{-7} *newton per metre of their length.*

Previously it was defined in terms of the chemical effect, and the choice of a force of 2×10^{-7} newton was made in the magnetic effect definition to keep the value of the new unit as near as possible to that of the old one.

Once the ampere has been defined, the value of μ_0 follows. Thus we have from the definition,

$$I_1 = I_2 = 1 \text{ A}$$

$$l = a = 1 \text{ m}$$

$$F = 2 \times 10^{-7} \text{ N}$$

Substituting in $F = \mu_0 I_1 I_2 l / (2\pi a)$, we get

$$2 \times 10^{-7} = \mu_0 \times 1 \times 1 \times 1/(2\pi \times 1)$$

$$\therefore \quad \mu_0 = 4\pi \times 10^{-7} \text{ H m}^{-1}$$

This is the value given previously.

(*c*) *Current balances and absolute measurement of current.* A current balance enables a current to be measured by weighing and uses the fact that adjacent current-carrying conductors, be they straight wires as considered in the definition of the ampere or coils, exert forces on each other. The measurement, when reduced to its essentials, involves finding a mass, a length and a time and is said to be an *absolute* one; no electrical quantities as such have to be measured, only mechanical ones. (Basically, a force is determined from

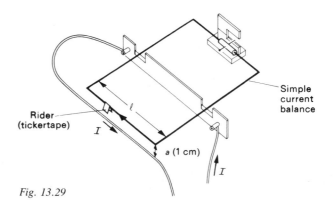

Rider (tickertape)

I

ℓ

a (1 cm)

I

Simple current balance

Fig. 13.29

the acceleration it produces in a known mass and has dimensions MLT^{-2}.) Current balances are used primarily to calibrate more convenient types of current-measuring instruments.

A simple arrangement is shown in Fig. 13.29 in which the current balance met earlier (p. 264) measures the current I (about 5 A) in a long, straight wire, close and parallel to its current-carrying arm of length l and in series with it. You should be able to show that $I^2 = mga/(2 \times 10^{-7} l)$, where m is the mass of the rider needed to counterpoise the balance. The accuracy of the result is poor because the force involved is very small. A larger force and greater accuracy are obtained if a wide, flat solenoid is used (see Fig. 13.10, p. 265) and

I calculated from $I^2 = mg/(\mu_0 nl)$ where n is the number of turns per metre on the solenoid.

The principle of a practical current balance such as is used in standardizing laboratories like the National Physical Laboratory (NPL) is shown in Fig. 13.30a. The six coils carry the same current and are connected in series so that the forces on the two movable coils tip the balance to the same side. A rider of known mass is moved on the beam of the balance to restore equilibrium and enables the forces between the coils to be found. Knowing this and the value of B (from the dimensions of the coils), the current can be calculated to an accuracy of a few parts in a million. The NPL current balance is shown in Fig. 13.30b.

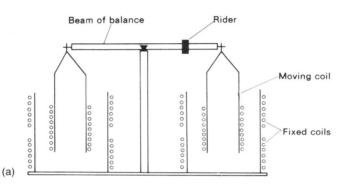

Fig. 13.30   (b)

Couple on a coil in a magnetic field

Current-carrying coils in magnetic fields are essential components of electric motors and meters of various kinds.

(a) *Expression for couple.* Consider a rectangular coil PQRS of N turns pivoted so that it can rotate about a vertical axis YY^1 which is at right angles to a *uniform* magnetic field of flux density B, Fig. 13.31a. Let the normal to the plane of the coil make an angle θ with the field, Fig. 13.31b.

When current I flows in the coil each side experiences a force (since all make some angle with B), acting perpendicularly to the plane containing the side and the direction of the field. The forces on the top and bottom (horizontal) sides are parallel to YY^1 and for the current direction shown, they lengthen the coil. The forces on the vertical sides, each of length l, are equal and opposite and have value F where

$$F = BIlN$$

Whatever the position of the coil, its vertical sides are at right angles to B and so F remains constant. The forces constitute a couple whose moment (or torque) C is given by

$C = $ one force × perpendicular distance between lines of action of the forces

$\quad = F \times \text{PT} = F \times b \sin \text{PQT}$ ($b = $ breadth of coil)

$\quad = Fb \sin \theta$

$\quad = BIlN \times b \sin \theta$

$\quad = BIAN \sin \theta$

where $A = $ area of face of coil $= l \times b$. The couple causes angular acceleration of the coil which rotates until its plane is perpendicular to the field (i.e. $\theta = 0$) and then $C = 0$.

The expression for C can be shown to hold for a coil of *any* shape of area A. Thus a circular coil carrying current I can be regarded as consisting of a large number of tiny rectangular coils, each with current I flowing in the same direction as in the circular coil, Fig. 13.32. The forces on all sides of the tiny coils cancel except where they lie on the circular coil itself.

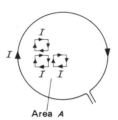

Circular coil

Fig. 13.32   Area A

(b) *Electromagnetic moment of a coil.* From the above expression we see that C depends on, among other things, I, A and N. It is convenient to write

$$m = IAN$$

where m is a property of the coil and the current it carries, and is called the *electromagnetic moment* of the coil. We then have

$$C = Bm \sin \theta$$

or $$m = C/(B \sin \theta)$$

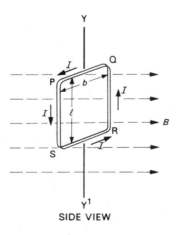

SIDE VIEW

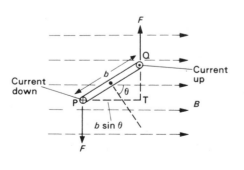

PLAN VIEW

Fig. 13.31   (a)   (b)

If $B = 1$ T and $\theta = 90°$ then $m = C$ and we can define the *electromagnetic moment* of a coil as *the moment of the couple (or torque) acting on it when it lies with its plane parallel to a magnetic field of unit flux density,* Fig. 13.33.

m is a vector whose direction is taken to be that of the flux density created along the axis of the coil by the current in it. We can therefore regard the couple which acts on a current-carrying coil as trying to align the electromagnetic moment of the coil with the direction of the flux density of the applied magnetic field.

PLAN VIEW

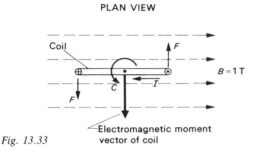

Fig. 13.33   Electromagnetic moment vector of coil

Moving-coil galvanometers

A galvanometer detects (or measures, if its scale is calibrated) small currents passing through it or small p.d.s across it: the addition of a shunt or multiplier converts it to an ammeter or a voltmeter. Most d.c. meters are of the moving-coil type.

(*a*) *Construction.* Basically a moving-coil galvanometer consists of a coil of fine, insulated copper wire which is able to rotate in a strong magnetic field. The field is produced in the narrow air gap between the concave pole pieces of a permanent magnet and a fixed soft iron cylinder and is *radial*, i.e. the field lines in the gap appear to radiate from the central axis of the cylinder and are always parallel to the plane of the coil, Fig. 13.34.

In the *pointer-type* meter shown, the coil is pivoted on jewelled bearings and its rotation is resisted by hair springs above and below it. The springs also lead the current in and out of the coil. In another common type of construction called *taut-ribbon suspension*, the coil is suspended and controlled not by pivots, jewelled bearings and hair springs but by two gold alloy ribbons held taut by springs above and below it. The ribbons conduct the current to and from the coil.

In the most sensitive galvanometers a small concave mirror (instead of a pointer) is fixed to the coil, as shown in the taut-ribbon instrument of Fig. 13.35*a*. The

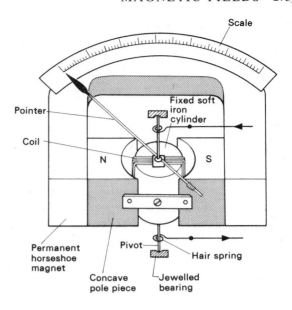

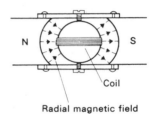

Fig. 13.34   Radial magnetic field

mirror throws an image of an illuminated hair line on to a scale via a return mirror, Fig. 13.35*b* and the light beam acts as a weightless pointer whose effective length is the distance from the coil to the scale. The angular deflection of the coil is magnified twice by this optical system. Why? A modern *light-beam* galvanometer with a millimetre scale is shown in Fig. 13.35*c*.

(*b*) *Theory.* The magnetic field is *radial* and so the plane of the coil is parallel to it, whatever the deflection. The forces acting on the vertical sides are therefore always perpendicular to the sides and the deflecting couple consequently has a maximum value for all positions of the coil (since $\theta = 90°$ and $\sin 90 = 1$). If the air gap is of constant width, the flux density B of the field is also nearly constant and the moment C of the deflecting couple due to current I in the coil is given by

$$C = BIAN$$

where A is the mean area of the coil and N is the number of turns on it.

The coil rotates until the resisting couple C^1 due to the suspension is equal and opposite to C. If the deflection is then α and k is the moment of the couple needed to produce unit angular deflection (in newton metres per radian) of the suspension, we have

$$C^1 = k\alpha$$

(assuming Hooke's law holds for the suspension). Hence, since at equilibrium $C = C^1$

$$BIAN = k\alpha$$

B, A, N and k are constants for a given meter and so

$$\alpha \propto I$$

The use of a radial field and a uniform air gap thus results in the deflection α of the coil being directly proportional to the current I in it; the galvanometer scale is therefore linear.

(c) *Sensitivity*. The *current sensitivity* of a galvanometer is defined as the *deflection per unit current* and equals α/I. It follows from above that

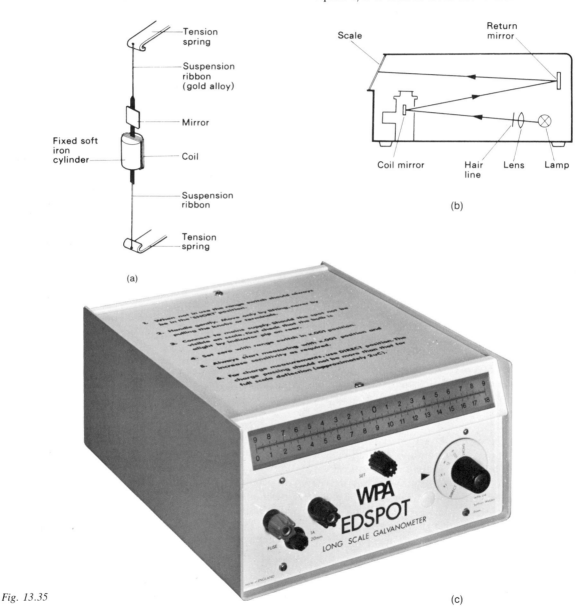

Fig. 13.35

$$\frac{\alpha}{I} = \frac{BAN}{k}$$

Maximum current sensitivity therefore requires

(*i*) *B* to be large in the air gap, i.e. the permanent magnet should be strong and the air gap narrow.

(*ii*) *A* to be as large as possible but if the coil is too large it swings about its deflected position before a reading can be taken.

(*iii*) *N* to be large but not at the expense of having to use a wide air gap.

(*iv*) *k* to be small but if the opposition of the suspension is too weak readings again take time.

The *voltage sensitivity* is defined as α/V where α is the deflection when the p.d. across the galvanometer is *V*. If its resistance is *R* then since $I = V/R$, we have

$$\frac{\alpha}{I} = \frac{BAN}{k}$$

$$\therefore \quad \frac{\alpha}{V} = \frac{BAN}{kR}$$

Sensitivities are usually expressed in mm per μA or mm per μV, this being the deflection produced on a mm scale by 1 μA or 1 μV.

(*d*) *Choice of galvanometer*. When choosing a galvanometer for a particular task the resistance of the rest of the circuit has to be considered. It can be shown that we should aim at transferring maximum power to the meter, and this occurs when the resistance of the meter equals the resistance of the rest of the circuit. If the latter is low, as it often is in potentiometer, Wheatstone bridge and thermocouple circuits, then a low resistance galvanometer is required. Such instruments have high voltage sensitivity. For high resistance circuits, galvanometers should have a high resistance and high current sensitivity.

Typical data for the general purpose galvanometer shown in Fig. 13.35*c* are given in Table 13.1 for the 'direct' position of the range switch.

Table 13.1

Galvanometer resistance Ω	Sensitivities		
	Current mm/μA	Voltage mm/μV	Charge mm/μC
14	25	1.8	75

(*e*) *Care and use of light-beam galvanometers.* Light-beam galvanometers should be handled with care. When not in use, being moved or when making connections, the range switch should be set at 'short' (see p. 287). The zero is usually set either at the end or the centre of the scale by turning the range switch to '× 0.001' and adjusting the set-zero control.

On 'direct' the instrument is most sensitive and the coil swings freely. It is used in this position for ballistic work to measure charge (see p. 304). The sensitivity is about the same on '× 1' but internal resistors (shunts) are then connected across the coil so that it is 'critically damped'. This means the coil reaches its steady deflection in the minimum of time, i.e. the movement is 'dead-beat' (see p. 287). The coil is critically damped for all settings of the range switch (except 'direct') unless the resistance of the external circuit is low compared with the critical damping resistance of the galvanometer (an external resistance of 120 Ω for the instrument of Fig. 13.35*c*).

The '× 0.1' range is 10 times less sensitive than the '× 1' setting. It is advisable to start on the '× 0.001' range and increase the sensitivity as required.

Moving-coil loudspeaker

Most speakers used today are of this type. The construction is shown in Fig. 13.36. Alternating current (p. 314) from the amplifier of a radio, record-player, etc., passes through a short cylindrical coil whose turns are at right angles to the magnetic field of a magnet

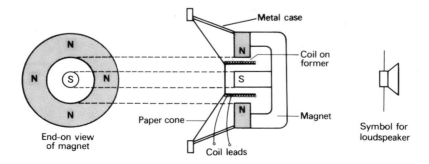

Fig. 13.36

with a central pole and a surrounding ring pole.

A force acts on the coil which, according to Fleming's left-hand rule, makes it move in and out. A paper cone attached to the former of the coil vibrates with it and sets up sound waves with the same frequency as the alternating current in the surrounding air.

Moving-coil microphones have a similar construction but they convert sound into alternating currents.

Relay

This is a switch worked by an electromagnet. It is useful if we want a small current in one circuit to control another circuit containing a device such as a lamp, electric bell or motor which requires a large current (p. 509).

The structure of a relay is shown in Fig. 13.37. When the controlling current flows through the coil, the soft iron core is magnetized and attracts the L-shaped soft iron armature. This rocks on its pivot and closes the electrical contacts in the circuit being controlled.

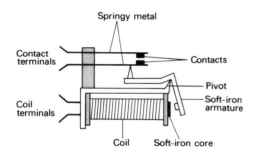

Fig. 13.37

QUESTIONS

1. In which of the four cases shown in Fig. 13.38 will the arm XY of the current balance experience a tilting force and will it be upwards or downwards?

2. A current of 5 A flows in a straight wire in a uniform flux density of 2×10^{-3} T. Calculate the force per unit length on the wire if it is (*i*) perpendicular to the field (*ii*) inclined at 30° to it.

3. A long straight vertical wire carrying 5.0 A in a downward direction passes through a horizontal board on which lines of force may be plotted. Draw a diagram of the lines of force to be expected and deduce a value for the horizontal component of the earth's magnetic field if there is a neutral point 5.0 cm from the wire. What would be the resultant horizontal field at a point an equal distance from the wire on a line through the wire at right angles to that joining the wire to the neutral point?

Permeability of free space (μ_0) = $4\pi \times 10^{-7}$ H m^{-1}.

(*J.M.B.*)

4. What current must be passed through a flat circular coil of 10 turns and radius 5.0 cm to produce a flux density of 2.0×10^{-4} T at its centre? ($\mu_0 = 4\pi \times 10^{-7}$ H m^{-1}).

5. Define the *magnitude* and *direction of a magnetic field*.

Describe experiments which would enable you to investigate how the magnetic field at the centre of a flat coil varies (*a*) with the current flowing in the coil, and (*b*) with the radius of the coil.

A flat circular coil of wire of 20 turns and of radius 10.0 cm is placed with its plane vertical and at 45° to the magnetic meridian. Calculate the current in the coil if a compass needle, free to move in a horizontal plane, points in the east–west direction when placed at the centre of the coil.

(Horizontal component of the earth's magnetic flux density = 2.0×10^{-5} T; $\mu_0 = 4\pi \times 10^{-7}$ H m^{-1}.) (*A.E.B.*)

6. Define the *ampere*.

Two long vertical wires, set in a plane at right angles to the magnetic meridian, carry equal currents flowing in opposite directions. Draw a diagram showing the pattern, in a horizontal plane, of the magnetic flux due to the currents alone—that is, neglecting for the moment the earth's magnetic field.

Next, taking into account the earth's magnetic field, discuss the various situations that can give rise to neutral points in the plane of the diagram.

Current balance

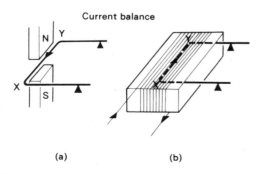

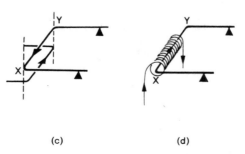

(a)   (b)   (c)   (d)

Fig. 13.38

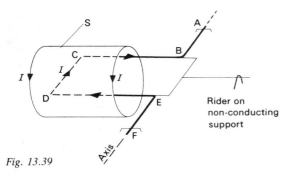

Fig. 13.39

Fig. 13.39 shows a simple form of current balance. The 'long' solenoid S, which has 2000 turns per metre, is in series with the horizontal rectangular copper loop ABCDEF, where BC = 10 cm and CD = 3.0 cm. The loop, which is freely pivoted on the axis AF, goes well inside the solenoid, and CD is perpendicular to the axis of the solenoid. When the current is switched on, a rider of mass 0.20 g placed 5.0 cm from the axis is needed to restore equilibrium. Calculate the value of the current, I.

($\mu_0 = 4\pi \times 10^{-7}$ H m^{-1}; $g = 10$ N kg^{-1}) (*O.*)

7. Draw a diagram to show the magnetic field (magnetizing force) due to a long solenoid carrying a current. Write down an equation for the flux density at its centre and at its ends, explaining your units. Describe an experiment you could perform to verify that the ratio of these flux densities is as predicted by your formulae.

A metal wire 10 m long lies east–west on a wooden table. What p.d. would have to be applied to the ends of the wire, and in what direction, in order to make the wire rise from the surface? Assume that the electrical connections to the wire cause no appreciable restraint.

(Density of the metal = 1.0×10^4 kg m^{-3}, resistivity of the metal = 2.0×10^{-8} Ω m, horizontal component of earth's field = 1.8×10^{-5} T, $g = 9.8$ N kg^{-1}.) (*C.*)

8. An electron of charge e and mass m describes a circular path of radius r when it is projected with velocity v into a uniform magnetic field of flux density B. Derive an expression for the frequency of revolution.

How many orbits per second are made by an electron in a fine beam tube (Fig. 13.23a, p. 269) if the flux density in the field due to the Helmholtz coils is 1.0×10^{-3} T? ($e = 1.6 \times 10^{-19}$ C; $m = 9.1 \times 10^{-31}$ kg.)

9. (*a*) In Fig. 13.40*a* electrons are shown drifting from right to left through a block of conductor. A flux density is applied as in Fig. 13.40*b* and the concentration of electrons increases near the front edge of the block. Why? If the current had consisted of positive charges drifting from right to left would they have been pushed towards the front?

(*b*) An electric field and a p.d. (the Hall p.d.) are created across the block and soon attain maximum values. Why?

(*c*) Fig. 13.40*c* shows an electron (charge e, mass m) drifting with velocity v near the middle of the conductor (width d) when the electric field and Hall p.d. have their maximum values, say E and V_H respectively. It is undeflected because two equal and opposite forces X and Y act on it. What are they? Derive an equation relating them (let B be the magnetic flux density).

(*d*) Hence calculate v if $V_H = 8$ μV, $B = 0.4$ T and $d = 2$ cm.

10. Calculate the magnetic flux density at a point 2.0 cm from a long straight wire carrying a current of 10 A. Hence calculate the force which would be exerted on a 50 cm length of another straight wire, parallel to the first and 2.0 cm away from it, if this second wire carried a current of 20 A. ($\mu_0 = 4\pi \times 10^{-7}$ H m^{-1}.) (*S.*)

11. Two long parallel wires in air with axes 50 cm apart carry currents of 100 A in opposite directions. Find (*a*) the magnetic field strength on the axis of one wire, due to the current in the other, and (*b*) the approximate value in newtons of the force per metre length on each wire.

Show clearly on a diagram the distribution of magnetic field around the wires and the direction of the force on one of them. Ignore the presence of the earth's magnetic field. ($\mu_0 = 4\pi \times 10^{-7}$ H m^{-1}.) (*L. part qn.*)

12. A rectangular coil 10 cm × 2.0 cm consisting of 100 turns is suspended vertically from the middle of a short side in a radial magnetic flux density of 2.0×10^{-2} T and supplied with current from a 25 V d.c. supply. Give a diagram of the arrangement showing the directions of the current, magnetic flux density (magnetic induction) and forces. If the resistance of the coil is 100 Ω, calculate the deflecting torque on the suspension. (*J.M.B.*)

13. Describe a simple experiment which demonstrates that a force is experienced by a current-carrying conductor in a magnetic field. State the factors that determine (*a*) the magni-

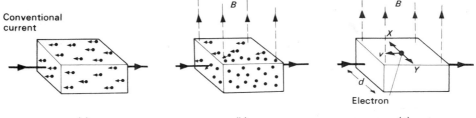

Fig. 13.40   **(a)**   **(b)**   **(c)**

tude of the force, and (b) its direction.

A rectangular coil of wire carrying a steady current is pivoted on an axis which is at right angles to a radial magnetic field. Obtain an expression for the torque experienced by the coil, and explain the relevance of this result to the design of moving-coil galvanometers.

A moving-coil galvanometer has a resistance of 25 Ω and gives a full-scale deflection when carrying a current of 4.4×10^{-6} A. What current will give a full-scale deflection when the galvanometer is shunted by a 0.10 Ω resistance? (O.)

14. Describe with the aid of diagrams the structure and mode of action of a moving-coil galvanometer having a linear scale and suitable for measuring small currents. If the coil is rectangular, derive an expression for the deflecting couple acting upon it when a current flows in it, and hence obtain an expression for the current sensitivity (defined as the deflection per unit current).

If the coil of a moving-coil galvanometer having 10 turns and of resistance 4.0 Ω is removed and replaced by a second coil having 100 turns and of resistance 160 Ω calculate

(a) the factor by which the current sensitivity changes, and

(b) the factor by which the voltage sensitivity changes.

Assume that all other features remain unaltered.   (J.M.B.)

14 Electromagnetic induction

Inducing e.m.f.s

An electric current creates a magnetic field, the reverse effect of producing electricity by magnetism was discovered independently in 1831 by Faraday in England and Henry in America and is called electromagnetic induction. Induced e.m.f.s can be generated in two ways.

(a) By relative movement (the dynamo effect). If a bar magnet is moved in and out of a stationary coil of wire connected to a centre-zero galvanometer, Fig. 14.1, a small current is recorded during the motion but not at other times. Movement of the coil towards or away from the stationary magnet has the same results. Relative motion between magnet and coil is necessary.

Observation shows that the direction of the induced current depends on the direction of relative motion. Also the magnitude of the current increases with the speed of motion, the number of turns on the coil and the strength of the magnet.

Although it is current we detect in this demonstration, an e.m.f. must be induced in the coil to cause the current. The induced e.m.f. is the more basic quantity and is always present even when the coil is not in a complete circuit. The value of the induced current depends on the resistance of the circuit as well as on the induced e.m.f.

We will be concerned here only with e.m.f.s induced in conductors but they can be produced in any

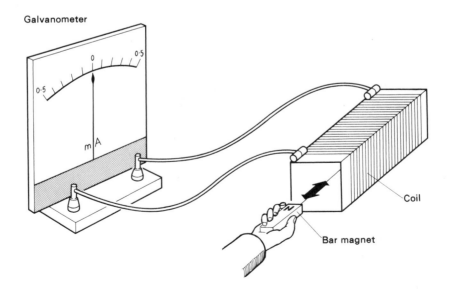

Fig. 14.1

medium—even a vacuum, where they play a basic role in the electromagnetic theory of radiation.

(*b*) *By changing a magnetic field* (the transformer effect). In this case two coils are arranged one inside the other as in Fig. 14.2*a*. One coil, called the *primary*, is in series with a 6 V d.c. supply, a tapping key and a rheostat. The other, called the *secondary*, is connected to a galvanometer. Switching the current on or off in the primary causes a pulse of e.m.f. and current to be induced in the secondary. Varying the primary current by quickly altering the value of the rheostat has the same effect. Electromagnetic induction thus occurs when there is any *change* in the primary current and so also in the magnetic field it produces.

Cases of electromagnetic induction in which current changes in one circuit cause induced e.m.f.s in a neighbouring circuit, not connected to the first, are examples of *mutual induction*—the transformer principle. (Our use of search coils when investigating magnetic fields due to various current-carrying conductors in the previous chapter depended on it.)

The induced e.m.f. is increased by having a soft iron rod in the coils or better still, by using coils wound on a complete iron ring. The iron ring and coils used by Faraday in his original experiment are shown in Fig. 14.2*b*. It is worth noting that the secondary current is in one direction when the primary current increases and in the opposite direction when it decreases.

An alternating current is continually increasing and decreasing first in one direction and then in the opposite direction. The magnetic field accompanying it

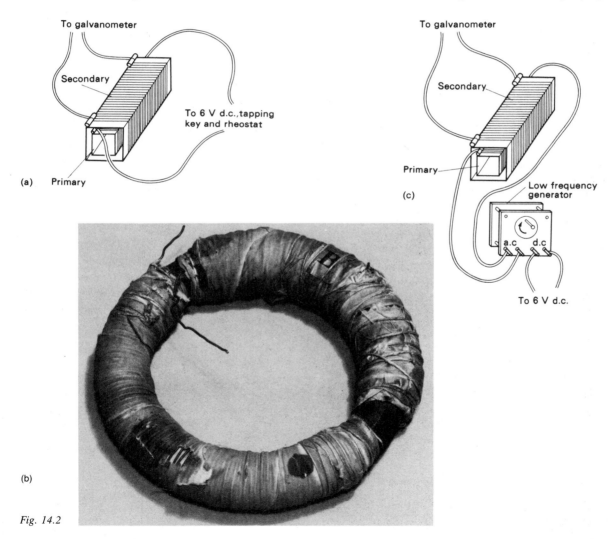

Fig. 14.2

changes similarly and if a.c. is applied to the primary coil we would expect an induced e.m.f. (also alternating?) to be induced in the secondary. A simple arrangement for investigating the effect is shown in Fig. 14.2c; very low frequency a.c. is produced when the generator is hand-operated.

Magnetic flux

Electromagnetic induction is one of those action-at-a-distance effects whose mechanism is not revealed to our senses. Consequently we have to invent a conceptual model or theory which enables us to picture and to 'explain' (i.e. describe in terms of the model) what might be happening.

Faraday suggested a model based on magnetic field lines. He proposed that an e.m.f. is induced in a conductor *either* when there is a change in the number of lines 'linking' it (i.e. passing through it) *or* when it 'cuts' across field lines. The two statements often come to the same thing, as can be seen by considering Fig. 14.3 (and later, the experiment described on p. 285). Thus if the coil moves towards the magnet from X to Y, the number of lines 'linking' or 'threading' it increases from three to five; alternatively we can say it 'cuts' two lines in moving from X to Y.

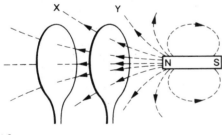

Fig. 14.3

Before expressing these ideas mathematically in the next section, we require some way of deciding how many lines link a coil or are cut by it. In the previous chapter (p. 261) it was suggested that magnetic field lines could be drawn so that the number per unit cross-sectional area represents the magnitude B of the flux density. It would then be reasonable to take the product, $B \times A$, as a measure of the number of lines linking a coil of cross-sectional area A.

This product is called the *magnetic flux* (Φ). It is a more useful concept than B alone for quantifying our electromagnetic induction model and it is defined by the equation

$$\Phi = BA$$

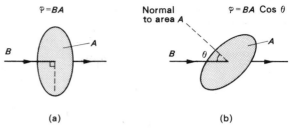

Fig. 14.4

where B is the flux density acting at right angles to and over an area A, Fig. 14.4a. In words, *the magnetic flux through a small plane surface is the product of the flux density normal to the surface and the area of the surface.* (Why a small surface?) If $B = 1$ tesla (T) and $A = 1$ square metre (m^2) then Φ is defined to be 1 weber (Wb).

Since $B = \Phi/A$, the reason for calling B the flux density will now be evident. Another unit of B is therefore the weber per square metre (Wb m^{-2}), i.e. 1 T $= 1$ Wb m^{-2}.

In general, if the normal to the area A makes an angle θ with B, Fig. 14.4b, then the flux is $BA \cos \theta$.

If Φ is the flux through the cross-sectional area A of a coil of N turns, the total flux through it, called the *flux-linkage*, is $N\Phi$ since the same flux Φ links each of the N turns.

Although the term *flux* suggests that something flows along the field lines, this is not so, but references to the flux 'entering', 'passing through' or 'leaving' a coil are in common use.

Faraday's law

(a) *Statement.* The law states that the *induced e.m.f. is directly proportional to the rate of change of flux-linkage or rate of flux-cutting.* In calculus notation it can be written

$$\mathscr{E} \propto \frac{d}{dt}(N\Phi)$$

or

$$\mathscr{E} = \text{constant} \times \frac{d}{dt}(N\Phi)$$

where $\mathscr{E}$ is the induced e.m.f. and $d(N\Phi)/dt$ is the rate of change of flux-linkage or the rate of flux-cutting. The law is found to be true for the dynamo and the transformer types of induction.

If instead of defining the weber in terms of the tesla (as we did in the previous section), we now redefine it as *the magnetic flux which induces in a one-turn coil an e.m.f. of 1 volt when the flux is reduced to zero in 1 second*, then the constant of proportionality in the above equation is 1. That is, if $d(N\Phi) = 1$ Wb when $\mathscr{E} = 1$ V and $dt = 1$ s, then we can write $1 = \text{constant} \times 1/1$. Hence

$$\mathscr{E} = \frac{d}{dt}(N\Phi)$$

where $\mathscr{E}$ is in volts, $d\Phi/dt$ in webers per second and N is a number of turns if we are considering a coil.

(*b*) *Numerical example.* Suppose a single-turn coil of cross-sectional area 5.0 cm^2 is at right angles to a flux density of 2.0×10^{-2} T, which is then reduced steadily to zero in 10 s. The flux-linkage change $d(N\Phi) = $ (number of turns) $\times$ (change in B) $\times$ (area of coil) $= (1) \times (2.0 \times 10^{-2}$ T$) \times (5.0 \times 10^{-4}$ m$^2) = 1.0 \times 10^{-5}$ Wb. The change occurs in time $dt = 10$ s, hence the e.m.f. $\mathscr{E}$ induced in the coil is given by

$$\mathscr{E} = \frac{d}{dt}(N\Phi) = \frac{1.0 \times 10^{-5}}{10}\frac{\text{Wb}}{\text{s}}$$

$$= 1.0 \times 10^{-6} \text{ V}$$

If the coil had 5000 turns, the *flux-linkage* would be 5000 times as great (i.e. $N\,d\Phi = 5.0 \times 10^{-2}$ Wb), hence

$$\mathscr{E} = 5000 \times 10^{-6} = 5.0 \times 10^{-3} \text{ V}$$

If the normal to the plane of this coil made an angle of 60° (instead of 0°) with the field then $N\,d\Phi = 5.0 \times 10^{-2} \cos 60 = 2.5 \times 10^{-2}$ Wb and

$$\mathscr{E} = 2.5 \times 10^{-3} \text{ V}$$

(*c*) *Experimental test of the law.* Two methods will be outlined, one for each type of induction.

(*i*) *Using a motor as a dynamo.* The arrangement of Fig. 14.5 uses a laboratory motor (12 V) with separate field and rotor terminals. When the hand drill is turned the rotor coil rotates and 'cuts' the flux due to the direct current (about 1 A) in the field coil.

To see if the e.m.f. induced in the rotor coil is proportional to the rate of flux-cutting, i.e. to the speed of rotation of the rotor, the drill is turned steadily and the number of turns made in, say, 10 seconds is counted. The reading (steady) on the d.c. voltmeter across the rotor is noted. The procedure is repeated for different drill speeds.

The effect on the induced e.m.f. of increasing the field current (and so also B and Φ) can be found.

(*ii*) *Using a.c., two coils and a CRO.* Alternating current from the low impedance output of a signal generator is passed through a solenoid and the peak-to-peak value of the e.m.f. induced (by mutual induction) in a ten-turn coil wound round the middle of the solenoid is measured on a CRO (used as a voltmeter and set on its most sensitive range, e.g. 0.1 V cm^{-1}) for different frequencies between 1 kHz and 3 kHz, Fig. 14.6. If the frequency of the a.c. in the solenoid is doubled, but the value of the current remains the same (as shown by the a.c. ammeter), the rate of change of the flux 'linking' the ten-turn coil doubles, as should the induced e.m.f.

(*d*) *Further experimental investigations.* These are concerned with properties of the coil itself which affect the e.m.f. induced in it. They help further with the understanding of Faraday's law.

(*i*) *Number of turns.* The effect of this can be investigated by comparing the vertical heights of the CRO traces due to ten-turn and five-turn coils round the middle of the solenoid, the same frequency of a.c. (say 2 kHz) being used in each case.

(*ii*) *Area.* In this case a.c. of frequency 2 kHz is passed through two solenoids in series, each having the

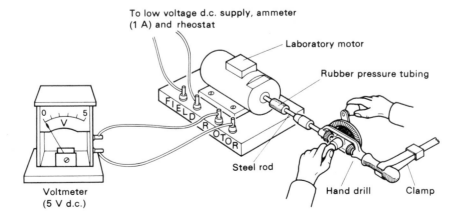

To low voltage d.c. supply, ammeter (1 A) and rheostat

Laboratory motor

Rubber pressure tubing

FIELD

ROTOR

Steel rod

Hand drill   Clamp

Voltmeter (5 V d.c.)

Fig. 14.5

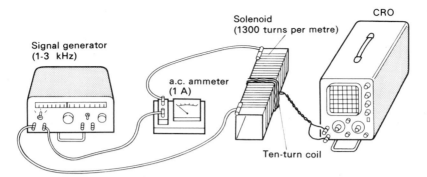

Fig. 14.6

same number of turns per metre, e.g. about 1300 (so that $B (= \mu_0 nI)$ is the same inside both), but one with twice the cross-sectional area of the other, Fig. 14.7. The flux linking ten-turn coils round the middle of each will be in the ratio of the areas (i.e. 2:1), as should be the ratio of the e.m.f.s induced in the coils and displayed on a CRO, preferably double beam.

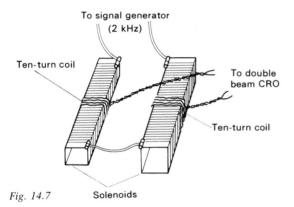

Fig. 14.7

(*iii*) *Orientation*. If the plane of the coil is at an angle (other than 90°) to the magnetic field, the e.m.f. induced in it is less, as can be shown by the apparatus of Fig. 14.8. What angle will the handle of the search coil make with the plane of the coil when the induced e.m.f. has half its maximum value?

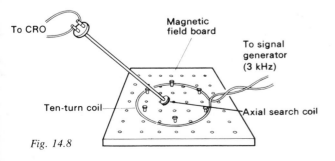

Fig. 14.8

(*e*) *Equivalence of flux cutting and flux linking*. This can be shown using the arrangement of Fig. 14.9. The outer solenoid is in series with a *smooth*, low voltage variable d.c. supply, an ammeter and a rheostat, and it carries a steady current of 1–2 A. The inner solenoid is connected to a light-beam galvanometer on 'direct' setting and is pulled out slowly so that the galvanometer deflection remains steady at, say, 50 mm. The time taken for the complete removal of the solenoid is noted.

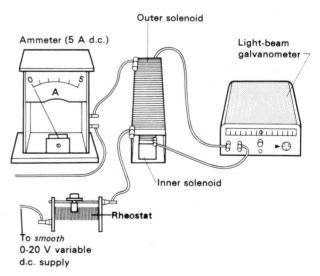

Fig. 14.9

The inner solenoid is reinserted and the current reduced to zero in the outer solenoid by turning down the output control on the low voltage supply. This is done at a rate which gives the same, as steady as possible, reading on the galvanometer as before. The time required is again measured.

In both cases the flux changes are the same and if the times are similar then the rate of flux-cutting equals the rate of change of flux-linkage.

The actions are apparently different, the first involves relative motion and the second a changing magnetic field. Nevertheless, the same law describes both.

Lenz's law

Whilst the magnitude of the induced e.m.f. is given by Faraday's law, its direction can be predicted by a law due to the Russian scientist Lenz. It may be stated as follows:

The direction of the induced e.m.f. is such that it tends to oppose the flux-change causing it, and does oppose it if induced current flows.

Thus in Fig. 14.10*a* a bar magnet is shown approaching the end of a coil, north pole first. If Lenz's law applies, the induced current should flow in a direction which makes the coil behave like a magnet with a north pole at the top. The downward motion of the coil and the accompanying flux-change will then be opposed. When the magnet is withdrawn, the top of the coil should behave like a south pole, Fig. 14.10*b*, and attract the north pole of the magnet, so hindering its removal and again opposing the flux-change. The induced current is therefore in the opposite direction to that when the magnet approaches. (Polarities may be checked using the right-hand screw rule if the direction of the windings on the coil are known and the current directions observed on the galvanometer.)

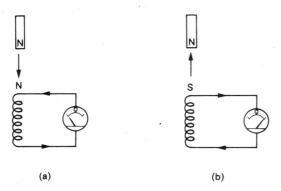

(a)　　　　　　　　　**(b)**

Fig. 14.10

Lenz's law is an example of the principle of conservation of energy; here, energy would be created from nothing if the e.m.f.s and currents acted differently. Thus if a south pole were produced at the top of the coil in Fig. 14.10*a*, attraction would occur between coil and magnet and the latter would, if it was allowed to, accelerate towards the coil, gaining kinetic energy as well as generating electrical energy. In practice, work has to be done to overcome the forces that arise, i.e. energy transfer occurs which in this case is from the mechanical to the electrical form.

For straight conductors moving at right angles to a magnetic field a more useful version of Lenz's law is *Fleming's right-hand rule* (also called the *dynamo rule*; his left-hand rule is often referred to as the *motor rule*, p. 262). It states that *if the thumb and first two fingers of the right hand are held so that each is at right angles to the other with the First finger pointing in the direction of the Field and the thuMb in the direction of Motion of the conductor, then the seCond finger indicates the direction (conventional) of the induced Current*, Fig. 14.11.

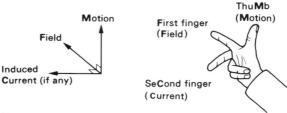

Fig. 14.11

Lenz's law is incorporated in the mathematical expression of Faraday's law by including a negative sign to show that current due to the induced e.m.f. produces an opposing flux-change, thus we write

$$\mathscr{E} = -\frac{\mathrm{d}}{\mathrm{d}t}(N\Phi)$$

Eddy currents

Any piece of metal moving in a magnetic field, or exposed to a changing one, has e.m.f.s induced in it, as we might expect. These can cause currents, called *eddy currents*, to flow inside the metal and they may be quite large because of the low resistance of the paths they follow. Their magnetic and heating effects are both helpful and troublesome.

(*a*) *Magnetic effect.* According to Lenz's law, eddy currents will circulate in directions such that the magnetic fields they create oppose the motion (or flux-change) producing them. This acts as a brake on the moving body and may be shown simply with the arrangement of Fig. 14.12. The solid copper cylinder quickly comes to rest if it is spun between the poles of the magnet. With the cylinder of coins there is very little braking because dirt on the coins increases the resistance of the cylinder as a whole, thereby reducing the eddy current flow.

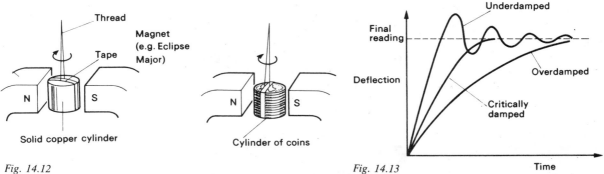

Fig. 14.12

Fig. 14.13

Use is made of the effect for the electromagnetic damping of moving-coil meters so that the coil takes up its deflected position quickly without overshooting and oscillating about its final reading. In most *pointer* instruments the coil is wound on a metal frame in which large eddy currents are induced and cause opposition to the motion of the coil as it cuts across the radial magnetic field of the permanent magnet. When oscillation is *just* prevented, the meter is said to be *critically damped* and its movement is 'dead-beat'. The curves in Fig. 14.13 show the effect of different degrees of damping.

In *light-beam* galvanometers the coil is not usually wound on a frame but is glued together and near-critical damping is achieved by having appropriate internal shunts across the coil on all ranges (except 'direct'). Suitable eddy currents then flow round the coil and shunt, whatever the current to be measured. On 'direct' setting, the electromagnetic damping is made a minimum (no internal shunts are connected) and the coil swings freely. Its first deflection can be shown to be proportional to the charge passing; the galvanometer is then said to be used *ballistically* (see p. 304). On 'short' the coil is short-circuited internally and the eddy currents induced in it bring it to rest quickly.

(*b*) *Heating effect.* In induction or eddy current heating, a coil carrying high frequency a.c. surrounds the material to be heated and the rapidly changing magnetic flux induces large eddy currents in the conducting parts of it. For example, in the zone refining of metals and semiconductors a narrow crucible containing the material is passed very slowly through the heating coil, Fig. 14.14*a*. The impurities tend to collect in the molten zone which moves to one end. After cooling this end is removed leaving a very pure, single crystal sample. In practice, multiple zone refining is employed as shown in Fig. 14.14*b* (for germanium).

Electric motors, dynamos and transformers contain iron that experiences flux changes when the device is in use. To minimize energy loss through eddy currents the iron parts consist of sheets, called *laminations*, insulated from each other by thin paper, varnish or some other insulator. The resistance of eddy current paths is thereby increased.

Calculation of induced e.m.f.s

(*a*) *Straight conductor.* In this case it is more helpful to consider flux-cutting. Suppose a conducting rod XY of length l moves sideways with steady velocity v through and at right angles to a uniform magnetic field of flux density B directed into the paper, Fig. 14.15*a*. The area swept out per second by XY is lv and therefore the flux cut per second is Blv. Assuming Faraday's law, we can say that the e.m.f. $\mathscr{E}$ induced in the rod is given numerically by

$$\mathscr{E} = \text{flux cut per second}$$

$$\therefore \quad \mathscr{E} = Blv$$

The e.m.f. induced in a rod cutting magnetic flux can be explained in terms of the forces acting on the charged particles in it and the above expression derived without recourse to Faraday's law. In a moving conductor both positive ions and electrons are carried along and both experience a force (magnetic) at right angles to the field and to the direction of motion of the conductor. However, only electrons are free to move inside the conductor and Fleming's left-hand (motor) rule indicates they will be forced to end X, making X negative and Y positive, Fig. 14.15*b*. (The conventional Current direction (seCond finger) will be opposite to the motion of the conductor since we are dealing with negative charges.) As a result of the charge separation and electron accumulation, an electric field is created inside the conductor which causes a repulsive electric

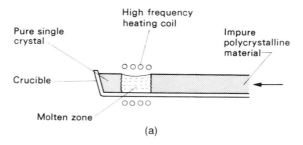

Fig. 14.14

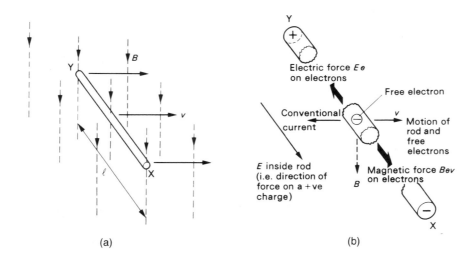

Fig. 14.15

force to be exerted on other electrons being urged towards X by the magnetic force.

These two forces act oppositely and when they become equal there is no further charge accumulation and we can say

$$Ee = Bev$$

where E is the equilibrium electric field strength, e the charge on an electron, B the magnetic flux density and v the velocity of the conductor. Hence

$$E = Bv$$

If V is the p.d. developed between the ends X and Y of the conductor and l is its length, then E = potential gradient = V/l and so $V = Blv$. The rod is on open circuit, therefore V equals the induced e.m.f.$\mathscr{E}$. We therefore get, as before,

$$\mathscr{E} = Blv$$

(b) *Spinning disc.* The first dynamo was made by Faraday and consisted of a metal disc rotated in a magnetic field, Fig. 14.16. A modern version of his apparatus is shown in Fig. 14.17. When the disc is driven at a steady speed by the motor a steady deflection is obtained on the light-beam galvanometer connected to two sliding contacts (e.g. 4 mm plugs) held one at the

Fig. 14.16

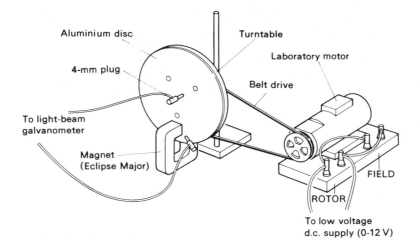

Fig. 14.17

To low voltage
d.c. supply (0-12 V)

centre and the other at the edge of the disc between the poles of the magnet. The effect of changing the speed of rotation and the position of the contacts may be investigated.

We can consider that an e.m.f. is induced in the circuit because the radius of the disc between the contacts at any instant is cutting the flux there, i.e. we

regard the disc as a many-spoked wheel. If the disc makes f revolutions per second and has radius r, the area swept out per second by a radius is $\pi r^2 f$. The flux cut per second = $B\pi r^2 f$ where B is the flux density (assumed uniform) between the contacts. Hence the induced e.m.f.$\mathscr{E}$ is, by Faraday's law,

$$\mathscr{E} = B\pi r^2 f$$

The 'homopolar generator' is a recent form of Faraday's disc and can deliver a very large direct current with very small e.m.f. for the production of powerful magnetic fields.

(c) *Rotating coil*. The coil in Fig. 14.18 has N turns each of area A and is being rotated about a horizontal axis in its own plane at right angles to a uniform magnetic field of flux density B. If the normal to the coil makes an angle θ with the field at time t (measured from the position where $\theta = 0$) then the flux Φ linking each turn is given by

$$\Phi = BA \cos \theta$$

But $\theta = \omega t$ where ω is the steady angular velocity of the coil, therefore

$$\Phi = BA \cos \omega t$$

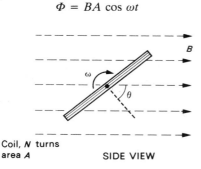

Coil, N turns
area A SIDE VIEW

Fig. 14.18

By Faraday's law, the e.m.f. $\mathscr{E}$ induced in N turns is

$$\mathscr{E} = -\frac{d}{dt}(N\Phi) = -N\frac{d}{dt}(BA \cos \omega t)$$

$$= -BAN\frac{d}{dt}(\cos \omega t)$$

$$= -BAN(-\omega \sin \omega t)$$

$$= BAN\omega \sin \omega t$$

The e.m.f. is thus an alternating one which varies sinusoidally with time and would cause a similar alternating current in an external circuit connected across the coil.

When the plane of the coil is parallel to B, $\theta = \omega t = 90°$, $\sin \omega t = 1$ and $\mathscr{E}$ has its maximum value $\mathscr{E}_0$ given by

$$\mathscr{E}_0 = BAN\omega$$

Hence we can write

$$\mathscr{E} = \mathscr{E}_0 \sin \omega t$$

When is $\mathscr{E}$ equal to (*i*) zero, and (*ii*) $\mathscr{E}_0/2$?

If a coil has area 1.0×10^{-2} m² (i.e. 100 cm²), 800 turns and makes 600 revolutions per minute in a magnetic field of flux density 5.0×10^{-2} T, then $\mathscr{E}_0$ is given by

$$\mathscr{E}_0 = BAN\omega$$

$$= (5.0 \times 10^{-2}\ \text{T}) \times (1.0 \times 10^{-2}\ \text{m}^2) \times (800) \times (2\pi \times 600/60\ \text{s}^{-1})$$

$$= 5.0 \times 10^{-2} \times 1.0 \times 10^{-2} \times 800 \times 20\pi\ \text{Wb s}^{-1}$$
$$(1\ \text{T} = 1\ \text{Wb m}^{-2})$$

$$= 25\ \text{V}$$

Generators (a.c. and d.c.)

A generator or dynamo produces electrical energy by electromagnetic induction. In principle it consists of a coil which is rotated between the poles of a magnet so that the flux-linkage changes, Fig. 14.19a. The flux Φ linking each turn of a coil of area A having N turns, rotating with angular velocity ω in a uniform flux density B, is given at time t (measured from the vertical position) by

$$\Phi = BA \cos \omega t$$

By Faraday's law, the induced e.m.f. $\mathscr{E}$ is, as we found in the previous section,

$$\mathscr{E} = -\frac{d}{dt}(N\Phi) = BAN\omega \sin \omega t$$

Both Φ and $\mathscr{E}$ alternate sinusoidally and their variation with the position of the coil is shown in Fig. 14.19b.

We see that although Φ is a maximum when the coil is vertical (i.e. perpendicular to the field), $\mathscr{E}$ is zero because the *rate of change* of Φ is zero *at that instant*, i.e. the tangent to the Φ-graph is parallel to the time axis and so has zero gradient—in calculus terms $d\Phi/dt = 0$. Also, when $\Phi = 0$, its rate of change is a maximum, therefore $\mathscr{E}$ is a maximum.

The expression for $\mathscr{E}$ shows that its instantaneous values increase with B, A, N and the angular velocity of the coil. If the coil makes one complete revolution, one cycle of alternating e.m.f. is generated, i.e. for a simple, single-coil generator the frequency of the supply equals the number of revolutions per second of the coil.

In an *a.c. generator* (or alternator) the alternating e.m.f. is taken off and applied to the external circuit by two spring-loaded graphite blocks (called 'brushes') which press against two copper *slip-rings*. These rotate with the axle, are insulated from one another and each is connected to one end of the coil, Fig. 14.20.

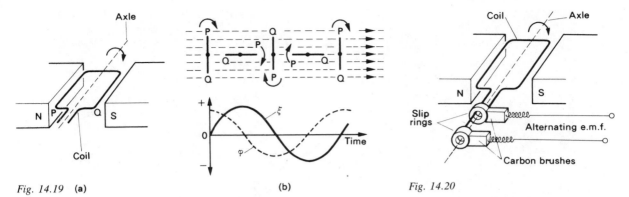

Fig. 14.19   (a)   (b)   Fig. 14.20

In a *d.c. generator* a *commutator* is used instead of slip-rings. This consists of a split-ring of copper, the two halves of which are insulated from each other and joined to the ends of the coil, Fig. 14.21a. The brushes are arranged so that the change-over of contact from one split ring to the other occurs when the coil is vertical. In this position the e.m.f. induced in the coil reverses and so one brush is always positive and the other negative. The graphs of Fig. 14.21b show the e.m.f.s in the coil and at the brushes; the latter, although varying, is unidirectional and produces d.c. in an external circuit.

In actual a.c. and d.c. generators, several coils are wound in uniformly spaced slots in a soft iron cylinder which is laminated to reduce eddy currents. The whole assembly is known as the *armature*. In the d.c. case the use of many coils and a correspondingly greater number of commutator segments gives a much steadier e.m.f. Also, in practical generators the magnetic field is produced by electromagnets (except in a cycle dynamo where a permanent magnet is used) and the coils which energize them are called *field* coils.

In power station alternators the armature coils and their iron cores are stationary (and are called the *stator*) whilst the field coils and their core (i.e. the electromagnets) rotate (and are called the *rotor*); Fig. 14.22 shows a simplified alternator. The advantage of this is that only the relatively small direct current needed for the field coils is fed through the rotating slip-rings. The large p.d.s and currents induced in the armature coils (25 kV and thousands of amperes in some modern alternators) are then led away through fixed connections. The rotor is driven either by a steam or water turbine which also powers a small d.c. dynamo (called the *exciter*) for supplying current to the field coils. In Fig. 14.23a a rotor is being inserted into the stator of an alternator and Fig. 14.23b shows a complete power station layout.

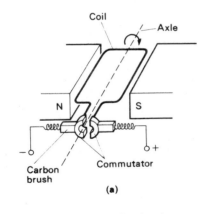

(a)

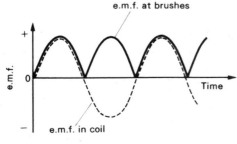

Fig. 14.21   (b)

Fig. 14.22

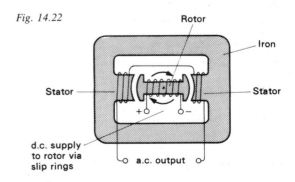

(a)

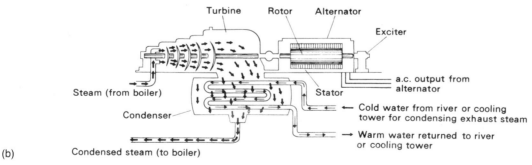

(b)

Fig. 14.23

Electric motors (d.c. and a.c.)

Electric motors form the heart of a whole host of devices ranging from domestic appliances such as vacuum cleaners and washing machines to electric locomotives and lifts. In a car the windscreen wipers are usually driven by one and the engine is started by another. There are many types, most of which work off a.c.; the principles involved in a few cases will be outlined.

(*a*) *d.c. motors.* Basically a d.c. motor consists of a coil on an axle, carrying a current in a magnetic field. The coil experiences a couple as in a moving-coil galvanometer (see pp. 274 and 275) which makes it rotate. When its plane is perpendicular to the field, a *split-ring commutator* reverses the current in the coil and, as Fig. 14.24 shows, ensures that the couple continues to act in the same direction thereby maintaining the rotation.

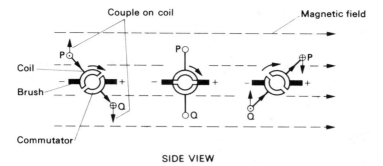

Fig. 14.24

SIDE VIEW

In practice several coils are wound in equally spaced slots in a laminated soft iron cylinder (the *armature* or *rotor*) and are connected to a commutator with many segments. Greater and steadier torque is thus obtained. Electromagnets with concave pole pieces frequently provide the magnetic field but the use of modern permanent magnets (e.g. of Magnadur) for this is increasing, especially in small motors. When electromagnets are used the field coils may be in series (a series-wound motor) or in parallel (a shunt-wound motor) with the armature, depending on what the motor is required to do. The construction of a d.c. motor is the same as that of a d.c. dynamo and in fact one can be used as the other (see p. 284).

When the armature coil in a motor rotates it cuts the magnetic flux of the field magnet and an e.m.f. $\mathscr{E}$, called the *back e.m.f.*, is induced in it (as in a dynamo) which, by Lenz's law, opposes the applied p.d. V causing current I in the coil. If r is the armature coil resistance, then

$$V - \mathscr{E} = Ir$$

Multiplying by I we get

$$VI = \mathscr{E}I + I^2r$$

Now VI is the power supplied to the motor and I^2r is the power dissipated as heat in the armature coil. The difference, $\mathscr{E}I$, must be the mechanical power output of the motor; it is also the rate of working against the induced e.m.f.

The armature resistance r of a d.c. motor is small (e.g. 1 Ω or less) to make I^2r small and give high efficiency. However, when the motor is started, the armature is at rest and the back e.m.f. $\mathscr{E}$ is zero. The armature current I then equals V/r and would be so large as to burn out the armature coils. This is prevented by connecting a 'starting' resistance in series with the motor and gradually reducing it as the motor speeds up. The back e.m.f. then limits the current and

will normally only be slightly less than the applied p.d. V.

(*b*) *a.c. motors.* A d.c. motor may be used on a.c. if the armature and field coils are in series. The current then reverses simultaneously in each and rotation in the same direction continues. (The torque developed in a shunt-wound motor on a.c. is very small due to inductive effects causing the armature and field currents to reach their maxima at different times, p. 318.)

The *induction motor* is widely used in industry and is the commonest type of a.c. motor. Its action depends on the fact that a moving magnetic field can set a neighbouring conductor into motion. The converse is also true, namely that a moving conductor can cause a magnetic field to move and is readily demonstrated with the arrangement in Fig. 14.25 in which the bar magnet starts spinning when the copper disc is rotated rapidly. Whether it be the conductor or the field that moves, eddy currents are induced in the conductor and these try to reduce the effect causing them, i.e. the relative motion between conductor and field. If the conductor is stationary it starts moving in the same direction as the field and tries to catch it up in an (unsuccessful) attempt to eliminate the relative motion between them.

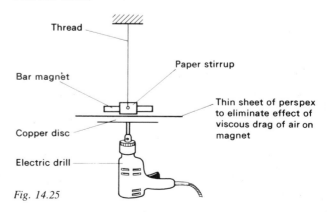

Fig. 14.25

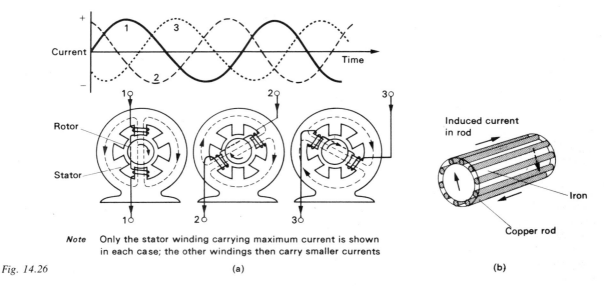

Note Only the stator winding carrying maximum current is shown
in each case; the other windings then carry smaller currents

Fig. 14.26   (a)   (b)

Moving magnetic fields are produced in various ways in actual induction motors. In large rotary machines three pairs of stationary electromagnets (the stator) are arranged at equal angles round a conductor (the rotor) and each pair is connected to one phase of a 3-phase a.c. supply, Fig. 14.26*a*; the graph shows how the phases reach their maximum values one after the other. The rotor is generally of the squirrel-cage pattern comprising a number of copper rods in an iron cylinder, Fig. 14.26*b*, and it 'interprets' the alternations of the magnetic field as a field sweeping round it, i.e. a rotating field.

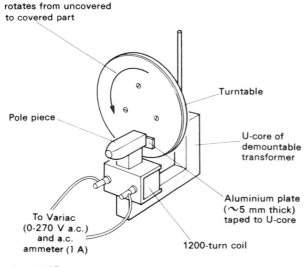

Fig. 14.27

The eddy currents induced in the copper rods set it into rotation as explained above. Linear induction motors operate on the same principle except that the field travels in a straight line; such motors are the subject of much research at present.

The *shaded-pole* induction motor, used in record players, produces a 'rotating' magnetic field by covering part of the pole of an electromagnet carrying a.c. with a thick conducting plate. The alternating field of the electromagnet induces eddy currents in the plate and these create another field, adjacent to the main electromagnet field. There is a phase difference between the two fields and a nearby metal disc regards this as a moving field and responds by rotating. A model shaded-pole motor is shown in Fig. 14.27. (*Note.* The small phase difference between the fields may be shown by holding two search coils over the 'shaded' and 'unshaded' parts of the pole and examining the traces on a double beam CRO.)

Transformers

(*a*) *Action.* A transformer changes, i.e. transforms, an alternating p.d. from one value to another of greater or smaller value using the mutual induction principle (p. 282).

Two coils, called the *primary* and *secondary* windings, which are not connected to one another in any way, are wound on a complete soft iron core, either one on top of the other as in Fig. 14.28*a* or on separate limbs of the core as in Fig. 14.28*b*. When an alternating p.d. is applied to the primary, the resulting current produces a

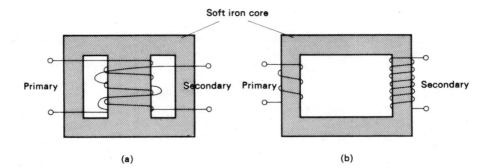

Fig. 14.28   **(a)**   **(b)**

large alternating magnetic flux which links the secondary and induces an e.m.f. in it. The value of this e.m.f. depends on the number of turns on the secondary and we will show shortly that under certain conditions it is approximately true to say

$$\frac{\text{e.m.f. induced in secondary}}{\text{p.d. applied to primary}} = \frac{\text{secondary turns}}{\text{primary turns}}$$

A 'step-up' transformer has more turns on the secondary than the primary and the e.m.f. induced in the secondary is greater than the p.d. applied to the primary, e.g. if the turns are stepped-up in the ratio 1:2, the secondary e.m.f. will be about twice the primary p.d. In a 'step-down' transformer the secondary e.m.f. is less than the primary p.d.

Three demonstrations to show the working of a

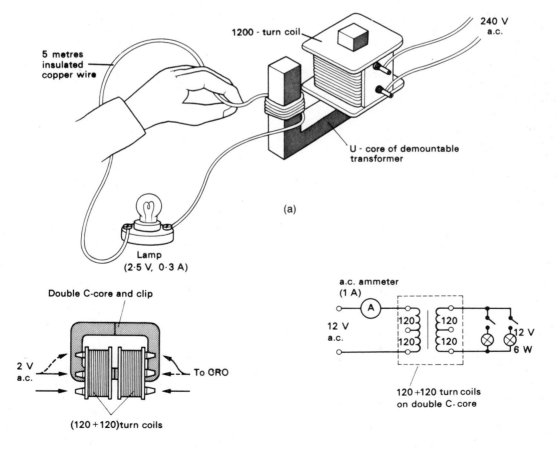

Fig. 14.29   **(b)**   **(c)**

transformer are illustrated in Fig. 14.29a, b, c. In the first the lamp lights up to full brightness when a sufficient number of secondary turns have been wound on. The effect of placing the iron yoke across the U-core can be investigated. In the second demonstration the relation between the turns and p.d. ratios may be studied, a CRO being used as a voltmeter to measure the secondary e.m.f. The third demonstration shows how the current in the primary depends on that in the secondary, the latter being increased by connecting lamps across it.

If the p.d. is stepped up by a transformer, the current is stepped down, roughly in the same ratio. This follows if conservation of energy is assumed, because, taking the transformer to be 100 per cent efficient (many approach this), if all the electrical energy supplied to the primary appears in the secondary, then

$$\text{power in primary} = \text{power in secondary}$$

that is,

primary p.d. × primary current =
secondary e.m.f. × secondary current

or,

$$\frac{\text{secondary current}}{\text{primary current}} = \frac{\text{primary p.d.}}{\text{secondary e.m.f.}}$$

The stepping up of current can be demonstrated effectively using the apparatus of Fig. 14.30a and b. In the first the iron nail melts spectacularly and in the second the water boils very quickly.

A transformer can have more than one secondary and may step up and down simultaneously. A transformer (with two secondaries) for a mains-operated h.t. power supply unit is shown diagrammatically in Fig. 14.31.

(b) *Energy losses.* Although transformers are very efficient devices, small energy losses do occur in them due to four main causes:

(i) *Resistance of windings.* The copper wire used for the windings has resistance and so ordinary (I^2R) heat losses occur. In high-current, low-p.d. windings these are minimized by using thick wire.

(ii) *Eddy currents.* The alternating magnetic flux induces eddy currents in the iron core and causes heating. The effect is reduced by having a laminated core (see p. 287).

(iii) *Hysteresis.* The magnetization of the core is repeatedly reversed by the alternating magnetic field. The resulting expenditure of energy in the core appears as heat and is kept to a minimum by using a magnetic material (such as mumetal) which has a low hysteresis loss (p. 307).

(iv) *Flux leakage.* The flux due to the primary may not all link the secondary if the core is badly designed or has air gaps in it.

Very large transformers like those in Fig. 14.32a and b (400 kV and 11 kV respectively) have to be oil-cooled to prevent overheating.

(c) *Theory.* The complete theory is complex and before we tackle even a simple version of it, consideration of the following will be helpful.

The primary winding of a mains transformer is found to have a resistance of 10 Ω, on a 240 V supply; the primary current should therefore be (by Ohm's law) 240/10 = 24 A. An a.c. ammeter connected in the primary records 0.10 A. The difference is very large and is due to the fact that the alternating flux (arising from the a.c. in the primary), which induces an e.m.f. in the secondary, induces an e.m.f. called a *back e.m.f.* in the primary as well. (The primary is said to have *self-inductance*, see p. 300). This e.m.f. opposes the applied p.d. and is nearly but not quite equal to it at every instant. The *net* e.m.f. in the primary is therefore quite small. We can now proceed to an approximate theoretical treatment.

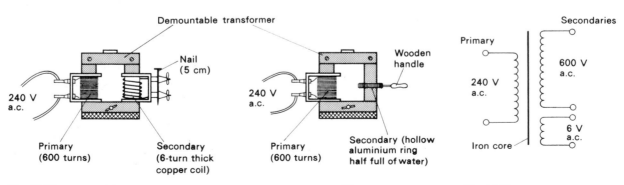

Fig. 14.30 **(a)** **(b)** *Fig. 14.31*

Fig. 14.32   (a)

Fig. 14.32   (b)

Consider an ideal transformer in which the primary has negligible resistance and all the flux in the core links both primary and secondary windings, Fig. 14.33. If Φ is the flux in the core at time t due to the current in the primary when a p.d. V_p is applied to it, then the back e.m.f. $\mathscr{E}_p$ induced in the primary of N_p turns (due to its self-inductance) is given by

$$\mathscr{E}_p = \frac{\mathrm{d}}{\mathrm{d}t}(N_p\Phi) = N_p \cdot \frac{\mathrm{d}\Phi}{\mathrm{d}t}$$

But, $\mathscr{E}_p = V_p$

If this were not so, the primary current would be infinite since the primary has zero resistance. Hence

$$V_p = N_p \cdot \frac{\mathrm{d}\Phi}{\mathrm{d}t} \qquad (1)$$

The e.m.f. $\mathscr{E}_s$ induced in the secondary (N_s turns) by the same flux in the core is

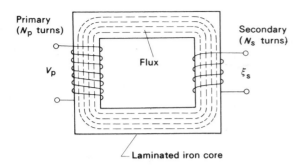

Fig. 14.33

$$\mathscr{E}_s = \frac{\mathrm{d}}{\mathrm{d}t}(N_s\Phi) = N_s \cdot \frac{\mathrm{d}\Phi}{\mathrm{d}t}$$

If the secondary is on open circuit or the current taken from it is small then, to a good approximation,

$$\mathscr{E}_s = V_s$$

where V_s is the p.d. across the secondary. Thus

$$V_s = N_s \cdot \frac{\mathrm{d}\Phi}{\mathrm{d}t} \qquad (2)$$

From (1) and (2),

$$\frac{V_s}{V_p} = \frac{N_s}{N_p}$$

This expression would be roughly true for an actual transformer if (*i*) the primary resistance and current were small, (*ii*) very little flux escaped from the core, and (*iii*) the secondary current was small.

When a greater load (i.e. a smaller resistance) is connected to the secondary it causes the secondary current to increase and this acts to reduce the flux in the core (since the secondary current opposes the change producing it). The back e.m.f. in the primary therefore falls and so the primary current increases. Eventually the flux is restored to its previous value and as a result the back e.m.f. rises and becomes nearly equal to the applied p.d. The net effect is an increase of primary current, i.e. more energy is drawn from the source connected to the primary.

Transmission of electrical power

(*a*) *Grid system.* The Grid system in Britain is a network of cables, most of it supported on pylons,

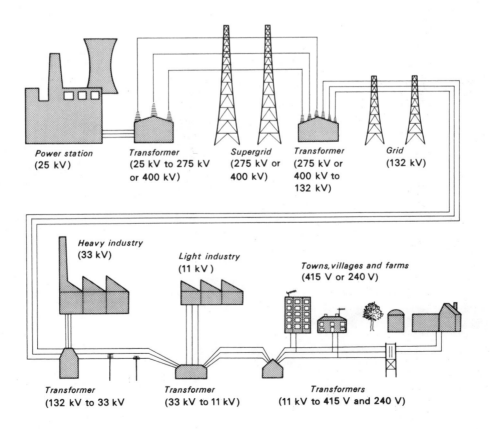

Fig. 14.34

which connects over 200 power stations, situated at convenient places throughout the country and carrying electrical energy from them to consumers.

In the largest modern power stations electricity is generated at about 25 kV (50 Hz) and stepped up in a transformer to 275 kV or 400 kV for transmission over long distances. The p.d. is subsequently reduced in sub-stations by other transformers for distribution to local users at suitable p.d.s—33 kV for heavy industry, 11 kV for light industry and 240 V for homes, schools, shops, farms, etc., Fig. 14.34. In the latest rail electri-fication schemes working at 25 kV, there are special sub-stations alongside the track taking their supply from the Grid system. This is fed to step-down transfor-mers in the electric locomotive and then rectified (see p. 328) for driving d.c. traction motors operating at about 900 V and 650 A.

To supervise the operation of the power stations, England and Wales are divided into eight areas, each with a Grid Control Centre. At these, engineers assess the demand, direct the flow and reroute it when breakdowns occur. In this way the electricity supply is made more reliable, requires less reserve plant to cover maintenance etc. and cuts costs by enabling smaller, less efficient power stations to be shut down at off-peak periods. All eight Grid Control Centres are in direct communication with the National Control Centre in London, Fig. 14.35.

(*b*) *Why high p.d.s are used.* Suppose electrical power P has to be delivered at a p.d. V by supply lines of total resistance R, Fig. 14.36. The current $I = P/V$ (since $P = IV$) and the power loss in the lines $= I^2R = (P/V)^2R$. Clearly, the greater V the smaller is the loss—in fact, doubling V quarters the loss. Electric-al power is thus transmitted more economically at high p.d.s but on the other hand they create insulation problems and raise installation costs. In the 400 kV Supergrid, currents of 2500 A are typical and the power loss is about 200 kW per kilometre of cable, i.e. a 0.02 per cent loss per kilometre.

The ease and efficiency with which alternating p.d.s are stepped up and down in a transformer and the fact that alternators produce much higher p.d.s than d.c. generators (25 kV compared with several thousand volts) are the main considerations influencing the use of high alternating, rather than direct, p.d.s in most situations. An exception to this is the cross-channel link between England and France where the underground cables favour a d.c. supply because of the high dielec-tric losses with a.c. in such cables.

The advantages of 'high' alternating p.d. power transmission may be shown with the model power line arrangements of Fig. 14.37*a* and *b*.

Self-inductance

The flux due to the current in a coil links that coil and if the current changes the resulting flux change induces an e.m.f. in the coil itself. This changing magnetic field type of electromagnetic induction is called *self-*

Fig. 14.35

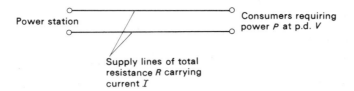

Fig. 14.36

induction, the coil is said to have *self-inductance* or simply *inductance* (symbol L) and is called an *inductor* (symbol ~~~ if air-cored and ⩘ if it has a core of magnetic material). The induced e.m.f. obeys Faraday's law like other induced e.m.f.s.

(*a*) *Some demonstrations.* From Lenz's law we would expect the induced e.m.f. to oppose the current change causing it. That it does so in both d.c. and a.c. circuits may be demonstrated.

In Fig. 14.38*a*, *L* is an iron-cored inductor and *R* is a

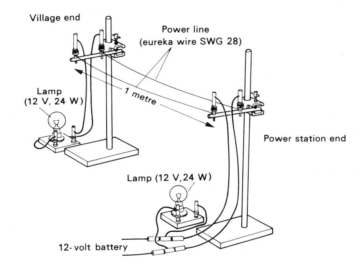

(a)

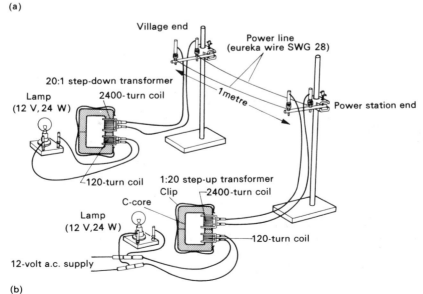

Fig. 14.37   (b)

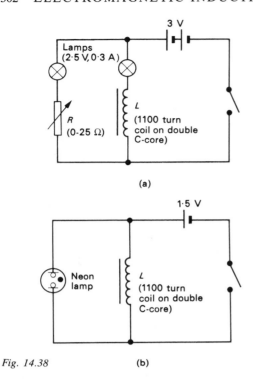

Fig. 14.38   (b)

variable resistor adjusted to have the same resistance as L. When the current is switched on, the lamp in series with L lights up a second or two *after* that in series with R. This can be attributed to the induced e.m.f. in L opposing the change and trying to drive a current against the increasing current due to the battery. The growth of the current to its steady value (when no self-induction occurs) is thus delayed.

The effect of the induced e.m.f. when the current is switched off is more striking and is shown with the circuit of Fig. 14.38*b*. Opening the switch causes the current to fall very rapidly to zero and the rate of change of flux is large. The induced e.m.f. is therefore large and in a direction which tries to maintain the current in its original direction. It is sufficiently great in this case (more than 100 V) to produce a brief flash of the neon lamp across L. When circuits carrying large currents in large inductors are switched off, the induced e.m.f. can cause sparking between the switch contacts and may even fuse them together. In the Grid network special circuit breakers are used.

Inductors are important components of a.c. circuits where, as we shall see later, the induced e.m.f. opposes the applied p.d. continuously. In Fig. 14.38*a* if the 3 V d.c. supply is replaced by a 3 V a.c. supply, the lamp in series with L does not light—unless the iron core is removed.

(*b*) *Definition and unit.* It would seem reasonable to say that a coil (or circuit) has a large inductance if a small rate of change of current in it induces a large back e.m.f. This is the basis of the following definition of inductance. If the e.m.f. induced in a coil is $\mathscr{E}$ when the rate of change of current in it is dI/dt, the inductance L of the coil is defined by the equation

$$L = -\frac{\mathscr{E}}{dI/dt}$$

The negative sign is inserted to make L a positive quantity since $\mathscr{E}$ and dI/dt act in opposite directions and are given opposite signs.

The unit of inductance is the *henry* (H), defined as *the inductance of a coil (or circuit) in which an e.m.f. of 1 volt is induced when the current changes at the rate of 1 ampere per second.* That is, $1\ \mathrm{H} = 1\ \mathrm{Vs\ A^{-1}}$.

(*c*) *Inductance of a solenoid.* Calculation of inductance is possible in certain cases, as it was of capacitance. Consider a long, air-cored solenoid of length l, cross-sectional area A having N turns and carrying current I. The flux density B is almost constant over A and, neglecting the ends, is given by

$$B = \mu_0 \frac{N}{l} I \qquad \text{(see p. 266)}$$

The flux Φ through each turn of the solenoid is BA and for the flux-linkage we have

$$N\Phi = BAN$$
$$= \left(\mu_0 \frac{N}{l} I\right) AN$$
$$= \frac{\mu_0 A N^2}{l} \cdot I$$

If the current changes by dI in time dt causing a flux-linkage change $d(N\Phi)$ then by Faraday's law the induced e.m.f. $\mathscr{E}$ is

$$\mathscr{E} = -\frac{d}{dt}(N\Phi)$$
$$= -\frac{\mu_0 A N^2}{l} \cdot \frac{dI}{dt}$$

If L is the inductance of the solenoid, then from the defining equation we get

$$\mathscr{E} = -L\frac{dI}{dt}$$

Comparing these two expressions it follows that

$$L = \frac{\mu_0 A N^2}{l}$$

L depends only on the geometry of the solenoid. If $N = 400$ turns, $l = 25$ cm $= 25 \times 10^{-2}$ m, $A = 50$ cm^2 $= 50 \times 10^{-4}$ m^2 and $\mu_0 = 4\pi \times 10^{-7}$ H m^{-1}, then $L = 4.0 \times 10^{-3}$ H $= 4.0$ mH.

A solenoid having a core of a magnetic material would have a much greater inductance but the value would vary depending on the current in the solenoid.

(d) Energy stored by an inductor. A current-carrying inductor stores energy in the magnetic field associated with it and it can be shown that for current I in an inductance L this equals $\frac{1}{2}LI^2$. Compare this with the analogous case of $\frac{1}{2}Q^2/C$ for the energy stored in the electric field of a capacitor.

Since every current produces a field, every circuit must have some self-inductance. On switching on any circuit some time is necessary to provide the energy in the magnetic field and so no current can be brought instantaneously to a non-zero value. Similarly on switching off any circuit, the energy of the magnetic field must be dissipated somehow, hence the spark. A capacitor across the switch can 'suppress' sparking.

Unit of μ_0

Reference was made to this on p. 265 and it is convenient to consider it now. The permeability of free space μ_0 was defined by the Biot-Savart law

$$\delta B = \frac{\mu_0 I \, \delta l \sin \theta}{4\pi r^2}$$

The unit of μ_0 from this equation is

$$\frac{(\text{Wb m}^{-2}) \times (\text{m}^2)}{(\text{A}) \times (\text{m})} \text{ or Wb A}^{-1} \text{ m}^{-1} \qquad (1)$$

From Faraday's law, $\mathscr{E} = \frac{\mathrm{d}}{\mathrm{d}t}(N\Phi)$, we can say that

$$1 \text{ Wb} = 1 \text{ V s} \qquad (2)$$

Also, from the inductance defining equation $L = \mathscr{E}/(\mathrm{d}I/\mathrm{d}t)$ we have

$$1 \text{ H} = 1 \text{ V s A}^{-1} \qquad (3)$$

From (2) and (3)

$$1 \text{ H} = 1 \text{ Wb A}^{-1}$$

Hence from (1), μ_0 can be expressed in

H m^{-1} (henry per metre)

This is the SI unit of μ_0 (and μ); it may be compared

with F m^{-1} (farad per metre) for the unit of ε_0 (and ε), the permittivity of free space.

Mutual inductance

(a) Definition and unit. In mutual induction, current changing in one coil or circuit (the primary) can induce an e.m.f. in a neighbouring coil or circuit (the secondary), as we saw earlier. The *mutual inductance*, M, of two coils, Fig. 14.39, is defined by the equation

$$M = -\frac{\mathscr{E}}{\mathrm{d}I_\mathrm{p}/\mathrm{d}t}$$

where $\mathscr{E}$ is the e.m.f. induced in the secondary when the rate of change of current in the primary is $\mathrm{d}I_\mathrm{p}/\mathrm{d}t$.

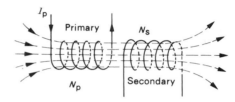

Fig. 14.39

It follows from the definition that M has the same unit as L, i.e. henry (H). *Two coils are said to have a mutual inductance of 1 henry if an e.m.f. of 1 volt is induced in the secondary when the primary current changes at the rate of 1 ampere per second.*

It can be shown that the mutual inductance of two coils is the same if current flows in the secondary and flux links the primary causing an induced e.m.f. when a flux-linkage change occurs.

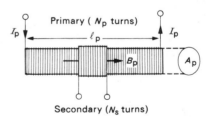

Fig. 14.40

(b) Mutual inductance of a solenoid and a coil. In Fig. 14.40 the long air-cored solenoid (the primary) with N_p turns, length l_p and cross-sectional area A_p carries current I_p. The flux density B_p in the centre of the solenoid is nearly constant over A_p and is given by

$$B_\mathrm{p} = \mu_0 \frac{N_\mathrm{p}}{l_\mathrm{p}} I_\mathrm{p} \qquad \text{(see p. 266)}$$

The flux Φ_s linking each of the N_s turns of the short coil (the secondary) round the middle of the solenoid is $B_p A_p$. The flux-linkage of the short coil is therefore

$$N_s \Phi_s = B_p A_p N_s$$
$$= \left(\mu_0 \frac{N_p}{l_p} I_p \right) A_p N_s$$

If the current in the primary changes by dI_p in time dt causing a flux-linkage change $d(N_s \Phi_s)$ in the secondary, the e.m.f. $\mathscr{E}_s$ induced in the secondary is

$$\mathscr{E}_s = - \frac{d}{dt} (N_s \Phi_s)$$
$$= - \frac{\mu_0 A_p N_p N_s}{l_p} \cdot \frac{dI_p}{dt}$$

If M is the mutual inductance, we can also say

$$\mathscr{E}_s = - M \frac{dI_p}{dt}$$

Hence
$$M = \frac{\mu_0 A_p N_p N_s}{l_p}$$

When the coils have a ferromagnetic core, the value of M is very much greater especially if the core is complete as in a transformer, but it does vary with the current.

Induced charge and flux change

When the flux linking or cutting a *complete* circuit changes, an e.m.f. is induced in it and current flows. A simple connection exists between the total charge circulation that constitutes the current and the flux change.

Consider a coil of N turns in a circuit of total resistance R in which the flux linking each turn is changing and has value Φ at time t. The induced e.m.f. at t will be (in magnitude)

$$\mathscr{E} = \frac{d}{dt} (N\Phi)$$

Also, the induced current I at time t will be

$$I = \frac{\mathscr{E}}{R} = \frac{1}{R} \cdot \frac{d}{dt} (N\Phi)$$

Now I = rate of flow of charge = dQ/dt

$$\therefore \quad \frac{dQ}{dt} = \frac{1}{R} \frac{d}{dt} (N\Phi) = \frac{N}{R} \frac{d\Phi}{dt}$$

If the flux changes from say Φ_1 to Φ_2 the total charge Q

that passes is

$$Q = \int_0^Q dQ = \frac{N}{R} \int_{\Phi_1}^{\Phi_2} d\Phi$$

$$\therefore \quad Q = \frac{N(\Phi_2 - \Phi_1)}{R}$$

$$= \frac{\text{flux-linkage change}}{R}$$

We see that Q does not depend on the time taken by the flux change.

Consider a numerical example. A search coil of average cross-sectional area 3.0 cm² has 400 turns and is in a circuit of total resistance 200 Ω. It is inserted into a magnetic field of flux density 2.5×10^{-3} T so as to produce the maximum flux change. We have, $N = 400$ turns, $A = 3.0 \times 10^{-4}$ m², $R = 200$ Ω, $B = 2.5 \times 10^{-3}$ T. Hence $\Phi_2 = BA = 2.5 \times 10^{-3} \times 3.0 \times 10^{-4}$ Wb and $\Phi_1 = 0$. The induced charge Q is given by

$$Q = \frac{N(\Phi_2 - \Phi_1)}{R}$$

$$= \frac{400 \times 2.5 \times 10^{-3} \times 3.0 \times 10^{-4}}{200} \text{ C}$$

$$= 1.5 \times 10^{-6} \text{ C} = 1.5 \text{ } \mu\text{C}$$

Measuring B by ballistic galvanometer and search coil

(a) *Ballistic galvanometer.* A moving-coil galvanometer will measure charge if (i) the period of oscillation of the movement is large (e.g. 2 seconds) so that all the charge passes through the coil before it moves appreciably, and (ii) the damping is very small. It is then called a ballistic galvanometer (since it is set into motion by an impulse, as is a projectile whose motion is under study in ballistics) and theory shows that the first deflection or 'throw' θ is proportional to the total charge Q that has passed. Hence

$$\theta \propto Q$$

or
$$\theta = bQ$$

where b is a constant called the *charge sensitivity* of the galvanometer. It is expressed in mm *per* μC and must either be known (from information supplied by the manufacturer) or found by a calibration experiment (see later).

Generally only light-beam galvanometers are suitable for ballistic use. Damping due to the air and the suspension are negligible and on 'direct' setting there are no internal shunts across the coil which, in modern instruments like that shown in Fig. 13.35c, is not wound on any kind of former. Electromagnetic damping is therefore due only to eddy currents in the coil. These depend solely on the resistance of the external circuit which should consequently be high enough to allow the coil to swing to and fro freely.

(b) *Measuring B.* A search coil in series with a ballistic galvanometer and a suitable high resistance (to reduce damping and adjust the sensitivity of the galvanometer) is placed in the magnetic field to be measured so that the flux links it normally, Fig. 14.41. It is then *quickly* removed (Why?) from the field and the first 'throw' θ produced by the flux change is noted. The charge Q driven through the coil (e.g. 1–2 μC) is proportional to θ.

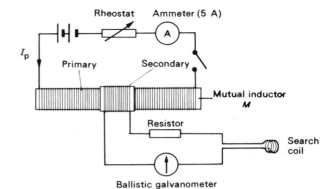

Fig. 14.42

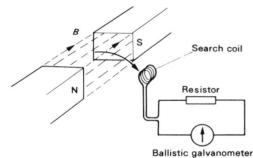

Fig. 14.41

If B is the flux density of the field and A is the cross-sectional area of the coil which has N turns, then

$$\text{flux-linkage change} = NAB$$

$$\therefore \quad Q = \frac{NAB}{R}$$

where R is the *total* resistance of the circuit. Hence

$$B = \frac{RQ}{NA}$$

R, N and A can be measured or are given. Q can be found from the charge sensitivity of the galvanometer—known or determined from (c)—and hence B calculated.

(c) *Calibrating a ballistic galvanometer.* A known current I_p is passed through the primary of a mutual inductance M in the circuit of Fig. 14.42. Let flux Φ_p link the secondary. I_p is switched off causing a flux-linkage change of $N_s\Phi_p$ with the secondary (N_s turns). Let the first 'throw' on the galvanometer be θ. The charge Q driven through the secondary is given by

$$Q = \frac{N_s\Phi_p}{R}$$

where R is the *total* resistance of the secondary circuit. If $\mathcal{E}_s$ is the e.m.f. induced in the secondary, then

$$\mathcal{E}_s = -\frac{d}{dt}(N_s\Phi_p) = -N_s\frac{d\Phi_p}{dt}$$

Also, $$\mathcal{E}_s = -M\frac{dI_p}{dt}$$

Therefore, $$N_s d\Phi_p = M dI_p$$

But here the change of current is I_p and the flux linkage change is $N_s\Phi_p$, and so

$$N_s\Phi_p = MI_p$$

$$\therefore \quad Q = \frac{MI_p}{R}$$

M can be calculated from

$$M = \frac{\mu_0 A_p N_p N_s}{l_p}$$

where the symbols have their previous meanings (p. 303). Hence Q and b (the charge sensitivity $= \theta/Q$) follow.

The search coil and the secondary of the mutual inductor are *both* in circuit during the calibration of the ballistic galvanometer; if they are so when the measurement of B is made, the damping is the same in both cases.

Absolute measurement of resistance

In an absolute method an electrical quantity is measured in terms of the basic mechanical quantities,

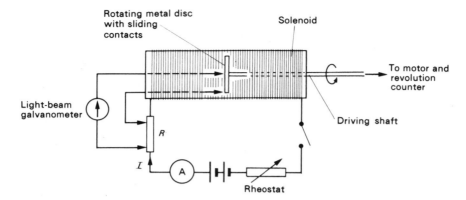

Fig. 14.43

i.e. mass, length and time and no electrical measurements are necessary. The absolute measurement of current was considered earlier (p. 272); the principle of a method for resistance, due to Lorenz and based on Faraday's disc dynamo (p. 289), is shown in Fig. 14.43.

A metal disc is rotated with its plane at right angles to the uniform flux density B at the centre of a long current-carrying solenoid having n turns per metre. The e.m.f. $\mathscr{E}$ induced between the centre and rim of the disc for the radius joining the sliding contacts is balanced against the p.d. across part of a low resistance R (a copper rod) which carries the same current I as flows through the solenoid. R is to be measured.

If the disc makes f revolutions per second at balance and has radius r, the area swept out per second by a radius is $\pi r^2 f$. The flux cut per second is $B\pi r^2 f$, hence when there is no deflection on the light-beam galvanometer,

$$\mathscr{E} = IR = B\pi r^2 f$$

$$\therefore \quad R = \frac{B\pi r^2 f}{I}$$

But at the centre of a long solenoid

$$B = \mu_0 n I \qquad \text{(p. 266)}$$

$$\therefore \quad R = \mu_0 n \pi r^2 f$$

R can thus be calculated if n, r and f are measured.

This method is used to measure the resistance of coils kept as standards in laboratories such as the NPL, but various modifications and precautions are necessary. Thermoelectric e.m.f.s, comparable with the small induced e.m.f., arise at the sliding contacts due to frictional heating and they must be allowed for. The earth's magnetic field has to be taken into account and allowance also made for the field inside the solenoid not being perfectly uniform over the disc.

Ferromagnetic materials

Iron, cobalt, nickel and substances containing them, are strongly magnetic and are called *ferromagnetic* materials. Many other materials exhibit large magnetic effects at very low temperatures.

(*a*) *Relative permeability*. The flux density in a coil increases many times when it has a ferromagnetic core because the core becomes magnetized and contributes flux. The relative permeability μ_r of a material is defined by the equation

$$\mu_r = \frac{B}{B_0}$$

where B_0 is the flux density in a current-carrying toroid (an end-less solenoid) containing air (strictly a vacuum) and B is the flux density when the same toroid is filled with the material, Fig. 14.44a. Since B and B_0 are both measured in teslas, μ_r has no units.

A toroid rather than a solenoid is specified so that the magnetization of the material is nearly uniform, if the difference between the external and internal radii is relatively small. A rod-shaped specimen in a solenoid would have poles at its ends which tend to demagnetize the rod near the ends (hence the use of keepers to store magnets and reduce self-demagnetization). Uniform magnetization, such as can be achieved all along the material under test when it forms a closed magnetic loop in a toroid, is impossible in a solenoid.

In a measurement of μ_r, B is found using a 'search' coil, connected to a calibrated ballistic galvanometer and wound round part of the toroid, Fig. 14.44b. B_0 is calculated since it can be shown that it equals $\mu_0 n I$ (the same as for the middle of a long solenoid) where n is the number of turns per metre on the toroid and I is the current in it.

(*b*) *Magnetization curve*. If B is measured for a

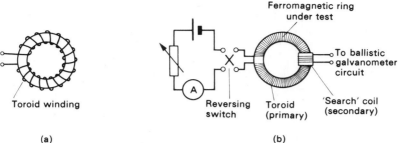

Fig. 14.44   **(a)**   **(b)**

ferromagnetic material as the magnetizing current is increased from zero, a magnetization curve of B against B_0 (which is proportional to the current) can be obtained and has the form of Fig. 14.45a.

Along OP the magnetization is small and reversible, i.e. it returns to zero when the magnetizing field B_0 is removed. Between P and Q the magnetization increases rapidly as B_0 increases and is irreversible, i.e. the specimen remains magnetized when the field is reduced to zero. For values of B_0 beyond Q very little increase of B occurs and the specimen is said to be approaching full magnetization or 'saturation' along QT.

The relative permeability μ_r ($= B/B_0$) at a point such as S is the gradient of the line joining O to S. Its value varies along the graph and is a maximum at Q. The value of B_0 at Q (i.e. OA) therefore gives the most efficient flux production and would be achieved by a correct choice of n and I for the magnetizing toroid. When values of μ_r are quoted (and they can be as high as 10^5) they usually refer to point Q on the magnetization curve.

(*c*) *Hysteresis loop.* When a specimen of a ferromagnetic material has reached saturation and the magnetizing field is reduced to zero, it remains quite strongly magnetized. The flux density B_r it retains is

called the *remanence* or *retentivity* of the material, OR in Fig. 14.45b. A reverse magnetizing field is required to demagnetize it completely and the value of B_0 which makes B zero is called the *coercive force* or *coercivity* of the material, OC in Fig. 14.45b. If the reverse field is increased more, the specimen becomes saturated in the reverse direction (D). Decreasing the field and again reversing to saturation in the first direction gives the rest of the loop, DEFT.

The curve in Fig. 14.45b is called a *hysteresis loop*. It shows that the magnetization of a material (i.e. B) lags behind the magnetizing field (i.e. B_0) when it is taken through a complete magnetization cycle, the effect is called *magnetic hysteresis*; the term *hysteresis* is derived from a Greek word meaning 'lagging behind'.

The shape of a hysteresis loop provides useful information to the designer of electrical equipment. For example, it can be shown that the area of the loop is proportional to the energy required to take unit volume of the material round one cycle of magnetization. This energy increases the internal energy of the specimen. It is called the *hysteresis loss* and is important when materials are subject to alternating fields which take them through many cycles of magnetization per second.

A hysteresis loop can be displayed on a CRO with an easily accessible tube from which any magnetic screen

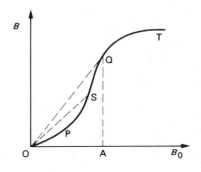

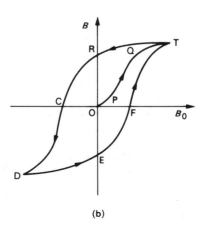

Fig. 14.45   **(a)**   **(b)**

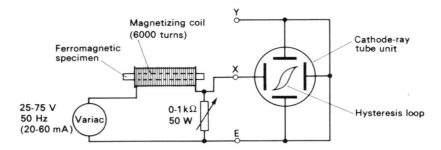

Fig. 14.46

must be removed. The specimen (e.g. a strip of soft iron tinplate or a length of steel clockspring) is inserted in a magnetizing coil set at right angles to the oscilloscope tube and close to the deflecting plates, Fig. 14.46. When current flows in the coil the specimen is magnetized and deflects the electron beam in the Y-direction (Fleming's left-hand rule). The Y-deflection is thus a measure of the flux density B of the specimen. The magnetizing coil current also passes through a variable resistor, and the p.d. across this is applied to the X-plates. The X-deflection is therefore proportional to the magnetizing current and so to B_0. With an a.c. input the specimen is taken through complete magnetization cycles and the spot produces a hysteresis loop.

(d) *Demagnetization.* A simple but effective way of demagnetizing a magnetic material is to insert it in a multi-turn coil carrying a.c. and then either to reduce the current to zero or to withdraw the specimen from the coil. In both cases the material is taken through a series of ever-diminishing hysteresis loops.

Properties and uses of magnetic materials

Ferromagnetic materials can be classified into two broad groups—'soft' and 'hard'. Soft magnetic materials are easily magnetized and demagnetized, hard materials require large magnetizing fields and retain

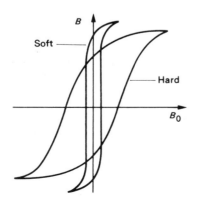

Fig. 14.47

their magnetization. Originally, the characteristics of the two groups were displayed by soft iron and hard steel but modern magnetic materials surpass these in performance. Typical hysteresis loops for soft and hard materials are shown in Fig. 14.47.

(a) *Permanent magnets* are made of hard magnetic materials with high remanence to give them 'strength' and high coercivity so that they are not easily demagnetized by stray magnetic fields or mechanical ill-treatment: Such materials are either (i) *alloys* containing, for example, iron, aluminium, nickel, copper and cobalt and having trade names like 'Ticonal', 'Alnico' and 'Alcomax', or (ii) *ceramics*, made by heat and pressure treatment from powders of iron oxide and barium oxide ($BaFe_{12}O_{19}$): they belong to the group of materials called *ferrites* and have hexagonal crystal structures; one is called 'Magnadur'.

Ceramic magnets are brittle, like china. The powder can be bonded with plastics and rubber to give flexible magnets of any shape. Very fine powder is used to coat tapes for tape recorders. In large fast computers many tiny ceramic ring magnets, each about 1 mm in diameter, Fig. 14.48a, are threaded together to act as memory stores. A close-up of a memory is shown in Fig. 14.48b.

(b) *Electromagnets* require a core of soft magnetic material which will give a strong but temporary magnet. Small coercivity is essential.

(c) *Transformer cores* are subject to many cycles of magnetization and must be 'soft' with a narrow hysteresis loop to prevent heating from hysteresis loss. They should also have high resistivity to reduce eddy current loss (p. 287) and must never be saturated when in normal use or the flux will not follow the changes in primary current. Silicon iron (e.g. 'Stalloy') and mumetal are used at mains frequencies and ferrite materials with cubic crystal structures (e.g. 'Ferroxcube'—general formula MFe_2O_4, where M is a divalent atom of copper, zinc, magnesium, manganese or

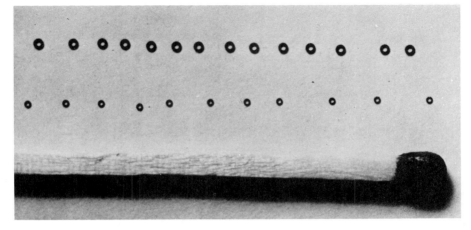

(a)

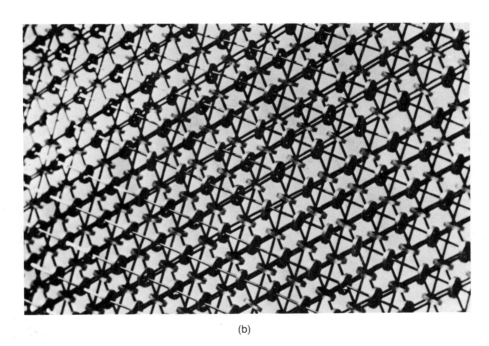

(b)

Fig. 14.48

nickel) and very high resistivities are suitable for high frequency applications in, for example, radio work.

Domain theory of ferromagnetism

(*a*) *Electrons, atoms and domains*. The magnetic field produced by a magnet can, in general, also be produced by a current in a suitably-shaped conductor. This suggests that possibly all magnetic effects, including permanent magnetism, may be due to electric currents.

In fact the magnetic properties of materials are attributed to the motion of electrons inside atoms and each electron may be regarded as a tiny 'current-carrying coil' having a magnetic field.

In the atoms of some materials the magnetic effects of different electrons cancel; in others they do not and each atom has a resultant magnetic field. With most of the latter materials, the vibratory motion of the atoms (due to their internal energy) causes their magnetic axes to have random orientations and no appreciable magnetization is shown by the material as a whole,

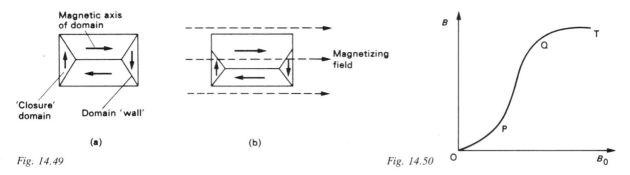

Fig. 14.49

Fig. 14.50

even when we attempt to align them all in the same direction by a strong applied field.

However, in ferromagnetic materials each atom has a resultant field and there is a force (explicable in terms of quantum mechanics) which causes *neighbouring* atoms to react on one another so that all their magnetic axes are lined up in the same direction even when there is no external magnetizing field. They do this in groups of about 10^{10} atoms to form regions called *domains* which behave like very small but very strong permanent magnets, each roughly 10^{-3} mm wide. The directions of alignment of the magnetic axes vary from one domain to another and in an unmagnetized specimen they form closed magnetic loops, Fig. 14.49*a*, with the 'closure' domains acting like the keepers on a pair of bar magnets. The magnetic fields of the domains thus neutralize one another and no detectable external magnetic effect is produced.

(*b*) *Explanation of magnetization and hysteresis.* The domain theory offers the following account. When a small field is applied to an unmagnetized specimen those domains whose magnetic axes are most in line with the field grow at the expense of others and a movement of domain 'walls' results, Fig. 14.49*b*. This first stage of magnetization, OP in the magnetization curve of Fig. 14.50, is almost wholly reversible and if the applied field is removed the walls return to their previous positions and the magnetization is again zero.

Larger magnetizing fields cause the magnetic axes of entire domains to 'jump' round quite suddenly in succession into alignment with the field and the magnetization increases sharply, PQ in Fig. 14.50. This stage is largely irreversible and the specimen retains its magnetization if the field is reduced to zero. When the field is great enough, more or less all domains are in line with the field and saturation occurs; QT on the curve. If a sufficiently strong reverse field is applied the domains can be re-aligned and saturation obtained in the opposite direction.

Hysteresis is considered to be due to domain walls being unable to move across grain boundaries and other defects in the polycrystalline specimen until a reverse field of sufficient strength is applied. The magnetization thus lags behind the magnetizing field.

(*c*) *Evidence for domains.* Various effects support their existence. Two will be considered briefly:

(*i*) *Bitter patterns.* These were first obtained by Bitter in 1931 when he allowed very fine iron powder in a colloidal suspension to settle on a single ferromagnetic crystal with a smooth surface. At the domain walls there is slight leakage of magnetic flux, the powder collects there and a 'maze' pattern like that in Fig. 14.51 is seen through a microscope, whether the specimen is magnetized or not.

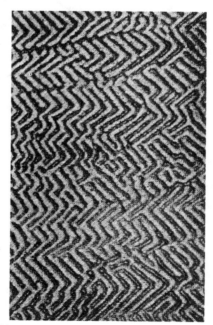

Fig. 14.51

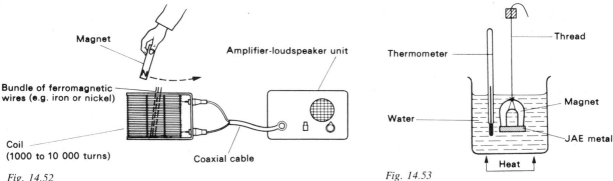

Fig. 14.52

Fig. 14.53

(*ii*) *Barkhausen effect*. This may be readily demonstrated using the apparatus of Fig. 14.52. When, say, the north pole of the magnet is drawn across and a little above the end of the bundle of ferromagnetic wires, a rushing noise is produced in the loudspeaker due to induced e.m.f.s in the coil arising from the succession of 'jumps' made by the magnetic axes of domains during magnetization. There is no repetition of the effect on subsequent transits of the magnet unless the south pole of the magnet is nearest the wires. Neither do copper wires give any effect.

(*d*) *Curie temperature*. At a certain temperature, called the *Curie point*, a ferromagnetic material loses its ferromagnetic properties. This is attributed to the internal energy and vibration of the atoms becoming so vigorous as to destroy the domain structure. The Curie point of iron is 770 °C, for the ferromagnetic alloy JAE metal (70 per cent Ni, 30 per cent Cu) it is about 70 °C. The latter can be used to show the effect, Fig. 14.53. The JAE metal drops off the magnet at the Curie point but becomes magnetic again below it.

QUESTIONS

1. Under what circumstances is an e.m.f. induced in a conductor? What factors govern the magnitude and direction of the induced e.m.f.? Describe a quantitative experiment which demonstrates how the magnitude of the induced e.m.f. depends on one of these factors.

By considering any simple case, show that if the induced e.m.f. acted in the opposite direction to that in which it does act the law of conservation of energy would be contravened.

(*S.*)

2. Write down an expression for the e.m.f. induced between the ends of a rod of length l moving with velocity v so as to cut a flux density B normally.

A straight wire of length 50 cm and resistance 10 Ω moves sideways with a velocity of 15 m s^{-1} at right angles to a uniform magnetic field of flux density 2.0×10^{-3} T. What

current would flow if its ends were connected by leads of negligible resistance?

3. State Lenz's law of electromagnetic induction and describe an experiment by which it may be demonstrated.

A circular disc of copper and a horseshoe magnet are mounted as shown in Fig. 14.54. The disc is free to rotate and the magnet can be rotated on the axle as shown. Describe and explain what happens when the magnet is set in rotation.

An aircraft is flying horizontally at 800 km h^{-1} at a point where the horizontal component of the earth's magnetic flux density is 2.0×10^{-5} T and the angle of dip is 60°. If the wing-span of the aircraft is 50 m calculate the potential difference in volts which is produced between the wing-tips of the aircraft.

(*A.E.B. part qn.*)

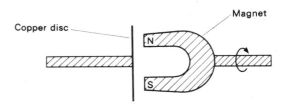

Fig. 14.54

4. A copper disc of radius 10 cm is situated in a uniform field of magnetic flux density 1.0×10^{-2} T with its plane perpendicular to the field.

The disc is rotated about an axis through its centre parallel to the field at 3.0×10^3 rev min^{-1}. Calculate the e.m.f. between the rim and centre of the disc.

Draw a circle to illustrate the disc. Show the direction of rotation as clockwise and consider the field directed into the plane of the diagram.

Explaining how you obtain your result, state the direction of the current flowing in a stationary wire whose ends touch the rim and centre of the disc.

(*J.M.B.*)

5. State an expression for the e.m.f. induced in a conductor moving in a magnetic field and show in a diagram the directional relations involved.

A rectangular coil 30.0 cm long and 20.0 cm wide has 25 turns. It rotates at the uniform rate of 3000 rev min^{-1} about an axis parallel to its long side and at right angles to a uniform magnetic field of flux density 5.00×10^{-2} T. Find (a) the frequency, and (b) the peak value of the induced e.m.f. in the coil.

Describe with the aid of diagrams how you would arrange for the rotating coil to supply to an external circuit (i) direct current, and (ii) alternating current.   (L.)

6. State the laws of electromagnetic induction and describe briefly experiments (one in each case) by which they may be demonstrated.

An electromagnet is in series with a 12 V battery and a switch across which is connected a 230 V neon lamp. Explain why, when the switch is closed, the neon remains unlit but, when the switch is opened, it flashes momentarily. Explain the importance of this observation in connection with large power switching.

A flat circular coil of 100 turns of mean radius 5.0 cm is lying on a horizontal surface and is turned over in 0.20 s. Calculate the mean e.m.f. induced if the vertical component of the earth's magnetic flux density is 4.0×10^{-5} T.   (A.E.B.)

7. The e.m.f. generated by a simple single-coil a.c. dynamo may be represented by the equation $E = E' \sin \omega t$.

(i) State the meanings of, and give the units for, the symbols employed.

(ii) Draw diagrams showing the relative position of the coil and the magnetic field (a) when $t = 0$, and (b) when $E = E'$.

(iii) Discuss the factors which, in practice, determine the maximum current which may be generated by such a dynamo.

(iv) Deduce a formula for the torque on the coil at the moment when the maximum current is flowing. Assume that the coil is rectangular, and state the units of any new symbols you employ.   (C.)

8. A designer has suggested the arrangement shown in the diagram for a heavy vehicle braking system to reduce fatigue by eliminating the need for large forces to be applied to the brake pedal.

A disc is mounted near each road wheel, as shown in Fig. 14.55. When the driver depresses the brake pedal, the effect is gradually to increase the current in the circuit and hence also the magnetic flux in the region of the disc.

Explain how braking is achieved and discuss its effectiveness: (a) at motorway speeds; (b) in city centre traffic

conditions; and (c) when the vehicle is stationary.

Select from the materials listed below the one you would use for each of the following, giving a reason for your choice in each case: (i) the disc, and (ii) the formers P on which the coils are wound.

Materials: copper, steel, cast iron, wood, rigid polythene.
          (J.M.B. Eng. Sc.)

9. Draw a simple diagram to illustrate the essential electrical connections of a shunt-wound d.c. motor.

The armature resistance of such a motor is 0.75 Ω and it runs from a 240 V d.c. supply. When the motor is running freely, i.e. under no applied load, the current in the armature is 4.0 A and the motor makes 400 revolutions per minute. What is the value of the back e.m.f. produced in the motor and what is the rate of working?

When a load is placed on the motor the armature current increases to 60 A. What is now the back e.m.f., the rate of working, and the speed of rotation? It may be assumed that the field current remains constant. (Hint. For a shunt-wound motor, torque ∝ armature current since field current and so flux-leakage is constant.)   (W.)

10. Describe the construction of a simple form of alternating current transformer.

If the secondary coil is on open circuit explain, without calculation, the effect on the current flowing in the primary of (a) a fall in the supply frequency, and (b) a reduction in the number of primary turns.

Calculate the current which flows in a resistance of 3 Ω connected to a secondary coil of 60 turns if the primary has 1200 turns and is connected to a 240 V a.c. supply, assuming that all the magnetic flux in the primary passes through the secondary and that there are no other losses.   (O. and C.)

11. By describing a suitable experiment using direct current explain what is meant by self inductance.

The current in a coil of inductance 0.10 H rises from zero to its maximum value at a mean rate of 2.0 A s^{-1}. Estimate the mean magnitude of the self-induced e.m.f. and state, giving your reason, its direction.   (J.M.B.)

12. A choke of large self inductance and small resistance, a battery and a switch are connected in series. Sketch and explain a graph illustrating how the current varies with time after the switch is closed. If the self inductance and resistance of the coil are 10 H and 5.0 Ω respectively and the battery has

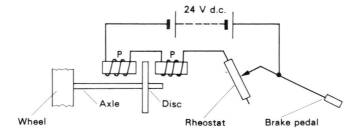

24 V d.c.

Fig. 14.55   Wheel   Axle   Disc   Rheostat   Brake pedal

an e.m.f. of 20 V and negligible resistance, what are the greatest values after the switch is closed of (a) the current, and (b) the rate of change of current? (J.M.B.)

13. The terminals of a moving-coil ballistic galvanometer are short-circuited after a charge has been passed through it. Explain why the oscillations of the coil are damped out and the coil returns slowly to its zero position.

A ballistic galvanometer is connected to a flat coil having 40 turns of mean area 3.0 cm^2 to form a circuit of total resistance 80 Ω. The coil, held between the poles of an electromagnet with its plane perpendicular to the field, is suddenly withdrawn from the field, producing a throw of 30 scale divisions. Find the magnetic induction (flux density) of the field at the place where the coil was held, assuming that the sensitivity of the galvanometer under the conditions of the experiment is 0.40 division per microcoulomb. (L. part qn.)

14. Explain the special features that are necessary in a moving-coil galvanometer intended for ballistic use.

A ballistic galvanometer is connected in series with a search coil and the secondary winding of a mutal inductance. When a current is reversed in the primary winding of the inductance a charge of 90 μC flows through the galvanometer. After switching off the primary current the search coil (which has 200 turns of mean diameter 1.00 cm) is placed in and perpendicular to a magnetic field. The deflection of the galvanometer caused by the rapid removal of the search coil from the magnetic field is the same as was observed when the primary current was reversed. The total resistance of the galvanometer circuit is 250 Ω. Calculate the magnetic induction (flux density) of the magnetic field. Draw a complete circuit diagram of the arrangement. (J.M.B.)

15. Explain what is meant by *cycle of magnetization,* and *hysteresis.*

Sketch on the same diagram the hysteresis loops for soft iron and hardened steel, indicating on the axes the physical quantities that have been plotted.

What information of practical importance can be obtained from a hysteresis loop? State, with reasons, which of the above metals would be suitable for (a) a permanent magnet, and (b) the core of a transformer.

Describe, with circuit diagram, an effective electrical method of demagnetizing a steel bar magnet and explain what is happening during the process of demagnetization. (L.)

16. Give a general account of the magnetization of iron. Show how the processes of magnetization and demagnetization, the phenomenon of hysteresis, and the existence of a Curie temperature can be explained in terms of elementary magnets and a domain structure.

What magnetic properties are desirable for the material of

(a) the core of a transformer,

(b) the core of a relay electromagnet,

(c) the tape of a magnetic tape recorder? (O.)

17. A student who was particularly interested in physics wrote at the end of his school course that he had found it interesting because 'you do a *few experiments* in the laboratory, you get out *a few rules and a few ideas about the sizes of some physical quantities,* and then presto—you find you can *understand and explain or predict* a whole range of phenomena and applications'.

Discuss this opinion illustrating your agreement or disagreement with it by choosing any *one* of the topics given in the list below or one stated topic of your own choice. You should explain, giving as many examples as possible, how each of the phrases given in italics applies to your study of the topic.

Topics:

The inverse square law for electric charges.

Electromagnetic induction.

The random behaviour of atoms and the energy they share.

(O. and C. Nuffield)

15 Alternating current

Introduction

(a) a.c. and d.c. In a direct current (d.c.) the drift velocity superimposed on the random motion of the charge carriers (e.g. electrons) is in one direction only. In an alternating current (a.c.) the direction of the drift velocity reverses, usually many times a second.

The effects of a.c. are essentially the same as those of d.c. Both are satisfactory for heating and lighting purposes. The magnetic field due to a.c. fluctuates with time and although, for example, a couple is exerted on a current-carrying coil, the inertia of the coil may prevent it responding, unless the frequency of the a.c. is very low. Thus a moving-coil meter gives a reading with d.c. but normally not with a.c. Chemical effects are observed in some cases. The electrolysis of acidulated water by mains frequency a.c. produces a hydrogen–oxygen mixture at both platinum electrodes. There is no resultant effect when a.c. passes through copper sulphate solution using copper electrodes.

As we have seen (p. 300) a.c. is more easily generated and distributed than d.c. and for this reason the mains supply is a.c. However, processes such as electroplating and battery charging require d.c., as does electronic equipment like radios and television receivers. When necessary a.c. can be rectified to give d.c.

(b) Terms. An alternating current or e.m.f. varies periodically with time in magnitude and direction. One complete alternation is called a *cycle* and the number of cycles occurring in one second is termed the *frequency* (*f*) of the alternating quantity. The unit of frequency is the *hertz* (Hz) and was previously the cycle per second. The frequency of the electricity supply in Britain is 50 Hz which means that the duration of one cycle, known as the *period* (*T*), is $1/50 = 0.02$ s. In general $f = 1/T$.

The simplest and most important alternating e.m.f. can be represented by a sine curve and is said to have a

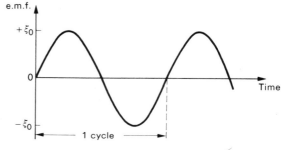

Fig. 15.1

sinusoidal waveform, Fig. 15.1. It can be expressed by the equation

$$\mathscr{E} = \mathscr{E}_0 \sin \omega t$$

where $\mathscr{E}$ is the e.m.f. at time t, $\mathscr{E}_0$ is the peak or maximum e.m.f. and ω is a constant which equals $2\pi f$ where f is the frequency of the e.m.f. Similarly, for a sinusoidal alternating current we may write

$$I = I_0 \sin \omega t$$

In the previous chapter we saw that a sinusoidal e.m.f. is induced in a coil rotating with *constant* speed in a *uniform* magnetic field. In that case ω was the angular velocity of the coil (in rad s^{-1}) and f equalled the number of complete revolutions of the coil per second. The mains supply is very nearly sinusoidal.

Alternating e.m.f.s and currents of many different waveforms can be produced and have their uses, Fig. 15.2. All, however irregular, can be shown to be combinations of sinusoidal e.m.f.s or currents.

Root-mean-square (r.m.s.) values

The value of an alternating current (and e.m.f.) varies from one instant to the next and the problem arises of

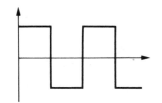
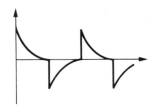
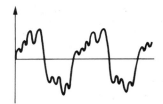

Fig. 15.2

what value we should take to measure it. The average value over a complete cycle is zero; the peak value is a possibility. However, the *root-mean-square* (r.m.s.) value is chosen because by using it many calculations can be done as they would be for direct currents.

The r.m.s. value of an alternating current (also called the *effective* value) is *the steady direct current which converts electrical energy to other forms of energy in a given resistance at the same rate as the a.c.*

Thus if the lamp in the circuit of Fig. 15.3 is lit first from a.c. and the brightness noted, then if 0.3 A d.c. produces the same brightness, the r.m.s. value of the a.c. is 0.3 A. A lamp designed to be fully lit by a current of 0.3 A d.c. will therefore be fully lit by an a.c. of r.m.s. value 0.3 A. Although the value (I) of the a.c. is varying, the *average* rate at which it supplies electrical energy to the lamp equals the *steady* rate of supply by the d.c. ($I_{d.c.}$) and in practice it is this aspect which is often important.

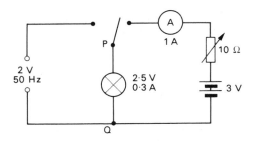

Fig. 15.3

In general, considering energy supplied to a resistance R we can say

$$I_{d.c.}^2 R = (\text{mean value of } I^2) \times R$$

$$\therefore \quad I_{d.c.} = \sqrt{\text{mean value of } I^2}$$

$$= \text{square } root \text{ of the } mean \text{ value of the}$$
$$\text{squares of the current}$$

$$= I_{r.m.s.}$$

If the a.c. is sinusoidal then

$$I = I_0 \sin \omega t$$

$$\therefore \quad I_{r.m.s.} = \sqrt{\text{mean value of } I_0^2 \sin^2 \omega t}$$

$$= I_0 \sqrt{\text{mean value of } \sin^2 \omega t}$$

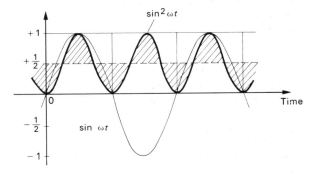

Fig. 15.4

From the graphs of $\sin \omega t$ and $\sin^2 \omega t$ in Fig. 15.4 it can be seen that $\sin^2 \omega t$ is always positive and varies between 0 to 1. The shaded areas above and below the dotted line are equal, there is symmetry and the mean value for $\sin^2 \omega t$ is therefore 1/2. Hence

$$I_{r.m.s.} = I_0 \sqrt{1/2} = \frac{I_0}{\sqrt{2}} = 0.707 \, I_0$$

The r.m.s. current is 0.707 times the peak current. Similar relationships hold for e.m.f.s. and p.d.s, and the r.m.s. value is the one usually quoted. Thus 240 V is the r.m.s value of the electricity supply and the peak value $\mathcal{E}_0 = \sqrt{2}\mathcal{E}_{r.m.s.} = \sqrt{2} \times 240 = 339$ V.

The circuit of Fig. 15.3 may be used to check roughly that the peak value of an alternating p.d. is 1.4 times its r.m.s. value. The lamp is adjusted to the same brightness on d.c. as on a.c. The CRO is then connected across PQ and is used as a voltmeter. It measures the r.m.s. value of the p.d. across the lamp when it is lit by d.c. and *twice* the peak value using the a.c. supply.

Most voltmeters and ammeters for a.c. use are calibrated to read r.m.s. values and give correct readings only if the waveform is sinusoidal.

Meters for a.c.

The deflection of an a.c. meter must not depend on the direction of the current.

(*a*) *Moving-iron meter.* The repulsion type consists of two soft iron rods P and Q mounted inside a solenoid S and parallel to its axis, Fig. 15.5. P is fixed and Q is carried by the pointer. Current passing either way through S magnetizes P and Q in the same direction and they repel each other. Q moves away from P until stopped by the restoring couple due, for example, to hair-springs. In many cases air damping is provided by attaching to the movement an aluminium pointer or vane which moves inside a curved cylinder.

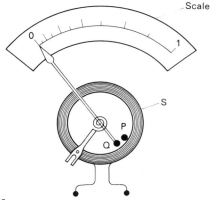

Fig. 15.5

The deflecting force is a function of the average value of the square of the current. Hence a moving-iron meter can be used to measure either d.c. or a.c. and in the latter case r.m.s. values are recorded. The scale is not divided uniformly being closed up for smaller currents.

A moving-iron voltmeter is a moving-iron milli-ammeter with a suitable (non-inductive) multiplier connected in series.

(*b*) *Thermocouple meter.* One junction of a thermo-couple (i.e. two wires of dissimilar metals) is joined to the centre of the wire XY carrying the current to be measured and is heated by it; the other junction is at room temperature, Fig. 15.6. When a.c. flows in XY, a

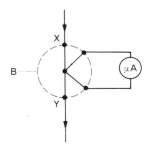

Fig. 15.6

thermoelectric e.m.f. is generated and produces a direct current that can be measured by a moving-coil microammeter (previously calibrated by passing known values of d.c. through XY). The hot junction is enclosed in an evacuated bulb B to shield it from draughts.

This type of meter relies on the heating effect of a current. It therefore measures r.m.s. values and can be used for alternating currents of high frequency (up to several megahertz) because of its low inductance and capacitance compared with other meters.

(*c*) *Rectifier meter.* A rectifier is a device with a low resistance to current flow in one direction and a high resistance for the reverse direction. When connected to an a.c. supply it allows pulses of varying but direct current to pass. In a rectifier-type meter the average value of these is measured by a moving-coil meter, Figs. 15.7 and 15.8. Rectification, i.e. the conversion of a.c. to d.c., thus occurs. There are various kinds of rectifier (p. 328); those in many instruments are semiconducting (germanium) diodes.

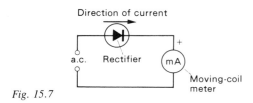

Fig. 15.7

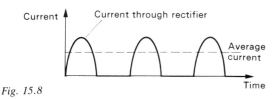

Fig. 15.8

Rectifier instruments, being based on the moving-coil meter, are much more sensitive than other a.c. meters and are used in multimeters that have a.c. as well as d.c. ranges. The scale of a rectifier meter is calibrated to read r.m.s. values of currents and p.d.s. with sinusoidal waveforms.

Capacitance in a.c. circuits

(a) *Flow of a.c. 'through' a capacitor.* If a 1000 μF capacitor is connected in series with a 2.5 V, 0.3 A lamp, and a 2 V d.c. supply, Fig. 15.9a, the lamp as expected, does not light. Is there *any* current flow? With a 2 V r.m.s. 50 Hz supply, Fig. 15.9b, it is nearly fully lit.

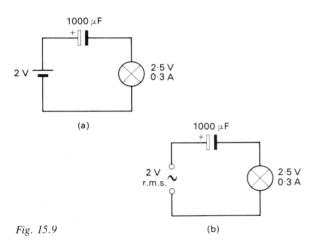

Fig. 15.9

(a)

(b)

The a.c. is apparently flowing through the capacitor. In fact, the capacitor is being charged, discharged, charged in the opposite direction and discharged again, fifty times per second (the frequency of the a.c.), and the charging and discharging currents flowing through the lamp light it. No current actually passes through the capacitor (since its plates are separated by an insulator) but it appears to do so and we talk as if it did. A current would certainly be recorded by an a.c. milliammeter.

When the 1000 μF capacitor is replaced by one of 100 μF, the charging and discharging currents are too small to light the lamp. Larger capacitances thus offer less 'opposition' to a.c. Increasing the frequency of the a.c. (at constant p.d. and capacitance) increases the current 'through' a capacitor since the same charge has to flow on and off the plates in a shorter time.

(b) *Phase relationships.* When a.c. flows through a resistor (having no capacitance or inductance) the current and p.d. reach their peak values at the same

instant, i.e. they are in phase. This is not so for a capacitor.

The circuit of Fig. 15.10a enables phase relationships to be studied using 'slow a.c.' of frequency less than 1 Hz. With a 2000 Ω resistor between X and Y the current (shown on the milliammeter) and the p.d. (shown on the voltmeter) rise and fall together. With a 250 μF capacitor replacing the resistor, the current through the capacitor is seen to lead the p.d. across it by one-quarter of a cycle, i.e. the current reaches its maximum value one-quarter of a cycle before the p.d. reaches its peak value, as shown in Fig. 15.10b by the cosine and sine curves.

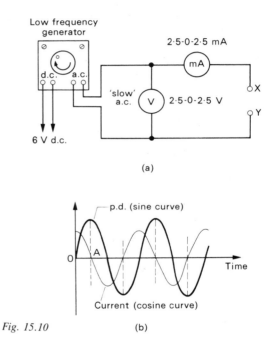

(a)

Fig. 15.10

(b)

The circuit for an alternative demonstration at 50 Hz using a double-beam CRO is given in Fig. 15.11. (A method of using an electronic beam splitter to convert a single beam CRO into a double beam one is given in Appendix 20.) The Y_2 trace is the p.d. across R and this gives the waveform of the current 'through' C

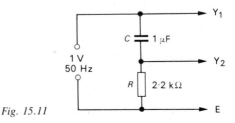

Fig. 15.11

since C and R are in series and the current in a resistor is in phase with the p.d. across it. The Y_1 trace is the p.d. across C and R in series and not just C, which accounts for the phase difference being less than a quarter of a cycle.

Current and applied p.d. are out of step because current flow is a maximum immediately an uncharged capacitor is connected to a supply be it d.c. (p. 239) or a.c. There is as yet no charge on the capacitor to oppose the arrival of charge. Thus at O the applied p.d. though momentarily zero, is increasing at its maximum rate (the slope of the tangent at O to the p.d. graph is a maximum) and so the rate of flow of charge—the current—is also a maximum. Between O and A the p.d. is increasing but at a decreasing rate, the charge on the capacitor is increasing but less quickly, which means that the charging current is less. At A the applied p.d. is a maximum and for a brief moment is constant. The charge on the capacitor will also be a maximum and constant. The rate of flow of charge is therefore zero, i.e. the current is zero. The phase difference between V and I can thus be explained.

(c) *Mathematical treatment.* Let a p.d. V be applied across a capacitance C and let its value at time t be given by

$$V = V_0 \sin \omega t$$

where V_0 is its peak value and $\omega = 2\pi f$ where f is the frequency of the supply. The charge Q on the capacitance at time t is

$$Q = VC$$

For the current I flowing 'through' the capacitor we can write

$$I = \text{rate of change of charge} = \frac{\mathrm{d}Q}{\mathrm{d}t}$$

$$= \frac{\mathrm{d}}{\mathrm{d}t}(VC) = C\frac{\mathrm{d}V}{\mathrm{d}t} = C\frac{\mathrm{d}}{\mathrm{d}t}(V_0 \sin \omega t)$$

$$= CV_0 \frac{\mathrm{d}}{\mathrm{d}t}(\sin \omega t)$$

$$\therefore \quad I = \omega C V_0 \cos \omega t$$

The current 'through' C (a cosine function) thus leads the applied p.d. (a sine function) by one quarter of a cycle or, as is often stated, by $\pi/2$ radians or 90° (1 cycle being regarded as 2π radians or 360°). This confirms the results of the demonstrations and the 'physical' explanation outlined above. We can also write

$$I = I_0 \cos \omega t$$

where I_0 is the peak current and is given by $I_0 = \omega C V_0$

$$\therefore \quad \frac{V_0}{I_0} = \frac{1}{\omega C}$$

But

$$\frac{V_{\text{r.m.s.}}}{I_{\text{r.m.s.}}} = \frac{V_0}{I_0}$$

$$\therefore \quad \frac{V_{\text{r.m.s.}}}{I_{\text{r.m.s.}}} = \frac{1}{\omega C} = \frac{1}{2\pi fC}$$

This expression resembles $V/I = R$ which defines resistance, $1/(2\pi fC)$ replacing R. The quantity $1/(2\pi fC)$ is taken as a measure of the opposition of a capacitor to a.c. and is called the *capacitive reactance* X_C. Hence

$$X_C = \frac{V_{\text{r.m.s.}}}{I_{\text{r.m.s.}}} = \frac{1}{2\pi fC}$$

The ohm is the unit of X_C since the unit of f is s^{-1} and that of C is C V^{-1}. The term $1/(fC)$ therefore has units $\text{V}/(\text{C s}^{-1}) = \text{V A}^{-1} = \Omega$. If f is in hertz and C in farads then X_C is in ohms. It is clear that X_C decreases as f and C increase. A 10 μF capacitor has a reactance of 320 Ω at 50 Hz. Check this. What will it be at 1 kHz?

Reactance is not to be confused with resistance, in the latter, electrical power is dissipated, it is not in a reactance as we shall see later.

Inductance in a.c. circuits

(a) *Phase relationships.* An inductor in an a.c. circuit behaves like a capacitor in that it causes a phase difference between the applied p.d. and the current. In this case, however, the current lags on the p.d. by one-quarter of a cycle (i.e. 90°).

The effect may be observed using 'slow a.c.' and the circuit of Fig. 15.10a (p. 317) with a 12 000 turn coil on a complete iron core (from a demountable transformer) as the inductor (500 H) connected to XY. Alternatively a double-beam CRO may be used with 50 Hz a.c. as in Fig. 15.12. In both cases the phase difference is less than 90° due to the resistance of the inductor and in the CRO demonstration because the Y_1 trace gives the p.d. across L and R in series; the graphs in Fig. 15.13 are for a pure inductor.

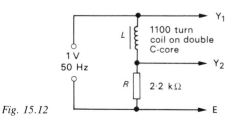

Fig. 15.12

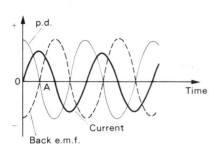

Fig. 15.13

We can explain the effects as follows. At O the current is zero but its rate of increase is a maximum (as given by the slope of the tangent to the current graph at O) which means, for an inductor of constant inductance (e.g. an air-cored coil), that the rate of change of flux is also a maximum. Therefore by Faraday's law the back e.m.f. is a maximum but, by Lenz's law, of negative sign since it acts to oppose the current change. At A the current and flux are momentarily a maximum and constant. Their rate of change is zero (slope of tangent to current graph is zero at A) and so the back e.m.f. is zero. If the inductor has negligible resistance, then at every instant the applied p.d. must be nearly equal and opposite to the back e.m.f. The applied p.d. curve is therefore as shown. The p.d. acts on the coil whilst the e.m.f. acts back upon the source, just like two forces acting on different bodies.

(b) *Mathematical treatment.* In this case it is simpler to start with the current. Consider an inductance L through which current I flows at time t where

$$I = I_0 \sin \omega t$$

I_0 is the peak current and $\omega = 2\pi f$ where f is the frequency of the a.c. The back e.m.f. $\mathscr{E}$ in the inductor due to the changing current is

$$\mathscr{E} = - L \frac{\mathrm{d}I}{\mathrm{d}t} \qquad \text{(p.302)}$$

$$\mathscr{E} = - L \frac{\mathrm{d}}{\mathrm{d}t}(I_0 \sin \omega t)$$

$$= - \omega L I_0 \cos \omega t$$

Assuming the inductor has zero resistance, then for current to flow the applied p.d. V must be equal and opposite to the back e.m.f., hence

$$V = - \mathscr{E} = \omega L I_0 \cos \omega t$$

The applied p.d. thus leads the current by 90°. We can also write

$$V = V_0 \cos \omega t$$

where V_0 is the peak value of the applied p.d. and is given by

$$V_0 = \omega L I_0$$

$$\therefore \quad \frac{V_0}{I_0} = \frac{V_{\text{r.m.s.}}}{I_{\text{r.m.s}}} = \omega L = 2\pi f L$$

The quantity $2\pi f L$ is called the *inductive reactance* X_L of the inductor and like X_C it is measured in ohms. Thus

$$X_L = \frac{V_{\text{r.m.s.}}}{I_{\text{r.m.s.}}} = 2\pi f L$$

X_L increases with f and L and is a measure of the opposition of the inductor to a.c. If $L = 10$ H for an inductor, X_L at 50 Hz is 3.1 kΩ and 63 MΩ at 1 MHz.

Vector diagrams

A sinusoidal alternating quantity can be represented by a *rotating vector* (often called a *phasor*). Suppose the graph, Fig. 15.14, represents an alternating quantity $y = Y_0 \sin \omega t$ where Y_0 is the peak value of the quantity and its frequency $f = \omega/2\pi$. If the line OP has length Y_0 and rotates in an anticlockwise direction about O with uniform angular velocity ω, the projection O'P' of OP on O'y at time t (measured from the time when OP passes through OO') is $Y_0 \sin \omega t$. If OP is directed as

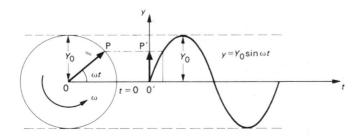

Fig. 15.14

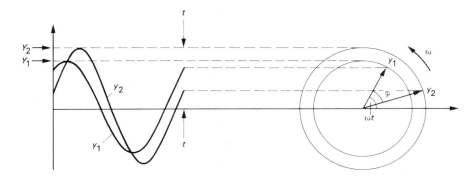

Fig. 15.15

shown by the arrow on it, then we can say that the projection on $O'y$ of the rotating vector OP, gives the value at any instant of the sinusoidal quantity y. (Simple harmonic motion, being sinusoidal, can be derived similarly from uniform motion in a circle.)

The method is very useful for representing two sinusoidal quantities which have the same frequency but are not in phase. In Fig. 15.15 the waveforms of two such quantities y_1 and y_2 and the corresponding vector diagram are shown for time t. The phase difference between them is ϕ, with y_2 lagging and this phase angle is maintained between them as the vectors rotate. Being vectors they can be added by the parallelogram law if they represent similar quantities, e.g. p.d.s.

Algebraically y_1 and y_2 are expressed by the equations

$$y_1 = Y_1 \sin \omega t$$

and

$$y_2 = Y_2 \sin (\omega t - \phi)$$

The vector diagram for a pure capacitance (i.e. infinite dielectric resistance) in an a.c. circuit is shown in Fig. 15.16a; the current I leads the applied p.d. V by 90°. That for a pure inductance (i.e. zero resistance) is given in Fig. 15.16b; in this case the current I lags on the applied p.d. V by 90°.

We shall see presently how vector diagrams are used

to solve problems involving capacitance, inductance and resistance in a.c. circuits.

Series circuits

When drawing vector diagrams a vector representing a quantity which is the same for all the circuit components should be drawn first. For a series circuit it would be the current vector. What would it be for a parallel circuit? This reference vector is drawn horizontal, directed to the right and the other vectors are then drawn so that their phases with respect to it are correct. The r.m.s. values of currents and p.d.s are used.

(*a*) *Resistance and capacitance.* Suppose an alternating p.d. V is applied across a resistance R and a capacitance C in series, Fig. 15.17a. The same current I flows through each component and so the reference vector will be that representing I. The p.d. V_R across R is in phase with I and V_C, that across C, lags on I by 90°. The vector diagram is as shown in Fig. 15.17b, V_C and V_R being drawn to scale.

The vector sum of V_R and V_C equals the applied p.d. V, hence

$$V^2 = V_R{}^2 + V_C{}^2$$

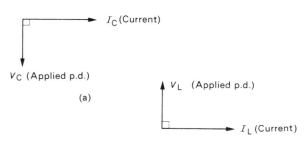

I_C (Current)

V_C (Applied p.d.)

(a)

V_L (Applied p.d.)

I_L (Current)

(b)

Fig. 15.16

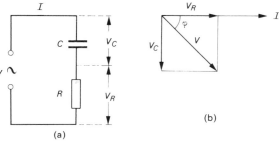

(a)

(b)

Fig. 15.17

But $V_R = IR$ and $V_C = IX_C$ where X_C is the reactance of C and equals $1/\omega C$, hence

$$V^2 = I^2(R^2 + X_C^2)$$

$$\therefore \quad V = I\sqrt{R^2 + X_C^2}$$

The quantity $\sqrt{R^2 + X_C^2}$ is called the *impedance Z* of the circuit and measures its opposition to a.c. It has resistive and reactive components and like both is measured in ohms. Hence

$$Z = \frac{V}{I} = \sqrt{R^2 + X_C^2}$$

Also, from the vector diagram we see that the current I leads V by a phase angle ϕ which is less than 90° and is given by

$$\tan \phi = \frac{V_C}{V_R} = \frac{IX_C}{IR} = \frac{X_C}{R}$$

(b) *Resistance and inductance.* The analysis is similar but in this case the p.d. V_L across L leads on the current I and the p.d. V_R across R is again in phase with I, Fig. 15.18a. As before the applied p.d. V equals the vector sum of V_L and V_R, Fig. 15.18b, and so

$$V^2 = V_R^2 + V_L^2$$

The phase angle ϕ by which I lags on V is given by

$$\tan \phi = \frac{V_L}{V_R} = \frac{IX_L}{IR} = \frac{X_L}{R}$$

(c) *Resistance, capacitance and inductance.* An *RCL* series circuit is shown in Fig. 15.19a; in practice R may be the resistance of the inductor. V_L leads the current (reference) vector I by 90°, V_C lags on it by 90° and V_R is in phase with it. V_L and V_C are therefore 180° (half a cycle) out of phase, i.e. in antiphase. If V_L is greater than V_C, their resultant $(V_L - V_C)$ is in the direction of V_L, Fig. 15.19b. The vector sum of $(V_L - V_C)$ and V_R equals the applied p.d. V, therefore

$$V^2 = V_R^2 + (V_L - V_C)^2$$

But $V_R = IR$, $V_L = IX_L$ and $V_C = IX_C$, hence

$$V^2 = I^2[R^2 + (X_L - X_C)^2]$$

$$\therefore \quad V = I\sqrt{R^2 + (X_L - X_C)^2}$$

The impedance Z is given by

$$Z = \frac{V}{I} = \sqrt{R^2 + (X_L - X_C)^2}$$

The phase angle ϕ by which I lags on V is given by

$$\tan \phi = \frac{V_L - V_C}{V_R} = \frac{X_L - X_C}{R}$$

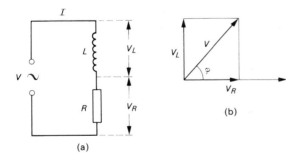

(a)

Fig. 15.18

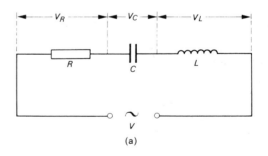

(a)

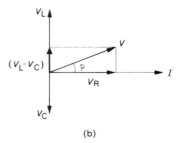

(b)

Fig. 15.19

But $V_R = IR$ and $V_L = IX_L$ where X_L is the reactance of L and equals ωL, hence

$$V^2 = I^2(R^2 + X_L^2)$$

$$\therefore \quad V = I\sqrt{R^2 + X_L^2}$$

Here the *impedance Z* is given by

$$Z = \frac{V}{I} = \sqrt{R^2 + X_L^2}$$

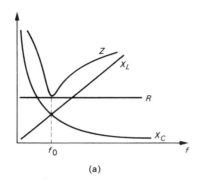

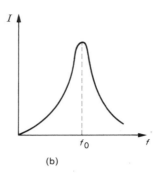

Fig. 15.20

Electrical resonance

(a) *Series resonance.* The expression just derived for the impedance Z of an RCL series circuit shows that Z varies with the frequency f of the applied p.d. since X_L and X_C both depend on f. $X_L = 2\pi f L$ and increases with f, $X_C = 1/(2\pi f C)$ and decreases with f, R is assumed to be independent of f (but it can vary). Fig. 15.20a shows how X_L, X_C, R and Z vary with f.

At a certain frequency f_0, called the *resonant frequency*, $X_L = X_C$ and Z has its minimum value, being equal to R. The circuit behaves as a pure resistance and the current I has a maximum value (given by $I = V/R$), Fig. 15.20b. The phase angle ϕ (given by $\tan \phi = (X_L - X_C)/R$) is zero, the applied p.d. V and the current I are in phase and there is said to be *resonance*. A series resonant circuit is called an *acceptor* circuit.

Series or current resonance may be shown using the arrangement of Fig. 15.21. As the frequency increases the milliammeter reading rises to a maximum and then falls.

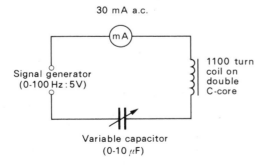

Fig. 15.21

An expression for f_0 is obtained from $X_L = X_C$, that is

$$2\pi f_0 L = \frac{1}{2\pi f_0 C}$$

or

$$4\pi^2 f_0^2 LC = 1$$

$$\therefore \quad f_0 = \frac{1}{2\pi \sqrt{LC}}$$

If L is in henrys and C in farads, f_0 will be in hertz.

At resonance V_L and V_C can both be very much greater than the total p.d. V applied across the whole circuit. Thus, if I is the resonance current, we have

$$I = \frac{V}{R}$$

$$\therefore \quad V_L = IX_L = \frac{V}{R} X_L$$

$$\therefore \quad \frac{V_L}{V} = \frac{X_L}{R}$$

Since R (which is mostly due in practice to the resistance of the inductor L) is usually very small compared with X_L, V_L (and V_C) can be large compared with V. The magnification or *Q-factor* (Q for quality) of the circuit at resonance is defined by

$$Q = \frac{V_L}{V} = \frac{X_L}{R}$$

In actual circuits Q-factors of over 200 are realized.

It may seem strange for a small applied p.d., V, to give rise to two large p.d.s, V_L and V_C. The explanation is that V_L and V_C are in antiphase (i.e. 180° out of phase) and so their vector sum can be quite small—zero, in fact, if they are equal in magnitude as they are at resonance, Fig. 15.22. The circuit of Fig. 15.23a may

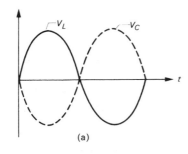

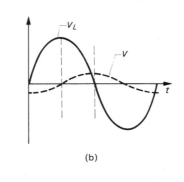

Fig. 15.22

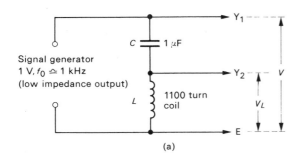

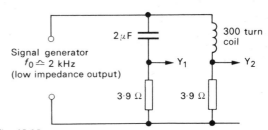

Fig. 15.23

be used to demonstrate on a double-beam CRO that at resonance V_L is much greater than V and leads it by nearly 90°, Fig. 15.23b. If C and L are interchanged, V_C is seen to be equal to V_L and to lag on V by almost 90°. V_L and V_C are thus approximately in antiphase.

(b) *Parallel resonance*. In a parallel *RCL* circuit resonance can also occur but in this case the impedance of the circuit becomes a maximum. The resonant frequency f_0 is also given to a good approximation by $f_0 = 1/(2\pi\sqrt{LC})$. Large *currents* I_C and I_L circulate to and fro *within* the circuit which are equal in magnitude but 180° out of phase, the supply current I is thus small. The p.d. across the circuit at resonance is large.

In the circuit of Fig. 15.24 the brightness of the lamps indicates the currents flowing in different parts of the circuit. Their behaviour as the frequency of the p.d.

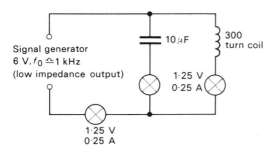

Fig. 15.24

applied from the signal generator increases (from about 0.5 kHz to 2 kHz) is worthy of study. In Fig. 15.25 the p.d.s across the 3.9 Ω resistors give the wave-forms of the currents in C and L and show that they have equal amplitude but are in antiphase at resonance.

Fig. 15.25

(c) *Tuned circuits*. The ability of a resonant circuit to select and amplify a p.d. of one particular frequency (strictly, a very narrow band of frequencies) is used in radio and television. For example in the aerial circuit of a radio receiver, Fig. 15.26, radio signals from different transmitting stations induce e.m.f.s of various frequencies in the aerial which cause currents to flow in the aerial coil. These induce currents of the same frequencies in coil L by mutual induction. If the capacitance C is adjusted (tuned) so that the resonant frequency of circuit LC equals the frequency of the wanted station, a large p.d. at that frequency (and no other) is developed

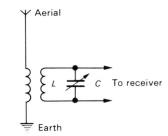

Aerial

L C To receiver

Fig. 15.26   Earth

across C. This p.d. is then applied to the next stage of the receiver.

(d) *Mechanical analogy.* The behaviour of a resonant circuit is similar to that of a vibrating system in mechanics. At one particular frequency it responds and stores a large amount of energy which passes to and fro between the electric field of the capacitor and the magnetic field of the inductor. Kinetic and potential energy behave similarly in an oscillating mass-spring system. Energy has then to be supplied only to compensate for that dissipated as heat in doing work against resistance—electrical or air.

Worked examples

1. A 1000 μF capacitor is joined in series with a 2.5 V, 0.30 A lamp and a 50 Hz supply. Calculate (a) the p.d. (r.m.s.) of the supply to light the lamp to its normal brightness, and (b) the p.d.s across the capacitor and the resistor respectively.

(a) Reactance X_C of capacitor $= 1/(2\pi fC)$

$$= 1/(2\pi \times 50 \times 10^3 \times 10^{-6})$$

$$= 10/\pi \ \Omega$$

Resistance R of lamp $= 2.5/0.30 = 8.3 \ \Omega$

Impedance Z of circuit $= \sqrt{R^2 + X_C^2}$

$$= \sqrt{(8.3)^2 + (10/\pi)^2}$$

$$= \sqrt{69 + 10}$$

$$= 8.9 \ \Omega$$

The applied r.m.s. p.d. V to cause an r.m.s. current of 0.30 A to flow round the circuit is given by

$$V = IZ$$

$$= 0.30 \times 8.9$$

$$= 2.7 \ V$$

(b) p.d. V_C across the capacitor $= IX_C$

$$= 0.30 \times 10/\pi = 0.96 \ V$$

p.d. V_R across the resistor $= IR = 2.5 \ V$

Note. $V_C + V_R = 0.96 + 2.5 \simeq 3.5$ V which is greater than the applied p.d. V of 2.7 V. This is due to V_C and V_R not being in phase. In fact $V^2 = V_C^2 + V_R^2$—as you can check from the above figures.

2. A 2.0 H inductor of resistance 80 Ω is connected in series with a 420 Ω resistor and a 240 V, 50 Hz supply. Find (a) the current in the circuit, and (b) the phase angle between the applied p.d. and the current.

(a) Reactance X_L of inductor $= 2\pi fL$

$$= 2\pi \times 50 \times 2$$

$$= 200\pi \ \Omega$$

Total resistance R of circuit $= 80 + 420$

$$= 500 \ \Omega$$

Impedance Z of circuit $= \sqrt{R^2 + X_L^2}$

$$= \sqrt{(500)^2 + (200\pi)^2}$$

$$= 800 \ \Omega$$

Current I in circuit $= V/Z$

$$= 240/800$$

$$= 0.30 \ A$$

(b) Phase angle ϕ between V and I is given by

$$\tan \phi = \frac{X_L}{R}$$

$$= \frac{200\pi}{500}$$

$$\therefore \quad \phi = 52°$$

V leads I by 52°.

3. A circuit consists of an inductor of 200 μH and resistance 10 Ω in series with a variable capacitor and a 0.10 V (r.m.s.), 1.0 MHz supply. Calculate (a) the capacitance to give resonance, (b) the p.d.s. across the inductor and the capacitor at resonance, and (c) the Q-factor of the circuit at resonance, Fig. 15.27.

(a) We have $L = 200 \times 10^{-6}$ H, $R = 10 \ \Omega$, $f_0 = 10^6$ Hz. Also,

$$f_0 = \frac{1}{2\pi\sqrt{LC}} \quad \text{or} \quad 4\pi^2 f_0^2 LC = 1$$

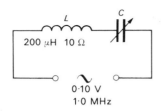

Fig. 15.27

$$\therefore \quad C = \frac{1}{4\pi^2 f_0^2 L}$$

$$= \frac{1}{4\pi^2 \times (10^6)^2 \times (200 \times 10^{-6})}$$

$$= 0.000\ 13\ \mu F$$

(*b*) At resonance the impedance $Z = R$ and if V is the applied p.d. (0.10 V), the current I is given by

$$I = \frac{V}{R} = \frac{0.10}{10} = 1.0 \times 10^{-2}\ A$$

If the inductor has reactance X_L, the p.d. V_L across it is

$$V_L = IX_L = I \times 2\pi f_0 L$$

$$= 1.0 \times 10^{-2} \times 2\pi \times 10^6 \times 200 \times 10^{-6}$$

$$= 4\pi$$

$$\simeq 13\ V$$

Since $V_C = V_L$, we have

$$V_C \simeq 13\ V$$

(*c*) The *Q-factor* at resonance is given by

$$Q = \frac{V_L}{V} = \frac{13}{0.10} = 130$$

Power in a.c. circuits

(*a*) *Resistance.* The general expression for the power absorbed by a device at any instant is IV where I and V are the instantaneous values of the current through it and the p.d. across it respectively.

In a resistor I and V are in phase and we can write $I = I_0 \sin \omega t$ and $V = V_0 \sin \omega t$, therefore the *instantaneous* power absorbed at time t is $I_0 V_0 \sin^2 \omega t$. The *mean* power P will be given by

$$P = \text{mean value of } I_0 V_0 \sin^2 \omega t$$

We saw previously (p. 315, Fig. 15.4) that the mean value of $\sin^2 \omega t$ is 1/2, thus

$$P = \frac{I_0 V_0}{2}$$

$$= \frac{I_0}{\sqrt{2}} \cdot \frac{V_0}{\sqrt{2}}$$

For sinusoidal quantities

$$I_{\text{r.m.s.}} = \frac{I_0}{\sqrt{2}} \quad \text{and} \quad V_{\text{r.m.s.}} = \frac{V_0}{\sqrt{2}}$$

$$\therefore \quad P = I_{\text{r.m.s.}} \times V_{\text{r.m.s.}}$$

For a resistance

$$V_{\text{r.m.s.}} = I_{\text{r.m.s.}} \times R$$

$$\therefore \quad P = I_{\text{r.m.s.}}^2 \times R = \frac{V_{\text{r.m.s.}}^2}{R}$$

The power varies sinusoidally at twice the frequency of either V or I, as shown in Fig. 15.28.

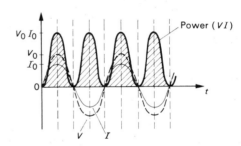

Fig. 15.28

(*b*) *Inductance.* In a pure inductor V leads I by 90° (or $\pi/2$ rad) and if $I = I_0 \sin \omega t$ then $V = V_0 \sin (\omega t + \pi/2) = V_0 \cos \omega t$. Hence the *instantaneous* power absorbed at time t is $I_0 V_0 \sin \omega t \cos \omega t$. But $\sin 2\omega t = 2 \sin \omega t \cos \omega t$, therefore

$$\text{instantaneous power absorbed} = \tfrac{1}{2} I_0 V_0 \sin 2\omega t$$

$$= I_{\text{r.m.s.}} V_{\text{r.m.s.}} \sin 2\omega t$$

This represents a sinusoidal variation with mean value zero (and frequency twice that of I and V), Fig. 15.29. It follows that the *power absorbed by an inductor in a cycle is zero*.

To explain this we consider that during the first quarter-cycle OA of current, power is drawn from the source and energy is stored in the *magnetic field* of the inductor. In the second quarter-cycle AB, the current and magnetic field decrease and the e.m.f. induced in the inductor causes it to act as a generator returning the energy stored in its magnetic field to the source. The

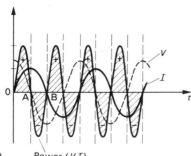

Fig. 15.29   Power (VI)

The *apparent power* absorbed in an a.c. circuit is, by comparison with the d.c. case, $I_{r.m.s.}$ $V_{r.m.s.}$. The *real power* is $I_{r.m.s.}$ $V_{r.m.s.}$ cos ϕ. The *power factor* of an a.c. circuit is defined by the equation

$$\text{power factor} = \frac{\text{real power}}{\text{apparent power}}$$

$$= \frac{I_{r.m.s.} V_{r.m.s.} \cos \phi}{I_{r.m.s.} V_{r.m.s.}} = \cos \phi$$

In a purely resistive circuit cos ϕ has its maximum value of 1.

power so restored is represented by the shaded area in Fig. 15.29. Thus, although the mean power taken over a cycle is zero, large amounts of energy flow in and out of the inductor every quarter-cycle. In practice an inductor has resistance and some energy is drawn from the source on this account and not returned.

(*c*) *Capacitance.* Similar reasoning shows that zero power is also taken by a pure capacitor over a cycle—since V and I are 90° out of phase. In this case energy taken from the source is stored in the *electric field* due to the p.d. between the plates of the charged capacitor. During the next quarter-cycle the capacitor discharges and the energy is returned to the source.

(*d*) *Formula for a.c. power.* Power is only absorbed by the *resistive* part of a circuit, i.e. in the expression IV for power, V is that part of the applied p.d. across the total resistance in the circuit; it is in phase with the current. Thus if $V_{r.m.s.}$ is the p.d. applied to an a.c. circuit and it leads the current $I_{r.m.s.}$ by angle ϕ, Fig. 15.30, the component of p.d. in phase with $I_{r.m.s.}$ is $V_{r.m.s.}$ cos ϕ. Hence the total power P expended in the circuit is given by

$$P = I_{r.m.s.} \ V_{r.m.s.} \ \cos \phi$$

The component $V_{r.m.s.}$ sin ϕ is that part of the applied p.d. across the total reactance of the circuit and is often called the 'wattless' component of the p.d.

Electrical oscillations

Electrical oscillators are used in radio and television transmitters and receivers, in signal generators, oscilloscopes and computers, to produce a.c. with waveforms which may be sinusoidal, square, sawtooth etc. and with frequencies from a few hertz up to millions of hertz.

(*a*) *Oscillatory circuit.* When a capacitor discharges through an inductor in a circuit of low resistance an a.c. flows. The circuit is said to oscillate at its *natural frequency* which, as we will show shortly, equals $1/(2\pi\sqrt{LC})$, i.e. its resonant frequency f_0. Electrical resonance thus occurs when the applied frequency equals the natural frequency—as it does in a mechanical system.

In Fig. 15.31*a* a charged capacitor C is shown connected across a coil L. C immediately starts to discharge, current flows and a magnetic field is created which induces an e.m.f. in L. This e.m.f. opposes the current. C cannot therefore discharge instantaneously and the greater the inductance of L the longer does the discharge take. When C is completely discharged the electrical energy originally stored in the electric field between its plates has been transferred to the magnetic field around L, Fig. 15.31*b*.

At this instant the magnetic field begins to collapse and a p.d. is induced in L which tries to maintain the

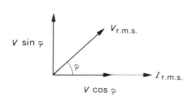

Fig. 15.30

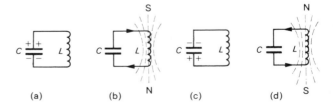

Fig. 15.31

field. Current therefore flows in the same direction as before and charges C so that the lower plate is positive. By the time the magnetic field has collapsed, the energy is again stored in C, Fig. 15.31c. Once more C starts to discharge but current now flows in the opposite direction, creating a magnetic field of opposite polarity, Fig. 15.31d. When this field has decayed, C is again charged with its upper plate positive and the same cycle is repeated.

In the absence of resistance in any part of the circuit, an undamped, sinusoidal a.c. would be obtained. In practice, energy is gradually dissipated by the resistance as heat and a damped oscillation is produced, Fig. 15.32.

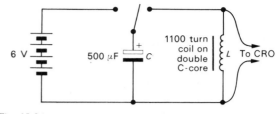

Fig. 15.34

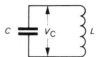

Fig. 15.35

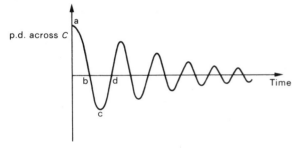

Fig. 15.32

(b) *Demonstrations.* Very slow damped electrical oscillations may be shown using the circuit of Fig. 15.33. Oscillations are started by charging C from a 20 V d.c. supply. Several cycles of 'slow' a.c. are indicated on the moving-coil milliammeter. If either L or C are reduced, the frequency of oscillation increases.

Slightly higher frequency oscillations (about 2 Hz) are obtained using the arrangement of Fig. 15.34 and a CRO on its slowest time base speed.

(c) *Frequency of the oscillations.* In Fig. 15.35 the capacitor has capacitance C and is in series with an inductor of inductance L and negligible resistance. If Q

is the charge on the capacitor at time t and I is the current flowing, then

$$\text{p.d. across the inductor} = V_L = -L\frac{dI}{dt}$$

$$\text{p.d. across capacitor} = V_C = \frac{Q}{C}$$

The net p.d. in the circuit is zero (since $R = 0$ and so there is no IR term), therefore $V_C = V_L$, that is,

$$\frac{Q}{C} = -L\frac{dI}{dt}$$

Also, current is rate of flow of charge, i.e. $I = dQ/dt$ and so

$$\frac{dI}{dt} = \frac{d}{dt}\left(\frac{dQ}{dt}\right) = \frac{d^2Q}{dt^2}$$

$$\therefore \quad \frac{Q}{C} = -L\frac{d^2Q}{dt^2}$$

$$\therefore \quad \frac{d^2Q}{dt^2} = -\frac{1}{LC}\cdot Q$$

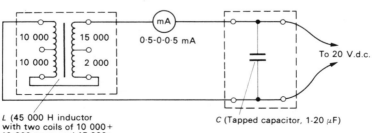

L (45 000 H inductor with two coils of 10 000 + 10 000 turns and 15 000 + 2000 turns on grain-orientated silicon steel core)

C (Tapped capacitor, 1-20 μF)

Fig. 15.33

This equation is of the same form as that which represents a s.h.m. (i.e. $d^2x/dt^2 = -\omega^2 x$) and indicates that the charge Q on the capacitor varies sinusoidally with time, having a period T given by $T = 2\pi/\omega = 2\pi\sqrt{LC}$. If f is the frequency of the oscillations, $f = 1/T$ and

$$f = \frac{1}{2\pi\sqrt{LC}}$$

As the resistance of an LC circuit increases, the oscillations decay more quickly, and when it is too large the capacitor discharge is unidirectional and no oscillations occur. To obtain undamped oscillations, energy has to be fed into the LC circuit in phase with its natural oscillations to compensate for the energy dissipated in the resistance of the circuit. This can be done with the help of a transistor in actual oscillators (p. 506).

Rectification of a.c.

Rectification is the conversion of a.c. to d.c. by a rectifier.

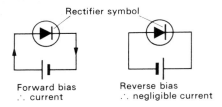

Fig. 15.36

(*a*) *Rectifiers.* Rectifiers have a low resistance to current flow in one direction, known as the *forward* direction, and a high resistance in the opposite or *reverse* direction. They are conductors which are largely unidirectional. When connection is made to a supply so that a rectifier conducts it is said to be *forward biased*; in the non-conducting state it is *reverse biased*, Fig. 15.36. The arrowhead on the symbol for a rectifier indicates the forward direction of conventional current flow. Three types of rectifier are

(*i*) the *thermionic diode valve*
(*ii*) the *selenium rectifier,* which is now being replaced by
(*iii*) the *semiconductor diode*, developed as a result of research aimed at producing rectifiers which could supply larger currents, withstand higher p.d.s and be made smaller than metal rectifiers, but yet be reliable and robust. The germanium diode rectifier was introduced in the early 1950s and shortly afterwards the

silicon rectifier. The latter has superseded germanium for power rectification: it can operate at a current density of 100 A cm^{-2} of rectifier surface (compared with 100 mA cm^{-2} for metal rectifiers), can withstand reverse p.d.s of several hundred volts and has a working temperature of about 150 °C. Fig. 15.37 is a full-size illustration of a typical silicon diode rectifier. The construction and action of semiconductor diodes will be considered later.

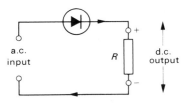

Fig. 15.37

(*b*) *Half-wave rectification.* The rectifying circuit of Fig. 15.38 consists of a rectifier in series with the a.c. input to be rectified and the 'load' requiring the d.c. output. For simplicity the 'load' is represented by a resistor R but it might be some piece of electronic equipment.

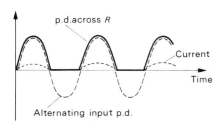

Fig. 15.38

In Fig. 15.39 the alternating input p.d. applied to the rectifier and load is shown. If the first half-cycle acts in the forward direction of the rectifier, a pulse of current flows round the circuit, creating a p.d. across R which will have almost the same value as the applied p.d. if the forward resistance of the rectifier is small compared with R. The second half-cycle reverse biases the rectifier, little or no current flows and the p.d. across R is zero. This is repeated for each cycle of a.c. input. The current pulses are unidirectional and so the p.d. across R is direct, for although it fluctuates it never changes direction.

Fig. 15.39

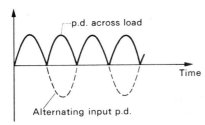

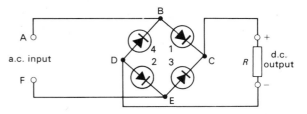

Fig. 15.40

Fig. 15.42

(c) *Full-wave rectification*. In this process both halves of every cycle of input p.d. produce current pulses and the p.d. developed across a load is like that shown in Fig. 15.40. There are two types of circuits.

(i) *Centre-tap full-wave rectifier*. Two rectifiers and a transformer with a centre-tapped secondary are used, Fig. 15.41. The centre tap O has a potential half-way between that of A and F and it is convenient to take it as a reference point having zero potential. If the first half-cycle of input makes A positive, rectifier B conducts, giving a current pulse in the circuit ABC, *R*, OA. During this half-cycle the other rectifier E is non-conducting since the p.d. across FO reverse biases it. On the other half of the same cycle F becomes positive with respect to O and A negative. Rectifier E conducts to give current in the circuit FEC, *R*, OF; rectifier B is now reverse biased.

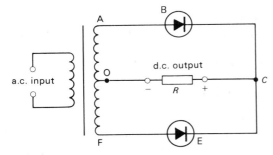

Fig. 15.41

In effect the circuit consists of two half-wave rectifiers working into the same load on alternate half-cycles of the applied p.d. The current through *R* is in the same direction during *both* half-cycles and a fluctuating direct p.d. is created across *R* like that in Fig. 15.40.

(ii) *Bridge full-wave rectifier*. Four rectifiers are arranged in a bridge network as in Fig. 15.42. If A is positive during the first half-cycle, rectifiers 1 and 2 conduct and current takes the path ABC, *R*, DEF. On the next half-cycle when F is positive, rectifiers 3 and 4 are forward biased and current follows the path FEC,

R, DBA. Once again current flow through *R* is unidirectional during both half-cycles of input p.d. and a d.c. output is obtained.

Smoothing circuits

To produce steady d.c. from the varying but unidirectional output from a half- or full-wave rectifier, smoothing is necessary.

(a) *Reservoir capacitor*. The simplest smoothing circuit consists of a large capacitor, 16 μF or more, called a *reservoir* capacitor, placed in parallel with the load *R*. In Fig. 15.43*a*, C_1 is the reservoir capacitor and its action can be followed from Fig. 15.43*b* where V_1 represents the p.d. developed across C_1 and *I* is the rectifier (full-wave) current.

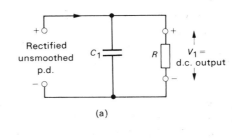

(a)

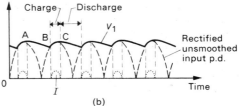

(b)

Fig. 15.43

Initially the rectifier input p.d. causes current to flow through *R* and at the same time C_1 becomes charged almost to the peak value of the input as shown by OA. At A, the input p.d. falls below V_1 and C_1 starts to discharge. It cannot do so through the rectifier since the polarity is wrong, but it does through the load and thus maintains current flow by its charge storing or reservoir

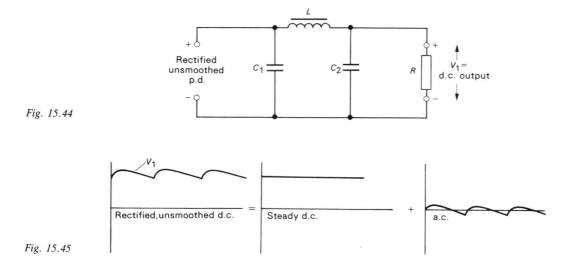

Fig. 15.44

Fig. 15.45

action. Along AB, V_1 falls. At B when the input p.d. equals the value to which V_1 has fallen, rectifier current I again flows to quickly recharge C_1 to the peak p.d., as shown by BC. The cycle of operations is then repeated. The d.c. output developed across R is V_1 and although it fluctuates at twice the frequency of the supply, the amplitude of the fluctuations is much less than when C_1 is absent.

The smoothing action of C_1 arises from its large capacitance making the time constant C_1R large so that the p.d. across it cannot follow the variations of input p.d. A very large value of C_1 would give better smoothing but initially the uncharged reservoir capacitor would act almost as a short-circuit and the resulting surge of current might damage the rectifier.

(*b*) *Capacitor-input filter.* A reservoir capacitor has a useful smoothing effect but it is usually supplemented by a filter circuit consisting of a choke L (i.e. an iron-cored inductor) having an inductance of about 15 H and a large capacitor C_2 arranged as in Fig. 15.44. The reservoir capacitor C_1 and the filter capacitor C_2 may be electrolytics enclosed in the same can.

C_1 behaves as explained previously and the p.d. across it is similar to V_1 in Fig. 15.43*b*. The action of the filter circuit $L - C_2$ can be understood if V_1 is resolved into a steady direct p.d. (the d.c. component) and an alternating p.d. (the a.c. component). This procedure is often used when dealing with a varying d.c. and is illustrated in Fig. 15.45.

By redrawing the smoothing circuit as in Fig. 15.46 we see that V_1 is applied across L and C_2 in series. L offers a much greater impedance to the a.c. component than

C_2 and most of the unwanted ripple p.d. appears across L. For the d.c. component, C_2 has infinite resistance and the whole of this component is developed across C_2 except for the small drop due to the resistance of the choke. The filter thus acts as a potential divider, separating d.c. from a.c. and giving a steady d.c. output across C_2.

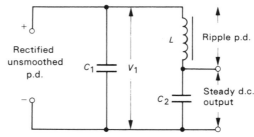

Fig. 15.46

(*c*) *Laboratory h.t. power supply unit* (a power pack). A typical circuit for a laboratory power pack is shown in Fig. 15.47; it employs a silicon diode or selenium bridge rectifier fed by a step-up mains transformer and has a capacitor-input smoothing filter. A steady d.c. output of 0–400 V at 100 mA is produced as well as a 6.3 V a.c. heater supply for thermionic devices such as valves and cathode-ray tubes. The 25 kΩ variable resistor acts as a potential divider giving a continuously variable output and also allowing the capacitors to discharge when the pack is switched off—this eliminates risk of shock should the output terminals be touched subsequently.

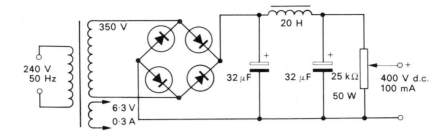

Fig. 15.47

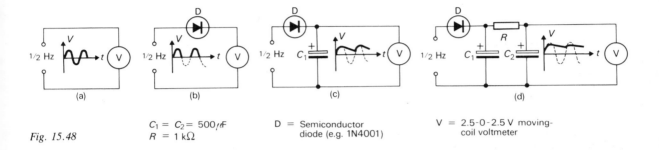

Fig. 15.48

$C_1 = C_2 = 500 \mu F$
$R = 1 k\Omega$

D = Semiconductor
diode (e.g. 1N4001)

V = 2.5-0-2.5 V moving-
coil voltmeter

It should be noted that r.m.s. p.d.s. are usually quoted for the secondary of a transformer but, depending on the current taken by the load, rectification and smoothing can result in a d.c. output which approaches the peak value, i.e. 1.4 times the r.m.s. value. In the circuit shown the nominal 350 V secondary gives a d.c. output of about 500 V at no load current and 400 V when 100 mA is supplied.

(d) *Demonstrations of rectification and smoothing using slow a.c.* The circuits are shown in Fig. 15.48. An a.c. of frequency about $\frac{1}{2}$ Hz is obtained by applying 6 V d.c. to the input of a low frequency generator which is rotated by hand. The action of the various stages of a rectifying (half-wave) and smoothing circuit can be followed from the voltmeter reading using each arrangement in turn. The voltmeter acts as the 'load'. A resistor is used instead of a choke but in practice the latter is preferred in power units where currents of more than a few milliamperes have to be supplied.

An alternative demonstration using a CRO is given in Appendix 11.

QUESTIONS

1. When a certain a.c. supply is connected to a lamp it lights with the same brightness as it does with a 12 V battery.

(a) What is the r.m.s. value of the a.c. supply?
(b) What is the peak p.d. of the a.c. supply?
(c) The 12 V battery is connected to the Y-plates of a CRO and the gain adjusted so that it deflects the spot by 1.0 cm. What will be the total length of the trace on the CRO screen when the a.c. supply replaces the battery?

2. Electrical energy is supplied to a distant consumer through a power line which has a total resistance of 5 Ω. Explain why an input potential difference of 100 kV r.m.s. would be more satisfactory than one of 10 kV r.m.s., and calculate in each case (a) the output voltage, and (b) the output power when the power input is 20 MW.

If the figure of 100 kV refers to the r.m.s. value of a sinusoidal a.c. what will be the maximum voltage for which the line must be insulated? (*O. and C. part qn.*)

3. A sinusoidal p.d. of r.m.s. value 10 V is applied across a 50 μF capacitor.
(a) What is the peak charge on the capacitor?
(b) Draw a graph of charge Q on the capacitor against time.
(c) When is the current flowing into the capacitor (*i*) a maximum, and (*ii*) a minimum?
(d) Draw a graph of current against time.
(e) If the a.c. supply has frequency of 50 Hz, calculate the r.m.s. current through the capacitor.

4. In Fig. 15.49 V_1 and V_2 are identical high resistance a.c. voltmeters. Explain why the sum of the r.m.s. p.d.s measured across XY and YZ by V_1 and V_2 exceeds the applied r.m.s. p.d. of 15.0 V. Calculate the value of C.

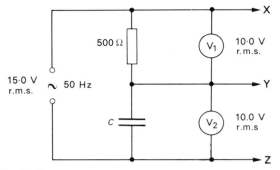

Fig. 15.49

5. As soon as the switch is closed in Fig. 15.50 what is the initial value of

(a) the p.d. across the capacitor
(b) the current through the resistor
(c) the rate at which the p.d. across the capacitor rises?

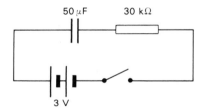

Fig. 15.50

6. What is meant by the *root-mean-square value* of a sinusoidal alternating current? Why is this a useful measure of alternating current?

A 3.0 Ω resistor is joined in series with a 10 mH inductor of negligible resistance, and a potential difference V (= 5.0 V r.m.s.) alternating at $200/\pi$ Hz is applied across the combination.

(a) Calculate the p.d. V_R across the resistor and V_L across the inductor.

(b) Showing clearly your procedure draw a vector diagram representing the relation between V, V_R and V_L.

(c) Determine the phase difference between V and V_L.

(d) How would you use a cathode-ray oscilloscope to show that there is a phase difference between V and V_L?   (*J.M.B.*)

7. What is the purpose of a *choke* in a circuit? Describe an experiment to demonstrate the action of a choke, explaining what happens.

When an impedance, consisting of an inductance L and a resistance R in series, is connected across a 12 V, 50 Hz supply a current of 0.050 A flows which differs in phase from that of the applied potential difference by 60°. Find the values of R and L.

Find the capacitance of the capacitor which, connected in series in the above circuit, has the effect of bringing the current into phase with the applied potential difference.   (*L.*)

8. A sinusoidal alternating current is represented by $I = I_0 \cos \omega t$, where I_0 is its peak value and ω its angular frequency in rad s^{-1}.

Derive an expression for the voltage necessary to send such a current through (a) a pure resistance R, (b) a pure inductance L, and (c) a pure capacitance C. Sketch the voltage waveforms in the three cases on the same axis as the current waveforms, and explain the terms *inductive* and *capacitive reactance, lag,* and *lead.*

What is the expression for the supply voltage if this current is to flow through a circuit consisting of R, L, and C, in series? What is meant by the impedance of the circuit?

If $R = 500$ Ω, $L = 0.50$ H, and $C = 1.0$ μF, draw the vector impedance diagram, for (i) $\omega = 1000$ rad s^{-1}, and (ii) $\omega = 2000$ rad s^{-1}. For what frequency is the impedance of the circuit smallest? If $I_0 = 10$ A, what would then be the peak voltage across (a) L, (b) C, and (c) the L-C combination? (*W.*)

9. Explain what is meant by *resonance* in an alternating current circuit containing inductance, resistance and capacitance in series. Give *one* practical application of this effect.

A variable capacitor is connected in series with a coil and a sinusoidal alternating supply of 20 V (r.m.s.) at a frequency of 50 Hz. When the capacitor has a value of 1.0 μF, the current in the circuit reaches a maximum value of 0.50 A (r.m.s.). Find (a) the resistance of the circuit, (b) the self-inductance of the coil, and (c) the potential difference across the capacitor.

(*L.*)

10. Define the *impedance* of an a.c. circuit.

A 2.5 μF capacitor is connected in series with a non-inductive resistor of 300 Ω across a source of p.d. of r.m.s. value 50 V alternating at $1000/2\pi$ Hz. Calculate (a) the r.m.s. values of the current in the circuit and the p.d. across the capacitor, and (b) the mean rate at which energy is supplied by the source.   (*J.M.B.*)

11. A simple alternator when rotating at 50 revolutions per second gives a 50 Hz alternating voltage of r.m.s. value 24 V. A 4.0 Ω resistance R and a 0.010 H inductance L are connected in series across its terminals.

(i) Assuming that the internal impedance of the generator can be neglected, find the r.m.s. current flowing, the power converted into heating, and the r.m.s. potential difference across each component.

(ii) Draw a vector diagram showing the relative phases of the applied voltage and the potential differences across R and L.   (*O. part qn.*)

12. An alarm bell is driven off a 20:1 step-down mains transformer from a primary voltage of 240 V at 50 Hz. If the

contact points fail to open, the coil presents an inductance of 20 mH and resistance of 10 Ω. For this fault condition, determine the current flowing through and the power dissipated in the coil. *(J.M.B. Eng. Sc.)*

13. Describe what will happen in the first few seconds after the switch is moved from position 1 to position 2 in the apparatus represented in Fig. 15.51. *(S.)*

14. Draw a circuit and explain the action of components to provide full-wave rectification of an alternating supply. Explain the action of a capacitor which could be used to smooth the output. *(A.E.B. part qn.)*

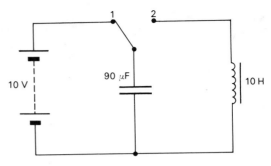

Fig. 15.51

Objective-type revision questions for Part 3

The first figure of a question number gives the relevant chapter, e.g. **13.2** is the second question for chapter 13.

Multiple choice

Select the response which you think is correct.

11.1. A sphere carrying a charge of Q coulombs and having a weight of W newtons falls under gravity between a pair of vertical plates distance d metres apart. When a potential difference of V volts is applied between the plates the path of the sphere changes as shown in Fig. 14, becoming linear along CD. The value of Q is

A $\dfrac{W}{V}$ **B** $\dfrac{W}{2V}$ **C** $\dfrac{Wd}{V}$ **D** $\dfrac{2Wd}{V}$ **E** $\dfrac{Wd}{2V}$ *(J.M.B.)*

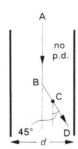

Fig. 14

12.1. A parallel-plate capacitor is to be made by inserting one of the sheets of dielectric material included below, between and in contact with two plates of copper.

	Relative permittivity	*Thickness (mm)*
Teflon	2	0.4
Quartz	3	0.8
Glass	4	1.0
Mica	5	1.2
Porcelain	6	1.3

The maximum capacitance will be obtained by using the sheet of

A Teflon   **B** Quartz   **C** Glass   **D** Mica   **E** Porcelain

(J.M.B.)

12.2. Fig. 15 represents a parallel-plate capacitor whose plate separation of 10 mm is very small compared with the size of the plates. If a potential difference of 5.0 kV is maintained across the plates, the intensity of the electric field in V m^{-1} at (*a*) X and (*b*) Y, is

A 50   **B** 2.5×10^2 **C** 5.0×10^2 **D** 2.5×10^5
E 5.0×10^5

Fig. 15

13.1. Two long solenoids, P and Q, have n_1 and n_2 turns per unit length respectively and carry currents I_1 and I_2 respectively. The magnetic flux density on the axis of P at a point near the middle is four times that at a corresponding point in Q. The value of I_1/I_2 is

A $\dfrac{2n_1}{n_2}$ **B** $\dfrac{2n_2}{n_1}$ **C** $\dfrac{4n_1}{n_2}$ **D** $\dfrac{4n_2}{n_1}$ **E** $\dfrac{16n_1}{n_2}$ *(J.M.B.)*

13.2. The wire X on Fig. 16 at right angles to the plane of the paper and carries a current into the paper. The magnetic flux density due to this current will be in the same direction as that of the horizontal component of the earth's field at a point on the diagram

A labelled 1   **B** labelled 2   **C** labelled 3   **D** labelled 4
E whose position is not deducible from the data without a knowledge of the value of the current.   *(J.M.B.)*

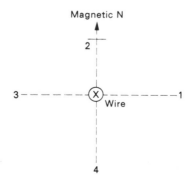

Fig. 16

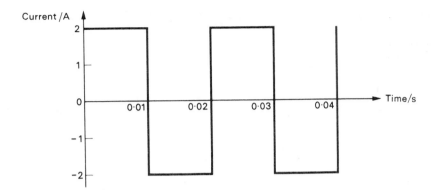

Fig. 17

14.1. The e.m.f. induced in a coil of wire, which is rotating in a magnetic field, does not depend on

A the angular speed of rotation   B the area of the coil
C the number of turns on the coil   D the resistance of
E the magnetic flux density.   the coil

14.2. When the speed of an electric motor is increased due to a decreasing load, the current flowing through it decreases. Which of the following is the best explanation of this?

A The resistance of the coil changes.
B Frictional forces increase as the speed increases.
C Frictional forces decrease as the speed increases.
D The induced back e.m.f. increases.
E At high speeds it is more difficult to feed current into the motor.

14.3. When an ammeter is well damped,

A large currents can be measured.
B small currents can be measured.
C readings can be taken quickly.
D it is very accurate.
E it is very robust.

14.4. Which feature of a moving-coil ammeter is of most importance in making it well damped?

A The strength of the hairsprings.
B The number of turns on the coil.
C The former on which the coil is wound.
D The size of the iron cylinder.
E The material used for the coil.

14.5. Alternating current is preferable to direct current for the transmission of power because

A it can be rectified.
B it is easier to generate.
C thinner conductors can be used.
D no question of polarity arises with equipment.
E it is safer.

15.1. The direct current which would give the same heating effect in an equal constant resistance as the current shown in Fig. 17, i.e. the r.m.s. current, is

A Zero   B $\sqrt{2}$ A   C 2 A   D $2\sqrt{2}$ A   E 4 A
(*J.M.B.*)

15.2. What value of L, in henries, will make the circuit in Fig. 18 resonate at 100 Hz?

A $\dfrac{1}{4\pi^2}$ B $\dfrac{1}{2\pi}$ C 1   D $4\pi^2$ (*J.M.B. Eng. Sc.*)

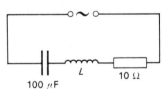

Fig. 18

15.3. The diagram in Fig. 19 shows a half-wave rectifier with reservoir capacitor and load resistance.
The time constant of C and R should be

A small compared with the time of one cycle.
B independent of the time of one cycle.
C large compared with the time of one cycle.
D the same as the time of one cycle.   (*J.M.B. Eng. Sc.*)

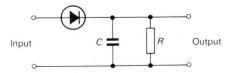

Fig. 19

Multiple selection

In each question one or more of the responses may be correct. Choose one letter from the answer code given.

*Answer **A** if (i), (ii) and (iii) are correct*
*Answer **B** if only (i) and (ii) are correct*
*Answer **C** if only (ii) and (iii) are correct*
*Answer **D** if (i) only is correct*
*Answer **E** if (iii) only is correct*

11.2. When a positive charge is given to an isolated hollow conducting sphere

   (*i*) a positively charged object anywhere inside the sphere experiences an electric force acting towards the centre of the sphere.
  (*ii*) the potential to which it is raised is proportional to its radius.
 (*iii*) the potential gradient outside the sphere is independent of its radius.

11.3. The intensity of an electric field

   (*i*) is a vector quantity.
  (*ii*) can be measured in N C^{-1}.
 (*iii*) can be measured in V m^{-1}.

12.3. A parallel-plate capacitor C is charged by connection to a battery using the switch S as shown in Fig. 20. S is opened and the plate separation is then increased. Increases occur in the following

   (*i*) the charge stored
  (*ii*) the potential difference between the plates
 (*iii*) the energy stored.

13.3. The magnetic flux density well inside a long uniformly wound solenoid depends on

   (*i*) the number of turns per unit length
  (*ii*) the area of cross-section
 (*iii*) the distance from the axis of the solenoid.

14.6. The circuit of Fig. 21 shows a cell of negligible internal resistance in series with a pure inductor and a pure resistor. When the switch is closed

   (*i*) the current rises initially at the rate of 6.0 A s^{-1}.
  (*ii*) the final value of the current is 1.5 A.
 (*iii*) the final energy stored in the space in and around the coil is 2.3 (2.25) J.

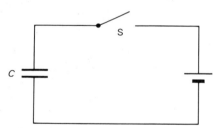

Fig. 20

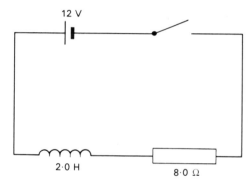

Fig. 21

Part 4 WAVES

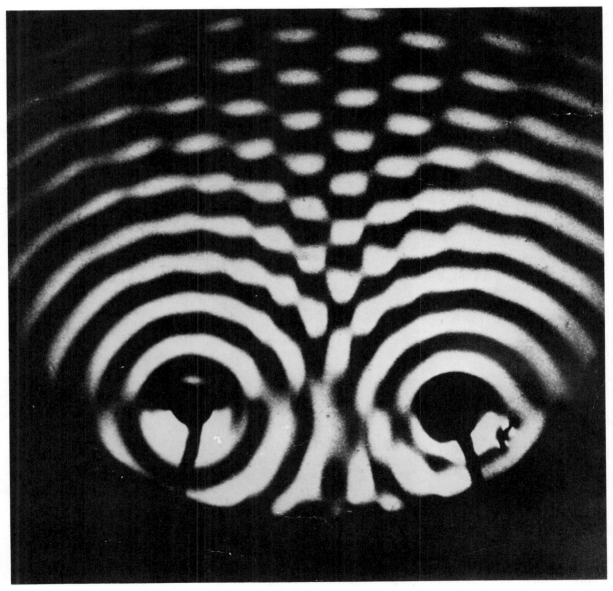

16 Wave motion

Progressive waves

The idea of a wave is useful for dealing with a wide range of phenomena and is one of the basic concepts of physics. A knowledge of wave behaviour is also important to engineers.

A *progressive* or travelling wave consists of a disturbance moving from a source to surrounding places as a result of which energy is transferred from one point to another.

There are two types of progressive wave. In the *transverse* type the direction associated with the disturbance is at right angles to the direction of travel of the wave, Fig. 16.1*a*. In the *longitudinal* type the disturbance is in the same direction as that of the wave, Fig. 16.1*b*. Transverse and longitudinal pulses (i.e. waves of short duration) can be sent along a Slinky spring, Fig. 16.2*a* and *b*. In both cases the disturbance generated by the hand is passed on from one coil of the spring to the next which performs the same motion but at a slightly later time. The pulse travels along the spring, the coils propagating the disturbance merely by vibrating to and fro (transversely or longitudinally) about their undisturbed positions. A succession of disturbances creates a continuous train of waves, i.e. a wave-train. The wave machine of Fig. 16.3*a* and *b* is useful for showing transverse and longitudinal waves.

Waves may be classified as *mechanical* or *electromagnetic*. Mechanical waves are produced by a disturbance (e.g. a vibrating body) in a material medium and are transmitted by the particles of the medium oscillating to and fro. Such waves can be seen or felt and include waves on a spring, water waves, waves on stretched strings (e.g. in musical instruments), sound waves in air and in other materials. Many of the properties of mechanical waves can be shown using water waves in a ripple tank. Such waves are less complex than sea waves and are transmitted by the surface layer, often being called *surface water waves*. However, the displacement of the water is not a simple up-and-down, transverse motion (although the effects we shall observe will be due to this motion); the water particles move in elliptical or circular paths in the direction of the wave at a crest and in the opposite direction in a trough.

Electromagnetic waves, as we shall see later (p. 410), consist of a disturbance in the form of varying electric and magnetic fields. No medium is necessary and they travel more easily in a vacuum than in matter. Radio signals, light and X-rays are examples of this type. In this chapter microwaves will be used to investigate those properties of electromagnetic radiation which are explained in terms of waves. Microwaves are radar-

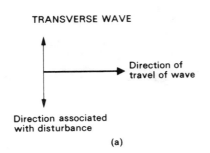

TRANSVERSE WAVE

Direction of travel of wave

Direction associated with disturbance

Fig. 16.1

(a)

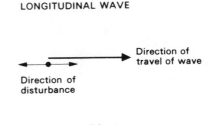

LONGITUDINAL WAVE

Direction of travel of wave

Direction of disturbance

(b)

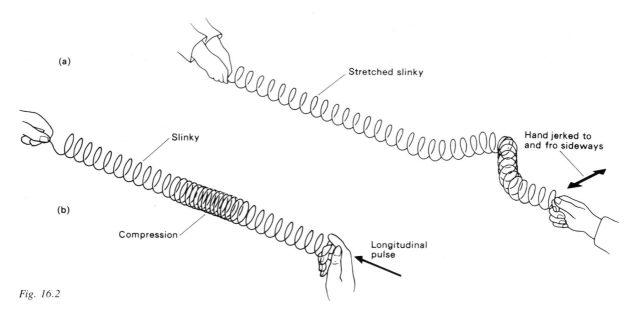

(a)

Stretched slinky

Slinky

(b)

Compression

Hand jerked to
and fro sideways

Longitudinal
pulse

Fig. 16.2

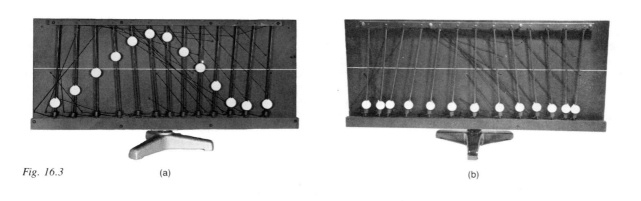

Fig. 16.3   (a)   (b)

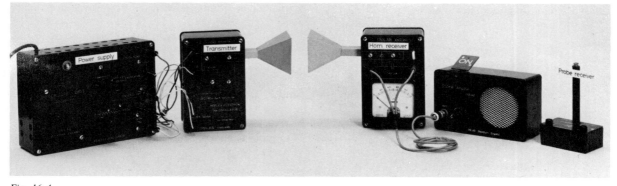

Fig. 16.4

type waves and are generated by an oscillator of very high frequency a.c. (10 000 MHz). In the transmitter of Fig. 16.4 the oscillator feeds a small aerial in a rectangular metal tube called a *wave guide* which opens into a horn at one end and is closed at the other. The radiation emitted has a wavelength of about 3 cm. The horn receiver has a wave guide and horn like those of the transmitter and contains a silicon diode mounted to 'detect' (see p. 527) the signal. The probe receiver is less sensitive and non-directional, and simply consists of a diode. The signal from the transmitter can be modulated so that when picked up by either receiver, a note is heard in an amplifier-loudspeaker unit. Otherwise it can produce current in a microammeter connected to the receiver.

Mechanical and electromagnetic waves give effects which are explicable by the same general principles as we shall see presently.

Describing waves

(*a*) *Graphical representation.* Two kinds of graph may be drawn. A *displacement-distance* graph for a transverse mechanical wave shows the displacements y of the vibrating particles of the transmitting medium at different distances x from the source *at a certain instant*. The dots in Fig. 16.5a show the positions of the particles at a particular time and the corresponding displacement–distance graph is given in Fig. 16.5b.

A longitudinal wave can also be represented by a transverse displacement–distance graph, the displacements in this case, however, being those of the vibrating particles in the line of travel of the wave. Thus y and x are in the same direction in the wave but at right angles to each other in the graph, Fig. 16.5c, i.e. the graph represents a longitudinal displacement as if it were transverse. Regions of high particle density are called *compressions* and regions of low particle density

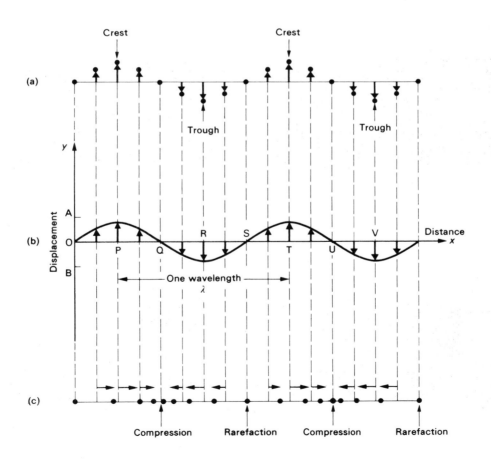

Fig. 16.5

are called *rarefactions*. Both move in the direction of travel of the wave whilst the particles of the medium vibrate to and fro about their undisturbed positions.

The maximum displacement of each particle from its undisturbed position is the *amplitude* of the wave; in Fig. 16.5b this is OA or OB. The *wavelength* λ of a wave is the distance between two consecutive points on it which are in step, i.e. have the same phase. Thus for a transverse wave it is the distance between two successive crests or two successive troughs; for a longitudinal wave it is the separation of consecutive compressions or consecutive rarefactions. In Fig. 16.5b, $\lambda = \text{OS} = \text{PT} = \text{RV}$.

Electromagnetic waves behave as transverse waves (p. 407) and may be represented by the same kind of displacement–distance graph, y being the instantaneous value of the electric field strength E or the magnetic flux density B at different distances from the source.

A *displacement–time* graph may also be drawn for a wave motion showing how the displacement of one particle (or the value of E or B) at a particular distance from the source varies with time. If this is a simple harmonic variation the graph is a sine curve.

(b) *Wavelength, frequency, speed.* If the source of a wave makes f vibrations per second, so too will the particles of the transmitting medium in the case of a mechanical wave and as will E or B in an electromagnetic wave. That is, the frequency of the wave equals the frequency of the source.

When the source makes one complete vibration, one wave is generated and the disturbance spreads out a distance λ from the source. If the source continues to vibrate with constant frequency f, then f waves will be produced per second and the wave advances a distance $f\lambda$ in one second. If v is the wave speed then

$$v = f\lambda$$

This relationship holds for all wave motions.

(c) *Wavefronts and rays.* A wavefront is a line or surface on which the disturbance has the same phase at all points. If the source is periodic it produces a succession of wavefronts, all of the same shape. A point source S generates circular wavefronts in two dimensions, Fig. 16.6a and spherical wavefronts in three dimensions. A line source S_1S_2 (e.g. the straight vibrator in a ripple tank) creates wavefronts that are straight in two dimensions, Fig. 16.6b, and cylindrical in three dimensions. Plane wavefronts are produced by a plane source or by *any* source at a distant point.

A line at right angles to a wavefront which shows its direction of travel is called a *ray*.

Huygens' construction

The construction proposed by Huygens (a contemporary of Newton) enables the new position of a wavefront to be found, knowing its position at some previous instant. It can be used to explain the reflection, refraction and dispersion of waves as we shall see presently.

According to Huygens, *every point on a wavefront may be regarded as a source of secondary spherical wavelets which spread out with the wave velocity. The new wavefront is the envelope of these secondary wavelets*, i.e. the surface which touches all the wavelets.

As a simple example, suppose AB is a straight

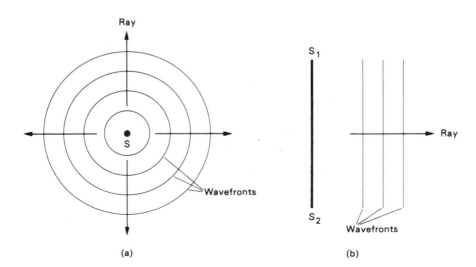

Fig. 16.6   (a)   (b)

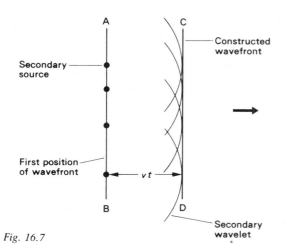

Secondary source

First position of wavefront

vt

Constructed wavefront

Secondary wavelet

Fig. 16.7

wavefront travelling from left to right, Fig. 16.7. The wavefront at time t later is the common tangent CD of the spherical wavelets of radius vt, drawn with centres every point on AB, where v is the speed of the waves in the medium.

The construction suggests that waves should also travel backwards to the left. This does not occur in practice and the assumption (which can be justified) has to be made that the amplitude of the secondary wavelets varies from a maximum in the direction of travel of the wave to zero in the opposite direction. This is indicated in Fig. 16.7 by drawing only arcs of the wavelets in the forward direction.

(a) Reflection. In Fig. 16.8*a* the wavefront AB is incident obliquely on the reflecting surface and A has just reached it. To find the new position of the wavefront when B is about to be reflected at B', we draw a

secondary wavelet with centre A and radius BB'. The tangent B'A' from B' to this wavelet is the required reflected wavefront.

In triangles AA'B' and ABB'

$$\text{angle AA'B'} = \text{angle ABB'} = 90°$$

$$\text{AB' is common}$$

$$\text{AA'} = \text{BB'} \text{ (by construction)}$$

Therefore the triangles are congruent (rt : h : s), hence

$$\text{angle BAB'} = \text{angle A'B'A}$$

These are the angles made by the incident and reflected wavefronts AB and A'B' respectively, with the reflecting surface. The incident and reflected rays, e.g. CA and B'D are at right angles to the wavefronts as are the normals to the surface (the dotted lines at A and B') and so the angle of incidence i equals the angle of reflection r. This is the law of reflection and Fig. 16.8*b* shows that it holds for the reflection of straight water waves in a ripple tank at a straight barrier. It may be tested for the reflection of microwaves from a metal plate with the apparatus of Fig. 16.9.

(b) Refraction. Fig. 16.10*a* shows the end A of a plane wavefront AB about to cross the boundary between media ① and ② in which its speeds are v_1 and v_2 respectively. The new position of the wavefront at time t later when B has travelled a distance BB' $= v_1 t$ and reached B' is found by drawing a secondary wavelet with centre A and radius AA' $= v_2 t$. The tangent B'A' from B' to this wavelet is the required wavefront and the ray AA', which is normal to the wavefront, is a refracted ray. If v_2 is less than v_1 the refracted ray is bent towards the normal in medium ②, as shown.

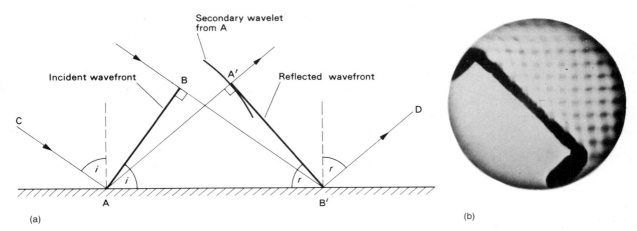

Secondary wavelet from A

Incident wavefront

B

A'

Reflected wavefront

C

D

i

i

r

r

A

B'

(a)

(b)

Fig. 16.8

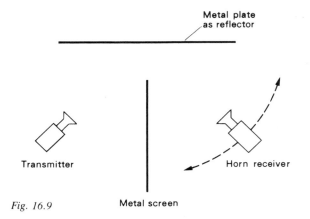

Metal plate as reflector

Transmitter

Horn receiver

Metal screen

Fig. 16.9

When v_2 is greater than v_1 refraction away from the normal occurs.

In triangles BAB′ and A′AB′

$$\frac{\sin i_1}{\sin i_2} = \frac{\sin \text{CAN}}{\sin \text{A}'\text{AN}'} = \frac{\sin \text{BAB}'}{\sin \text{A}'\text{B}'\text{A}}$$

$$= \frac{\text{BB}'/\text{AB}'}{\text{AA}'/\text{AB}'} = \frac{\text{BB}'}{\text{AA}'} = \frac{v_1 t}{v_2 t} = \frac{v_1}{v_2}$$

But v_1 and v_2 are constants for given media and a particular wavelength

$$\therefore \quad \frac{\sin i_1}{\sin i_2} = \text{a constant}$$

This is Snell's law of refraction and it holds for electromagnetic, sound and water waves. The constant is called the *refractive index*, $_1n_2$ for waves passing from medium ① to medium ②. Hence we can write

$$_1n_2 = \frac{v_1}{v_2} = \frac{\sin i_1}{\sin i_2} = \frac{n_2}{n_1}$$

where n_1 and n_2 are the *absolute* refractive indices of media ① and ②.

Huygens' construction thus explains the refraction of, for example, light, on the assumption that the speed of light decreases when it passes into a medium in which the refracted ray is bent towards the normal, i.e. when it enters an optically denser medium. This assumption is verified by experiment.

Refraction occurs when a wave passes from one medium into another in which it has a different speed. The speed of surface waves depends mostly on the depth in shallow liquids, i.e. the depth is not large compared with the wavelength (in deep liquids it depends on wavelength, not depth). Water ripples travel more slowly in shallow than in deeper water and are refracted on passing from one to the other. In Fig. 16.10*b* a glass plate on the bottom lower half of a ripple tank reduces the depth. Note the shorter wavelength (and hence smaller speed since the frequency is un-

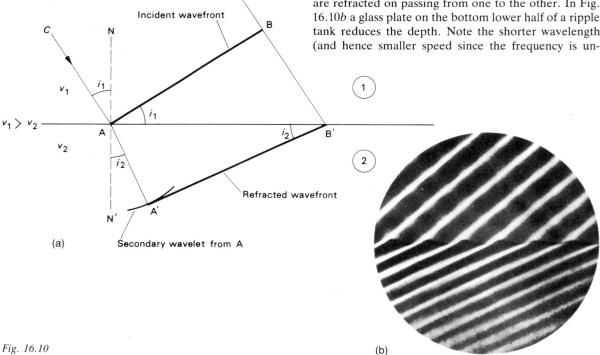

Incident wavefront

C   N   B

v_1 i_1

     i_1

$v_1 > v_2$ ─────────────────────────── A i_2 B′

v_2

i_2

     N′   A′

Refracted wavefront

① ②

(a)

Secondary wavelet from A

Fig. 16.10   (b)

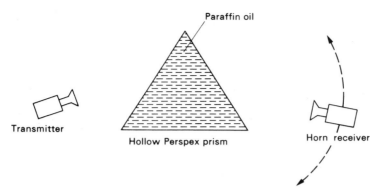

Fig. 16.11

changed—how would you check that it is so?) as well as the change of direction of the refracted wavefront in the shallower water. The refraction of microwaves may be demonstrated as in Fig. 16.11.

(*c*) *Dispersion*. Huygens' construction predicts that if the speed of waves in a given medium depends on the frequency (and therefore wavelength) of the waves, then when refraction occurs its extent will also depend on the frequency. In Fig. 16.12*a* the waves of frequencies f_1 and f_2 travel with the same speed in medium ① but in medium ②, v_1 (the speed of wave with frequency f_1) is greater than v_2 and so f_2 suffers greater deviation than f_1. Dispersion occurs; medium ① is called a *non-dispersive* medium and medium ② a *dispersive* medium. Thus, red light travels faster than blue light in glass and if white light travels from air to glass, the blue wavefront is refracted more than the red one, i.e. dispersion is observed, Fig. 16.12*b*.

Principle of superposition

What happens when waves meet? Is their motion changed as it is when solid objects collide? Answers to these questions may be obtained by producing pulses on a long narrow spring.[1]

The drawings in Fig. 16.13*a* show two transverse pulses (of slightly different shapes) on a spring approaching each other. Where they cross (diagram 3) the resultant displacement of the spring is evidently the sum of the displacements which each would have caused at that point, i.e. the pulses superpose. After crossing, each pulse travels along the spring as if nothing had happened and it has its original shape. The result is always the same whatever shapes the pulses have. Fig. 16.13*b* shows the superposition of two equal and opposite pulses. In this case, when the pulses meet (diagram 3) they cancel out and the net displacement of the spring is zero.

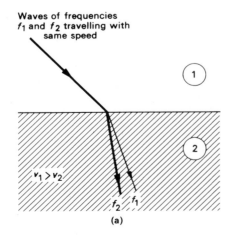

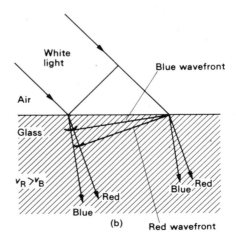

Fig. 16.12

[1] A PSSC-type spring is better than a Slinky for this.

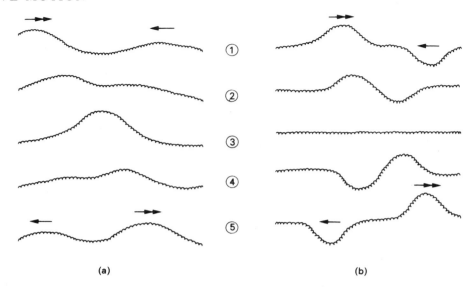

Fig. 16.13   (a)   (b)

In general we can conclude that pulses (and waves), unlike particles, pass through each other unaffected and *where they cross the total displacement is the vector sum of the individual displacements due to each pulse at that point.* This statement is called the *principle of superposition* and can be used to explain many wave effects. It is applied in Fig. 16.14 where the resultant, ①, of two waves, ② and ③, of different amplitudes and frequencies is obtained by adding the displacements; ③ has half the amplitude of ② and three times its frequency. The shape of the resultant is often quite different from those of its components, as here.

Interference

In a region where wave-trains from *coherent* sources cross, superposition occurs giving reinforcement of the waves at some points and cancellation at others. The resulting effect is called an *interference pattern* or a *system of fringes.*

Coherent sources have a constant phase difference which means they must have the same frequency and for complete cancellation to occur the amplitudes of the superposing waves they produce must be about equal. If their phase difference is not constant, at a certain

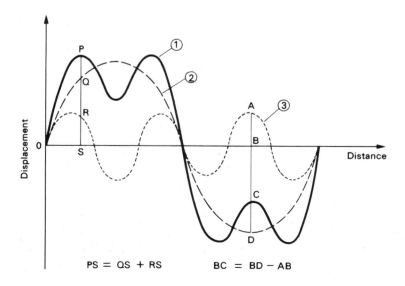

Fig. 16.14   PS = QS + RS   BC = BD − AB

point there may be reinforcement at one instant and cancellation at the next. If these variations follow one another rapidly, the effect at the point is to produce uniformity, i.e. the interference pattern changes so quickly with the continuously changing phase difference between the sources that the detector—for example, the eye—cannot follow the alterations and records an average effect.

In practice coherent sources are derived from a single source.

(*a*) *Mechanical waves*. Interference is readily shown by the transverse component of surface water waves in a ripple tank using two small spheres attached to the same vibrating bar, Fig. 16.15*a*. Circular waves are produced and give a pattern like that of Fig. 16.15*b*.

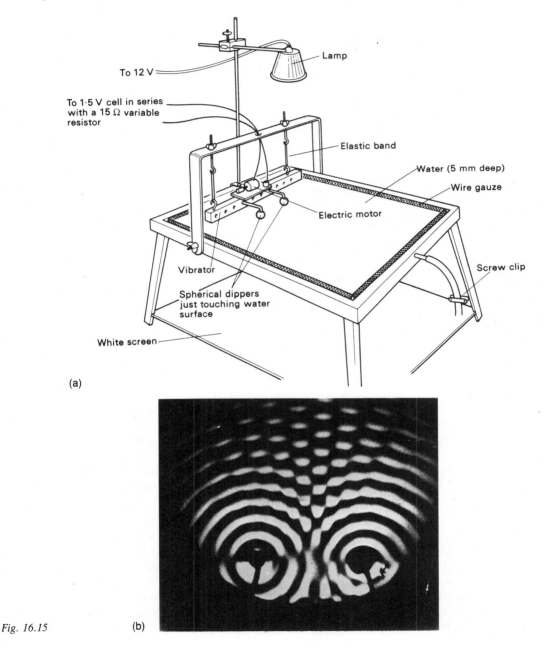

Fig. 16.15

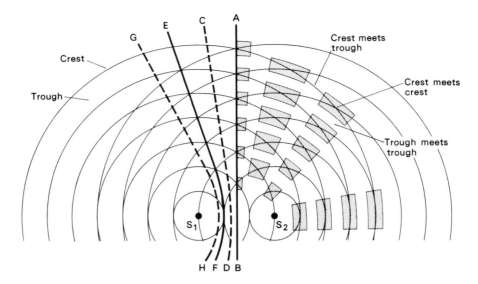

Fig. 16.16

To explain this interference pattern, consider Fig. 16.16. All points on AB are equidistant from the sources S_1 and S_2 and since the vibrations of these are in phase, crests (or troughs) from S_1 arrive at the same time as crests (or troughs) from S_2. Hence along AB reinforcement occurs by superposition and a wave of double amplitude is formed, Fig. 16.17a. Points on CD are half a wavelength (a wavelength being the distance between the centres of successive bright bands in Fig. 16.15b) nearer to S_1 than to S_2, i.e. there is a path difference of half a wavelength. Therefore crests (or troughs) from S_1 arrive simultaneously with troughs (or crests) from S_2 and the waves cancel, Fig. 16.17b.

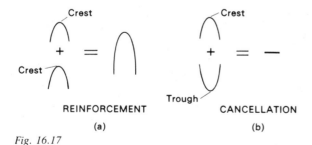

REINFORCEMENT   CANCELLATION

(a)   (b)

Fig. 16.17

Along EF the difference of the distances from S_1 and S_2 to any point is one wavelength. EF is a line of reinforcement, i.e. constructive interference occurs. GH is a line of cancellation where destructive interference occurs since at every point on GH the distance to

S_1 is one and a half wavelengths less than to S_2. Lines such as CD and GH are called *nodal* lines: AB and EF are *antinodal* lines, both sets are hyperbolic curves.

It can be shown that the separation of nodal (and antinodal) lines increases

   (i) as the distance from S_1 and S_2 increases

   (ii) the smaller the separation of S_1 and S_2

   (iii) as the wavelength increases (i.e. as the frequency decreases).

(b) *Electromagnetic waves.* Two arrangements for showing the interference of microwaves are given in Fig. 16.18a and b. In (a) interference occurs between the two wave-trains emerging from the 3 cm wide slits which act as two coherent sources. The receiver detects the maxima and minima of the fringe pattern as it is moved round; if it is a minimum and the waves from one slit are cut off by a metal plate, the signal *increases*—clear evidence of destructive interference.

In (b), the glass plate partially reflects and partially transmits the microwaves. The transmitted wave-train traverses the air 'film' to the metal plate where it is reflected back across the film and again transmitted (partially) by the glass. Two wave-trains (derived from the same source) thus reach the receiver and the thickness of the air 'film' decides the path difference between them (i.e. how much farther one has travelled than the other). If 'crests' of both arrive simultaneously at the receiver, there is reinforcement but if a 'crest' arrives with a 'trough' there is cancellation. Changing the path difference by moving the metal plate towards or away from the glass, produces maxima and minima

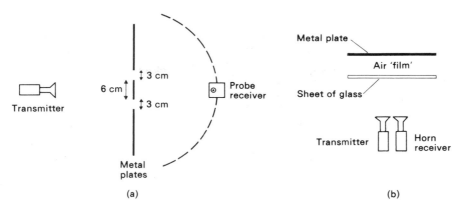

Fig. 16.18

(a)

(b)

at the receiver. What is the minimum change in the thickness of the air 'film' that will cause a maximum to be replaced by another maximum?

The Decca system of navigation is based on the interference of radio waves, the position of the aircraft or ship being plotted automatically. Fig. 16.19 shows a typical layout on a small boat with the track plotter on the extreme left.

Summing up, effects in which disturbances when added produce no disturbance can be regarded as being due to some kind of wave motion. Thus radio signals and light are treated as wave motions because radio signals sometimes 'fade' and light plus light can result in darkness. However, no energy disappears: it is redistributed (p. 366). In the case of invisible waves where the nature of the disturbance is not revealed directly to our

Fig. 16.19

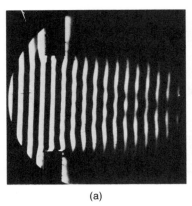

(a)

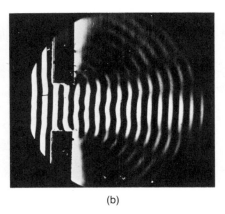

(b)

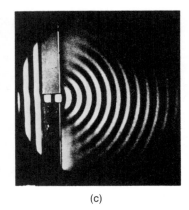

(c)

Fig. 16.20

senses we are really using a wave model, based on our experience of visible waves, to enable us to explain their behaviour.

(c) *Rectilinear propagation.* Interference shows the mechanism by which Huygens' secondary wavelets superpose to form the new wavefront. Rectilinear propagation for an infinite plane wave is then explained by the destructive interference of out-of-phase waves in all directions except forwards.

Diffraction

The spreading of waves when they pass through an opening or round an obstacle into regions where we would not expect them is called *diffraction*.

(a) *Mechanical waves.* The diffraction of water ripples at an opening in a barrier may be studied in a ripple tank, a vibrating bar being used to produce straight ripples.

(a)

(b)

Fig. 16.21

In Fig. 16.20a, the opening is wide and the incident waves emerge almost unchanged (i.e. straight). A fairly sharp 'shadow' of the opening is obtained and diffraction is not marked. Most of the energy propagated through the opening is in the same direction as the incident waves. In Fig. 16.20c, the opening is narrow and the emerging waves are circular, i.e. the opening behaves like a point source. In this case diffraction is appreciable and the energy of the diffracted waves is more or less equally distributed over 180°. In Fig. 16.20b some diffraction is evident at the 'medium-sized' gap. Diffraction effects are thus greatest when the width of the opening is comparable with the wavelength of the waves.

Huygens' procedure for predicting the future position of a wavefront by replacing it by point sources is able to account for diffraction, as well as for reflection and refraction. Thus the patterns obtained can be considered to arise from the interference (superposition) of the secondary wavelets produced by the point sources imagined to exist at the unrestricted part of the wavefront which falls on the opening. The two ripple tank photographs of Fig. 16.21a and b support this view. The diffraction pattern of straight waves passing through a slit is shown in (a); (b) is the interference pattern of a line of equally-spaced point sources occupying the same position and width as the slit and vibrating together with the same frequency as that of the waves in (a). The two patterns are very similar except near the sources in (b).

(b) *Electromagnetic waves.* Diffraction of microwaves at a slit may be shown with the apparatus of Fig. 16.22a. Spreading into the regions of geometrical shadow behind the slit is greatest when the slit width is comparable with the wavelength of the microwaves, i.e. 3 cm. (It should now be evident why it is possible to demonstrate the interference of microwaves using the double slit arrangement of Fig. 16.18a.)

It is also instructive to position a single metal plate as in Fig. 16.22b so that it just cuts off the signal to the receiver, placed in the shadow of the plate. When a second plate is slid up (along the dotted line) to make a slit 3–6 cm wide with the first plate, the received signal increases showing that a *smaller* wavefront (less energy) can produce a *stronger* signal. This is due to destructive interference between different parts of the larger wavefront and by removing some of them (with the second plate) the resultant signal increases in this particular direction.

If a circular metal disc (about 15 cm diameter) is used as an 'obstacle', a signal is received along the axis in the centre of the shadow, indicating appreciable diffraction round the edges of the disc.

Stationary waves

(a) *Mechanical waves.* If one end of a long, narrow, stretched spring (or a Slinky) is fixed and the other end is moved continuously from side to side, a progressive transverse wave is generated. At the fixed end it is reflected, travels back to the vibrating end and repeated reflection occurs. Two progressive trains of waves travel along the spring in opposite directions. If the shaking frequency is slowly increased, at certain frequencies one or more vibrating loops of large amplitude are formed in the spring. A *stationary* or *standing wave* is said to have been produced since the waveform does not seem to be travelling along the spring in either direction.

A more effective demonstration, called Melde's experiment, is shown in Fig. 16.23a using a $\frac{1}{2}$ m length of rubber cord (3 mm square section) stretched to 1 m.

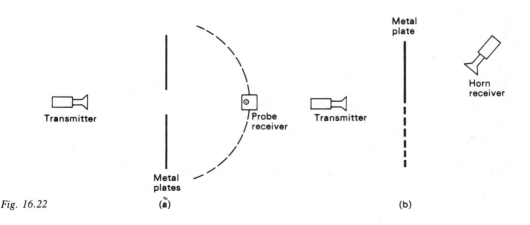

Fig. 16.22   (a)   (b)

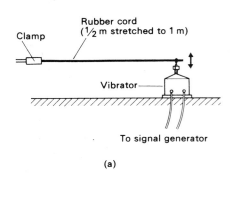

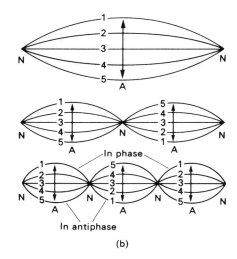

Fig. 16.23

As the frequency of the a.c. from the low impedance output of the signal generator is increased from 10 Hz to about 100 Hz, standing waves with one, two, three or four loops are obtained. If the cord is illuminated in the dark by a lamp stroboscope and the flashing rate adjusted to be nearly equal to that of the vibrator, the cord can be seen moving up and down slowly, the crests of the standing wave becoming troughs, then crests again and so on, Fig. 16.23*b*. Note that:

(*i*) There are points such as N, called *nodes* of the stationary wave, where the displacement is always zero.

(*ii*) Within one loop all particles oscillate in phase but with different amplitudes (how does this compare with the behaviour of a progressive wave?) and so all points (except nodes) have their maximum displacement simultaneously; points such as A with the greatest amplitude are called *antinodes*.

(*iii*) The oscillations in one loop are in antiphase with those in an adjacent loop.

(*iv*) The frequency of the particle vibration in both the standing wave and the progressive waves is the same, and the wavelength of the stationary wave is *twice* the distance between successive nodes or successive antinodes and equals the wavelength of either of the progressive waves. This offers a solution to the otherwise difficult problem of measuring the wavelength of a progressive wave. We simply produce standing waves from the progressive wave and measure the wavelength of the former directly.

(*b*) *Explanation of stationary waves.* Stationary waves result from the superposition of two trains of progressive waves of equal amplitude and frequency travelling with the same speed in opposite directions. In

Fig. 16.24 the dotted and broken curves are the displacement–distance graphs of two such waves at successive equal time intervals. The continuous curve in each case is their resultant at these instants, formed by superposition. The formation of the stationary wave loops of Fig. 16.23*b* can now be understood.

In the two previous demonstrations one wave-train is obtained by reflection of the other and if, in the time for a wave to travel along the spring or cord and back again, the vibrator is about to send off the second wave, the latter will reinforce the first. This will also be true if the vibrator produces exactly two, three or any integral number of waves in the time for a wave to make a return journey on the spring or cord. The amplitude of the waves builds up since they are returned in phase to the source, i.e. there is resonance. The coupling is thus improved between the source and the medium, producing 'stimulated emission'. If the stationary wave has one loop at frequency *f* then a vibrator frequency of 2*f* (called the *second harmonic*) will give two loops and so on.

We can regard the stretched spring or cord as having a number of natural frequencies of vibration and when the applied frequency from the vibrator is near one of these, there is a large-amplitude standing-wave vibration, i.e. resonance occurs. A stretched spring thus has a number of natural frequencies. By contrast, a mass hanging from the end of a spring, for example, has only one natural frequency. (We will see later, p. 355, that a system can only vibrate if it has mass and elasticity. The difference between the two systems arises from the fact that in the mass-spring system, the mass is concentrated in one part of the system and the elasticity property is in another part; in the stretched spring or cord system all

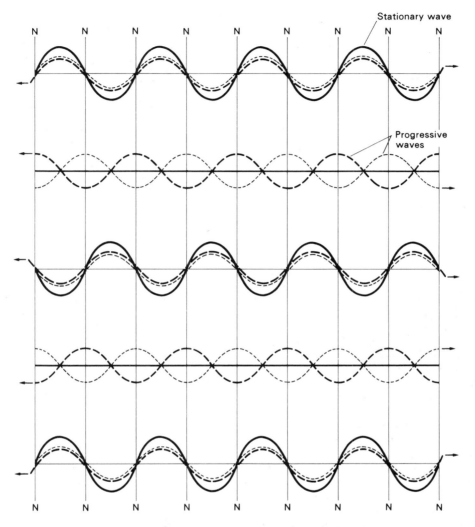

Fig. 16.24

parts have both properties. The first is said to have *lumped* elements and the second *distributed* elements.)

Standing-wave resonance can occur only in systems with 'boundaries' (e.g. the ends of a spring or cord) that restrict wave travel. Progressive waves reflected from the 'boundaries' interfere with progressive waves travelling to the 'boundaries' and the resulting standing waves have to 'fit' into the system, e.g. have nodes at the fixed ends of the cord in Fig. 16.23*a*, otherwise the pattern is not fixed in form. Large amounts of energy are stored locally in standing waves and become trapped with the waves; there is no energy transmission as with progressive waves. This is an important difference between standing and progressive waves. In 'unbounded' systems the waves are not confined and travel on unrestricted.

Some other demonstrations of standing waves in mechanical systems are illustrated in Figs. 16.25*a*, *b*, *c* and *d*. The first two should be viewed in the dark in stroboscopic light.

(*c*) *Electromagnetic waves.* Standing microwave patterns can be obtained with the arrangement of Fig. 16.26. Waves from the transmitter are superposed on those reflected from the metal plate. If the latter is moved slowly towards or away from the transmitter, the signal at the receiver varies and the distance moved by the reflector between two consecutive minima (nodes) equals half the wavelength of the microwaves. Why is the signal not quite zero at a node?

(*d*) *Importance of standing waves.* Standing waves are a feature of many physical phenomena. The mech-

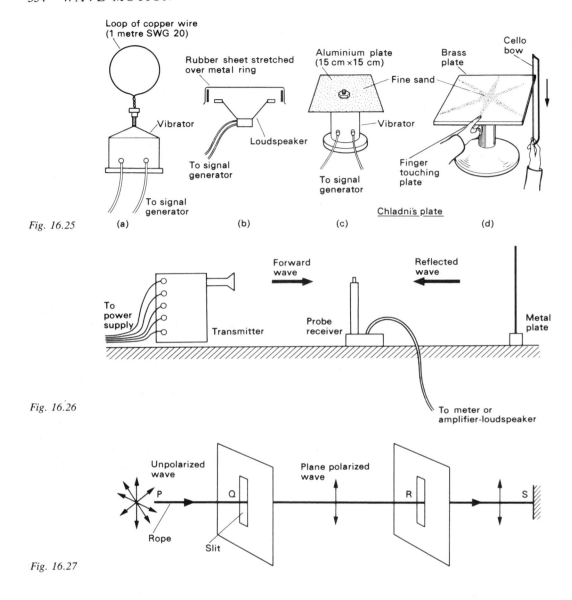

Fig. 16.25

Fig. 16.26

Fig. 16.27

anical type are produced on the strings and air columns of musical instruments when sounding a note (pp. 370 and 372). Electromagnetic examples occur in radio and television aerial systems and, as we shall see later (p. 485), the idea that an electron in an atom behaves like a standing wave is used to account for the fact that atoms have definite energy levels (p. 479).

Engineers are often confronted with standing-wave problems when dealing with systems such as turbines, propellers, aircraft wings, car bodies, etc., which can vibrate and have edges.

Polarization

This effect occurs only with transverse waves.

(*a*) *Mechanical waves*. In Fig. 16.27 the rope PQRS is fixed at end S and passes through two vertical slits at Q and R. If end P is moved to and fro in all directions (as shown by the short arrowed lines), vibrations of the rope occur in every plane and transverse waves travel towards Q. At Q only waves due to vibrations in a vertical plane can emerge from the slit and the wave between Q and R is said to be *plane polarized* (in the

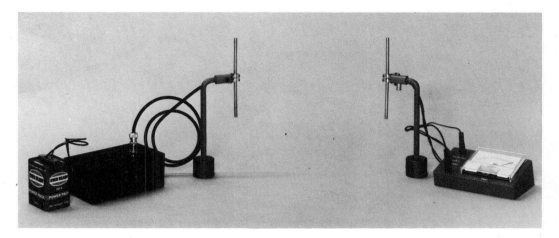

Fig. 16.28

vertical plane containing the slit at Q) in contrast to that between P and Q which is unpolarized. If the slit at R is vertical, as shown, the wave travels on, but if it is horizontal the wave is stopped and the slits are 'crossed'.

(b) *Electromagnetic waves.* Electromagnetic radiation from radio, television and radar aerials is plane polarized. This may be shown for 3 cm microwaves using the microwave transmitter (10 000 MHz = 10 GHz) and receiver or with 30 cm 'television-type' waves using the transmitter (1000 MHz = 1 GHz) and receiver of Fig. 16.28. When the receivers are rotated through 90° in a vertical plane from the maximum signal position, the signal decreases to a minimum. Polarization provides the evidence for the transverse nature of electromagnetic waves.

Consideration of the way in which a transmitting aerial produces radio and other types of waves, shows that the electric field comes off parallel to the aerial and the magnetic field at right angles to it. If the transmitting aerial is vertical, the signal in the receiving aerial is a maximum when it too is vertical. In this case the radio waves are said to be *vertically polarized*, the direction of polarization being given by the direction of the electric field. This follows the practice adopted with light waves where experiments show that the coupling of light with matter is more often through the electric field. The polarization of light will be considered later.

Speed of mechanical waves

(a) *Factors affecting.* A mechanical wave is transmitted by the vibration of particles of the propagating medium. A system can vibrate only if it has *mass* and *elasticity*.

Consider a load hung from the lower end of a spring. When the load is pulled down and released, the spring—due to its elasticity—pulls the load up again. However, owing to the load having mass it acquires kinetic energy and overshoots its original position. The opposition of the spring to compression gradually brings it to rest and it starts to fall. It again overshoots its original position and stretches the spring. The cycle of events is then repeated. Proper vibrations of the spring would not occur if it was very light (i.e. had no mass on the end) or was inelastic.

Waves can therefore only be transmitted by a medium if it has mass and elasticity and it is these two factors which determine the speed of a wave. This last point may be demonstrated using a wave model in which trolleys represent particles of the transmitting medium and springs (slightly stretched) represent the bonds between the particles. The time is found for a transverse pulse to travel along to the fixed end and back again of a line of twelve trolleys linked as in Fig. 16.29.

Doubling the mass of each trolley (by adding a load, usually 1 kg) *decreases* the wave speed since it takes longer for the larger mass to acquire a certain speed. Doubling the tension (by connecting another spring in parallel with each one in the model) *increases* the wave speed because the larger force causes the next trolley to respond more quickly. If mass and tension are both doubled, the wave speed has its original value since both quantities change the speed by the same factor.

Note how the energy of the vibrating 'particles' (trolleys) changes from k.e. to p.e. to k.e. and so on as the wave progresses.

(b) *Expressions for wave speeds.* These may be

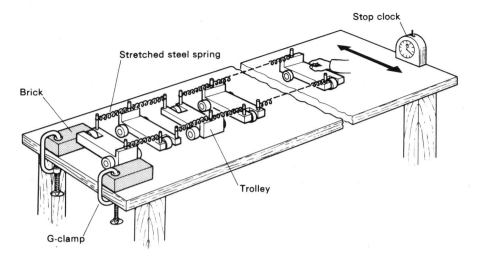

Fig. 16.29

---

derived for different types of waves using basic mechanical ideas. Some are quoted below.

Transverse waves on a taut string or spring

$$v = \sqrt{T/\mu}$$

T = tension
μ = mass per unit length

Longitudinal waves along masses (e.g. trolleys) linked by springs

$$v = x\sqrt{k/m}$$

x = spacing between mass centres
k = spring constant
m = one mass

Short wavelength ripples on deep water

$$v = \sqrt{2\pi\gamma/(\lambda\rho)}$$

γ = surface tension
λ = wavelength
ρ = density

It should be noted that in each case the numerator contains an 'elasticity' term and the denominator a 'mass' term.

Reflection and phase changes

(*a*) *Mechanical waves.* The behaviour of a wave at a boundary can be studied by sending pulses along a long narrow spring.

In Fig. 16.30*a*, the left-hand end of the spring is fixed and a transverse 'upward' pulse travelling towards it is reflected as a trough. A phase change of 180° or π rad has occurred and there is a phase difference of half a wavelength ($\lambda/2$) between the incident and reflected pulses. In Fig. 16.30*b* the left-hand end of the spring is attached to a heavier spring and at the boundary the pulse is partly transmitted and partly reflected, the reflected pulse being inverted.

In Fig. 16.30*c* a pulse passes from a heavy spring on the left to a light spring. Partial reflection and transmission again occur but the reflected pulse is not turned upside down. In Fig. 16.30*d* the left-hand end of the long narrow spring is fastened to a length of thin string and is in effect 'free'. Here almost the whole of the incident pulse is reflected the right way up, i.e. a crest is reflected as a crest and no phase change occurs.

These results may be summarized by saying that when a transverse wave on a spring is reflected at a 'denser' medium (e.g. a fixed end or a heavier spring) there is a phase change of 180° (or π rad or $\lambda/2$). The phase change occurs in the case of the spring with one end fixed for example, because there can be no displacement of the fixed end, it must be a node. The incident and reflected waves therefore cause equal and opposite displacements at the fixed end so that they superpose to give resultant zero displacement as shown in Fig. 16.31.

Phase changes also occur when longitudinal waves are reflected, as can be shown by sending pulses along a Slinky spring to 'denser' and 'less dense' boundaries, i.e. to fixed and free ends. At a fixed end a compression is reflected as a compression, at a free end it is reflected as a rarefaction. Similar effects are obtained when sound waves are reflected in pipes with closed and open ends; a compression is reflected as a compression at a closed end and as a rarefaction at an open end. In the latter case the air in the compression is able to expand

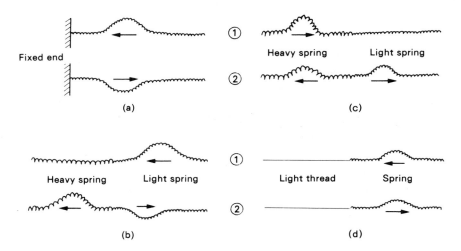

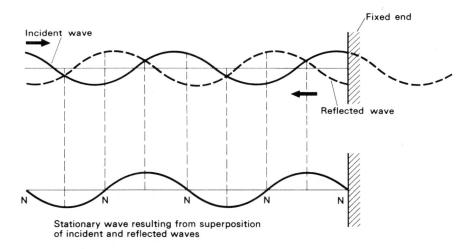

Fig. 16.30

Fig. 16.31   Stationary wave resulting from superposition of incident and reflected waves

outwards suddenly at the end of the pipe and a rarefaction travels back along the pipe.

(b) *Electromagnetic waves.* The half-wavelength ($\lambda/2$) phase change suffered by microwaves when they are reflected at a metal plate can be shown as in Fig. 16.32. Interference occurs between the waves reaching the probe receiver directly and those arriving at it by reflection from the metal plate. Maxima and minima are detected when the receiver is moved along PQ. At Q the geometrical path difference between the direct and reflected waves is zero and the two sets of waves should reinforce to give a maximum. In fact a *minimum* is obtained due to the $\lambda/2$ phase change of the reflected wave.

Using reflected waves

Various techniques have been developed for locating the positions of 'objects' by reflecting from them waves of known speed.

In *radar* (*ra*dio *d*etection *a*nd *r*anging), radio waves (e.g. 3 cm microwaves) are emitted in short pulses by a transmitter and picked up after reflection from the 'object', Fig. 16.33. The received pulse is displayed on a CRO with a calibrated time base which is triggered to start by the transmitted pulse. The time for the waves to travel twice the distance from the transmitting station to the 'object' is thus found (the interval between transmitted pulses must be greater than this to avoid confusion). Military 'objects' include aircraft and

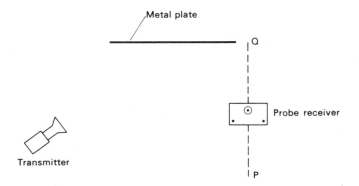

Fig. 16.32

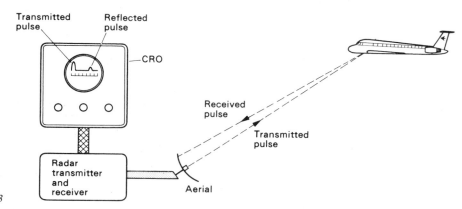

Fig. 16.33

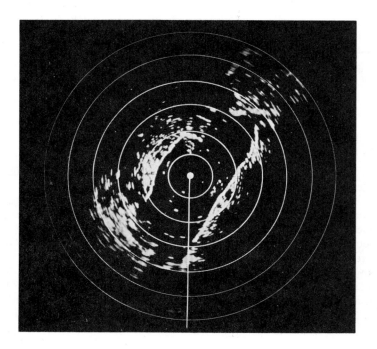

Fig. 16.34

missiles. In ships equipped with radar to assist navigation in fog and at night, a narrow microwave beam is swept continuously through 360° by a rotating aerial. The pulses reflected from land, other ships and buoys are shown on a CRO, called a *plan position indicator* (PPI), which has the time base origin in the centre of the screen and represents the ship; Fig. 16.34 shows Southampton Water on a yacht radar screen. Radar also helps to control aircraft waiting to land.

Sonar or sound navigation and ranging is similar to radar but employs ultrasonic waves, i.e. waves with frequencies above the maximum audible frequency of about 20 kHz. It is used to measure the depth of the sea (i.e. in echo sounding) and to detect shoals of fish.

In the non-destructive testing of materials ultrasonic waves can detect flaws. If three pulses are obtained on the display CRO they are due to the transmitted pulse A, the pulse B reflected from the flaw and pulse C reflected from the boundary of the specimen, Fig. 16.35.

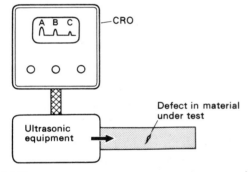

Fig. 16.35

Wave equations

(*a*) *Progressive wave.* Suppose the oscillation of the particle at O in Fig. 16.36 is simple harmonic of frequency f. Its displacement with time t will be given by $a \sin \omega t$ where a is the amplitude of the oscillation and $\omega = 2\pi f$. If the wave generated travels from left to right a particle at P, distance x from O, will lag behind the particle at O, say by phase angle ϕ. For the displacement y of the particle at P we can write

$$y = a \sin (\omega t - \phi) \qquad \text{(p. 320)}$$

But $\phi/2\pi = x/\lambda$ since at Q, distance λ from O, the phase difference between the motion of particles at O and Q is 2π (rad). Hence substituting for ϕ in the above equation we get

$$y = a \sin (\omega t - 2\pi x/\lambda)$$
$$= a \sin (\omega t - kx)$$

where $k = 2\pi/\lambda$.

If the wave travelled in the opposite direction, the vibration at P would lead on that at O by ϕ and the displacement y at P would be written

$$y = a \sin (\omega t + kx)$$

This is the equation for a wave travelling from *right to left*.

(*b*) *Stationary wave.* Let the two progressive waves of equal amplitude a and frequency f ($\omega = 2\pi f$) travelling in opposite directions be represented by

$$y_1 = a \sin (\omega t + kx) \quad \text{(to the left)}$$

and $\quad y_2 = a \sin (\omega t - kx) \quad \text{(to the right)}$

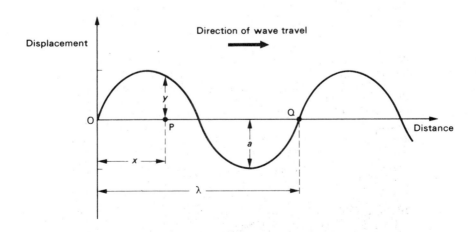

Fig. 16.36

By the principle of superposition, the resultant displacement y is given by

$$y = y_1 + y_2$$
$$= a \sin (\omega t + kx) + a \sin (\omega t - kx)$$

Using the trigonometrical transformation for converting the sum of two sines to a product

$$\left(\text{i.e. } \sin \alpha + \sin \beta = 2 \sin \frac{\alpha + \beta}{2} \cos \frac{\alpha - \beta}{2} \right)$$

we get

$$y = 2 a \sin \omega t \cos kx$$

This is the equation of the stationary wave. It can be written

$$y = A \sin \omega t$$

where $A = 2a \cos kx = 2a \cos (2\pi x/\lambda)$ is the amplitude of oscillation of the various particles. We see that when $x = 0, \lambda/2$ etc. A is a maximum and equal to $2a$. These are the antinodes. Nodes occur midway between antinodes since $A = 0$ when $x = \lambda/4, 3\lambda/4, 5\lambda/4$ etc.

QUESTIONS

1. A plane wavefront of monochromatic light is incident normally on one face of a glass prism of refracting angle 30°, and is transmitted. Using Huygens' construction trace the course of the wavefront. Explain your diagram and find the angle through which the wavefront is deviated. (Refractive index of glass = 1.5.) (*J.M.B.*)

2. State Snell's law of refraction and define refractive index.

   Show how refraction of light at a plane interface can be explained on the basis of the wave theory of light.

   Light travelling through a pool of water in a parallel beam is incident on the horizontal surface. Its speed in water is 2.2×10^{10} cm s^{-1}. Calculate the maximum angle which the beam can make with the vertical if light is to escape into the air where its speed is 3.0×10^{10} cm s^{-1}.

   At this angle in water, how will the path of the beam be affected if a thick layer of oil, of refractive index 1.5, is floated on to the surface of the water? (*O. and C.*)

3. S_1 and S_2 in Fig. 16.37 are two sources of circular water waves of the same frequency, phase and amplitude. There is a maximum disturbance at P, a minimum at Q and a maximum at R.

   (*a*) Write *two* expressions for the wavelength λ of the ripples.

   (*b*) How does the pattern change if (*i*) S_1 and S_2 are moved farther apart, and (*ii*) the frequency of the waves decreases?

Fig. 16.37

4. (*a*) What is meant by (*i*) diffraction, and (*ii*) superposition of waves? Describe **one** phenomenon to illustrate each in the case of *sound waves*.

   (*b*) The floats of two men fishing in a lake from boats are 21 m apart. A disturbance at a point in line with the floats sends out a train of waves along the surface of the water, so that the floats bob up and down 20 times per minute. A man in a third boat observes that when the float of one of his colleagues is on the crest of a wave that of the other is in a trough, and that there is then one crest between them. What is the velocity of the waves? (*O. and C.*)

5. The range of sound frequencies which can be detected by a certain person is 17.0 Hz to 20.0 kHz. What is the corresponding range of wavelengths if the speed of sound in air is 340 m s^{-1}?

6. (*a*) BBC Radio 4 broadcasts on a wavelength of 1.50 km. What is its frequency?

   (*b*) What are the wavelengths of a television station which transmits vision on 500 MHz and sound on 505 MHz?

   (*c*) What is the frequency of yellow light of wavelength $0.6\ \mu m$?

   (Speed of electromagnetic waves in free space = 3×10^8 m s^{-1}; $1\ \mu m$ (micrometre) = 10^{-6} m.)

7. The map in Fig. 16.38 shows part of a coastline, with two land-based radio navigation stations A and B. Both stations transmit continuous sinusoidal radio waves with the same amplitude and same wavelength (200 m).

   A ship X, exactly midway between A and B, detects a signal whose amplitude is twice that of either station alone.

   (*a*) What can be said about the signals from the two stations?

   (*b*) The ship X travels to a new position by sailing 100 m in the direction shown by the arrow. What signal will it now detect? Explain this.

   (*c*) A ship Y also starts at a position equidistant from A and B and then travels in the direction shown by the arrow. Exactly the same changes to the signal received were observed as in the case of ship X in (*b*).

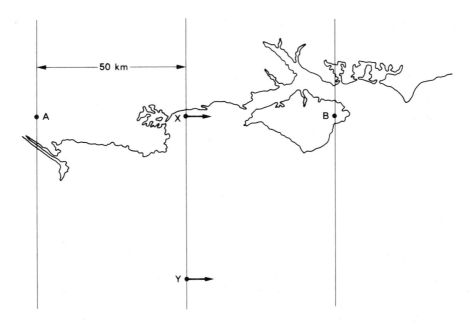

Fig. 16.38

Explain whether Y has sailed 100 m, more than 100 m, or less than 100 m. (*O. and C. Nuffield*)

8. The velocity v of waves of wavelength λ on the surface of a pool of liquid, whose surface tension and density are σ and ρ respectively, is given by

$$v^2 = \frac{\lambda g}{2\pi} + \frac{2\pi\sigma}{\lambda\rho}$$

where g is the acceleration due to gravity.

Show that the equation is dimensionally correct.

A vibrator of frequency 480 ± 1 Hz produces, on the surface of water, waves whose wavelength is 0.125 ± 0.001 cm. Assuming that for this wavelength the first term on the right-hand side of the equation is negligible, calculate the value which these results give for the surface tension of water.

Discuss whether the assumption is justified. (*O. and C.*)

9. The equation $y = a \sin (\omega t - kx)$ represents a plane wave travelling in a medium along the x-direction, y being the displacement at the point x at time t. Deduce whether the wave is travelling in the positive x-direction or in the negative x-direction.

If $a = 1.0 \times 10^{-7}$ m, $\omega = 6.6 \times 10^3$ s^{-1} and $k = 20$ m^{-1}, calculate (*a*) the speed of the wave, and (*b*) the maximum speed of a particle of the medium due to the wave. (*J.M.B.*)

17 Sound

Sound waves

Sound exhibits all the properties of waves discussed in the previous chapter, except polarization. This suggests it is a longitudinal wave motion and since it cannot travel in a vacuum but requires a material medium it must be of the mechanical type.

A sound wave is produced by a vibrating object which superimposes, on any movement the particles of the transmitting medium have, an oscillatory to-and-fro motion along the direction of travel of the wave. The frequency of these oscillations—i.e. of the sound wave—equals that of the vibrating source and to be audible to human beings must be roughly in the range 20 Hz to 20 000 Hz. The speed of sound in air is about 330 m s^{-1} and the corresponding wavelength limits are therefore (using $v = f\lambda$) 17 m and 17 mm respectively.

A progressive sound wave carries energy from the source to the receiver where, in the case of the ear, the succession of compressions and rarefactions causes the ear-drum to vibrate. The sensation of what we *also* call 'sound' is then produced by impulses sent to the brain.

It is worthwhile connecting a loudspeaker to a signal generator, Fig. 17.1, varying the frequency from a very low to a very high value (e.g. 10 Hz to 30 kHz) and observing the effect on (*i*) the cone of the loudspeaker, (*ii*) the ear and (*iii*) a CRO with a microphone feeding its Y input. At the lowest frequencies the vibrations of the cone can be seen and felt with the finger, but not heard (except for a weak thudding). As the frequency increases, the pitch of the note rises from that of a hum to a whistle and then to a hiss before it becomes inaudible. Above a few hundred hertz the movement of the speaker cone cannot be felt. The CRO responds throughout.

Sound waves with frequencies greater than 20 kHz or so are called *ultrasonic* waves. They are used by bats as a kind of navigational 'radar' for night-flying. A correctly-cut quartz crystal generates ultrasonic waves when an alternating p.d. (of ultrasonic frequency) is applied across its faces. Their use in the non-destructive testing of materials has already been mentioned (p. 359) and Fig. 17.2 shows the shades of street lamps being cleaned by immersion in a tank of water which has an ultrasonic vibrator in the base.

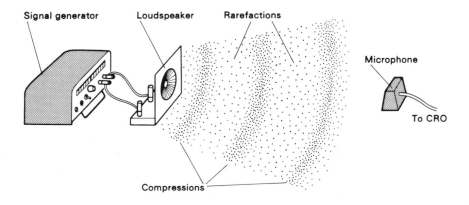

Fig. 17.1

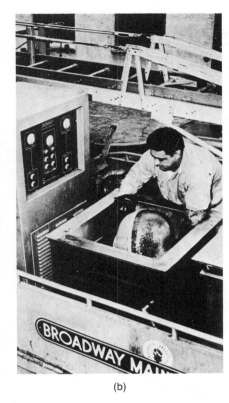

Fig. 17.2   (a)   (b)

Graphical representation

In Fig. 17.3a the circled figures ①, ②, ③ etc. represent parts of the air in which a progressive longitudinal wave is generated when the blade P is plucked. It can be seen that as P and the air oscillate to and fro about their undisturbed positions, compressions (C) and rarefactions (R) travel to the right. The following conclusions may be drawn.

(*i*) *Displacement.* Air at the centres of compressions and rarefactions has zero displacement. A displacement–*distance* graph for the instant of time corresponding to the last line of Fig. 17.3a can be drawn as in Fig. 17.3b where the longitudinal displacements are represented as if they were transverse and displacements to the right are taken as positive. The amplitude of vibration of the air is much exaggerated in the graph; even for a very loud sound it is only of the order of 10^{-2} mm. The velocities of the central parts are a maximum but they are directed away from the source in a compression and towards it in a rarefaction, i.e. the air moves forward in a compression and backwards in a rarefaction. The variation with time of the displacement of the air is thus 90° ($\pi/2$ rad) out of phase with its

variation of velocity with time. Draw two graphs to show this for a sinusoidal wave.

(*ii*) *Pressure excess.* The crowding together of the air in a compression causes the density and pressure of the air to be greater there than normal; both are a maximum at the centre of the compression. Conversely, density and pressure have minimum values at the centre of a rarefaction. Fig. 17.3c shows how the pressure (or density) varies with *distance* in a progressive longitudinal wave. The variations of the graph are exaggerated, actual variations are very small, 30 Pa being typical for a very loud sound (normal atmospheric pressure is about 10^5 Pa). The ear responds to pressure changes rather than displacements and is able to detect variations of as little as 2×10^{-5} Pa.

Reflection and refraction of sound

(*a*) *Reflection.* Sound waves obey the laws of reflection. Regular reflection occurs at a surface if it treats all parts of the incident wavefront similarly. To do this it must be 'flat' to within a fraction of the wavelength of the waves falling on it. Speech sound waves have wavelengths of several metres and so it is not surprising

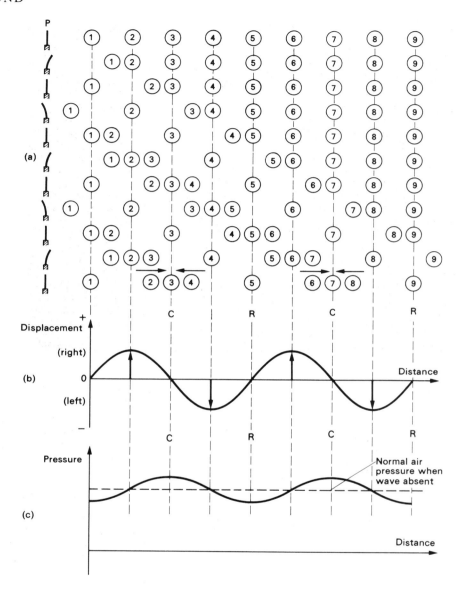

Fig. 17.3

that surfaces as 'rough' as cliffs, large buildings, etc., reflect sound regularly to give *echoes*, i.e. sound images. (Light requires a highly polished surface for regular reflection. Why?)

In a hall or large room sound reaches a listener by many paths, some much longer than others, as a result of reflection at walls, ceilings, etc. *Reverberation* occurs when the sound produced at one instant by the source persists at the listener for some time afterwards. Too short a reverberation time makes a room sound 'dead',

but if it is too long, 'confusion' results. The best value depends on the function of the building; for speech about 0.5 s is acceptable, whereas for music between one and two seconds is required.

In a modern concert hall such as the Royal Festival Hall in London, Fig. 17.4, the reverberation time is made the same irrespective of the size of the audience by upholstering the seats with material having similar absorption to clothing. Also, to distribute the sound as equally as possible, walls, etc., tend to be built up from

Fig. 17.4

many small, flat surfaces at various inclinations. Curved surfaces, which might focus the sound strongly in one place, are avoided.

(b) Refraction. When the speed of sound waves changes at a boundary between two media, refraction occurs, i.e. the direction of travel changes as does the wavelength (since the frequency is fixed by the source). Snell's law (p. 344) is obeyed and the refractive index $_1n_2$ for waves of a given frequency, having speed v_1 in medium ① and v_2 in medium ② is defined by

$$_1n_2 = \frac{v_1}{v_2}$$

Sounds are more audible at night than during the day because the speed of sound in warm air exceeds that in cold air (p. 376) and refraction occurs. At night the air is usually colder near the ground than it is higher up and refraction towards the earth occurs, Fig. 17.5a. During the day the reverse is usually true, Fig. 17.5b.

Interference and diffraction of sound

(a) Interference. The interference of sound waves may be demonstrated with two loudspeakers connected in parallel to the low impedance output of a signal generator, set at 4 kHz, Fig. 17.6. The fringes are

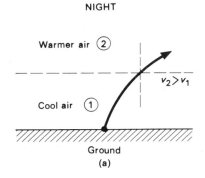

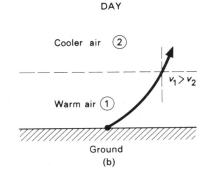

Fig. 17.5

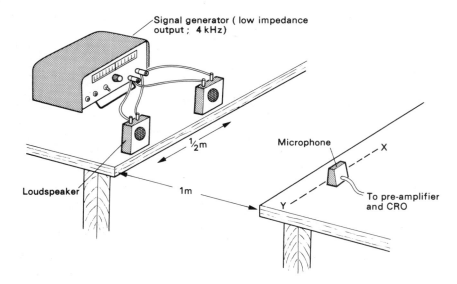

Fig. 17.6

detected by moving a microphone feeding a CRO, preferably via a pre-amplifier, along XY. (The effect of unwanted reflections is minimized by working between two benches or, better still, out of doors.)

When the microphone is at the central maximum of the interference pattern (i.e. along AB in Fig. 16.16, p. 348) and one loudspeaker is covered up, the output p.d. from the microphone (as shown on the CRO) falls to *half* of the value it had with both speakers uncovered. If the CRO were replaced by a resistor, the current through it would also be halved on blocking off the sound from one speaker and so the rate of energy conversion in it would be a *quarter* of its value with two speakers (since power = p.d. × current). Conversely, the energy at a central maximum will be *four* times that

emitted in a given time by one of the speakers. We must conclude that since there is zero energy at minima, the energy from them goes to the maxima when two waves superpose to produce interference fringes. That is, energy is redistributed.

A tuning fork, Fig. 17.7a, may also be used to show interference of sound. If it is struck and held close to the ear with the prongs vertical whilst being slowly rotated about a vertical axis, no sound is heard four times in each revolution. When the prongs PP are moving outwards, a rarefaction is produced in the air between them and travels along OY and OY'. Simultaneously, compressions are generated outside the prongs and move along OX and OX', Fig. 17.7b. The compressions and rarefactions spread out as spherical

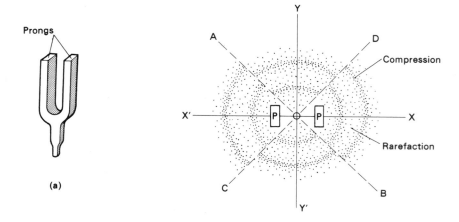

Fig. 17.7

wavefronts and destructive interference occurs along AOB and COD where compressions and rarefactions superpose.

(*b*) *Diffraction*. The spreading of sound waves round the corners of an open doorway is an everyday event and suggests that the width of such an aperture (say, one metre) is of the same order as the wavelength of sound. A note of frequency 300 Hz has a wavelength in air of about one metre.

Beats

When two notes of slightly different frequencies but similar amplitudes are sounded together, the loudness increases and decreases periodically and *beats* are said to be heard. The effect may be demonstrated with two signal generators, each feeding a loudspeaker. Alternatively, two tuning forks of the same frequency can be used, one having a little Plasticine stuck on to one prong to lower its frequency slightly. They are struck simultaneously and the stems pressed against the bench top. If the difference between the frequencies of the two sources is increased, the beat frequency also rises until it is too high to be counted.

The production of beats is a wave effect explained by the principle of superposition. The displacement–time graphs for the wave-trains from two sources of nearly equal frequency are shown in Fig. 17.8*a* for a certain observer. At an instant such as A the waves from the sources arrive in phase and reinforce to produce a loud sound. The phase difference then increases until a compression (or rarefaction) from one source arrives at the same time as a rarefaction (or compression) from the other. The observer hears little or nothing, point B. Later the waves are in phase again (point C) and a loud note is heard. Fig. 17.8*b* gives the resultant variation of amplitude. Beats are thus due to interference but because the sources are not coherent (frequencies different) there is sometimes reinforcement at a given point and at other times cancellation.

We will now show that the *beat frequency equals the difference of the two almost equal frequencies*. Suppose the beat period (i.e. the time between two successive maxima) is T and that one wave-train of frequency f_1 makes one cycle more than the other of frequency f_2 then

number of cycles of frequency f_1 in time $T = f_1 T$

and,

number of cycles of frequency f_2 in time $T = f_2 T$

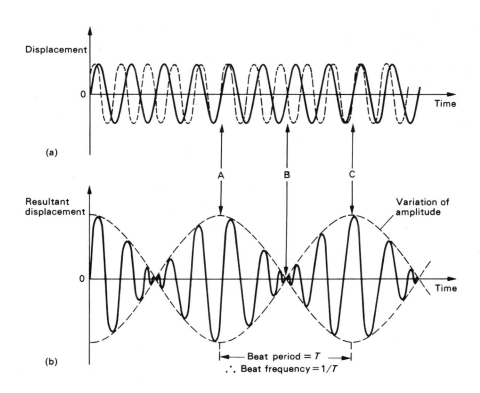

Fig. 17.8

(a)

Displacement

0

Time

A   B   C

(b)

Resultant displacement

0

Variation of amplitude

Time

Beat period = T

∴ Beat frequency = $1/T$

$$\therefore \quad f_1 T - f_2 T = 1$$

$$\therefore \quad f_1 - f_2 = 1/T$$

But one beat has occurred in time T and so $1/T$ is the beat frequency, hence

$$beat\ frequency = f_1 - f_2$$

Stationary sound waves

(a) *Production*. The production of stationary or standing waves by the superposition of two trains of progressive waves of equal amplitude and frequency, travelling with the same speed in opposite directions, was considered previously for transverse waves. We saw that nodes and antinodes were formed. At a node the displacement of the vibrating particle was always zero but at an antinode it varied periodically from zero to maxima in opposite directions (see Fig. 16.23b, p. 352). Also, the wavelength of the stationary wave equalled that of either progressive wave and was twice the distance between two consecutive nodes or antinodes.

Stationary sound waves can be obtained in air using apparatus called *Kundt's tube*, Fig. 17.9. The speaker, driven by a signal generator, produces progressive longitudinal waves which travel through the air to the end of the cylinder where they are reflected to interfere with the incident waves. The frequency of the sound is varied and resonance occurs when it equals one of the natural frequencies of vibration of the air column. Stationary waves are then formed and are revealed by the lycopodium powder which swirls away from the antinodes (where the air is vibrating strongly) and after a short time forms into heaps at the nodes.

(b) *Displacement and pressure graphs*. Earlier we discussed the pressure variation in a progressive sound wave (p. 363), we shall now consider how it varies in a stationary sound wave. In Fig. 17.10a, curves ① and ② are displacement–distance graphs for a stationary wave for times when the displacements are greatest. As before (Fig. 17.3b) displacements to the right are taken to be positive. Consider ①. Air to the left of node N_1 is displaced towards N_1 since this displacement is positive; air to the right has a negative displacement but is also displaced towards N_1. At N_1 the air is thus compressed and the pressure greater than normal. At node N_2 the air is again displaced in opposite directions but away from N_2. In this case the air is less dense and the pressure less than normal. Thus on either side of a node the air is vibrating in antiphase. Approach causes pressure increase and separation makes it decrease. Parts of the air on either side of antinode A_2, that are equidistant from it, have the same displacement to the left, i.e. they vibrate in phase and the air pressure is normal. The pressure–distance graph corresponding to displacement–distance graph ① is therefore given by ① in Fig. 17.10b. By treating displacement graph ② in the same way we obtain pressure–distance graph ②.

From Fig. 17.10b we can conclude that in a stationary longitudinal wave the pressure *variation* is a maximum at a displacement node and is always zero at a displacement antinode. It is sometimes stated that displacement nodes and pressure antinodes coincide as do displacement antinodes and pressure nodes. Adjacent nodes and antinodes are always a quarter of a wavelength apart.

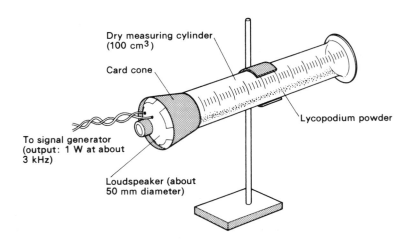

Dry measuring cylinder (100 cm³)

Card cone

To signal generator (output: 1 W at about 3 kHz)

Loudspeaker (about 50 mm diameter)

Lycopodium powder

Fig. 17.9

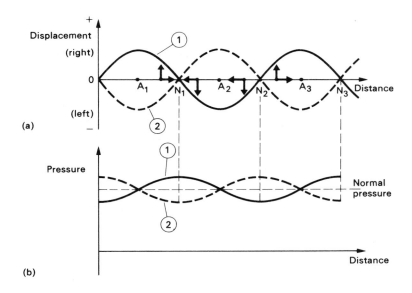

Fig. 17.10

Musical notes

A musical note is produced by vibrations that are regular and repeating, i.e. by periodic motion. Non-periodic motion creates noise which is not pleasant to the ear.

A musical note has three characteristics.

(*a*) *Loudness.* Loudness is a subjective sensation (i.e. it depends on the listener) which is determined by the intensity of the sound and by the sensitivity of the observer's ear. *Intensity,* unlike loudness, is a physical quantity and is defined as the *rate of flow of energy through unit area perpendicular to the direction of travel of the sound at the place in question.* It is measured in watts per square metre $(\mathrm{W\,m^{-2}})$ and if P is the total energy emitted per second equally in all directions by a point source, then neglecting absorption in the surrounding medium, the intensity I at a distance r from the source is given by

$$I = \frac{P}{4\pi r^2}$$

This is an *inverse square* relationship and it holds for the intensity at a point due to any source of spherical waves of any type.

The energy of a particle describing simple harmonic motion (s.h.m.) is directly proportional to the square of the amplitude of vibration if other factors are held constant (p. 184). Now any periodic motion which is not simple harmonic can be resolved into a number of such motions and can conversely be thought of as formed from these separate s.h.m.s. Sound energy is carried by periodic vibrations of the air and so the *intensity (and loudness) of a sound must also be directly proportional to the square of the amplitude of vibration of the air* and, in turn, of the source.

Amplitudes are alternately positive and negative and their average value is zero. The squares of amplitudes are always positive and their average is not zero, which further confirms their use to measure the rate at which a wave transports energy, i.e. its intensity. The demonstration described on p. 365 using two loudspeakers, a microphone and CRO to show interference with sound waves offers some experimental evidence for this point. An analogous situation occurs with alternating currents where the rate of delivery of energy is proportional to the square of the peak current.

(*b*) *Pitch.* This is also a sensation experienced by a listener. It depends mainly on the frequency of vibration of the air, which of course is the same as that of the source of sound. Frequency is a physical quantity and is measured in hertz (Hz), one vibration per second being one hertz. A high frequency produces a high-pitched note and a low frequency gives a low-pitched note. A note of frequency 500 Hz sounded in front of a microphone connected to a CRO gives twice as many waves (for the same time base setting) as one of frequency 250 Hz. The notes are said to differ in pitch by one *octave.* Pitch is analogous to colour in light.

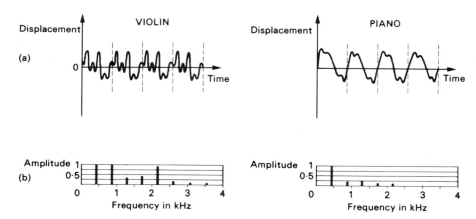

Fig. 17.11

(c) *Quality or timbre*. The same note played on two different instruments does not sound the same and different waveforms are obtained on a CRO. The notes are said to have different *quality* or *timbre*. This is due to the fact that with one exception (a tuning fork), sounds are never 'pure notes', i.e. of one frequency; their waveforms are not sinusoidal. They consist of a main or *fundamental* note—which usually predominates—and others, generally with smaller intensities, called *overtones*. The fundamental is the component of lowest frequency and the overtones have frequencies that are multiples of the fundamental frequency. *The number and intensities of the overtones determines the quality of a sound* and depends mostly on the instrument producing the sound. The fundamental is also called the *first harmonic* and if it has frequency f, the overtones with frequencies $2f$, $3f$, etc., are the second, third, etc., harmonics. As we shall see shortly (p. 371), with some instruments certain harmonics are not present as overtones in a note.

Waveforms for the same note of fundamental frequency 440 Hz played on a violin and a piano are shown in Fig. 17.11a. They are very different and provide visual evidence that the notes are of different 'quality'. Using a mathematical technique known as *Fourier analysis* the frequencies and amplitudes of the harmonics present as overtones can be worked out in each case and sound spectra obtained, Fig. 17.11b. Loud higher harmonics (notably the second and fifth) are present in the violin spectrum. The analysis depends on the fact that a periodic waveform, however complex, can be resolved into a number of sine waveforms with frequencies that are multiples of the fundamental. For example, a waveform like that of graph ① in Fig. 16.14 (p. 346) has two components—the fundamental ② and an overtone

③ (the third harmonic) with three times the frequency and half the amplitude of the fundamental.

Vibrating air columns

(a) *Wind instruments*. Stationary longitudinal waves in a column of air in a pipe are the source of sound in wind instruments. To set the air into vibration a disturbance is created at one end of the pipe, the other end can be open or closed (stopped). In an organ pipe, Fig. 17.12, a jet of air strikes a sharp edge, in a recorder the player blows across a hole, while in an oboe a reed vibrates when blown. Progressive sound waves travel

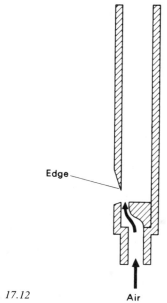

Fig. 17.12

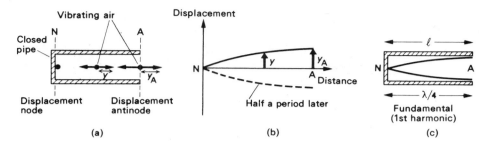

Fig. 17.13   (a)   (b)   (c)

from the source to the end of the pipe where they are reflected and interfere with the incident waves. The wavelength of some waves will be such that the length of the pipe produces resonance, i.e. a compression in the reflected wave arrives back at the source just as it is producing another compression. The amplitude of these waves will be large and a stationary longitudinal wave is formed.

The possible modes of vibration of the air in closed and open pipes will now be considered.

(*b*) *Closed pipe.* In the simplest stationary wave vibration possible, there will be a displacement node N at the closed end of the pipe, since the air there must be at rest, and a displacement antinode A at the open end where the air can vibrate freely, Fig. 17.13*a*. (Where will the pressure (*i*) node, and (*ii*) antinode, be formed?) Fig. 17.13*b* shows the displacement–distance graph for this vibration and from Fig. 17.13*c* we see that the length *l* of the pipe equals the distance between a node and the next antinode. That is NA = $\lambda/4$ where λ is the wavelength of the stationary wave. Hence

$$l = \frac{\lambda}{4}$$

$$\therefore \quad \lambda = 4l$$

The frequency *f* of the note is given by $v = f\lambda$ where v is the speed of sound in air. Therefore

$$f = \frac{v}{\lambda} = \frac{v}{4l}$$

This is the lowest frequency produced by the pipe, i.e. the fundamental frequency *f* or first harmonic.

The next two simplest ways in which the air may vibrate in the same pipe are shown in Fig. 17.14*a* and *b*. In both cases there are displacement nodes at the closed ends and displacement antinodes at the open ends. In (*a*), if the wavelength is λ_1, we have

$$l = \frac{3\lambda_1}{4}$$

The frequency f_1 is therefore given by

$$f_1 = \frac{3v}{4l}$$

This is the first overtone and since $f_1 = 3f$ it is the third harmonic. Similarly from (*b*) we find that the frequency f_2 of the second overtone is $5f$, i.e. it is the fifth harmonic and in general a *closed pipe gives only odd-numbered harmonics.*

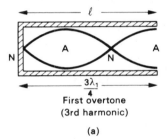

First overtone
(3rd harmonic)

(a)

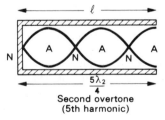

Second overtone
(5th harmonic)

Fig. 17.14   (b)

(*c*) *Open pipe.* Here, both ends of the pipe are open and are displacement antinodes A. The simplest stationary wave is shown in Fig. 17.15*a*. There is a displacement node N midway between the two antinodes, hence

$$l = \frac{\lambda}{2}$$

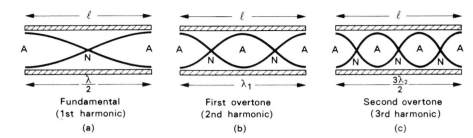

Fundamental
(1st harmonic)

First overtone
(2nd harmonic)

Second overtone
(3rd harmonic)

Fig. 17.15   (a)   (b)   (c)

The fundamental frequency f is given by

$$f = \frac{v}{\lambda} = \frac{v}{2l}$$

The next simplest mode of vibration, Fig. 17.15b gives the first overtone and we have

$$l = \lambda_1$$

$$\therefore \quad f_1 = \frac{v}{l}$$

This is the second harmonic since $f_1 = 2f$. The second overtone is obtained from Fig. 17.15c and is the third harmonic. In an open pipe all harmonics are possible as overtones.

(d) *Further points.*

(i) The fundamental frequency of an open pipe is *twice* that of a closed pipe of the same length.

(ii) The note from an open pipe is richer than that from a closed pipe due to the presence of extra overtones.

(iii) Higher overtones are encouraged by blowing harder.

(iv) The actual vibration of the air in a pipe is the resultant, by superposition, of the various modes that occur.

(v) In practice the air just outside the open end of a pipe is set into vibration and the displacement antinode of a stationary wave occurs a distance c—called the *end-correction*—beyond the end. The effective length of the air column is therefore greater than the length of the pipe. Hence for the fundamental of a closed pipe, Fig. 17.16a, we should write $\lambda/4 = l + c$ and for an open pipe, Fig. 17.16b, we have $\lambda/2 = l + 2c$. The value of c is generally taken to be $0.6\,r$ where r is the radius of the pipe.

The impressive array of pipes of the Festival Hall organ is shown in Fig. 17.17.

Vibrating strings

(a) *String instruments.* The 'string' is a tightly-stretched wire or length of gut. When it is struck, bowed or plucked, progressive transverse waves travel to both ends, which are fixed, where they are reflected to meet the incident waves. A stationary wave pattern is formed for waves whose wavelengths 'fit' into the length of the string, i.e. resonance occurs. A progressive sound wave (i.e. a longitudinal wave) is produced in the surrounding air with a frequency equal to that of the stationary transverse wave on the string.

The weak sound produced by the vibrating strings is transmitted to a sounding board in a piano and to the hollow body of instruments such as the violin. A larger mass of air is thereby set vibrating and the loudness of the sound increased.

(b) *Modes of vibration.* The fixed ends must be displacement nodes N. If the string is plucked in the middle the simplest mode of vibration is shown in Fig. 17.18a, A being a displacement antinode. This vibration

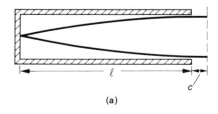

(a)

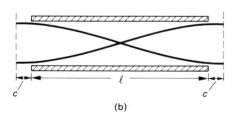

(b)

Fig. 17.16

Fig. 17.17

creates the fundamental note of frequency *f,* and

$$l = \frac{\lambda}{2}$$

where *l* is the length of the string. The frequency *f* is therefore

$$f = \frac{v}{\lambda} = \frac{v}{2l}$$

where *v* is the speed of the transverse wave along the string.

By plucking the string at a point a quarter of its length from one end, it can vibrate in two segments,

Fig. 17.18*b*. Then, if λ_1 is the wavelength of the resulting stationary wave,

$$l = \lambda_1$$

$$\therefore \quad f_1 = \frac{v}{\lambda_1} = \frac{v}{l} = 2f$$

This is the first overtone or second harmonic. If the string vibrates in three segments, Fig. 17.18*c*, the second overtone or third harmonic is obtained.

A string can vibrate in several modes simultaneously, depending on where it is plucked, etc., and the frequencies and relative intensities of the overtones produced determine the quality of the note emitted.

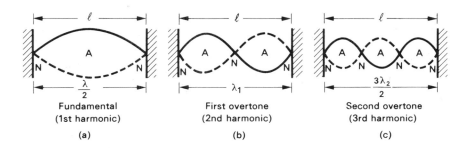

Fig. 17.18

Melde's experiment (p. 351) shows the various stationary wave patterns of a vibrating string (or rubber cord). It will be seen that, rather unexpectedly, the driven point of the string is nearly a node; certainly the amplitude of vibration is much less than at an antinode. The motion there is no more than is required for the vibrator to make good the energy dissipated in the system. Also, the frequency of the wave is the same as that of the vibrator.

(c) *Laws of vibration of stretched strings*. It was stated earlier (p. 356) that the speed v of a transverse wave travelling along a stretched string is given by

$$v = \sqrt{\frac{T}{\mu}} \tag{1}$$

where T is the tension in the string and μ is its mass per unit length. If the string has length l, the frequency f of the fundamental note emitted by it has just been shown to be

$$f = \frac{v}{2l} \tag{2}$$

Hence from (1) and (2),

$$f = \frac{1}{2l}\sqrt{\frac{T}{\mu}} \tag{3}$$

From (3) it follows that

(i) $f \propto 1/l$ if T and μ are constant
(ii) $f \propto \sqrt{T}$ if l and μ are constant
(iii) $f \propto 1/\sqrt{\mu}$ if l and T are constant.

These three statements are known as the *laws of vibration of stretched strings*. They indicate that short, thin wires under high tension emit high notes. The laws may be verified experimentally using a *sonometer*. This consists of a wire under tension, arranged on a hollow wooden box as in Fig. 17.19. The vibrations of the wire are passed by the movable bridges to the box and then to the air inside it.

To investigate the relationship between f and l, the bridges are moved so that different lengths between them emit their fundamental frequencies when plucked at the centre. The frequencies may be found by one or other of the methods outlined on p. 382 or by taking lengths of wire which vibrate in unison (as judged by the ear) with tuning forks of known frequencies. The product $f \times l$ should be constant within the limits of experimental error.

To check that $f \propto \sqrt{T}$, the same length of wire is subjected to different tensions by changing the hanging mass and the frequencies found as before.

The relation between f and $1/\sqrt{\mu}$ requires the use of wires of different diameters and materials but of the same vibrating length and under the same tension.

(d) *Measuring the frequency of an a.c. supply*, Fig. 17.20. The wire carrying the a.c. experiences an alternating transverse force when the magnet is placed so that the wire passes between its poles. Bridge B_1 is arranged near B_2 at the fixed end of the wire and then slowly separated from it, keeping the magnet approximately midway between them, until the wire vibrates in one loop with maximum amplitude, the current being reduced to a low value. The frequency f of the a.c. is given by

$$f = \frac{1}{2l}\sqrt{\frac{T}{\mu}}$$

where $l = B_1B_2$, $T = (0.20\,\text{kg}) \times (10\,\text{N}\,\text{kg}^{-1}) = 2.0\,\text{N}$ and μ is found by weighing a two-metre length of the SWG 32 copper wire. If T is kept constant and the magnet left in position, doubling the distance between the bridges should produce two loops and a second value of f may be calculated (in this case $l = B_1B_2/2$).

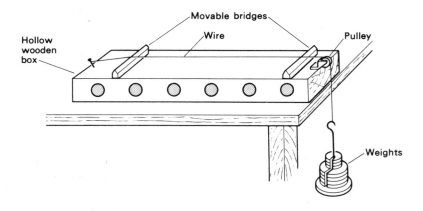

Fig. 17.19

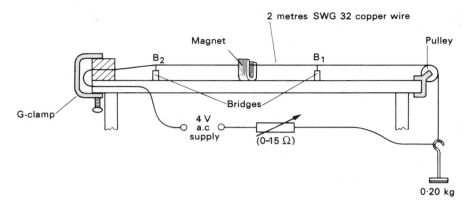

Fig. 17.20

A striking variation of the experiment can be shown in a darkened room with a suitable length of nichrome wire connected directly across the 240 V a.c. mains supply—*but great care is necessary not to touch the wire.* If the current is sufficient and several loops are obtained, the nodes become red-hot and show up clearly. (With a ten metre length of SWG 28 nichrome wire the current is satisfactory and if three metres are allowed to vibrate, four loops are obtained with a load of 0.5 kg.)

Speed of sound

It was shown on p. 355 that the speed of a mechanical wave in a medium depends on 'elasticity' and 'mass' factors and formulae for the speeds of different waves were quoted. An expression for the speed of sound in *any* medium was derived by Newton and may be written

$$\text{speed} = \sqrt{\frac{\text{elastic modulus of medium}}{\text{density of medium}}}$$

The elastic modulus (i.e. stress/strain) concerned depends on the type of strain produced by the wave in the medium.

(*a*) *Solids.* In a solid rod of small diameter compared with the sound wavelength, the compressions and rarefactions of the sound wave cause a change of length strain and the Young modulus E is appropriate in the above expression. Hence for a solid rod of density ρ in which the speed of sound is v, we have

$$v = \sqrt{\frac{E}{\rho}}$$

When E is in N m^{-2} (i.e. Pa) and ρ in kg m^{-3}, v is in

m s^{-1}. For steel $E = 2.0 \times 10^{11}$ Pa and $\rho = 7.8 \times 10^3$ kg m^{-3} giving $v = 5.1 \times 10^3$ m s^{-1}.

The formula for the speed of sound in a solid is derived in Appendix 12.

(*b*) *Gases.* Volume changes occur in the case of gases (and liquids) and the bulk modulus K of the gas is the relevant elastic modulus. It can be shown that under the conditions in which a sound wave travels in a gas, $K = \gamma p$ where p is the pressure of the gas and γ is the ratio of its principal heat capacities, i.e. C_p/C_v (see p. 444). We then have for the speed of sound v in a gas of density ρ

$$v = \sqrt{\frac{\gamma p}{\rho}}$$

For air at s.t.p., $\rho = 1.3$ kg mg^{-3}, $\gamma = 1.4$ and $p = 1.0 \times 10^5$ Pa, therefore $v = 3.3 \times 10^2$ m s^{-1}, which agrees well with the experimental value.

(*c*) *Effect of pressure, temperature and humidity.* For one mole of an ideal gas having volume V and pressure p at temperature T, we can write (p. 431).

$$pV = RT$$

where R is a constant. If M is the mass of gas

$$\rho = \frac{M}{V} = \frac{Mp}{RT}$$

$$\therefore \quad v = \sqrt{\frac{\gamma p}{\rho}} = \sqrt{\frac{\gamma RT}{M}}$$

The final expression does not contain p and so *the speed of sound in a gas is independent of pressure.* This is not unexpected since changes in density are proportional to changes of pressure, i.e. p/ρ is a constant.

Also, since R has the same value for all gases and γ

and M are constants for a particular gas

$$v \propto \sqrt{T}$$

That is, *the speed of sound in a gas is directly proportional to the square root of the absolute temperature of the gas* (provided γ is independent of temperature).

For moist air ρ is less than for dry air and γ is slightly greater. The net result is that *the speed of sound increases with humidity*.

(*d*) *Effect of wind.* If the air carrying sound waves is itself moving, i.e. there is a wind, the wind velocity has to be added vectorially to that of the sound to obtain the velocity of the latter with respect to the ground. And since wind velocities usually increase with height, wavefronts become distorted as in Fig. 17.21, resulting in greater audibility down-wind than up-wind.

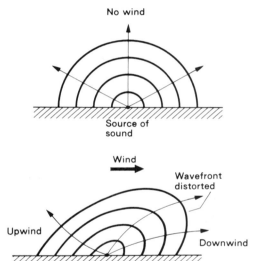

Fig. 17.21

Measuring the speed of sound

(*a*) *In air by progressive waves using a CRO.* The CRO must have an output terminal from its time base and the latter must be calibrated. The procedure described here refers to the Telequipment S51E (Fig. 21.3*b*, p. 489).

The VARIABLE control (fine time base) is rotated fully clockwise to the CAL position and the X GAIN turned fully anticlockwise to its minimum setting. These two adjustments ensure that the time base will run at the speed indicated by TIME/CM control. The latter is set to 100 ms. The VOLTS/CM knob is set at the maximum Y-amplifier gain of 0.1 V and the STABILITY and TRIG LEVEL controls rotated to their maximum clockwise positions. The CRO is then switched on at the BRIGHTNESS control and X-SHIFT adjusted so that the start of the trace at the left of the screen is visible.

The SWEEP OUTPUT and E terminals on the CRO are connected to the input of an amplifier-loudspeaker, Fig. 17.22*a*. The method uses the fact that as the time base starts off the trace at the left of the screen, a pulse comes from the sweep output and produces a noise in the loudspeaker. That this is so should be checked.

To make a measurement the time-base speed is increased to 1 ms/cm and a microphone, set close to the loudspeaker, is connected to the INPUT and E terminals on the CRO, Fig. 17.22*b*. The signal received by the microphone from the loudspeaker causes a 'wavy' trace on the CRO. (The volume control on the amplifier may have to be turned up and the position of the trace adjusted by the X-SHIFT.) When the microphone is moved away from the speaker the 'wavy' trace moves to the right on the screen. If we find the distance the microphone has to move to cause the trace to move

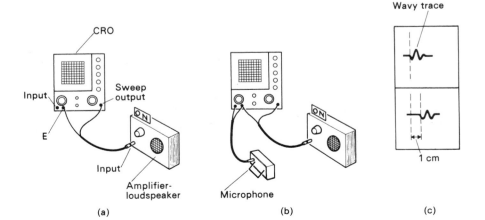

Fig. 17.22   (a)   (b)   (c)

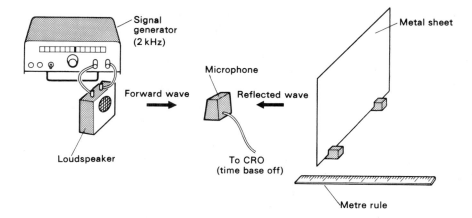

Fig. 17.23

one centimetre to the right, Fig. 17.22*c*, then we can say sound travels that distance in one millisecond. The speed in air can then be calculated.

(*b*) *In air by stationary waves.* The apparatus is arranged as in Fig. 17.23. The signal generator delivers a note of known frequency *f* (2 kHz is suitable) to a loudspeaker which is directed towards a metal sheet. Interference occurs between the forward and reflected progressive sound waves and a stationary wave pattern with nodes and antinodes is established. If the reflector is moved slowly towards or away from the microphone the vertical trace on the CRO varies from a maximum to a minimum and the distance moved by the reflector between two consecutive maxima or minima equals $\lambda/2$ where λ is the wavelength of the sound wave (progressive or stationary). The speed *v* is obtained from $v = f\lambda$. In practice the reflector is moved through several maxima and minima so that a greater distance is measured.

The stationary wave pattern produced in Kundt's tube (p. 368) and indicated by the lycopodium powder, may also be used to find the speed of sound in air or in different gases.

(*c*) *In air using a resonance tube.* The arrangement is shown in Fig. 17.24*a*. The tuning fork is struck and held over the top of the tube whose position in the water is raised or lowered until the note is at its loudest. The fundamental frequency of the air column then equals the frequency of the fork, i.e. there is resonance. A stationary wave now exists in the tube with a displacement node N at the closed end and a displacement antinode A near the open end. We then have

$$l_1 + c = \frac{\lambda}{4} \tag{1}$$

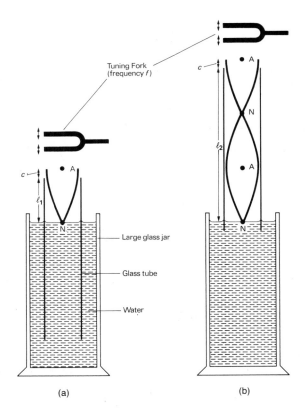

(a)   (b)

Fig. 17.24

where l_1 is the length from the water level to the top of the tube and c is the end correction.

A second weaker resonance can be obtained with the same fork by slowly raising the tube out of the water. The air column of greater length l_2 is then producing its first overtone (which equals the frequency of the fork) and from the stationary wave pattern of Fig. 17.24b we have

$$l_2 + c = \frac{3\lambda}{4} \qquad (2)$$

Subtracting (1) from (2) eliminates c and we get

$$l_2 - l_1 = \frac{\lambda}{2}$$

Knowing the frequency f of the tuning fork, v can be calculated from $v = f\lambda$.

Alternatively using several forks of different, known frequencies f the fundamental resonance lengths l can be found for each and we have

$$l + c = \frac{\lambda}{4} = \frac{v}{4f}$$

$$\therefore \quad l = \frac{v}{4f} - c$$

A graph of l against $1/f$ should be a straight line of slope $v/4$ (and intercept $-c$ on the $1/f$ axis).

(d) *In a metal rod.* This method gives an approximate value but allows a rough check to be made on the expression $v = \sqrt{E/\rho}$ (p. 375) for, say, steel. In Fig. 17.25a

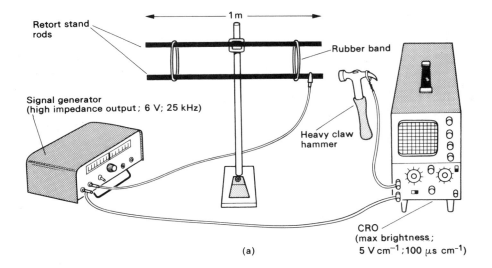

(a)

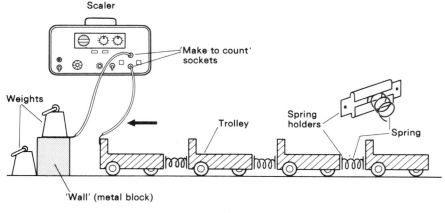

Fig. 17.25

(b)

the output from the signal generator is applied to the input of the CRO when the near end of the suspended rod is hit with the hammer. A compression pulse travels to the far end and is reflected as an expansion (rarefaction) pulse to the near end where *it breaks the contact between the hammer and the rod*. The time of contact is calculated from the length of the trace on the CRO (using its calibrated time scale of $100\ \mu s\ cm^{-1}$) and is the time for sound to travel *twice* the length of the rod. (*Note*. The stability control on the CRO should be turned only as far anticlockwise as is needed to give a trace when contact is made.)

A demonstration in slow motion with the apparatus of Fig. 17.25b may help to reinforce the idea that the reflected expansion pulse breaks the contact between the hammer and the rod. The four trolleys linked by springs are *all pushed together* to the left. The front trolley stops on striking the 'wall' and a compression pulse travels along the row as each trolley stops in turn. The rear trolley then moves to the right, starting an expansion pulse which travels left to the front trolley and on reaching it contact with the 'wall' is broken. The disturbance thus travels twice along the row of trolleys during the time the front trolley is in contact with the 'wall'. The scaler records the time of contact and if the speed of the pulses is required the distance from the front of the first trolley to the front of the rear one has to be measured.

Doppler effect

The pitch of the note from the siren of a fast-travelling ambulance or police car appears to a stationary observer to drop suddenly as it passes. This *apparent* change in the frequency of a wave motion when there is relative motion between the source and the observer is called the *Doppler effect*. It occurs with electromagnetic waves as well as with sound. The expressions derived below apply to the latter and only to the former if the relative speed of source and observer is small compared with the speed of electromagnetic waves (otherwise relativistic effects have to be considered). The microwave Doppler effect is used in police radar speed checks and in tracking satellites.

(*a*) *Source moving*. In Fig. 17.26a S is the source of waves of frequency f and velocity v and O is the stationary observer. If S were at rest, the f waves emitted per second would occupy a distance v and the wavelength would be v/f. When S is moving towards O with velocity u_s, f waves are now compressed into the smaller distance $(v - u_s)$ since S moves a distance u_s towards O per second, Fig. 17.26b. To O the effect thus appears to be a *decrease of wavelength* to a value λ_0 given by

$$\lambda_0 = \frac{(v - u_s)}{f}$$

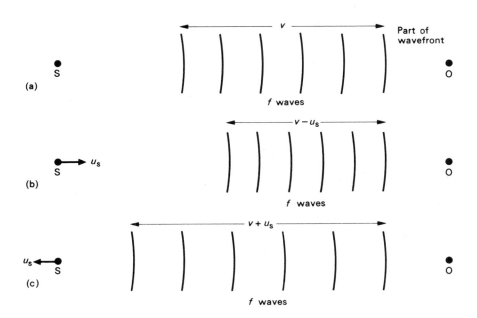

Fig. 17.26

Hence if f_0 is the apparent frequency we have

$$f_0 = \frac{\text{velocity of waves}}{\text{apparent wavelength}}$$

$$= \frac{v}{(v - u_s)/f}$$

$$\therefore \quad f_0 = \left(\frac{v}{v - u_s}\right)f$$

The apparent frequency is therefore greater than the true frequency since $(v - u_s) < v$.

If S is moving away from the stationary observer O, the apparent wavelength $\lambda_0 = (v + u_s)/f$, Fig. 17.26c, and the apparent frequency f_0 is

$$f_0 = \left(\frac{v}{v + u_s}\right)f$$

In this case $f_0 < f$ since $(v + u_s) > v$.

It should be noted that if, for example, a source of sound is approaching an observer with *constant velocity*, the apparent pitch of the note heard does not increase as the source gets nearer; it is constant but higher than the true pitch. Similarly, if the source recedes with constant velocity the observer hears a note of constant but lower pitch than the true pitch. The apparent change heard by the observer occurs *suddenly as the source passes*. What expression gives the value of the apparent change?

(*b*) *Observer moving*. In this case the wavelength is unchanged and is given by v/f since the f waves sent out per second by the stationary source S occupy a distance v, Fig. 17.27. If the observer O has velocity u_0 towards S, the velocity of the waves relative to O is $(v + u_0)$. The apparent frequency f_0 is given by

$$f_0 = \frac{\text{relative velocity of waves}}{\text{wavelength}}$$

$$= \frac{v + u_0}{v/f}$$

$$\therefore \quad f_0 = \left(\frac{v + u_0}{v}\right)f$$

Thus $f_0 > f$ since $(v + u_0) > v$.

If O is moving away from the stationary source S, the velocity of the waves relative to O is $(v - u_0)$ and the apparent frequency f_0 is

$$f_0 = \left(\frac{v - u_0}{v}\right)f$$

Here $f_0 < f$.

(*c*) *Source and observer moving*. Motion of the source affects the *apparent wavelength* and motion of the observer affects the *velocity of the waves* he receives. If the source and observer are approaching each other with velocities u_s and u_0 respectively then, as before, we can say

velocity of waves relative to O $= v_0 = v + u_0$

apparent wavelength $\lambda_0 \qquad = (v - u_s)/f$

For the apparent frequency f_0 we therefore have

$$f_0 = \frac{v_0}{\lambda_0} = \frac{v + u_0}{(v - u_s)/f}$$

$$\therefore \quad f_0 = \left(\frac{v + u_0}{v - u_s}\right)f$$

When source and observer are moving away from one another

$$\lambda_0 = \frac{v + u_s}{f} \quad \text{and} \quad v_0 = v - u_0$$

Hence $\qquad f_0 = \frac{v_0}{\lambda_0} = \frac{v - u_0}{(v + u_s)/f}$

$$\therefore \quad f_0 = \left(\frac{v - u_0}{v + u_s}\right)f$$

In general $\qquad f_0 = \left(\frac{v \pm u_0}{v \mp u_s}\right)f$

where the upper signs apply to approach and the lower signs to separation.

(*d*) *Doppler effect in light*. Because of the very great

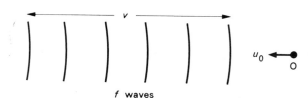

Fig. 17.27

speed o. ht the effect is negligible for most terrestrial sources but it has been used to measure the speeds of stars relative to the earth. The positions of certain wavelengths (i.e. spectral lines), due to an identifiable element in the star's spectrum are compared with their positions in a laboratory produced spectrum of the element. Red shift indicates recession of the star from the earth and from its size the speed may be calculated.

The speed of rotation of the sun has also been found from the apparent difference in wavelength between the Fraunhöfer lines (p. 404) in the spectra of the western and eastern edges of its disc; the former is moving towards the earth and the latter is receding. The value obtained agrees with that deduced from observations of sunspots.

Doppler effect red shift measurements on the light from galaxies (each comprising many millions of stars) suggest that the universe is expanding, with each galaxy retreating from every other at speeds of up to one-third that of light.

One cause of the broadening of spectral lines is due to the Doppler effect. The molecules of a gas or vapour which is emitting light are moving at different angles towards and away from the observer with various very high speeds. As a result the wavelength of a particular spectral line has a range of apparent values. How will the molecules be moving that are responsible for the edges of the line?

Sound calculations

1. What are the first two successive resonance lengths of a closed pipe containing air at 27 °C for a tuning fork of frequency 341 Hz? Take the speed of sound in air at 0 °C to be 330 m s⁻¹.

The speed of sound in a gas is directly proportional to the square root of the absolute temperature

$$\therefore \quad \frac{v_1}{v_0} = \sqrt{\frac{273 + 27}{273}} = \sqrt{\frac{300}{273}}$$

where v_1 and v_0 are the speeds of sound at 27 °C and 0 °C respectively,

$$\therefore \quad v_1 = 330\sqrt{\frac{300}{273}} = 346 \text{ m s}^{-1}$$

For the first resonance length l_1, the closed pipe emits its fundamental frequency f_1 where $f_1 = 341$ Hz. The wavelength of the fundamental note is given by

$$\lambda_1 = \frac{v_1}{f_1} = \frac{346}{341} = 1.02 \text{ m}$$

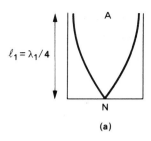

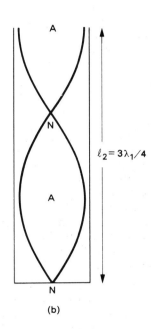

Fig. 17.28

The stationary wave pattern in the tube is then as in Fig. 17.28a, hence

$$l_1 = \frac{\lambda_1}{4} = \frac{1.02}{4} = 0.255 \text{ m}$$

At the second resonance length l_2 the stationary wave is as in Fig. 17.28b. The resonating note has frequency f_1 and wavelength λ_1 as before, but it is now the first overtone for length l_2. Therefore

$$l_2 = \tfrac{3}{4}\lambda_1 = \frac{3 \times 1.02}{4} = 0.765 \text{ m}$$

2. A sonometer wire of length 0.50 m and mass per unit length 1.0 × 10⁻³ kg m⁻¹ is stretched by a load of 4.0 kg. If it is plucked at its mid-point, what will be (a) the wavelength and (b) the frequency, of the note emitted? Take g = 10 N kg⁻¹.

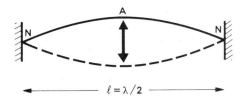

Fig. 17.29

(a) The wire vibrates as in Fig. 17.29 and emits its fundamental frequency f of wavelength λ. If l is the length of the wire then

$$\therefore \quad l = \frac{\lambda}{2}$$

$$\therefore \quad \lambda = 2l = 2 \times 0.50 = 1.0 \text{ m}$$

(b) The fundamental frequency f emitted by a wire of length l, mass per unit length μ and under tension T is given by

$$f = \frac{1}{2l} \sqrt{\frac{T}{\mu}}$$

Now $T = 4.0 \text{ kg} \times 10 \text{ N kg}^{-1} = 40 \text{ N}$, therefore

$$f = \frac{1}{2 \times 0.50} \sqrt{\frac{40}{1.0 \times 10^{-3}}} \text{ Hz}$$

$$= 2.0 \times 10^2 \text{ Hz}$$

3. *Calculate the frequency of the beats heard by a stationary observer when a source of sound of frequency 100 Hz moves directly away from him with a speed of 10.0 m s⁻¹ towards a vertical wall. (Speed of sound in air = 340 m s⁻¹.)*

The observer hears beats because of interference between the sound coming to him directly from the source and that reflected from the wall. The apparent frequency of the former is less than 100 Hz since the source is moving away from him, while the apparent frequency of the latter is greater than 100 Hz. The motion of the source towards the wall causes the waves incident on it to have a shorter wavelength than they would if the source were at rest and so, to the observer, the reflected waves appear to come from a source moving towards him.

Apparent frequency of direct sound

$$= \left(\frac{340}{340 + 10}\right) 100$$

$$= 97.1 \text{ Hz}$$

Apparent frequency of reflected sound

$$= \left(\frac{340}{340 - 10}\right) 100$$

$$= 103 \text{ Hz}$$

$$\therefore \quad \text{Number of beats per second} = 5.90$$

Measuring the frequency of a source of sound

Three methods will be outlined with instruments that are assumed to be correctly calibrated.

(a) *Using a signal generator and a loudspeaker.* The loudspeaker is fed from a signal generator whose frequency is altered until the notes from the speaker and the source of sound are judged to have the same pitch.

Fig. 17.30   (a)   (b)

(b) *Using a microphone and a CRO.* The sound is received by the microphone and fed to the CRO (if need be via a pre-amplifier) which is set on a suitable, known time base range. By noting the number of cycles of a.c. on a certain length of the time scale, the frequency can be worked out.

(c) *Using a stroboscope.* A stroboscope makes an object which is moving appear to be at rest. A lamp stroboscope, Fig. 17.30, consists of a lamp (containing xenon gas) which emits brief but intense flashes of light from about 2 to 250 times a second according to the setting of the speed control. Thus if, for example, 200 is the *highest* flashing speed which makes a sonometer wire appear to be at rest and in one position, then 200 Hz is the frequency of the stationary wave on the wire and also of the progressive sound wave emitted into the surrounding air by it. What would be seen if the flashing speed were (i) 100 Hz, and (ii) 400 Hz?

QUESTIONS

1. A point source A emits spherical sound waves. State how the intensity of sound varies with position around the source, assuming that there is no absorption in the medium.

A second identical source B is placed near A, the two sources being in phase. Explain why there will be positions of maximum and minimum intensity near the sources.

If the wavelength for each source is 0.40 m, and the rate at which sound energy is emitted by A is 1.44 times the rate of that emitted by B, explain why the sound intensity is found to be zero at a point 13.2 m from A and 11.0 m from B. (*J.M.B.*)

2. Explain what is meant by the statement that 'sound is propagated in air as longitudinal progressive waves', and outline the experimental evidence in favour of this statement. Discuss how changes in atmospheric conditions of pressure, temperature and humidity might be expected to affect the speed of sound.

Two loudspeakers face each other at a separation of about 100 m and are connected to the same oscillator, which gives a signal of frequency 110 Hz. Describe and explain the variation of sound intensity along the line joining the speakers. A man walks along the line with a uniform speed of 2.0 m s^{-1}. What does he hear? (Speed of sound = 330 m s^{-1}.) (*C.*)

3. Draw a diagram showing the positions of nodes and antinodes in the vibrations of an air column in a pipe closed at one end when it is giving the third harmonic of its fundamental note. Indicate on the diagram the directions and relative magnitudes of motion of the air at several significant points in the pipe at an instant when these motions are at maximum velocity. Calculate the frequency of this third harmonic if the effective length of the closed pipe is 72.0 cm. Speed of sound in air = 330 m s^{-1}. (*S.*)

4. Distinguish between *free vibrations* and *forced vibrations*, and explain the term *resonance*. Discuss the meanings of these terms by considering a specific vibrating system chosen from any branch of physics other than sound.

A small loudspeaker, actuated by a variable frequency oscillator, is sounded continuously over the open end of a vertical tube 40 cm long and closed at its lower end. At what frequencies will resonance occur as the frequency of the note emitted by the loudspeaker is increased from 200 Hz to 1200 Hz, given that the velocity of sound in air is 3.44×10^4 cm s^{-1}? Neglect the end effect of the tube. (*L.*)

5. Two open organ pipes of lengths 50 and 51 cm respectively, give beats of frequency 6.0 Hz when sounding their fundamental notes together. Neglecting end corrections what value does this give for the velocity of sound in air?

(*A.E.B. part qn.*)

6. Two sonometer wires, A of diameter 7.0×10^{-4} m and B of diameter 6.0×10^{-4} m, of the same material, are stretched side by side under the same tension. They vibrate at the same fundamental frequency of 256 Hz. If the length of B is 0.91 m, find the length of A. Calculate also the number of beats per second which will occur if the length of B is reduced to 0.90 m. (*O. part qn.*)

7. What is meant by a *wave motion*? Define the terms *wavelength* and *frequency* and derive the relationship between them.

Given that the velocity v of transverse waves along a stretched string is related to the tension T and the mass μ per unit length by the equation

$$v = \sqrt{\frac{T}{\mu}}$$

derive an expression for the natural frequencies of a string of length l when fixed at both ends.

Explain how the vibration of a string in a musical instrument produces sound and how this sound reaches the ear. Discuss the factors which determine the quality of the sound heard by the listener. (*O. and C.*)

8. Explain why the note emitted by a stretched string can easily be distinguished from that of a tuning fork with which it is in unison. How would you justify your answer by experiment?

Describe how, using a set of standard forks, you would verify experimentally the relationship between the frequency of the note emitted by a string of fixed length and tension and the mass per unit length of the wire.

A sonometer wire of length 1.0 m emits the same fundamental frequency as a given tuning fork. The wire is shortened by 0.05 m, the tension remaining unaltered, and 10 beats per second are heard when the wire and fork are sounded together. What is the frequency of the fork? If the mass per unit length of the wire is 1.4×10^{-3} kg m^{-1}, what is the tension? (*A.E.B.*)

9. Describe the motion of the particles of a string under constant tension and fixed at both ends when the string executes transverse vibrations of (a) its fundamental frequency, and (b) the first overtone (second harmonic). Illustrate your answer with suitable diagrams.

A horizontal sonometer wire of fixed length 0.50 m and mass 4.5×10^{-3} kg is under a fixed tension of 1.2×10^2 N. The poles of a horse-shoe magnet are arranged to produce a horizontal transverse magnetic field at the midpoint of the wire, and an alternating sinusoidal current passes through the wire. State and explain what happens when the frequency of the current is progressively increased from 100 to 200 Hz. Support your explanation by performing a suitable calculation. Indicate how you would use such an apparatus to measure the fixed frequency of an alternating current.

(J.M.B.)

10. A source emitting a note of certain frequency f approaches a stationary observer at a constant speed of one-tenth the speed of sound in air. The source is then maintained stationary and the observer moves towards it at the same constant speed. Determine from first principles the frequency of the note heard by the observer in each case.

(J.M.B.)

11. Show that when a source emitting sound waves of frequency f moves towards a stationary observer with velocity u, the observer hears a note of frequency $fv/(v - u)$, where v is the velocity of sound.

Describe the effect of a steady wind blowing with velocity w directly from source to observer (a) in the case above, and (b) if the source and the observer are both at rest.

A model aircraft on a control line travels in a horizontal circle of 10 m radius, making 1 revolution in 3.0 s. It emits a note of frequency 300 Hz. Calculate the maximum and minimum frequencies of the note heard at a point 20 m from the centre of the circle and in the plane of the path, and find the time interval between a maximum and the minimum that next succeeds it. (Take the velocity of sound in air to be 330 m s^{-1}.)

(O.)

12. Show that two identical progressive wave-trains travelling along the same straight line in opposite directions in a given medium set up a system of stationary waves. Compare the properties of a stationary wave system in air with those of a progressive wave-train in respect of (a) amplitude, (b) phase, and (c) pressure variation.

An observer moving between two identical sources of sound along the straight line joining them, hears beats at the rate of 4.0 s^{-1}. At what velocity is he moving if the frequency of each source is 500 Hz and the velocity of sound when he makes the observations is 3.40×10^4 cm s^{-1}?

(L.)

13. A car travelling normally towards a cliff at a speed of 30 m s^{-1} sounds its horn which emits a note of frequency 100 Hz. What is the apparent frequency of the echo as heard by the driver? Speed of sound in air = 330 m s^{-1}.

14. The velocity of sound v in a gas is given by $v = \sqrt{\dfrac{\gamma p}{\rho}}$, where γ is the ratio of the principal specific heats of the gas, p is its pressure, and ρ is its density. If v is found to be 400 m s^{-1} for a certain gas under particular conditions, what would be the new value of v if (a) the pressure were reduced by 4 per cent, and (b) the absolute temperature were increased by 4 per cent? Explain how you arrive at your answers.

(S.)

15. Compare the mode of propagation of sound with that of light.

How is the speed of sound in a gas affected by (a) an increase in the temperature of the gas, and (b) a decrease in the pressure of the gas?

A generator of ultrasonic waves of frequency 1.00 MHz is set up in a rectangular tank of paraffin facing a detector, which is connected to an amplifier and oscilloscope. It is found that the amplitude of the oscilloscope trace varies periodically as the detector is moved away from the generator. A series of consecutive maxima occurs at 2.99 mm, 3.65 mm, 4.33 mm, 5.00 mm and 5.66 mm from the generator. Explain the variation of the oscilloscope trace, and calculate the speed of sound in paraffin.

(C.)

16. Explain the formation of 'beats' and show that if two notes of frequencies f_1 and f_2 are sounded together the frequency of the beats is given by $f_1 \sim f_2$. How would you determine, other than by ear, which frequency is the higher?

Explain with the aid of diagrams how sound waves may be refracted by (a) a wind gradient, and (b) a temperature gradient. What is the effect of these refractions on the audibility of a sound?

A man standing close to an iron railing consisting of evenly-spaced uprights makes a sharp sound and hears a note of frequency 640 Hz. Calculate the spacing between the uprights. (Speed of sound in air = 330 m s^{-1} at 0 °C. Ambient temperature = 17.0 °C.)

(A.E.B.)

17. A vibrating tuning fork, observed by a light which flashes at regular intervals, has the same appearance as it has when at rest. A circular disc with 70 equally-spaced radii drawn on it is rotated from rest with gradually increasing speed. When viewed in the same light the disc first appears at rest when it rotates at one revolution per second. What are the possible values of the frequency of the fork?

(S.)

18 Physical optics

Nature of light

Two apparently contradictory theories of the nature of light were advanced in the seventeenth century.

The *corpuscular theory* regarded light as a stream of tiny particles or corpuscles travelling at high speed in straight lines. Newton supported this view which did account for rectilinear propagation, reflection and re-fraction—the latter by assuming that on entering an optically denser medium the corpuscles are attracted, thereby causing bending towards the normal. On the other hand, the *wave theory* proposed by Huygens about 1680 considered light to travel as waves. As we have seen (p. 343), a wave model can account satisfactorily for reflection and refraction.

Opponents of the wave theory argued that waves require a transmitting medium and there did not appear to be one for light which was able to travel in a vacuum. Subsequently a medium, called the *ether*, was invented but it defied all attempts at detection. An apparently crucial difference was that whereas the corpuscular theory required light to have a greater speed in a material than in air, the wave theory predicted a lower speed. It was not until 1862 that the wave theory prediction was confirmed when Foucault found the speed of light in water to be less than that in air. In the meantime, interference and diffraction effects had been discovered which were more readily explicable in terms of waves than in terms of corpuscles.

The need for the ether disappeared when Maxwell suggested in 1864 that light was an electromagnetic 'wave' and consisted of a fluctuating electric field coupled with a fluctuating magnetic field (p. 410). By the end of the nineteenth century Maxwell's electro-magnetic wave theory of radiation had established itself as one of the great intellectual pillars of physics, a unifying principle linking electricity, magnetism and light. However, as we shall see later, it did not on its own give a completely satisfactory explanation of all the properties of electromagnetic radiation.

In this chapter we will deal with physical optics—the subject which is concerned with those effects that make sense if we regard light as having a wave-like nature: such effects are interference, diffraction and polariza-tion.

Speed of light

A knowledge of the speed of light is important for several reasons. First, if the experimental value agrees with the theoretical value predicted by Maxwell for electromagnetic waves (see p. 411) then it is reasonable to assume that (*i*) light is an electromagnetic wave and (*ii*) the electromagnetic theory of radiation is valid. Second, it occurs in certain basic expressions of atomic and nuclear physics, such as Einstein's mass–energy equation $E = mc^2$ (p. 548). Third, the measurement of distance by radar techniques becomes possible (p. 357).

The rotating mirror method is based on one due to Foucault (1862) and in the school laboratory version of the apparatus, Fig. 18.1*a*, an attempt is made to esti-mate, in effect, the time taken by light to travel a distance of about four metres. This is of the order of 10^{-8} second, consequently the result obtained is subject to a large error.

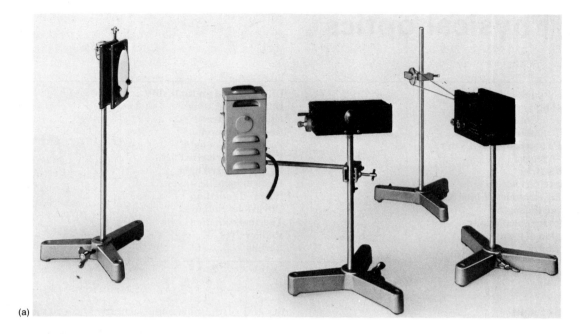

(a)

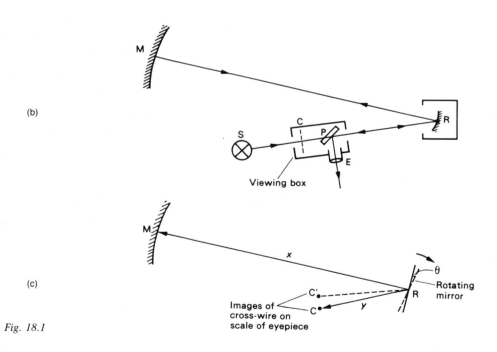

(b)

(c)

Fig. 18.1

The principle of the method is shown by Fig. 18.1*b*. Light from a source S illuminates a cross-wire C in the viewing box and travels on through the glass plate P to the rotating mirror R where it is reflected to the large fixed concave mirror M, placed so that R is at its centre of curvature. M reflects the light back to R and into the viewing box where it is partially reflected by P into the eyepiece E. An image of C is formed on the eyepiece scale. When R is rotated at high speed (by a small electric motor), the image of C is displaced by a very small distance to C'.

To obtain an expression for the speed of light c, consider the simplified diagram Fig. 18.1c. Let $MR = x$, $CR = y$ and when the rotating mirror makes n revolutions per second, suppose the displacement of the image of C is $CC' = d$. Then

distance travelled by light between reflections at R
$= 2x$

time for rotating mirror to make one revolution
$= 1/n$

time for mirror to rotate angle θ (in rad)
$= \theta/(2\pi n)$

Hence

$$c = \frac{\text{distance travelled by light}}{\text{time taken}} = \frac{2x}{\theta/(2\pi n)}$$

$$= \frac{4\pi nx}{\theta}$$

Also, since the reflected ray turns through twice the angle of rotation θ of the mirror,

angular rotation of image of C (i.e. angle CRC')
$= 2\theta = d/y$

Substituting for θ,

$$c = \frac{8\pi nxy}{d}$$

In practice d is extremely small (about 0.05 mm) and the error in measuring it is reduced by noting the displacement $2d$ of the cross-wire when the rotation of the mirror is reversed. Typical values for x and y are 2 m and 0.6 m respectively.

The speed of rotation n can be estimated in various ways. A signal generator connected to a loudspeaker can be adjusted until the note from the latter is judged to be the same frequency as that produced by the rotating mirror. Alternatively, a length of thread is tied to form a loop—say exactly one metre (i.e. one metre from the knot round the loop and back to the knot). It is then placed round the boss of the idler pulley in the rotating mirror unit and run round another pulley on a stand. By finding the time of revolution of the knot in the loop and knowing the diameters of the idler pulley boss, the idler pulley and the mirror shaft, Fig. 18.2, the speed of the mirror can be calculated.

Measurements of the speed of other members of the electromagnetic family confirm that they all—whatever their wavelength—travel in a vacuum with the *same* speed of 2.998×10^8 m s^{-1}. In other media the speed varies with the wavelength; for example, red light travels faster in glass than does blue light.

Interference of light

We saw earlier (p. 346) that interference occurs when waves from two coherent sources cross. Such sources produce waves having the same frequency, equal or comparable amplitudes and a phase difference that does not alter with time.

The wavelength of light must be very small otherwise the diffraction effects it gives would be more evident than they are. It therefore follows from the experiments with water waves in a ripple tank (p. 347) that to obtain nodal and antinodal lines (called *interference fringes*) sufficiently far apart to be seen, we must have

(*i*) the sources very close together
(*ii*) the screen (or eyepiece) as far as possible from the sources.

Also, if the sources are to be coherent they must be derived, in practice, from the same source. If an attempt is made to produce interference with two separate light sources, uniform illumination results instead of regions of light and dark. This is due to the fact that in a light source the phase is constantly

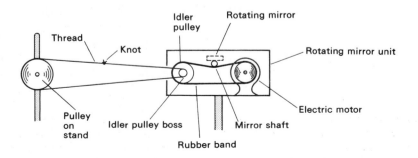

Fig. 18.2

changing because light is emitted in short bursts that last about 10^{-9} s when electrons in individual atoms suffer energy changes that occur very quickly and randomly. Phase changes happen abruptly as different atoms come into action and the eye is unable to follow the rapidly-changing interference pattern.

This is also true for light coming from different parts of the same source (with one exception). What is necessary then is that two wave-trains arriving at a given point, say by different paths, should have come from the *same point* of the *same source*. Any phase change occurring in the source then occurs in both wave-trains and stationary interference effects result. To obtain two coherent wave-trains from a point of a single source one of two methods is adopted: (*i*) *division of wavefront*, as is done in Young's double slit, Fresnel's biprism and Lloyd's mirror, and (*ii*) *division of amplitude*, usually by partial reflection and transmission at a boundary, as occurs in wedge fringes (p. 391) and Newton's rings (p. 392).

The exception mentioned above is the *laser* which does produce coherent light because the atoms are made to act in unison and all undergo energy changes simultaneously. Thus if a screen with two small holes is placed in a laser beam so that different parts of the source are used, an interference pattern is obtained.

Interference accounts for the colours of soap bubbles and of thin films of oil on a wet road. It also has practical applications (p. 394).

Young's double slit

One of the first to demonstrate the interference of light was Thomas Young in 1801.

(*a*) *Principle.* The principle of his method is shown in Fig. 18.3a. Monochromatic light (i.e. of one colour) from a narrow vertical slit S falls on two other narrow slits S_1 and S_2 which are very close together and parallel to S. S_1 and S_2 act as two coherent sources (both being derived from S) and if they (as well as S) are narrow enough, diffraction causes the emerging beams to spread into the region beyond the slits. Superposition occurs in the shaded area of Fig. 18.3a where the diffracted beams overlap. Alternate bright and dark equally-spaced vertical bands (interference fringes) can be observed on a screen or at the cross-wires of an eyepiece, Fig. 18.3b. If either S_1 or S_2 is covered the bands disappear.

(*b*) *Theory.* An expression for the separation of two bright (or dark) fringes can be obtained from Fig. 18.4a.

The path difference between waves reaching O from S_1 and S_2 is zero, i.e. $S_1O = S_2O$, they therefore arrive in phase and so there is a bright fringe at O, in the centre of the pattern. At P, distance x_1 from O, there will be a bright fringe if the path difference is a whole number of wavelengths, that is, if

$$S_2P - S_1P = n\lambda$$

where n is an integer (or zero) and λ is the wavelength of the light. We say the nth bright fringe is formed at P.

In Fig. 18.4a, d is the distance from the screen or cross-wire to the double slit and a is the slit separation. Hence

$$S_2P^2 = d^2 + (x_1 + a/2)^2 = d^2 + x_1^2 + ax_1 + a^2/4$$
$$S_1P^2 = d^2 + (x_1 - a/2)^2 = d^2 + x_1^2 - ax_1 + a^2/4$$

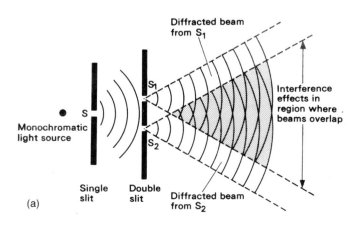

Monochromatic light source

Single slit

Double slit

Diffracted beam from S_1

Diffracted beam from S_2

Interference effects in region where beams overlap

(a)

Fig. 18.3

(b)

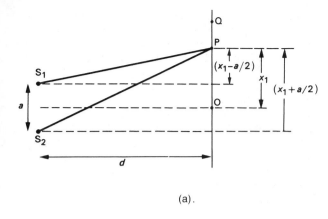

(a).

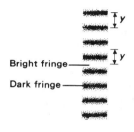

Bright fringe ———

Dark fringe ———

Fig. 18.4   (b)

$$\therefore \quad S_2P^2 - S_1P^2 = 2ax_1$$

But $S_2P^2 - S_1P^2 = (S_2P - S_1P)(S_2P + S_1P)$

In practice a is very small (e.g. 0.5 mm) compared with d (e.g. 1 m) and if P is near O then S_2P and S_1P are each just greater than d. Therefore we can say $(S_2P + S_1P) = 2d$. It follows that

$$(S_2P - S_1P)2d = 2ax_1$$

$$\therefore \quad S_2P - S_1P = ax_1/d$$

For the nth bright fringe at P we have

$$n\lambda = ax_1/d \qquad (1)$$

If the next bright fringe, i.e. the $(n + 1)$th, is formed at Q where $OQ = x_2$ then

$$S_2Q - S_1Q = (n + 1)\lambda$$

$$(n + 1)\lambda = ax_2/d \qquad (2)$$

Subtracting (1) from (2) we get

$$\lambda = a(x_2 - x_1)/d$$

If y is the distance between two adjacent bright (or dark) fringes, called the *fringe spacing*, Fig. 18.4*b*, then $y = x_2 - x_1$ and so

$$\lambda = \frac{ay}{d}$$

We see that (*i*) $y \propto 1/a$ if λ and d constant (therefore the slit separation should be small), (*ii*) $y \propto d$ if λ and a constant (therefore the fringes should be viewed from a distance) and (*iii*) $y \propto \lambda$ if a and d constant.

If a dark fringe were formed at P then $(S_2P - S_1P)$ would equal an odd number of half wavelengths.

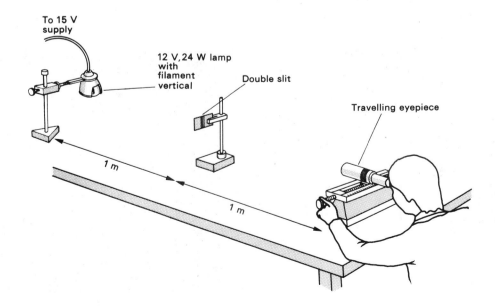

Fig. 18.5

(*c*) *Measurement of* λ. One experimental arrangement is shown in Fig. 18.5 in which the lamp filament acts as the single slit. (The double slit may be made by ruling two slits 0.5 mm apart on a microscope slide coated with Aquadag, using a blunt needle as described in Appendix 13, or it can be purchased ready-made from a laboratory supplier.) The fringes are viewed in a blacked-out room at the cross-wire of a travelling eyepiece (e.g. a travelling microscope with the objective removed). Filters are used to obtain coloured light from a white light source.

The average fringe spacing y is found by measuring across as many fringes as possible with the travelling eyepiece. A metre rule is used to measure d and the slit separation a is measured directly with a travelling microscope. The value of λ obtained is approximate; for violet light it is about 4×10^{-7} m (0.4 μm) and for red light about 7×10^{-7} m (0.7 μm).

(*d*) *Further points.*

(*i*) The overlapping beams which interfere are produced by diffraction at S_1 and S_2. Since the fringes can be observed anywhere in the overlapping region they are called *non-localized* fringes.

(*ii*) Point sources would give the same fringe system but slits (i.e. a line of point sources) give brighter fringes by reinforcing the pattern.

(*iii*) The fringes are really the intersections with a vertical plane of hyperboloids of revolution having S_1 and S_2 as foci. (These correspond in three dimensions to the two-dimensional hyperbolic nodal and antinodal lines, see Fig. 16.16, p. 348.)

(*iv*) The interference is incomplete because for all fringes except the central bright one, the amplitudes of the two wave-trains are not exactly equal. Why?

(*v*) Using white light fewer fringes are seen and each colour produces its own set of fringes which overlap. Only the central one is white, its position being the only one where the path difference is zero for all colours.

The first coloured fringe is bluish near the central fringe and red at its far side, Fig. 18.6. Why?

(*vi*) The fringe spacing for red light is greater than for blue light. Red light must therefore have a greater wavelength than blue light since $y \propto \lambda$ (if a and d are constant).

(*vii*) The number of fringes obtained depends on the amount of diffraction occurring at the slits and this in turn depends on their width (see pp. 398 and 572). The narrower the slits, the greater will be the number of fringes due to the increased diffraction but the fainter will they be, since less light gets through. In practice, to give easily seen fringes, the slits have to be many wavelengths wide; the case is similar to that shown for water waves in Fig. 16.20*b* (p. 350).

Fresnel's biprism: Lloyd's mirror

Two other ways of producing interference fringes from two coherent sources that are derived by division of the wavefront from a single source will now be outlined.

(*a*) *Fresnel's biprism.* Monochromatic light from a narrow slit S falls on a double glass prism arranged as in Fig. 18.7. Two virtual images S_1 and S_2 are formed of S, one by refraction at each half of the prism, and these act as coherent sources which are close together because of the small refracting angles (about 0.5°) of the prisms. An interference pattern, similar to that given by the double slit but brighter, is obtained in the shaded region where the two refracted beams overlap. The fringes can be observed as before using a travelling eyepiece focused on the fringes at its cross-wire.

The theory and the expression for the fringe spacing y are the same as for Young's method, i.e. $y = d\lambda/a$. To obtain a, a convex lens is moved between the biprism and the eyepiece until a real magnified image of the virtual 'objects' S_1 and S_2 is in focus. The distance b between the images of S_1 and S_2 (on the cross-wire of

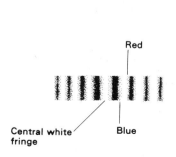

Fig. 18.6

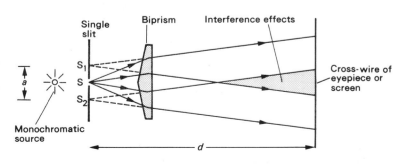

Fig. 18.7

the eyepiece) is measured, then knowing the object and image distances u and v respectively we can say that the magnification m is given by $m = v/u = b/a$. The wavelength λ can then be calculated if y and d are also determined, but this is not one of the most accurate ways of measuring the wavelength of light.

(b) *Lloyd's mirror*. A plane glass plate (acting as a mirror) is illuminated at almost grazing incidence by light from a slit S_1, parallel to the plate, Fig. 18.8. A virtual image S_2 of S_1 is formed close to S_1 by reflection and these two act as coherent sources. In the shaded area direct waves from S_1 cross reflected waves which appear to come from S_2 and interference fringes can be seen.

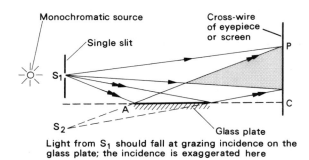

Monochromatic source
Single slit
Cross-wire of eyepiece or screen
S_1
S_2
A
C
P
Glass plate
Light from S_1 should fall at grazing incidence on the glass plate; the incidence is exaggerated here

Fig. 18.8

The expression giving the fringe spacing is the same as for the double slit and the biprism experiments, but the fringe system differs in one important respect. In Lloyd's mirror, if the point P, for example, is such that the path difference $(S_1A + AP) - S_1P$ (or $S_2P - S_1P$) is a whole number of wavelengths, the fringe at P is *dark*, not bright. This is due to the 180° phase change which occurs when light is reflected at a rare–dense boundary. This is equivalent to adding an extra half-wavelength to the path of the reflected wave and is similar to the phase change a pulse on a spring undergoes at a fixed end, or to that when microwaves are reflected at a metal plate (p. 357). At grazing incidence a fringe is formed at C, where the geometrical path difference between the direct and reflected waves is zero and it follows that it will be dark rather than bright.

Wedge fringes

Interference fringes are produced by a thin wedge-shaped film of air, the thickness of which gradually increases from zero along its length. The wedge can be formed from two microscope slides clamped at one end and separated by a thin piece of paper at the other so that the wedge angle is very small. In the arrangement of Fig. 18.9a monochromatic light from an extended source (e.g. a sodium lamp or flame) is partially reflected vertically downwards by the glass plate G. When the microscope is focused on the wedge, light and dark equally-spaced fringes are seen, parallel to the edge of contact of the wedge, Fig. 18.9b.

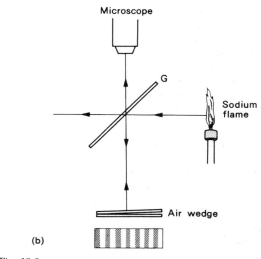

Microscope
G
Sodium flame
Air wedge

(b)

Fig. 18.9

Some of the light falling on the wedge is reflected upwards from the bottom surface of the top slide and some of the rest, which is transmitted through the air wedge, is reflected upwards from the top surface of the bottom slide. Both wave-trains have arisen from the same point P, Fig. 18.10, by *division of the amplitude*. They are therefore coherent and when brought together (by the eye or a microscope) they can interfere.

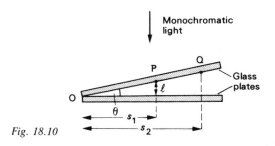

Monochromatic light
P
Q
Glass plates
O
θ
l
s_1
s_2

Fig. 18.10

If l is the thickness of the air wedge at P then, since the incidence is nearly normal, the path difference between the rays at P is $2l$. At O where the path difference is zero, we would expect a bright band but a *dark* band is observed. This is due to the 180° phase change which occurs when the wave-train *in* the air wedge is reflected at the top surface of the bottom slide, i.e. at a denser medium. The phase change, in effect, adds an extra path of half a wavelength (i.e. a crest is reflected as a trough). The path difference between the two wave-trains at P is thus $(2l + \lambda/2)$ where λ is the wavelength of the light. A bright fringe is formed at P if

$$2l + \lambda/2 = n\lambda$$
or
$$2l = (n - \tfrac{1}{2})\lambda$$

where n = 1 gives the first bright fringe, $n = 2$ gives the second bright fringe, etc. ($n = 0$ is impossible). A dark fringe is formed at P if

$$2l = n\lambda$$

where $n = 0$ gives the first dark fringe, $n = 1$ gives the second dark fringe, $n = 2$ gives the third dark fringe, etc.

The microscope (or the unaided eye) has, for normal incidence, to be focused on the top surface of the air wedge. This ensures that superposition of the two interfering wave-trains then occurs at a particular point in the retina. The fringes in this case are called *localized* fringes. Each fringe is the locus of points of equal air wedge thickness (i.e. same path difference) and they are often referred to as 'fringes of equal thickness'. There are also reflections from the other surfaces; these are usually out of focus and since they involve large path differences they do not spoil the fringes.

The angle θ of the wedge can be found if the reading on a travelling microscope is taken when the cross-wire is on a dark fringe at P, say the nth from O. We then have from Fig. 18.10 that

$$2l = n\lambda$$

But $l = s_1\theta$ (if θ is in radians), hence

$$2s_1\theta = n\lambda \tag{1}$$

If the microscope is now moved until it is on the $(n + k)$th dark fringe, say at Q, and the reading again taken, then

$$2s_2\theta = (n + k)\lambda \tag{2}$$

Subtracting (1) from (2)

$$2\theta(s_2 - s_1) = k\lambda$$

$$\therefore \quad \theta = \frac{k\lambda}{2(s_2 - s_1)}$$

where $(s_2 - s_1)$ is the distance moved by the microscope and λ is the wavelength of the light.

Newton's rings

This system of interference fringes, also produced by division of amplitude, was discovered by Newton but it was Young who gave a satisfactory explanation of their formation in terms of waves.

(*a*) *Principle.* The arrangement is shown in Fig. 18.11. Monochromatic light (e.g. from a sodium lamp or flame) is reflected by the glass plate G so that it falls normally on the air film formed between the convex lens of long focal length (about one metre) and the flat glass plate. The thickness of the air film gradually increases outwards from zero at the point of contact B but is the same at all points on any circle with centre B.

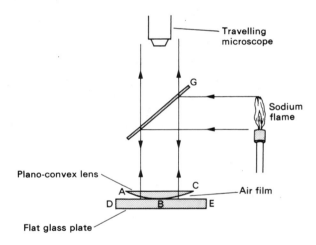

Fig. 18.11

Interference occurs between light reflected from the lower surface ABC of the lens and the upper surface DBE of the plate. A series of bright and dark rings is seen through G when a travelling microscope is focused *on the air film*, Fig. 18.12. The rings are fringes of 'equal thickness', 'localized' in the air film (like those formed by a wedge) and as their radii increase, the separation decreases. At the centre B of the fringe system where the geometrical path difference between the two wave-trains is zero, there is a *dark* spot. As with wedge fringes, this is due to the 180° phase change which occurs when light is reflected at an optically denser medium. In effect, the path difference at B is not zero but half a wavelength.

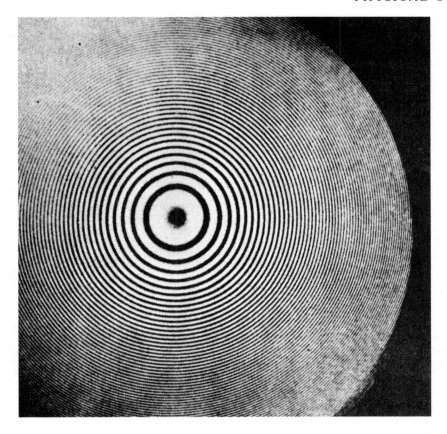

Fig. 18.12

(b) *Radius of a ring*. In Fig. 18.13 a complete circular section of a sphere of radius R is shown, R being the radius of curvature of the lower surface ABC of the lens. Let r_n be the radius OP of the ring at P, where the thickness PQ of the film is l. If BO is produced to meet the circle at L then BOL is a diameter. By the theorem of intersecting chords we have

$$SO \cdot OP = LO \cdot OB$$

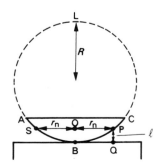

Fig. 18.13

$$\therefore \qquad r_n \cdot r_n = (2R - l)l$$
$$\text{(since OB} = \text{PQ} = l)$$
$$\therefore \qquad r_n{}^2 = 2Rl - l^2$$

But l^2 is very small compared with $2Rl$ since R is large, hence

$$r_n{}^2 = 2Rl$$

The path difference between the two interfering wave-trains at P is $2l$ for light incident normally on the air film. If a dark ring is formed at P then

$$2l = n\lambda$$

where n is an integer and λ is the wavelength of the light. Even though the geometrical path difference is a whole number of wavelengths, the $\lambda/2$ phase change at Q results in destructive interference. Hence from the two previous expressions we can write for the radius r_n of a *dark* fringe

$$r_n{}^2 = Rn\lambda$$

Thus $n = 0$ gives the central dark spot, $n = 1$ gives the

first dark ring, $n = 2$ gives the second dark ring and so on.

A *bright* ring would be formed at P if $2l$ equalled $\frac{1}{2}\lambda$, or $1\frac{1}{2}\lambda$, or $2\frac{1}{2}\lambda$, etc., since the phase change at the glass plate in effect increases the path length of the light reflected there by $\lambda/2$. Hence if the nth bright ring is formed at P, then

$$2l = (n - \tfrac{1}{2})\lambda$$

The radius r_n of a *bright* ring is therefore given by

$$r_n^2 = R(n - \tfrac{1}{2})\lambda$$

where $n = 1$ gives the first bright ring, $n = 2$ gives the second bright ring and so on.

(*c*) *Measurement of* λ. It is better to measure the diameters of rings rather than their radii because of the uncertainty of the position of the centre of the system. Hence if d_n is the diameter of the nth *dark* ring we have

$$r_n^2 = (d_n/2)^2 = Rn\lambda$$

$$\therefore \quad d_n^2 = 4Rn\lambda$$

A graph of d_n^2 against n is thus a straight line of slope $4R\lambda$. If it does not pass through the origin, n may have been miscounted each time *or* the contact between lens and plate is poor (and may give a central bright spot). Neither effect, however, affects the slope of the graph. A travelling microscope is used to measure d_n and R can be found by Boys' method.

Using interference

(*a*) *Testing of optical surfaces.* Fringes of equal thickness are useful for testing optical components. For example, in the making of optical 'flats', the plate under test is made to form an air wedge with a standard plane glass surface. Any uneven parts of the surface which require more grinding will show up as irregularities in what should be a parallel, equally-spaced, straight set of fringes.

The grinding of a lens surface may be checked if it is placed on an optical flat and Newton's rings observed in monochromatic light. The rings should be exactly circular if the lens is spherical.

(*b*) *Non-reflecting glass.* In optical instruments containing lenses or prisms light is lost by reflection at each refracting surface and results in reduced brightness of the final image. There will also be a loss of contrast against such a background of stray light.

The amount of light reflected at a surface can be appreciably reduced by coating it (by evaporation in a

vacuum) with a film of transparent material (e.g. magnesium fluoride) to a thickness of *one quarter of a wavelength of light in the film*, Fig. 18.14. Light reflected from the top (ray 1) and bottom (ray 2) surfaces of the film then interfere destructively since the latter has to travel twice the thickness of the film, i.e. the path difference is $2l$. The refractive index of the film is less than that of glass and so each reflection, being at a rare-dense boundary, suffers a 180° phase change. The net effect of the phase changes on the path difference is thus zero. The refractive index of the film should be as nearly as possible the mean of the refractive indices for air and glass so that the amounts of light reflected at the two surfaces are almost equal.

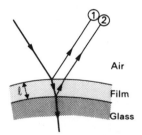

Air

Film

Glass

Fig. 18.14

For light of wavelength λ_a in air and λ_f in the film, the condition for destructive interference, at normal incidence, is

$$2l = \tfrac{1}{2}\lambda_f$$

$$\therefore \quad l = \tfrac{1}{4}\lambda_f$$

When light passes from one medium into another, its speed and wavelength change but not its frequency f. Therefore if v_a and v_f are the speeds of light in air and the film respectively, we have

$$v_a = f\lambda_a \quad \text{and} \quad v_f = f\lambda_f$$

Also if $_an_f$ is the refractive index of the material of the film relative to that of air then

$$_an_f = \frac{v_a}{v_f} \qquad \text{(see p. 344)}$$

$$= \frac{\lambda_a}{\lambda_f}$$

Hence

$$\lambda_f = \frac{\lambda_a}{_an_f}$$

The previous expression for l may then be written more

conveniently as

$$l = \frac{\lambda_a}{4 \, _a n_f}$$

The interference is complete for one wavelength only, usually taken to be that at the centre of the visible range (i.e. yellow-green). For red and blue light the reflection is weakened but not eliminated and a coated or 'bloomed' lens appears purple in white light. Energy which would have been wasted as reflected light increases the amount of transmitted light when destructive interference occurs.

If $\lambda_a = 550$ nm and $_a n_f = 1.38$, then $l = 100$ nm.

(c) *Measurement of length.* The SI unit of length, the metre, is defined as a certain number of wavelengths (1 650 763.73) in a vacuum of a particular line in the spectrum of an isotope of the inert gas krypton. The measurement required to set up this standard involves forming an interference pattern between two mirrors and counting the number of fringes which cross the field of view as one mirror is moved to the other. The movement of one-tenth of a fringe can just be detected and so the accuracy attainable is about 1 part in 10^7.

A similar technique is used to measure the very small expansion of a crystal when it is heated, the motion of interference fringes again being observed.

Everyday examples of interference

(a) *Colours of oil films on water.* Interference occurs between two wave-trains—one reflected from the surface of the oil and the other from the oil-water interface, Fig. 18.15. When the path difference gives constructive interference for light of one wavelength, the corresponding colour is seen in the film. The path difference varies with the thickness of the film and the angle of viewing, both of which affects the colour produced at any one part.

(b) *Pulsing of the picture on a television receiver.* This occurs when an aircraft passes low overhead. The signal travelling directly from the transmitting to the receiving aerial interferes with that reflected from the aircraft, Fig. 18.16. Usually the reflected wave is much weaker and so the interference is never completely destructive.

Optical path length

We can show that a length l in a medium of refractive index n is equivalent to a length nl in a vacuum. Let PQ in Fig. 18.17 be a plane wavefront falling obliquely on the surface separating a vacuum and the medium, in which the speeds of light are c and v respectively. Let t be the time for Q to reach R, i.e. QR $= ct$. In this time suppose P travels a distance PS in the medium, then PS $= vt$. If n is the absolute refractive index of the medium then

$$n = \frac{c}{v} \quad \text{and} \quad v = \frac{c}{n}$$

Hence

$$PS = \frac{ct}{n} = \frac{QR}{n}$$

$$\therefore \quad QR = n \, . \, PS$$

That is, if light travels a distance PS in the medium then in the same time it would travel a distance n. PS in a vacuum. In general,

length l in medium of refractive index n is optically equivalent to length nl in a vacuum.

nl is called the *optical path length* of distance l in the medium. It follows that if a thickness l of transparent material of refractive index n is placed in the path of a beam of light, the path difference between the new and the previous paths is $nl - l = (n - 1)l$.

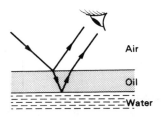

Fig. 18.15

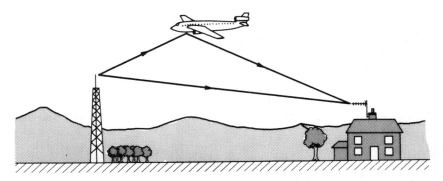

Fig. 18.16

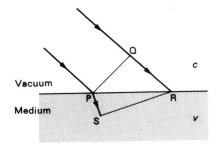

Fig. 18.17

Interference calculations

1. In a Young's double slit experiment the distance between the slits and the screen is 1.60 m and using light of wavelength 5.89×10^{-7} m the distance between the centre of the interference pattern and the fourth bright fringe on either side is 16.0 mm. What is the slit separation?

From $\lambda = ay/d$ (see p. 389) we have

$$a = \frac{d\lambda}{y}$$

where $\lambda = 5.89 \times 10^{-7}$ m, $d = 1.60$ m and $y =$ fringe spacing $= 16.0/4 = 4.0$ mm $= 4.0 \times 10^{-3}$ m. Hence

$$a = \frac{1.60 \times 5.89 \times 10^{-7}}{4.0 \times 10^{-3}} \frac{\text{m} \times \text{m}}{\text{m}}$$

$$= 0.236 \times 10^{-3} \text{ m}$$

$$= 0.236 \text{ mm}$$

2. An air wedge is formed between two glass plates which are in contact at one end and separated by a piece of thin metal foil at the other end. Calculate the thickness of the foil if 30 dark fringes are observed between the ends when light of wavelength 6.0×10^{-7} m is incident normally on the wedge, Fig. 18.18.

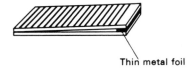

Thin metal foil

Fig. 18.18

The *first* dark fringe is formed at the end where the plates are in contact and the thickness *l* of the air wedge is zero.

The *second* dark fringe occurs where $l = \lambda/2$; the geometrical path difference is then $2l = \lambda$ and the $\lambda/2$ phase change at the bottom plate causes destructive interference.

The *third* dark fringe occurs where $l = 2(\lambda/2)$, the *fourth* where $l = 3(\lambda/2)$ and the *thirtieth* where $l = 29(\lambda/2)$.

$$\therefore \quad \text{Thickness of foil} = 29 \times \frac{6.0 \times 10^{-7}}{2} \text{ m}$$

$$= 8.7 \times 10^{-3} \text{ mm}$$

3. Calculate the radius of curvature of a plano-convex lens used to produce Newton's rings with a flat glass plate if the diameter of the tenth dark ring is 4.48 mm, viewed by normally reflected light of wavelength 5.00×10^{-7} m. What is the diameter of the twentieth bright ring?

For the *n*th dark ring of radius r_n

$$r_n^2 = Rn\lambda \qquad \text{(see p. 393)}$$

where R is the radius of curvature of the lens surface and λ is the wavelength of the light

$$\therefore \quad R = \frac{r_n^2}{n\lambda}$$

$$= \frac{(4.48 \times 10^{-3}/2)^2}{10 \times 5.00 \times 10^{-7}} \frac{\text{m}^2}{\text{m}}$$

since $n = 10$ for the tenth dark ring

$$\therefore \quad R = 1.00 \text{ m}$$

The radius r_n of the twentieth bright ring is given by

$$r_n^2 = R(n - \tfrac{1}{2})\lambda$$

where $n = 20$ and $R = 1.00$ m. Hence

$$r_n^2 = 1(20 - \tfrac{1}{2})5.00 \times 10^{-7}$$

$$\therefore \quad r_n = \sqrt{9.75} \times 10^{-3} \text{ m}$$

$$= 3.12 \text{ mm}$$

$$\therefore \quad d_n = 6.24 \text{ mm}$$

Diffraction of light

Light can spread round obstacles into regions that would be in shadow if it travelled exactly in straight lines, i.e. it exhibits the typical wave-like property of diffraction. Thus the edges of shadows are not sharp. Behind the obstacles or apertures at which diffraction occurs, a diffraction pattern of dark and bright fringes is formed which, under the right conditions, may be seen on a screen or at the cross-wire of an eyepiece.

A striking example of a diffraction pattern is shown in Fig. 18.19 and was produced by placing a razor blade midway between an illuminated pinhole and a photographic film. In most cases, small sources (e.g. a pinhole or a slit) are needed to observe diffraction effects and these are more evident if obstacles and apertures have linear dimensions comparable with the wavelength of light. With a large source each point gives rise to a diffraction pattern and there is uniform illumination when these overlap.

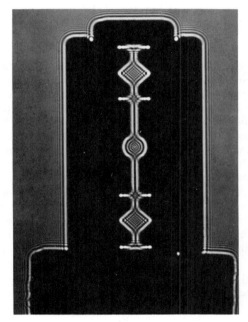

Fig. 18.19

Diffraction is regarded as being due to the superposition of secondary wavelets from coherent sources on the unrestricted part of a wavefront that has been obstructed by an obstacle or aperture (p. 351). Thus, whereas interference involves the superposition of

waves on two different wavefronts, in diffraction there is superposition of waves from different parts of the *same* wavefront.

Before considering the important case of diffraction at a single slit, we shall briefly explain in general terms the diffraction patterns due to various other obstacles.

(*a*) *Straight edge* (e.g. a razor blade). It can be shown that the fringes at points such as P in Fig. 18.20, which are close to the region of geometrical shadow, are due to the superposition of secondary wavelets from point sources on the unrestricted wavefront near R.

(*b*) *Circular obstacle* (e.g. a ball-bearing). In this case there is, rather surprisingly, a *bright* spot at the centre of the geometrical shadow, Fig. 18.21.

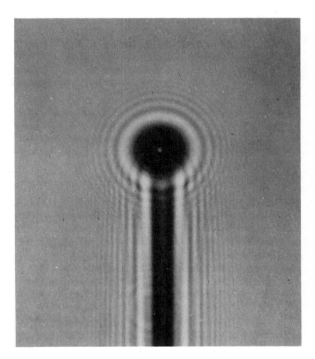

Fig. 18.21

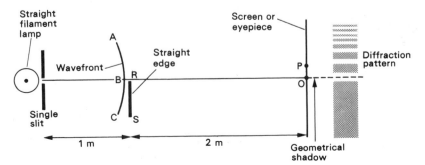

Fig. 18.20

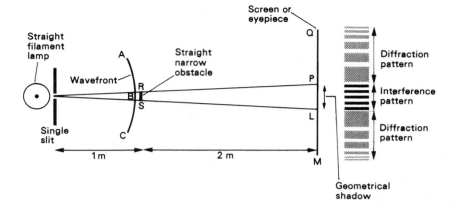

Fig. 18.22

(c) *Straight narrow obstacle* (e.g. a pin). Between P and Q in Fig. 18.22 the fringes arise largely from the superposition of secondary wavelets from points near R, i.e. the pattern is a diffraction one due to coherent sources on the same wavefront. Similarly, secondary wavelets from points near S cause the pattern between L and M. The evenly-spaced fringes inside the geometrical shadow LP are formed by the superposition of secondary wavelets from points around R and S acting as *two* coherent sources; this is an interference pattern similar to that obtained with Young's double slit.

Diffraction at a single slit

(a) *Experimental arrangement*, Fig. 18.23. The lamp has a straight filament, arranged to be vertical and the slit, whose width can be adjusted by a screw, is mounted with its length parallel to the filament. Initially the slit is opened wide and the lens moved to give a sharp image of the filament on the translucent screen. The shield and the large stop cut out stray light.

The diffraction pattern depends on the slit width and is observed from behind the screen, preferably through a magnifying glass. For widths of a few millimetres a

rectangle of light is produced, the result of near-rectilinear propagation. As the slit is narrowed, a pattern is obtained with a white central band having dark bands either side, fringed with colour. Inserting a red filter between lamp and screen gives red and black bands and the effect is as in Fig. 18.24a. In the diagrammatic representation of Fig. 18.24b, bright red bands are formed at A, B, C, D and E. With blue light the bands are closer together. Just before the slit closes the pattern dims and the central bright band widens causing illumination well into the geometrical shadow of the slit, i.e. the diffraction is very marked as the width of the slit approaches the wavelength of light and the slit acts like a secondary source.

(b) *Theory*. To account for the maxima and minima of the diffraction pattern we use Huygens' construction and consider each point of the slit as a source of secondary wavelets. The slit is imagined to consist of strips of equal width, parallel to the length of the slit. The total effect in a particular direction is then found by adding the wavelets emitted in that direction by all the strips, using the superposition principle. In practice this operation presents mathematical difficulties too

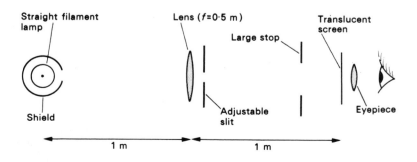

Fig. 18.23

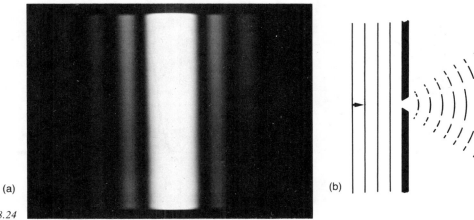

(a)

(b)

Fig. 18.24

complex to be dealt with here, but a simplified two-dimensional treatment based on dividing the slit into strips is possible.

Suppose plane waves (e.g. from a distant light source) fall normally on a narrow rectangular slit of width a, Fig. 18.25. Consider the first dark band (i.e. the first minimum) where there is no light. It will be formed at an angle θ to the incident beam if the path difference for the secondary wavelets from the strip just below A and the strip just below C (the midpoint of the slit) is $\lambda/2$, where λ is the wavelength of the light. Destructive interference will then occur for wavelets from this pair of strips since a crest from one strip reaches the observer with a trough from the other. This happens for all pairs of *corresponding* strips in AC and CB because the same path difference of $\lambda/2$ exists. Hence there is no light in direction θ when

$$CD = \lambda/2 \quad (\text{or } BE = \lambda)$$

But $\sin \theta = \sin CAD = CD/(a/2)$, therefore

$$CD = \frac{a \sin \theta}{2} \quad (\text{or } BE = a \sin \theta)$$

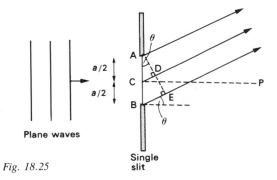

Plane waves

Single slit

Fig. 18.25

That is, the direction of the first minimum is given by

$$a \frac{\sin \theta}{2} = \frac{\lambda}{2} \quad (\text{or } a \sin \theta = \lambda)$$

$$\therefore \quad \sin \theta = \frac{\lambda}{a}$$

It can be shown by similar 'pairing' processes that other minima occur when

$$\sin \theta = \frac{n\lambda}{a}$$

where $n = \pm 1, \pm 2, \pm 3$, etc.; the $\pm$ signs indicate that, for example, there are two 'first order' minima, one on each side of the original direction of the incident beam. If θ is small and in radians we can write $\theta \simeq n\lambda/a$.

At P in Fig. 18.25, where the central bright band is formed, wavelets from all the imaginary strips in the slit arrive in phase since they have the same path length and the intensity of light is greatest. Other maxima occur roughly half-way between the minima at angles such that $\sin \theta$ has values $\pm 3\lambda/2a$, i.e. $(\lambda/a + 2\lambda/a)/2$, $\pm 5\lambda/2a$, etc. The first maximum is explained if the slit is divided into three equal parts and a direction considered in which the path differences between their ends are $\lambda/2$, Fig. 18.26. Wavelets from strips in two adjacent parts then cancel (as above), leaving only wavelets from one part to give a much less bright band. A graph of the relative intensity distribution for a single slit diffraction pattern is shown in Fig. 18.27.

(c) *Further points.* From $\sin \theta = \lambda/a$ we see that if the slit is wide, λ is much less than a and so $\sin \theta$, and therefore θ, are very small. The directions of the first (and all other) minima are thus extremely close to the

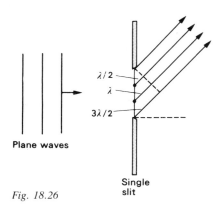

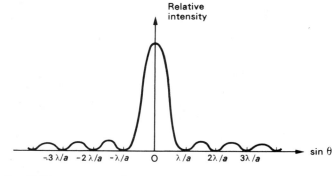

Fig. 18.26

Fig. 18.27

middle of the central maximum. Most of the light emerging from the slit is in the direction of the incident light and there is little diffraction. Propagation is almost rectilinear and the laws of geometrical optics are applicable. On the other hand, if the slit is one wavelength wide, $\lambda = a$, $\sin \theta = \sin 90 = 1$ and the central bright band spreads completely into the geometrical shadow. In this case the behaviour of light requires a wave model.

The width of the central bright band can be seen

from Fig. 18.27 to be twice that of any other bright band. (Also see Fig. 18.24a.)

Diffraction at multiple slits

Effects similar to those given by Young's double slit are obtained if more than two slits are used. They may be produced with the apparatus of Fig. 18.28, set up as for observing diffraction at a single slit (p. 398). It can be seen that as the number of slits is increased, the bright

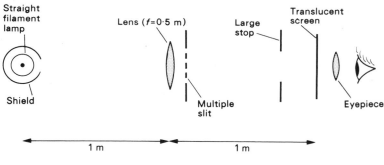

Fig. 18.28

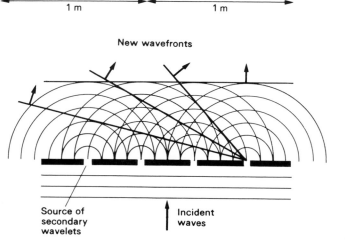

Fig. 18.29

bands become both brighter and sharper. With equally-spaced slits each pattern is in effect the same as a two-slit one.

Diffraction occurs at the slits and these, being very narrow, act as sources of secondary wavelets (semicircular in two dimensions) which superpose beyond the slits, Fig. 18.29. In certain directions the wavelets interfere constructively to form a new, straight wavefront (a bright band), whilst in others they interfere destructively (a dark band).

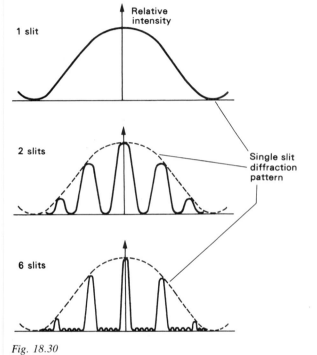

Fig. 18.30

As well as interference occurring between secondary wavelets from *different* slits, it also occurs between secondary wavelets from the *same* slit. Each slit therefore produces its own diffraction pattern; these are similar, in the same direction and will coincide exactly if focused by a lens. The diffraction pattern due to a single slit is thus superimposed on the interference pattern and determines the variations of intensity of the bright bands in the latter, as shown by the dotted curves in Fig. 18.30 and by the double slit interference fringes in the photograph of Fig. 18.3*b* (p. 381).

Diffraction grating

A diffraction grating consists of a large number of fine, equidistant, closely-spaced parallel lines of equal width, ruled on glass or polished metal by a diamond point. In *transmission* gratings glass is used; the lines scatter the incident light and are more or less opaque while the spaces between them transmit light and act as slits. Such gratings are very expensive and cheaper plastic replicas are made. In *reflection* gratings the lines, ruled on metal, are again opaque but the unruled parts reflect light regularly. This type has the advantage that radiation absorbed by transmission grating material can be studied and if it is ruled on a concave spherical surface it focuses the radiation as well as diffracting it and no lenses are needed.

Diffraction gratings are used to produce spectra and for measuring wavelengths accurately. They have replaced the prism in much modern spectroscopy. Their usefulness arises from the fact that they give very sharp spectra, most of the incident light being concentrated in certain directions.

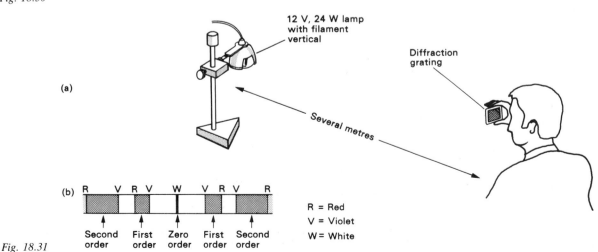

Fig. 18.31

(*a*) *'Fine' and 'coarse' gratings.* A low-voltage lamp with its filament vertical is viewed at a distance of a few metres through a transmission grating held near the eye and with its lines parallel to the filament, Fig. 18.31*a*. A 'fine' grating (e.g. 300 lines per mm) and a 'coarse' grating (e.g. 100 lines per mm) should be tried in turn.

A typical pattern for a 'fine' grating is shown in Fig. 18.31*b*. The central bright band, called the *zero order image*, is white (W) but on either side of it are brilliant bands of colour, called *first* and *second order spectra*, like those given by a prism but having red light (R) deviated more than violet (V) and the dispersion increasing with order. Each spectrum consists of a series of adjacent images of the filament formed by the constituent colours of white light. A 'coarse' grating forms many more orders of spectra, closer together. The effect of allowing a narrower band of wavelengths to fall on the grating can be observed by placing red and green filters in turn in front of the lamp.

A 'very coarse' grating (e.g. 10 lines per mm) gives a pattern that resembles Young's fringes.

The result is very striking and beautiful if white light from the lamp is viewed through two 'coarse' gratings (100 lines per mm) with their rulings 'crossed'.

An arrangement for projecting grating spectra on to a screen is shown in Fig. 18.32.

(*b*) *Theory.* Suppose plane waves of monochromatic light of wavelength λ fall on a transmission grating in which the slit separation (called the *grating spacing*) is d, Fig. 18.33. Consider wavelets coming from corresponding points A and B on two successive slits and travelling at an angle θ to the direction of the incident beam. The path difference AC between the wavelets is

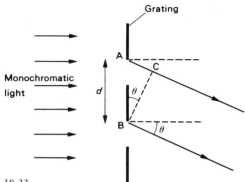

Fig. 18.33

$d \sin \theta$, as it is for all pairs of wavelets from other corresponding points in these two slits and in all pairs of slits in the grating. Hence if

$$d \sin \theta = n\lambda$$

where n is an integer giving the order of the spectrum, then reinforcement of the diffracted wavelets occurs in direction θ and a maximum will be obtained when the wavelets are brought to a focus by a lens.

When $n = 0$, $\theta = 0$ and we observe in the direction of the incident light the central bright maximum, i.e. the zero order image, for which the path difference of diffracted wavelets is zero. First, second, etc., order spectra are given by $n = 1, 2$, etc., but these are much less bright than the $n = 0$ maximum.

As an example, consider a grating with 500 lines per mm on which yellow light of wavelength 6×10^{-7} m falls normally. Since there are 500 lines and 500 spaces per mm of the grating, the grating spacing (i.e. 1 line + 1 space) $d = 1/500$ mm $= 10^{-3}/500$ m $= 2 \times 10^{-6}$ m

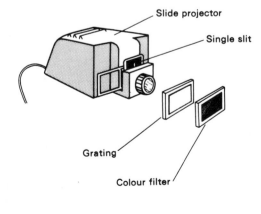

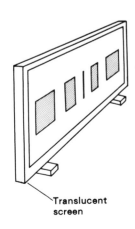

Fig. 18.32

(this is about three times the wavelength of yellow light). For the first order, $n = 1$

$$\therefore \quad \sin \theta_1 = \frac{\lambda}{d} = \frac{6 \times 10^{-7}}{2 \times 10^{-6}}$$

$$= 0.3$$

$$\therefore \quad \theta_1 = 17°$$

For the second order, $n = 2$

$$\therefore \quad \sin \theta_2 = \frac{2\lambda}{d} = 0.6$$

$$\therefore \quad \theta_2 = 37°$$

With this grating a third order spectrum is obtained, Fig. 18.34, but not a fourth. Why?

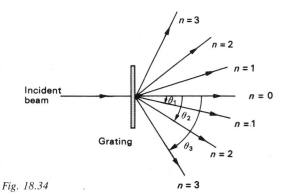

Fig. 18.34

(c) *Measurement of wavelength.* The wavelength of monochromatic light, e.g. from a sodium lamp or flame, can be measured to four-figure accuracy using a spectrometer and a transmission diffraction grating having 600 lines per mm (which is typical of many gratings).

The usual adjustments are first made on (i) cross-wires, (ii) telescope and (iii) collimator of the spectrometer[1] so that the collimator produces parallel light and the telescope focuses it at the cross-wires.

The telescope is then turned through 90° exactly, from the position T_1 in which it is directly opposite the illuminated collimator slit, to position T_2, Fig. 18.35a.

The grating is placed on the spectrometer table, at right angles to the line joining two of the levelling screws, say A and B, and the table turned until the image of the slit reflected from the grating is in the centre of the field of view. Adjustment of A or B may

[1] See Chapter 5.

be necessary to achieve this. The plane of the grating is now parallel to the axis of rotation of the telescope.

The table (and grating) are next turned through 45° exactly so that the incident light falls on the grating normally. The telescope is rotated, say to T_3, Fig. 18.35b, where the first order image is seen. If the lines of the grating are parallel to the axis of rotation of the telescope the image will be central, otherwise levelling screw C will need altering. The first order reading on the other side of the normal, at T_4, should also be taken. Half the angle between these two telescope settings gives θ, whence λ can be calculated from $\lambda = d \sin \theta$ where d is the grating spacing.

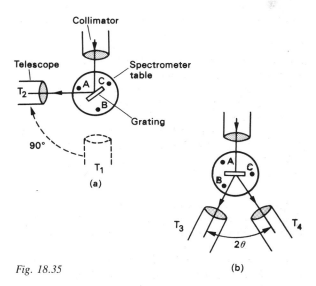

Fig. 18.35

Optical spectra

Optical spectra fall into two basic groups, as do the spectra of all types of electromagnetic radiation when analysed by an appropriate spectrometer. As we shall see later (p. 480) the study of spectra, known as *spectroscopy*, provides information about the structure of atoms.

(a) *Emission spectra.* These are obtained when the light from a luminous source undergoes dispersion (formerly by a prism, nowadays often by a diffraction grating) and is observed directly. There are three types.

(i) *Line spectra* consist of quite separate bright lines of definite wavelengths on a dark background and are given by luminous gases and vapours at low pressure. Sodium vapour emits two bright yellow lines which are very close together (wavelengths 0.5896 and 0.5890 μm), Fig. 18.36a. Hydrogen emits red, blue-green and violet lines, Fig. 18.36b. Each line is an image of the slit (on the collimator) of the spectrometer on which the light falls. No two elements give the same line spectrum and spectroscopic methods are so sensitive that they can identify and reveal the presence of the most minute quantities of materials. In general line spectra are due to the individual *atoms* concerned since only in a gas (especially at low pressure) are the atoms far enough apart not to interact. They are also called *atomic* spectra.

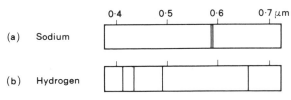

Fig. 18.36

A convenient source for producing line spectra is the discharge tube. This comprises a glass tube containing a gas (e.g. hydrogen, helium or neon) or vapour (e.g. mercury or sodium) at low pressure and two metal electrodes, across which a p.d. of several thousand volts is applied. The gas or vapour conducts and a luminous discharge occurs—an effect used in neon advertising signs and certain types of street lighting.

(*ii*) *Band spectra* have several well-defined groups or bands of lines. The lines are close together at one side of each band, making this side sharper and brighter than the other, Fig. 18.37. Band spectra are complex and are obtained from the molecules of glowing gases or vapours, heated or excited electrically. They arise from interaction between atoms in each molecule. The blue inner cone of a Bunsen burner flame gives a band spectrum as do nitrogen and oxygen in a discharge tube.

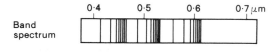

Fig. 18.37

(*iii*) *Continuous spectra* are emitted by hot solids and liquids and also hot gases at high pressures. The atoms are then so close that interaction is inevitable, and all wavelengths are emitted, Fig. 18.38. A continuous spectrum is not characteristic of the source and is conveniently produced by the white-hot tungsten filament of an electric lamp.

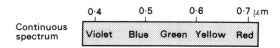

Fig. 18.38

(*b*) *Absorption spectra.* These form the second basic group of spectra and are observed when part of the radiation emitted is absorbed by a material between the source and observer. Line, band and continuous spectra are again obtained.

A line absorption spectrum occurs when white light passes through a cooler gas or vapour. Dark lines occur, against the continuous spectrum of white light, exactly at some of those wavelengths which are present in the line emission spectrum of the gas or vapour. The absorption spectrum of an element is thus the same as its emission spectrum except that the latter consists of bright lines on a dark background and the former of dark lines on a bright background. The atoms of the cooler gas absorb light of the wavelengths which they can emit, and then re-radiate the same wavelengths almost immediately but in all directions. Consequently, the parts of the spectrum corresponding to those wavelengths appear dark by comparison with other wavelengths not absorbed. The production of the line absorption spectrum of iodine vapour is described in Appendix 16. The presence of a layer of relatively cooler gas round the sun causes the so-called *Fraunhöfer* lines in the solar spectrum, which is thus an example of a line absorption spectrum. The lines indicate the presence of hydrogen, helium, sodium, etc., in the sun's atmosphere (i.e. the chromosphere).

Band absorption spectra are formed when a continuous emission spectrum is observed through a material which itself can emit a band spectrum. In a continuous absorption spectrum the brightness of one region of the spectrum may be reduced but not another.

Resolving power

This is the ability of an optical system to form separate images of objects that are very close together and so reveal detail.

(*a*) *Factors affecting it.* It can be shown that resolving power depends on the size of the slit or hole through which the objects are viewed and on the colour, i.e. wavelength, of the light used. Thus if the multiple light source of Fig. 18.39, with a green filter in front of it, is observed through an adjustable slit it will be found that if the slit is narrow or the viewing distance large the

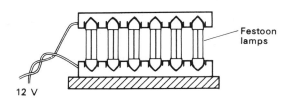

Fig. 18.39

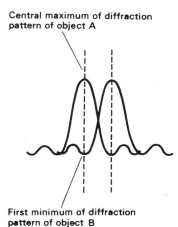

Central maximum of diffraction pattern of object A

Fig. 18.41   First minimum of diffraction pattern of object B

lamps are not seen separately but appear to be continuous.

If the distance and the slit width are arranged so that the lamps can *just* be resolved, replacing the green filter by a blue one improves the resolution whilst a red one worsens it. Evidently the shorter the wavelength of the light the easier it is to see detail.

(*b*) *Resolving power of the eye.* The smallest angle θ which two points can subtend at the eye and still be seen as separate is taken as a measure of the resolving power of the eye. For example, if two well-lit black lines 2 mm apart on a card can just be distinguished separately by a certain observer at a distance of 5 m, then from Fig. 18.40 and since θ will be small, we have $\theta = a/d = 2 \times 10^{-3}$ m/5 m $= 4 \times 10^{-4}$ rad. The resolving power of his eye is 4×10^{-4} rad. The smaller θ is, the *greater* the resolving power.

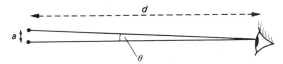

Fig. 18.40

(*c*) *Resolving power and diffraction.* A limit on the resolving power of an optical system is set by diffraction at the slit or hole through which observation occurs. A point object does not give a point image (even in an aberration-free system as we assume in geometrical optics); instead a diffraction pattern is obtained with its centre where the point image would be formed.

Rayleigh suggested that we should consider objects to be *just* resolved when the first minimum of the diffraction pattern of one falls on the central maximum of the other, Fig. 18.41. In the case of a slit of width a, the angular separation of these two fringes is λ/a (see p. 399) and in general we can take this as being roughly true for other diffraction apertures of width a, e.g. a circle of diameter a.

Assuming the pupil of the eye has a diameter of two millimetres and that the average wavelength of light is 6×10^{-7} m, we might expect the eye to have a resolving power of $6 \times 10^{-7}/(2 \times 10^{-3}) = 3 \times 10^{-4}$ rad. This agrees with experimental values—see (*b*) above.

However, it must be pointed out that in many cameras, for example, the resolving power is limited more by aberrations than by diffraction and so the lens system has to be 'stopped down' for finer resolution.

(*d*) *Resolving power and magnifying power of optical instruments.* When a lens forms an image of a small object, it acts as a circular aperture and the image is a diffraction pattern like that in Fig. 18.42. From (*c*) it follows that the resolving power (λ/a) of a lens is improved by increasing its diameter a or by decreasing the wavelength λ of the light used. By having a large diameter, a telescope objective gathers a large amount of light and forms a small diffraction pattern of a distant star, i.e. the image is bright, sharp and detailed. Some microscopes obtain greater resolution by using ultra-violet radiation, which has a smaller wavelength than light. In an electron microscope moving electrons behave as waves with wavelengths about 10^5 times less than those of light and so extremely small objects can be resolved.

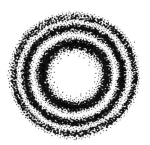

Fig. 18.42

The resolving power of an instrument sets a limit on its useful magnifying power. For example, the objective of the Mount Palomar telescope has a diameter of 5 m, giving an approximate resolving power of $6 \times 10^{-7}/5 = 10^{-7}$ rad. The resolving power of the eye is about 3×10^{-4} rad, therefore the magnifying power required to increase the angular separation from 10^{-7} to 3×10^{-4} rad is given by

$$\frac{\text{resolving power of the eye}}{\text{resolving power of telescope}} = \frac{3 \times 10^{-4}}{10^{-7}} = 3 \times 10^3$$

Any magnification beyond this reveals no further detail and is like stretching a rubber sheet on which there is a picture. The size of the picture increases but not the resolution.

In an electron microscope with a resolving power of 10^{-12} rad, a magnifying power of the order of 10^5 could be usefully employed. In any type of microscope the magnifying (and the resolving) power is limited basically by the wavelength of the 'illumination' used in the instrument.

Polarization of light

Interference and diffraction require a wave model of light but they do not show whether the waves are longitudinal or transverse. Polarization suggests they are transverse in character.

If a lamp is viewed through a piece of Polaroid (used for example in sunglasses), apart from it appearing slightly less bright there is no effect when the Polaroid is rotated about an axis perpendicular to itself. However, using two pieces, one of which is kept at rest and the other rotated slowly, Fig. 18.43, the light is cut off

more or less completely in one position. The Polaroids are then said to be 'crossed'. Rotation through a further 90° allows maximum light transmission.

This experiment is analogous to that shown in Fig. 18.44 where transverse waves are sent along a spring to two slots B and C. The waves are generated by moving end A of the spring to and fro in *all* directions perpendicular to the direction of travel. Slot B passes only waves due to vibrations in a vertical plane. The transmitted wave is said to be *plane polarized* in a vertical plane and is unable to pass through the horizontal slot C. In this position slots B and C are 'crossed'. A longitudinal wave would emerge from both slots whatever their relative position, i.e. it cannot be polarized.

The optical experiment can be explained if light is regarded as a transverse wave motion which the first Polaroid plane polarizes by transmitting only the light 'vibrations' in one particular plane and absorbing those in a plane at right angles. The second Polaroid transmits or absorbs the plane polarized light incident on it depending on its orientation with respect to the first Polaroid. It thus acts as a detector of polarized light.

Light from the sun and other sources is unpolarized and consists of 'vibrations' in every plane perpendicular to the direction of travel. Fig. 18.45*a* is an end-on representation of unpolarized light, the arrowed lines show some of the directions in which 'vibrations' may occur. At one point and time the vibration is in just one direction and is not the same for different points at that time. It also changes rapidly (about 10^9 times a second) and randomly at each point. Any vibration can be resolved into two perpendicular components and another end-on picture of unpolarized light is given in Fig. 18.45*b*. A side view of an unpolarized ray is shown

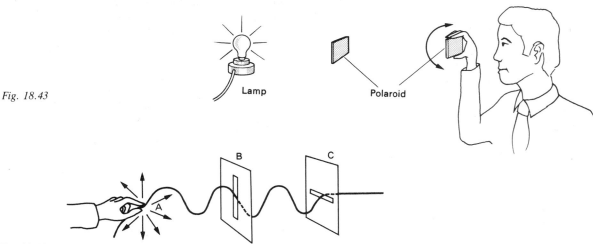

Fig. 18.43   Lamp   Polaroid

Fig. 18.44

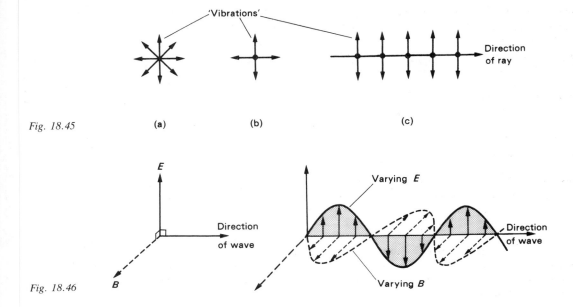

Fig. 18.45   (a)   (b)   (c)

Fig. 18.46

in Fig. 18.45c, the dots representing 'vibrations' at right angles to the paper and the arrowed lines 'vibrations' in the plane of the paper.

Like other forms of electromagnetic radiation light is regarded as a varying electric field (E) coupled with a varying magnetic field (B), at right angles to each other and to the direction of travel. Fig. 18.46 is an attempt to indicate diagrammatically an electromagnetic wave. Experiment shows that the coupling of light with matter is more often through the electric field, e.g. affecting a photographic film. For this reason light 'vibrations' are taken to be the variations of the electric field strength E. If light is vertically polarized we mean that the plane containing E and the direction of travel of the wave, called the *plane of vibration*, is vertical.

Producing polarized light

(a) *By Polaroid.* Polaroid is made from tiny crystals of quinine iodosulphate all lined up in the same direction in a sheet of nitrocellulose. Crystals which transmit light vibrations, i.e. electric field variations, in one particular plane and absorb those in a mutually perpendicular plane are said to be *dichroic*. They produce polarization by *selective absorption*. The effect is analogous to the absorption of vertically polarized microwaves by a grid of vertical wires, Fig. 18.47a, and transmission when the wires are horizontal, Fig. 18.47b.

(b) *By reflection.* When unpolarized light falls on glass, water and some other materials the reflected light is, in general, partially plane polarized. But at one particular angle of incidence, called the *polarizing angle* i_p, the polarization is complete. At this angle, the reflected ray and the refracted ray in a transparent medium are found to be at right angles to each other. The vibrations in the reflected ray are parallel to the surface.

Applying Snell's law to Fig. 18.48 we have

$$n_1 \sin i_p = n_2 \sin r$$

where n_1 and n_2 are the absolute refractive indices of media ① and ②. Also

$$i_p + 90° + r = 180°$$

$$\therefore \quad r = 90 - i_p$$

Hence
$$n_1 \sin i_p = n_2 \sin (90 - i_p)$$

$$= n_2 \cos i_p$$

$$\therefore \quad \frac{n_2}{n_1} = \frac{\sin i_p}{\cos i_p} = \tan i_p$$

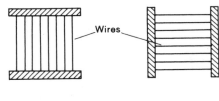

Fig. 18.47   (a)   (b)

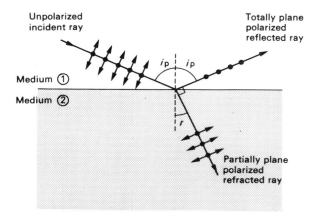

Fig. 18.48

If medium ① is air or a vacuum $n_1 = 1$ and so

$$n_2 = \tan i_p$$

This is *Brewster's law*. For glass of refractive index 1.5, $i_p = 57°$.

(*c*) *By double refraction.* If a crystal of calcite (a form of calcium carbonate and also called 'Iceland spar') is placed over, say, a triangle drawn on a piece of paper, two images are seen, Fig. 18.49. Such crystals exhibit double refraction and an incident unpolarized ray is split into two rays, called the *ordinary* and *extraordinary* rays, which are plane polarized in directions at right angles to each other. The former obeys Snell's law, but the latter does not because it travels with different speeds in different directions in the crystal.

Fig. 18.49

In a *Nicol prism*, Fig. 18.50, which is made by cutting a calcite crystal in a certain way, only the extraordinary ray (E) is transmitted. The prism may therefore be used to produce plane polarized light, as Polaroid does. If the angle of incidence exceeds the critical angle, total internal reflection of the ordinary ray (O) occurs at the transparent layer of Canada balsam cement because it is optically less dense than the crystal for the ordinary ray but not for the extraordinary ray.

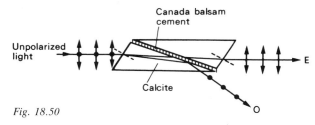

Fig. 18.50

(*d*) *By scattering.* This may be shown by sending a narrow beam of unpolarized light through a tank of water in which scattering particles are obtained by adding one drop of milk. Polarization of the scattered light is detected by holding a piece of Polaroid in different positions, shown by the dotted lines in Fig. 18.51.

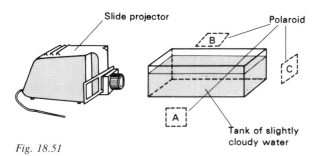

Fig. 18.51

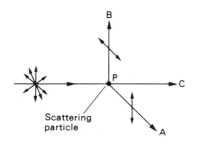

Fig. 18.52

The polarization arises from the transverse wave nature of light. That scattered along PA in Fig. 18.52 must be polarized in a direction parallel to PB (otherwise light would be longitudinal in character) and along PB the direction of polarization must be parallel to PA. What will happen if a piece of Polaroid is placed between the tank and the lamp?

Using polarized light

(*a*) *Reducing glare.* Glare caused by light reflected from a smooth surface can be reduced by using polarizing materials since the reflected light is partially or completely polarized. Thus Polaroid discs, suitably orientated, are used in sunglasses and also in photography as 'filters' where they are placed in front of the camera lens, thereby enabling detail to be seen that would otherwise be hidden by glare. Blue light from the sky, in a direction at right angles to the sun, is polarized and the brightness of the sky may be reduced in a colour photograph by a polarizing 'filter'.

(*b*) *Optical activity.* Certain crystals (e.g. quartz) and liquids (e.g. sugar solutions) rotate the plane of vibration of polarized light passing through them and are said to be optically active. For a solution the angle of rotation depends on its concentration and in an instrument known as a *polarimeter* this is used to measure concentration.

(*c*) *Stress analysis.* When glass, Perspex, polythene and some other plastics are under stress (e.g. by bending, twisting or uneven heating) they become doubly refracting and if viewed in white light between two 'crossed' Polaroids, coloured fringes are seen round the regions of strain.[1] The effect is called *photoelasticity* and is used to analyse stresses in plastic models of various structures. A modern polariscope for such work is shown in Fig. 18.53a and Fig. 18.53b is a photograph of the strain pattern for the model of the crane-hook under stress in the polariscope.

In a more recent development of the technique, actual structures are coated with a photoelastic plastic to which strains created in the structure are transmitted directly. In this case a reflection polariscope is used.

(*d*) *Liquid crystal displays (LCD).* These are used as numerical indicators, especially in digital watches where their very small current requirement makes them preferable to LED displays (p. 496). Normally they transmit light but when a voltage is applied the light is cut off and the crystal goes 'dark'.

[1] See Fig. 2.17a, page 26.

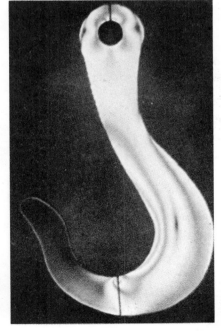

Fig. 18.53   (a)   (b)

Liquid crystals are organic compounds (such as cyano-biphenyls) which exhibit both solid and liquid properties. Their molecules, which are long, narrow, stiff and colourless, are arranged regularly as in a crystal (tending to align with each other) but the material in bulk flows like a liquid. Their use in displays depends on two effects. First, when a voltage is applied, the molecules line up end-on with the resulting electric field (like the grass seeds in Fig. 11.8). And second, they can rotate the plane of vibration of polarized light passing through them.

In a display, a thin liquid crystal cell is used in which the molecules have been arranged so that their alignment gradually changes through 90° from one face of the cell to the other. (One of several ways of achieving this is to gently rub a piece of glass so that fine scratches, invisible to the naked eye, are produced on it all in the same direction; the molecules of a liquid crystal poured on to the glass then tend to lie along the scratches—like matches in the grooves of corrugated paper. Two such pieces of glass with a very thin gap between them, having their scratch lines at right angles, produce a 90° twist in the alignment of the molecules.)

If the cell is set between two *crossed* polarizing films, light passes through due to the 90° twist produced by the liquid crystal, Fig. 18.54a. The glass faces of the cell each have a very thin, transparent, conducting coating and when a p.d. is applied across them, the direction of the resulting electric field is such that it alters the twist of the liquid crystal molecules which no longer rotate the polarized light, and the display goes from light to dark. When the p.d. (and field) is removed, the molecules return to their twisted state. The operation can be repeated almost indefinitely.

The pattern of the conductive coating in a 7-segment liquid crystal display for producing numbers 0 to 9 is shown in Fig. 18.54b: only the liquid crystal under those segments of the coating to which the voltage is applied is untwisted by the electric field. The display has a silvered background which reflects back incident light; it is continuously visible (except at night when it has to be illuminated); by contrast an LED display is lit only when required, to save the battery.

Electromagnetic waves

The idea that fields of force exist in the space surrounding electric charges, current-carrying conductors and magnets is useful for 'explaining' electrical and magnetic effects. The flow of current in a conductor may also be considered to be due to the existence of an electric field in the conductor and since current can be induced by a changing magnetic field it follows that *a changing magnetic field creates an electric field* in the medium where the change occurs (whether it be free space or a material medium). This effect is electromagnetic induction and was discovered by Faraday in 1831 as we have seen (p. 281).

The converse effect, i.e. *a changing electric field sets up a magnetic field*, was proposed by Maxwell in 1864 on the grounds that if this assumption is made, the mathematical equations expressing the laws of electricity and magnetism (in terms of Faraday's ideas of fields of force) become both simple and symmetrical. There is no easy way of demonstrating Maxwell's assumption directly on account of the difficulty of detecting a weak magnetic field, but one consequence is that any electric field produced by a changing magnetic field must inevitably create a magnetic field and vice versa, i.e. one effect is coupled with the other.

On this basis Maxwell predicted that when a moving electric charge is oscillating it should radiate an electro-

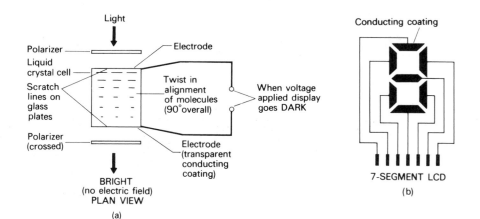

Fig. 18.54

magnetic wave consisting of a fluctuating electric field accompanied by a fluctuating magnetic field of the same frequency and phase, the fields being at right angles to each other and to the direction of travel of the wave. The intensities (E and B) of the fields thus vary periodically with time *like* the amplitude of a wave motion of the transverse type but, in fact, no medium seems to be involved in the transmission. A diagrammatic representation of an electromagnetic wave was given in Fig. 18.46 (p. 407).

Maxwell also showed that the speed c of all electromagnetic waves in free space, irrespective of their wavelength or origin, should be given by

$$c = \frac{1}{\sqrt{\mu_0 \varepsilon_0}}$$

where μ_0 and ε_0 are the permeability and permittivity respectively of free space.

Substituting numerical values in this expression gives $c = 3.0 \times 10^8$ m s^{-1}, i.e. the speed of light. This led Maxwell to suggest that light might be one form of electromagnetic wave motion.

The search for other forms of electromagnetic radiation was taken up and resulted in the discovery by H. Hertz in 1887 of waves having the same speed as light but with wavelengths of several metres compared with a very small fraction of a millimetre for light. These waves, now called *radio waves*, were generated by Hertz using a spark gap transmitter which produced bursts of high frequency a.c.

A whole family of electromagnetic radiations is now known, extending from gamma rays of very short wavelength to very long radio waves. Fig. 18.55 shows the wavelength ranges of the various types but there is no rapid change of properties from one to the next. Although methods of production and detection differ from member to member, all exhibit reflection, refraction, interference, diffraction and polarization. When they fall on a body partial reflection, transmission or absorption occurs; the part absorbed becomes internal energy.

For any wave motion, $v = f\lambda$. For all electromagnetic waves in a vacuum $v = c = 3.0 \times 10^8$ m s^{-1} and so if f or λ for a particular radiation is known, the other can be calculated. Thus for violet light of $\lambda = 0.40$ μm $= 4.0 \times 10^{-7}$ m we have $f = c/\lambda = 3.0 \times 10^8/(4.0 \times 10^{-7}) = 7.5 \times 10^{14}$ Hz. By contrast a 1 MHz radio signal has a wavelength of 300 m. Electromagnetic waves may be distinguished either by their wavelength or frequency but the latter is more fundamental since unlike wavelength (and speed) it does not change when the wave travels from one medium to another. The production of various types of electromagnetic radiation will be considered later.

Whilst a wave model accounts for many of the properties of electromagnetic radiation it cannot explain some other aspects of its behaviour (see p. 465).

Infrared radiation

The wavelength of light varies from 0.4 μm for violet light to about 0.7 μm for red light. This is the 'visible' region of the electromagnetic spectrum. Longer wavelength radiation, extending from 0.7 μm to about one millimetre, is called *infrared* and is not detected by the eye. It has all the properties of electromagnetic waves.

(*a*) *Sources.* The surfaces of all bodies emit infrared radiation in a continuous range of wavelengths. The relative amount of each wavelength depends mainly on the temperature but also on the nature of the surface of the body. At low temperatures, long wavelength infrared is emitted; as the temperature of the body rises the emission is more copious and shorter and shorter wavelengths are present. At about 500 °C, red light is emitted as well as long and short wave infrared, i.e the body is red-hot. Increasing the temperature adds orange, yellow, green, blue and violet light in turn and at around 1000 °C the body becomes white-hot. Eventually ultraviolet radiation (p. 413) is emitted as well.

The presence of infrared in the radiation from a white-hot filament lamp may be shown with the

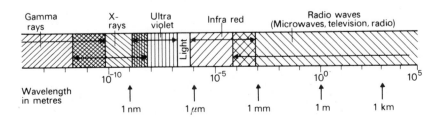

Fig. 18.55

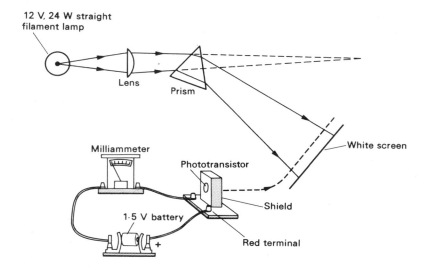

Fig. 18.56

arrangement of Fig. 18.56 in which a phototransistor is used as a detector. When this is moved slowly through the spectrum produced by the glass prism, the milliammeter gives a reading beyond the visible red region, i.e. in the infrared.

(*b*) *Absorption.* When absorbed by matter, infrared—like all other types of electromagnetic radiation—causes an increase of internal energy which usually results in a temperature rise. Thus, infrared falling on the skin produces the sensation of warmth. Infrared lamps are used for the treatment of muscular complaints and to dry the paint on cars during manufacture. Infrared cooking is also practised.

Except for wavelengths near to that of red light, infrared is absorbed by glass but transmitted by rock salt. The action of a greenhouse depends on the fact that sunlight passes through the glass, is absorbed by and warms up the plants, soil, etc., which reradiate long wavelength infrared (because of their comparatively low temperature). Most of this infrared cannot penetrate the glass and is trapped in the greenhouse.

Water vapour and carbon dioxide in the lower layers of the atmosphere exhibit the same 'selective absorption' effect and prevent infrared emitted by the earth from escaping. It has been estimated that if the average temperature of the earth rose by 3.5 °C due, for example, to the combustion of fossil fuels and the resulting increase of carbon dioxide in the atmosphere, dramatic climatic and geographical changes could occur. Some scientists believe that 'thermal' pollution is as great a threat as any other form.

Fig. 18.57

(*c*) *Detection*. There are three types of detector—photographic, photoelectric (e.g. phototransistors) and thermal.

Special *photographic films* sensitive to infrared may be used. They enable pictures to be taken in the dark (like that of Fig. 18.57 of a car; the white parts are hottest) or in hazy conditions since infrared is scattered less than light by particles in the atmosphere (because of its longer wavelength).

Very sensitive *photoelectric devices* have been developed which enable the infrared emitted by, for example, a distant rocket to be detected and early warning of its launching obtained. Similar detectors are used in weather satellites to obtain cloud formation patterns over the whole of the earth's surface and are especially useful for detecting hurricanes.

Thermal detectors include the thermometer, the thermopile and the bolometer.

A *thermopile* consists of many thermocouples in series. In a thermocouple an e.m.f. is produced between the junction of two different metals or semiconductors if one junction is at a higher temperature than the other, Fig. 18.58*a*. The hot junctions of a thermopile, Fig. 18.58*b*, are blackened to make them good absorbers of the incident radiation to be measured whilst the cold junctions are shielded from it. The hot junction temperature rises until the rate of gain of heat equals the rate of loss of heat to the surroundings and then the e.m.f. is measured with a potentiometer. Alternatively it may be enough to observe the reading produced on a galvanometer connected across the thermopile.

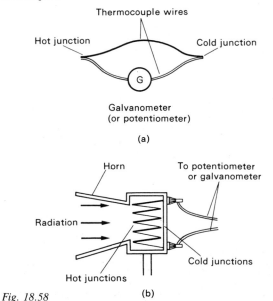

Fig. 18.58

A *bolometer* is a device such as a thermistor in which a small temperature change causes a large change of electrical resistance. In practice two thermistors are used in a Wheatstone bridge circuit, Fig. 18.59, radiation being allowed to fall on the 'active' one but not on the 'compensating' one. The resulting temperature change creates an out-of-balance p.d. and the bridge has to be rebalanced. Greater sensitivity is achieved if the incident radiation is 'chopped', i.e. interrupted by a rotating disc with slots, so that the output from the bridge is an alternating p.d. which can be amplified.

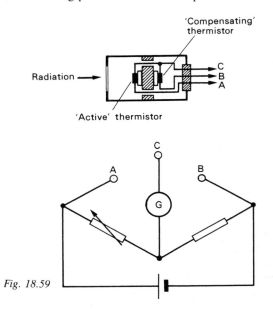

Fig. 18.59

Ultraviolet radiation

Shortly after infrared was discovered in 1800, radiation beyond the violet end of the sun's visible spectrum was found. This radiation, which has typical electromagnetic wave properties, has a shorter wavelength than light and covers the range 0.4 μm to about 1nm.

The existence of ultraviolet in the spectrum formed from an overrun filament lamp may be shown by

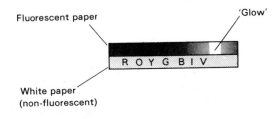

Fig. 18.60

allowing it to fall on strips of fluorescent paper and non-fluorescent white paper, one above the other, Fig. 18.60, in a darkened room. The fluorescent paper glows outside (as well as at) the violet of the visible spectrum, the latter being revealed by the non-fluorescent white paper. If an ultraviolet filter is inserted it cuts off most of the visible light.

The sun emits ultraviolet but much of it is absorbed by a layer of ozone in the earth's atmosphere. This is fortunate, for although ultraviolet produces vitamins in the skin and causes 'sun-tan', an overdose can be harmful, particularly if it is short-wavelength radiation. The eyes are especially vulnerable and *an ultra-violet lamp should never be viewed directly.* Such a lamp contains mercury vapour and is a convenient source of the radiation. It usually has a quartz bulb or window— not glass, which would absorb much of the ultraviolet.

Fluorescent materials absorb 'invisible' ultraviolet radiation and reradiate 'visible' light. Fluorescent lamps contain mercury vapour and their inner surfaces are coated with fluorescent powders which emit light of a characteristic colour when struck by ultraviolet. Fluorescent powders are also used in poster paints for advertising and in washing powders that add 'brightness to cleanness and whiteness'. Why?

Photoelectric devices and photographic films can also detect ultraviolet.

Thermal radiation

Thermal radiation is energy which travels as electro-magnetic waves, having been produced by a source because of its temperature. Its chief component is usually infrared radiation but light and ultraviolet may also be present. Being energy in the process of transfer it can be regarded as heat. (See Chapter 4.)

At the start of the twentieth century certain aspects of the emission of thermal radiation were responsible, along with the photoelectric effect to be discussed later (p. 464), for a revolutionary change of view about the nature of light and other types of electromagnetic radiation. This resulted in the emergence of the *quantum theory*. For the present we shall deal mainly with the facts concerning the emission and absorption of thermal radiation and with other related ideas.

(*a*) *Prévost's theory of exchanges.* In Fig. 18.61 the small body A, at temperature T_A, is suspended by a non-conducting thread inside the box B whose walls are at a constant, different temperature T_B. If B is then evacuated, energy exchange between A and B can occur only by radiation. If $T_A > T_B$, A's temperature falls until it is also T_B, but if $T_A < T_B$, it rises to T_B. In

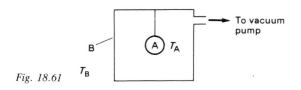

Fig. 18.61

either case A acquires B's temperature and it might appear that energy exchange then ceases.

Prévost suggested in 1792 that, on the contrary, *when a body is at the same temperature as its surroundings its rate of emission of radiation to the surroundings equals its rate of absorption of radiation from the surroundings.* That is, there is dynamic equilibrium, and energy exchange continues and at a rate depending on the temperature. This view is now generally accepted.

If follows that a body which is a good absorber of radiation must also be a good emitter of radiation otherwise its temperature would rise above that of its surroundings. Conversely a good emitter must be a good absorber. Experiments confirm these conclusions and indicate that a dull, black surface is the best absorber and, as we might anticipate, is also the best emitter.

(*b*) *Black body radiation.* The theoretical concept of a perfect absorber, called a *black body, which absorbs all the radiation of every wavelength falling on it,* is, like the idea of an ideal gas in kinetic theory (p. 431), a useful standard for judging the performance of other bodies. At normal temperatures it must appear black since it does not reflect light—hence the term 'black body'. We would expect a black body to be the best possible emitter at any given temperature. The radiation emitted by it is called *black body radiation,* or *full radiation* or *temperature radiation*; the latter term is used because the relative intensities of the various wavelengths present depend only on the temperature of the black body.

In practice an almost perfect black body consists of an enclosure, such as a cylinder, with dull black interior walls and a small hole, Fig. 18.62. Radiation entering the

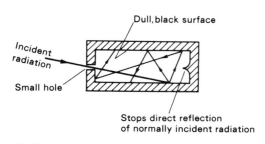

Fig. 18.62

hole from outside has little chance of escaping since any energy not absorbed when the wall is first struck, will be subsequently. The *hole* in the enclosure thus acts as a black body because it absorbs all the radiation falling on it.

When the enclosure is heated to a certain temperature, say by an electric heating coil wrapped round it, *black body* or *cavity* radiation emerges from the hole (which nonetheless may appear red or yellow or even white if the temperature is high enough). A spectrum of the radiation can be formed if it falls on the slit of a suitable spectrometer and in general infrared, light and ultraviolet will be present. All three have a heating effect when absorbed and can therefore be detected by a thermopile or a bolometer. If a narrow slit is placed in front of one of these instruments and moved along the spectrum, the energy carried by the small band of wavelengths passing through the slit at different mean wavelengths can be obtained. Repeating this for different temperatures enables a family of curves to be drawn showing how the energy in the spectrum of a black body is distributed among the various wavelengths.

The curves have the form shown in Fig. 18.63; the following points emerge.

(*i*) As the temperature rises the energy emitted in each band of wavelengths increases, i.e. the body becomes 'brighter'.

(*ii*) Even at 1000 K only a small fraction of the radiation is light. (The maximum does not lie in the visible region until about 4000 K.)

(*iii*) At each temperature T the energy radiated is a maximum for a certain wavelength λ_{max} which decreases with rising temperature; in fact $\lambda_{max} \propto 1/T$, i.e. $\lambda_{max}T = $ constant—a statement known as *Wien's displacement law*. This explains why a body appears successively red-hot, yellow-hot and white-hot, etc. Sirius (the Dog Star) looks blue, not white. Why?

It is interesting to compare the spectral emission curve of a tungsten filament lamp (a non-black body) with that of a black body at the same temperature (2000 K), Fig. 18.64.

(*c*) *Stefan's law.* This states that if E is the *total* energy radiated of all wavelengths per unit area per unit time by a *black body* at thermodynamic temperature T (see page 430), then

$$E \propto T^4$$
or
$$E = \sigma T^4$$

where σ is a constant, called *Stefan's constant*. If E is in W m^{-2} (i.e. J s^{-1} m^{-2}) and T is in K (kelvins) then σ has units W m^{-2} K^{-4}. Its value is

$$\sigma = 5.7 \times 10^{-8} \text{ W m}^{-2} \text{ K}^{-4}$$

The area under each spectral emission curve, Fig. 18.64, represents the total energy radiated per unit area per unit time at temperature T and is found to be proportional to T^4—agreeing with Stefan's law.

A black body at temperature T in an enclosure at a lower temperature T_0 loses energy by emission but also gains some from the enclosure. The net loss of energy E

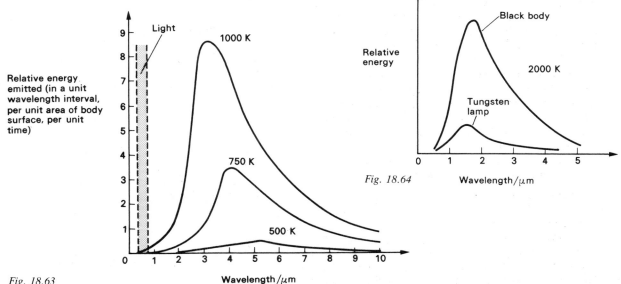

Fig. 18.63

Relative energy emitted (in a unit wavelength interval, per unit area of body surface, per unit time)

Light

1000 K

750 K

500 K

Wavelength/μm

Fig. 18.64

Relative energy

Black body

2000 K

Tungsten lamp

Wavelength/μm

from the black body per unit area per unit time is thus given by

$$E = \sigma T^4 - \sigma T_0^4$$
$$= \sigma(T^4 - T_0^4)$$

If $T \gg T_0$ then T_0^4 can be neglected and we have

$$E = \sigma T^4$$

If $(T - T_0)$ is small then we can show that

$$E \simeq 4\sigma T_0^3(T - T_0)$$

Thus, let $x = T - T_0$ and so $E = \sigma(T^4 - T_0^4) = \sigma[(T_0 + x)^4 - T_0^4] = \sigma[T_0^4 + 4T_0^3x + 6T_0^2x^2 + 4T_0x^3 + x^4 - T_0^4] \simeq 4\sigma T_0^3x$ since x^2, x^3 and x^4 are negligible. The rate of loss of energy by radiation is therefore proportional to the temperature excess of the body over its surroundings if this is small.

For a non-black body Stefan's law can be applied in the form

$$E = \varepsilon\sigma T^4$$

where ε is called the *total emissivity* of the body and has a value between 0 and 1.

(d) *Optical pyrometer.* A pyrometer is an instrument for measuring high temperatures and certain types use all or part of the thermal radiation emitted by the hot source. The principle of the *disappearing filament* optical pyrometer, which responds to light only, is shown in Fig. 18.65. Light from the source and the tungsten filament lamp passes through a red filter of a known wavelength range before reaching the eye; both appear red. The current through the filament is adjusted until it appears as bright as the background of light from the source. The temperature is then read off from the ammeter, previously calibrated in K.

Calibration may be done against a gas thermometer up to 1800 K. The range of a pyrometer can be extended up to about 3000 K by having a sectored disc rotating in front of the source which cuts down the radiation in a known proportion.

Pyrometers are used to measure the temperatures of the interiors of furnaces—without getting too close— and of the surfaces of molten metals.

Radio waves

The radiation from a transmitting aerial can reach the receiving aerial by three different paths, Fig. 18.66.

(i) The *ground wave* travels along the ground following the curvature of the earth.

(ii) The *sky wave* leaves the aerial at an angle exceeding that between the aerial and the horizon and travels skywards.

(iii) The *space wave* takes a straight line path and strikes the ground between the aerial and the horizon.

Which one of the three is most effective in reaching a particular location depends on the frequency of the radiated signal as well as on the power of the transmitter. A common classification of frequencies is given in Table 18.1.

The range of the ground wave is limited mainly by the extent to which energy is absorbed from it by the

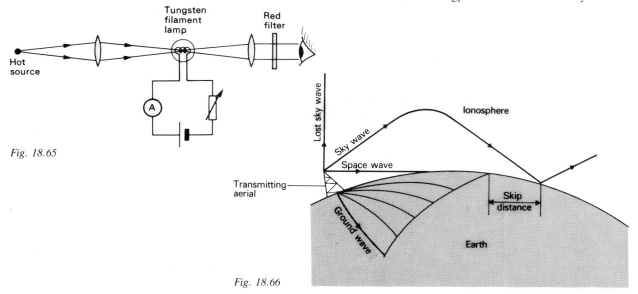

Fig. 18.65

Fig. 18.66

Table 18.1

	Long Waves	Medium Waves	Short Waves	v.h.f. (very high frequency)	u.h.f. (ultra high frequency)	Microwaves
	30 kHz–300 kHz	300 kHz–3 MHz	3 MHz–30 MHz	30 MHz–300 MHz	300 MHz–3000 MHz	above 3000 MHz
Ground wave	Medium range communication	Local sound broadcasts				
Sky wave	Long range communication	Distant sound broadcasts	Distant sound broadcasts and communication			
Space wave				F.M. sound broadcasts T.V. (B.B.C.1 and I.T.A.1)	T.V. (B.B.C.2 and I.T.A.2)	Radar, Radioastronomy, Satellite communication, Telephone links

ground: thus poor conductors such as sand absorb more strongly than water. This absorption attenuates the wave and the higher the frequency the greater the attenuation. The range may be about 1500 km at low frequencies (long waves) but only a few kilometres for v.h.f. waves.

The sky wave, so long as it is below a certain *critical frequency*, is returned to earth by layers of ionized gas, collectively called the *ionosphere*. This extends from a height of about 80 km above the earth to 500 km and it changes the direction of travel of the sky wave by gradual refraction. On striking the earth the sky wave bounces back to the ionosphere where it is once again directed to earth, and the process continues until the wave is completely attenuated. Since sky waves leave a transmitting aerial at many angles, the waves returning from the ionosphere cover quite large areas of the earth's surface. Those radiated at too steep an angle pass into space due to the refraction being insufficient to direct the wave earthwards. As a result no sky waves reach the area around the aerial and the distance between the limit of the ground wave and the point where the first sky wave returns to earth is called the 'skip distance' and receives no signals. The critical frequency (i.e. the frequency above which no sky wave, whatever its angle of radiation, will return to earth) varies with the time of day, the seasons and the eleven-year sun-spot cycle. Whereas in the long and medium wavebands the sky wave can be used for long-distance communications over several thousand kilometres, in the short waveband it is erratic and

unreliable. If both the ground wave and the sky wave from a particular transmitter are received at the same place, interference can occur if the two waves are out of phase. When the phase difference varies the signal 'fades'.

Above the critical frequency the ground wave is rapidly attenuated and the sky wave passes through the ionosphere. Therefore for v.h.f., u.h.f. and microwave transmissions only the space wave, travelling in a direct line from transmitter to receiver, is effective. When the transmitting aerial is at the top of a tall mast standing on high ground a range of up to 150 km or so is possible. Microwave frequencies approach those of light waves and in radar and radioastronomy where microwaves are used, the aerial is often in the form of a huge curved metal bowl which collects and reflects the incoming signal to a focus, just as an optical aerial, i.e. a telescope objective, focuses light.

Artificial satellites for world-wide communication are not only more dependable than the ionosphere but they contain amplifiers to boost the microwave signals used.

Table 18.1 shows some applications of the various frequency bands.

Aerials

Consideration of the way in which a transmitting aerial produces radio waves shows that the electric field emerges parallel to the aerial and the magnetic field at right angles to it, as we will see shortly. If the aerial is

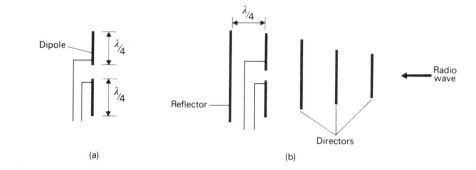

Fig. 18.67

(a)　　　　　　　　　　　　　(b)

vertical the waves are said to be vertically polarized, the direction of polarization being given by the direction of the electric field, so following the practice adopted with light waves (p. 407).

Television reception aerials are *dipoles* consisting of two conducting rods, each having a length of one-quarter of the wavelength (i.e. $\lambda/4$) to be received, Fig. 18.67*a*. Another conducting rod, called a *reflector*, placed $\lambda/4$ behind the dipole, reflects waves that reach it back to the dipole. These reflected waves travel an extra distance of $2 \times \lambda/4 = \lambda/2$ and they also suffer a $\lambda/2$ phase change on reflection (p. 357). They therefore arrive back at the dipole in phase with the main wave and so reinforce it to produce a larger signal. Several conducting rods in front of the dipole, called *directors*, Fig. 18.67*b*, add to the directional property that the reflector gives to the aerial.

In a portable medium/long wave radio, the aerial is a coil of wire wrapped on a ferrite rod, which for strongest reception, should lie parallel to the direction of the magnetic field.

A very simple but plausible *model* of the way an aerial might radiate electromagnetic waves is represented pictorially in Fig. 18.68. Suppose a positively charged particle is moved rapidly to and fro between X and Y. When it is *stationary* at either of these points it is surrounded by a radial electric field, parts of which are shown for each point; being at rest, it does not create a magnetic field. Whilst it is *moving*, say from X to Y, it behaves like an electric current and produces a magnetic field. It also 'twists' the electric field lines which become almost parallel to direction XY. At the 'twist' we have both an electric field and a magnetic field (the latter shown by dotted lines acting at right angles to the plane of the page). We can think of the 'twist' as a pulse of electromagnetic radiation produced by the moving charge and radiated with the speed of light.

If the charged particle is an electron accelerating and slowing down in an aerial, a continuous train of electromagnetic pulses will be radiated. As indicated previously (p. 410), we can represent this as a varying electric field at right angles to a varying magnetic field, both being perpendicular to the direction of travel, Fig. 18.46 (p. 407).

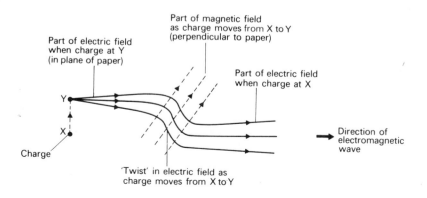

Fig. 18.68

QUESTIONS

Speed of light: interference

1. A plane mirror rotating at 35 revolutions per second reflects a narrow beam of light to a stationary mirror 200 m away. The stationary mirror reflects the light normally so that it is again reflected from the rotating mirror. The light now makes an angle of 2.0 minutes of arc with the path it would travel if both mirrors were stationary. Calculate the velocity of light.

Give *two* reasons why it is important that an accurate value of the velocity of light should be known.   (*J.M.B.*)

2. A beam of monochromatic light of wavelength 6.0×10^{-7} m in air passes into glass of refractive index 1.5. If the speed of light in air is 3.0×10^{8} m s^{-1}, calculate (*a*) the speed of the light in glass, (*b*) the frequency of the light, and (*c*) the wavelength of the light in glass.

3. Describe an experiment by means of which both the wave nature of light can be demonstrated and the wavelength of the light can be determined. Draw and label a diagram of the apparatus and give the theory of the experiment.

Plane waves in air are incident obliquely on a plane boundary between the air and glass. Use Huygens' construction to derive the sine law of refraction.

What test of the wave theory is suggested by the result?
          (*J.M.B.*)

4. Monochromatic light illuminates a narrow slit which is 4.0 m away from a screen. Two very narrow parallel slits 0.50 mm apart are placed midway between the single slit and the screen so that interference fringes are obtained. If the spacing of five fringes is 10 mm, calculate the wavelength of the light.

What will be the effect of (*a*) halving the distance between the double slit and the screen, (*b*) halving the slit separation, (*c*) covering one of the double slits, and (*d*) using white light?

Discuss whether interference, diffraction or both are involved in the formation of the fringes.

5. Describe the experimental arrangement for the production of interference fringes by Young's two-slit method. Describe and explain the appearance of the fringes if a white light source is used.

Explain how you would take the necessary measurements in order to determine the wavelength of light from a monochromatic source. Give the theory which enables you to calculate your result.

What would be the effect of introducing a microscope slide in the path of one of the interfering beams?   (*A.E.B.*)

6. Draw and label diagrams to illustrate the apparatus you would use to demonstrate optical interference giving:
   (*a*) straight line fringes, and
   (*b*) circular fringes.

In a Young's double-slit experiment the fringe separation observed using yellow light was found to be 0.275 mm. The yellow lamp, giving a wavelength 5.50×10^{-7} m is replaced by a purple one giving wavelengths of 4.00×10^{-7} m in the violet and 6.00×10^{-7} m in the red. The remainder of the apparatus is undisturbed. Calculate

   (*i*) the distance between the fringes formed by the violet light,

   (*ii*) the distance between the fringes formed by the red light, and

   (*iii*) the distance from the purple fringe on the axis to the next purple fringe observed.

Hence draw a diagram of the appearance of the new fringe system, indicating the colours, and extending as far as 1 mm from the axis.   (*C.*)

7. Explain what is meant by the term *path-difference* with reference to the interference of two wave-motions.

Why is it not possible to see interference where the light beams from the headlamps of a car overlap?

Interference fringes were produced by the Young's slits method, the wavelength of the light being 6.0×10^{-5} cm. When a film of material 3.6×10^{-3} cm thick was placed over one of the slits, the fringe pattern was displaced by a distance equal to 30 times that between two adjacent fringes. Calculate the refractive index of the material. To which side are the fringes displaced?

(When a layer of transparent material whose refractive index is *n* and whose thickness is *d* is placed in the path of a beam of light, it introduces a path difference equal to $(n - 1)d$.)   (*O. and C.*)

8. Give an account of the evidence that leads you to believe that light is propagated as a transverse wave motion.

In an interferometer, two coherent beams are made to follow separate paths, one traversing the length of a glass tube *A*, the other the length of a glass tube *B*, before being brought together to interfere. The tubes, each 20 cm long, are closed by plane glass plates, and initially contain air at atmospheric pressure. Fringes are obtained with light of wavelength 5.6×10^{-5} cm. Explain why the fringes move across the field of view while one of the tubes is being evacuated, and calculate the number which cross a fixed point in the field of view before a good vacuum is finally reached.

(Take the refractive index of air to be 1.000 28 at atmospheric pressure.)   (*O.*)

9. State the conditions necessary for the effects of interference in optics to be observed.

Two microscope slides, each 7.5 cm in length are placed with two of their faces in contact. A cover glass is then inserted between them at one extreme end to enclose a wedge-shaped layer of air. When illuminated normally by sodium light and the reflected light viewed, parallel interference bands are seen at a uniform distance apart of 0.11 mm. Explain the formation of these bands and calculate the thickness of the cover glass. (Wavelength of sodium light may be taken as 5.9×10^{-5} cm.)
          (*L.*)

10. Draw a diagram to illustrate how you would set up apparatus to view Newton's rings, formed by normal reflection of sodium light. Explain the formation of the rings.

If the diameter of the fifth bright ring is 2.55×10^{-1} cm and the diameter of the fifteenth bright ring is 4.27×10^{-1} cm, calculate an approximate value for the wavelength of the sodium light. The radius of curvature of the lens employed is 50.0 cm. (W.)

11. Describe how you would set up the necessary apparatus to observe and measure Newton's rings. Derive an expression for the difference in the squares of the radii of two consecutive bright rings in the pattern.

A set of Newton's rings was produced between one surface of a biconvex lens and a glass plate, using green light of wavelength 5.46×10^{-5} cm. The diameters of two particular bright rings, of orders of interference p and $p + 10$, were found to be 5.72 mm and 8.10 mm respectively. When the space between the lens surface and the plate was filled with liquid the corresponding values were 4.95 mm and 7.02 mm. Determine the radius of curvature of the lens surface and the refractive index of the liquid. (L.)

Diffraction: spectra: polarization

12. Calculate the angular separation of the red and violet rays of the first order spectrum when a parallel beam of white light is incident normally on a diffraction grating with 5000 lines per cm. Take the wavelengths of red and violet light to be 7×10^{-7} m and 4×10^{-7} m respectively.

13. When a source of monochromatic light of wavelength 6.0×10^{-7} m is viewed 2.0 m away through a diffraction grating two similar lamps are 'seen' one on each side of it, each 0.30 m away from it. Calculate the number of lines in the grating.

14. Explain the action of a plane diffraction grating. Giving the necessary theory and experimental detail, explain how it may be used to measure the wavelength of light from a monochromatic source. It may be assumed that the initial adjustments to the spectrometer have been made.

With a given grating used at normal incidence a yellow line, $\lambda = 6.0 \times 10^{-7}$ m, in one spectrum coincides with a blue line, $\lambda = 4.8 \times 10^{-7}$ m, in the next order spectrum. If the angle of diffraction for these two lines is 30° calculate the spacing between the grating lines. (A.E.B.)

15. Describe *two* experiments to illustrate the diffraction of light. How, in general, do the effects produced depend on the size of the obstacle or aperture relative to the wavelength of the light used?

A diffraction grating consists of a number of fine equidistant wires each of diameter 1.00×10^{-3} cm with spaces of width 0.75×10^{-3} cm between them. A parallel beam of infrared radiation of wavelength 3.00×10^{-4} cm falls normally on the grating, behind which is a converging lens made of material transparent to the infrared. Find the angular positions of the second order maxima in the diffraction pattern. (L.)

16. Draw a diagram to show plane monochromatic waves falling normally on a grating and being diffracted. Using the diagram, explain in which directions diffracted maxima will be seen.

A diffraction grating is fitted on a spectrometer table with monochromatic light incident normally on it. The instrument is adjusted to observe the spectra produced. State two ways in which the spectra would be affected if a grating of greater spacing were used.

The slit of the collimator is now illuminated with light of wavelengths 5.89×10^{-5} cm and 6.15×10^{-5} cm. The grating has 6.00×10^3 lines per cm and the telescope objective has a focal length of 20.0 cm. Calculate (*a*) the angle between the first-order diffracted waves leaving the grating, and (*b*) the separation between the centres of the two first-order lines formed in the focal plane of the objective. (J.M.B.)

17. Distinguish between emission spectra and absorption spectra. Describe the spectrum of the light emitted by (*i*) the sun, (*ii*) a car headlamp fitted with yellow glass, and (*iii*) a sodium vapour street lamp.

What are the approximate wavelength limits of the visible spectrum? How would you demonstrate the existence of radiations whose wavelengths lie just outside these limits? (O. and C.)

18. Describe and explain the action of one device for producing a beam of plane polarized light.

How would you make an experimental determination of the refractive index of a polished sheet of black glass using this device?

Mention two practical applications of polarized light. (L. part qn.)

19. Explain what is meant by *plane polarized light.*

Describe how a beam of plane polarized light may be produced by (*a*) reflection, and (*b*) double refraction. Give an account of two uses of polarized light.

Why is it not possible to polarize sound waves? (A.E.B.)

Thermal radiation

20. The cylindrical element of a 1.0 kW electric fire is 30.0 cm long and 1.0 cm in diameter. Taking the temperature of the surroundings to be 20 °C and Stefan's constant to be 5.7×10^{-8} W m^{-2} K^{-4}, estimate the working temperature of the element.

Give *two* reasons why the actual temperature is likely to differ from your estimate. (J.M.B.)

21. State *Stefan's* law and explain what is meant by *Prévost's theory of exchanges.*

What is meant by a *black body radiator?* Draw labelled curves which show how the energy in the spectrum of the radiation from such a body varies with wavelength at various temperatures.

A copper sphere, at a temperature of 727 °C, is suspended in an evacuated enclosure the walls of which are maintained at 27 °C. If the radius of the sphere is 5.0 cm, and if conditions are such that Stefan's law is obeyed, calculate the initial rate of fall of temperature of the sphere. (Density of copper $= 9.0 \times 10^3$ kg m^{-3}; specific heat capacity of copper = 400 J kg^{-1} K^{-1}; Stefan's constant = 5.8×10^{-8} W m^{-2} K^{-4}.)

(*A.E.B.*)

22. What do you understand by a *black body*? In what respects does the radiation from a black body at 2000 K differ from that from a black body at 1000 K? How would you devise a black body to radiate at 1000 K?

A blackened sphere, of radius 2.0 cm, is contained within a hollow evacuated enclosure the walls of which are maintained at 27 °C. Assuming that the sphere radiates like a black body and that Stefan's constant is 5.7×10^{-8} W m^{-2} K^{-4}, calculate the rate at which the sphere loses heat when its temperature is 227 °C.

(*J.M.B.*)

23. Discuss the nature of the processes by which a hot body may lose heat to its surroundings.

A blackened platinum strip of area 0.20 cm^2 is placed at a distance of 200 cm from a white-hot iron sphere of diameter 1.0 cm, so that the radiation falls normally on the strip. The radiation causes the temperature, and hence the resistance, of the platinum to increase. It is found that the same increase in resistance can be produced under similar conditions, but in the absence of radiation, when a current of 3.0 mA is passed through the platinum strip, the potential difference between its ends being 24 mV. Estimate the temperature of the iron sphere.

(Stefan's constant = 5.7×10^{-8} W m^{-2} K^{-4}.)   (*C.*)

Objective-type revision questions for Part 4

The first figure of a question number gives the relevant chapter, e.g. **16.2** is the second question for chapter 16.

Multiple choice

Select the response which you think is correct.

16.1. Which one of the following statements is *not* correct? For interference to occur between two sets of waves

A each set must have a constant wavelength.
B the two sets must have the same wavelength.
C the two sets must not be polarized or must have corresponding polarizations.
D the waves must be transverse.
E the waves must have similar amplitudes.

16.2. Which of the graphs in Fig. 22 best represents the variation of the frequency of waves with wavelength if their speed remains constant?

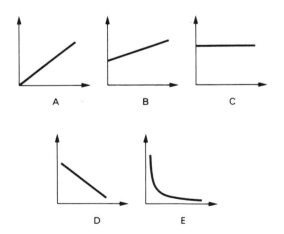

Fig. 22

16.3. Fig. 23 shows two coherent sources of waves vibrating in phase and separated by a distance of 8.0 cm. P is the nearest point to the axis at which constructive interference occurs in a plane 24 cm from the line joining the sources. If S_1P is 26 cm, the wavelength of the waves in cm is

A 4.0 **B** 8.0 **C** 10 **D** 16 **E** 18

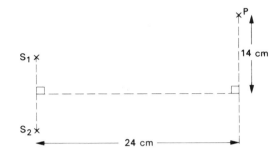

Fig. 23

16.4. Two waves each of amplitude 1.5 mm and frequency 10 Hz are travelling in opposite directions with velocity 20 mm s^{-1}. The distance in mm between adjacent nodes is

A 1.0 **B** 1.5 **C** 2.0 **D** 5.0 **E** 10

17.1. A steel piano wire 0.5 m long has a total mass of 0.01 kg and is stretched with a tension of 800 N. The frequency when it vibrates in its fundamental mode is

A 2 Hz **B** 4 Hz **C** 100 Hz **D** 200 Hz
E 20 000 Hz *(J.M.B.)*

17.2. A closed pipe resonates at its fundamental frequency of 300 Hz. Which one of the following statements is *not* correct?

A If the pressure rises the fundamental frequency increases.
B If the temperature rises the fundamental frequency increases.
C The first overtone is of frequency 900 Hz.
D An open pipe with the same fundamental frequency has twice the length.
E If the pipe is filled with a gas of lower density the fundamental frequency increases.

18.1. For a monochromatic light wave passing from air to glass which one of the following statements is true?

A Both frequency and wavelength decrease.
B The frequency increases and wavelength decreases in the same proportion.
C Frequency stays the same but wavelength increases.
D Frequency stays the same but wavelength decreases.
E Frequency and wavelength are unchanged. *(J.M.B.)*

18.2. In a Young's double-slit interference experiment using green light the fringe width was observed to be 0.20 mm. If red light replaces green light the fringe width becomes

  A 0.31 mm   **B** 0.25 mm   **C** 0.20 mm   **D** 0.16 mm
  E 0.13 mm

(Wavelength of green light = 5.2×10^{-7} m; of red light = 6.5×10^{-7} m.)   (*J.M.B.*)

18.3. Newton's rings are formed in the air gap between a plane glass surface and the convex surface of a lens which is in contact with it. If the first bright ring has a radius of 2.0×10^{-4} m, the radius of the second bright ring in m is

  A 1.4×10^{-4} **B** 2.8×10^{-4} **C** 3.5×10^{-4}
  D 4.0×10^{-4} **E** 6.0×10^{-4}

Part 5 ATOMS

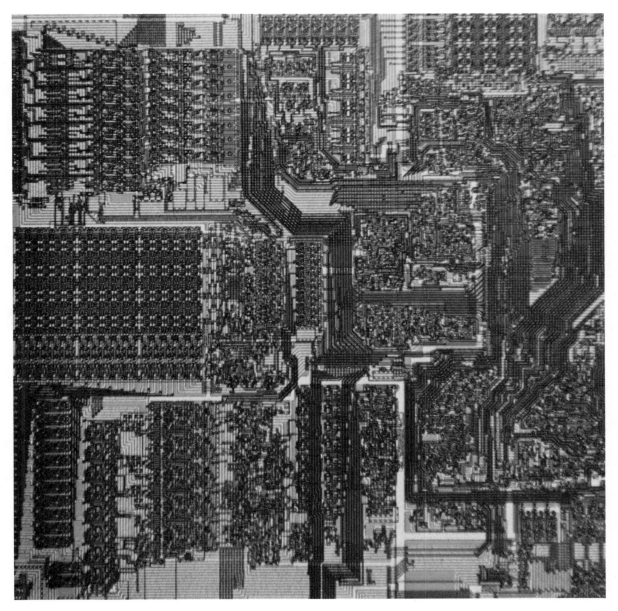

19 Kinetic theory: thermodynamics

Introduction

The behaviour of matter in bulk can be described in terms of quantities such as density, pressure and temperature which are measurable and are often capable of perception by the senses. One of the aims of modern science is to relate these macroscopic (i.e. large-scale) properties of matter to the masses, speeds, energies, etc., of the constituent atoms and molecules that cannot be perceived directly. An attempt is made to explain the macroscopic in terms of the microscopic. If the atomic model of matter is valid this should be possible since the same thing is being considered from two different points of view. We thereby hope to obtain deeper understanding of the phenomenon under investigation.

Here we will pursue this approach by studying *thermodynamics* and *kinetic theory* (a branch of statistical mechanics). In thermodynamics, thermal effects are considered using macroscopic quantities like pressure, temperature, volume and internal energy. Kinetic theory covers similar ground but assumes the existence of atoms and molecules and applies the laws of mech-anics to large numbers of them, using simple averaging techniques.

We will be concerned mainly with gases. In these, conditions are simpler and the mathematical problems less difficult as a result.

Gas laws

Experiments have established three laws describing the thermal behaviour of gases.

(*a*) *Boyle's law*. This relates the pressure and volume of a gas and arose from work done by Boyle about 1660.

The pressure of a fixed mass of gas is inversely proportional to its volume if the temperature is constant.

In symbols, if p is the pressure and V the volume then

$$p \propto \frac{1}{V} \quad \text{or} \quad p = \frac{\text{constant}}{V}$$

$$\therefore \quad pV = \text{constant}$$

The value of the constant depends on the mass of gas

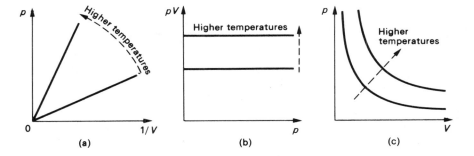

Fig. 19.1   (a)   (b)   (c)

and the temperature. Graphically the law may be represented by plotting (*i*) p against $1/V$, Fig. 19.1*a*, which is a straight line through the origins, (*ii*) pV against p, Fig. 19.1*b*, giving a straight line parallel to the *p*-axis, or (*iii*) p against V, Fig. 19.1*c*, to obtain a rectangular hyperbola. If a fixed mass of gas is investigated at different temperatures, a series of graphs, each called an *isothermal* (i.e. a curve of constant temperature), can be drawn as shown.

Measurements made since Boyle's time have indicated that the law is obeyed only when the density of the gas is low. Thus, gases such as oxygen, nitrogen, hydrogen and helium (known as 'permanent' gases) follow the law well at normal temperatures and pressures but they deviate from it at high pressures (e.g. above several hundred atmospheres) when their density is high. Their *p-V* graphs become more like that for a liquid and the volume decrease is quite small for large pressure increases, as shown by the dotted curve in Fig. 19.2. At high pressures carbon dioxide and some other gases and vapours are close enough to being liquid at ordinary temperatures to show marked deviations from the law.

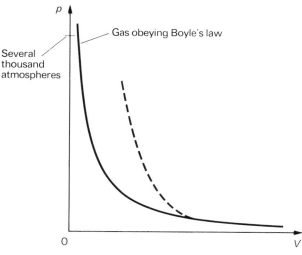

p

Gas obeying Boyle's law

Several thousand atmospheres

O

V

Fig. 19.2

(*b*) *Charles' law.* The connection between the volume change with temperature of a fixed mass of gas kept at constant pressure was published by Charles in 1787 (and independently by Gay-Lussac in 1802). It states that the volume V of the gas, of fixed mass and pressure, measured at temperature θ by, say, a mer-

cury-in-glass thermometer is related approximately to its volume V_0 at the ice point (0 °C) by

$$V = V_0(1 + \alpha\theta)$$

where α is the *cubic expansivity of the gas at constant pressure* and has roughly the same value of 0.00366 °C^{-1} (1/273) for all gases at low pressure. This is a linear relationship and a graph of V against θ would be a straight line but not passing through the volume origin. How would you find α from such a graph?

(*c*) *Pressure law.* A similar relationship is found to exist between the pressure p of a fixed mass of a gas, kept at constant volume, and the temperature θ, again measured by the mercury-in-glass thermometer. It is

$$p = p_0(1 + \beta\theta)$$

where p_0 is the pressure at the ice point and β is a constant, known as the *pressure coefficient* of the gas. It has practically the same value as α.

Temperature scales

(*a*) *Disagreement between thermometers.* The early experimenters measured temperatures with mercury thermometers and often found that different instruments gave different readings for the same temperature. Later, when thermometers based on other temperature-dependent properties of matter were used (e.g. electrical resistance of a wire, pressure of a gas at constant volume) there was disagreement between the different kinds as well (except at the fixed points where they had to agree by definition). (See page 65.)

It became clear that one type of thermometer based on one scale of temperature would have to be chosen as a standard. Gas thermometers, either constant volume Fig. 19.3, or constant pressure, showed the closest agreement over a wide range of temperatures, especially at low pressures. They were also very sensitive, accurate and highly reproducible. Although the readings depend on the nature of the gas used and on its pressure, all indicated the same reading as the pressure was lowered and approached zero.

(*b*) *Ideal gas scale.* Kelvin suggested that the standard scale of temperature should be based on an imaginary ideal gas with properties that were those of real gases at very low pressure, i.e. it obeyed Boyle's law. He proposed that the product of the pressure and the volume of this ideal gas be used as the thermometric property and the simplest procedure is then to say that if p_1 and p_2 are the pressures and V_1 and V_2 are the corresponding volumes of a fixed mass of the gas at

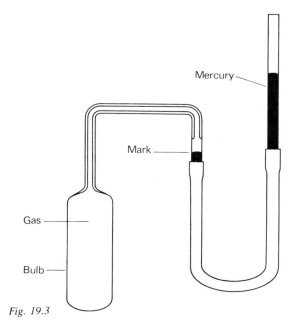

Mercury

Mark

Gas

Bulb

Fig. 19.3

calibrated. This is done by choosing two fixed points and assigning numbers to them. Before 1954 the ice and steam points were used and on the Celsius method of numbering, given the values 0 °C and 100 °C respectively.

For practical reasons (among which was the extreme sensitivity of the steam point to pressure change) the upper fixed point is now taken as the only temperature at which ice, liquid water and water vapour coexist in equilibrium and is called the *triple point* of water. It is given the value of 273.16 K (which is nearly 0.01 °C), K being the symbol for the post-1954 unit of temperature, i.e. the kelvin. This number was chosen so that there were still 100 kelvins between the ice and steam points. The temperatures of the ice, triple and steam points in °C and in K are given below.

	°C	K
Steam point	100	373.15
Triple point	0.01	273.16
Ice point	0	273.15

The lower fixed point, to which a value 0 K is assigned, is called *absolute zero* and more will be said about it presently (0 K = −273.15 °C).

If we use a constant volume gas thermometer, and in the equation $T_1/T_2 = p_1/p_2$ we put

$T_1 = T$ (the unknown temperature)
$T_2 = 273.16$ (triple point)
$p_1 = p$ (pressure of the ideal gas in the constant volume thermometer at temperature T)

and $\quad p_2 = p_{tr}$ (pressure at the triple point)

we get

$$T = 273.16 \left(\frac{p}{p_{tr}}\right)$$

(*d*) *Measuring temperature on the ideal gas scale.* To find an unknown temperature T using a constant volume gas thermometer, the pressure p_{tr} at the triple point has to be found when the bulb is surrounded by water at 273.16 K in a triple point cell, Fig. 19.4. The pressure p at the unknown temperature must also be determined (keeping the volume constant) and then the value calculated of 273.16 (p/p_{tr}).

Some of the gas in the thermometer is then removed so reducing the pressure and the new values for p_{tr} and p measured. The value of 273.16 (p/p_{tr}) is again obtained and will be different. This process is repeated for smaller and smaller amounts of gas in the bulb and a graph of the calculated value of 273.16 (p/p_{tr}) plotted against p_{tr}.

two temperatures T_1 and T_2 on this scale, then these temperatures are defined by

$$\frac{T_1}{T_2} = \frac{p_1 V_1}{p_2 V_2}$$

A linear relationship between pV and temperature is thus assumed and means we have *chosen* to make equal changes of pV represent equal changes of temperature. That is, the ideal gas scale of temperature has been *defined* so that pV for a fixed mass of ideal gas is directly proportional to the temperature of the gas measured on this scale. The measurements can be imagined to be made either by a constant volume or a constant pressure gas thermometer or by an experiment in which neither pressure nor volume remain constant. In all cases the ratio T_1/T_2 will be the same.

If the volume of the gas is kept constant we have

$$\frac{p_1}{p_2} = \frac{T_1}{T_2}$$

which is the pressure law for an ideal gas and is a consequence of the way temperature is defined on the ideal gas scale. Similarly, at constant pressure we can say

$$\frac{V_1}{V_2} = \frac{T_1}{T_2}$$

(*c*) *Fixed points.* The unit of temperature on this scale has to be defined so that thermometers can be

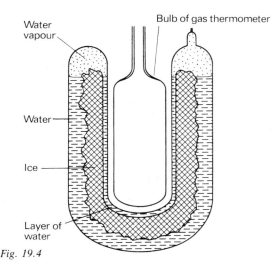

Fig. 19.4

Graphs for different gases at the temperature of steam in equilibrium with water at standard pressure are shown in Fig. 19.5. When extrapolated to $p_{tr} = 0$, they all indicate the same value, which is the temperazure T of the steam point on the ideal gas scale. A constant volume thermometer containing a *real* gas, e.g. helium, hydrogen or nitrogen, can thus be used to measure temperatures on the ideal gas scale.

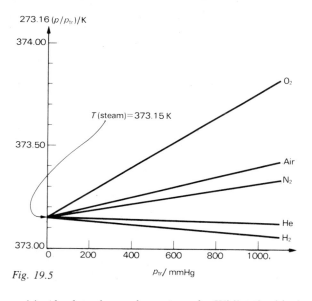

Fig. 19.5

(e) *Absolute thermodynamic scale.* Whilst the ideal gas scale is independent of the properties of any particular gas, it does depend on the properties of gases in general. The absolute thermodynamic or Kelvin scale is quite independent of the properties of any

substance and is based on the efficiency of an ideal reversible heat engine (p. 448). It is a theoretical scale and the SI unit of temperature, i.e. the kelvin, is defined in terms of it, as is absolute zero (0 K).

The ideal gas scale and the Kelvin scale can be shown in a more advanced treatment, to be theoretically identical at all temperatures. We are therefore justified in writing 'K' after an ideal gas scale temperature as we have done. The symbol for temperatures in kelvins is T.

Theory also predicts that temperatures below absolute zero do not exist and certainly so far all attempts to reach 0 K have failed.

(f) *International Practical Temperature Scale* (IPTS). Measuring ideal gas scale temperatures is a tedious task and the IPTS has been adopted as a more practical scale for general use. It consists of eleven primary fixed points (ranging from the triple point of hydrogen at 13.81 K to the freezing point of gold at 1337.58 K), which have been accurately determined on the ideal gas scale by a constant volume gas thermometer, and also some other secondary fixed points. Particular types of thermometer are specified for measuring temperatures over a certain range using agreed formulae. At the fixed points there is agreement between the IPTS and the Kelvin scale; the differences at intervening temperatures are usually negligible.

Ideal gas equation

The equation defining temperature on the ideal gas scale is (p. 429),

$$\frac{T_1}{T_2} = \frac{p_1 V_1}{p_2 V_2}$$

or

$$\frac{p_1 V_1}{T_1} = \frac{p_2 V_2}{T_2}$$

Hence for an *ideal* gas we can say

$$\frac{pV}{T} = \text{constant}$$

where p and V represent any pair of simultaneous values of the pressure and volume of a given mass of ideal gas at the ideal gas scale temperature T.

This equation is *not* based on experiment but incorporates *two definitions*. The first is of an ideal gas as one which obeys Boyle's law ($pV = \text{constant}$) and the second of temperature on the ideal gas scale as a quantity proportional to pV.

The value of the constant depends on the mass of gas and experiments with real gases at low enough pressures show that it has the same value R for all gases if one mole is considered. R is called the *universal molar gas constant* and has a value 8.31 J mol^{-1} K^{-1}. (Check that the units are correct.) We can now write for one mole

$$pV = RT$$

This is called the *equation of state for an ideal gas*, the state being determined by the values of pressure and temperature. (In view of this use of the term 'state', confusion is avoided if we refer on other occasions to the three 'phases' of matter, i.e. solid, liquid and gas.) The ideal gas does not exist of course, but since real gases behave almost ideally at low pressures the concept is sufficiently close to actual fact to be useful.

Notes. 1. The *mole* is the amount of substance which contains the same number of elementary entities as there are atoms in 12 grams of carbon 12. Experiment shows this to be about 6.02×10^{23}—a value denoted by L and called the *Avogadro constant*, i.e. $L = 6.02 \times 10^{23}$ particles per mole where the particles may be atoms, molecules, electrons, etc.

2. The *relative molecular mass* M_r (known previously as the *molecular weight*) of a substance is defined by

$$M_r = \frac{\text{mass of a molecule of substance}}{\text{mass of carbon 12 atom}} \times 12$$

The values for hydrogen, carbon and oxygen are 2, 12 (by definition) and 32 respectively.

3. The *molar mass* M_m of a substance is the mass of one mole and if expressed in kg mol^{-1} it is related (in SI units) to M_r by the equation

$$M_m = M_r \times 10^{-3}$$

M_m may thus be the mass of a substance (in kg) containing L molecules; its value in SI units for oxygen is 32×10^{-3} kg mol^{-1}. ('Molar' means 'per unit amount of substance' and terms such as 'gram-molecule' are now obsolescent.)

4. In $pV = RT$, V is the molar volume given by

$$V = \frac{\text{molar mass (in kg mol}^{-1})}{\text{density (in kg m}^{-3})}$$

$$= \frac{M_m}{\rho} \text{ (in m}^3 \text{ mol}^{-1})$$

Hence we can also write $\quad R = \dfrac{pM_m}{T\rho}$

$$= \frac{pM_r}{T\rho} \times 10^{-3}$$

Also for n moles of an ideal gas

$$pV = nRT$$

where n = mass of gas (in kg)/molar mass (in kg mol^{-1}).

5. One mole of an ideal gas at s.t.p. has a volume of 22.4×10^{-3} m^3 as can be shown by substituting $p = 1.01 \times 10^5$ Pa, $R = 8.31$ J mol^{-1} K^{-1}, $T = 273$ K and $n = 1$ in $pV = nRT$. $(1 \times 10^{-3}$ m^3 = 1 litre.)

Kinetic theory of an ideal gas

So far, the ideal gas has been considered from the macroscopic point of view and treated in terms of quantities such as pressure, volume and temperature, which are properties of the gas as a whole. The kinetic theory, on the other hand, attempts to relate the macroscopic behaviour of an ideal gas with the microscopic properties (e.g. speed, mass) of its molecules. A theoretical model is constructed by attributing certain properties to the molecules and the laws of mechanics are applied to it. Any results obtained must then be studied to see if the theory throws any light on the macroscopic behaviour of ideal and real gases.

(*a*) *Assumptions.* The main assumption, which in effect defines an ideal gas on the microscopic scale, is that the range of intermolecular forces (both attractive and repulsive) is small compared with the average distance between molecules. From this it follows that

(*i*) the intermolecular forces are negligible except during a collision.

(*ii*) the volume of the molecules themselves can be neglected compared with the volume occupied by the gas (see p. 439),

(*iii*) the time occupied by a collision is negligible compared with the time spent by a molecule between collisions, and

(*iv*) between collisions a molecule moves with uniform velocity.

We also assume that there is a large number of molecules even in a small volume and that a large number of collisions occurs in a small time.

(*b*) *Calculation of pressure.* An expression for the pressure exerted by an ideal gas due to molecular bombardment can be derived and the model made quantitative.

Suppose a gas in a cubical box of side l contains N molecules each of mass m. Consider one molecule moving with velocity c as shown in Fig. 19.6. We can resolve c into components u, v and w in the three directions Ox, Oy and Oz respectively, parallel to the edges of the box. First consider motion along Ox. The molecule has momentum mu to the right and on striking the shaded wall X of the box, we can assume for mathematical simplicity that on average (since the system is in equilibrium) it rebounds with the same velocity reversed. Its momentum is now mu to the left, i.e. $-mu$ and so, by providing an impulse (i.e. a force for a short time), the wall has caused a change of momentum of

$$mu - (-mu) = 2\,mu$$

The momentum components mv and mw have no effect on the momentum exchanges at wall X since they are directed at right angles to Ox.

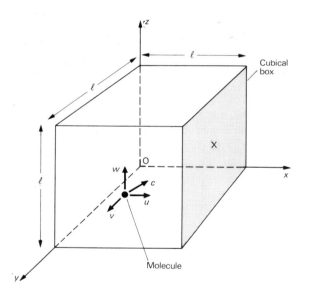

Fig. 19.6

If the molecule travels to the wall opposite X and rebounds back to X again without striking any other molecule on the way, it covers a distance $2l$ in time $2l/u$. This is the time interval between successive collisions of the molecule with wall X. Hence

rate of change of momentum at X =

$$\frac{\text{change of momentum}}{\text{time}}$$

$$= \frac{2mu}{2l/u} = \frac{mu^2}{l}$$

Newton's second law of motion states that force equals rate of change of momentum. Therefore the

total force on wall X due to impacts by all N molecules is given by

$$\text{force on X} = \frac{m}{l}\,(u_1^2 + u_2^2 + \ldots + u_N^2)$$

where u_1, u_2, etc., are the different Ox components of the velocities of molecules 1, 2, etc.

Since pressure is force per unit area, we can say that the pressure p on wall X of area l^2 is given by

$$p = \frac{m}{l^3}\,(u_1^2 + u_2^2 + \ldots + u_N^2)$$

If $\overline{u^2}$ represents the mean value of the squares of all the velocity components in the Ox direction, then

$$\overline{u^2} = \frac{u_1^2 + u_2^2 + \ldots + u_N^2}{N}$$

and

$$N\overline{u^2} = u_1^2 + u_2^2 + \ldots + u_N^2$$

$$\therefore \quad p = \frac{Nm\overline{u^2}}{l^3}$$

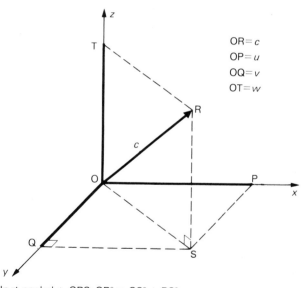

$$OR = c$$
$$OP = u$$
$$OQ = v$$
$$OT = w$$

In rt. angled $\triangle$ ORS: $OR^2 = OS^2 + RS^2$
In rt. angled $\triangle$ OQS: $OS^2 = QS^2 + OQ^2$
$\therefore OR^2 = QS^2 + OQ^2 + RS^2 = OP^2 + OQ^2 + OT^2$
$\therefore c^2 = u^2 + v^2 + w^2$

Fig. 19.7

For any molecule, the application of Pythagoras' theorem twice to Fig. 19.7 shows that

$$c^2 = u^2 + v^2 + w^2$$

This will also hold for the mean square values

$$\therefore \quad \overline{c^2} = \overline{u^2} + \overline{v^2} + \overline{w^2}$$

But since N is large and the molecules move randomly (i.e. they show no preference for moving parallel to any one edge of the cube, see (c) (v) below), it follows that the mean values of u^2, v^2 and w^2 are equal. Thus

$$\overline{c^2} = 3\overline{u^2}$$

$$\therefore \quad \overline{u^2} = \frac{\overline{c^2}}{3}$$

Hence

$$p = \frac{Nm\overline{c^2}}{3l^3}$$

But $l^3 =$ volume of the cube = volume of gas = V and so

$$pV = \frac{1}{3}Nm\overline{c^2}$$

This important result links the macroscopic property pressure with the number, mass and speed of the molecules. We will see presently what further information it yields.

Alternatively, if ρ is the density of the gas we can write

$$p = \frac{1}{3}\rho\overline{c^2}$$

since $\rho = Nm/V$, Nm being the total mass of gas.

(c) *Further points*. The following points about the above derivation should be noted.

(i) A container of any shape could have been chosen but a cube simplifies matters.

(ii) Intermolecular collisions were ignored but these would not affect the result because the average momentum of the molecules on striking the walls is unaltered by their collisions with each other.

(iii) The mean-square speed $\overline{c^2}$ is not the same as the square of the mean speed. Thus if five molecules have speeds of 1, 2, 3, 4, 5 units, their mean speed is $(1 + 2 + 3 + 4 + 5)/5 = 3$ and its square is 9. The mean-square speed on the other hand is $(1^2 + 2^2 + 3^2 + 4^2 + 5^2)/5 = 55/5 = 11$.

(iv) The pressure on one wall only was found but it is the same on all sides (neglecting the weight of the gas).

(v) Fig. 19.8 shows a horizontal table with lots of small holes through which air is blown and forms a cushion on which lightweight pucks can move almost without friction. A vibrating wire around the 'air table' keeps the pucks moving randomly and compensates for the inelastic collisions they have with each other and with

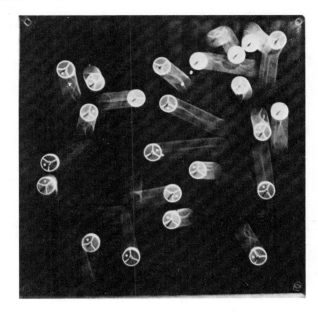

Fig. 19.8

the sides. The apparatus is a two-dimensional model of a gas.

The photograph was taken from above by opening a camera shutter about one second before the air supply was cut off and closing it one second after. The pucks stop immediately the air supply stops and 'tracks' of their direction of motion are obtained. We see from them that *roughly* half the pucks are moving east or west if we include those that are going more nearly east or west than north or south. The other half are going more nearly north or south.

(vi) The derivation is a simplified version of a more complex piece of theory which makes similar assumptions and arrives at the same result.

Temperature and kinetic theory

We have found from the kinetic theory that for a mole of an ideal gas,

$$pV = \frac{1}{3}Lm\overline{c^2}$$

where L is the Avogadro constant. This may be rewritten as

$$pV = \frac{2}{3}L(\tfrac{1}{2}m\overline{c^2})$$

The ideal gas equation of state, also for a mole, is

$$pV = RT$$

where R is the universal molar gas constant. If we combine these last two equations we get

$$\frac{2}{3}L(\tfrac{1}{2}m\overline{c^2}) = RT$$

$$\therefore \quad \tfrac{1}{2}m\overline{c^2} = \frac{3}{2}\frac{R}{L}T$$

Now $\tfrac{1}{2}m\overline{c^2}$ is the average kinetic energy (k.e.) of translational motion[1] per molecule and from the above equation we see, since R and L are constants, that *for an ideal gas, it is proportional to the thermodynamic temperature T.*

The kinetic theory thus offers a microscopic basis for the macroscopic quantity of temperature. It suggests that an increase in the temperature of a gas is a manifestation of an increase of molecular speed and so of translational k.e.

The ratio R/L is called *Boltzmann's constant* and is denoted by k. It is the gas constant per molecule and has the value 1.38×10^{-23} J K^{-1}. Hence we can say

$$\tfrac{1}{2}m\overline{c^2} = \frac{3}{2}kT$$

The total translational k.e. per mole of an ideal gas is given by

$$\tfrac{1}{2}Lm\overline{c^2} = \frac{3}{2}RT$$

Whilst the above expression indicates that at absolute zero the molecular k.e. of an *ideal gas* is zero, this is not so for real substances. In their case, the molecular k.e. can be shown to have a *minimum* value, called the 'zero-point energy' and it is wrong to think that all molecular motion stops at 0 K.

Deductions from kinetic theory

Three well-established laws relating to real gases will be deduced.

(*a*) *Avogadro's law (or hypothesis).* Consider two ideal gases 1 and 2. We can write

$$p_1V_1 = \frac{1}{3}N_1m_1\overline{c_1^2}$$

$$p_2V_2 = \frac{1}{3}N_2m_2\overline{c_2^2}$$

---

[1] This is motion in which every point of the body has the same instantaneous velocity and there is no rotational or vibrational motion.

If their pressures, volumes and temperatures are the same, then

$$p_1 = p_2$$

$$V_1 = V_2$$

$$\tfrac{1}{2}m_1\overline{c_1^2} = \tfrac{1}{2}m_2\overline{c_2^2}$$

$$\therefore \quad N_1 = N_2$$

Hence, *equal volumes of ideal gases existing under the same conditions of temperature and pressure contain equal numbers of molecules.*

This statement is Avogadro's law. It cannot be proved directly for real gases but has to be assumed to explain the volume relationships which exist between combining gases (and are expressed by Gay-Lussac's law).

(*b*) *Dalton's law of partial pressures.* The kinetic theory attributes gas pressure to bombardment of the walls of the containing vessel by molecules. In a mixture of ideal gases we might therefore expect the total pressure to be the sum of the partial pressures due to each gas. This statement is Dalton's law of partial pressures, which holds for real gases when they behave ideally. Fig. 19.9 illustrates the law.

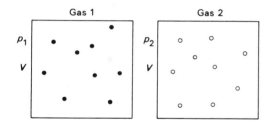

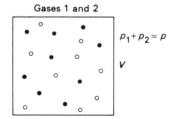

Fig. 19.9

(*c*) *Graham's law of diffusion.* It is reasonable to assume that in a closed container, the rate of diffusion of a gas through a small hole is directly proportional to the rate at which each molecule hits the side with the hole, i.e. the rate is proportional to $\sqrt{\overline{c^2}}/2l$, i.e. to $\sqrt{\overline{c^2}}$, where l is the length of the container. Hence if two gases, 1 and 2, diffuse through the same hole under the same conditions of pressure and temperature,

$$\frac{\text{rate of diffusion of 1}}{\text{rate of diffusion of 2}} = \sqrt{\frac{\overline{c_1^2}}{\overline{c_2^2}}}$$

But, from $p = \frac{1}{3}\rho\overline{c^2}$ we can say

$$\rho_1\overline{c_1^2} = \rho_2\overline{c_2^2}$$

$$\therefore \quad \frac{\overline{c_1^2}}{\overline{c_2^2}} = \frac{\rho_2}{\rho_1}$$

Hence
$$\frac{\text{rate of diffusion of 1}}{\text{rate of diffusion of 2}} = \frac{\sqrt{\rho_2}}{\sqrt{\rho_1}}$$

That is, the *rate of diffusion of a gas is inversely proportional to the square root of its density*. This is the law which Graham discovered by experiment.

Molecular magnitudes

Information about gas molecules can be obtained from the kinetic theory if use is also made of knowledge obtained from macroscopic measurements on gases.

(*a*) *Speed of air molecules.* For a gas of density ρ, exerting pressure p we have

$$p = \frac{1}{3}\rho\overline{c^2}$$

where $\overline{c^2}$ is the mean-square speed of the molecules. Assuming we can apply this equation to air, for which $\rho = 1.29$ kg m^{-3} at s.t.p. (i.e. 273 K and 1.01×10^5 Pa) then

$$\overline{c^2} = \frac{3p}{\rho} = \frac{3 \times 1.01 \times 10^5}{1.29} \frac{\text{Pa}}{\text{kg m}^{-3}}$$

$$= 2.35 \times 10^5 \text{ m}^2 \text{ s}^{-2}$$

The square root of $\overline{c^2}$ is called the *root-mean-square* (r.m.s.) *speed*. It is denoted by $\sqrt{\overline{c^2}}$ or $c_{\text{r.m.s.}}$ (is greater than the mean speed, see p. 433) and for air has the value

$$\sqrt{\overline{c^2}} = \sqrt{2.35 \times 10^5} = 485 \text{ m s}^{-1}$$

This is a surprisingly high speed. Some molecules will have greater speeds, others smaller, and these will change due to energy exchanges in intermolecular collisions.

The speeds of gas molecules have been measured directly and the results are in good agreement with calculated values. The fact that the speed of sound in a gas is of the same order as the speed of its molecules offers some confirmation. We would not expect the speed with which the energy of a sound wave is passed on from one air molecule to the next to be greater than the speed of the air molecules themselves. In fact, it could be less due to the random nature of molecular motion. A 'message' carried by a 'runner' can only be transmitted as fast as he can 'run' and may even travel more slowly if there is 'traffic' congestion. The speed of sound in air at 0 °C is about 330 m s^{-1}, which compares with the 485 m s^{-1} we calculated for the r.m.s. speed of air molecules at 0 °C.

The very high speed with which molecules of brown bromine vapour travel into a vacuum may be demonstrated, but *care is necessary as the liquid and vapour are both dangerous.* (For precautions see Appendix 14.) The air is removed from the tube of Fig. 19.10*a*,

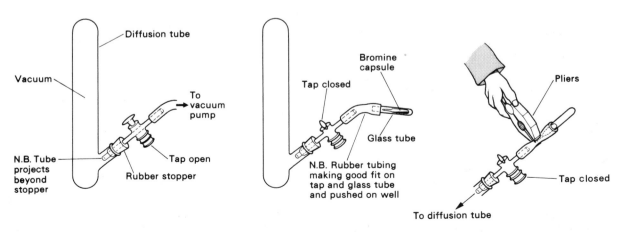

Fig. 19.10   (a)   (b)   (c)

the tap closed and another tube containing a bromine capsule connected, Fig. 19.10*b*. The neck of the capsule is broken by squeezing the rubber tubing with pliers, Fig. 19.10*c*, and the tap opened.

(*b*) *Mean free path.* Despite their high average molecular speeds, gases diffuse very slowly into air as we know from the fact that a 'smell' (even when greatly aided by convection) takes time to travel. This suggests that gas molecules, because of their finite size, suffer frequent collisions with other molecules. At each collision the direction of motion changes and the path followed is a succession of zig-zag steps—called a *random walk*. The progress of a molecule is slow, although the sum of all the separate steps it has taken between collisions is a very large distance.

There is a statistical rule (which you can verify by a game, see Appendix 15) which states that the average distance travelled in a 'random walk' of *n* equal steps taken in succession in random directions will be $\sqrt{n}$ steps. Thus in Fig. 19.11, if A and B are the starting and finishing points respectively and 25 equal random steps are taken, then the distance from A to B as the crow flies is five steps.

Now consider the diffusion of bromine into air. Using the same apparatus as before the average distance bromine has diffused up the tube (containing air) in say, 500 seconds can be estimated by guessing where the vapour looks 'half-brown', Fig. 19.12. At the bottom it is 'full-brown' and at the top it is almost clear. Suppose the 'half-brown' level is 0.1 m up from the original, fairly distinct bromine-air boundary. In 500 seconds the average progress of a bromine molecule is, from start to finish, 0.1 m. If the average length of a step between collisions (called the *mean free path*) in this random walk of, say, *n* steps is λ then, applying the statistical rule, we have

$$0.1 = \sqrt{n} \, . \, \lambda \qquad (1)$$

But it is also known from calculations like the one we did for air molecules that bromine molecules have an r.m.s. speed of 200 m s^{-1} and so in 500 seconds they will travel $200 \times 500 = 10^5$ m. This will be the length of the *straightened-out* track of a bromine molecule and since it makes *n* steps (and therefore *n* collisions) in 500 seconds, we can say that the mean free path λ is also given by

$$\lambda = \frac{10^5}{n} \qquad (2)$$

Substituting for *n* in (1),

$$0.1 = \sqrt{\frac{10^5}{\lambda}} \, . \, \lambda = \sqrt{10^5 \lambda}$$

Squaring both sides gives

$$\lambda = 10^{-7} \text{ m}$$

How many collisions does a bromine molecule make (mostly with air molecules) in 500 seconds?

The mean free path of an air molecule in air at atmospheric pressure is also about 10^{-7} m = 100×10^{-9} m = 100 nm.

Mean free path λ is not to be confused with the average separation *D* of the gas molecules. At atmospheric pressure λ is appreciably greater than *D*, Fig. 19.13, as we shall see presently.

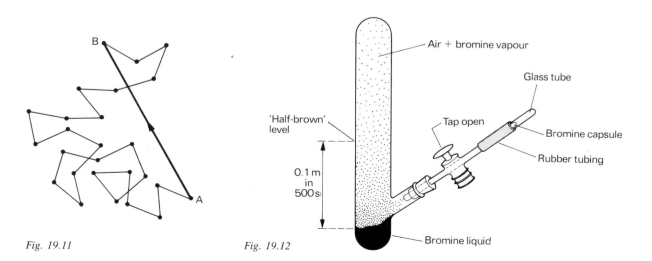

Fig. 19.11

'Half-brown' level

0.1 m in 500 s

Air + bromine vapour

Glass tube

Tap open

Bromine capsule

Rubber tubing

Bromine liquid

Fig. 19.12

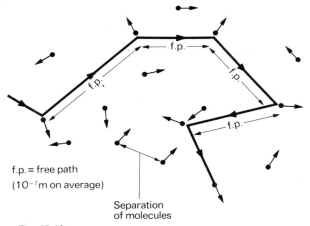

f.p. = free path
(10⁻⁷m on average)

Separation
of molecules

Fig. 19.13

(*c*) *Size of an air molecule.* The volume swept out on average by a molecule between two consecutive collisions is the volume of a cylinder of radius d and length λ where λ is the mean free path and d is the *diameter* of a molecule. This can be seen from Fig. 19.14; a collision will occur when the centres of the two molecules involved are up to distance d apart. In other words there is, on average, one gas molecule in a volume $\pi d^2 \lambda$ and this is therefore the volume 'taken up' by a gas molecule.

When liquid air changes to atmospheric air the volume increases about 750 times and so we may assume that the mean volume per gas molecule is 750 times the mean volume per liquid molecule. Assuming that the latter is d^3 (i.e. we suppose liquid molecules are close-packed like spheres in a box), then the former will be $750\,d^3$. Hence

$$750\,d^3 = \pi d^2 \lambda$$

$$\therefore \quad d = \frac{\pi \lambda}{750}$$

Substituting $\lambda = 10^{-7}$ m, we get

$$d \simeq 4 \times 10^{-10} \text{ m}$$

The diameter of an air molecule is thus 0.4 nm, approximately. More precise measurements can be made but there is no point in trying to obtain too exact values since molecules do not have hard, definite surfaces. The value obtained depends to some extent on the method of measurement and there will be an element of doubt when this involves collisions between molecules.

(*d*) *Separation of air molecules.* Gas molecules lack any kind of regular spacing such as exists in the case of the molecules of a solid. However, we can obtain a value for their 'average separation' D for air using the fact that in the change from liquid air to atmospheric air, the volume increases 750 times. We assume as above that each liquid molecule occupies a volume d^3, where d is the diameter of a molecule (in whatever phase). If we *imagine* a given volume of gaseous air as divided up into tiny cubes each of side D and each having one molecule at its centre, then the volume of a gaseous air molecule will be D^3. Hence

$$750\,d^3 = D^3$$

$$\therefore \quad \frac{D}{d} = \sqrt[3]{750} \simeq 9$$

The average separation of molecules in air is roughly 9 molecular diameters. Taking $d = 0.4$ nm, $D \simeq 3.6$ nm which is much smaller than the mean free path.

We now have the following *rough* estimates for air molecules at s.t.p.

Diameter (d)	Separation (D)	Mean free path (λ)	Speed $\sqrt{\overline{c^2}}$ or $c_{\text{r.m.s.}}$
0.4 nm	4 nm	100 nm	500 m s⁻¹

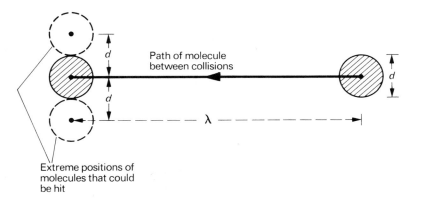

Path of molecule
between collisions

λ

Extreme positions of
molecules that could
be hit

Fig. 19.14

Properties of vapours

(*a*) *Saturated and unsaturated vapours.* If a very small quantity of a volatile liquid like ether is introduced at the bottom of a barometer, Fig. 19.15*a*, it rises to the top of the mercury column. Here it evaporates into the vacuum and the pressure exerted by the vapour causes the mercury level to fall, Fig. 19.15*b*. When more ether is injected, the level drops further but stops as soon as liquid ether appears on top of the column. The ether vapour is then said to be *saturated* and exerting its *saturation vapour pressure* (s.v.p.) at the temperature of the surroundings, Fig. 19.15*c*. Before liquid ether gathers above the mercury, the ether vapour is *unsaturated*.

The s.v.p. of a substance is the pressure exerted by the vapour in equilibrium with the liquid, e.g. the vapour is in a closed vessel above the liquid. If the space above the mercury in Fig. 19.15*a* had contained some air initially, the s.v.p. would be the same but the evaporation would take place more slowly. At 15 °C the s.v.p. of water is 1.3 cmHg and of ether 35 cmHg.

(*b*) *Vapours and the gas laws.* Unsaturated vapours obey Boyle's law roughly up to near saturation point, i.e. along AB in the *p*-*V* graph of Fig. 19.16*a*. At B condensation of the vapour starts, liquid and saturated vapour exist together along BC and, since the mass of vapour is changing, Boyle's law is no longer relevant. The pressure along BC is the s.v.p. at the temperature concerned and we see that it does not change as the volume of saturated vapour decreases. This may be shown by tilting a barometer tube containing saturated vapour as in Fig. 19.15*d*; the vertical depth of the mercury surface below its level in Fig. 19.15*a* indicates the s.v.p. and remains the same as the volume of saturated vapour diminishes.

Unsaturated vapours obey Charles' law and the pressure law approximately: saturated vapours do not. The s.v.p. of a vapour does increase with temperature but much more rapidly than is required by Charles' law, as Fig. 19.16*b* shows for water.

A mixture of gas and unsaturated vapour obeys the gas laws fairly well, each exerting the same pressure as it would exert if it alone occupied the total volume (Dalton's law of partial pressures, p. 434). A graph of *p* against 1/*V* for such a mixture is shown in Fig. 19.16*c*. What occurs at B? What does OD represent?

(*c*) *Kinetic theory explanations.* In an unsaturated vapour the rate at which molecules leave the liquid surface exceeds that at which they enter it from the vapour, i.e. evaporation occurs more rapidly than condensation. In a saturated vapour the rates are equal and *dynamic equilibrium* exists. The rate of leaving depends on the average k.e. of the molecules and so on the temperature; the rate of entering is determined by the temperature and density of the vapour.

The s.v.p. increases when the temperature of the liquid is raised because the average k.e. of the molecules increases and more are able to escape from the surface. The rate of evaporation becomes greater, thereby increasing the density of the vapour and so also the rate of condensation. Eventually equilibrium and

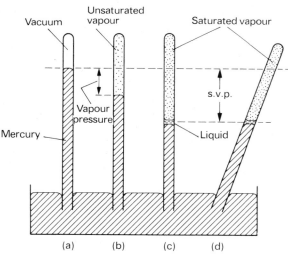

Fig. 19.15

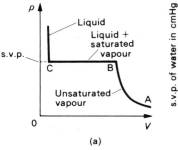

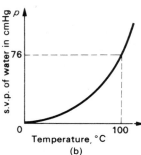

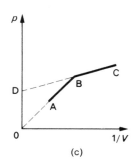

Fig. 19.16

saturation are re-established at a greater s.v.p. than before.

The s.v.p. is not affected by changes of volume (at a constant temperature). If the volume available to the vapour decreases, its density momentarily increases and more molecules return to the liquid in a given time than previously. The rate at which molecules leave the surface remains steady, however, and so the rate of condensation now exceeds the rate of evaporation until the density of the vapour falls to its original value and dynamic equilibrium is restored once more, with the s.v.p. having its initial value. What happens if the volume of the vapour increases?

(*d*) *Boiling.* Whereas evaporation occurs from the surface of a liquid at all temperatures, boiling takes place at a temperature determined by the external pressure and consists in the formation of bubbles of vapour throughout the liquid. The pressure inside such bubbles must at least equal the pressure in the surrounding liquid. The pressure inside is the s.v.p. at the temperature of the boiling point (since the vapour is in contact with liquid), that outside may be taken as practically equal to atmospheric pressure for a liquid in an open vessel (if we neglect the hydrostatic pressure of the liquid itself and surface tension effects). Hence, from this qualitative treatment it would seem that *a liquid boils when its s.v.p. equals the external pressure.*

(*e*) *Measuring s.v.p.* The apparatus for the 'dynamic method' is shown in Fig. 19.17 and is suitable for s.v.p.s above 50 mmHg (for water from about 50 °C upwards). The boiling point of the liquid is found at different external pressures and these equal the s.v.p.s.

The vacuum pump or compressor enables the pressure to be set to any desired value below or above atmospheric pressure. With cold water circulating through the condenser, the liquid in the flask is heated until it boils. The temperature of the saturated vapour is then given by the thermometer and the difference between the pressure inside and outside the apparatus is registered by the manometer. The reservoir helps to prevent large pressure fluctuations by acting as a buffer.

Van der Waals' equation

The equation of state for an ideal gas $pV = RT$ holds fairly well for real gases (and unsaturated vapours) at low pressures but not at moderate to high pressures. An accurate knowledge of the deviations of real gases from ideal behaviour is important for two reasons. First, they are required to make the necessary corrections to constant volume gas thermometer readings so that temperatures can be expressed on the thermodynamic scale. Second, they provide information about the nature of intermolecular forces and the structure of molecules.

The ideal gas equation is consistent with the kinetic theory. As a step towards obtaining an equation of state for real gases and explaining their deviations, we might consider if any of the assumptions of the theory are invalid for real gases. Van der Waals assumed that when real gases are *not* at low pressure, the range of intermolecular forces is not small compared with the average distance between the molecules (see assumptions, p. 431). There are then two points to consider.

(*i*) *Effect of repulsive intermolecular forces.* These act at very short range, i.e. when molecules approach each other very closely and in effect cause a reduction in the volume in which the molecules can move. This is often

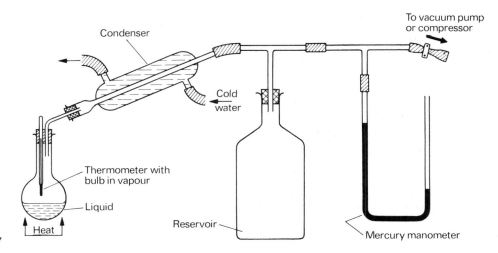

Fig. 19.17

expressed by saying that the molecules themselves have a certain volume which is not negligible in relation to the volume V occupied by the gas. A factor called the 'co-volume', denoted by b is introduced, making the 'free-volume' $(V - b)$.

(*ii*) *Effect of attractive intermolecular forces.* These act over slightly greater distances and may cause two or more molecules to form loose associations or 'complexes', thereby reducing the number of particles in the gas. The observed pressure p is less than it would be if this did not happen. (The 'complexes' do not have greater momentum than single molecules because, being in thermal equilibrium with the latter they have the same average translational k.e.) It can be shown in a more advanced treatment that to allow for this 'pressure defect', p, which still represents the pressure, should be replaced by $(p + a/V^2)$ where a, like b, is a 'constant' that varies from gas to gas and is found by experiment.

Van der Waals' equation of state for a mole of a real gas is

$$\left(p + \frac{a}{V^2}\right)(V - b) = RT$$

The quantities a/V^2 and b become important at high pressures when the molecules are close together.

The isothermals (i.e. graphs of p against V at different, constant temperatures) predicted by van der Waals' equation for three temperatures are shown in Fig. 19.18*a*. Those in Fig. 19.18*b* were obtained experimentally by Andrews (in 1863) for carbon dioxide. The top curve is a rectangular hyperbola, the middle one (which represents a lower temperature) has a point of inflexion C and occurs at the *critical temperature*, i.e. the temperature above which a gas cannot be liquefied by increasing the pressure (see *Note* below). The upper

two van der Waals curves are very similar to the corresponding Andrews curves but the third one has a curved section PQRST whereas the lowest Andrews curve has a straight section PRT (which represents the change from liquid to vapour), although parts PQ and TS have been realized in practice.

The van der Waals equation thus describes fairly well the behaviour of real gases above, but not below, their critical temperature. Evidently the modifications we have made to the kinetic theory are still oversimplified. Other equations of state have been tried but there is no simple one which applies to all gases under all conditions.

Note. The critical temperature of carbon dioxide is 304 K (31 °C) and of oxygen 154 K (−119 °C). Gases such as oxygen were wrongly called 'permanent' gases because attempts to liquefy them by compression without cooling were unsuccessful. A distinction is sometimes made between a gas and a vapour; the former is above its critical temperature and the latter below.

Thermodynamics, heat and work

Thermodynamics deals with processes which cause energy changes as a result of heat flow to or from a system and/or of work done on or by a system. A thermodynamic system consists of a fixed mass of matter, often a gas, separated from its surroundings, perhaps by a cylinder and a piston. Heat engines such as a petrol engine, a steam turbine and a jet engine all contain thermodynamic systems designed to convert heat into mechanical work. Heat pumps and refrigerators are thermodynamic devices for transferring heat from a cold body to a hotter one.

Heat and work are terms used to describe *energy in the process of transfer*.

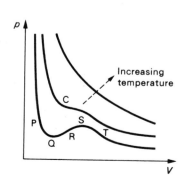

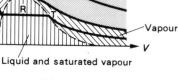

Fig. 19.18   (a)   (b)

Heat is energy that flows by conduction, convection or radiation from one body to another because of a temperature difference between them.

The energy is only heat as it flows. The notion that heat is something in a body is contrary to experimental fact. For example, an unlimited amount of heat can be obtained by rubbing two surfaces together and the system remains the same during the process. No definite meaning can therefore be given to the expression 'the heat in the system'.

Work is energy that is transferred from one system to another by a force moving its point of application in its own direction.

The force may have a mechanical, electrical, magnetic, gravitational, etc., origin but no temperature difference is involved. Again, experience shows that an indefinite amount of work can be put into a system without it changing and so the phrase 'work in a body' is also meaningless.

Energy may be transferred to or from a system as heat or work and afterwards it is impossible to tell which form it took. Thus if the air in a bicycle pump is hot, it could either have been compressed by the piston or held near a body at a higher temperature. Again there is no point in talking about the 'heat in the air' since heat may not have been transmitted to it. Unfortunately everyday usage militates against this use of the term 'heat'.

The energy in a system, whether transferred to it as heat or work, is called *internal energy* and will be considered shortly.

Laws of thermodynamics

(*a*) *Zeroth law and thermal equilibrium.* This basic law was proposed after the first law had been stated. It stems from the fact that all energy exchange appears to stop after a time when hot and cold bodies are brought into contact. The bodies are then said to be in 'thermal equilibrium' and we call the common property which they have *temperature*.

The zeroth law states that if bodies A and B are each separately in thermal equilibrium with body C, then A and B are in thermal equilibrium with each other.

For example, if *C* is a thermometer and reads the same when in contact with two bodies *A* and *B*, then *A* and *B* are at the same temperature. This apparently simple fact is not altogether obvious since in the realm of human relations Black and White may both know Brown but they may not know each other. In effect, the law says that there exists a useful quantity called 'temperature' and if it was not true, taking thermometer readings would be a pointless exercise.

(*b*) *First law and internal energy.* Heat supplied to a gas (or a liquid or a solid) may (*i*) raise its *internal energy*, and (*ii*) enable it to expand and thereby do *external work* by pushing back the atmosphere or if it is in a cylinder, by moving a piston against a force.

In general, the internal energy of a gas consists of two components.

(*i*) Kinetic energy due to the translational, rotational and vibrational motion of the molecules, all of which depend only on the temperature. (In a monatomic gas there is only translational motion.)

(*ii*) Potential energy due to the intermolecular forces; this depends on the separation of the molecules, i.e. the volume of the gas.

In an ideal gas only the kinetic component is present (Why?) and the kinetic theory shows that the translational part of it equals $3RT/2$ per mole. For real gases, both components are present with the kinetic form predominating and their internal energy depends on the temperature and the volume of the gas. (In a solid k.e. and p.e. are present in roughly equal amounts.)

If δQ is the heat supplied to a mass of gas and if δW is the external work done by it then the increase of internal energy δU equals $(\delta Q - \delta W)$ if energy is conserved. Hence

$$\delta Q = \delta U + \delta W$$

This equation is taken as the first law of thermodynamics and is a particular case of the principle of conservation of energy. δQ is taken as positive if heat is supplied to the gas and negative if heat is transferred from it. δW is positive when external work is done by the gas (expansion) and negative if work is done on it (compression).

(*c*) *Second law, heat engines and heat pumps.* Experience shows that in a heat engine, of the total heat Q_1 absorbed at the high-temperature reservoir (at T_1) only part is converted into work W, the rest Q_2 being rejected at the low-temperature reservoir (at T_2) or exhaust of the engine, Fig. 19.19a. A reservoir in thermodynamics is a source or sink of infinite heat capacity whose temperature remains constant however much heat it supplies or receives. In practice a source may take the form of a continuous combustion of fuel and a sink may be the atmosphere. The ratio W/Q_1 for one cycle of operations is defined as the *efficiency* of the engine and it can be shown that it increases as the temperature ratio T_1/T_2 increases.

In a modern power station steam enters the turbines at about 830 K (560 °C) but, even so, the efficiency attained is just over 30 per cent which means that about

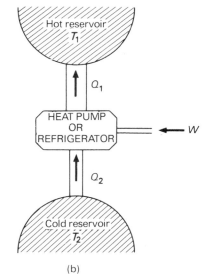

$$W = Q_1 - Q_2$$

Fig. 19.19   (a)   (b)

two-thirds of the heat absorbed is rejected to the atmosphere usually via the cooling towers. In a jet engine the turbine blades, Fig. 19.20, 'glow' during operation.

The heat engine version of the second law may be stated as follows.

No continually working heat engine can take heat from a source and convert it completely into work.

Whilst all the work done on a system may become heat, the conversion of heat to work can only occur to a limited extent no matter how good the engine design may be.

The action of heat pumps and refrigerators is the reverse of that occurring in a heat engine. In these devices work W is done on the system which enables it to transfer heat Q_2 from a low-temperature reservoir at T_2 to a high-temperature reservoir at T_1, Fig. 19.19b. The function of a heat pump is to supply as much heat as possible to the hot body, i.e. to make Q_1 large; its *coefficient of performance* is measured by Q_1/W (which

Fig. 19.20

is greater than 1). A refrigerator, on the other hand, is designed to remove heat from the cold body, i.e. to make Q_2 large; its *coefficient of performance* is given by Q_2/W.

Heat naturally flows from a higher temperature to a lower one and the heat pump-refrigerator statement of the second law may be expressed in the following way.

Heat cannot be transferred continually from one body to another at a higher temperature unless work is done by an external agent.

The two statements of the second law are equivalent and state the same thing in different ways. They cannot be proved directly but their consequences can. The law has been applied to a very wide range of phenomena.

(*d*) *Degradation of energy.* All energy ultimately becomes internal energy of the surroundings through being used to overcome friction. This is the most unavailable form of energy and is useless unless we have something cooler to which to transfer it. Whilst energy is always conserved, it does become inaccessible and we say it is *degraded* as internal energy. The oceans of the world are an enormous energy reservoir but their use requires another reservoir at a lower temperature and this we unfortunately do not have.

Work done by an expanding gas

(*a*) *Expression for work done.* Consider a mass of gas enclosed in a cylinder by a frictionless piston of cross-section area A which is in equilibrium under the action of an external force F acting to the left and a force due to the pressure p of the gas acting to the right, Fig. 19.21. We have

$$F = pA$$

Let the gas expand moving the piston outwards through a distance δx which is so small that p remains practically constant during the expansion. The external work done δW by the gas against F will be

$$\delta W = F\delta x$$
$$= pA . \delta x$$
$$= p . \delta V$$

where δV ($= A\delta x$) is the increase in volume of the gas. The work done by the gas in this small expansion equals its pressure multiplied by the increase of volume.

The total work done W by the gas in a finite expansion from V_1 to V_2 is given, in calculus notation, by

$$W = \int dW = \int_{V_1}^{V_2} p \, dV$$

To evaluate this integral and obtain a value for W requires the relation between p and V to be known. If p is constant during the expansion from V_1 to V_2, then

$$W = p(V_2 - V_1)$$

When a gas is compressed, work is done on it and is also given by $\int_{V_1}^{V_2} p \, dV$.

(*b*) *Reversible changes.* To calculate the work done from the above expressions the pressure of the gas must have a determinable value. This will only be so if the changes of volume occur infinitely slowly, as we can now show. Suppose the external force F in Fig. 19.21 is reduced rapidly during an expansion, there is then not enough time for the gas pressure to be uniform at any stage of the expansion. It will have different values throughout the cylinder, being least near the piston and no proper value can be assigned to it.

Hence if the quantities (p, V, T) defining the state of the system are to be expressed at every stage of the change the system should always be in equilibrium. In other words, we have to regard it as passing through an infinite series of states of equilibrium. Such a change is called a *reversible* one because an infinitesimal change of the controlling conditions will reverse the direction of the change at every stage. In Fig. 19.21 this may be done by increasing F very slightly. The energy transformations must also be reversed exactly in a reversible change and so there must be no energy 'losses'. This means that as well as proceeding infinitely slowly, friction and heat losses must not occur.

In practice the perfect reversible change does not exist but the idea is a useful theoretical standard for judging real changes. A change which is nearly reversible is the very small alternating volume changes in a sound wave, the departures from equilibrium being

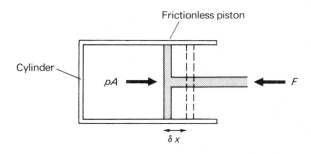

Fig. 19.21

very small. The reactions in some electrical cells can also be approximately reversible. In other cases, whilst the initial and final states are equilibrium ones, which can be expressed in terms of p, V and T, the intermediate ones are not, especially if rapid changes occur.

(c) *Indicator diagram.* This is a graph showing how the pressure p of a gas varies with its volume V during a change. The work done in any particular case can be derived from it.

In Fig. 19.22a if the pressure is $p = AB$ at the start of a very small expansion $\delta V = BC$, the work done, $p\delta V$, is represented to a good approximation by the area of the shaded strip ABCD. The total work done *by* the gas in a large expansion from V_1 to V_2 is therefore the sum of the areas of all such strips, that is, area $p_1 p_2 V_2 V_1$ and clearly depends on the shape of the p-V graph. If the graph had represented a compression of the gas from V_2 to V_1, the work done on it would be represented by the same area. When the gas suffers changes which eventually return it to its initial state, it is said to have undergone a *cycle* of operations and the indicator diagram is a closed loop like that in Fig. 19.22b. The net work done by the gas in this case is represented by the shaded area.

Indicator diagrams play a very important part in the theory of heat engines. Those in Fig. 19.23 show the theoretical relations between p and V for different positions of the piston during one complete cycle of operations (known as an *Otto* cycle after the inventor of the internal combustion engine) of a four-stroke petrol engine. A practical p-V diagram for a complete cycle is similar to that in Fig. 19.23f but the sharp corners tend to be rounded, because neither the combustion of the fuel nor the opening and closing of the valves is instantaneous. They are produced while the engine is working by an electronic or mechanical device called an *engine indicator*.

Principal heat capacities of a gas

(a) *Definitions.* Heat supplied to a solid, liquid or gas increases its internal energy and, especially in the case of a gas, may enable it to expand and do external work. If the kinetic energy component of the internal energy increases, so also will the temperature but the temperature rise produced by a given amount of heat in a particular case will depend on just how much external work the gas is allowed to do. A gas therefore has an infinite number of heat capacities but only the two simplest are important. These relate to constant volume and constant pressure conditions and are called the *principal* heat capacities.

The molar heat capacity at constant volume (C_V) is the heat required to produce unit rise of temperature in one mole of the gas when the volume is kept constant.

The molar heat capacity at constant pressure (C_p) is the heat required to produce unit rise of temperature in one mole of gas when its pressure remains constant.

It follows that C_p is greater than C_V for a given gas, as Fig. 19.24 helps to show. The cylinders X and Y each contain one mole of gas at the same temperature T and pressure (and so volume). The piston in X is fixed and that in Y can move but has a constant force applied to it. If heat is supplied to each until the temperature has risen by one kelvin, the increase of internal energy must be the same in each case (since the temperature rise is the same). All the heat supplied to X is used to increase the internal energy of the gas. In Y, however, the gas expands and work is done by it on the piston; the heat supplied in this case equals the increase of internal energy plus the (heat equivalent of the) work done.

Note that the definitions refer to *reversible* changes, i.e. those that occur infinitely slowly, otherwise pressure, volume and temperature are not determinable properties (see p. 443).

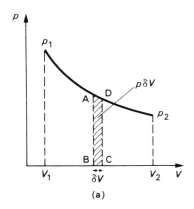

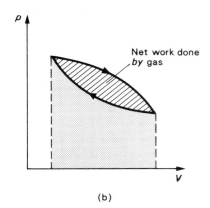

Fig. 19.22   (a)   (b)

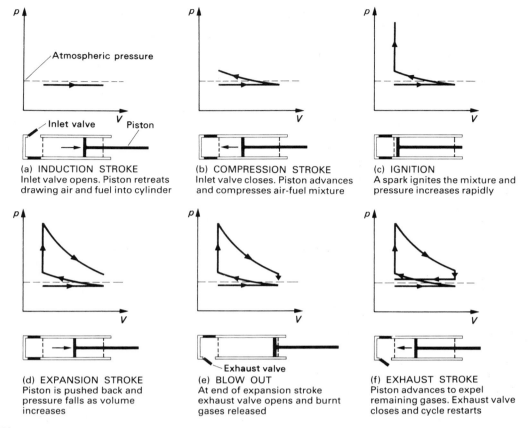

Fig. 19.23

If unit mass of gas is considered instead of one mole we use the term *specific heat capacity*, denoted by c_p or c_V according to whether the pressure or the volume is kept constant.

(*b*) *Relation between C_p and C_V for an ideal gas.* Consider one mole of an *ideal gas* in a cylinder with a frictionless piston, Fig. 19.25. Let a quantity of heat δQ be given to the gas which is allowed to expand *reversibly* at constant pressure p, i.e. the external force on the piston does not change during the expansion. Suppose the volume of the gas increases from V to $V + \delta V$ and the temperature from T to $T + \delta T$.

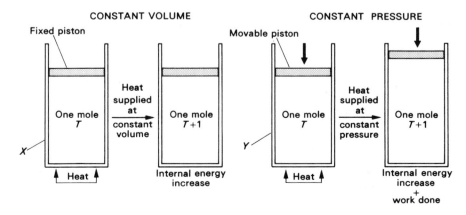

Fig. 19.24

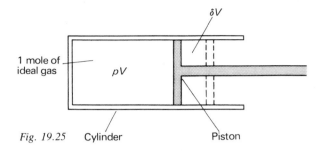

Fig. 19.25   Cylinder   Piston

From the first law of thermodynamics we have,

$$\delta Q = \delta U + \delta W \qquad (1)$$

where δQ is the heat to raise the temperature of one mole of gas by δT at constant pressure. Hence, from the definition of molar heat capacity C_p we have

$$\delta Q = \text{number of moles} \times C_p \times \text{temperature rise}$$

$$= C_p \delta T$$

Also, the increase of internal energy δU is, for an ideal gas, the heat needed to raise the temperature by δT at constant volume. Therefore

$$\delta U = C_V \delta T$$

Further, the external work δW done by the gas is, because the pressure is constant and the change reversible,

$$\delta W = p \delta V$$

Substituting in (1) for δQ, δU and δW, we get

$$C_p \delta T = C_V \delta T + p \delta V \qquad (2)$$

Applying the ideal gas equation to the initial and final stages, we have for 1 mole,

$$pV = RT$$

$$p(V + \delta V) = R(T + \delta T)$$

Subtracting $p \delta V = R \delta T$

Hence from (2) $C_p \delta T = C_V \delta T + R \delta T$

$$\therefore \quad C_p - C_V = R$$

This equation is approximately true for real gases and (as with solids and liquids) the values of C_p and C_V may vary with temperature.

(c) *Calculation of C_p and C_V for an ideal monatomic gas.* The increase of internal energy δU of a mole of an ideal gas for a temperature rise δT is given by

$$\delta U = C_V \delta T \qquad (3)$$

where C_V is the molar heat capacity at constant volume. The internal energy of an ideal *monatomic* gas consists entirely of translational kinetic energy, i.e. $U = \frac{1}{2} L m \overline{c^2}$ for a mole. From the kinetic theory we have that this also equals $3RT/2$ (p. 434), hence

$$U = \frac{3}{2} RT \qquad (4)$$

If the temperature increases by δT and the internal energy by δU then

$$U + \delta U = \frac{3}{2} R(T + \delta T) \qquad (5)$$

Subtracting (4) from (5)

$$\delta U = \frac{3}{2} R \delta T \qquad (6)$$

From (3) and (6)

$$C_V = \frac{3}{2} R$$

Taking $R = 8.31$ J mol^{-1} K^{-1}

$$C_V = 12.5 \text{ J mol}^{-1} \text{ K}^{-1}$$

But

$$C_p = C_V + R$$

$$\therefore \quad C_p = \frac{3}{2} R + R = \frac{5}{2} R$$

$$= 20.8 \text{ J mol}^{-1} \text{ K}^{-1}$$

These values agree well with those for real monatomic gases.

(d) *Importance of γ.* The ratio of the two principal heat capacities of a gas is denoted by γ, that is

$$\gamma = \frac{c_p}{c_V} = \frac{C_p}{C_V}$$

It is a ratio which appears in the equation for the speed of sound in a gas (p. 375) and in that for a reversible adiabatic change, to be considered in the next section. It can also provide information about the atomicity of gases. Experiments give the following approximate values which do, however, generally decrease with increasing temperature.

Atomicity	γ
monatomic	1.67
diatomic	1.40
polyatomic	1.30

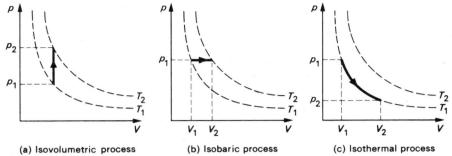

Fig. 19.26   **(a) Isovolumetric process**   **(b) Isobaric process**   **(c) Isothermal process**

Isothermal, adiabatic and other processes

The processes used to change the state (i.e. the values of p, V and T) of a gas in a heat engine affect its performance in converting heat into mechanical work. In general they are not simple. Here we will consider in terms of the first law of thermodynamics four special cases performed *reversibly* on a mole of an ideal gas. (*Note: isos* is the Greek for 'same'.)

(*a*) *Isovolumetric process.* This is a constant volume change and one is represented by the continuous line on the *p-V* graphs of Fig. 19.26*a*. All the heat entering the system becomes internal energy. No work is done, the temperature rises from T_1 to T_2 and the pressure from p_1 to p_2. Thus $\delta W = 0$ and

$$\delta Q = \delta U = C_V(T_2 - T_1)$$

(*b*) *Isobaric process.* Pressure remains constant and, of the heat received by the gas, some becomes internal energy as the temperature rises from T_1 to T_2: the rest is used to do work. Hence for the expansion in Fig. 19.26*b*

$$\delta Q = \delta U + \delta W$$

$$\therefore \quad C_p(T_2 - T_1) = C_V(T_2 - T_1) + p_1(V_2 - V_1)$$

(*c*) *Isothermal process.* The change occurs at constant temperature. Work is done at the same rate as heat is supplied so there is no increase of internal energy (for an ideal gas). Hence $\delta U = 0$, $\delta Q = \delta W$. The expansion curve follows a Boyle's law *p-V* graph, Fig. 19.26*c*, and the equation relating p and V is

$$pV = \text{constant}$$

A reversible isothermal process is an ideal one which requires the gas to be contained, for example, in a cylinder with thin, good-conducting walls having a frictionless piston, surrounded by a constant temperature reservoir. The changes must also occur infinitely

slowly. These conditions are necessary to keep the temperature constant at all times, otherwise say in an expansion, external work is done by drawing on the internal energy of the gas and its temperature falls.

(*d*) *Adiabatic process.* This case is different from those discussed so far in that the expansion (or contraction) of the gas takes place without it receiving (or rejecting) heat, i.e. no heat enters or leaves the gas and so $\delta Q = 0$. Hence

$$0 = \delta U + \delta W \qquad \therefore \quad \delta W = -\delta U$$

In an adiabatic expansion all the work is done at the expense of the internal energy of the gas which therefore cools. Conversely, in an adiabatic compression the work done on the gas by an external agent increases the internal energy and the temperature of the gas rises.

In Fig. 19.27 two *p-V* isothermals are given for a mass of ideal gas. If the gas has initially a temperature T_1 and volume V_1, its state (i.e. p_1 V_1 T_1) is represented by

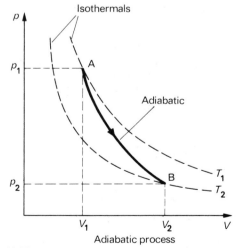

Fig. 19.27

point A on the T_1 isothermal. If it then expands adiabatically to volume V_2 so that its temperature falls to T_2, its state is now represented by point B (i.e. $p_2 V_2 T_2$) on isothermal T_2. The curve through A and B relates the pressure and volume of the mass of gas for this adiabatic change and is called an 'adiabatic'. We see that it is steeper than the 'isothermals' through either A or B. Its equation can be shown to be

$$pV^\gamma = \text{constant}$$

where γ is the ratio of the two principal heat capacities of the gas. It applies to a *reversible adiabatic* change for an ideal gas having a constant value of γ.

A reversible adiabatic process is also an ideal one and requires the gas to be in a thick-walled, poorly-conducting cylinder and piston and to undergo a very rapid, but very small, change of volume to minimize the escape of heat. Very few actual processes can be regarded as adiabatic; those which are approximately so occur (*i*) in insulated vessels, (*ii*) very rapidly, e.g. as in a sound wave, or (*iii*) in a large mass of material. Practical processes fall somewhere between isothermal and adiabatic.

Expressions for the temperature change during a reversible adiabatic process can also be obtained. We have, considering a mole of ideal gas,

$$p_1 V_1^\gamma = p_2 V_2^\gamma \qquad (1)$$
$$\text{(from } pV^\gamma = \text{constant)}$$

For *any* type of change

$$\frac{p_1 V_1}{T_1} = \frac{p_2 V_2}{T_2} \qquad (2)$$
$$\text{(from } pV = RT)$$

Dividing (1) by (2)

$$T_1 V_1^{\gamma-1} = T_2 V_2^{\gamma-1} \qquad (3)$$
$$\therefore \quad TV^{\gamma-1} = \text{constant}$$

This gives a relation between T and V. To obtain one between T and P we raise equation (2) to the power γ,

$$\therefore \quad \left(\frac{p_1 V_1}{T_1}\right)^\gamma = \left(\frac{p_2 V_2}{T_2}\right)^\gamma \qquad (4)$$

Dividing (4) by (1), we get

$$\frac{p_1^{\gamma-1}}{T_1^\gamma} = \frac{p_2^{\gamma-1}}{T_2^\gamma}$$

$$\therefore \quad \frac{p^{\gamma-1}}{T^\gamma} = \text{constant}$$

It can be shown that a gas does more work in a reversible isothermal expansion than in a reversible adiabatic one.

Carnot's ideal heat engine

In 1824 the French scientist Carnot took the first steps towards developing a scientific theory of heat engines. He imagined an ideal engine, not of any particular type, but free from all imperfections such as friction, in which the working substance was taken reversibly through a cycle (called the *Carnot* cycle) consisting of two isothermal and two adiabatic processes, Fig. 19.28.

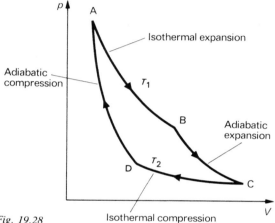

Fig. 19.28

Along AB the substance expands isothermally absorbing heat Q_1 from a source at temperature T_1 and doing external work. Along BC there is an adiabatic expansion, more work is done by it and the temperature ultimately falls to T_2, CD represents an isothermal compression during which work is done on the substance and heat is rejected to a sink at a temperature T_2. Finally along DA, adiabatic compression occurs and more work is done on the substance. The net external work W done by the substance during the cycle equals $(Q_1 - Q_2)$ and we can say

$$\text{efficiency of engine} = \frac{\text{external work done in one cycle}}{\text{heat received from source}}$$

$$= \frac{W}{Q_1} = \frac{Q_1 - Q_2}{Q_1}$$

It can be shown in an advanced treatment that using an ideal gas as the working substance, a heat engine working in a Carnot cycle obeys the relation

$$\frac{Q_1 - Q_2}{Q_1} = \frac{T_1 - T_2}{T_1}$$

where T_1 and T_2 are the temperatures of source and sink respectively. Hence,

$$\text{efficiency} = \frac{T_1 - T_2}{T_1}$$

Reversible processes are ideal and the most efficient imaginable and so there can be no more efficient engine than a Carnot one. An upper *theoretical* limit thus exists for the efficiency, which practical factors further reduce.

The theoretical efficiency of a steam turbine driven by steam at 830 K (560 °C) and exhausting it directly into the air at 373 K (100 °C) is $(830 - 373)/830$, i.e. 55 per cent which compares with the actual efficiency of just over 30 per cent. The efficiency of a petrol engine is about 20 per cent and of a large Diesel engine about 40 per cent.

There are similar expressions involving temperature for the coefficients of performance (p. 442) of heat pumps and refrigerators working in Carnot cycles.

Some calculations

1. Calculate the root-mean-square speed of the molecules of hydrogen at (a) 273 K, and (b) 373 K. Density of hydrogen at s.t.p. = 9.00×10^{-2} kg m^{-3} and one standard atmosphere = 1.01×10^5 Pa.

(a) At 273 K
For an ideal gas of density ρ and pressure p we have from the kinetic theory

$$p = \frac{1}{3}\rho\overline{c^2}$$

$$\therefore \quad \sqrt{\overline{c^2}} = \sqrt{\frac{3p}{\rho}}$$

where $\sqrt{\overline{c^2}}$ is the root-mean-square molecular speed. Assuming this expression can be applied to hydrogen at s.t.p., for which $\rho = 9.00 \times 10^{-2}$ kg m^{-3} and $p = 1.01 \times 10^5$ Pa, we have

$$\sqrt{\overline{c^2}}_{273} = \sqrt{\frac{3 \times 1.01 \times 10^5}{9.00 \times 10^{-2}}}$$

$$= 1.84 \times 10^3 \text{ m s}^{-1} \ (\approxeq 1.2 \text{ miles per second})$$

(b) At 373 K
For an ideal gas the kinetic theory suggests that $\overline{c^2} \propto T$, hence

$$\frac{\sqrt{\overline{c^2}}_{373}}{\sqrt{\overline{c^2}}_{273}} = \sqrt{\frac{373}{273}}$$

$$\therefore \quad \sqrt{\overline{c^2}}_{373} = \sqrt{\frac{373}{273}} \times 1.84 \times 10^3 \text{ m s}^{-1}$$

$$= 2.15 \times 10^3 \text{ m s}^{-1}$$

2. Calculate the two principal molar heat capacities of oxygen if their ratio is 1.40. Density of oxygen at s.t.p. = 1.43 kg m^{-3}, one standard atmosphere = 1.01×10^5 Pa and molar mass of oxygen = 32.0×10^{-3} kg mol^{-1}.

For 1 mole of an ideal gas

$$pV = RT$$

where $V = V_m$ = molar volume = molar mass/density = M_m/ρ

$$\therefore \quad R = \frac{pV_m}{T} = \frac{pM_m}{T\rho}$$

Assuming we can apply this equation to oxygen

$$R = \frac{1.01 \times 10^5 \times 32.0 \times 10^{-3}}{273 \times 1.43} \frac{\text{N m}^{-2} \text{ kg mol}^{-1}}{\text{K kg m}^{-3}}$$

$$= 8.31 \text{ J mol}^{-1} \text{ K}^{-1}$$

Also for an ideal gas

$$C_p - C_V = R$$

$$\therefore \quad C_p - C_V = 8.31 \qquad (1)$$

But $\quad \dfrac{C_p}{C_V} = 1.40 \quad \therefore \quad C_p = 1.40\, C_V \qquad (2)$

Substituting (2) in (1)

$$1.40\, C_V - C_V = 8.31$$

$$\therefore \quad C_V = \frac{8.31}{0.40} = 20.8 \text{ J mol}^{-1} \text{ K}^{-1}$$

Hence $\quad C_p = 20.8 + R = 29.1 \text{ J mol}^{-1} \text{ K}^{-1}$

3. A mass of an ideal gas of volume 400 cm^3 at 288 K expands adiabatically and its temperature falls to 273 K. (a) What is the new volume if $\gamma = 1.40$? (b) If it is then compressed isothermally until the pressure returns to its original value, calculate the final volume of the gas.

(a) For the reversible adiabatic expansion of an ideal gas we have (p. 448),

$$T_1 V_1^{\gamma-1} = T_2 V_2^{\gamma-1}$$

where $V_1 = 400$ cm^3, $T_1 = 288$ K, $T_2 = 273$ K and $\gamma = 1.40$

$$\therefore \quad V_2^{0.40} = \frac{288}{273} \times 400^{0.40}$$

Taking logs,

$$0.40 \log V_2 = \log 288 + 0.40 \log 400 - \log 273$$

$$= 1.0640$$

$$\therefore \quad \log V_2 = 1.0640/0.40 = 2.66$$

$$\therefore \quad V_2 = 457 \text{ cm}^3$$

This is the new volume.

(b) If p_1 and p_2 are the pressures of the gas before and after the adiabatic expansion,

$$p_1 V_1{}^\gamma = p_2 V_2{}^\gamma$$

$$\therefore \quad p_2 = \left(\frac{V_1}{V_2}\right)^\gamma p_1 = \left(\frac{400}{457}\right)^{1.40} p_1$$

After the isothermal compression, the pressure is p_1 and let the final volume be V_3; then from Boyle's law

$$\therefore \quad p_2 V_2 = p_1 V_3$$

$$\therefore \quad V_3 = \left(\frac{p_2}{p_1}\right) V_2 = \left(\frac{400}{457}\right)^{1.40} \times 457$$

$$= \frac{400^{1.40}}{457^{0.40}}$$

Taking logs,

$$\log V_3 = 1.40 \log 400 - 0.40 \log 457$$

$$= 2.58$$

$$\therefore \quad V_3 = 380 \text{ cm}^3$$

Thermodynamics and chance

Why does heat flow from a hot to a cold body and never spontaneously in the reverse direction? The highly mathematical subject of statistical mechanics gives insight into the average, but not the individual, behaviour of molecules in such one-way processes. Our treatment will be much simpler and aided by analogies. As a first step we will consider the number of ways in which a situation can arise when things are left to chance, as they are in random changes.

(a) *How many ways?* Suppose we wish to know the number of ways in which six numbered objects can be distributed between two compartments of a box. We see from Fig. 19.29a that one object can be placed in 2^1 ways; from Fig. 19.29b, two objects can be arranged in $4 = 2^2$ ways; from Fig. 19.29c, three objects can be placed in $8 = 2^3$ ways and so on. In general n objects can be distributed in $W = 2^n$ ways.

We would expect that any given arrangement would occur once in every W arrangements if these were produced *randomly*. It is worthwhile testing this by starting with all the objects in one compartment and throwing a die to obtain the number of the object to be moved. On average *all* the objects might be in, say, the left-hand compartment once in every $2^6 = 64$ throws but it would not be too common an occurrence even with only six objects.

With a very large number of objects the likelihood of them all ever being in the same compartment together becomes very remote. Thus if we have a gas jar of bromine vapour containing, say, 10^{22} molecules and we place another jar over the first, the chance of seeing all the bromine molecules back in the first jar at some future instant is so unlikely as to be impossible. Diffusion is evidently a one-way process, the result of chance when large numbers of particles are involved. By spreading, more arrangements become possible, which is what chance seems to favour.

Many mixing processes such as diffusion, the addition of milk to a cup of tea etc. occur in one direction only; they appear ridiculous if seen in a film run backwards.[1] Can you think of any *unmixing* processes which occur spontaneously?

[1] Three such 8 mm film loops made for the Nuffield Advanced Physics course are worth viewing. Penguin XX1668, XX1669, XX1670.

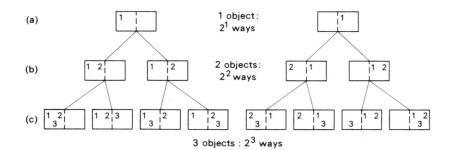

Fig. 19.29

3 objects : 2^3 ways

(b) *Thermal equilibrium*.[1] This is attained when there is no net flow of energy from one body to another. A picture, at the atomic level, of how it might occur in solids and of the difference between hot and cold bodies can be obtained using a model proposed by Einstein and which we will develop by means of a game that simulates the model.

In Einstein's solid the atoms vibrate more or less independently about their mean positions in the crystal lattice with internal energies (k.e. + p.e.) which, as we shall see later (p. 491), can only have certain values. If we assume the oscillation is simple harmonic, the values are simple integral multiples of a certain minimum called a *quantum* of energy. It is helpful to think of a ladder whose *equally-spaced* rungs represent the definite energy values or levels, any one of which an atom can occupy. When energy is gained the atom 'jumps' from a lower to a higher level and vice versa when energy is lost. A simple experiment supporting this idea is described in Appendix 16.

If we assume quanta can move from atom to atom and do so randomly, the next step is to find out what numbers of atoms in the solid have any particular energy, i.e. what are the relative numbers of atoms in each energy level. An idea of whether there is any pattern in this *distribution* can be obtained from the following game.

Each of the 36 spaces in the grid of Fig. 19.30a represents an atom whose energy will be indicated by

[1] The material of this section is covered by the computer-made film *Change and chance: a model of thermal equilibrium in a solid* made for the Nuffield Advanced Physics course. Penguin XX1673.

the number of counters in the space. An atom in its lowest energy level has no counters; one in its first level has one counter; in its second level two counters, and so on. We will start with each atom having one counter, as shown. Two dice are then thrown to give a random choice of the atom which is to lose a quantum. If the numbers obtained are 2 and 3 then the atom indicated may be taken as the second from the left and third up from the bottom. The counter on that space (atom) is then moved to the space (atom) given by the next throw of the two dice. This is repeated at least 100 times.

A distribution curve (in the form of a histogram) can be drawn but when chance is involved the results are most reliable if large numbers are concerned. Fig. 19.30b shows the curve that would be obtained by programming a computer to make 10 000 moves with 900 spaces and 900 quanta. A further 10 000 moves make little difference and the result is similar if we start with a different initial distribution or if 400 atoms and 400 quanta are used. The distribution thus depends on the ratio of the number of atoms to the number of quanta and not on their actual numbers.

We see from Fig. 19.30b that the number of atoms with no quanta is twice that with one quantum and the number with one quantum is twice that with two quanta and so on. The curve is an exponential one (dotted) and decreases by a constant fraction as the number of quanta increases by one. It represents a *Boltzmann* distribution.

The curve for an Einstein-type solid with 300 quanta shared among 900 atoms is shown in Fig. 19.30c. The shape is still exponential but steeper, and the ratio of the numbers of atoms with a certain number of quanta to those with one more quantum is still constant, but

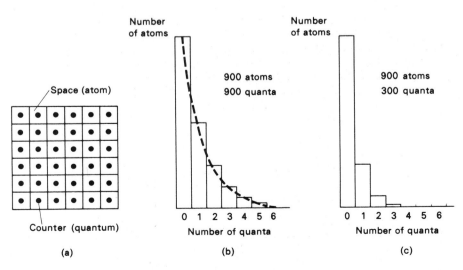

Fig. 19.30   (a)   (b)   (c)

greater (four compared with two). If we regard temperature as the quantity which determines the likelihood of energy flowing from one body to another, the 900-atom solid with 900 quanta is hotter than the 900-atom solid with 300 quanta (it has more energy per atom), and so it is at a higher temperature.

Temperature can thus be associated with the distribution of atoms among energy levels, i.e. with the steepness of the distribution curve. Theory shows that in general if q quanta are shared among N atoms, the distribution curve ratio is $(1 + N/q)$. (Check this for the cases considered.) The *larger* the value of $(1 + N/q)$ the greater the steepness and the *lower* the temperature of the solid and vice versa. In other words, the temperature is high if the proportion of atoms in higher energy levels is high.

On this interpretation it would be reasonable to suppose that heat flows from an Einstein-type solid with a small distribution curve slope to one with a large slope until both slopes are the same. When this has happened thermal equilibrium will have been reached, although quanta will still be exchanged. The computer, suitably programmed, does in fact show that the ratio of the resultant distribution curve is greater than that of the 'hot' solid but less than that of the 'cold' one.

Attaining thermal equilibrium is thus a one-way process, like diffusion, and happens inevitably if left to chance. Heat flows due to the random sharing out of internal energy among the atoms of the bodies involved.

It must be remembered that Einstein's model is a simplification. In fact the atoms in a solid do not vibrate independently and consequently the energy levels of one atom affect those of others so that to talk of 'the' energy levels of 'an' atom is not quite realistic. Neither are the atoms generally harmonic oscillators. Nonetheless many of the results derived hold for more complex and 'life-like' models which require advanced mathematical treatment. The use of a model to represent a real system is a common technique in science.

Entropy

The fundamental, difficult, but very useful statistical concept of entropy can be obtained by combining ideas about 'numbers of ways' and the 'distribution curve steepness ratio'.

(*a*) *Effect of more quanta on W.* When energy is added to a substance the number of ways W of arranging the quanta among the atoms increases. Thus in Fig. 19.31 we see that there are 3 ways of sharing 1 quantum among 3 atoms and 6 ways of sharing 2 quanta between them. It can be shown for an Einstein solid that, in general, if there are W ways in which q quanta can be distributed among N atoms, the addition of one more

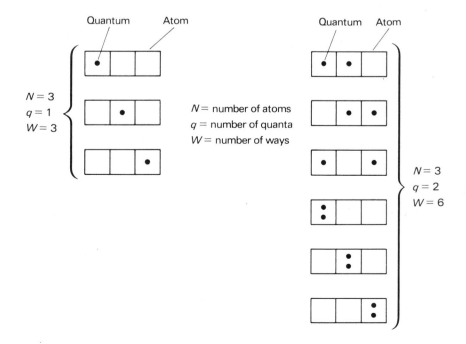

N = number of atoms
q = number of quanta
W = number of ways

Fig. 19.31

quantum increases the number of ways to W_1 where

$$\frac{W_1}{W} = \frac{N + q}{q + 1}$$

If N and q are both large (as they are in practice) then

$$\frac{W_1}{W} = \frac{N + q}{q} = 1 + \frac{N}{q}$$

We met the factor $(1 + N/q)$ in the previous section as a measure of the steepness of the Boltzmann distribution curve and we linked it with temperature. We now see that it, and therefore temperature also, are related to the *change* in the number of ways of distributing quanta when energy is added (or removed). Presently we will derive a relation between temperature and change in the number of ways.

Suppose that in a certain solid A, $q = N$, then $W_1/W = 2$ so that the loss (or gain) of one quantum would *halve* (or double) the number of ways of distributing the remaining quanta. On the other hand, in solid B having $q = N/3$, $W_1/W = 4$ and in this case the gain (or loss) of one quantum *quadruples* (or quarters) the number of ways of arranging the quanta.

The transfer of a quantum has a greater effect on W in solid B than in solid A. Now, as we saw with diffusion, chance favours those events which lead to an increase in W and so we would expect that bringing A and B together would lead to the net transfer of quanta from A (the hotter, since q/N is greater) to B. Thermal equilibrium would be established when exchange of quanta between A and B has the same effect on W for each and the process can be regarded as one which must happen inevitably since chance and large numbers are involved.

(b) *Temperature and change in number of ways.* From what has been said it follows that every quantum added to an Einstein solid increases W by the same factor, say a, where $a = (1 + N/q)$ and depends only on the temperature of the solid. Then if W_n is the number of ways after adding n quanta

$$\frac{W_n}{W} = a \times a \times a \times \ldots \text{ (} n \text{ times)}$$

This is conveniently and usually expressed in natural logarithms (i.e. to base e) and if the extra quanta do not materially affect the ratio (i.e. the temperature) we have, writing $\log_e = \ln$,

$$\ln (W_n/W) = \ln W_n - \ln W = \ln a + \ln a +$$
$$\ln a + \ldots \text{ (} n \text{ times)}$$

The number of terms on the right-hand side equals the number of quanta added and the right-hand side is therefore proportional to the change δU in the internal energy of the solid. Writing $\delta \ln W$ for $\ln W_n - \ln W$,

$$\delta \ln W \propto \delta U$$

the volume being constant and the temperature nearly so.

Earlier we had associated a *high* temperature T with a *small* change in W and so if T is to be incorporated in the above expression we must write

$$\delta \ln W \propto \frac{\delta U}{T}$$

Introducing a constant k, the relation between W and T is, under the conditions given above,

$$\delta \ln W = \frac{\delta U}{kT}$$

$$\therefore \quad k \, \delta \ln W = \frac{\delta U}{T}$$

This expression, which is applicable to solids, liquids and gases as well as Einstein solids, defines temperature in terms of energy and number of ways and could, if desired, be used as the basis of a scale of temperature. However, as we have seen, T is defined on the Kelvin scale and then k is found to have the value 1.38×10^{-23} J K^{-1}—which is *Boltzmann's* constant (p. 434).

(c) *Entropy and the direction of a process.* The quantity $k \, \delta \ln W$ is called the *change of entropy*, δS, hence

$$\delta S = k \, \delta \ln W = \frac{\delta U}{T}$$

It can be calculated by measuring δU and T and gives us information about the increase in the number of ways resulting from adding energy δU to matter (in any phase) at temperature T and constant volume.

Absolute zero is taken as the arbitrary zero of *entropy* S and we can write

$$S = k \ln W$$

Knowing S we then have information about the direction processes will take, for we have seen that *chance favours those for which W tends to increase.* If W increases so does the entropy $k \ln W$ and an alternative but more basic statement of the second law of thermodynamics than those given earlier (pp. 442 and 443) is as follows.

The direction of a process is such as to increase the total entropy.

Hence any physical or chemical change only occurs if the number of ways W and the entropy S do not diminish as a result. To take a very simple example, ice and water at s.t.p. have entropies of 41 J K^{-1} mol^{-1} and 63 J K^{-1} mol^{-1} respectively and so when heat is supplied to melt ice, the entropy increases.

(d) *Entropy and heat engines.* The efficiency of a heat engine, previously discussed in terms of the Carnot cycle (p. 448), can also be viewed from the stand-point of entropy. Suppose a steam turbine receives Q_1 joules of energy per second from steam at temperature T_1. The entropy *decrease* δS_1 of the steam for the *removal* of this amount of energy is

$$\delta S_1 = \frac{\delta U}{T_1} = \frac{Q_1}{T_1}$$

The entropy in the *whole* process must increase and does so as a result of the smaller amount of energy, say Q_2 per second, which is *supplied* to the atmosphere as 'exhausted' steam at temperature T_2. The entropy *increase* δS_2 of the atmosphere for this part must be greater than δS_1 and is

$$\delta S_2 = \frac{Q_2}{T_2}$$

The *net* change of entropy δS is thus

$$\delta S = \frac{Q_2}{T_2} - \frac{Q_1}{T_1}$$

If δS is to be positive then

$$\frac{Q_2}{T_2} > \frac{Q_1}{T_1}$$

That is, for the turbine to operate without a net decrease in entropy, T_1 must be high and T_2 low so that even though $Q_1 > Q_2$, the term Q_1/T_1 is smaller than Q_2/T_2. The difference $(Q_1 - Q_2)$ is used by the turbine to do external work, e.g. to drive a dynamo and we are again led to conclude that an engine must necessarily have a maximum efficiency, i.e. it must be inefficient to some extent.

QUESTIONS

Kinetic theory: gas laws

1. If a mole of oxygen molecules occupies 22.4×10^{-3} m^3 at s.t.p. (i.e. 273 K and 1.00×10^5 Pa), calculate the value of the molar gas constant R in J mol^{-1} K^{-1}.

2. (a) Assuming the equation of state for an ideal gas, show that the number of molecules in a volume V of such a gas at pressure p and temperature T is $pVL/(RT)$ where L is the Avogadro constant and R is the molar gas constant.

(b) Hence find the number of molecules in 1.00 m^3 of an ideal gas at s.t.p. (One standard atmosphere = 1.00×10^5 Pa.)

3. Distinguish between a *saturated* and an *unsaturated* vapour. Describe how the variation of the saturation vapour pressure of water vapour with temperature may be investigated over a range from about 30 °C to 100 °C.

A closed vessel of fixed volume contains air and water. The pressures in the vessel at 20 °C and 75 °C are respectively 737.5 mm and 1144 mm of mercury and some of the water remains liquid at 75 °C. If the saturation vapour pressure of water at 20 °C is 17.5 mm of mercury, find its value at 75 °C.

(L.)

4. Explain what is meant by *a saturated vapour.* Sketch curves which show the relationship between pressure and temperature for (a) an ideal gas, (b) saturated water vapour, and (c) a mixture of an ideal gas and saturated water vapour. The temperature axis should extend from 0–100 °C and the pressure axis from 0–1000 mmHg.

A flask contains a mixture of air and unsaturated water vapour at a temperature of 50 °C and a pressure of 8.0×10^2 mmHg. The mixture is cooled and when the temperature reaches 20 °C water begins to condense out of the mixture. Given that the s.v.p. of water vapour at 20 °C is 18 mmHg and at 5.0 °C is 7.0 mmHg, calculate the pressure of the mixture if it is cooled to 5.0 °C. You may assume that the unsaturated vapour obeys the gas laws up to the point of saturation.

(A.E.B. part qn.)

5. If the density of nitrogen at s.t.p. is 1.25 kg m^{-3}, calculate the root-mean-square speed of nitrogen molecules at 227 °C. (One standard atmosphere = 1.00×10^5 Pa.)

6. A closed vessel contains hydrogen which exerts a pressure of 20.0 mmHg at a temperature of 50.0 K. At what temperature will it exert a pressure of 180 mmHg? If the r.m.s. velocity of the hydrogen molecules at 50.0 K was 800 m s^{-1}, what will be their r.m.s. velocity at this new temperature? Assume that there is no change in volume of the vessel.   (S.)

7. Define *pressure* and explain in qualitative terms how the pressure exerted by a gas may be interpreted in terms of the momenta of the gas molecules.

List the postulates of the simple kinetic theory which are used in the derivation of the expression $p = \frac{1}{3}\rho\overline{c^2}$ for the pressure p exerted by a gas of density ρ whose molecules have a r.m.s. speed $\sqrt{\overline{c^2}}$.

Establish the relation between the temperature of an ideal gas and its molecular kinetic energy.

Give a descriptive account of the arguments which lead to the introduction of the terms a/V^2 and b in van der Waals' equation $(p + a/V^2)(V - b) = RT$.   (J.M.B.)

8. A vessel of volume 50 cm³ contains hydrogen at a pressure of 1.0 Pa and at a temperature of 27 °C. Estimate (a) the number of molecules in the vessel, (b) their distance apart, on the average, and (c) their root-mean-square speed.

($R = 8.3$. J mol⁻¹ K⁻¹; Avogadro constant = 6.0×10^{23} mol⁻¹; mass of 1 mole of hydrogen molecules = 2.0×10^{-3} kg mol⁻¹.)

Thermodynamics

9. The graph in Fig. 19.32 shows three curves relating the pressure and the volume of a fixed mass of a perfect gas at three different temperatures. Curve 2 is for 0 °C.

(i) Name and state the law which any one of these three curves represents.

(ii) What relationship exists between all the points on the three curves?

(iii) Name the law which connects all points on the line through A and B and express it mathematically.

(iv) Deduce the pressure at the point C, explaining your reasoning. (Do not attempt to read it from the graph.)

(v) What temperatures do the curves 1 and 3 refer to?

(vi) How much external work (in joules) would have to be done to take the gas from the state represented at A to that at B?

(vii) What other energy would have to be supplied during this process?

(viii) Under what conditions does an ordinary gas behave as a perfect gas?

(ix) What properties must be postulated for the molecules of an ordinary gas to account for its deviations, under other conditions, from perfect gas behaviour?

(Density of mercury = 13.6×10^3 kg m⁻³; $g = 10$ N kg⁻¹.)

(S.)

10. Derive an expression for the total translational kinetic energy of the molecules of 1.0 g of helium at T K and calculate the principal specific heats of the gas. You may assume that helium is an ideal monatomic gas of relative molecular mass 4.0 and that the molar (or 'universal') gas constant R is 8.3 J K⁻¹ mol⁻¹. (J.M.B. part qn.)

11. State the first law of thermodynamics. What is the evidence for its validity? When applied to a fixed mass of gas this law can be written in the form

$$dU = dQ - p\, dv$$

Explain the meaning of each of the three terms.

Explain how the equation is applied when the gas is (a) heated at constant volume, and (b) compressed adiabatically.

For water at constant pressure of one standard atmosphere the specific latent heat of vaporization is 2.26×10^6 J kg⁻¹. During the transformation to vapour the increase in volume of 1.00 kg of water is 1.67 m³. Calculate the work done against the external pressure during this process. What has happened to the remainder of the heat supplied during the evaporation?

Density of water = 1.00×10^3 kg m⁻³; one standard atmosphere = 1.00×10^5 Pa. (J.M.B.)

12. Explain why, when quoting the specific heat capacity of a gas, it is necessary to specify the conditions under which the change of temperature occurs. What conditions are normally specified?

A vessel of capacity 10 litres contains 1.3×10^2 grams of a gas at 20 °C and 10 atmospheres pressure. 8.0×10^3 joules of heat energy are suddenly released in the gas and raise the pressure to 14 atmospheres. Assuming no loss of heat to the vessel, and ideal gas behaviour, calculate the specific heat capacity of the gas under these conditions.

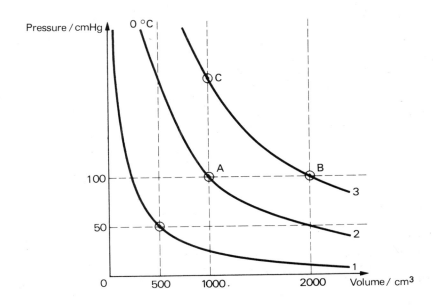

Fig. 19.32

In a second experiment the same mass of gas, under the same initial conditions, is heated through the same rise in temperature while it is allowed to expand slowly so that the pressure remains constant. What fraction of the heat energy supplied in this case is used in doing external work? Take 1 atmosphere = 1.0×10^5 pascals.   (*O. and C.*)

13. 'In a reversible adiabatic change in a gas, the pressure p and the volume V obey the relation pV^γ = constant.' Explain the terms *adiabatic* and *reversible* and discuss how the value of the index γ depends on the atomicity of the gas.

Air initially at atmospheric pressure is compressed adiabatically and reversibly to a pressure of 4.00 atmospheres and is then allowed to expand isothermally and reversibly to its original volume. Find the final pressure. Sketch on a *p-V* diagram the curves representing the changes. State, with reasons, which is the greater: the work done on the air during compression or that done by the air in expanding. (γ for air = 1.40.)   (*L.*)

14. Define isothermal and adiabatic changes and give the equation relating the pressure and volume of an ideal gas for each type of change.

Why has it been concluded that the pressure and volume changes accompanying the passage of sound waves through a gas are adiabatic?

Air occupying a volume of 10 litres at 0.0 °C and atmospheric pressure is compressed isothermally to a volume of 2.0 litres and is then allowed to expand adiabatically to a volume of 10 litres. Show the process on a *p-V* diagram and calculate the final pressure and temperature of the air. At what volume was the pressure momentarily atmospheric? Mark this point on your diagram.

Assume $\gamma = 1.4$ for air (and that all changes are reversible).   (*W.*)

15. Sketch on a *p-V* diagram the theoretical pressure and volume changes which take place in the cylinder of a four-stroke engine working on the Otto cycle. Label each part of the diagram.

Sketch the shape of the indicator diagram which is obtained in practice and account for the differences between the two diagrams.   (*J.M.B. Eng. Sc.*)

16. Use the data given below to make *quantitative* predictions, estimates or comparisons concerning man's use of the fuel resources (coal, oil, etc.) stored in the earth.

In particular, you should consider the following problems:

(*a*) The possibility of using up these fuel resources, and any need there may be to find alternative sources of energy.

(*b*) Any differences between groups of people in the amount of energy they use.

Energy reaching earth from the sun: 5×10^{24} joules per year.

Maximum energy likely to be available from hydroelectric power if all such sources were used: 5×10^{19} joules per year.

Total energy stored in all known reserves of fuels (coal, oil, etc.) 10^{23} joules.

Total energy stored in all known reserves of fuels (coal, oil, etc.) that can be extracted economically at present: 10^{22} joules.

Total energy stored in known oil deposits that can be extracted economically at present: 10^{21} joules.

Time taken to produce the earth's store of coal, oil and other fuels: 10^8 years. Energy used by one car in a year: 3×10^{10} joules.

Work energy one labourer can produce in one year: 10^{10} joules.

Total energy used in 1964		Population
World	10^{20} joules	3000×10^6
North America	4×10^{19} joules	200×10^6
Western Europe	2×10^{19} joules	300×10^6
Middle East and Africa	0.3×10^{19} joules	300×10^6
Asia	2×10^{19} joules	1700×10^6

(*O. and C. Nuffield*)

20 Atomic physics

Thermionic emission

In a metal each atom has a few loosely-attached outer electrons which move randomly through the material as a whole. The atoms thus exist as positive ions in a 'sea' of free electrons. If one of these electrons near the surface of the metal tries to escape, it experiences an attractive inward force from the resultant positive charge left behind. The surface cannot be penetrated by an electron unless an external source does work against the attractive force and thereby increases the kinetic energy of the electron. If this is done by heating the metal, the process is called *thermionic emission*.[1]

The work function Φ of a metal is the energy which must be supplied to enable an electron to escape from its surface. It is conveniently expressed in electron-volts (p. 232).

---

[1] It will be seen in Chapter 21 that modern solid-state theory views conditions in a metal from a slightly different standpoint—that of energy levels; here the above picture is adequate.

The smaller the work function of a metal the lower the temperature of thermionic emission; in most cases the temperature has to be too near the melting point. Two materials used are (*i*) thoriated tungsten having $\Phi = 2.6$ eV and giving good emission at about 2000 K and, in most cases, (*ii*) a metal coated with barium oxide for which $\Phi = 1$ eV, copious emission occurring at 1200 K.

In many thermionic devices a plate, called the *anode*, is at a positive potential with respect to the heated metal and attracts electrons from it. The latter is then called a *hot cathode*. Hot cathodes are heated electrically either directly or indirectly. In direct heating current passes through the cathode (or filament) itself which is in the form of a wire, Fig. 20.1*a*. An indirectly heated cathode consists of a thin, hollow metal tube with a fine wire, called the *heater*, inside and separated from it by an electrical insulator, Fig. 20.1*b*. Indirect heating is most common since it allows a.c. to be used without the potential of the cathode continually varying. A typical heater supply for many thermionic devices is 6.3 V a.c., 0.3 A.

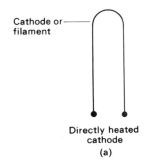

Cathode or filament

Directly heated cathode

(a)

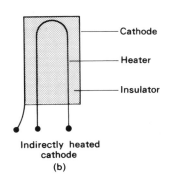

Cathode

Heater

Insulator

Indirectly heated cathode

(b)

Fig. 20.1

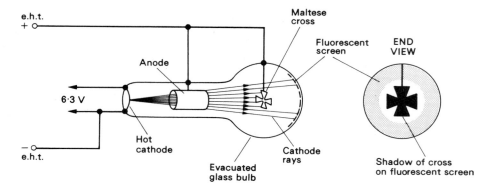

Fig. 20.2

Cathode rays and the electron

(*a*) *Properties*. Streams of electrons moving at high speed are called *cathode rays*. They exhibit several important and useful effects, some of which can be demonstrated with the Maltese cross tube of Fig. 20.2. It consists of a hot cathode and a hollow cylindrical anode enclosed in an evacuated glass envelope having a coating of fluorescent material on the inside of the bulb. The anode is connected to the positive of an e.h.t. supply of 2–3 kV so that electrons from the cathode are accelerated along the tube in a divergent beam. Most bypass the anode and a dark shadow of the cross appears on the screen against a blue or green fluorescent background. This suggests that the rays are travelling in straight lines from the cathode and those not intercepted by the cross cause the screen to fluoresce.

When a magnet is brought near the side of the tube level with the anode, the beam is deflected vertically and the shadow can be made partially or wholly to disappear. Using Fleming's left-hand rule the direction of the deflection shows that the rays behave like a flow of negative charge, travelling from cathode to anode.

The beam also carries energy since the end of the tube struck by it becomes warm.

These and other properties of cathode rays may be summarized as follows:

(*i*) They travel from the cathode in straight lines.
(*ii*) They cause certain substances to fluoresce.
(*iii*) They possess kinetic energy.
(*iv*) They can be deflected by a magnetic field.
(*v*) They can be deflected by an electric field (p. 459).
(*vi*) They produce X-rays on striking matter (p. 468).

(*b*) *Discovery of the electron*. The first evidence to establish the existence of the electron is usually considered to be provided by J. J. Thomson's experiment in 1897 in which he measured the speeds and the charge to mass ratio (e/m) called the *specific charge*, of cathode

rays. The view that cathode rays were not electromagnetic waves was based on the fact that (*i*) the speeds were typically one-tenth that of light, and (*ii*) they could be deflected by electric and magnetic fields.

The value of e/m obtained by Thomson was always the same, whatever the source or method of production of the cathode rays. This suggested that electrons are all alike, universal constituents of matter, and by assuming that the charge carried was equal to that on a monovalent ion in electrolysis, Thomson estimated the mass m of an electron, knowing e/m. Using modern values m is 9.11×10^{-31} kg, i.e. it is 1837 times smaller than the mass of the hydrogen atom. Other interpretations of the value of e/m are possible; it could be that the electron has the same mass as a hydrogen ion but a much greater charge. However, there is now no doubt that an electron carries the fundamental unit of electric charge.

For most purposes the electron can be regarded as a sub-atomic particle having a negative charge of value e, the electronic charge, and a very small mass (since force is required to accelerate it). The value of the mass quoted above is known as the 'rest mass'—it cannot be measured directly. The term has arisen because it has been found, as predicted by Einstein in the theory of relativity, that the mass of a particle accelerated to a speed approaching that of electromagnetic waves in vacuo increases with speed.

Ordinary particles are characterized by size and shape but such properties cannot be stated precisely for sub-atomic particles. To some extent the size of the electron depends on the method of determination and while some measurements indicate that it is a sphere of diameter 10^{-15} m, it is also satisfactory on occasions to regard it as a dimensionless point. By contrast its charge and mass can be uniquely specified. More will be said about the nature of the electron at the end of this chapter.

Electron dynamics

If cathode rays are assumed to consist of particles (electrons), to which the laws of mechanics apply, we can obtain information about their speed and specific charge from their behaviour in electric and magnetic fields.

(*a*) *Speed of electrons.* Consider an electron of charge *e* and mass *m* which is emitted from a hot cathode and then accelerated by an electric field towards an anode. It experiences a force due to the field and work is done on it. The system (of field and electron) loses electrical potential energy and the electron gains kinetic energy. Let *V* be the p.d. between anode and cathode responsible for the field. If the electron starts from the cathode with zero speed and moves in a vacuum attaining speed *v* as it reaches the anode, the energy change *W* is given by

$$W = eV \qquad \text{(p. 231)}$$

But

$$W = \tfrac{1}{2}mv^2$$

$$\therefore \quad eV = \tfrac{1}{2}mv^2$$

From this 'energy equation' it follows that

$$v = \sqrt{\frac{2eV}{m}}$$

Substitution of numerical values for *e*/*m* and *V* shows that *v* is about 1.9×10^7 m s^{-1} (one-sixteenth of the speed of light) when *V* = 1000 V.

(*b*) *Deflection of electrons by a magnetic field.* It was shown earlier (p. 269) that the force *F* on charge *Q* moving with speed *v* at right angles to a uniform magnetic field of flux density *B* is

$$F = BQv$$

For an electron *Q* = *e* and is negative, hence

$$F = Bev$$

The direction of the force is given by Fleming's left-hand rule (remembering that a negative charge moving one way is equivalent to conventional current in the opposite direction) and is always at right angles to the field and to the direction of motion. Therefore at X in Fig. 20.3 the speed of the electron remains unaltered but it is deflected from its original path to, say, Y. Here the force acting on it still has the same value, *Bev*, and since the direction of motion and the field continue to be mutually perpendicular, the force is perpendicular to the new direction. The force thus only changes the direction of motion but not the speed and so the

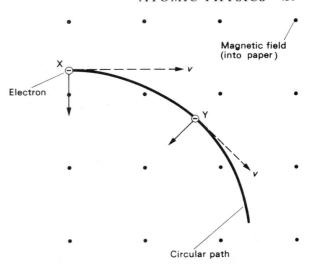

Fig. 20.3

electron of mass *m* describes a *circular arc* of radius *r*. The constant radial force *Bev* is the centripetal force and so

$$Bev = \frac{mv^2}{r}$$

This equation describes the path of the electron in the magnetic field.

(*c*) *Deflection of electrons by an electric field.* If electrons enter an electric field acting at right angles to their direction of motion, they are deflected from their original path. In Fig. 20.4 a p.d. applied between plates P and Q of length *l* creates a uniform electric field of strength *E*, non-uniformities at the edges of the plates being ignored.

Consider an electron of charge *e*, mass *m* and horizontal speed *v* on entering the field. If the upper plate is positive the electron experiences a force *Ee* and an acceleration *Ee*/*m* (second law of motion) both acting vertically upwards. Since the field is uniform, the acceleration is uniform and combines with the initial horizontal speed *v*, which the electron retains during the whole of its journey in the vacuum between the plates, to give a path which we shall show is a *parabola*. The behaviour of the electron is similar to that of a projectile fired horizontally; its path (neglecting air resistance) is also a parabola, the resultant of a uniform horizontal velocity and the vertical acceleration due to gravity.

The vertical displacement *y* of the electron at any time *t* is given by

$$y = \tfrac{1}{2}at^2$$

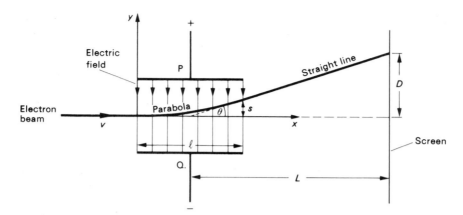

Fig. 20.4

But $a = Ee/m$,

$$\therefore \quad y = \tfrac{1}{2} \cdot \frac{Ee}{m} \cdot t^2$$

The corresponding horizontal displacement x is given by

$$x = vt$$

Eliminating t,

$$y = \left(\frac{Ee}{2mv^2}\right) \cdot x^2$$

This equation is of the form $y = kx^2$ (where $k = Ee/(2mv^2) =$ a constant) and so the path of the electron between the plates is a parabola.

The deflection D of the electron (i.e. its displacement from the original direction) on a screen distance L from the centre of the plates, can be obtained using the fact that it continues in a straight line after leaving the field. From Fig. 20.4, $\tan \theta \simeq D/L$, where $\tan \theta$ is the slope of the tangent at the end of the parabolic path. The slope equals, in calculus terms, the differential coefficient ds/dx when $x = l$. Differentiating $s = kx^2$ we get $ds/dx = 2kx$, therefore the slope is $2kl$. Hence $D/L = 2kl$ giving $D = 2klL$. Substituting for k we get $D = EelL/(mv^2)$. If V is the p.d. which has accelerated the electron to speed v then $eV = \tfrac{1}{2}mv^2$ and so

$$D = \frac{ElL}{2V}$$

Thus D is proportional to E if V is constant and inversely proportional to V if E is constant.

(*d*) *Worked example.* (*i*) *An electron emitted from a hot cathode in an evacuated tube is accelerated by a p.d. of 1.0×10^3 V. Calculate the kinetic energy and*

speed acquired by the electron. ($e = 1.6 \times 10^{-19}$ C, $m = 9.1 \times 10^{-31}$ kg).

We have

$$\tfrac{1}{2}mv^2 = eV$$
$$= (1.6 \times 10^{-19} \text{ C}) (1.0 \times 10^3 \text{ J C}^{-1})$$
$$= 1.6 \times 10^{-16} \text{ J}$$
$$\therefore \quad v = \sqrt{\frac{2 \times 1.6 \times 10^{-16}}{9.1 \times 10^{-31}}}$$
$$= 1.8 \times 10^7 \text{ m s}^{-1}$$

(*ii*) *The electron now enters at right angles a uniform magnetic field of flux density 1.0×10^{-3} T. Determine its path.*

The magnetic force Bev, on the electron makes it describe a circular path of radius r given by

$$Bev = \frac{mv^2}{r}$$

Hence

$$r = \frac{mv}{Be}$$

where $B = 1.0 \times 10^{-3}$ T, $m = 9.1 \times 10^{-31}$ kg, $e = 1.6 \times 10^{-19}$ C and from (*i*), $v = 1.8 \times 10^7$ m s^{-1}.

$$\therefore \quad r = \frac{(9.1 \times 10^{-31} \text{ kg}) (1.8 \times 10^7 \text{ m s}^{-1})}{(1.0 \times 10^{-3} \text{ T}) (1.6 \times 10^{-19} \text{ C})}$$
$$= 0.10 \text{ m}$$

(*iii*) *Find the intensity of the uniform electric field which, when applied perpendicular to the previous magnetic field so as to be co-terminous with it, compensates for the magnetic deflection. If the electric field plates are 2.0×10^{-2} m apart what is the p.d. between them?*

For the electric and magnetic forces to balance

$$Ee = Bev$$

Hence $\quad E = Bv$

$$= (1.0 \times 10^{-3} \text{ T}) (1.8 \times 10^7 \text{ m s}^{-1})$$

$$= 1.8 \times 10^4 \text{ V m}^{-1}$$

If V_1 is the p.d. between the plates and d their separation, then $E = V_1/d$ and so

$$V_1 = Ed$$

$$= (1.8 \times 10^4 \text{ V m}^{-1}) (2.0 \times 10^{-2} \text{ m})$$

$$= 3.6 \times 10^2 \text{ V}$$

Specific charge of electron

Two methods that can be performed in a school laboratory will be outlined. A modern value of the specific charge of the electron is

$$\frac{e}{m} = 1.76 \times 10^{11} \text{ C kg}^{-1}$$

(a) *Cathode-ray tube method using crossed fields.* This method is similar in principle to that developed by J. J. Thomson in which an electron beam is subjected simultaneously to mutually perpendicular, i.e. crossed, electric and magnetic fields. Here a vacuum-type cathode-ray tube is used, Fig. 20.5, having a hot cathode C and an anode A with a horizontal collimating slit from which the electrons emerge in a flat beam. The beam produces a narrow luminous trace when it hits a vertical fluorescent screen S, marked in squares and set at an angle. S is supported by two parallel deflecting plates Y_1, Y_2, across which a p.d. (about 3 kV) is applied, thereby creating an electric field between them.

Helmholtz coils X_1, X_2 (p. 266) mounted on opposite sides of the bulb produce a magnetic field between the plates, at right angles to both the direction of travel of the beam and the electric field. The coils are separated by a distance equal to their radius and when connected in series—so that the current has the same direction in each—they give an almost uniform field for a short distance along their common axis midway between them.

Consider an electron of charge e and mass m which emerges from the slit in the anode having been accelerated to speed v. Let E be the electric field strength between Y_1 and Y_2 and B the magnetic flux density along the axis of X_1 and X_2. When E and B are such that the electron suffers no deflection, the electric force Ee on it must be equal and opposite to the magnetic force Bev. Hence

$$Ee = Bev \qquad (1)$$

If the electron is emitted from the cathode with zero speed and moves in a good vacuum, its kinetic energy $\frac{1}{2}mv^2$ is given by

$$\tfrac{1}{2}mv^2 = eV \qquad (2)$$

where V is the accelerating p.d. between anode and cathode. Eliminating v from (1) and (2)

$$\frac{e}{m} = \frac{E^2}{2B^2V}$$

If the p.d. between Y_1 and Y_2 creating the electric field equals the accelerating p.d., then $E = V/d$ (i.e. field strength = potential gradient) where d is the separation of Y_1 and Y_2. Hence

$$\frac{e}{m} = \frac{V}{2B^2d^2}$$

Thus e/m can be found if V, B and d are known. B may be determined experimentally by removing the tube

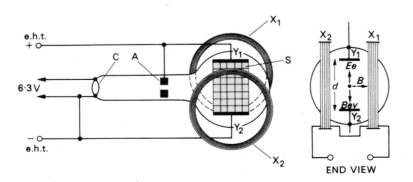

Fig. 20.5

END VIEW

and investigating the region between the coils with a current balance (p. 272) or it can be calculated from the expression $B \simeq 0.72 \, \mu_0 NI/r$ (p. 267) where $\mu_0 = 4\pi \times 10^{-7}$ H m^{-1}, N is the number of turns on one coil, I is the current and r the radius of the coil.

In the above simple treatment the fields are assumed to be uniform and coterminous, i.e. to extend over the same length of the electron beam. In practice such conditions are not achieved and this partly accounts for only an approximate value of e/m being obtained.

(b) *Fine beam tube method.* The fine beam tube, Fig. 20.6, is a special type of cathode-ray tube containing a small quantity of gas (often hydrogen) at a pressure of about 10^{-2} mmHg. Electrons from a hot cathode emerge as a narrow beam from a small hole at the apex of a conical-shaped anode and collide with atoms of the gas in the tube. As a result, the latter may lose electrons (i.e. inelastic collisions occur) and form positive gas ions.

The electrons created by ionization are easily scattered but the relatively heavy gas ions form a line of positive charge along the path of the beam which attracts the fast electrons from the anode, focusing them into a 'fine' beam. It will be seen later (p. 479) that a gas atom which has lost an electron can emit light when it recaptures an electron. The gas, therefore, not only focuses the beam but also reveals its path.

Helmholtz coils are arranged one on each side of the tube and produce a fairly uniform magnetic field at right angles to the beam. If the field is sufficiently strong the electrons are deflected into a circular orbit and a luminous circle of low intensity appears when the tube is viewed in the dark. The diameter of the circle may be measured by placing a large mirror with a scale behind the tube so that the observer sees the circle, its image and the corresponding marks on the scale all in line. The circle diameter is altered by varying either the anode voltage or the current in the coils.

If B is the magnetic flux density between the coils in the region of the tube, r the radius of the luminous circle and e, m and v are the charge, mass and orbital speed of the electron then the circular motion equation gives

$$Bev = \frac{mv^2}{r} \qquad (3)$$

We shall assume electrons are emitted from the cathode with zero speed and that their orbital speed v is constant and equal to that with which they emerge from the anode after being accelerated by the p.d. V between anode and cathode. The energy equation then gives

$$\tfrac{1}{2}mv^2 = eV \qquad (4)$$

Eliminating v from (3) and (4),

$$\frac{e}{m} = \frac{2V}{B^2 r^2}$$

As in the cathode-ray tube method, B is obtained experimentally or by calculation.

Electronic charge

(a) *Electrolysis and the Faraday.* During the latter half of the nineteenth century it was suggested that electricity, like matter, was atomic and that a natural

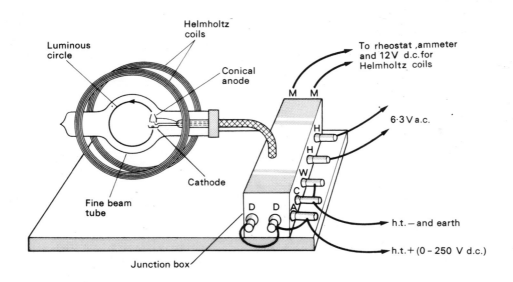

Luminous circle

Helmholtz coils

Conical anode

To rheostat ,ammeter and 12V d.c.for Helmholtz coils

6·3 V a.c.

Cathode

Fine beam tube

h.t. – and earth

h.t. + (0 – 250 V d.c.)

Junction box

Fig. 20.6

unit of electric charge existed. The basis for this belief was Faraday's work on electrolysis (1831–34) which may be summarized by the statement *a mole of monovalent ions of any substance is liberated by 9.65×10^4 coulombs*. This electric charge is called the *Faraday constant* and is denoted by *F*.

The number of atoms (ions) in a mole of any substance is 6.02×10^{23} and if we assume every atom is associated with the same charge during electrolysis, it follows that éach monovalent ion carries a charge of $9.65 \times 10^4/(6.02 \times 10^{23}) = 1.60 \times 10^{-19}$ C. It would appear that the natural unit of charge, called the *electronic charge*, has this value. For a monovalent ion we therefore have

$$F = Le$$

where *L* is the Avogadro constant (6.02×10^{23} mol^{-1}) and *e* is the electronic charge. Nowadays, the charge indicated by the Faraday is referred to as a 'mole of electrons'. A mole of divalent ions is liberated by two 'mồles of electrons' (i.e. $2 \times 9.65 \times 10^4$ C) and of trivalent ions by three 'moles of electrons' (i.e. $3 \times 9.65 \times 10^4$ C).

The above expression provides one of the most accurate ways of measuring *e*. *F* is found by electrolysis and *L* from X-ray crystallography measurements.

(*b*) *Millikan's oil drop experiment*. In 1909 Millikan started a series of experiments lasting many years which supplied evidence for the atomic nature of electricity and provided a value for the magnitude of the electronic charge. The principle of his method is to observe very small oil drops, charged either positively or negatively, falling in air under gravity and then either rising or being held stationary by an electric field.

The essential features of the apparatus are shown in Fig. 20.7. A spray of oil drops is formed above a tiny hole in the upper of two parallel metal plates and some find their way into the space between them. The drops are strongly illuminated and appear as bright specks on a dark background when viewed through a microscope.

With no electric field between the plates, one drop is selected and its velocity of fall found by timing it over a convenient number of divisions on a scale in the eyepiece of the microscope. (To find the actual distance fallen, the eyepiece scale is calibrated by viewing a millimetre scale through the microscope and comparing it with the eyepiece scale.) For a spherical drop of radius *r*, moving with uniform velocity *v* through a homogeneous medium having coefficient of viscosity η, Stokes' law states that the viscous force retarding its motion is $6\pi\eta rv$. In falling, the drop attains its terminal velocity almost at once because it is so small. The retarding force acting up then equals its weight, given by $\frac{4}{3}\pi r^3 \rho g$, where ρ is the density of the oil and *g* the acceleration due to gravity. If *v* is the terminal velocity and the small upthrust of the air is neglected

$$6\pi\eta rv = \frac{4}{3}\pi r^3 \rho g \qquad (1)$$

From this the radius *r* of the drop can be found.

Some of the drops become charged either by friction in the process of spraying or from ions in the air. Suppose the drop under observation has a negative charge *Q*. When a p.d. is applied to the plates so that the top one is positive, an electric field is created which exerts an upward force on the drop. If *V* is the p.d. and *E* is the intensity of the field required to keep the drop at rest, then the electric force experienced by it is *EQ* (since *E* is the force per unit charge). The electric force on the drop then equals its weight and so

$$EQ = \frac{4}{3}\pi r^3 \rho g \qquad (2)$$

$E = V/d$ where *d* is the distance between the plates, *r* is known from (1) and hence *Q* can be calculated.

Certain measures were adopted by Millikan to improve accuracy.

(*i*) He used non-volatile oil to prevent evaporation altering the mass of the drop.

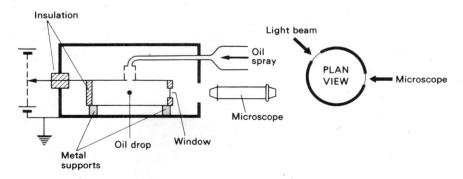

Fig. 20.7

(*ii*) Convection currents between the plates and variation of the viscosity of air due to temperature change were eliminated by enclosing the apparatus in a constant-temperature oil bath.

(*iii*) Stokes' law assumes fall in a homogeneous medium. The air consists of molecules and, as Millikan put it, very small drops 'fall freely through the holes in the medium'. He investigated this effect and corrected the law to allow for it.

(*iv*) To simplify the theory we have considered the drop held at rest by the electric field but Millikan reversed the motion and found the upward velocity of the drop.

Millikan found that the charge on an oil drop, whether positive or negative, was always an integral multiple of a basic charge. He studied drops having charges many times the basic charge and by using X-rays he was able to change the charge on a drop. The same minimum charge, equal to that on a monovalent ion, was always involved. The value of the 'atom' of electric charge, i.e. the electronic charge e, is

$$e = 1.60 \times 10^{-19} \text{ C}$$

Photoelectric emission

In photoelectric emission electrons are ejected from metal surfaces when electromagnetic radiation of high enough frequency falls on them. The effect is given by zinc when exposed to X-rays or ultraviolet. Sodium gives emission with X-rays, ultraviolet and all colours of light except orange and red, while preparations

containing caesium respond to infrared as well as to high frequency radiation.

(*a*) *Simple demonstration of photoelectric effect.* Ultraviolet from a mercury vapour lamp is allowed to fall on a small sheet of zinc, freshly cleaned with emery cloth and connected to an electroscope as in Fig. 20.8*a*. If the electroscope is given a positive charge the pointer is unaffected by the ultraviolet. When negatively charged, however, the electroscope discharges quite rapidly when the zinc is illuminated with ultraviolet. A sheet of glass between the lamp and the zinc plate halts the discharge.

The photoelectric effect was discovered in 1887 but was not explained until the electron had been 'discovered' by J. J. Thomson. When the zinc plate is positively charged the electrons ejected from it by the ultraviolet fail to escape, being attracted back to the plate, Fig. 20.8*b*. A negative charge on the zinc repels the emitted electrons and both the zinc and the electroscope lose negative charge, Fig. 20.8*c*. The insertion of the sheet of glass cuts off much of the ultraviolet but allows the passage of violet light, also emitted by the lamp, and shows that violet light does not produce the effect with zinc.

Experiment shows that 'photoelectrons' have the same specific charge as any other electrons.

(*b*) *Laws of photoelectric emission.* An experimental study of the photoelectric effect yields some surprising results which may be summarized as follows:

Law I. The *number* of photoelectrons emitted

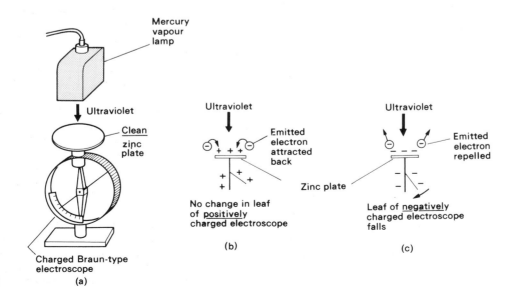

Fig. 20.8

(a) Mercury vapour lamp / Ultraviolet / Clean zinc plate / Charged Braun-type electroscope

(b) Ultraviolet / Emitted electron attracted back / No change in leaf of *positively* charged electroscope

(c) Ultraviolet / Emitted electron repelled / Zinc plate / Leaf of *negatively* charged electroscope falls

per second is proportional to the *intensity* of the incident radiation.

Law II. The photoelectrons are emitted with a range of kinetic energies from zero up to a *maximum* which increases as the *frequency* of the radiation increases and is independent of the intensity of the radiation. (Thus a faint blue light produces electrons with a greater maximum kinetic energy than those produced by a bright red light, but the latter releases a greater number.)

Law III. For a given metal there is a certain minimum frequency of radiation, called the *threshold frequency*, below which no emission occurs irrespective of the intensity of the radiation. For zinc, the threshold frequency is in the ultraviolet.

Most of these facts appear to be inexplicable on a wave theory of electromagnetic radiation. Law I can be vindicated in terms of waves because if the radiation has greater intensity, more energy is absorbed by the metal and it is possible for more electrons to escape. Also, it is reasonable to suppose that the range of emission speeds (and kinetic energies) from zero to a maximum is due to electrons having a range of possible kinetic energies inside the metal. Those with the highest kinetic energy are emitted with the maximum speeds. However, we would expect a certain number of photoelectrons to be ejected with greater speeds when the radiation intensity increases but this is not so according to Law II.

The increase of maximum kinetic energy with frequency and the existence of the threshold frequency are even more enigmatic. Furthermore, according to the wave theory, radiation energy is spread over the wavefront and since the amount incident on any one electron would be extremely small, some time would elapse before an electron gathered enough energy to escape. No such time lag between the start of radiation and the start of emission is observed, even when the radiation is weak.

Quantum theory

By the end of the nineteenth century the wave theory, despite its earlier notable successes, was unable to account for most of the known facts concerning the interaction of electromagnetic radiation with matter. One of these, as we have just seen, was photoelectric emission and another was black body radiation, which is considered in Chapter 18.

(*a*) *Planck's theory.* In 1900 Planck tackled the problem of finding a theory which would fit the facts of black body radiation. Whereas others had considered the radiation to be emitted continuously, Planck supposed this to occur intermittently in integral multiples of an 'atom' or *quantum* of energy, the size of which depended on the frequency of the oscillator producing the radiation. A body would thus emit one, two, three etc. quanta of energy but no fractional amounts.

According to Planck the quantum E of energy for radiation of frequency f is given by

$$E = hf$$

where h is a constant, now called *Planck's constant.* For electromagnetic radiation of wavelength λ, $c = f\lambda$, where c is its speed in vacuo and so we also have $E = hc/\lambda$. The energy of a quantum is thus inversely proportional to the wavelength of the radiation but directly proportional to the frequency. It is convenient to express many quantum energies in electron-volts; the quantum for red light has energy of about 2 eV and for blue light of about 4 eV.

Using the equation $E = hf$, Planck derived an expression for the variation of energy with wavelength for a black body which agreed with the experimental curves (Fig. 18.63, p. 415), at all wavelengths and temperatures. At the time the quantum theory was too revolutionary for most scientists and little attention was paid to it. Nevertheless the interpretation of black body radiation was the first of its many successes.

(*b*) *Einstein's photoelectric equation.* Einstein extended Planck's ideas in 1905 by deriving an equation which explained in a completely satisfactory way the laws of photoelectric emission. He assumed that not only were light and other forms of electromagnetic radiation emitted in whole numbers of quanta but that they were also absorbed as quanta, called *photons*. This implied that electromagnetic radiation could exhibit particle-like behaviour when being emitted and absorbed and led to the idea that light etc. has a dual nature; under some circumstances it behaves as waves and under others as particles. Wave-particle duality will be considered later in the chapter.

When dealing with thermionic emission it was explained that to liberate an electron from the surface of a metal a quantity of energy, called the *work function*, Φ, which is characteristic of the metal, has to be supplied. In photoelectric emission, Einstein proposed that a photon of energy hf causes an electron to be emitted if $hf \geqslant \Phi$. The excess energy $(hf - \Phi)$ appears as kinetic energy of the emitted electron which escapes with a speed having any value up to a maximum v_{max}. The actual value depends on how much energy the electron has inside the metal. Hence

$$hf - \Phi = \tfrac{1}{2}mv^2_{max}$$

This is Einstein's photoelectric equation.

If the photon has only just enough energy to liberate an electron, the electron gains no more kinetic energy. It follows that since Φ is constant for a given metal, there is a minimum frequency, the threshold frequency f_0, below which no photoelectric emission is possible. It is given by

$$hf_0 = \Phi$$

Einstein's equation can also be written as

$$h(f - f_0) = \tfrac{1}{2}mv^2_{max}$$

The increase of maximum emission speed with higher frequency radiation can now be seen to be due to the greater photon energy of such radiation. It should also be noted that it is assumed that a photon imparts all its energy to one electron and then no longer exists.

(c) *Millikan's experiment on photoelectric emission.* In 1916 Millikan verified Einstein's photoelectric equation experimentally and provided irrefutable evidence for the photon model of light. He was the first to obtain photoelectrically a value for Planck's constant which agreed with values from other methods. His apparatus is shown in simplified form in Fig. 20.9a.

Monochromatic light from a spectrometer entered the window of a vacuum chamber and fell on a metal X mounted on a turntable R controlled from outside the chamber. The photoelectrons emitted were collected by an electrode C and detected by a sensitive current measuring device E. The minimum positive potential, called the *stopping potential*, which had to be applied to X to prevent the most energetic photoelectrons reaching C and causing current flow, was found for different frequencies of the incident radiation. The procedure was repeated with Y and then Z opposite C. X, Y and Z were made from the alkali metals lithium, sodium and potassium, since these emit photoelectrons with light and each one can therefore be studied over a wide range of frequencies. Immediately before taking a set of readings R was rotated and the knife K adjusted so that a fresh surface was cut on the metal, thus eliminating the effects of surface oxidation.

In such experiments the p.d. between the electrodes, as measured by any form of voltmeter, is not the same as the p.d. in the space between them unless their work functions are equal. The error is called the *contact p.d.* In Millikan's experiment this was a few volts, that is, the correction was as large as the effect measured. One feature of his experiment was an ingenious device (not shown in Fig. 20.9a) by which an accurate correction was made.

The relation between the stopping potential V_s and the maximum kinetic energy of the photoelectrons is given by the energy equation (p. 459), $eV_s = \tfrac{1}{2}mv^2_{max}$. Einstein's equation may then be written

$$\tfrac{1}{2}mv^2_{max} = eV_s = hf - \Phi$$

$$\therefore \quad V_s = \frac{h}{e} \cdot f - \frac{\Phi}{e}$$

The graph of V_s against f should be a straight line, a fact which Millikan's results confirmed. Einstein's relation was thus verified. Furthermore, the stopping potential for a given frequency of light was independent of the intensity of the light. One of Millikan's graphs is shown in Fig. 20.9b. He found that whatever the metal all graphs had the same slope and from the above

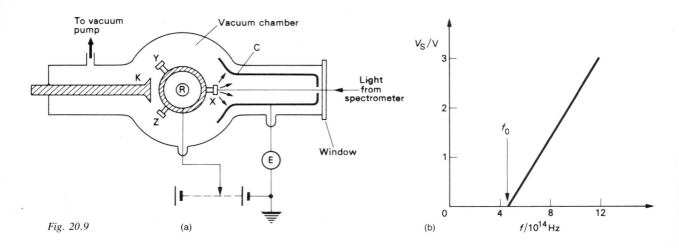

Fig. 20.9   (a)   (b)

equation this is seen to be h/e. Knowing e, h can be calculated and the value obtained agrees with that found from black body radiation experiments. The threshold frequency f_0 and the work function Φ are characteristic for each metal and may also be deduced from the graph. The values of Φ are in good agreement with those found from thermionic emission.

Planck's constant h is a fundamental physical constant and occurs in many formulae in atomic physics. Its value to three figures is

$$h = 6.63 \times 10^{-34} \text{ J s}$$

It is because of this smallness that quantum effects are not normally apparent.

Because of the need for very clean surfaces, quantitative photoelectric experiments are difficult with simple apparatus in a school laboratory but one which gives fair results is outlined in Appendix 17.

(d) *Worked example. If a photoemissive surface has a threshold wavelength of 0.65 μm, calculate (i) its threshold frequency, (ii) its work function in electron volts, and (iii) the maximum speed of the electrons emitted by violet light of wavelength 0.40 μm.* (Speed of light $c = 3.0 \times 10^8$ m s^{-1}, $h = 6.6 \times 10^{-34}$ J s, $e = 1.6 \times 10^{-19}$ C and mass of electron $m = 9.1 \times 10^{-31}$ kg.)

(i) $\qquad \lambda_0 = 0.65 \, \mu\text{m} = 6.5 \times 10^{-7}$ m

$$\therefore \quad f_0 = c/\lambda_0$$

$$= \frac{3.0 \times 10^8 \text{ m s}^{-1}}{6.5 \times 10^{-7} \text{ m}}$$

$$= 4.6 \times 10^{14} \text{ Hz}$$

(ii) We have $\qquad \Phi = hf_0$

$$= (6.6 \times 10^{-34} \text{ J s})$$
$$(4.6 \times 10^{14} \text{ s}^{-1})$$

$$= 6.6 \times 10^{-34} \times 4.6 \times 10^{14} \text{ J}$$

But $\qquad 1 \text{ eV} = 1.6 \times 10^{-19}$ J

$$\therefore \quad \Phi = \frac{6.6 \times 4.6 \times 10^{-20} \text{ eV}}{1.6 \times 10^{-19}}$$

$$= 1.9 \text{ eV}$$

(iii) For violet light $\quad \lambda = 0.40 \, \mu\text{m} = 4.0 \times 10^{-7}$ m

$$f = c/\lambda$$

$$= \frac{3.0 \times 10^{-8} \text{ m s}^{-1}}{4.0 \times 10^{-7} \text{ m}}$$

$$= 7.5 \times 10^{14} \text{ Hz}$$

From the photoelectric equation $\frac{1}{2}mv_{\text{max}}^2 = hf - \Phi$

$$\frac{1}{2}mv_{\text{max}}^2 = (6.6 \times 10^{-34} \times 7.5 \times 10^{14} - 1.9 \times 1.6 \times 10^{-19}) \text{ J}$$

$$= 1.9 \times 10^{-19} \text{ J}$$

$$\therefore \quad v_{\text{max}} = \sqrt{\frac{2 \times 1.9 \times 10^{-19}}{9.1 \times 10^{-31}}} \text{ m s}^{-1}$$

$$= 6.5 \times 10^5 \text{ m s}^{-1}$$

Photocells and their uses

There are several kinds of photocell.

(a) *Photoemissive cell.* The construction of a typical cell and its symbol are shown in Fig. 20.10a. Two electrodes are enclosed in a glass bulb which may be evacuated or contain an inert gas at low pressure. The cathode, often called the *photocathode*, is a curved metal plate having an emissive surface facing the anode, here shown as a single metal rod. When electromagnetic radiation falls on the cathode, photoelectrons are emitted and are attracted to the anode if it is at a suitable positive potential. A current of a few microamperes flows and increases with the intensity of the incident radiation. An inert gas in the cell gives greater current but causes a time lag in the response of the cell to very rapid changes of radiation which may make it unsuitable for some purposes.

The choice of material for the cathode surface depends on the frequency range over which the cell is to operate, and should be such that a good proportion of the incident photons yield electrons. Ideally, every photon should release one electron but in practice the yield is much lower. Pure metals are rarely used because of their high reflecting power; most photocathode surfaces are composite. Caesium on oxidized silver has a peak response near the red end of the spectrum and a threshold wavelength of 12 μm which makes it suitable for use with infrared.

In the reproduction of sound from a film, light from an exciter lamp, Fig. 20.10b, is focused on the 'sound track' at the side of the moving film and then falls on a photocell. The sound track varies the intensity of the light passing through it so that the photocell creates a varying current which is a replica of that obtained in the recording microphone when the film was made. The fluctuating p.d. developed across a load R is amplified and the output converted to sound by a loudspeaker. In the variable area sound track the width of the 'white' part varies in the same way as the amplitude of the original sound wave. The track is produced by allowing the electrical variations from the recording microphone

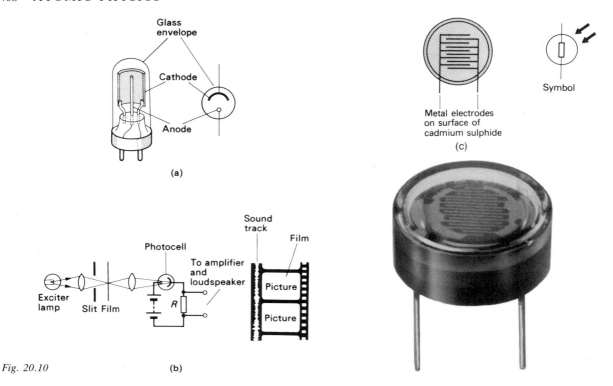

Fig. 20.10   (b)

to control the width of a narrow slit through which light passes on its way to the side of the moving film.

(b) *Photoconductive cell or light dependent resistor (l.d.r.)*. The resistance of certain semiconductors such as cadmium sulphide decreases as the intensity of the light falling on them increases. The effect is due to light photons setting free electrons in the semiconductor (p. 493), so increasing its conductivity, i.e. reducing its resistance.

A popular l.d.r. is the ORP12 shown in Fig. 20.10c. There is a 'window' over the grid-like metal structure to allow light to fall on a thin layer of cadmium sulphide. Its resistance varies from about 10 MΩ in the dark to 1 kΩ or so in daylight.

An alarm circuit using an l.d.r. is described later (p. 508).

X-rays

X-rays, so named because their nature was at first unknown, were discovered in 1895 by Röntgen. They are produced whenever cathode rays (electrons) are brought to rest by matter.

(a) *Production*. A modern X-ray tube is highly evacuated and contains an anode and a tungsten filament connected to a cathode, Fig. 20.11. Electrons are obtained from the filament by thermionic emission and are accelerated to the anode by a p.d., typically up to 100 kV. The anode is a copper block inclined to the electron stream and having a small target of tungsten, or another high-melting-point metal, on which electrons are focused by the concave cathode. The tube has a lead shield with a small window to allow the passage of the X-ray beam.

Less than $\frac{1}{2}$ per cent of the kinetic energy of the electrons is converted into X-rays—the type of electromagnetic radiation produced in cases where electrons are rapidly brought to rest by striking matter (see p. 472). The rest of the kinetic energy becomes internal energy of the anode which has to be kept cool by circulating oil or water through channels in it or by the use of cooling fins.

The *intensity* of the X-ray beam increases when the number of electrons hitting the target increases and this is controlled by the filament current. The *quality* or penetrating power of the X-rays is determined by the speed attained by the electrons and increases with the p.d. across the tube. 'Soft' X-rays only penetrate such objects as flesh, 'hard' X-rays can penetrate much more solid matter.

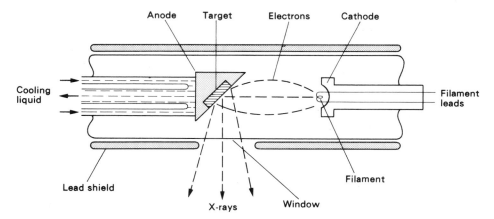

Fig. 20.11

The p.d. required to operate an X-ray tube may be obtained from a half-wave rectifying circuit containing a step-up transformer and in which the X-ray tube itself acts as the rectifier.

(b) *Properties*. These may be summarized as follows:
(i) They travel in straight lines.
(ii) They readily penetrate matter; penetration is least in materials containing elements of high density and high atomic number. Thus, while sheets of cardboard, wood and some metals fail to stop them, all but the most penetrating are absorbed by a sheet of lead 1 mm thick. Lead glass is a much better absorber than ordinary soda glass.
(iii) They are not deflected by electric or magnetic fields.
(iv) They eject electrons from matter by the photoelectric effect and other mechanisms. The ejected electrons are responsible for the next three effects.
(v) They ionize a gas, permitting it to conduct. An electrified body such as an electroscope charged positively or negatively is discharged when the surrounding air is irradiated by X-rays.
(vi) They cause certain substances to fluoresce, e.g. barium platinocyanide.
(vii) They affect a photographic emulsion in a similar manner to light.

(c) *Nature: von Laue's experiment*. X-rays cannot be charged particles since they are not deflected by electric and magnetic fields. The vital experiment which established their electromagnetic nature was initiated by von Laue in 1912. After unsuccessful attempts had been made to obtain X-ray diffraction patterns with apparatus similar to that used for light, it was realized that the failure might be due to X-rays having much smaller wavelengths.

Von Laue's suggestion was that if the regular spacing of atoms in a crystal is of the same order as the wavelength of the X-rays, the crystal should act as a

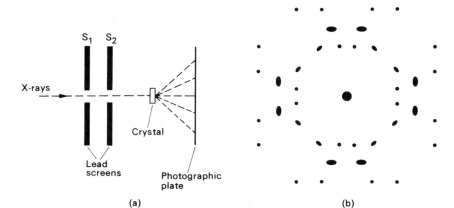

Fig. 20.12   (a)   (b)

three-dimensional diffraction grating. This idea was put to the test with conspicuous success by Friedrich and Knipping, two of von Laue's colleagues. The arrangement of their apparatus is shown in Fig. 20.12a. A narrow beam of X-rays from two slits S_1 and S_2 fell on a thin crystal in front of a photographic plate. After a long exposure the plate (on developing) revealed that most of the radiation passed straight through to give a large central spot, but some of it produced a distinct pattern of fainter spots around the central one, Fig. 20.12b.

The electromagnetic wave behaviour of X-rays was thus demonstrated and analysis of diffraction patterns confirmed they had wavelengths of about 10^{-10} m (0.1 nm). They are considered to be electromagnetic waves because (i) their method of production involves accelerated charges, (ii) they eject electrons from matter (implying strong electric fields), (iii) they give line spectra similar in general character to the optical spectrum of hydrogen (p. 404).

(d) Uses of X-rays. The usefulness of X-rays is largely due to their penetrating power.

(i) Medicine. Here, radiographs or X-ray photographs are used for a variety of purposes. Since X-rays can damage healthy cells of the human body, great care is taken to avoid unnecessary exposure. In radiography the X-ray film is sandwiched between two screens which fluoresce when subjected to a small amount of X-radiation. The film is affected by the fluorescent light rather than by the X-rays; formerly X-rays were used to sensitize the film directly and much longer exposures were required.

Suspected bone fractures can be investigated since X-rays of a certain hardness can penetrate flesh but not bone. In the detection of lung tuberculosis by mass radiography use is made of the fact that diseased tissue is denser than healthy lung tissue which consists of air sacs and so the former absorbs X-rays more strongly. When an organ is being X-rayed whose absorptive power is similar to that of the surrounding tissue, a 'contrast' medium is given to the patient, orally or by injection. This is less easily penetrated by the X-rays and enables a shadow of the organ to be obtained on a radiograph.

In the treatment of cancer by radiotherapy, very hard X-rays are used to destroy the cancer cells whose rapid multiplication causes malignant growth.

(ii) Industry. Castings and welded joints can be inspected for internal imperfections using X-rays. A complete machine may also be examined from a radiograph without having to be dismantled.

(iii) X-ray crystallography. The study of crystal structure by X-rays is now a powerful method of scientific research. The first crystals to be analysed were of simple compounds such as sodium chloride but in recent years the structure of very complex organic molecules has been unravelled.

X-ray diffraction

(a) Bragg's law. In von Laue's method of producing X-ray diffraction, the crystal is used as a transmission grating and interpretation of the patterns is not easy. Sir William Bragg and his son Sir Lawrence developed a simpler technique in which the crystal acts as a reflection diffraction grating.

When X-rays fall on a single plane of atoms in a crystal, each atom scatters a small fraction of the incident beam and may be regarded as the source of a weak secondary wavelet of X-rays. In most directions destructive interference of the wavelets occurs but in the direction for which the angle of incidence equals the angle of 'reflection', there is reinforcement and a weak reflected beam is obtained. The X-rays thus behave as if they were weakly 'reflected' by the layer of atoms, Fig. 20.13a.

Other planes of atoms, such as p, q and r in Fig. 20.13b, to which the X-rays penetrate, behave similarly. The reflected beams from all the planes involved

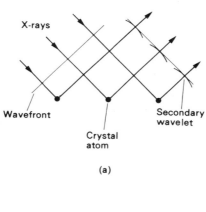

X-rays

Wavefront

Crystal atom

Secondary wavelet

(a)

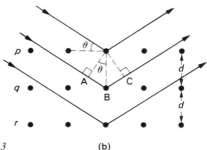

Fig. 20.13   (b)

interfere and the resultant reflected beam is only strong if the path difference between successive planes is a whole number of wavelengths of the incident X-radiation. Thus reinforcement only occurs for planes p and q when

$$AB + BC = n\lambda$$

where n is an integer and λ is the wavelength of the X-rays. If d is the distance between planes of atoms and θ is the angle between the X-ray beam and the crystal surface, called the *glancing angle*, then $AB + BC = 2d \sin \theta$ and the reflected beam has maximum intensity when

$$2d \sin \theta = n\lambda$$

This equation is Bragg's law; note that it refers to the glancing angle and not to the angle of incidence. Intensity maxima occur for several glancing angles. The smallest angle is given by $n = 1$ and is the first-order reflection; $n = 2$ gives the second order, and so on.

This process is sometimes termed X-ray 'reflection' and although it may appear that reflection occurs it is in fact a diffraction effect, since interference takes place between X-rays from secondary sources, i.e. crystal atoms, on the *same* wavefront.

The atoms in a crystal can be considered to be arranged in several different sets of parallel planes, from all of which strong reflections may be obtained to give a pattern of spots characteristic of the particular structure.

A microwave analogue of X-ray 'reflection' from the planes of 'atoms' in a polystyrene ball crystal can be demonstrated. (See Chapter 1.)

(b) *X-ray spectrometer.* The Bragg X-ray spectrometer was developed to measure (i) X-ray wavelengths, and (ii) the spacing of atoms in crystals. The principle

of the instrument is shown in Fig. 20.14. X-rays from the target of an X-ray tube are collimated by two slits S_1 and S_2 (made in lead sheets) and the narrow beam so formed falls on a crystal C set on the table T of the spectrometer. The reflected beam passes through a third slit S_3 into an ionization chamber I (p. 533) where it creates an ionization current which is a measure of the intensity of the reflected radiation.

As the crystal and the ionization chamber are rotated, the angle of reflection always being kept equal to the angle of incidence, the ionization current is found. Strong reflection occurs for glancing angles satisfying Bragg's law $2d \sin \theta = n\lambda$. Knowing either d or λ, the other can be calculated.

X-ray wavelengths are now measured directly using mechanically ruled diffraction gratings (similar to optical gratings) provided the X-rays strike the grating at a glancing angle much less than 1°.

(c) *X-ray powder photography.* If instead of a single crystal, a polycrystalline specimen or a crystalline powder is used, many planes are involved at once in X-ray 'reflection' and thousands of spots are produced from

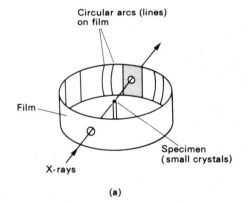

(a)

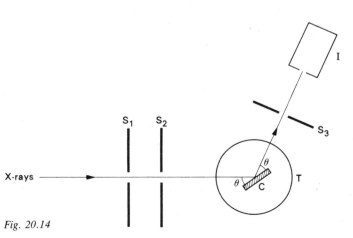

Fig. 20.14

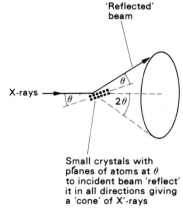

Fig. 20.15　　　　　(b)

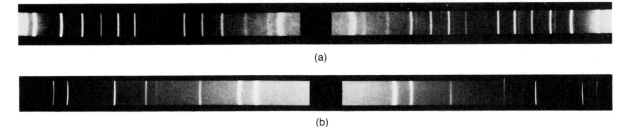

(a)

(b)

Fig. 20.16

'reflection' at all possible angles. As a result, circles or circular arcs are obtained on a suitably placed X-ray film. Fig. 20.15a shows the arrangement in the X-ray powder (or polycrystalline) technique using a strip of film and Fig. 20.15b indicates how the 'lines' are formed. Fig. 20.16a and b are photographs for sodium chloride powder and a polycrystalline copper wire.

X-ray spectra and the quantum theory

(a) *Continuous and line spectra.* The radiation from an X-ray tube can be analysed with a spectrometer and an intensity wavelength graph obtained to show the spectral distribution. A typical X-ray spectrum is given in Fig. 20.17. It has two parts.

(i) A *continuous* spectrum which has a definite lower wavelength limit, increases to a maximum and then decreases gradually in the longer wavelengths. All targets emit this type of radiation.

(ii) A *line* spectrum consisting of groups or series of two or three peaks of high-intensity radiation superimposed on the continuous spectrum. The series are denoted by the letters K, L, M, etc., in order of

increasing wavelength, and the peaks by α, β, γ. The wavelengths of the peaks are characteristic of the target element; all the series are not normally given by one element.

(b) *Quantum theory and the continuous spectrum.* Certain features of the continuous spectrum are readily explained by the quantum theory. Thus the existence of a definite *minimum* wavelength can be justified if we assume that this radiation consists of X-ray photons produced by electrons which have given up all their kinetic energy in a single encounter with a target atom. If such an electron has mass m and speed v on striking the target, the energy hf of the photon is given by

$$hf = \tfrac{1}{2}mv^2 \tag{1}$$

h being Planck's constant and f the frequency of the radiation. With a p.d. V across the X-ray tube, an electron of charge e has work eV done on it by the electric field and so

$$eV = \tfrac{1}{2}mv^2 \tag{2}$$

From (1) and (2)

$$hf = eV \tag{3}$$

The value of f given by (3) is the maximum frequency of the X-rays emitted at p.d. V, since all the energy of the electron is converted to the photon. The corresponding wavelength will have a minimum value, and if this is λ_{min} then $c = f\lambda_{min}$ where c is the speed of travel of X-rays. It follows that

$$\lambda_{min} = \frac{hc}{eV} \tag{4}$$

As V increases we see from (4) that λ_{min} decreases, i.e. X-rays of higher frequency and greater penetrating power are emitted. The values of λ_{min} calculated from this equation agree with those found experimentally.

Most of the electrons responsible for the X-radiation

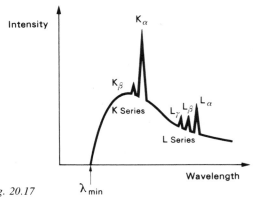

Fig. 20.17

usually have more than one encounter before losing all their energy. Several photons are produced with smaller frequencies than f and therefore with greater wavelengths than λ_{min}. Different electrons lose different amounts of energy and so a continuous spectrum covering a range of wavelengths is obtained. The great majority of electrons however lose their kinetic energy too gradually for X-rays to be emitted and merely increase the internal energy of the target.

(c) *Explanation of line spectra.* This will be considered later (p. 480).

(d) *Worked example. An X-ray tube operates at 30 kV and the current through it is 2.0 mA. Calculate (i) the electrical power input, (ii) the number of electrons striking the target per second, (iii) the speed of the electrons when they hit the target, and (iv) the lower wavelength limit of the X-rays emitted.*

(*i*) If V is the p.d. across the tube and I the tube current then

$$\text{power input} = VI$$
$$= (30 \times 10^3 \text{ V}) (2.0 \times 10^{-3} \text{ A})$$
$$= 60 \text{ W}$$

(*ii*) Current through the tube is given by $I = ne$, where n is the number of electrons striking the target per second and e is the electronic charge (i.e. 1.6×10^{-19} C).

$$\therefore \quad n = \frac{I}{e} = \frac{2.0 \times 10^{-3} \text{ A}}{1.6 \times 10^{-19} \text{ C}}$$
$$= 1.3 \times 10^{16}$$

(*iii*) If m is the mass of an electron (i.e. 9.0×10^{-31} kg) and v its speed at the target, then from equation (2) above

$$\tfrac{1}{2}mv^2 = eV$$
$$\therefore \quad v = \sqrt{\frac{2eV}{m}}$$
$$= \sqrt{\frac{2 \times 1.60 \times 10^{-19} \times 30 \times 10^3}{9.0 \times 10^{-31}}}$$
$$= 1.0 \times 10^8 \text{ m s}^{-1}$$

(*iv*) From equation (4) above, the lowest X-ray wavelength emitted is given by

$$\lambda_{min} = \frac{hc}{eV}$$

where $h = 6.6 \times 10^{-34}$ J s

and $c = 3.0 \times 10^8$ m s^{-1}

$$\therefore \quad \lambda_{min} = \frac{(6.6 \times 10^{-34} \text{ J s}) (3.0 \times 10^8 \text{ m s}^{-1})}{(1.6 \times 10^{-19} \text{ C}) (30 \times 10^3 \text{ V})}$$
$$= 0.41 \times 10^{-10} \text{ m}$$

Electrical conduction in gases

(*a*) *Ionized gases.* Gases at s.t.p. are very good electrical insulators and about 30 kV is required to produce a spark discharge between two rounded electrodes 1 cm apart in air; the value for pointed electrodes 1 cm apart is 12 kV. To conduct, a gas must be ionized.

Ionization occurs when an electron is removed from an atom (or molecule) by supplying a certain amount of energy to overcome the attractive force securing the electron to the atom. The two resulting charged particles form an *ion-pair*, the electron being a negative ion and the atom, now deficient of an electron, a positive ion. When a p.d. is applied between two electrodes in the ionized gas, positive ions move towards the cathode and electrons towards the anode. The ions thus act as charge carriers and current flows in the gas. If an atom gains an electron it becomes a heavy negative ion but these are generally few in number.

Various agents can ionize a gas and these include a flame, sufficiently energetic electrons, ultraviolet, X-rays and the radiation from radioactive substances, i.e. α-, β- and γ-rays. The few ions always present in the atmosphere are caused by cosmic rays from outer space and radioactive minerals in the earth. They are responsible for a charged, insulated body such as an electroscope, gradually discharging.

(*b*) *Current-p.d. relationship.* The gas between two parallel plates P and Q, Fig. 20.18a, is ionized by a beam of ionizing radiation. When a p.d. is applied the resulting ionization current is recorded by a sensitive current detector, e.g. a d.c. amplifier as described in Chapter 22. If the p.d. is varied while the intensity of the radiation remains constant, the current variation is shown by the curve OABCD in Fig. 20.18b.

The shape of the curve is explained as follows. Small p.d.s cause the electrons and positive ions to move slowly to the anode P and cathode Q respectively. On the way some ions recombine to form neutral atoms. As the p.d. increases, the ions travel more quickly and there is less opportunity for recombination; the ionization current increases. At B all the ions produced by the radiation reach the electrodes, no recombination occurs and further increase of p.d. between B and C does not affect the current. Along BC the current has

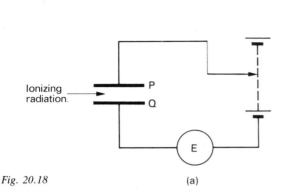

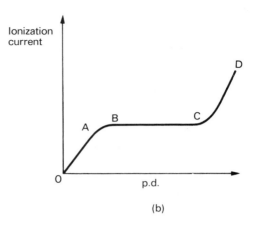

Fig. 20.18   (a)   (b)

its *saturation value* and is independent of the applied p.d.

Beyond C the current rises rapidly with p.d. and indicates that a new source of ion-pairs has become operative. The original ions due to the radiation are now accelerated sufficiently by the large p.d. to form new electrons and new positive ions by collision with neutral gas molecules between the plates. Thus along CD each original ion-pair creates several other ion-pairs, the process being known as *ionization by collision*. It can be shown that electrons are mainly responsible for this.

Two points should be noted:

(*i*) Ohm's law is obeyed between O and A

(*ii*) the saturation current is proportional to the rate of production of ions by the radiation and so measures the intensity of the radiation.

Terms and definitions

(*a*) *Atomic or proton number Z.* After it had been established that the negatively charged electron was one of the basic constituents of all atoms, the search began for a positively charged counterpart.

Experiments on the bombardment of atoms by suitable high-speed particles revealed that in certain cases, a particle with the properties of a hydrogen ion is ejected from the nucleus of an atom (p. 545). This provided fairly conclusive evidence for the belief that such particles are one of the fundamental particles from which atoms are made. Being the lightest and simplest positively charged particles then known, they were called *protons* (from the Greek *protos* meaning first).

The proton, denoted by the symbol p, has a charge equal in magnitude but opposite in sign to that on an electron and its rest mass is 1836 times the rest mass of the electron.

The number allotted to an element in the periodic table was called its *atomic number*. The significance of this term was not apparent until the nuclear theory of the atom had been proposed and evidence obtained (p. 478) that it was the *number of protons in the nucleus of the atom of the element*—hence the alternative term, *proton number*.

An atom is normally electrically neutral and so the atomic number is also the number of electrons in a neutral atom of the element. Hydrogen with $Z = 1$ has 1 proton and 1 electron while uranium with $Z = 92$ has 92 protons and 92 electrons.

(*b*) *Mass or nucleon number A.* Since the atomic number of an element is about half its relative atomic mass, if follows that if (as the nuclear theory supposes, p. 476) the mass of an atom is due to its nucleus, then there must be other constituents besides protons in the nucleus. The possibility of the existence of an electrically neutral particle, having a similar mass to that of the proton, was suggested in 1920. This particle, named the *neutron*, symbol n, remained undiscovered until 1932 (p. 547), partly on account of the difficulty of detecting a particle which, being uncharged, is not deflected by electric or magnetic fields and produces no appreciable ionization in its path.

Neutrons can also be expelled from certain nuclei by bombardment and there is now no doubt that they are basic constituents of matter. Each does not consist of a proton in close association with an electron but is an entity in itself. The rest mass of the neutron is 1839 times that of the electron. Protons and neutrons are called *nucleons*.

The mass or nucleon number A of an atom is the

number of nucleons in the nucleus.

The simplest nucleus is that of hydrogen which consists of 1 p; in symbolic notation it is written $_1^1\text{H}$, where the superscript gives the nucleon number and the subscript the proton number. Helium has 2 p and 2 n, giving $A = 4$ and $Z = 2$ and symbol $_2^4\text{He}$. Lithium $_3^7\text{Li}$ has $A = 7$ and $Z = 3$ and its atom has 3 p and 4 n. In general, atom X is represented by $_Z^A\text{X}$. The neutron is written $_0^1\text{n}$ since it has $A = 1$ and zero charge, i.e. $Z = 0$; the proton can be written $_1^1\text{p}$ and the electron $_{-1}^0\text{e}$. Sometimes the proton number is omitted and the name or symbol of the element is given followed by the mass number, e.g., lithium 7 or Li 7.

(*c*) *Isotopes.* If two atoms have the same number of protons but different numbers of neutrons, their atomic (proton) numbers are equal but not their mass (nucleon) numbers. Each atom is said to be an *isotope* of the other. They are chemically indistinguishable (since they have the same number of electrons) and occupy the same place in the periodic table.

Few elements consist of identical atoms: most are isotopic mixtures. Hydrogen has three forms: $_1^1\text{H}$ with 1 p, heavy hydrogen or deuterium $_1^2\text{D}$ with 1 p and 1 n and tritium $_1^3\text{T}$ with 1 p and 2 n. Ordinary hydrogen contains 99.99 per cent of $_1^1\text{H}$ atoms. Isotopes are not as a rule given separate names and symbols; exception is made in the case of hydrogen because there is an appreciable difference in the physical properties of the three forms. Water made from deuterium is called *heavy water* and is denoted by D_2O; it has density 1.108 g cm^{-3}, a freezing point of 3.82 °C and a boiling point of 101.42 °C. The nucleus of the deuterium atom is called a *deuteron*.

Isotopes account for fractional atomic masses. For example, chlorine with atomic mass 35.5 has two forms: $_{17}^{35}\text{Cl}$ and $_{17}^{37}\text{Cl}$ and these are present in ordinary chlorine in the approximate ratio of three atoms of the former to one of the latter. Chemically they are identical but one is slightly denser than the other.

Isotopes were discovered among the radioactive elements in 1906 but their nature was not understood. Although there are only 104 elements (89 in nature and 15 man-made), each one has isotopes and the total number known at present is about 1500. Of these about 300 occur naturally and the rest are artificial; all of the latter and some of the former are radioactive.

The term *nuclide* is used to specify an atom with a particular proton-neutron combination. $_3^6\text{Li}$ and $_3^7\text{Li}$ are isotopes and nuclides, $_4^9\text{Be}$ and $_5^{10}\text{B}$ are nuclides (they have the same number of neutrons, i.e. 5, but different numbers of protons).

Atomic mass: mass spectrograph

(*a*) *Atomic mass* (more correctly, *relative atomic mass*). This is denoted by A_r and was previously called the *atomic weight*. It is defined as the ratio

$$\frac{mass\ of\ an\ atom}{1/12\ mass\ of\ _6^{12}C\ atom}$$

Formerly atomic masses were referred to the hydrogen atom as 1 and then to the oxygen isotope $_8^{16}\text{O}$ as 16, but in 1960 physicists and chemists agreed to adopt the carbon 12 scale. This was chosen because in measuring atomic masses by mass spectroscopy (to be explained shortly), carbon 12 is a convenient standard for comparison since it forms many compounds. Furthermore, there are only two isotopes of carbon and their proportions vary very little in naturally-occurring carbon.

Atomic masses were first measured to a high degree of accuracy by Aston who established between 1919 and 1927 that (*i*) most elements exhibit isotopy, and (*ii*) most isotopic masses are very nearly but not quite whole numbers.

In Aston's apparatus the deflecting electric and magnetic fields were arranged so that all particles of the same mass, irrespective of their speed, were brought to a line focus. When ions of different masses were present, a series of lines, i.e. a mass spectrum, Fig. 20.19, was obtained on a photographic film. The relative intensities of the lines enabled an estimate to be made of the relative amounts of isotopes. Aston called his instrument a *mass spectrograph*.

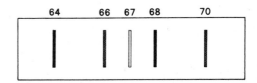

Fig. 20.19

(*b*) *Mass spectrograph.* Many types have been constructed for the accurate determination of atomic mass; the essential features of that due to Bainbridge are shown in Fig. 20.20.

A stream of positive ions, called positive rays is produced by passing electrons from a hot cathode into the gas or vapour to be investigated. The positive ions so formed are directed through slits S_1 and S_2 and emerge as a narrow beam with a range of speeds and specific charges. In the region between S_2 and S_3 crossed uniform electric and magnetic fields are applied. The electric field between P_1 and P_2 exerts a

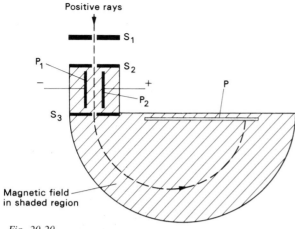

Positive rays

S_1

P_1

S_2

P_2

P

S_3

Magnetic field
in shaded region

Fig. 20.20

force acting to the left on the positive ions. The magnetic field, which acts normal to and into the plane of the diagram, tends to deflect the ions to the right. If Q is the charge of an ion and E the electric field strength, the electric force is EQ. The magnetic force is BQv where v is the velocity of the ion and B the flux density. When the forces are equal

$$BQv = EQ$$

$$\therefore \quad v = \frac{E}{B} \tag{1}$$

The ion will be undeflected and will emerge from the *selector system*, as S_2–S_3 is called, if its velocity equals the ratio E/B. All ions leaving S_3 thus have the same velocity v whatever their specific charge and *velocity selection* is said to have occurred.

Beyond S_3 only the magnetic field acts, the ions describe circular arcs and strike the photographic plate P. For particles of mass M, the radius r of the path is given by

$$BQv = \frac{Mv^2}{r}$$

$$\therefore \quad r = \frac{Mv}{BQ} \tag{2}$$

From (1) and (2)

$$r = \frac{M}{Q} \cdot \frac{E}{B^2}$$

If B and E are constant, r is directly proportional to M (assuming Q is the same for all ions). When ions with different masses are present each set produces a definite line and from their positions the masses can be found. In a mass spectrograph the masses of individual atoms are being measured; by contrast chemical methods give the average atomic mass for a large number of atoms.

The mass spectrograph, which uses photographic detection, is employed primarily for precise mass determinations and can measure the very small differences of mass occurring in nuclear reactions when one nuclide changes to another. The mass spectrometer uses an electrometer as a detector and gives the exact relative abundances, as in gas analysis. Such an instrument is shown in Fig. 20.21. The focusing magnet (here an electromagnet) is on the right surrounding the analyser tube from which cables can be seen going to the various control and recording instruments on the left. In this case the results of the analysis are displayed on a meter or a chart recorder.

Nuclear model of the atom

(a) *Rutherford's nuclear atom*. The nuclear atom is the basis of the modern theory of atomic structure and was proposed by Rutherford in 1911. He had observed that the passage of alpha particles through a very thin metal foil was accompanied by some scattering of the particles from their original direction. Two of his assistants, Geiger and Marsden made a more detailed study of this effect. They directed a narrow beam of alpha particles on to gold foil about 1 μm thick and found that while most of the particles passed straight through, some were scattered appreciably and a very few—about 1 in 8000—suffered deflections of more than 90°. In effect they were reflected back towards the radioactive source.

To account for this very surprising result Rutherford suggested that *all the positive charge and nearly all the mass were concentrated in a very small volume or nucleus at the centre of the atom*. The large-angle scattering of alpha particles would then be explained by the strong electrostatic repulsion to which the alpha particles (also positively charged) are subjected on approaching closely enough to the tiny nucleus; the closer the approach the greater the scattering. We now believe that protons are responsible for the positive charge on the nucleus and protons and neutrons together for the nuclear mass.

Rutherford considered the electrons to be outside the nucleus and at relatively large distances from it so that their negative charge did not act as a shield to the positive nuclear charge when an alpha particle penetrated the atom. The electrons were supposed to move in circular orbits round the nucleus (like planets round

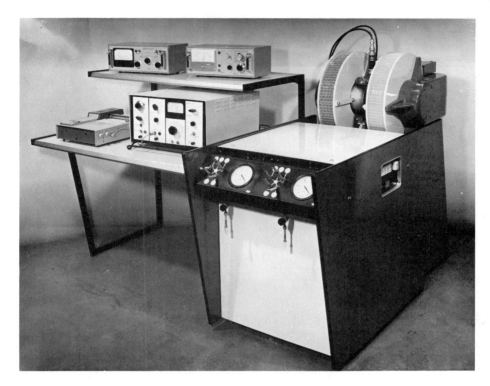

Fig. 20.21

the sun), the electrostatic attraction between the two opposite charges being the required centripetal force for such motion.

On this planetary model it would be reasonable to expect that many physical and chemical properties of atoms could be explained in terms of the number and arrangement of the electrons on account of their greater accessibility.

(*b*) *Geiger-Marsden scattering experiment.* To test his theory Rutherford derived an expression for the number of alpha particles deflected through various angles.

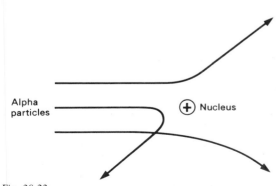

Alpha particles ⊕ Nucleus

Fig. 20.22

The derivation was complex but involved, among other factors, the charge on the nucleus of the scattering atom, the thickness of the foil, the charge, mass and speed of the bombarding alpha particles and was based on the assumption that the repulsive force between the two positive charges obeys an inverse square law. The path predicted for the scattered alpha particle was a hyperbola, Fig. 20.22.

The test was performed by Geiger and Marsden using the apparatus shown in Fig. 20.23. A fine beam of alpha particles from a radioactive source fell on a thin foil of gold, platinum or other metal in an evacuated box. The angular deflection of the particles was measured by using a microscope to observe the scintillations (flashes of light) on a glass screen coated with zinc sulphide. The screen and microscope could be rotated together relative to the foil and source, which were fixed.

Geiger and Marsden spent many hours in a darkened room counting the scintillations for a wide range of angles. Their results completely confirmed Rutherford's deductions and vindicated the use of the inverse square law. (We will see later however (p. 547), that in certain cases where the alpha particle approaches extremely closely to the nucleus the inverse square law no longer holds.)

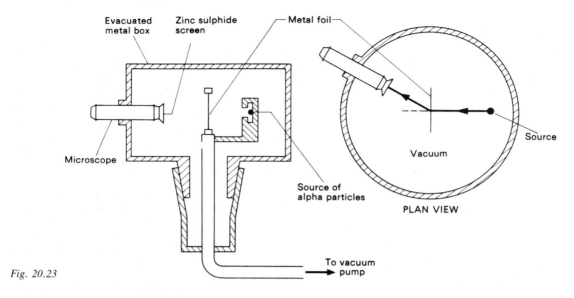

Fig. 20.23

This experiment represents one of the great landmarks in physics. As well as putting the nuclear model on a sound footing, it inaugurated the technique of using high-speed particles as atomic probes. The subsequent exploitation of the technique was responsible for profound discoveries in nuclear physics.

(c) *Nuclear size.* The maximum angle of scattering will occur when the distance between the centres of an alpha particle and the atomic nucleus involved in the encounter is a minimum. This distance gives an upper limit for the sum of the radii of an alpha particle and the nucleus. It is about 10^{-14} m, therefore the radius of the nucleus must be of the order of 10^{-15} m. With the reservation mentioned earlier about the *size* of atomic particles (p. 458), if the radius of the nucleus is compared with that of the atom (about 10^{-10} m), it is evident, since electrons are similar in size to nuclei, that most of the atom is empty. The volume of the nucleus and electrons in an atom is roughly 10^{-12} of the total volume of the atom. The penetration of thin foil, with negligible deflection, by most alpha particles is not surprising; close approaches to the nucleus are rare.

(d) *Moseley and X-ray spectra.* The determination of the charge on the nucleus was a vital problem in the development of models of the atom. It was felt that the position of an element in the periodic table, i.e. its atomic number Z, was equal to the number of protons in the nucleus and so also to the number of extra-nuclear electrons. The measurements made by Geiger and Marsden and later improved upon by Chadwick in 1920 showed that this was approximately true. How-

ever, in 1913 Moseley, a young physicist who was also working with Rutherford, obtained convincing evidence using a different technique.

He carried out a detailed study of X-ray spectra using a Bragg X-ray spectrograph (i.e. a spectrometer with a photographic plate instead of an ionization chamber to detect the X-rays). It was explained earlier (p. 472) that in the production of X-rays, a characteristic line spectrum is superimposed on a continuous spectrum. Moseley measured the frequencies of the lines in the line spectra of nearly 40 elements and found that the frequency of any particular line, such as the K_α of Fig. 20.17 (p. 472), increased progressively from one element to the next in the periodic table. If f is the frequency of this line for a certain element, it is related to a number Z which always works out to be the atomic number, by the expression

$$\sqrt{f} = a(Z - b)$$

where a and b are constants for this particular line.

Moseley wrote, 'We have here a proof that there is in the atom a fundamental quantity which increases by regular steps from one element to the next.' He supported his belief by theoretical arguments that this fundamental quantity was the positive charge on the nucleus.

The Bohr atom

Rutherford's model of the atom, although strongly supported by evidence for the nucleus, is inconsistent with classical physics. An electron moving in a circular orbit round a nucleus is accelerating and according to electromagnetic theory it should therefore emit radiation continuously and thereby lose energy. If this happened the radius of the orbit would decrease and the electron would spiral into the nucleus. Evidently either this model of the atom or the classical theory of radiation requires modification.

In 1913, in an effort to overcome this paradox, Bohr, drawing inspiration from the success of the quantum theory in solving other problems involving radiation and atoms, made two revolutionary suggestions.

(*i*) Electrons can revolve round the nucleus only in certain *allowed orbits* and while they are in these orbits they do not emit radiation. An electron in an orbit has a definite amount of energy. It possesses kinetic energy because of its motion and potential energy on account of the attraction of the nucleus. Each allowed orbit is therefore associated with a certain quantity of energy, called the *energy of the orbit*, which equals the total energy of an electron in it.

(*ii*) An electron can 'jump' from one orbit of energy E_2 to another of lower energy E_1 and the energy difference is emitted as one quantum of radiation of frequency f given by Planck's equation $E_2 - E_1 = hf$.

By choosing the allowed orbits correctly Bohr was able to explain quantitatively why particular wavelengths appeared in the line spectrum of atomic hydrogen and this provided evidence for his ideas concerning how electromagnetic radiation originates in an atom.

Despite its considerable achievements the Bohr atom had certain shortcomings. First, it could not interpret the details of the optical spectra of atoms containing more than one electron. Second, the very arbitrary method of selecting allowed orbits had no theoretical basis and third, it involved quantities, such as the radius of an orbit, which could not be checked experimentally. Nevertheless great credit is due to Bohr for linking spectroscopy and atomic structure and for introducing quantum ideas into atomic theory.

Bohr's model of the atom has been superseded by a new theory based on *wave mechanics* in which there is no need to make assumptions because they give correct results. But whereas the Bohr atom was easily visualized and involved fairly simple mathematics, its successor is abstract and the mathematics more difficult.

For the present it is sufficient to say that while wave mechanics preserves the general idea of a hollow, nuclear atom it discards the Bohr picture of electrons moving in allowed orbits. However, the essential characteristic of Bohr's orbits, i.e. definite energy values, is retained.

Energy levels in atoms

Wave mechanics permits the electrons in an atom to have only certain energy values. These values are called the *energy levels* of the atom. They are not something we can observe in the usual sense but, as we will see shortly, there is fairly direct experimental evidence to support our belief in their existence. Any theory of atomic structure must be able to explain how they arise; the wave mechanical justification will be given later.

The levels can be represented by horizontal lines, arranged one above the other to form an energy level diagram (or a ladder, of unequally-spaced rungs), each line indicating by its position a particular energy value. Every *atom* has a characteristic set of energy levels whose values can be found experimentally or calculated using wave mechanics. Whilst an electron is permitted to pass from one level to another by gaining or losing energy, it is not allowed to have an amount of energy that would put it between two levels. An atom can thus only accept 'parcels' of energy of certain definite sizes, i.e. its energy is *quantized*.

All levels have negative energy values because the energy of an electron at rest outside an atom is taken as zero and when the electron 'falls' into the atom, energy is lost as electromagnetic radiation (compare the loss of p.e. when a body falls in the earth's attractive gravitational field). In effect the electron is passing from a higher to a lower energy level and the lower the level the larger the negative energy value. The most stable or *ground state* of the atom is the condition in which every electron is in the lowest energy state available. An atom is in an *excited state* when an electron is in a state of energy above that of a state which is unoccupied.

Energy level diagrams can be drawn for every atom—that for atomic hydrogen is shown in Fig. 20.24*a*. In this case the lowest level has energy—13.6 eV and is the one normally occupied by the single electron of a hydrogen atom. Above this state are the excited states to which the hydrogen atom may be raised by absorbing the correct amount of energy. If the energy absorbed is sufficient to allow the electron to escape from the atom, the latter becomes ionized; for hydrogen the ionization energy is 13.6 eV $(21.8 \times 10^{-19} \text{ J})$.

Electrons in the lower energy levels are held strongly by the nucleus. Those in higher levels need less energy to escape and are more loosely held in the atom. We may think of them as being near the outside of the electron cloud which surrounds the nucleus and partly screened from the attraction of the positive nuclear charge by inner, lower energy electrons.

Detailed study of the periodic variation of the chem-

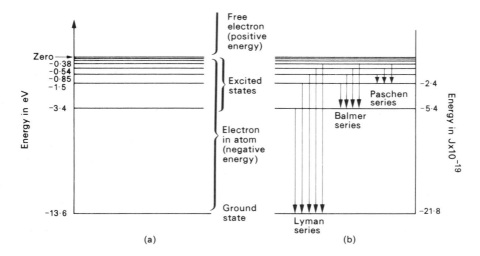

Fig. 20.24   (a)   (b)

ical properties and the ionization energies (Fig. 20.27, p. 482) of the elements and of their optical and X-ray spectra suggests that the electrons in an atom fall into groups—called *shells*. All the energy states which the electrons in one group can occupy have *about* the same value. Each shell is given a *principal quantum number n* and they are labelled in ascending order of energy by the letters K, L, M, etc.

Wave mechanics indicates that the number of electrons that can be accommodated in a shell is $2(n)^2$. Thus the K-shell ($n = 1$) can take at most $2(1)^2 = 2$ electrons, the L-shell ($n = 2$) can take a maximum of $2(2)^2 = 8$ electrons and so on. For the first five shells the numbers are 2, 8, 18, 32, 50. Normally the electrons occupy the lowest energy shells first, i.e. starting with K, but this is not always so. Thus whilst argon with 18 electrons has its full complement of 2 and 8 electrons in the K- and L-shells respectively and 8 electrons in the incomplete M-shell, in potassium (atomic number 19) the electronic configuration is 2–8–8–1, i.e. the extra electron is in the N-shell although the M-shell can take 10 more electrons.

Evidence of energy levels

(*a*) *Optical line spectra.* A line in an optical emission spectrum indicates the presence of a particular frequency (and wavelength) of light and is considered to arise from the loss of energy which occurs in an excited atom when an electron 'jumps', directly or in stages, from a higher to a lower level. The frequency *f* of the quantum of electromagnetic radiation emitted in a transition between levels of energies E_1 and E_2 ($E_2 > E_1$) is given

by

$$E_2 - E_1 = hf$$

where *h* is Planck's constant. The transitions for some of the lines in the visible spectrum of atomic hydrogen, called the *Balmer series*, are shown in Fig. 20.24*b*. An electronic transition from the -2.4×10^{-19} J level to the -5.4×10^{-19} J level represents an energy loss to the atom of $[-2.4 - (-5.4)] \times 10^{-19} = 3.0 \times 10^{-19}$ J. The wavelength λ of the radiation emitted is given by

$$3.0 \times 10^{-19} = hf = \frac{hc}{\lambda}$$

where *c* is the speed of light = 3.0×10^8 m s⁻¹. Hence

$$\lambda = \frac{hc}{3.0 \times 10^{-19}} = \frac{6.6 \times 10^{-34} \times 3.0 \times 10^8}{3.0 \times 10^{-19}} \text{ m}$$

$$= 6.6 \times 10^{-7} \text{ m}$$

$$= 0.66 \ \mu\text{m}$$

The spectrum of atomic hydrogen contains a line of this wavelength (Fig. 18.36, p. 404). The lines of the *Lyman series* are in the ultraviolet and of the *Paschen* in the infrared. Thus, assuming the energy levels for hydrogen have been calculated from wave mechanics, the optical spectrum of hydrogen can be explained.

(*b*) *X-ray line spectra.* The energy of a light photon is a few electron-volts, that of an X-ray photon is several kiloelectron-volts. This suggests that whereas optical line spectra are due to transitions of loosely held electrons in higher energy levels, X-ray line spectra originate from electronic transitions in the lowest energy levels which have their full complement of electrons. They are only produced when materials are bombarded by high-energy electron beams (of the order of thousands of electron-volts) which are able to penetrate deep into atoms and displace electrons from very 'deep' energy levels. The subsequent fall of an electron from a higher level into one of these gaps in an otherwise complete, but appreciably lower energy level, causes the emission of a high-energy X-ray photon.

The K-series of X-ray lines is produced when an electron is knocked out of the lowest or K-shell. The return of the same or another electron to the gap in the K-shell causes the emission of a line of the K-series. If the electron falls from the L-shell the K_α line occurs, when from the M-shell we have the K_β line and so on, Fig. 20.25. Similarly the L-series arises when electrons return to vacancies in the L-shell. This series is excited by smaller energies than the K-series because the L-electrons are less strongly held; the X-rays emitted have lower frequencies and longer wavelengths and are less penetrating.

The progressive increase of frequency with increasing atomic number, in say the K_α line, is due to small energy differences in the K- and L-shells of different atoms. This arises from the increasing positive charge on the nucleus and it makes possible the determination of atomic numbers from a study of X-ray line spectra (p. 478; Moseley).

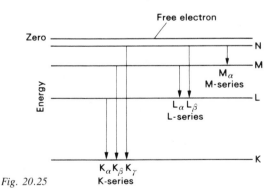

Fig. 20.25

(c) *Electron collision experiments.* Direct evidence for the existence of energy levels is provided by experiments, first successfully performed in 1914 by Franck and G. Hertz (nephew of H. Hertz), in which electrons have collisions with the atoms of a gas at low pressure. Whilst a free electron completely detached from an atom can be accelerated to any energy, these experiments lead to the conclusion that the electrons *in an*

atom can have only *certain* values, i.e. those permitted by its energy levels.

When an electron has an encounter with a gas atom one of three things can happen:

(*i*) an elastic collision occurs in which the bombarding electron, being much less massive than the gas atom, suffers only a slight loss of kinetic energy (see p. 557, Qn. 10),

(*ii*) an inelastic collision occurs in which an electron in a gas atom gains *exactly* the amount of energy it requires to reach a higher energy level. Excitation has then occurred and the gas atom is in an excited state. The collision is inelastic since the kinetic energy lost by the bombarding electron does not reappear as kinetic energy of the gas atom but is emitted as electromagnetic radiation when the gas atom returns to a lower state or the ground state, usually after about 10^{-9} s. The p.d. through which the bombarding electron has to be accelerated from rest to cause excitation is called the *excitation potential* and since every atom has many energy levels, there are numerous excitation potentials characteristic of a particular atom,

(*iii*) an inelastic collision occurs in which an electron in a gas atom gains enough energy to escape from the atom, so causing ionization. The accelerating p.d. is then the *ionization potential* of the atom.

The principle of a Franck-Hertz type of experiment is illustrated in Fig. 20.26a. Electrons emitted by the hot cathode C in a tube containing gas at a low pressure are accelerated by a positive potential V_1 on the wire mesh G, called the grid. When V_1 just exceeds the small negative potential V_2 which the anode A has with respect to G, electrons reach A and a small current is indicated on the galvanometer. As V_1 is increased the current increases. During this phase, PQ in Fig. 20.26b, the bombarding electron energies are small and all collisions between electrons and gas atoms are elastic. At a certain value of V_1, called the *first excitation potential*, some electrons have inelastic collisions with gas atoms near G and raise them to their first excited energy level. The negative potential V_2 ensures that the electrons, having lost their kinetic energy, are carried back to G. The anode current therefore falls abruptly, shown by QR. Further increase of V_1 allows even those electrons which have inelastic collisions to overcome V_2 and reach A. The galvanometer reading increases again, RS.

Other effects may be obtained at higher values of V_1 depending on the tube and circuit conditions. Thus it is possible to find the ionization potential of the gas, i.e. the potential at which the bombarding electrons have enough energy to cause gas molecules to be ionized.

Collision experiments, as just outlined, indicate that

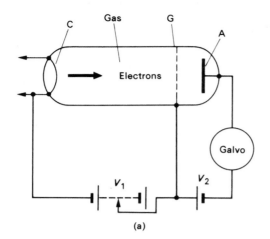

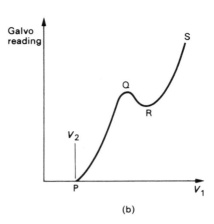

Fig. 20.26   (a)   (b)

an atom cannot absorb any amount of energy but only certain definite amounts which are determined by its energy levels.

Franck-Hertz type tubes that give reliable results are difficult to make and so are expensive. An experiment with a commercial tube containing the inert gas xenon is described in Appendix 18) and gives rough values of the ionization potential and one excitation potential.

The ionization potentials and energies of many gases and vapours have been investigated. The graph of Fig. 20.27 shows the variation of ionization energy with atomic number for the first thirty elements in the periodic table. A periodicity, like that displayed by other properties, is evident. The small ionization energies of the alkali metals suggests they have a loosely-held electron in a higher energy level. By contrast the inert gases must have very stable electronic structures since they require most energy for ionization.

(d) *Spectrum of mercury vapour.* Electron collision experiments with mercury vapour show that there is an energy difference of 4.9 eV (7.8×19^{-19} J) between two of the energy levels in an isolated mercury atom. We might therefore expect photons with this energy to be emitted by a mercury vapour lamp. The wavelength of such radiation is 2.5×10^{-7} m, i.e. 0.25 μm (check this); it is in the ultraviolet region as may be shown using the arrangement of Fig. 20.28a (*in which the lamp should be well-screened from observers*). The ultraviolet lines A and B in Fig. 20.28b appear only on the fluorescent paper (the visible lines are also on the white paper) and are cut off if a piece of glass is held in front of the lamp. Assuming the wavelength of the green line C is 5.5×10^{-7} m (it can be measured readily using a transmission diffraction grating of known spacing), the wavelength λ of the ultraviolet line A is given approximately by

$$\frac{\lambda}{5.5 \times 10^{-7}} = \frac{AA}{CC}$$

If *cold* mercury vapour is puffed *gently* into the air from a polythene bottle held in front of the reflection grating, both A and B fade or disappear. What does this suggest about B? *Note. Mercury vapour is highly poisonous and on no account must the mercury be warmed.*

Wave-particle duality of matter

The wave-particle nature of electromagnetic radiation discussed previously (p. 465), led de Broglie (pronounced 'de Broy') to suggest in 1923 that matter might also exhibit this duality and have wave properties. His

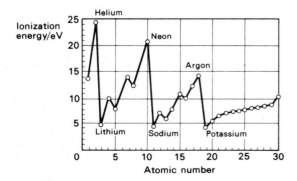

Fig. 20.27

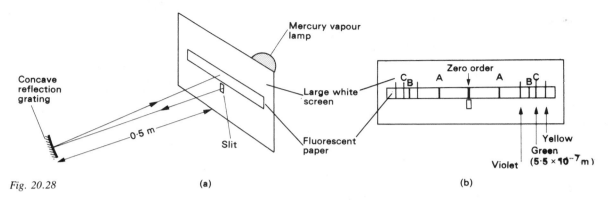

Fig. 20.28   (a)   (b)

ideas can be expressed quantitatively by first considering radiation. A photon of frequency f and wavelength λ has, according to the quantum theory, energy $E = hf = hc/\lambda$ where c is the speed of light and h is Planck's constant. By Einstein's energy-mass relation (p. 548) the equivalent mass m of the photon is given by $E = mc^2$. Hence

$$\frac{hc}{\lambda} = mc^2$$

$$\therefore \quad \lambda = \frac{h}{mc}$$

By analogy de Broglie suggested that a particle of mass m moving with speed v behaves in some ways like waves of wavelength λ given by

$$\lambda = \frac{h}{mv}$$

Calculation shows that electrons accelerated through a p.d. of about 100 V should be associated with *de*

Broglie or *matter waves*, as they are called, having a wavelength of the order of 10^{-10} m. This is about the same as for X-rays and it was suggested that the conditions required to reveal the wave nature of X-rays might also lead to the detection of electron waves. At first de Broglie's proposal was no more than speculation, but within a few years a variety of experiments proved beyond dispute that moving particles of matter had wave-like properties associated with them.

Interference and diffraction patterns can be obtained with electrons. Fig. 20.29 shows interference fringes produced by Young's double-slit type experiments (*a*) with light and (*b*) with a stream of electrons. An arrangement for producing electron interference is given in Fig. 20.30*a*. The stream of electrons is split into two by a very thin wire between two metal plates. When the wire is made a few volts positive with respect to the plates, the electron streams passing on each side of the wire are brought closer together and overlap on the photographic film as if they had come from two sources, Fig. 20.30*b*. The interference pattern is so

Fig. 20.29

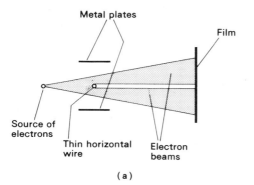

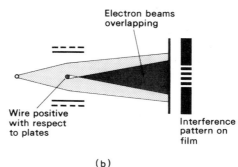

Fig. 20.30

(a)

(b)

small that it has to be magnified by an electron microscope and then by an optical enlarger to give fringes of the width shown in Fig. 20.29*b*.

Electron diffraction may be shown in a school laboratory using the arrangement of Fig. 20.31 in which a beam of electrons strikes a thin film of graphite on a metal grid just beyond a hole in the anode. Diffraction effects have been obtained with streams of protons, neutrons and alpha particles, but it is evident from de Broglie's equation that the greater the mass of the moving particle, the smaller is the associated wavelength and so the more difficult does detection become. Ordinary objects have extremely small wavelengths.

The idea of matter waves is useful when we consider the various kinds of microscopes. The resolving power of any microscope increases as the wavelength of whatever is used to 'illuminate' the object decreases. Thus electron waves have a smaller wavelength than light waves and so an electron microscope reveals much more detail. The field ion microscope gives even greater resolution because the waves associated with the helium ions used have an even shorter wavelength.

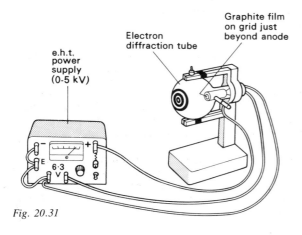

Fig. 20.31

Wave mechanics

(*a*) *Wave-particle duality.* In the early decades of the present century physicists faced an apparently puzzling situation. How could matter and radiation have *both* wave-like and particle-like properties? It seemed to be a complete contradiction.

However, to some extent the dilemma was of man's own making for originally an attempt was made to explain this dual behaviour in terms of the real waves and real particles of everyday life. This was a perfectly reasonable thing to do when faced with phenomena which appeared to have similarities. But perhaps it was not altogether surprising that ideas which helped to deal with the large-scale world of water waves and cricket balls were not completely suited to the atomic world of photons and electrons. At any rate it came to be realized that part of the dilemma, at least, arose from the use of models based on our experience of the macroscopic and that such analogies might not be valid when pushed too far.

It is now appreciated that a scientific model is an aid to understanding and not necessarily a true description. In the same way a map showing only towns and roads is no more a complete representation of a country than is one showing only physical features; each describes one aspect and neither is wrong. Both models are necessary for an adequate description of the behaviour of matter and radiation: they are complementary and not contradictory. The sensible thing to do is to use one or other model when it is appropriate and helpful.

Such arguments help to make the duality dilemma more acceptable but they do not explain the nature of the 'particles' and 'waves'. The present view is that the 'waves' are not waves of moving matter or varying fields but are probability waves. Thus the wave-like behaviour of a moving electron is considered to be due to the fact that the chance of it being found in an element of volume depends on the intensity (i.e. the square of the amplitude) at the point of the electron wave associated with it. Where this is high we are likely to have a high electron density. In the same way an electromagnetic wave, although undoubtedly connected with electric and magnetic fields, is regarded as consisting of photons (i.e. bundles of energy) whose probable locations are given by the intensity of the wave. This outlook forms the basis of wave mechanics which, as the name implies, considers that the motion of a 'particle' is determined by a wave equation. For 'particles' such as electrons and photons it predicts quite a different behaviour from that given by the particle mechanics of Newton but agrees with the latter for macroscopic bodies. It deals in probabilities and not certainties.

(b) *Wave-mechanical model of the atom.* The real objection to the Bohr model of the atom is that it pin-points an electron in a definite orbit and takes no account of its wave-like aspect. The wave-mechanical model requires advanced mathematics but we can attempt to explain it by making analogies with real waves.

In a vibrating string fixed at both ends, waves are reflected to and fro and a stationary wave system is set up having nodes and antinodes. In its simplest mode of vibration the string has a node at each end and an antinode in the centre but it can vibrate with two, three or any integral number of loops. The vibration is restricted to those modes having a complete number of loops and each mode produces a note of different frequency.

Similarly we can imagine an electron, confined to an atom by the attractive force exerted on it by the nucleus, being associated with electron waves which are reflected when they reach the boundary of the atom. A stationary wave system is established inside the atom and can vibrate in various modes. Each mode corresponds to a particular frequency and, therefore, to a particular energy level of the atom. When an electron wave changes its mode of vibration, the energy difference between the levels is emitted (or absorbed) as radiation. This argument, although much too realistic, enables us to see how wave mechanics can account for the existence of energy levels.

QUESTIONS

Electrons

1. What is an *electron-volt*? Assuming that the charge on an electron is -1.60×10^{-19} C express one electron-volt in terms of another unit.

Calculate the kinetic energy and velocity of protons after being accelerated from rest through a potential difference of 2.00×10^5 V. (Assume that the mass of a proton = 1.67×10^{-27} kg.) (*J.M.B.*)

2. How may cathode rays be produced? What are their chief properties?

An electron starts from rest and moves freely in an electric field whose intensity is 2.4×10^3 V m^{-1}. Find (a) the force on the electron, (b) its acceleration and (c) the velocity acquired in moving through a potential difference of 90 V. (The charge on an electron = 1.6×10^{-19} C and the mass of an electron = 9.1×10^{-31} kg.) (*W.*)

3. What do you understand by an *electron*?

Electrons in a certain cathode-ray tube are accelerated through a potential difference of 2.0 kV between the cathode and the screen. Calculate the velocity with which they strike the screen. Assuming they lose all their energy on impact and given that 10^{12} electrons pass per second, calculate the power dissipation.
(Charge on the electron = 1.6×10^{-19} C; mass of electron = 9.1×10^{-31} kg.) (*O. and C. part qn.*)

4. Describe a method for measuring the ratio of the charge to mass (e/m) for an electron.

Calculate (a) the speed achieved by an electron accelerated in a vacuum through a p.d. of 2.00×10^3 V and (b) the magnetic flux density required to make an electron travelling with speed 8.00×10^6 m s^{-1} traverse a circular path of diameter 10.0×10^{-2} m. Take e/m for an electron as 1.76×10^{11} C kg^{-1}.

5. Give an account of a method by which the charge associated with an electron has been measured.

Taking this electronic charge to be -1.60×10^{-19} C, calculate the potential difference in volts necessary to be maintained between two horizontal conducting plates, one 5.00×10^{-3} m above the other, so that a small oil drop, of mass 1.31×10^{-14} kg with two electrons attached to it, remains in equilibrium between them. Which plate would be at the positive potential? ($g = 9.81$ N kg^{-1}) (*L.*)

6. Two plane parallel conducting plates 1.50×10^{-2} m apart are held horizontally, one above the other, in air. The upper plate is maintained at a positive potential of 1.50×10^3 V while the lower plate is earthed. Calculate the number of electrons which must be attached to a small oil drop of mass 4.90×10^{-15} kg if it remains stationary in the air between the plates. (Assume that the density of air is negligible in comparison with that of oil.)

If the potential of the upper plate is suddenly changed to -1.50×10^3 V what is the initial acceleration of the charged drop? Indicate, giving reasons, how the acceleration will change. (The charge of an electron is 1.60×10^{-19} C and g is 9.81 N kg^{-1}.) (*J.M.B.*)

Photoelectric effect

Speed of light $= 3.0 \times 10^8$ m s^{-1}
Planck's constant $= 6.6 \times 10^{-34}$ J s
Mass of electron $= 9.1 \times 10^{-31}$ kg
Electronic charge $= 1.6 \times 10^{-19}$ C

7. (*a*) If a surface has a work function of 3.0 eV, find the longest wavelength light which will cause the emission of photoelectrons from it.
(*b*) What is the maximum velocity of the photoelectrons liberated from a surface having a work function of 4.0 eV by ultraviolet radiation of wavelength 0.20 μm?

8. When light of wavelength 0.50 μm falls on a surface it ejects photoelectrons with a maximum velocity of 6.0×10^5 m s^{-1}. Calculate (*a*) the work function in electron-volts, and (*b*) the threshold frequency for the surface.

9. What do you understand by the *quantum theory*? Describe the evidence provided by experiments on the photoelectric effect in favour of this theory.
Calculate the energy of the photons associated with light of wavelength 3.0×10^{-7} m. Give your answer in joules. (Planck's constant $= 6.6 \times 10^{-34}$ J s; speed of light $= 3.0 \times 10^8$ m s^{-1}.) (*C.*)

10. When light is incident on a metal plate electrons are emitted only when the frequency of the light exceeds a certain value. How has this been explained?
The maximum kinetic energy of the electrons emitted from a metallic surface is 1.6×10^{-19} J when the frequency of the incident radiation is 7.5×10^{14} Hz. Calculate the minimum frequency of radiation for which electrons will be emitted. Assume that Planck's constant $= 6.6 \times 10^{-34}$ J s. (*J.M.B.*)

X-rays

11. If the p.d. applied to an X-ray tube is 20 kV, what is the speed with which electrons strike the target? Take the specific charge of the electron as 1.8×10^{11} C kg^{-1}.

12. Describe the modern hot cathode X-ray tube, and give a diagram of a circuit suitable for its operation.
Discuss briefly the energy conversions that take place in the tube.
If the p.d. across the tube is 1.5×10^3 V, and the current 1.0×10^{-3} A, find (*a*) the number of electrons crossing the tube per second, and (*b*) the kinetic energy gained by an electron traversing the tube without collisions. (Take the electronic charge *e* to be 1.6×10^{-19} C.) (*O.*)

13. What is the minimum wavelength of the X-rays produced when electrons are accelerated through a potential difference of 1.0×10^5 V in an X-ray tube? Why is there a minimum wavelength? (*e* $= 1.6 \times 10^{-19}$ C; *h* $= 6.6 \times 10^{-34}$ J s; *c* $= 3.0 \times 10^8$ m s^{-1}.)

Structure of atom

14. The atomic nucleus may be considered to be a sphere of positive charge with a diameter very much less than that of the atom. Discuss the experimental evidence which supports this view. Describe briefly how this experimental evidence was obtained. (*J.M.B.*)

15. How are X-rays produced, and how are their wavelengths determined?
Discuss briefly the origin of the lines in the spectrum produced by an X-ray tube that are characteristic of the target metal.
Give a brief account of Moseley's work and the part it played in establishing the idea of atomic number. (*O.*)

16. What are the chief characteristics of a line spectrum? Explain how line spectra are used in analysis for the identification of elements.
Fig. 20.32 which represents the lowest energy levels of the electron in the hydrogen atom, specifies the value of the principal quantum number *n* associated with each state and the corresponding value of the energy of the level, measured in electron-volts. Work out the wavelengths of the lines associated with the transitions A, B, C and D marked in the figure. Show that the other transitions that can occur give rise to lines that are either in the ultraviolet or the infrared regions of the spectrum. (Take 1 eV to be 1.6×10^{-19} J; Planck's constant *h* to be 6.5×10^{-34} J s; and *c*, the velocity of light in vacuo, to be 3.0×10^8 m s^{-1}.) (*O.*)

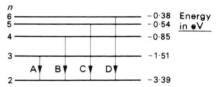

Fig. 20.32

17. Some of the energy levels of the mercury atom are shown in Fig. 20.33.
(*a*) How much energy in electron-volts is required to raise an electron from the ground state to each of the levels shown?
(*b*) What is the ionization energy of mercury?
(*c*) If a mercury vapour atom in the ground state has a collision with an electron of energy (i) 5 eV and (ii) 10 eV, how much energy might be retained by the electron in each case?
(*d*) What would happen to a photon of energy (i) 4.9 eV and (ii) 8 eV, which has a collision with a mercury atom?

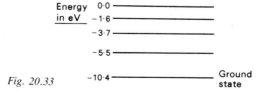

Fig. 20.33

21 Electronics

Introduction

The striking advances in electronics during recent years have been due to the development of semiconductor devices such as junction diodes, transistors and integrated circuits. On account of their small size, low power requirements and very long life they have largely replaced thermionic valves in modern electronic equipment. They are, however, damaged by too high temperatures and p.d.s that are too great or of the wrong polarity.

First we shall consider the cathode-ray oscilloscope—an important tool in electronics.

Cathode-ray oscilloscope

The cathode-ray oscilloscope (CRO) consists of a cathode-ray tube and associated circuits. It has four main parts, as described in (*a*) to (*d*) below.

(*a*) *Electron gun.* This is an electrode assembly for producing a narrow beam of cathode rays. A typical CRO tube is shown in Fig. 21.1 connected to a potential divider which supplies appropriate voltages from an e.h.t. power supply.

The gun comprises an indirectly heated cathode C, a cylinder G called the *grid* and two anodes A_1 and A_2. G

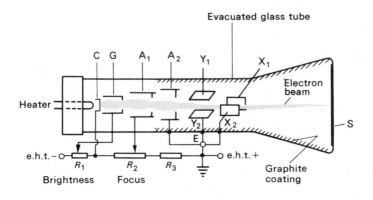

Fig. 21.1

has a negative potential (or bias) with respect to C and controls the number of electrons reaching A_1 from C. R_1 determines the potential of G and thus acts as a *brightness* control.

A_1 and A_2 are metal discs or cylinders with central holes and both have positive potentials relative to C, A_2 more so than A_1. They accelerate the electrons to a high speed down the tube (which is highly evacuated) and their shapes and potentials are such that the electric fields between them focus the beam to a small spot on the fluorescent screen S at the end of the tube. Adjustment of the potential of A_1 by R_2 provides a *focus* control. A_1 and A_2 are an 'electron lens' system. Typical potentials for a small tube are 1000 V for A_2, 200 to 300 V for A_1 and -50 V to zero for G.

There must be a return path to A_2 (and hence e.h.t.+) for electrons reaching S. Its provision depends on an effect known as *secondary emission* in which electrons, called *secondary electrons,* are emitted from a surface struck by high-speed electrons. The secondary electrons from S are collected by a coating of graphite on the inside of the tube which is connected to A_2. S gains bombarding electrons but loses about the same number of secondary ones and has a potential roughly equal to that of A_2. The electron beam therefore travels with *constant speed* between A_2 and S. In addition, the coating shields the beam from external electric fields. A mumetal screen round the tube protects it from stray magnetic fields.

To prevent earthed objects (e.g. the experimenter) near the screen affecting the beam, A_2 (and the coating) are usually earthed. This means e.h.t.+ is also at earth potential, Fig. 21.1, and the other electrodes become negative with respect to A_2. The electrons are still accelerated through the same p.d. between A_2 and C.

(b) Deflecting system. The beam from A_2 passes first between a pair of horizontal metal plates, the Y-plates whose electric field causes vertical deflection and then between two vertical plates, the X-plates that cause horizontal deflections when a p.d. is applied to them.

To minimize the effect of electric fields between the deflector plates and other parts of the tube (which might cause defocusing), one of each pair of plates is at the same (earth) potential as A_2. In practice X_2, Y_2 and A_2 are connected internally and brought out to a single terminal marked E (for earth). Deflecting p.d.s are then applied to Y_1 (marked Y or input) and E or to X_1 (marked X) and E.

The deflection sensitivity for p.d.s applied directly to the deflector plates is typically 50 V cm^{-1}, which is not high. X and Y deflection amplifiers are usually built into the CRO to amplify p.d.s that are too small to give measurable deflections before they are connected to the plates.

X and Y shift controls are used to move the spot 'manually' in the X or Y directions respectively. They apply a positive or negative voltage to one of the deflecting plates according to the shift required.

(c) Fluorescent screen. The inside of the wide end of the tube is coated with a *phosphor* which emits light when struck by fast-moving particles. This may occur as fluorescence, i.e. the emission stops with the bombardment, and phosphorescence, when there is 'afterglow', possibly for a few seconds. Zinc sulphide is commonly used in cathode-ray tubes; it emits blue light and has no afterglow.

(d) Time base. When the variation of a quantity with time is to be studied, an alternating p.d. representing the quantity is applied to the Y-plates via the vertical or Y-amplifier and the X-plates are connected via the horizontal or X-amplifier to a circuit in the CRO, called the *time base,* which generates a sawtooth p.d. like that in Fig. 21.2. The 'sweep' AB must be linear so that the deflection of the beam is proportional to time and causes the spot to travel across the screen from left to right at a speed which is steady but which can be varied by the time base control(s) to suit the frequency of the alternating input p.d.

To maintain a stable trace on the screen each horizontal sweep must start at the same point on the waveform being displayed. This is done by feeding part of the input signal to a trigger circuit that gives a pulse (at a chosen point on the input signal selected by the TRIG LEVEL control) which is used to start the sweep of the time base sawtooth, i.e. it 'triggers' the time base and initiates the horizontal motion of the spot at the left-hand side of the screen. However, automatic triggering by the input can be obtained (and is used for most applications) by selecting the AUTO position on the TRIG LEVEL control.

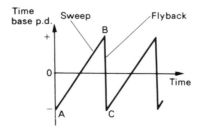

Fig. 21.2

A trace appears only during the sweep part of the

sawtooth p.d. since when the time base is not triggered by an input signal, the electron beam, in one arrangement, is deflected so that it does not reach the screen. The deflection is achieved by a set of plates (not shown in Fig. 21.1) at suitable potentials immediately beyond A_1. When the time base is triggered a positive 'unblanking' pulse, derived from the time base, is applied to the set of plates and allows the electrons to travel along the tube to the screen.

(e) *Mono and PDA tubes.* Two desirable properties of a CRO are high light output from the phosphor (especially for viewing high frequency signals) and good deflection sensitivity. The former requires the electrons to be accelerated through a large p.d. so that they strike the screen at high speed. The latter is attained when the electrons travel slowly between the deflecting plates (see p. 460).

The cathode-ray tube described above is a mono-accelerator type (*mono* for short) in which the electrons are accelerated to their final high speed in the electron gun, i.e. *before* they reach the deflection plates; this is not conducive to high deflection sensitivity. In the post-deflection acceleration tube (PDA for short), the electrons leave the gun and enter the deflection region at a comparatively low speed and the main acceleration is achieved *after* deflection. The deflection factor is, as a result, affected less adversely than in a mono tube with the same overall accelerating p.d.

A mono tube usually has a conducting graphite coating (as stated earlier) on the inside of the tube from the deflection plates to the screen. In some early PDA tubes this coating was split into bands by insulating gaps, and each band was at a different potential. In many present-day PDA tubes a continuous helically-wound resistive material is used instead of the graphite bands at the final accelerating 'anode'.

Fig. 21.3a is a block diagram of a CRO; Fig. 21.3b and c show single and double beam CROs respectively.

Some uses of the CRO

(a) *As a voltmeter.* The CRO can be used as a d.c./a.c. voltmeter by connecting the p.d. to be measured across the Y-plates, with the time base off and

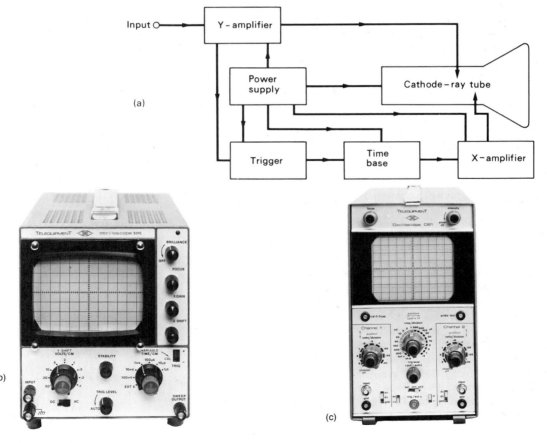

Fig. 21.3

the X-plates earthed. The spot is deflected by d.c.; a.c. causes it to move up and down and if the motion is fast enough (e.g. at 50 Hz) a vertical line is observed.

In the CRO of Fig. 21.3b the 'gain' control (marked VOLTS/CM) of the Y-deflection amplifier is calibrated. If it is on the '50' setting a deflection of 1 cm would be given by a 50 V d.c. input; a line 1 cm long would be produced by an a.c. input of 50 V peak-to-peak (i.e. peak voltage $V_0 = 25$ V and r.m.s. voltage $V_{r.m.s.} = V_0/\sqrt{2} = 18$ V). On the '0.1' setting the gain of the amplifier is greatest and small p.d.s can be measured, 0.1 V causing a deflection of 1 cm. (The calibration can be checked by applying known input p.d.s.) The use of the CRO as a voltmeter is shown in Fig. 14.6 (p. 285) and Fig. 14.29 (p. 295).

Among the advantages of the CRO as a voltmeter are:

(i) the electron beam behaves as a pointer of negligible inertia, responding instantaneously and having a perfect 'dead-beat' action,

(ii) direct and alternating p.d.s can be measured, the latter at frequencies of several megahertz,

(iii) it has an almost infinite resistance to d.c. and a very high impedance to a.c. and so the circuit to which connection is made is little affected, and

(iv) it is not damaged by overloading.

(b) *Displaying waveforms.* The signal to be examined is connected to the Y-plates and the time base to the X-plates. As the spot is drawn horizontally across the screen by the time base, it is also deflected vertically by the alternating signal p.d. which is therefore 'spread out' on a time axis and its waveform displayed. The waveform will be a faithful representation, free from distortion, only if the time base has a linear sweep. When the time base has the same frequency as the input, one complete wave is formed on the screen; if it is half that of the input, two waves are displayed. The CRO is used to display waveforms in the circuits of Fig. 15.34 (p. 327) and Fig. A11.1 (p. 570).

(c) *Measuring time intervals.* This may be done if the CRO has a calibrated time base like that in Fig. 21.3b. The coarse time-base control is marked TIME/CM and gives six pre-set sweep speeds. The calibration only holds when the fine time-base control (marked VARIABLE) is in the CAL position and the X-GAIN control is fully anticlockwise.

If the TIME/CM control is on 10 ms, the spot takes 10 ms to move 1 cm and so travels 10 cm on the screen graticule in 100 ms (0.1 s). How could the calibration of the time base be checked using a 50 Hz a.c. mains signal?

A CRO was used to estimate a very small time interval in the measurement of the speed of sound in air and in a metal rod (pp. 376 and 378).

(d) *Measurement of phase relationships.* If two sinusoidal p.d.s of the same frequency and amplitude are applied simultaneously to the X- and Y-plates (time base off), the electron beam is subjected to two mutually perpendicular simple harmonic motions and the trace produced on the screen by the spot depends on the phase difference between the p.d.s. Fig. 21.4 shows the traces obtained using 50 Hz supplies (of a few volts) from, say, a signal generator and a step-down transformer. In general, the resultant motion of the spot is an ellipse except when the phase difference is 0°, 90° or 180°.

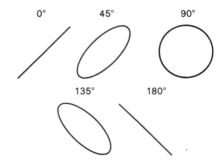

Fig. 21.4

The method can be used to find the phase relationships between p.d.s and currents in a.c. circuits if a double beam CRO is not available.

(e) *Comparison of frequencies.* When two p.d.s of different frequencies f_x and f_y are applied to the X- and Y-plates (time base off), more complex figures than those in Fig. 21.4 (known as *Lissajous' figures*) are obtained. Some are shown in Fig. 21.5. In any particular case the frequency ratio can be found from inspection by imagining a horizontal and a vertical line to be drawn at the top and side of the trace. It is given by

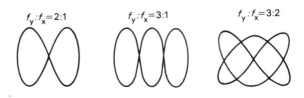

Fig. 21.5

$$\frac{f_y}{f_x} = \frac{\text{number of loops touching horizontal line}}{\text{number of loops touching vertical line}}$$

The pattern is stationary when f_y/f_x is a ratio of whole numbers and the phase difference is constant.

When f_y/f_x is large, comparison by Lissajous' figures is difficult and in such cases a *circular time base* method can be used in which one alternating p.d. of known frequency (say 50 Hz) applied to a 'phase-splitting' R-C series circuit, produces a circular trace on the CRO. (For a 2 V, 50 Hz input and a CRO gain setting of 1 V, convenient values of R and C are 10 kΩ and 0.47 μF respectively.) Three connections from the R-C circuit to the Y-, E- and X-terminals are necessary, Fig. 21.6a. If an unknown, much higher frequency (from a signal generator), is applied to the Z- and E-terminals on the CRO the brightness of the trace varies and a broken circle is obtained, Fig. 21.6b. Here the frequency ratio is 8:1. Why? (The Z-input on the CRO is joined to the grid of the tube via an isolating capacitor, and an external p.d. applied to it brightens or blacks-out the trace, depending on whether it is positive or negative. The beam is thus 'intensity modulated'.)

The method can be used to check the calibration of a signal generator at frequencies up to 2 kHz against the 50 Hz mains supply. If the generator has a square wave output it should be used in preference to the sine wave output. Why?

energy. The present theory of conduction, which is based on wave mechanics and is called the *band theory*, permits them to have only certain ranges of energies.

We saw previously that an *isolated* atom, as in a gas, has a characteristic set of well-defined energy levels that electrons can occupy. When two similar atoms are brought together, the electric fields due to their charges overlap and mutual interaction occurs. As a result, according to the band theory, each of the higher energy levels is slightly altered in value, one becoming a little higher and the other a little lower than it was in the isolated atoms. (Why are the lower levels less likely to be affected?) The closer the atoms are together, the greater is the energy difference between the two levels, Fig. 21.7a.

In a solid, many atoms are closely packed and for each original higher level in an isolated atom, there will be many levels. Since the spacing between atoms varies, the levels will be altered by different amounts and will, in effect, form a band of extremely closely spaced levels corresponding to one particular level in the isolated atoms, Fig. 21.7b. A solid can thus be considered to have energy bands that are shared by all atoms. The bands, like the levels in an isolated atom, are separated by gaps which are forbidden to the electrons.

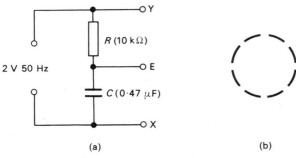

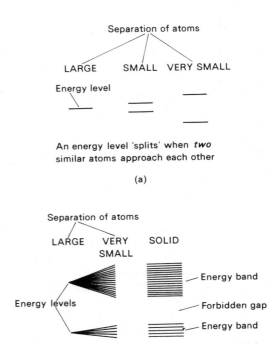

Fig. 21.6

An energy level 'splits' when *two* similar atoms approach each other

(a)

Energy bands, conductors and insulators

(*a*) *Band theory*. Charge carriers are necessary for electrical conduction. The 'free' electron theory, described earlier for metals (p. 457), suggests that solids possessing 'free' electrons are conductors and those which do not are insulators. This theory was developed at the start of the present century and whilst it deals adequately with certain aspects of conduction, it cannot satisfactorily account for others. One of its assumptions, is that the 'free' electrons can have any

Fig. 21.7   (b)

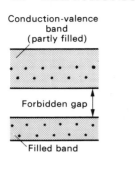

Conduction-valence band (partly filled)

Forbidden gap

Filled band

(a) Conductor

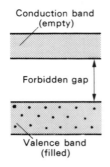

Conduction band (empty)

Forbidden gap

Valence band (filled)

(b) Insulator

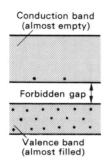

Conduction band (almost empty)

Forbidden gap

Valence band (almost filled)

(c) Semiconductor at room temperature

Fig. 21.8

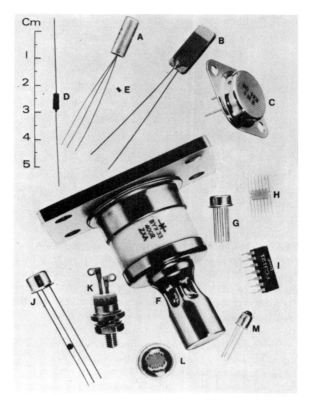

A, J	Low power transistors
B	Plate thermistor
C	Power transistor
D	Low power diode
E	Leadless inverted device (LID)
F	250 A rectifier diode
G, H, I	Integrated circuits
K	Thyristor
L	Photoconductive cell
M	Phototransistor

Fig. 21.9

(b) *Conductors.* The electrical conductivity of a solid depends on the spacing of its energy bands and the extent to which these are occupied by electrons. In a conductor such as copper the lower bands are full but the highest occupied band, called the *conduction band,* is only half-full. An electron in it can accept from an electric field (due to the p.d. applied to it) the very small amount of energy it requires to be raised to an empty, slightly higher energy level in the band where it is so loosely held that it can drift through the solid under the action of the applied field (as a 'free' electron) thus creating an electric current.

It is the conduction band electrons in copper which form chemical bonds with other atoms and so they are also called 'valence' electrons and the band the '*conduction-valence*' band. Conductors in general are characterized by having either a partly filled conduction-valence band, like copper, Fig. 21.8*a,* or an empty conduction band overlapping a filled valence band, like magnesium, so that there is in effect one continuous partly-filled band.

(c) *Insulators.* An insulator is considered on this model to have neither partly-filled bands nor overlapping bands. The highest occupied band, the valence band, is full but the band above it, the conduction band, is empty Fig. 21.8*b.* There is, however, a substantial energy gap—the forbidden gap—between the two bands. The energy required to raise a valence electron into the conduction band is about 5 eV, which is more than can be supplied by the electric field of an ordinary source of p.d.

Semiconductors

A semiconductor may be defined as a material whose conductivity lies between that of a good conductor and a good insulator. Silicon and germanium are the best known semiconductors because they are used to make the most common devices, i.e. transistors and diodes, the present trend being towards silicon devices. Other semiconductor materials are in common use such as cadmium sulphide, lead sulphide, gallium arsenide. These tend to be used in specialist devices such as photoelectric cells (p. 467). A selection of semiconductor devices is shown in Fig. 21.9.

Intrinsic semiconductors

When semiconductors such as germanium and silicon are extremely pure, they resemble insulators but their forbidden gap is narrower, about 1 eV. At absolute zero the valence band is completely filled and the conduction band is empty. At room temperatures a few valence electrons gain enough energy from the vibration of atoms in the crystal lattice of the material (due to their internal energy) to reach the conduction band, Fig. 21.8c. There they behave like the conduction electrons in a conductor, and when a p.d. is applied they move under the action of the electric field to form a current. There is also a second type of charge carrier which arises as follows.

When a valence electron enters the conduction band, it leaves in the valence band an empty energy level. A position in the crystal which was previously electrically neutral is now deficient of an electron and behaves like a positive charge. Such positive vacancies are called 'holes'. Under an applied electric field a nearby electron in the valence band can occupy the hole but in doing so creates another positive hole. Thus, whilst the actual particle that moves is a valence electron, it seems that a hole with a positive electronic charge has moved in the opposite direction. The very small current obtainable in a *very pure* semiconductor therefore consists of electrons in the nearly empty conduction band moving one way and an equal number of positively charged holes in the almost full valence band drifting the opposite way, Fig. 21.10. Such a semiconductor is said to exhibit *intrinsic* conduction, i.e. the charge carriers have their origin inside the material, arising in this case from the transfer of electrons from the valence to the conduction band.

It is possible for a hole to be filled by a conduction electron thereby reducing the number of charge carriers. But at any temperature the rate at which holes are filled equals the rate at which they are created when

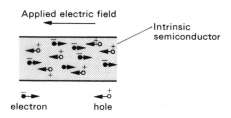

Fig. 21.10

a valence electron gains enough energy from the lattice to move to the conduction band.

The behaviour of valence electrons and holes in a semiconductor can be likened to a row of chairs, all occupied except one at the end. If everyone moves one place towards the vacant seat, the vacancy appears at the other end. The vacancy (hole) seems to have moved along the row but the motion has really been of the occupants (electrons) of the chairs in the opposite direction.

The conductivity of a pure semiconductor at room temperature is very small but may be increased greatly by raising the temperature so that many more valence electrons can reach the conduction band. This fact is used in thermistors. Above 100 °C for germanium and 150 °C for silicon the conductivity is no longer under control and precautions have to be taken to limit their working temperatures.

Extrinsic semiconductors

The use of semiconducting materials in devices such as the transistor depends on increasing their conductivity by introducing tiny, but controlled, amounts of certain 'impurities' into very pure (i.e. intrinsic) semiconductors. The process is known as 'doping' and the material obtained is an *extrinsic* semiconductor because the impurity introduces charge carriers, additional to the intrinsic ones. The impurity atom must have a similar size to that of the semiconducting atom so that it can occupy a position in the crystal lattice of the semiconductor without distorting it.

The valency of germanium and silicon is four, i.e. each atom has four electrons in the valence band. Each valence electron forms a covalent bond with a valence electron of four neighbouring atoms. The impurity atoms must be either pentavalent or trivalent.

(*i*) n-*type.* Suppose a pentavalent atom, such as phosphorus, arsenic or antimony, is introduced into the lattice of a pure silicon crystal. The impurity atom has five valence electrons but only four are required to form covalent bonds with adjacent tetravalent silicon atoms. One electron from every pentavalent atom added is spare and being loosely held it can take part in conduction. The impurity atom is called a *donor* and the 'impure' silicon is known as an n-type semiconductor because the main charge carriers are *n*egative electrons. However, note that the overall charge in the crystal remains zero because each atom present is electrically neutral. In terms of energy bands, the donor atom provides *filled* energy levels in the forbidden gap of silicon, just below its conduction band, Fig. 21.11. An impurity electron in one of these donor levels can easily acquire from the lattice vibrations, the small amount of energy it needs to reach the empty conduction band in silicon.

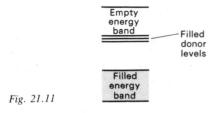

Fig. 21.11

(*ii*) p-*type*. A trivalent impurity such as boron, aluminium, or indium, also increases the conductivity of a pure semiconductor but in a different way. In this case the impurity atom has three valence electrons and can only form covalent bonds with three of the four surrounding silicon atoms in the crystal lattice. One bond is incomplete and, as before, the position of the missing electron behaves like a positive hole. This can accept a valence electron from an adjacent silicon atom (when it receives just a small amount of energy) and thereby form another hole. The impurity atoms are therefore called *acceptors* and since the conduction is now chiefly by *positive* charge carriers (i.e. holes), the 'impure' silicon is said to be a p-type semiconductor. Note again that the semiconductor is electrically neutral. In the energy band diagram of Fig. 21.12, the acceptor atom is shown as supplying *empty* energy levels in the forbidden gap just above the filled valence band of silicon. A valence electron can easily occupy one of these slightly higher levels and create a vacancy in the valence band which makes increased conduction possible.

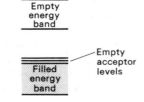

Fig. 21.12

To sum up, the *majority carriers* in an n-type semiconductor are electrons, positive holes being very much *minority carriers*. In p-type material, the reverse is true. The sign and concentration of the majority carriers can be found from the Hall effect (p. 270). In both types of material a temperature rise increases the proportion of minority carriers (due to increased intrinsic conduction) and upsets the semiconductor device whose action depends entirely on its extrinsic charge carriers.

In the manufacture of n- and p-type semiconductors, the semiconducting material is first purified (e.g. by zone refining, p. 287) to 1 part in 10^{10} and then impurity atoms added to produce the required conductivity.

Junction diode

A crystal of silicon or germanium with a junction of p- and n-type materials can act as a rectifier. The junction must be formed in the *same* continuous crystal lattice.

(*a*) *The* p-n *junction.* A p-n junction is represented in Fig. 21.13*a*. As soon as the junction is produced free electrons near the junction in the n-type material move (by diffusion) across the junction into the p-type material where they fill holes. (Diffusion occurs because the concentration of electrons in n-type material is large and in p-type material it is small. The electrons behave

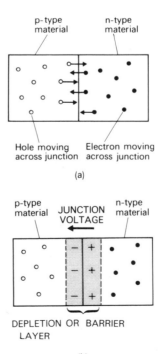

Fig. 21.13

like scent molecules when a scent bottle is opened—they diffuse from the vapour in the bottle, where their concentration is high, into the air where it is initially zero.)

As a result the n-type material near the junction becomes positively charged and the p-type material negatively charged (both previously being neutral). At the same time, and for a similar reason, holes diffuse from p-type to n-type, capturing electrons there. The exchange of charge soon stops because negative charge on the p-type material opposes the further flow of electrons and the positive charge on the n-type opposes the further flow of holes. The region on either side of the junction becomes fairly free of majority charge carriers and is called the *depletion* or *barrier layer*; it is less than one-millionth of a metre wide and is, in effect, an insulator, Fig. 21.13*b*.

The situation is just as if there was a battery across the junction with a voltage of a few tenths of a volt, called the *junction voltage,* acting from n- to p-type.

If a battery is connected across a p-n junction with its positive terminal joined to the n-type side and its negative terminal to the p-type side, it helps the junction voltage and the junction is said to be *reverse biased*. Electrons and holes are repelled further from the junction and the depletion layer widens, Fig. 21.14*a*. Only a few minority carriers (produced by bonds breaking at ordinary temperatures in both p- and n-type materials) cross the junction and a tiny current, called the *leakage* or *reverse* current, flows.

If a battery is connected so as to oppose the junction voltage, the depletion layer narrows. When the battery voltage exceeds the junction voltage, appreciable current flows because majority carriers are able to cross the junction, electrons from the n- to the p-side and holes in the opposite direction, Fig. 21.14*b*. The junction is then *forward biased*, i.e. the p-type side is connected to the *positive* of the battery and the n-type to the *negative*. As before a small leakage current flows.

(*b*) *Construction.* A junction diode consists of a p-n junction with one connection to the p-side (the *anode* A) and another to the n-side (the *cathode* K). A simplified section of a silicon diode with 'planar-type' construction is shown in Fig. 21.15. (In practice the boundaries are neither straight nor well-defined.) The manufacturing process involves first soldering a thin slice of n-type silicon to a metal base from which the cathode lead K is taken. A film of highly insulating silicon oxide is then formed on the surface of the slice by heating it in steam at about 1100 °C. A 'window' is next etched chemically in the oxide film and vapour of the appropriate impurity allowed to diffuse through it so converting the top of the slice into p-type silicon. Aluminium is evaporated on to the p-type region to allow the anode lead A to be soldered. The diode is sealed in a case to exclude moisture and light.

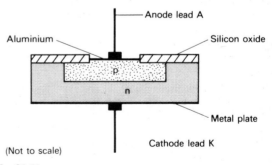

Fig. 21.15

(*c*) *Characteristic.* A typical characteristic for a silicon diode is shown in Fig. 21.16. You can see from it that the forward current I_F is small until the forward voltage V_F is about 0.6 V, thereafter a very small change in V_F causes a large increase in I_F. The reverse current I_R is negligible (the scales on the graph have been changed to show it) and remains so as the reverse voltage V_R is increased. At a certain reverse voltage

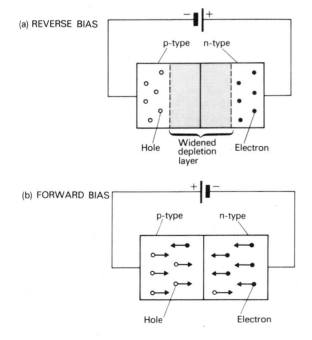

Fig. 21.14

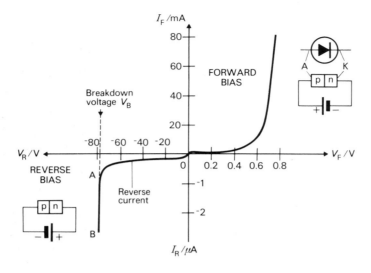

Fig. 21.16

V_B, called the *breakdown voltage*, the insulation of the barrier layer breaks down and I_R increases suddenly and rapidly and damage may occur by overheating. V_B can have any value from a few volts up to 1000 V depending on the construction of the diode and the level of doping.

The characteristic of a germanium junction diode is similar to the one in Fig. 21.16 but I_F 'turns on' when V_F is about 0.2 V, I_R is greater and V_B has a maximum value of about 100 V.

As rectifiers, silicon junction diodes are preferred to germanium types because their much lower reverse current makes them more efficient (i.e. more complete conversion of a.c. to d.c. occurs). Silicon also has a higher breakdown voltage and can work at higher temperatures.

Other junction diodes

(*a*) *Light-emitting diode (LED)*. An LED, shown in Fig. 21.17 with its symbol, is a junction diode made from the semiconducting compound gallium arsenide phosphide. When forward biased it emits red, yellow or green light depending on its exact composition.

As we have seen, at a forward biased p-n junction, recombination of electrons and holes occurs. Every recombination results in the release of a quantum of energy which, in most semiconductors, causes an increase of internal energy and a temperature rise in the material. In gallium arsenide phosphide however, many of these quanta are emitted as light which gets out of the LED because the junction is formed very close to the surface.

LEDs are used as indicator lamps, especially in digital electronic circuits to show whether outputs are 'high' or 'low'. Unless they are the constant-current type (incorporating an integrated circuit regulator for use on a 2 to 18 V d.c. or a.c. supply), they *must have an external resistor* connected in series to limit the forward current (typically 2 to 25 mA). On a 5 V d.c. supply 270 Ω would be suitable for a red LED and 120 Ω for green and yellow LEDs.

Many electronic calculators, clocks and measuring instruments have seven-segment LED displays as numerical indicators, Fig. 21.18a. Each segment is an LED and depending on which are energized, the display lights up the numbers 0 to 9, Fig. 21.18b and c. Such displays are usually designed to work on a 5 V supply. Each segment needs a separate current-limiting resistor and all the cathodes (or anodes) are joined together to form a common connection.

The advantages of LEDs (over, for example, filament lamps) is their small size, reliability, long life and high operating speed.

(*b*) *Zener diode*. This is used to stabilize (i.e. keep steady) the voltage output of a power supply. It looks

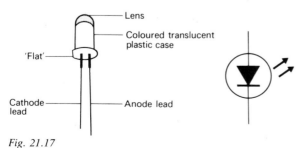

Fig. 21.17

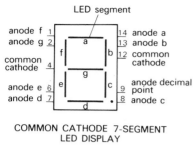

LED segment

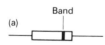

anode f 1   14 anode a
anode g 2   13 anode b
common cathode 4   12 common cathode
anode e 6   9 anode decimal point
anode d 7   8 anode c

COMMON CATHODE 7-SEGMENT
LED DISPLAY

(a)   (typical pin connections)

(b)

(c)

Fig. 21.18

like a rectifier diode, Fig. 21.19*a*; the symbol for one is given in Fig. 21.19*b*.

Band

(a)

(b)
Anode   Cathode

Fig. 21.19

When the reverse p.d. applied to a silicon junction diode reaches the breakdown voltage V_B, the reverse current I_R increases suddenly and rapidly as Fig. 21.16 shows. Damage may then occur to the diode by over-heating unless the current is limited by a series resistor. If this is connected, *the p.d. across the diode remains almost constant at V_B* (also called the *Zener voltage*) *over a wide range of reverse currents*, i.e. part AB of the characteristic is nearly at right angles to the V_R axis. It is this property of a Zener diode which makes it useful in stabilized power supplies.

The value of the current-limiting resistor should ensure that the power rating of the diode is not exceeded. For example, a 400 mW (0.4 W) Zener diode for which $V_B = 10$ V can pass a maximum current I_{max} given by

$$I_{max} = \frac{power}{voltage} = \frac{0.4}{10} = 0.04 \text{ A} = 40 \text{ mA}$$

Zener diodes with specified Zener voltages are made e.g. 3.0 V, 3.9 V, 5.1 V, 6.2 V, 9.1 V, 15 V up to 200 V—by controlling the doping.

A circuit for demonstrating the action of a Zener diode is given in Fig. 21.20; note that the diode is *reverse biased*. If the voltage of the d.c. supply is gradually increased from zero, at the Zener voltage (6.2 V), the diode starts to conduct. The current shown on the milliammeter increases with the supply voltage but the voltmeter connected across the diode gives a constant reading of 6.2 V (i.e. is stabilized) even when the supply voltage is greater.

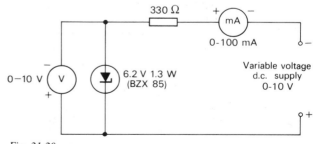

Fig. 21.20

(*c*) *Photodiode.* It consists of a normal p-n junction in a case with a transparent 'window' through which light can enter. Reverse bias is applied and the reverse current increases in proportion to the amount of light falling on the junction. The effect is due to the light energy breaking bonds in the crystal lattice of the semiconductor to produce electrons and holes.

Photodiodes are used as fast 'counters' which generate a pulse of current every time a beam of light is interrupted. The symbol for one is shown in Fig. 21.21.

Fig. 21.21

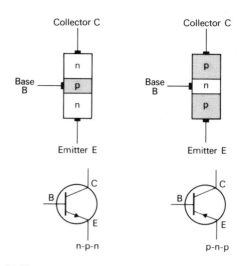

Fig. 21.22

Transistors

Transistors are the most important device in electronics today. Not only are they made as discrete (separate) components but integrated circuits (ICs) may contain several thousands on a tiny slice of silicon.

Transistors are three-terminal devices used as *amplifiers* and as *switches*. There are two basic types. They are

(*i*) the *bipolar* or *junction transistor* (usually called *the* transistor); its operation depends on the flow of both majority and minority carriers and it has two p-n junctions, and

(*ii*) the *unipolar* or *field effect transistor* (called the FET) in which current is due to majority carriers only (either electrons or holes) and there is just one p-n junction.

We will consider (*i*) now and (*ii*) later (p. 509).

(*a*) *Construction.* A junction transistor consists of two p-n junctions in the same crystal. A very thin wafer of lightly doped p- or n-type semiconductor (the *base* B) is sandwiched between two thicker, heavily doped materials of the opposite type (the *collector* C and *emitter* E). The two possible arrangements are shown diagrammatically in Fig. 21.22. The arrows on the symbols give the direction in which conventional (positive) current flows; in the n-p-n type it points from B to E and in the p-n-p type from E to B.

As with diodes, silicon transistors are in general preferred to germanium ones because they withstand higher temperatures and voltages and have lower leakage currents. Silicon n-p-n types are more easily mass-produced than p-n-p types, the opposite is true of germanium. A simplified section of an n-p-n silicon

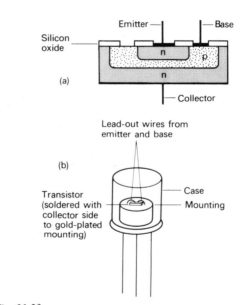

Fig. 21.23

transistor made by the *planar* process (outlined on p. 495) is shown in Fig. 21.23*a*; Fig. 21.23*b* shows a transistor complete with case (called the encapsulation) and three wire leads.

(*b*) *Action.* In Fig. 21.24 an n-p-n silicon transistor is connected in a *common-emitter* circuit, i.e. the emitter is joined (via batteries B_1 and B_2) to both the base and the collector. For transistor action to occur the base-emitter junction must be forward biased, i.e. positive terminal of B_1 to p-type base, and the collector-base

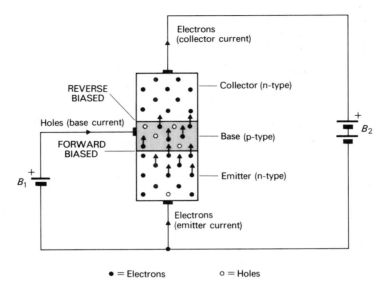

REVERSE BIASED

Holes (base current)

FORWARD BIASED

B_1

Electrons (collector current)

Collector (n-type)

Base (p-type)

Emitter (n-type)

B_2

Electrons (emitter current)

Fig. 21.24   ● = Electrons   ○ = Holes

junction reverse biased, i.e. positive terminal of B_2 to n-type collector.

When the base-emitter bias is about +0.6 V, electrons (the majority carriers in the heavily doped n-type emitter) cross the junction (as they would in any junction diode) into the base. Their loss is made good by electrons entering the emitter from the external circuit to form the *emitter current*. At the same time holes flow from the base to the emitter but, since the p-type base is lightly doped, this is small compared with the electron flow in the opposite direction, i.e. electrons are the majority carriers in an n-p-n transistor.

In the base, only a small proportion (about 1 per cent) of the electrons from the emitter combine with holes in the base because the base is very thin (less than a millionth of a metre) and is lightly doped. Most of the electrons are swept through the base due to being attracted by the positive voltage on the collector and cross the base-collector junction to become the *collector current* in the external circuit.

The small amount of electron-hole recombination which occurs in the base gives it a momentary negative charge which is immediately compensated by battery B_1 supplying it with (positive) holes. The flow of holes to the base from the external circuit creates a small *base current*. This keeps the base-emitter junction forward-biased and so maintains the larger collector current.

Transistor action is the turning on (and controlling) of a large current through the high-resistance (reverse biased) collector-base junction by a small current through the low-resistance (forward biased) base-emitter junction. The term transistor refers to this effect and comes from the two words '*trans*fer—re*sistor*'.

The behaviour of a p-n-p transistor is similar to that of the n-p-n type but it is holes that are the majority carriers which flow from the emitter to the collector and electrons are injected into the base to compensate for recombination. To obtain correct biasing the polarities of both batteries must be reversed. *Wrong battery connection can seriously damage transistors.*

Transistor as a current amplifier

(*a*) *d.c. current gain*. In the circuit of Fig. 21.25 when the base-emitter junction of the n-p-n silicon transistor is forward biased to about 0.6 V (0.2 V for a germanium transistor), a small base current I_B flows and 'turns on' a larger collector current I_C. That is, I_C is zero until I_B flows (but see p. 502). A junction transistor is therefore a *current-operated* device.

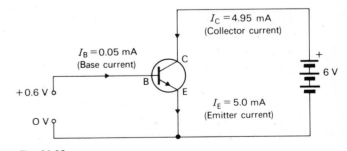

$I_C = 4.95$ mA (Collector current)

$I_B = 0.05$ mA (Base current)

+0.6 V

0 V

C

B

E

6 V

$I_E = 5.0$ mA (Emitter current)

Fig. 21.25

Typically I_C may be 10 to 1000 times greater than I_B depending on the transistor. If we look upon I_B as the input current to the transistor and I_C as the output current from it, then it is basically a *current amplifier*. The d.c. current gain h_{FE} (previously β) is an important property of a transistor and is defined by

$$h_{FE} = \frac{I_C}{I_B}$$

For the transistor in Fig. 21.25, where the arrows show the direction of conventional current flow, $I_C = 4.95$ mA and $I_B = 0.05$ mA (50 μA)

$$\therefore \quad h_{FE} = \frac{4.95}{0.05} = \frac{495}{5} = 99$$

Also note that since the current flowing out of a transistor must equal the current flowing into it, then

$$I_E = I_B + I_C$$

In the above example $I_E = 0.05 + 4.95 = 5.0$ mA, i.e. $I_C \simeq I_E$.

In the symbol h_{FE}, F indicates that we are considering forward currents and E that the transistor is connected in the common-emitter mode. Two other less usual methods of connection are 'common-base' and 'common-collector'.

(*b*) *Demonstration.* A simple circuit to show current amplification by a transistor is given in Fig. 21.26. The forward bias for the base-emitter junction and the reverse bias for the collector-base junction are both obtained from the same battery. The base current I_B flows through lamp L_1 and resistor R and 'turns on' the transistor. L_2 lights up but not L_1 showing that I_C is much greater than I_B. If L_1 is removed so that the base circuit is broken, I_B becomes zero as does I_C and L_2 goes out.

The value of R is chosen so that I_B causes an I_C that is large enough (greater than about 0.04 A) to light L_2. (See *Worked example*, p. 503.)

I_C and I_B can be measured by connecting a 0-100 mA meter in series with L_2 and a 0-100 μA meter in series with L_1 and an exact value found for h_{FE}.

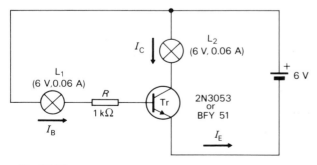

Fig. 21.26

Transistor characteristics

These are graphs found by experiment which show the relationships between various currents and voltages and enable us to see how best to use a transistor. A module[1] and circuit for investigating an n-p-n transistor (e.g. BFY51 or 2N3053) in common-emitter connection are shown in Fig. 21.27a and b.

The voltmeter for measuring V_{BE} must have a very high resistance (e.g. an electronic type with a resistance of the order of 1 MΩ or more); if it is a moving-coil type, allowance has to be made for the current it takes when finding (*a*). R_2 protects the transistor from excessive base currents. Three characteristics are important.

(*a*) *Input (base) characteristic* (I_B—V_{BE}). The collector-emitter voltage V_{CE} is kept constant (e.g. at the battery voltage of 6 V) and the base-emitter voltage V_{BE} measured for different values of the base current I_B, obtained by varying R_1. (The procedure when using a 'low resistance' voltmeter for V_{BE} is to open switch S, adjust R_1 to make $I_B = 5$ μA say, record I_C, close S, readjust R_1 until I_C is the same as before, record V_{BE}.)

A typical graph for a silicon transistor is given in Fig. 21.28a. Note that I_B is negligibly small until V_{BE} exceeds about 0.6 V and thereafter small changes in V_{BE} cause large changes in I_B. The *input resistance* r_i is defined as the ratio $\Delta V_{BE}/\Delta I_B$ where ΔI_B is the change in I_B due to a change of ΔV_{BE} in V_{BE}. ('Δ' means 'a small increase in'.) Since the input characteristic is non-linear, r_i varies but is of the order of 1 to 5 kΩ.

(*b*) *Output (collector) characteristic* (I_C—V_{CE}). I_B is fixed at a low value, e.g. 10 μA and I_C measured as V_{CE} is increased in stages by varying R_3. This is repeated for different values of I_B to give a family of curves as in Fig. 21.28b.

You can see that I_C depends almost entirely on I_B and hardly at all on V_{CE} (except when V_{CE} is less than about 0.5 V). As an *amplifier*, a transistor operates well to the right of the sharp bend or 'knee' of the characteristic, i.e. where I_C varies linearly with V_{CE} for a given I_B. The small slope of this part of the characteristic shows that the *output resistance* r_0 of the transistor is fairly high, of the order 10 kΩ to 50 kΩ; it is given by $r_0 = \Delta V_{CE}/\Delta I_C$ where ΔI_C is the change in I_C caused by a change of ΔV_{CE} in V_{CE}. As a *switch*, a transistor operates in the shaded parts of Fig. 21.28b and changes over rapidly from the 'off' state in which $I_C = 0$ (cut-off) to the 'on' state in which I_C is a maximum (saturation).

[1] Unilab 'Blue Chip': 511.008.

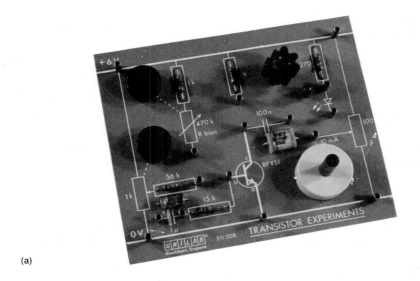

(a)

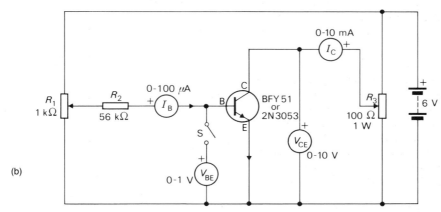

(b)

Fig. 21.27

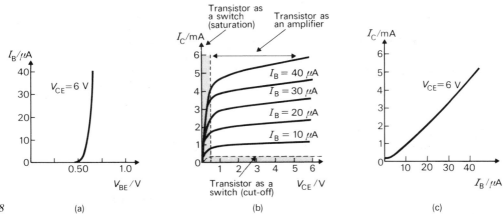

Fig. 21.28   (a)   (b)   (c)

(c) Transfer characteristic (I_C—I_B). V_{CE} is kept fixed and I_C measured for different values of I_B by varying R_1. The graph, Fig. 21.28c, is almost a straight line showing that I_C is directly proportional to I_B, i.e. the relation between I_C and I_B is linear.

The a.c. current gain h_{fe} (previously β) is defined by

$$h_{fe} = \frac{\Delta I_C}{\Delta I_B}$$

where ΔI_C is the change in I_C produced by a change of ΔI_B in I_B. For most purposes h_{fe} and h_{FE} (the d.c. current gain = I_C/I_B) can be considered as equal.

The characteristic also shows that when I_B is zero, I_C has a small value (about 0.01 μA for silicon and 2 μA for germanium at 15 °C). This is called the *leakage current I_{CEO}* and is due to minority carriers (holes for an n-p-n transistor and electrons for a p-n-p type) crossing the reverse biased collector-base junction from collector to base to emitter. Since minority carriers are produced by heat breaking bonds in the crystal lattice, the leakage current increases with temperature rise (much more so for germanium than silicon) and upsets the working of the transistor (p. 494).

Summing up, in a silicon junction transistor

    (i) I_C is zero until I_B flows,

    (ii) I_B is zero until V_{BE} is about 0.5 V, and

    (iii) V_{BE} remains close to 0.6 V for a wide range of values of I_B.

Transistor as a voltage amplifier

Amplifiers are necessary in many types of electronic equipment such as radios, oscilloscopes and record players. Often it is a small alternating voltage that has to be amplified. A junction transistor in the common-emitter mode can act as a voltage amplifier if a suitable resistor called the *load*, is connected in the collector circuit.

The small alternating voltage, the *input v_i*, is applied to the base-emitter circuit and causes small changes of base current which produce large changes in the collector current flowing through the load. The load converts these current changes into voltage changes which form the alternating *output* voltage v_o, v_o being much greater than v_i. (Note the use of small italic letters to represent instantaneous values of alternating quantities.)

A simple circuit is shown in Fig. 21.29. To see just why voltage amplification occurs, consider first the situation when there is no input, i.e. $v_i = 0$, called the quiescent (quiet) state.

(a) Quiescent state. For transistor action to take place the base-emitter junction must always be forward biased (even when v_i is applied and goes negative). A simple way of ensuring this is to connect a resistor R_B, called the *base bias resistor*, as shown. A steady (d.c.) base current I_B, flows from battery +, through R_B into the base and back to battery −, via the emitter. The value of R_B can be calculated (see *Worked example*, p. 503) once the value of I_B for the best amplifier performance has been decided.

If V_{CC} is the battery voltage and V_{BE} is the base-emitter junction voltage (always about +0.6 V for an n-p-n silicon transistor), then for the *base-emitter* circuit, since d.c. voltages add up, we can write

$$V_{CC} = I_B \times R_B + V_{BE} \tag{1}$$

I_B causes a much larger collector current I_C which produces a voltage drop $I_C \times R_L$ across the load R_L.

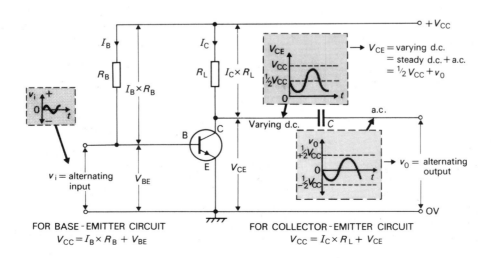

Fig. 21.29

FOR BASE-EMITTER CIRCUIT   FOR COLLECTOR-EMITTER CIRCUIT

$V_{CC} = I_B \times R_B + V_{BE}$ $V_{CC} = I_C \times R_L + V_{CE}$

The voltage at the end of R_L joined to battery + is fixed and so the voltage drop must be at the end connected to the collector. If V_{CE} is the collector-emitter voltage, then for the *collector-emitter* circuit

$$V_{CC} = I_C \times R_L + V_{CE} \qquad (2)$$

Component values are chosen so that the quiescent collector-emitter voltage V_{CE} is about *half the battery voltage V_{CC}*. As you will see shortly (p. 505), this ensures the best working conditions for the amplifier.

(*b*) *Input applied.* When v_i is applied and goes positive it increases the base-emitter voltage slightly (e.g. from +0.60 V to +0.61 V). When v_i swings negative the base-emitter voltage decreases slightly (e.g. from +0.60 V to +0.59 V). As a result a small alternating current is superimposed on the quiescent base current I_B which in effect becomes a varying d.c.

When the base current *increases*, large proportionate increases occur in the collector current. From equation (2) it follows that there is a corresponding large *decrease* in the collector-emitter voltage (since V_{CC} is fixed). A decrease of base current causes a large increase of collector-emitter voltage. In practice positive and negative swings of a few millivolts in v_i can result in a fall or rise of several volts in the voltage across R_L and so also in the collector-emitter voltage.

The collector-emitter voltage is a varying direct voltage and may be regarded as an alternating voltage superimposed on a steady direct voltage, i.e. on the quiescent value of V_{CE}. Only the alternating part is wanted and capacitor C blocks the direct part but allows the alternating part, i.e. the output v_o, to pass.

(*c*) *Further points.* The transistor *and* load together bring about voltage amplification.

The output v_o is 180° out of phase with the input v_i, i.e. when v_i has its maximum positive value, v_o has its maximum negative value (see graphs in Fig. 21.29).

The emitter is common to the input, output and battery circuits and is usually taken as the reference point for all voltages, i.e. 0 V. It is called 'common' or 'ground' (symbol 𝝿𝝿) or 'earth' (symbol ⟱) if connected to earth.

Worked example

A silicon transistor in the simple voltage amplifier circuit of Fig. 21.29 operates satisfactorily on a quiescent (no a.c. input) collector current (I_C) of 3 mA. If the battery supply (V_{CC}) is 6 V, what must be the value of (a) the load resistor (R_L) and (b) the base bias resistor (R_B), for the quiescent collector-emitter voltage (V_{CE}) to be half

the battery voltage? The transistor d.c. current gain (h_{FE}) is 100.

(*i*) The collector-emitter circuit equation is

$$V_{CC} = I_C \times R_L + V_{CE}$$

Rearranging we get

$$I_C \times R_L = V_{CC} - V_{CE}$$

That is

$$R_L = (V_{CC} - V_{CE})/I_C$$

Substituting $V_{CC} = 6$ V, $V_{CE} = \frac{1}{2} V_{CC} = 3$ V and $I_C = 3$ mA gives

$$R_L = (6 - 3)\text{V}/3 \text{ mA} = 1 \text{ k}\Omega$$

(*ii*) The d.c. current gain is given by

$$h_{FE} = \frac{I_C}{I_B}$$

where I_B is the quiescent base current to produce the quiescent collector current I_C.
Rearranging

$$I_B = \frac{I_C}{h_{FE}}$$

Substituting $I_C = 3$ mA and $h_{FE} = 100$, we get

$$I_B = 3 \text{ mA}/100 = 0.03 \text{ mA } (30 \text{ } \mu\text{A})$$

The base-emitter circuit equation is

$$V_{CC} = I_B \times R_B + V_{BE}$$

where V_{BE} is the base-emitter voltage.
Rearranging gives

$$R_B = (V_{CC} - V_{BE})/I_B$$

Substituting $V_{CC} = 6$ V, $V_{BE} = 0.6$ V (for a silicon transistor) and $I_B = 0.03$ mA gives

$$R_B = (6 - 0.6) \text{ V}/0.03 \text{ mA} = 5.4 \text{ V}/0.03 \text{ mA}$$

$$= 540/3 = 180 \text{ k}\Omega$$

Voltage gain of an amplifier

The voltage gain A of an amplifier is the ratio of the output voltage v_o to the input voltage v_i.

$$A = \frac{v_o}{v_i}$$

(*a*) *Measurement.* A circuit for finding A ex-

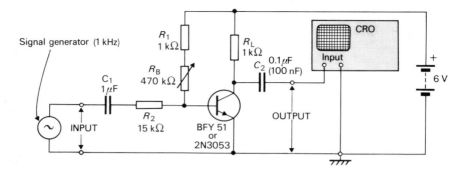

Fig. 21.30

perimentally using the module of Fig. 21.27a is shown in Fig. 21.30. Variable resistor R_B controls the base current and initially is turned fully anticlockwise to its maximum resistance. This sets the quiescent collector voltage at 2.5 to 3.0 V, i.e. about half the d.c. supply voltage. R_1 prevents damage to the transistor by limiting the base current should R_B be made zero.

A sine-wave input of several millivolts at 1 kHz is applied via C_1 and R_2 from a signal generator. C_1 stops the resistance of the generator upsetting the base-emitter bias. R_2 protects the transistor in the input circuit.

The peak-to-peak value of v_o is noted from the CRO (in volts if it is calibrated, otherwise in scale divisions). To measure v_i the CRO input connection is transferred to the input terminals on the amplifier module.

The gain (v_o/v_i) may be only in the range 5 to 10. This is due to most of the input being 'lost' across R_2 and just a fraction of it applied to the base-emitter. If R_2 is shortcircuited (but the input still applied to C_1) and v_o and v_i remeasured, the value obtained for A is much greater, e.g. 70 to 100.

(b) *Theoretical value.* Suppose that the input voltage v_i causes a change ΔI_B in I_B then $\Delta I_B = v_i/($resistance of input circuit$)$, that is

$$\Delta I_B = \frac{v_i}{R_2 + r_i}$$

where r_i is the input resistance of the transistor, the resistance of the signal generator being neglected. If the resulting change of I_C is ΔI_C, the voltage change across R_L, i.e. v_o, is given by

$$v_o = \Delta I_C \times R_L$$

But

$$\Delta I_C = h_{fe} \times \Delta I_B = h_{fe} \times \frac{v_i}{(R_2 + r_i)}$$

Hence

$$v_o = h_{fe} \times \frac{R_L}{(R_2 + r_i)} \times v_i$$

Therefore

$$A = \frac{v_o}{v_i} = h_{fe} \times \frac{R_L}{(R_2 + r_i)}$$

In the circuit of Fig. 21.30, $R_L = 1$ kΩ and $R_2 = 15$ kΩ. Transistor characteristics like those in Fig. 21.28 give typical values for r_i and h_{fe} of 1 kΩ and 100 respectively. Substituting in the equation for A we get

$$A = \frac{100 \times 1}{(15 + 1)} = \frac{100}{16} = 6 \text{ (approx)}$$

If $R_2 = 0$ then $A = 100$. These values for A are approximate since h_{fe} has been taken from characteristics obtained at a constant collector-emitter voltage. In practice, in a voltage amplifier this voltage changes when the input is applied. A better way of calculating A is given in the next section.

Load lines: operating point

When designing a voltage amplifier the aim is to obtain
   (i) the desired voltage gain,
   (ii) minimum distortion of the output so that it is a good copy of the input, and
   (iii) operation within the current, voltage and power limits for the transistor.
The choice of the quiescent (d.c.) operating point, i.e. the values of I_C and V_{CE}, determines whether these requirements will be met. This is made by constructing a *load line*.

The output characteristics of a transistor (p. 501, Fig. 21.28b) show the relation between V_{CE} and I_C with *no load* in the collector circuit. With a load R_L, the equation connecting them is (p. 503)

$$V_{CC} = I_C \times R_L + V_{CE}$$

where V_{CC} is the battery voltage.

Rearranging we get

$$V_{CE} = V_{CC} - I_C \times R_L \qquad (3)$$

Knowing V_{CC} and R_L this equation enables us to calculate V_{CE} for different values of I_C. If a graph of I_C (as the y-axis) is plotted against V_{CE} (as the x-axis) we get a straight line, called a load line. If we accept that (3) is the equation of a straight line (you can see this by rewriting (3) as

$$I_C = \frac{-1}{R_L} \times V_{CE} + \frac{V_{CC}}{R_L}$$

which is of the form $y = mx + c$), the line can be drawn if we know just two points. The easiest to find are the end points A and B where the line cuts the V_{CE} and I_C axes respectively.

For A we put $I_C = 0$ in (3) and get $V_{CE} = V_{CC} = 6$ V (say).

For B we put $V_{CE} = 0$ in (3) and get $I_C = V_{CC}/R_L$. If $R_L = 1$ kΩ, then $I_C = 6$ V/1 kΩ = 6 mA.

In Fig. 21.31, AB is the load line for $V_{CC} = 6$ V and $R_L = 1$ kΩ. It is shown superimposed on the output characteristics of the transistor used in the circuit of Fig. 21.29. We can regard a load line as the output characteristic of the *transistor and load* for particular values of V_{CC} and R_L. Different values of either give a different load line; for example, a smaller value of R_L gives a steeper line.

The choice of load line and d.c. operating point affects the shape and size of the output waveform. Choosing a line which cuts the characteristics where they are not linear (straight) or where they are not equally spaced can cause distortion. Selecting an operating point too near either the I_C or the V_{CE} axis can have the same effect. The best position for the d.c. operating point is *near the middle of the chosen load line*, e.g. at Q on Fig. 21.31. The 'swing' capability of the output is then a maximum (from near V_{CC} to near 0 V) and distortion a minimum.

Having chosen Q, the quiescent (d.c.) values of V_{CE} and I_C can be read off. In Fig. 21.31, they are $V_{CE} = 3$ V (i.e. half V_{CC}) and $I_C = 3$ mA. The value of I_B which gives these values is obtained from the transistor output characteristics passing through Q (since I_C and V_{CE} have to satisfy *at the same time* both the characteristic and the load line). Here it is the 30 μA characteristic. R_B can then be calculated as on p. 503.

The *voltage gain* can be obtained from the load line by noting that when the input causes I_B to vary from 10

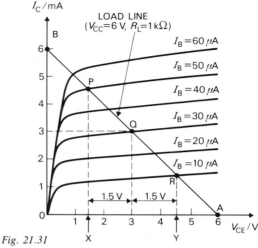

Fig. 21.31

to 50 μA (from R to P), V_{CE} varies from 4.5 to 1.5 V (from Y to X). From the input characteristic of the transistor (Fig. 21.28a, p. 501) we can find the change in V_{BE} to cause this change of 40 μA in I_B. If it is, say 40 mV (0.04 V), the voltage gain $A = (4.5 - 1.5)/0.04 = 3.0/0.04 = 75$.

Stability and bias

(a) *Thermal runaway*. If the temperature of a transistor rises, there is greater vibration of the semiconductor atoms resulting in the production of more free electrons and holes. The collector current increases causing further heating of the transistor and so on until it is damaged or destroyed. The initial temperature rise may be due to slight overloading of the transistor, to an increase in the surrounding temperature or to the replacement of one transistor by another of greater h_{FE}.

To stop this 'thermal runaway' effect and stabilize the d.c. operating point, special bias circuits have been designed which automatically compensate for variations of collector current. The simplest of these will now be considered.

(b) *Collector-to-base bias*. The basic circuit of Fig. 21.29 for a voltage amplifier can be adequately stabilized for many applications by *halving* the value of R_B and connecting it between the collector and base as in Fig. 21.32, rather than between battery + and base.

For this circuit we can write

$$V_{CC} = I_C \times R_L + V_{CE} \qquad (4)$$

where

$$V_{CE} = I_B \times R_B + V_{BE} \qquad (5)$$

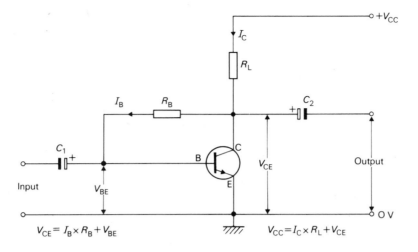

Fig. 21.32

From (4) you can see that if I_C increases for any reason, V_{CE} decreases since V_{CC} is fixed. From (5) it therefore follows that since V_{BE} is constant (0.6 V or so), I_B must also decrease and in so doing tends to bring back I_C to its original value.

Taking the quiescent conditions to be (as in the *Worked example* on p. 503) $V_{CE} = \frac{1}{2}V_{CC} = 3.0$ V, $I_C = 3.0$ mA and $I_B = 0.03$ mA, the value of R_B in Fig. 21.32 is found by rearranging equation (5) to be

$$R_B = \frac{V_{CE} - V_{BE}}{I_B} = \frac{(3.0 - 0.6)\ \text{V}}{0.03\ \text{mA}} = \frac{2.4\ \text{V}}{0.03\ \text{mA}}$$

$$= 80\ \text{k}\Omega$$

This is about half the value of 180 kΩ for R_B in the unstabilized circuit of Fig. 21.29, as calculated in the *Worked example* of p. 503.

Transistor as an oscillator

An oscillator is a generator of alternating current and in essence consists of an amplifier which feeds back a small part of its output to its input. If the feedback is positive, i.e. in phase with the input and sufficient to compensate for resistive energy losses, then undamped oscillations are obtained.

A simple tuned oscillator is shown in Fig. 21.33. The *L-C* circuit is connected in the collector (as the load) and oscillations start in it when the supply is switched on (p. 326). Left to themselves these would decay but changes of current in L are fed back by mutual induction to the base-emitter (input) circuit by coil L_1, arranged close to L. The frequency f of the oscillations is given by $f = 1/(2\pi\sqrt{LC})$, (p. 328), i.e. the natural

frequency of the *LC* circuit; the transistor merely ensures that energy is fed back at the correct instant from the battery.

The current bias for the base of the transistor is obtained through R; C_1 allows the a.c. component of the base-emitter current, at the frequency of the oscillations, to pass whilst not short-circuiting the d.c. bias.

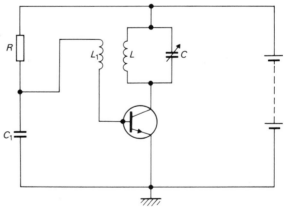

Fig. 21.33

Transistor as a switch

Transistors are used as switches in many important electronic circuits.

(*a*) *Action.* The switching action can be investigated with the circuit of Fig. 21.34 (using the module of Fig. 21.27*a*) in which an n-p-n silicon transistor (e.g. BFY51 or 2N3053) is connected in common-emitter mode with

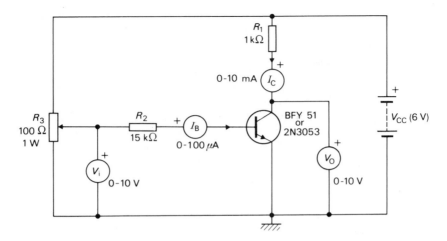

Fig. 21.34

a resistor R_1 in its collector (output) circuit. By adjusting R_3 the input voltage V_i can be varied and we find that

(*i*) when V_i increases from 0 to about 0.5 V, the base current I_B and the collector current I_C are both zero and the output voltage V_o remains almost equal to the battery voltage V_{CC} (6 V),

(*ii*) when V_i increases from about 0.5 V to 1 V, I_B and I_C increase rapidly from zero while V_o falls rapidly, and

(*iii*) when V_i increases from 1 V to 6 V, I_B goes on increasing but soon I_C reaches a maximum and V_o falls to nearly zero.

The graph of V_o against V_i is shown in Fig. 21.35; the dotted curve shows the effect of increasing the current-limiting resistor R_2. The table below summarizes the behaviour of V_i and V_o.

V_i	V_o
'low' (< 0.5 V)	'high' (6 V)
'high' (> 1 V)	'low' (0 V)

Depending on the value of V_i, V_o has one of two 'levels'—either a 'high' positive voltage V_{CC} (here, +6 V) or a 'low' voltage (almost zero).

(*b*) *Explanation.* The basic transistor switching circuit is shown in Fig. 21.36. R_2 is a current limiting resistor to keep I_B below the maximum allowed value.

V_{CC} is applied across R_1 and the transistor in series, therefore

$$V_{CC} = I_C \times R_1 + V_o$$

or
$$V_o = V_{CC} - I_C \times R_1 \qquad (6)$$

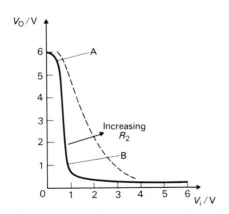

Fig. 21.35

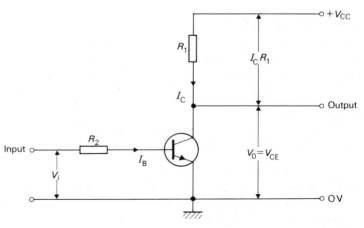

Fig. 21.36

When V_i is 'low' (e.g. by connecting the input terminal to 0 V), $I_C = 0$ and from (6) $V_o = V_{CC}$, i.e. all the battery voltage is dropped across the transistor which behaves like a very large resistor and is said to be *cut-off*.

When V_i is 'high' (e.g. by connecting the input terminal to $+ V_{CC}$), it makes I_C and therefore $I_C \times R_1$ large. From (6), the greatest value $I_C \times R_1$ can have is V_{CC}. Then $V_o = 0$ and all the battery voltage is dropped across R_1 and none across the transistor which behaves as if it had zero resistance. I_C has its maximum possible value, given by $I_C = V_{CC}/R_1$ and any further increase in I_B does not increase I_C. In this case the transistor is said to be *saturated* or *bottomed*.

As an amplifier a transistor acts *linearly* (i.e. the output is more or less directly proportional to the input) because it is biased to work on the straight part AB of Fig. 21.35. As a switch it operates between cut-off and saturation, its behaviour is *non-linear*, the output having one of two values.

(c) *Power considerations.* The power P used by a transistor as a switch should be as small as possible. It is given by $P = V_o \times I_C$. At cut-off $I_C = 0$ and at saturation $V_o \simeq 0$ (typically 0.2 V; for the ideal switching transistor it should be zero) therefore $P \simeq 0$ in both cases. To ensure saturation the transistor must be driven 'hard' by increasing I_B so that I_C/I_B is about *five* times less than h_{FE}.

For example, if $V_{CC} = 6$ V, $R_1 = 1$ kΩ and $h_{FE} = 100$, then I_C (max) $= 6$ V/1 kΩ $= 6$ mA. Hence since 20 is one-fifth of 100, if $I_C/I_B = 20$, i.e. $I_B = I_C/20 = 6/20 = 0.3$ mA, then 'hard' bottoming (satisfactory saturation) occurs.

Power is also used if the switch-over from cut-off to saturation takes place slowly. The larger R_2 is, the slower is the switching rate and the greater the value of V_i at which it happens (as the dotted curve in Fig. 21.35 shows). Fast-switching is desirable to avoid overheating and damage to the transistor and for this reason n-p-n types are preferred because their majority carriers (electrons) travel faster than the majority carriers (holes) in p-n-p types.

(d) *Advantages of transistor switching.* It occurs electrically and may be done by taking the input from another circuit. Also, it can occur millions of times a second.

(e) *Comparison of transistor as an amplifier and as a switch.* In both cases a 'load' (e.g. a resistor) is required in the collector circuit but the base bias is different. As an amplifier, the bias has to cause a collector current which makes the quiescent value of $V_{CE} \simeq \frac{1}{2} V_{CC}$. As a

switch, the bias has to make the collector current either zero or a maximum. In the first case the transistor is cut-off and $V_{CE} = V_{CC}$, in the second case the transistor is saturated and $V_{CE} \simeq 0$. The table sums up these facts.

Use	V_{CE}
Amplifier	$\frac{1}{2} V_{CC}$
Switch-on -off	0 V_{CC}

Alarm circuits

The two alarm circuits to be described use transistors as switches. The first is controlled by a photoconductive cell (p. 468) or light-dependent resistor (l.d.r.) and the second by a thermistor (p. 48).

(a) *Light-operated.* A simple circuit which switches on a lamp L when it gets dark is shown in Fig. 21.37. The resistor R and the l.d.r. form a potential divider across the supply. The input, applied between the base and emitter of the transistor, is the voltage across the l.d.r. and depends on its resistance.

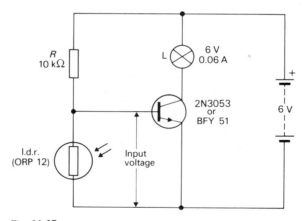

Fig. 21.37

In bright light, the resistance of the l.d.r. is low (e.g. 1 kΩ) compared with that of R (10 kΩ). Most of the supply voltage is dropped across R and the input is too small to switch on the transistor (for silicon about 0.6 V is needed). In the dark the l.d.r. has a much greater resistance (e.g. 10 MΩ) and more of the supply voltage is dropped across it and less across R. The voltage across the l.d.r. is now enough to switch on the transistor and produce a collector current sufficient to

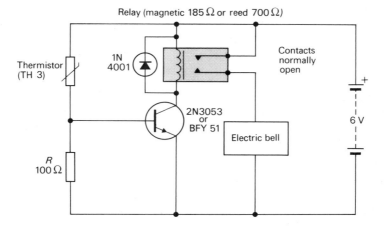

Fig. 21.38

light L. If R is replaced by a variable resistor the light level at which L comes on can be adjusted.

When R and the l.d.r. are interchanged, L is on in the light and off in the dark.

(*b*) *Temperature-operated.* In the circuit of Fig. 21.38, a thermistor and resistor R form a potential divider across the supply and the input to the transistor is the voltage across R. When the temperature of the thermistor rises (e.g. by heating with a match), its resistance decreases, causing less of the supply voltage to be dropped across the thermistor and more across R. If the latter exceeds 0.6 V (for silicon), the transistor is switched on and collector current (too small to ring the bell directly) flows through the coil of the relay (p. 278). As a result the 'normally open' contacts close enabling the bell to obtain the larger current it requires directly from the supply.

The diode across the relay coil protects the transistor from damage by the large e.m.f. induced in the coil when the current through it is switched off (p. 301). It is connected in reverse bias and offers an easy path for the current the e.m.f. produces.

How would you modify the circuit to make it a low-temperature alarm?

Field effect transistor (FET)

The junction (bipolar) transistor is a current-controlled amplifying device. In the field effect (unipolar) transistor it is the input *voltage* which controls the output current; the input current is usually negligible. This is a big advantage when whatever supplies the input cannot give much current, e.g. the crystal pick-up of a record player.

In a FET, a narrow channel of doped semiconductor connects two metal electrodes called the *drain* D and the *source* S. The voltage (or more correctly the electric field it produces) applied to a third electrode known as the *gate* G and located between S and D, determines the current which, as the terms suggest, flows from S to D.

The type of FET to be considered here is called a *junction-gate* FET (or JUGFET). Another type is the *metal-oxide semiconductor* FET (or MOSFET).

An n-channel JUGFET is shown diagrammatically in Fig. 21.39a. Its action depends on the formation and control of the depletion layer (see p. 495) at the p-n junction, to which *reverse* bias is applied. With respect to the source, the voltage of the drain V_{DS} is positive and that of the gate V_{GS} is negative. When V_{GS} is made more negative, the depletion layer (an insulator) widens and hence narrows the channel (a conductor). This reduces the electron flow, i.e. the drain current I_D, from S to D.

Normally only the *transfer* and *output characteristics* are plotted for FETs because the gate current is negligible. Those for an n-channel JUGFET such as the general purpose 2N3819 can be found using the circuit of Fig. 21.39b. Since a FET is voltage-operated, a transfer characteristic, Fig. 21.40a, shows the relation between the gate voltage V_{GS} and I_D (for fixed V_{DS}); it is almost linear.

The performance of a FET is measured by its *transconductance* g_m, defined by

$$g_m = \frac{\Delta I_D}{\Delta V_{GS}}$$

where ΔI_D is the change in I_D caused by a change of ΔV_{GS} in V_{GS}. Its value can be found from a transfer characteristic and is roughly in the range 1 to

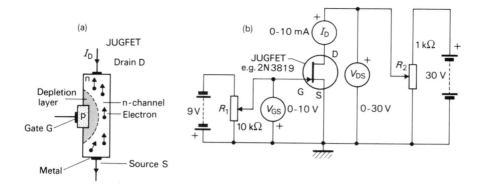

Fig. 21.39

10 mA V^{-1}. g_m corresponds to h_{fe} of a junction transistor.

The output characteristics, Fig. 21.40b, are similar to those of a junction transistor (Fig. 21.28), but their slope to the right of the 'knee' is less, indicating a higher *output resistance* r_o (50 kΩ to 1 MΩ).

The *input resistance* r_i is very high ($>10^9$ Ω). The reason is the very small capacitance (a few pF) of the capacitor formed by the metal electrode of the gate as one plate and the channel as the other plate with the depletion layer as the insulator between them.

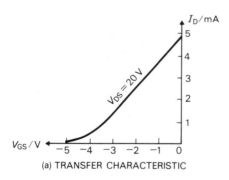

(a) TRANSFER CHARACTERISTIC

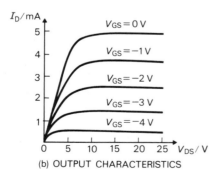

(b) OUTPUT CHARACTERISTICS

Fig. 21.40

FET as a voltage amplifier

In a FET voltage amplifier, changes in the gate (input) *voltage* cause changes in the drain current which are converted into larger voltage changes by a load resistor in the drain (output) circuit. The load line and d.c. operating point are selected as for a junction transistor voltage amplifier (p. 504). The chosen operating point is realized in practice by applying the correct quiescent bias voltage to the gate.

The circuit for an n-channel JUGFET voltage amplifier in common-source connection is shown in Fig. 21.41. The load resistor R_L and the battery voltage V_{DD} (note the symbol: V_{CC} is used for the supply voltage to a junction transistor) are both higher than for a bipolar transistor to obtain a reasonable gain. Negative (quiescent) bias is required for the gate. It is provided as follows.

(*a*) R_S *and* R_G. In the quiescent state the source current I_S ($= I_D + I_G \simeq I_D$ since $I_G \simeq 0$) is steady and causes a voltage drop across resistor R_S. The source end of R_S is therefore positive with respect to the other end connected by the high resistor R_G to the gate. R_G ensures that the gate has the same potential as the lower end of R_S. This is so because there is negligible current through R_G and hence practically no voltage across it. Both ends of R_G are at the same potential, namely that of the lower end of R_S, i.e. 'ground' or 0 V. The source is therefore at a higher potential than the gate, i.e. the gate is negative with respect to the source.

The circuit automatically compensates for any change of I_S and helps to stabilize the d.c. operating point because any increase in I_S increases the voltage V_S ($= I_S \times R_S$) across R_S. The potential of the source end of R_S (and so of the source) rises and since the lower end of R_S and the gate are tied to 0 V, the gate-source voltage V_{GS} must go more negative, tending to reduce I_S to its previous value.

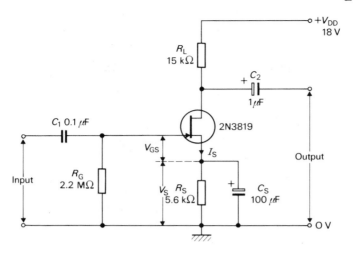

Fig. 21.41

(b) C_S. The large *decoupling* capacitor C_S provides a bypass (i.e. a low impedance) round R_S for the a.c. part of the source current (which becomes a varying d.c.) when an alternating input is applied. Otherwise the varying voltage developed across R_S would cause unwanted changes in the value (quiescent) of V_{GS} required to give the chosen operating point.

(c) C_1. Capacitor C_1 blocks any d.c. voltage from the input which would affect the operating point. With R_G it forms a voltage divider across the input. The alternating voltage developed across R_G is applied to the gate for amplification and for this voltage to be as large as possible, R_G should be large compared with the reactance of C_1. Since R_G is usually in the range 2.2 to 10 MΩ, C_1, can be small (e.g. 0.1 μF).

FET voltage amplifiers give lower gains than junction transistor types (typically 10 compared with up to 1000). It can be shown that the voltage gain A is given approximately by $A = g_m \times R_L$ where g_m is the transconductance of the FET. However their much greater input impedances make them better for certain applications, e.g. as impedance matching devices and as r.f. amplifiers.

Logic gates

Logic gates are circuits in which transistors, junction or field effect, act as high-speed switches. They are used in for example, pocket calculators, computers and industrial control systems. Their output depends on the input(s) and is either 'high' (e.g. the supply voltage V_{CC}, say 6 V) or 'low' (near 0 V).

(a) *NOT gate or inverter*. This is the most basic logic gate. Its circuit is the same as that of the simple

transistor switch in Fig. 21.36 and is given again in Fig. 21.42, along with the symbol most often used for it.

As explained before (p. 506), its action is to produce an output which is 'high' if the input is 'low' and vice-versa. The behaviour is summarized by its *truth table* where the digit '1' represents a 'high' input or output and the digit '0' indicates a 'low' one. We see that the output is 'high' only if the input is *not* 'high'.

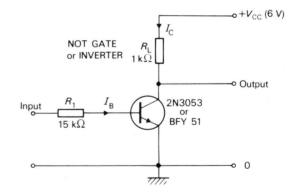

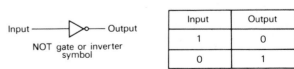

NOT gate or inverter symbol

Input	Output
1	0
0	1

Fig. 21.42

(b) *NOR gate*. The circuit is similar to that of the NOT gate but it has two (or more) inputs A and B, Fig. 21.43.

When both A and B are set to zero volts (or left disconnected) $V_i = 0$, i.e. the input is 'low' or '0',

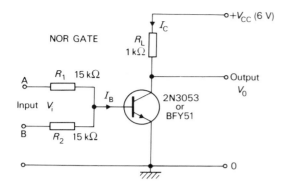

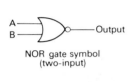

NOR gate symbol
(two-input)

A	B	Output
0	0	1
1	0	0
0	1	0
1	1	0

Fig. 21.43

therefore $I_B = 0$ and so $I_C = 0$. The transistor is cut-off, making $V_0 = V_{CC}$, i.e. the output is 'high' or '1'.

If either or both A and B are set 'high' by, for example, connection to $+V_{CC}$ through suitable values of R_1 and R_2, then $V_i = V_{CC}$. I_B will be large enough to saturate the transistor, i.e. I_C has its maximum value given by $I_C \times R_L \simeq V_{CC}$ and so $V_o \simeq 0$, giving a 'low' output.

These results are given in the truth table and show that the output is 'high' if neither A *nor* B is high. With more inputs the same happens, i.e. if any input is 'high', the output is 'low'.

A NOT gate is a NOR gate with one input.

Other gates can be made by combining NOT and NOR gates.

(*c*) *OR gate.* This is a NOR gate followed by a NOT gate; it is shown in Fig. 21.44 in circuit and symbol form. The truth table can be worked out from that of the NOR gate by changing '0's to '1's and '1's to '0's in the output of the NOR gate.

The output is '1' when either A *or* B *or* both is a '1', i.e. if any of the inputs is 'high' the output is 'high'.

(*d*) *AND gate.* It consists of a NOR gate with a NOT gate in each input and is shown in symbol form in Fig. 21.45. The output is a '1' only if input A *and* input B are also '1's.

(*e*) *NAND gate.* In this case an AND gate feeds a NOT gate, Fig. 21.46. The truth table can be derived from that of the AND gate by 'inverting' the outputs. We see that the output is a '1' if either or both inputs are not '1'.

Note. Other sets of symbols are used for logic gates. Those given here are favoured by electronic component manufacturers and most journals.

(*f*) *Experiment.* The truth tables for all the above logic gates can be derived or checked using one or more Transistor Units[1] like that in Fig. 21.47 which also shows an Indicator Unit (containing a 6 V, 60 mA lamp) for finding the level of the output voltage. If the

[1] Designed for the Nuffield Advanced Physics Course and called Basic Units.

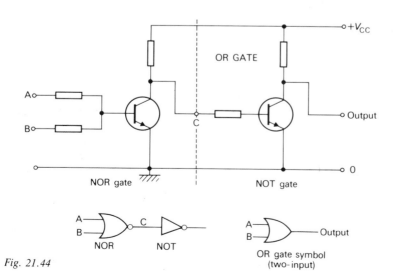

Fig. 21.44

A	B	C	Output
0	0	1	0
1	0	0	1
0	1	0	1
1	1	0	1

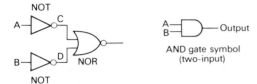

A	B	C	D	Output
0	0	1	1	0
1	0	0	1	0
0	1	1	0	0
1	1	0	0	1

A	B	E	Output
0	0	0	1
1	0	0	1
0	1	0	1
1	1	1	0

NOT

A — C

B — D

NOR

NOT

AND gate symbol
(two-input)

A
B — Output

AND   E   NOT

A
B

NAND gate symbol
(two-input)

A
B — Output

Fig. 21.45

Fig. 21.46

Fig. 21.47

lamp lights, the output is a '1', if it does not, the output is a '0'.

Multivibrators

Multivibrators are two-stage transistor switching circuits in which the output of each stage is fed to the input of the other by coupling resistors or capacitors. As a result the transistors are driven alternately into saturation and cut-off and whilst the output from one is 'high' the other is 'low'; we say their outputs are *complementary*.

The switch-over in each transistor from one output level or state to the other is so rapid that the collector voltage waveforms are almost 'square'. The term 'multivibrator', refers to this since a square wave can be analysed into a large number of sine waves with frequencies that are multiples (harmonics) of the fundamental.

Multivibrators are of three types.

(*a*) *The bistable or 'flip-flop'* has two stable states. In one, the output of the first transistor is 'high' and of the second 'low'. In the other state the opposite is the case. It will remain in either state until an external trigger pulse makes it switch. Bistables are used as electrical *memories* or *registers* in computers to store the binary digits '0' ('low' output) and '1' ('high' output). They are also at the heart of *binary counters* (p. 519).

(*b*) *The astable or 'free-running multivibrator'* has no stable states. It switches from one state to the other automatically at a rate determined by the circuit components. Consequently it generates a continuous stream of almost square-wave pulses, i.e. it is a square-wave oscillator or pulse generator. It has many uses including producing timing pulses to keep the various parts of a computer in step (it is then known as the 'clock') and generating musical notes in an electronic organ.

(*c*) *The monostable or 'one-shot'* has one stable state and one unstable state. Normally it rests in its stable state but can be switched to the other state by applying an external trigger pulse, where it stays for a certain time before returning to its stable state. Monostables are used to produce a square pulse of a certain height (voltage) and length (time) which may act as a 'gate' for another circuit and allow a number of timing pulses to pass for a certain time.

Multivibrators can be constructed using junction or field effect transistors, logic gates and operational amplifiers (p. 522).

Bistable

The basic circuit is shown in Fig. 21.48. The collector of each transistor is coupled to the base of the other by a resistor R_1 or R_2. The output of Tr_1 is thus fed to the input of Tr_2 and vice versa.

(*a*) *Action.* When the supply is first connected both transistors draw base current but, because of slight differences (e.g. in h_{FE}), one conducts more than the other. A cumulative effect occurs and as a result one, say Tr_1, rapidly saturates while the other, Tr_2, is driven to cut-off.

The collector voltage of Tr_1 is therefore 'low' (most of V_{CC} being dropped across R_3) and so no current can flow through R_1 into the base of Tr_2 (since about 0.6 V is required for this in a silicon transistor). Tr_2 remains off, its collector voltage is thus 'high' ($+V_{CC}$) and causes current to flow via R_2 into the base of Tr_1, thus reinforcing Tr_1's saturated condition. The feedback is positive and the circuit is in a stable state which it can maintain indefinitely with output Q = '1' (since Tr_2 is cut-off) and its complementary output $\bar{Q}$ = '0' (since Tr_1 is saturated). $\bar{Q}$ is pronounced 'not Q'.

The state can be changed by applying a *positive* pulse ($> + 0.6$ V) to the base of Tr_2 via R_6 (e.g. by temporarily connecting input 2 to $+V_{CC}$). Tr_2 now draws base current (through R_6) which is large enough to drive it into saturation. Its collector voltage falls from $+V_{CC}$ to near zero, so cutting off the base current to Tr_1. Tr_1 switches off, its collector voltage rises from near zero to $+V_{CC}$ and is fed via R_1 to the base of Tr_2 to keep it saturated. The circuit is in its second stable state but with Tr_1 off ($\bar{Q}$ = '1') and Tr_2 saturated (Q = '0').

The collector voltage (the output) of each transistor can thus be made to 'flip' to $+V_{CC}$ or 'flop' to zero by an appropriate trigger pulse. A *negative* pulse applied to the base of the transistor that is saturated also causes a change of state.

(*b*) *Further points.* To ensure the transistors are

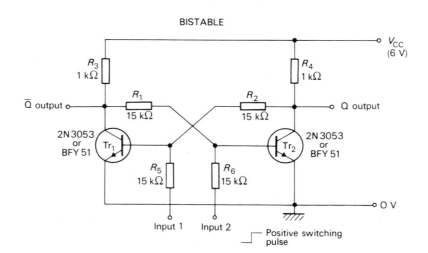

BISTABLE

Fig. 21.48

driven into cut-off and saturation (thereby keeping the power dissipation to a minimum during switching, p. 508), R_1/R_3 and R_2/R_4 should be less than h_{FE} for the transistors, i.e. in Fig. 21.48 we require $h_{FE} > 15$.

If the input 2 is made 'high' (e.g. at +6 V) and input 1 is 'low' (e.g. at 0 V), Q = '0', i.e. the bistable has been 'reset'; if input 1 is 'high' and input 2 'low', Q = '1', i.e. the bistable is 'set'.

(c) *Demonstration*. Two Transistor Units and two Indicator Units arranged as in Fig. 21.49 can be used to show bistable action. If input 1 is connected to +6 V, lamp L_1 is off and L_2 on. When input 1 is removed from +6 V and input 2 connected instead, L_1 comes on and L_2 goes off. The process can be repeated indefinitely and always when the output from one transistor goes 'high' the other goes 'low', but each state persists until an external pulse is applied.

(d) *Bistable as a memory*. A bistable can 'store' a single *binary digit* (called a 'bit') by staying in one of its stable states until there is a switching pulse. Taking the output of the bistable as the collector voltage of Tr_2 (i.e. the Q output in Fig. 21.48), when Q = 1, the bistable stores (remembers) the bit '1' as a 'high' voltage level. It is an electrical memory which stores

data presented to its input and makes that data available at its output. When Q = 0, the bit stored is '0'.

A separate bistable is needed to store each bit. For example to store the number 9 (1001 in binary), four bistables are required, the Q outputs of the first and fourth being 'high', i.e. a '1', whilst those of the second and third are 'low', i.e. a '0'.

Astable

The circuit of Fig. 21.50 is similar to that of the bistable (Fig. 21.48) but the transistors are coupled by capacitors C_1 and C_2, not by resistors.

(a) *Action*. When the supply is connected one transistor quickly saturates and the other cuts-off (as with the bistable). Each then switches automatically to its other state, then back to its first state and so on. As a result, the output voltage, which can be taken from the collector of either transistor is alternately 'high' (+6 V) and 'low' (near 0 V) and is a series of almost square pulses.

To see how these are produced suppose that Tr_2 was saturated (i.e. on) and has just cut off, while Tr_1 was off and has just saturated. Plate L of C_1 was at +6 V, i.e.

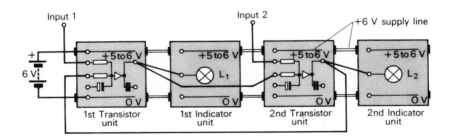

Fig. 21.49

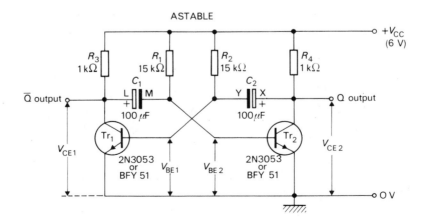

Fig. 21.50

the collector voltage of Tr_1 when it was off; plate M was at $+0.6$ V, i.e. the base voltage of Tr_2 when it was saturated. C_1 was therefore charged with a p.d. between its plates of $(6 V - 0.6 V) = +5.4$ V. At the instant when Tr_1 suddenly saturates, the voltage of the collector of Tr_1 and so also of plate L, falls to 0 V (nearly). But since C_1 has not had time to discharge, there is still $+5.4$ V between its plates and therefore the voltage of plate M must fall to -5.4 V. This negative voltage is applied to the base of Tr_2 and turns it off.

C_1 starts to charge up via R_1 (the p.d. across which is now 11.4 V), aiming to get to $+6$ V but when it reaches $+0.6$ V, it turns on Tr_2. Meanwhile plate X of C_2 has been at $+6$ V and plate Y at $+0.6$ V (i.e. p.d. across it is 5.4 V), therefore when Tr_2 turns on, X falls to 0 V and Y to -5.4 V and so turns Tr_1 off. C_2 now starts to charge through R_2 and when Y reaches $+0.6$ V, Tr_1 turns on again. The circuit thus switches continuously between its two states.

(b) *Voltage waveforms.* The voltage changes at the base and collector of each transistor are shown graphically in Fig. 21.51. We see that V_{CE1} (collector voltage of Tr_1) and V_{CE2} (collector voltage of Tr_2) are complementary (i.e. when one is 'high' the other is 'low'). Also, they are not quite square but have a rounded

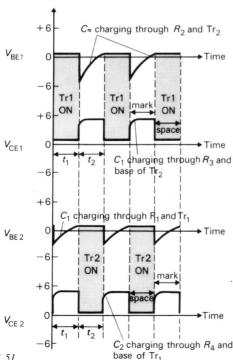

Fig. 21.51

rising edge. This happens because as each transistor switches off and V_{CE} rises from near 0 V to $+V_{CC}$, the capacitor (C_1 or C_2) has to be charged (via the collector load R_3 or R_4). In doing so it draws current and causes a small, temporary voltage drop across the collector load which prevents V_{CE} rising 'vertically'.

(c) *Frequency of the square wave.* The time t_1 for which Tr_1 is *on* (i.e. saturated with $V_{CE1} \simeq 0$) depends on how long C_1 takes to charge up through R_1 from -5.4 V to $+0.6$ V (i.e. 6.0 V $= V_{CC}$) and switch on Tr_2. It can be shown that

$$t_1 = 0.7\, C_1 R_1$$

That is, it depends on the time constant $C_1 R_1$ (p. 251). If $C_1 R_1$ is reduced, Tr_1 is on for a shorter time.

Similarly the time t_2 for which Tr_2 is *on* (i.e. the time for which Tr_1 is off) is given by

$$t_2 = 0.7\, C_2 R_2$$

The frequency f of the square wave is therefore

$$f = \frac{1}{t_1 + t_2} = \frac{1}{0.7(C_1 R_1 + C_2 R_2)}$$

If the circuit is symmetrical $C_1 = C_2$ and $R_1 = R_2$ and $t_1 = t_2$, i.e. the transistors are on and off for equal times. Their *mark-to-space ratio* is said to be 1. The frequency f in hertz (Hz) is therefore

$$f = \frac{1}{1.4\, C_1 R_1} = \frac{0.7}{C_1 R_1}$$

where C_1 is in farads (F) and R_1 in ohms (Ω). For example, if $C_1 = C_2 = 100\ \mu F = 100 \times 10^{-6} F = 10^{-4} F$ and $R_1 = R_2 = 15\ k\Omega = 1.5 \times 10^4\ \Omega$ then $f = 0.7/(10^{-4} \times 1.5 \times 10^4) = 0.7/1.5 \simeq 0.5$ Hz.

The mark-to-space ratio of V_{CE} (the output from either transistor) can be varied by choosing different values for the time constants $C_1 R_1$ and $C_2 R_2$. However, the base resistors R_1 and R_2 should be low enough to saturate the transistors. The condition for satisfactory saturation and cut-off is (as for the bistable) R_1/R_3 and $R_2/R_4 < h_{FE}$. In Fig. 21.50, the transistors must have $h_{FE} > 15$.

(d) *Demonstration.* The 'oscillations' of an astable can be shown in slow motion, by flashing lamps using two Transistor and two Indicator Units as in Fig. 21.52. The mark-to-space ratio and frequency of flashing can be changed by the variable resistors R_1 and R_2.

Monostable

The circuit is shown in Fig. 21.53; one stage is coupled

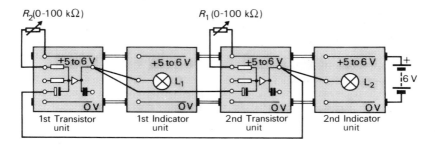

Fig. 21.52

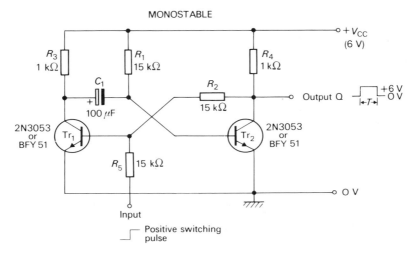

Fig. 21.53

by a capacitor C_1, and the other by a resistor R_2.

(*a*) *Action.* When the supply is connected, the circuit settles in its one stable state with Tr_1 off and Tr_2 held on (saturated) by R_1. The output voltage is therefore zero (i.e. Q = '0').

A positive pulse applied via R_5 (e.g. by connecting it momentarily to +6 V) to the base of Tr_1 switches it on. The potential of the right-hand plate of C_1 therefore falls rapidly from +0.6 V to −5.4 V (as explained for the bistable on p. 514) and switches off Tr_2, making Q go 'high' (+6 V). The monostable is now in its second state but only until the right-hand plate of C_1 charges

up through R_1 to +0.6 V. Then Tr_2 is switched on again and Q goes 'low' (0 V).

The time T of the square output pulse is the time for which Q is 'high' and is given approximately by $T = 0.7\ C_1 R_1$.

(*b*) *Demonstration.* If the circuit of Fig. 21.54 is used, a positive input pulse can be applied from the Switch Unit. With the switch first connected to 0 V as shown, the input to the first Transistor Unit is 'low', giving a 'high' output (L_1 on). This 'high' output is the input to the second Transistor Unit whose output is therefore 'low' (L_2 off), and, being fed back to the

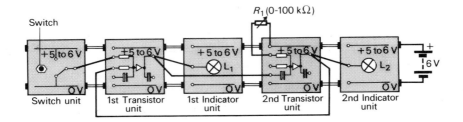

Fig. 21.54

input of the first Unit, it holds the output of that Unit 'high'. This is the one stable state of the circuit.

Pressing the switch makes the input to the first Transistor Unit 'high', its output goes 'low', L_1 goes off and the second Transistor Unit produces a pulse which brings on L_2 while the pulse lasts. On releasing the switch L_1 comes on again, L_2 goes off and the circuit returns to its stable state.

Varying R_1 varies the length of the output pulse.

Triggered bistable

Triggered bistables are the building blocks of binary counting circuits. They are similar to the basic bistable of Fig. 21.48, with extra components that enable successive pulses, applied to an input called the *trigger*, to make the bistable switch to and fro (or 'toggle') from one state to the other.

(*a*) *Action.* In Fig. 21.55 the pulse 'steering' is done by connecting between the collector and base of each transistor, a diode (D_1 or D_2) and a resistor (R_5 or R_6), their junctions going via capacitors (C_1 or C_2) to the trigger input. The action is as follows. Suppose Tr_1 is on (saturated) and Tr_2 off. The base of Tr_1 (point N) is therefore at +0.6 V and its collector (point O) near 0 V, so D_1 is on the verge of conducting since it is almost forward biased (the p.d. between points N and M being close to 0.6 V). D_2 on the other hand is heavily reverse biased because the collector of Tr_2 (point Z) is at +6 V (and so also is point X because no current flows through R_6) and its base (point Y) is at 0 V.

When square-wave pulses (6 V amplitude) are applied to the trigger input, the rapidly rising, *positive-going edge* of the first pulse (i.e. AB) raises the potential of both L *and* M to +6 V and of W *and* X to +6 V and +12 V respectively since the charges on C_1 and C_2 and therefore the p.d.s across them, cannot change instantly. As a result the potential of the lower ends of D_1 and D_2 (points M and X) are raised by 6 V, forcing D_1 into reverse bias and D_2 even more so than previously. The state of the bistable stays the same (Tr_1 on and Tr_2 off) and the charges on C_1 and C_2 quickly adjust to restore the potential at M and X to 0 V and 6 V respectively.

The rapidly falling *negative-going edge* of the first pulse (i.e. CD) suddenly lowers the potential of L from 6 V to 0 V, M from 0 V to −6 V and W and X both from 6 V to 0 V. This means that while D_2 is now on the point of conducting, D_1 becomes forward biased, conducts heavily and in effect short-circuits the base-emitter of Tr_1, diverting the current into C_1 which becomes charged. Tr_1 is cut off, its collector current falls, thereby causing its collector voltage to rise and turn on Tr_2. This switches the bistable into its second state (Tr_1 off and Tr_2 on).

The switching process is repeated and the bistable returns to its first state (Tr_1 on and Tr_2 off) when the *negative-going edge* (GH) of the second input pulse arrives and so on.

The table below shows that each output is a '1' once for every two trigger input pulses. That is, the circuit *divides by two* and this is the basis of binary counting.

Trigger input pulse no.	1	2	3	4
Q output	0	1	0	1
$\overline{Q}$ output	1	0	1	0

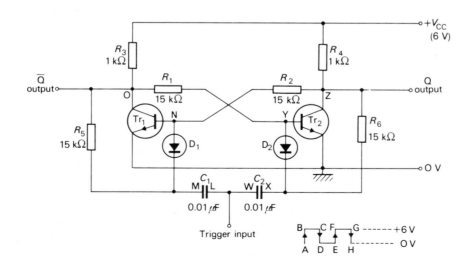

Fig. 21.55

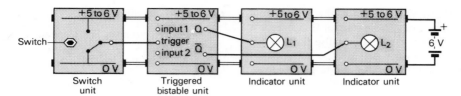

Fig. 21.56

(*b*) *Demonstration*. The divide-by-two action can be shown using modules connected as in Fig. 21.56. Pressing and releasing the switch on the Switch Unit once, applies to the trigger input a positive pulse which goes from 0 V to +6 V to 0 V. When it falls from +6 V to 0 V (i.e. when the switch is released) the bistable switches (so long as inputs 1 and 2 are both unconnected or at 0 V). L_1 and L_2 show that (*i*) the Q and $\bar{Q}$ outputs are *complementary* (i.e. when L_1 is on, L_2 is off and vice versa) and (*ii*) *half* as many pulses appear at Q (or $\bar{Q}$) as are applied to the trigger.

Binary counter

Binary counters are used in devices such as digital watches and computers. They count electrical pulses in the binary system using '0's and '1's and can be made from triggered bistables.

(*a*) *Action*. In the block diagram of Fig. 21.57*a* the three bistables change state on the *falling* (negative-going) edges of pulses. All outputs are set initially to zero and each is connected to the input of the next bistable. The pulses to be counted are applied to the trigger input of the first bistable BS1.

When the first pulse arrives, the output from BS1 switches from 'low' to 'high' on edge ab, making $Q_1 = '1'$, Fig. 21.57*b*. This rising (positive-going) edge AB of the output pulse from BS1 goes to the input of BS2 but because it is not a falling edge, no change in the state of BS2 occurs, leaving Q_2 and Q_3 as '0's. The count in binary is 001.

The falling edge cd of the second pulse switches the output of BS1 again, this time from 'high' to 'low', making $Q_1 = '0'$. The falling edge CD of the output pulse from BS1 switches BS2 whose output goes from 'low' to 'high', i.e. $Q_2 = '1'$. The positive-going change

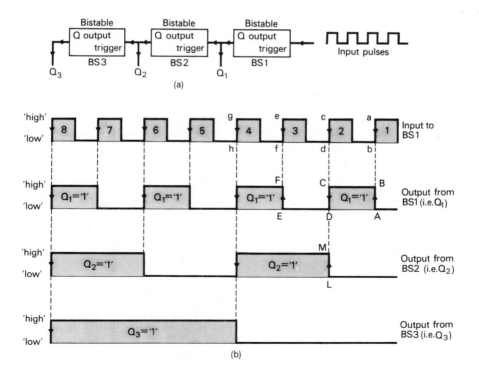

Fig. 21.57

LM in the output of BS2 is applied to the trigger input of BS3 but it does not change the state of BS3 since it is not a falling edge. After the second pulse the count is 010.

The edge ef of the third input pulse to BS1 changes the output of BS1 from 'low' to 'high', i.e. $Q_1 = $ '1'. However, this output change EF is a rising edge and does not switch BS2, so leaving $Q_2 = $ '1' and $Q_3 = $ '0'. The count is now 011.

The changes caused by the falling edge gh of the fourth input pulse to BS1 are shown in Fig. 21.57b.

You can see that (i) each bistable produces one output pulse for every two pulses applied to its trigger input, i.e. it divides by two, and (ii) the bistable outputs represent powers of 2; for example Q_1 gives 2^0 ($= 1$) and is the only 'high' output after the first pulse, Q_2 gives 2^1 ($= 2$) and is the only 'high' output after the second pulse, Q_3 gives 2^2 ($= 4$) and is the only 'high' output after the fourth pulse and so on. If Q_1, Q_2 and Q_3 are all 'high' (i.e. '1's), the count is 7 (111 in binary).

By using large numbers of bistables (as is done in integrated circuits), large numbers of pulses can be counted. In general n bistables count to $(2^n - 1)$, e.g. if $n = 12$, 4095 pulses can be counted.

(b) *Demonstration.* A 3-bit binary up-counter to count to 7 can be made using three Bistable Units each with an Indicator Unit connected to its Q output, Fig. 21.58. The counter is first set to zero by briefly connecting to +6 V the input 1 of any bistable whose Q output is 'high', i.e. lamp lit.

Pulses are fed in by pressing and releasing the switch on the Switch Unit. It will be seen that the changes of state occur on 'release' when the input pulse *falls* from +6 V to 0 V. The truth table is given below; L_1 gives the least significant digit.

Pulse no.	L_3	L_2	L_1
1	0	0	1
2	0	1	0
3	0	1	1
4	1	0	0
5	1	0	1
6	1	1	0
7	1	1	1

A down-counter is made by connecting the Indicator Units to $\bar{Q}$, not Q; the Bistables still go from Q to trigger.

Integrated circuits (ICs)

An integrated circuit is an electronic circuit on a 'chip' of silicon about 5 mm square, consisting in some cases, of hundreds of thousands of transistors and perhaps diodes, resistors and capacitors along with interconnections. Integrated circuits are the building blocks of microelectronics.

(a) *Manufacture.* Silicon containing no more than 1 in 10^{10} parts of impurity is produced chemically from silicon dioxide, the main constituent of sand. It is then melted in an inert atmosphere and as a small crystal of pure silicon (a 'seed') is inserted and slowly withdrawn from it, crystallization starts. A cylindrical bar, up to 10 cm in diameter and 1 metre or so long, is formed as a single, near-perfect crystal (i.e. its atoms are arranged almost perfectly regularly throughout). It is cut into $\frac{1}{4}$ to $\frac{1}{2}$ mm thick wafers whose surfaces are ground and highly polished, Fig. 21.59a.

Depending on the size of the chip, up to five hundred identical ones are formed side by side on the surface of one wafer by the *planar* process. This process first involves depositing an insulating layer of silicon oxide on the wafer and then using a pattern of photographic masks, designed from a very large drawing of one chip, to create 'windows' in the oxide by exposure to ultraviolet, developing and etching, Fig. 21.59b.

Doping then occurs by exposing the wafer at high temperature to the vapour of either boron or phosphorus, so that their atoms diffuse through the 'windows' (guided by masks) into the silicon. The p- and n-type regions so produced for the various components are next interconnected to give the required circuit by depositing aluminium, again using masks, Fig. 21.59c. Several layers can be built in this way, one on top of the other.

Before the wafers are cut into separate chips, each is tested and faulty ones discarded, Fig. 21.59d and e (up to 70 per cent can fail). Each chip is enclosed in a plastic case and connected (automatically) by gold wires to the pins on the case, Fig. 21.59f.

Fig. 21.59g shows an IC with the pins arranged in two lines, called the *dual-in-line* (d.i.l.) arrangement.

The complete process, which can require up to three months, must be done in a controlled, absolutely clean environment.

(b) *Types.* ICs fall into two broad groups, *linear* or *analogue ICs* which contain amplifying circuits and *digital ICs* containing switching circuits.

The output of a *linear IC* changes more or less uniformly as the input changes at a constant rate, i.e. the output is directly proportional to the input (it varies

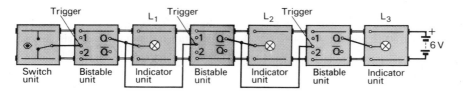

Fig. 21.58

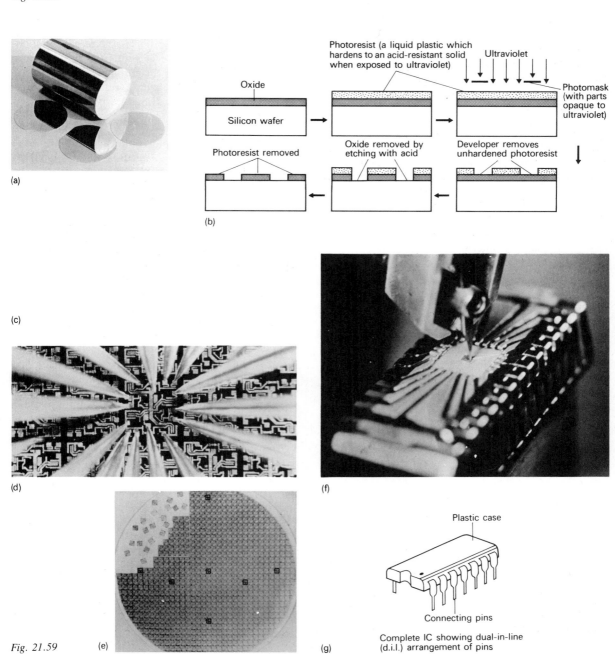

Fig. 21.59

linearly with the input). A wide variety of linear ICs is available, for example, as a.f. amplifiers (p. 526), r.f. amplifiers (p. 526) and most importantly as operational amplifiers.

The output of a *digital IC* jumps suddenly from one value to another when the input changes, and not smoothly as in a linear I.C. They are made as complete systems containing many logic gates, multivibrators, counters, memories, etc. There are two main families— TTL (transistor-transistor logic) and CMOS (complementary metal-oxide-semiconductor logic). The first (e.g. the 74 series) uses junction transistors as well as other components, the second (e.g. the 4000B series) is built almost entirely from field effect transistors (MOSFETS).

TTL was the first to be developed and although it gives faster switching than CMOS it requires much larger quiescent currents. Also, whereas CMOS works from any unstabilized voltage between 3 and 15 V, TTL requires a stabilized supply of 5 V ± 0.25 V. Greater component density is achieved with CMOS, so making large-scale integration (LSI) easier and accounting for their use in digital watches, electronic calculators and microprocessors for controlling computers, robots and machinery.

Operational amplifier

An operational amplifier (op amp) is so-called because it can perform electronically, mathematical operations such as addition, multiplication and integration. These operations form the basis of analogue computing (in which mathematical equations representing physical systems, e.g. the forces on a bridge, are solved) and it was for this that the op amp was designed originally. Nowadays it is also used widely as a low power amplifier.

The first op amps were made from discrete (separate) components, they are now available in integrated circuit form and belong to the linear (analogue) group, although they can perform non-linear (i.e. digital) operations. One might contain twenty or so transistors as well as resistors and small capacitors, all on the same tiny silicon chip.

(*a*) *Properties.* The chief properties of an op amp are:
(*i*) *a very high voltage gain*, called the *open-loop gain* A_0, which typically is 10^5 for d.c. and low frequencies but decreases with frequency, Fig. 21.60a,
(*ii*) *a very high input resistance* r_i, typically 10^{12} Ω: it therefore draws a minute current from the device or circuit supplying its input (or putting it another way, it does not alter the value of the voltage applied to its input, i.e. it behaves like a high-resistance voltmeter) and
(*iii*) *a very low output resistance* r_o typically 100 Ω, which means that its output voltage can be transferred with little loss to a load greater than a few kilohms.

(*b*) *Connections.* A knowledge of the internal circuit of an op amp is not necessary to use it for a particular job, but we have to know how to connect it. The symbol for any amplifier is shown in Fig. 21.60b with the five main connections for an op amp added.

A d.c. power supply is needed with a centre tap which is used as the reference level for input and output voltages, i.e. 'ground' or 0 V. A supply in the range ±5 V to ±15 V is suitable in most cases. To simplify circuit diagrams the power supply connections are often omitted.

There are two separate inputs, the *inverting* (marked −) and the *non-inverting* (marked +). The signs should not be confused with supply voltage polarities; they arise as follows. If a *positive* voltage V_1 is applied to the inverting input, an amplified *negative* voltage $-A_0V_1$ is produced between the output terminal and 0 V; if V_1 is negative the output is positive, i.e. input and output voltages are in antiphase, hence the − sign. A *positive* voltage V_2 at the non-inverting input

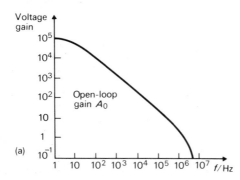

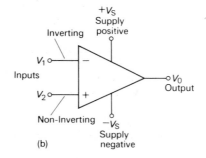

Fig. 21.60

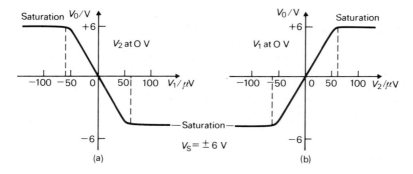

Fig. 21.61   (a)   (b)

gives a *positive* output voltage $+A_0V_2$; if V_2 is negative, the output is negative, i.e. input and output voltages are in phase, hence the + sign.

(*c*) *Characteristics.* Typical input-output voltage characteristics for both inverting and non-inverting inputs are given in Fig. 21.61*a* and *b*. They are called *transfer characteristics* and show that there is a very small range of input voltages over which the output is directly proportional to the input, i.e. when the op amp behaves more or less linearly with minimum distortion of the amplified output. Outside this input range, whatever the input, the output voltage does not change but equals approximately either the positive or the negative supply voltage. The op amp is then *saturated*, acts non-linearly and a suitable alternating input voltage would make its output 'switch' from a 'high' voltage (near supply +) to a 'low' one (near supply −).

The limited linear behaviour of an op amp is due to its very high gain. Thus if $A_0 = 10^5$ and the supply is ±6 V, the maximum output voltage swing is 12 V, giving a maximum input voltage swing (for linear operation) of 12 V/10^5 = 120 μV (±60 μV).

Op amp as a voltage amplifier

The very high gain of an op amp would, for many purposes, give much greater amplification than is required. Also, over its linear region an op amp tends to be thermally unstable and sensitive to electrical pickup. In practice, it is almost always used with *negative feedback*. This reduces the gain of any amplifier to which it is applied (not just an op amp), but it has the following considerable advantages in all cases:

(*i*) the gain is predictable and less dependent on the properties of the IC (or transistor),

(*ii*) stability is greater,

(*iii*) response is more linear, i.e. less distortion of the output, and

(*iv*) more uniform amplification of a wider band of frequencies.

(*a*) *Theory.* Fig. 21.62 shows the basic circuit of an op amp voltage amplifier with a similar performance to the junction and FET amplifiers described earlier (Figs. 21.32 and 21.41). The input voltage V_i (d.c. or a.c.) is applied to the inverting (−) terminal *via* a resistor R_i; the non-inverting (+) terminal goes to 0 V. R_f, called the *feedback resistor*, feeds back a certain fraction of the output voltage V_o to the inverting input and since the output is in antiphase with the input, the input to the inverting terminal, and therefore the output and gain, are reduced, i.e. the feedback is negative.

The potential at the inverting input (point P) can never be far from zero (a maximum of ±60 μV from it in the case considered at the end of the previous section) because of the high value of A_0. Therefore the p.d. across $R_i \simeq V_i$ and that across $R_f \simeq V_o$. When V_i is positive, current I flows as shown through R_i and then through R_f, *only a negligible fraction of it enters the inverting input* of the op amp.[1] And so

$$I = \frac{V_i}{R_i} = -\frac{V_o}{R_f}$$

The minus sign shows V_i and V_o are in antiphase. Hence the gain, called the *closed-loop gain A*, is given by

$$A = \frac{V_o}{V_i} = -\frac{R_f}{R_i}$$

[1] This assumption can be justified in a more rigorous treatment such as is given in an article in *Physics Education* Vol. 15 No. 2 of March 1980 by M. K. Summers, entitled 'Operational amplifiers—some misconceptions'. The essence of the argument is, if V_p is the small potential at P, the p.d. across R_f is $V_p - (-A_0V_p) = V_p(1 + A_0)$, while the p.d. across the input resistance r_i of the op amp is only V_p. Therefore the current through R_f will always be $(1 + A_0)r_i/R_f$ times greater than the current i_{in} flowing into the op amp and since $r_i \simeq 10^{12}$ Ω and $A_0 \simeq 10^5$, the assumption is justified for all practical purposes.

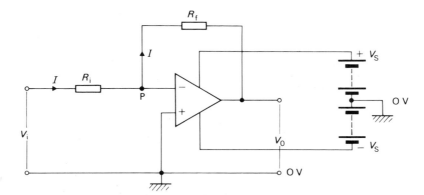

Fig. 21.62

For example, if $R_f = 1\ M\Omega$ and $R_i = 10\ k\Omega$, $A = -100$ exactly and an input of 0.01 V will cause an output change of 1 V. If the Texas op amp TL081C is used with these resistors, this value of A is maintained over a frequency range of 0 to 30 kHz.

The gain of the amplifier would therefore seem to depend on the values chosen for the discrete resistors R_f and R_i (which can be made very accurately) and is independent of the characteristics of the op amp (which cannot be made so accurately). Compare this expression with that for the gain of a transistor voltage amplifier (p. 504).

(b) *Demonstration.* A module[1] and circuit for demonstrating the amplification of an alternating voltage by an op amp (the TL081C, which has the same pin connections as the 741 but greatly improved characteristics) in inverting mode are shown in Fig. 21.63a and b. The signal generator, op amp module and CRO are set up with the values indicated. The amplitude of the input from the generator is slowly increased until the sine wave output waveform on the CRO starts to be 'clipped' (i.e. the crests and troughs are 'squared'). The input is then slightly reduced to give a pure sine wave output and the Y-gain on the CRO adjusted until the sine wave just occupies ten vertical divisions of the screen.

If the Y input 'high' lead from the op amp output (point A) is transferred to the inverting input on the op amp (point B), the input waveforms should occupy about one vertical division, confirming that $A = V_o / V_i = R_f / R_a = 100/10 = 10$ (numerically). Other values of R_f and R_a can be tried.

The same arrangement may be used to show that the gain falls off rapidly above a certain frequency. The signal generator is set to 10 kHz and 25 mV (e.g. 2.5 V

[1] Unilab 'Blue Chip': 511.006.

with −40 dB attenuation) and the CRO to 1 V/div. and 0.1 ms/div. The Y-gain on the CRO is adjusted until the trace just occupies 10 vertical divisions. If the generator frequency is now raised to 100 kHz, the amplitude of the output waveform decreases appreciably.

The pin connections for the TL081C (and 741) are shown in Fig. 21.63c. The offset connections are used in analogue computing applications to ensure that the output is zero when the inputs are zero.

Op amp as a voltage comparator

If both inputs of an op amp are used simultaneously the output voltage V_o is given by

$$V_o = A_o(V_2 - V_1)$$

where V_1 is the inverting input, V_2 the non-inverting input and A_0 the open-loop gain. The voltage difference between the input terminals i.e. $(V_2 - V_1)$, is amplified and appears at the output.

However, A_0 is so large that if $(V_2 - V_1)$ exceeds about $100\ \mu V$, the op amp saturates. When $V_2 > V_1$, V_o rises to a steady value close to the positive supply voltage $+V_s$. V_o is therefore either 'high' or 'low' and the circuit can be used as a voltage comparator to indicate whether V_2 is greater or less than V_1. The op amp is then behaving as a digital (non-linear) device.

The action may be demonstrated with the circuit of Fig. 21.64, again using the module of Fig. 21.63a. The signal generator, op amp and CRO are adjusted as shown and V_1 set to zero. The waveforms that will be observed, either on a double beam CRO or by connecting the Y input 'high' lead of a single beam CRO to A and B in turn, are shown in Fig. 21.65a. We see that when V_2 is positive, $V_o \simeq +6$ V and when V_2 is negative, $V_o \simeq -6$ V. V_o thus switches between supply voltage + and −, generating a train of square pulses.

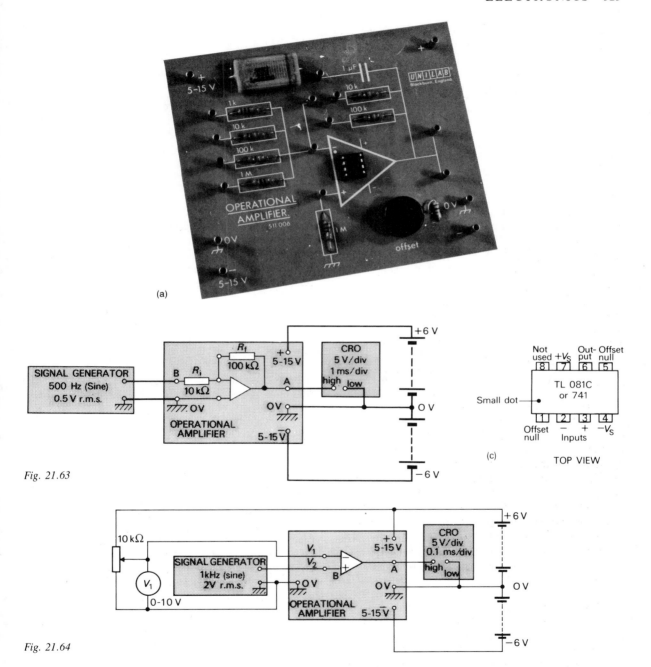

Fig. 21.63

Fig. 21.64

(a)

(c)

TOP VIEW

If V_1 is increased from zero, the switching occurs when $V_2 \simeq V_1$, as shown in Fig. 21.65*b*, giving a mark-to-space ratio which is no longer 1.

An op amp voltage comparator is used in a digital voltmeter (p. 527).

Electronic systems

No matter how complex an electronic system such as a radio receiver, a digital voltmeter, a CRO or a computer may seem, it can be regarded as consisting of a

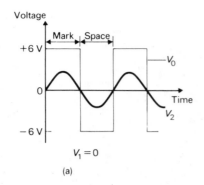

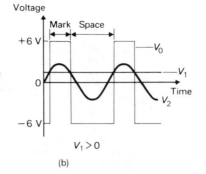

Fig. 21.65   (a)   (b)

number of basic building blocks or *modules* (often in the form of ICs), each performing a certain task. This may be to amplify, switch, count or store a signal.

The job of the electronics engineer today is to know what different modules can and cannot do and how to assemble the minimum number to achieve a particular end. An understanding of the system as a whole is required and not necessarily a detailed knowledge of how individual circuits work.

In this so-called *systems approach*, the block diagram rather than the circuit diagram is used to simplify matters and give us a broad but working appreciation of the system overall. Its use will be illustrated for a radio communication system (an analogue system) and a digital voltmeter (a digital system).

Radio transmitter and receiver

The radiation from an aerial is appreciable only when the length of the aerial is comparable with the wavelength of the electromagnetic wave produced by the a.c. flowing in it. A 50 Hz alternating current corresponds to a wavelength of 6×10^6 m and so in practice radio frequency (r.f.) currents (i.e. frequency > 20 kHz) must be supplied. However, since speech and music generate audio frequency (a.f.)

currents (i.e. frequencies 20 Hz to 20 kHz), some means of combining the two is necessary if sound is to be conveyed over a distance.

In a transmitter an *oscillator* generates r.f. current which if applied to the aerial would produce an electromagnetic wave, called the *carrier wave*, of constant amplitude and having the same frequency as the r.f. current. If a normal receiver picked up such a signal nothing would be heard. The r.f. signal can be modified or *modulated* in various ways so that it carries the a.f. intelligence.

In *amplitude modulation* (a.m.) the amplitude of the r.f. is varied so that it depends on the a.f. current from the microphone, the process occurring in a *modulator*. This type of transmission is used for medium and long-wave broadcasting in Britain. Fig. 21.66a shows a simplified block diagram for an a.m. transmitter. In *frequency modulation* (f.m.) which is used in v.h.f. broadcasts, the frequency of the carrier is altered at a rate equal to the frequency of the a.f. signal but the amplitude remains constant. Frequency modulated signals are relatively free from various kinds of electrical interference.

A block diagram of a simple receiver of a.m. signals is given in Fig. 21.66b. The *tuning circuit* selects the wanted signal from the *aerial* before it is amplified by

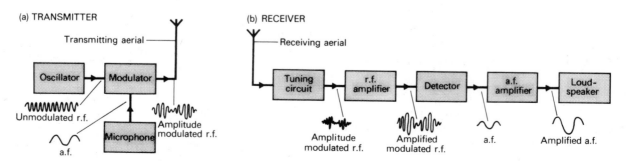

Fig. 21.66

the *r.f. amplifier*. The *detector* (or demodulator) separates the a.f. intelligence (speech or music) from the r.f. carrier. The *a.f. amplifier* then boosts the a.f. so that an audible sound is produced in the *loudspeaker*.

Digital voltmeter

An electronic digital voltmeter, Fig. 21.67, gives a reading on a numerical display (e.g. LED). It eliminates errors due to parallax that can arise in instruments requiring the position of a pointer on a scale to be estimated and also has a very high input resistance (e.g. 10 MΩ).

The simplified block diagram in Fig. 21.68 and the waveforms below it, help us to follow the action. The d.c. voltage to be measured is fed to one input of a voltage *comparator* (p. 524). The other input of the comparator is supplied by a *ramp generator* which produces a repeating sawtooth waveform. The output from the comparator is 'high' (a '1') until the ramp voltage equals the input voltage when it goes 'low' (a '0').

The comparator output is applied to one input of an *AND gate*, the other input of the gate being fed by a steady train of pulses from a *pulse generator*. When both these inputs are 'high', the gate opens (p. 512) and gives a 'high' output, i.e. a pulse. The number of output pulses so obtained from the *AND gate* depends on the length of the comparator output pulse, i.e. on the time taken by the ramp voltage to reach the value of the input voltage. If the ramp is linear, this time is proportional to the input voltage.

The output pulses from the *AND gate* are recorded by a *binary counter* and then converted into decimal form by a *decoder* before being passed on to the *display*. The voltmeter is thus sampling the input voltage at regular intervals.

The whole process commences when the voltmeter is switched on and a pulse from a trigger circuit starts the ramp generator and sets the counter to zero. When the input voltage is of the order of millivolts it is amplified before being measured. With some additional circuitry the voltmeter can be adapted for use as a *multimeter* to measure a.c. voltages, current and also resistance.

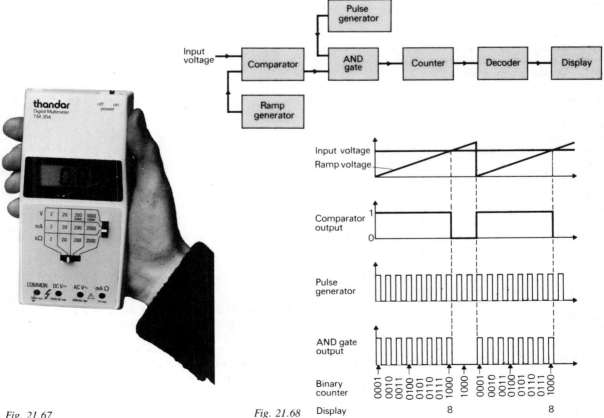

Fig. 21.67

Fig. 21.68

QUESTIONS

1. Draw a sketch to show the essential parts of a cathode-ray oscilloscope having electrostatic deflection.

With the help of your sketch explain how in a cathode-ray oscilloscope: (*a*) the electrons are produced, (*b*) the electrons are focused, (*c*) the spot is made visible, and (*d*) the brightness of the spot is controlled.

What is meant by stating that a cathode-ray oscilloscope is fitted with a linear time base of variable frequency? (*J.M.B.*)

2. (*a*) Draw a block diagram for an oscilloscope.

(*b*) Sketch and explain the forms of the traces seen on an oscilloscope screen when a p.d. alternating at 50 Hz is connected across the Y-plates if the time base is linear and has a frequency of (*i*) 10 Hz, and (*ii*) 100 Hz.

(*c*) What is the frequency of an alternating p.d. which is applied to the Y-plates of an oscilloscope and produces five complete waves on a 10 cm length of the screen when the time base setting is 10 ms cm^{-1}?

3. The gain control of an oscilloscope is set on 1 V cm^{-1}. What is (*i*) the peak value, and (*ii*) the r.m.s. value of an alternating p.d. which produces a vertical line trace 2 cm long when the time base is off?

4. With the help of a simple diagram discuss qualitatively the factors which determine the deflection of the spot on the screen of a cathode-ray oscilloscope when a given p.d. is applied between the Y plates.

What is meant by the statement that a CRO has a sensitivity of 40 V cm^{-1} for Y deflection?

A generator is believed to supply a sinusoidal voltage of 80 V r.m.s. at 200 Hz. Describe how you would use a CRO having the above sensitivity to test this. You may assume that a sinusoidal 50 Hz a.c. mains supply at any desired voltage is available. (*J.M.B.*)

5. One sinusoidal voltage alternating at 50 Hz is connected across the X plates of a cathode-ray oscilloscope and another 50 Hz sinusoidal alternating voltage of approximately the same amplitude is connected across the Y plates. Sketch what you would expect to observe on the screen if the phase difference between the voltage is (*a*) zero, (*b*) π/2, and (*c*) π/4.

If the voltage on the X plates is replaced by a 100 Hz sinusoidal alternating voltage of similar amplitude, sketch what you would observe on the screen.

Explain briefly why figures of this type are useful in the study of alternating voltages. (*J.M.B.*)

6. (*a*) What is meant by (*i*) intrinsic and (*ii*) extrinsic conductivity?

(*b*) Explain the terms p-type and n-type semiconductors.

(*c*) Describe a p-n junction diode and draw a graph to show how the current through it varies with the p.d. across it.

7. Write a brief essay on the transistor and explain how transistors may be used to produce a simple amplifying circuit. (*W.*)

8. The resistance of a slice of semiconductor material falls when exposed to light of sufficiently short wavelength; explain briefly why this occurs.

The resistance, *R*, of the CdS photoconductive cell in Fig. 21.69 falls when the cell is illuminated. Find the approximate value of *R* at which the relay will close when the illumination is increased.

Relay closes at 20 mA. Current gain of the transistor = 80. Assume that p.d. between base and emitter is negligible. (*J.M.B. Eng. Sc.*)

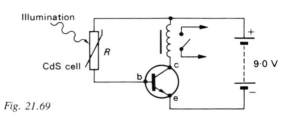

Fig. 21.69

9. In the junction transistor voltage amplifier circuit of Fig. 21.70 if $R_1 = 100$ kΩ, $R_2 = 1$ kΩ, $V_{CC} = 6.0$ V and $V_{BE} = 0.6$ V, calculate
    (*i*) the voltage across R_1,
    (*ii*) I_B,
    (*iii*) I_C if $h_{FE} = 60$,
    (*iv*) the voltage across R_2, and
    (*v*) the voltage across the collector-emitter.

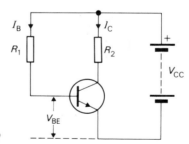

Fig. 21.70

10. The output characteristics of a junction transistor in common-emitter connection are shown in Fig. 21.71*a*. The transistor is used in an amplifier with a 9 V supply and a load resistor of 1.8 kΩ.

(*a*) On a separate graph draw the load line.

(*b*) Choose a suitable d.c. operating point and read off the quiescent values of I_C, I_B and V_{CE}.

(*c*) What is the quiescent power consumption of the amplifier?

(*d*) If an alternating input voltage varies the base current by

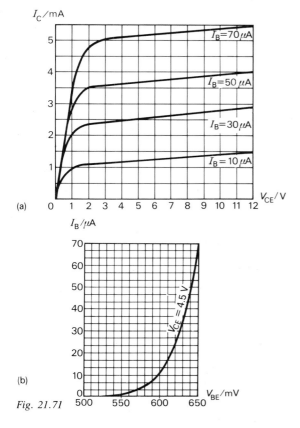

Fig. 21.71

$\pm 20\ \mu A$ about its quiescent value, what is (*i*) the variation in the collector-emitter voltage and (*ii*) the peak output voltage?

(*e*) An input characteristic of the transistor is given in Fig. 21.71*b*. Use it to find the base-emitter voltage variation which causes a change of $\pm 20\ \mu A$ in the base current.

(*f*) Using your answers from (*d*) and (*e*) find the voltage gain of the amplifier.

(*g*) If the amplifier uses the collector-to-base bias circuit of Fig. 21.32 (p. 506), calculate the value of R_B to give the quiescent value of I_B. (Assume $V_{BE} = 0.6$ V.)

11. (*a*) Explain briefly the action of a JUGFET.
(*b*) If the transconductance g_m of a FET is 5 mA/V, what does this mean?
(*c*) In the FET voltage amplifier circuit of Fig. 21.41 (p. 511), state the purpose of (*i*) R_L, (*ii*) R_S, (*iii*) R_G and (*iv*) C_S.

12. State what each of the *three* types of multivibrator does and give one use for each.

13. (*a*) State *three* important properties of an op amp and say why they are important.
(*b*) Explain the term *negative feedback* and state *four* advantages of using it in an op amp (or any type of) voltage amplifier.
(*c*) Define *closed-loop gain A* and derive an expression for it for an inverting op amp voltage amplifier with an input resistor R_1 and a feedback resistor R_2. Use the expression to calculate A if $R_1 = 10$ kΩ and $R_2 = 100$ kΩ.

14. To find whether there was a minimum time for which a sound must persist to be heard a psychologist required an electronic device to produce single bursts of high-pitched sound, each lasting for a short time.
Here is a block diagram of the arrangement used, Fig. 21.72.
(*a*) What is the function of A?
(*b*) What is the function of B?
(*c*) Suggest an addition or additions to the arrangement to enable the time for which each burst lasts to be timed to an accuracy of at least 1 millisecond. You may add to the sketch as well as explaining your idea below. (*O. and C. Nuffield*)

15. The diagram, Fig. 21.73, is of a very simple radio receiver which can be used for broadcasts from one station only.
(*a*) The tuning circuit selects one station (one frequency). Is the output of energy of the selected signal coming from the

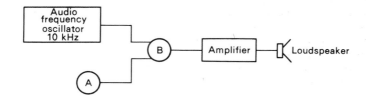

Fig. 21.72

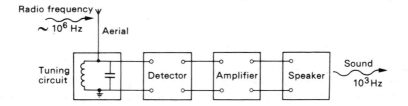

Fig. 21.73

tuning circuit greater than the energy of the signal collected by the aerial? Explain your answer briefly.

(*b*) For which of the other three boxes is the energy of the output signal of the box larger than the energy of the input signal to the box? Give below the name of the box or boxes for which there is energy increase.

(*c*) How could the tuning circuit be altered so that it could select other stations?

(*d*) Which box or boxes is/are designed to transform energy from one form to another?

(*e*) The frequency of the signal received at the aerial is about 10^6 Hz, and the frequency of the speaker output is about 10^3 Hz. Would it matter if the amplifier could only amplify signals of frequency less than 10^5 Hz and so failed to amplify signals of 10^6 Hz? Explain your answer.

(*O. and C. Nuffield*)

16. A NOR gate 'opens' and gives an output only if *both* its inputs are 'low', whilst an OR gate 'closes'. An AND gate 'opens' only if both inputs are 'high', but a NAND gate 'closes'.

Copy and complete the truth table for each gate. 'High' is represented by '1' and 'low' by '0'.

Input 1	Input 2	NOR output	OR output	AND output	NAND output
0	0				
1	1				
1	0				
0	1				

17. The system in Fig. 21.74*a* makes three lamps flash in the order of British traffic signals. The amber lamp is connected to the output of the slow astable, the red one to one of the outputs of the bistable and the green one to the output of the NOR gate.

In Fig. 21.74*b* the output pulses from the astable are shown. Copy it and add the corresponding outputs from the bistable and the NOR gate. The bistable switches only when its input goes from 'high' to 'low' and a NOR gate 'opens' only when both its inputs are 'low'.

Also show which lamp(s) is (are) on during intervals 1, 2, 3, 4 and 5. Write out the truth table for the system.

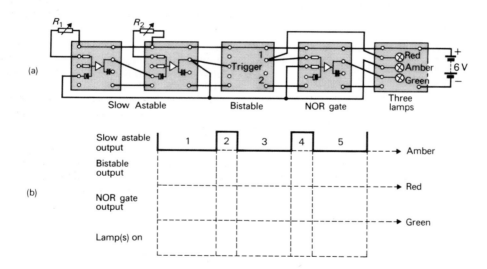

Fig. 21.74

22 Nuclear physics

Radioactivity

In 1896 the French scientist Becquerel found that uranium compounds emitted radiation which affected a photographic plate wrapped in black paper and, like X-rays, ionized a gas. The search for other *radioactive* substances was taken up by Marie Curie, who extracted from the ore pitchblende two new radioactive elements which she named polonium and radium.

(*a*) *Alpha, beta and gamma rays.* One or more of three types of radiation may be emitted which can be identified by their different penetrating power, ionizing ability and behaviour in a magnetic field.

Alpha (*α*) rays have a range of a few centimetres in air at s.t.p. and are stopped by a thick sheet of paper. They produce intense ionization in a gas and are deflected by a *strong* magnetic field in a direction which suggests they are relatively heavy, positively charged *particles*.

Beta (*β*) rays are usually more penetrating, having ranges which can be as high as several metres of air at s.t.p. or a few millimetres of aluminium. They cause much less intense ionization than alpha particles but are more easily deviated by a magnetic field and in a direction which indicates they are negatively charged *particles* of small mass. The magnetic deflection of beta particles may be demonstrated using the apparatus of Fig. 22.1. With the Geiger-Müller (G-M) tube in position A and without the magnet, the count-rate produced by the beam of beta particles is observed on the ratemeter. When the magnet is inserted, the count-rate decreases but rises again when the G-M tube is moved to some position such as B.

Gamma (*γ*) rays of high energy can penetrate several centimetres of lead. They ionize a gas weakly and are not deflected in a magnetic field. Their behaviour is not that of charged particles.

Alpha, beta and gamma rays are termed *nuclear radiation*, since, as we shall see later, they originate in atomic nuclei.

(*b*) *Nature of rays.* The specific charges of alpha and beta particles can be deduced from measurements of their deflections in electric and magnetic fields and give information about their nature.

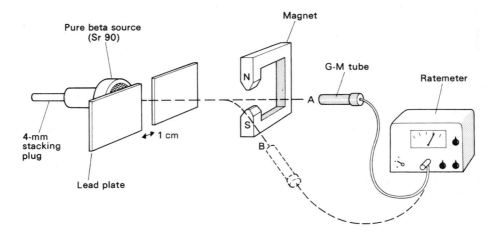

Fig. 22.1

Alpha particles have a specific charge which suggests they might be helium atoms which have lost two electrons, i.e. helium ions with a double positive charge—a helium nucleus $_2^4$He. In 1909 Rutherford and Royds confirmed this by compressing some of the radioactive gas radon in a tube A, Fig. 22.2, whose walls were thin enough to allow the alpha particles emitted by radon to escape into the evacuated space enclosed by the thicker-walled tube B. After a week the mercury level was raised so that gas which had collected in B was forced into C. On passing a current through C the line spectrum of helium was observed. Each alpha particle penetrating A had collected two electrons and changed into a helium atom.

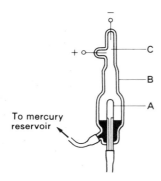

Fig. 22.2

Specific charge measurements for beta particles show they are high-speed electrons. (Allowance has to be made for the increase of mass with speed as predicted by relativity since in some cases beta particles are emitted with speeds (v) very close to that of light (c) and if the relativistic mass m is not used, the specific charge decreases with increasing speed: $m = m_0(1 - v^2/c^2)^{-\frac{1}{2}}$ where m_0 is the rest mass of an electron.)

Gamma rays were the subject of controversy until they were shown to be diffracted by a crystal, thus establishing their wave-like nature. Their wavelengths are those of very short X-rays, and like X-rays they are a form of electromagnetic radiation travelling with the speed of light (since among other things they give the photoelectric effect). Whilst diffraction gives the most accurate method of finding gamma ray wavelengths, it is difficult and rarely done. Nowadays the rays are usually absorbed by a solid state detector calibrated by rays of known energy.

(c) *Energy and speed of emission.* The energies of alpha and beta particles are found from measurements of their paths in magnetic fields and in the case of gamma rays as explained above.

(i) Alpha particles. In many cases the alpha particles emitted by a particular nuclide all have the same energy and are said to be monoenergetic. Energies vary from 4 to 10 MeV, corresponding to emission speeds of 5 to 7 per cent of the speed of light.

(ii) Beta particles. They exhibit quite a different behaviour. Their energy spectrum is a continuous one in which all energies are present from quite small values up to a certain maximum as shown by Fig. 22.3. The maxima are characteristic of the nuclide and vary from 0.025 to 3.2 MeV for natural radioactive sources. The highest represent beta particle emission speeds of 99 per cent of the speed of light.

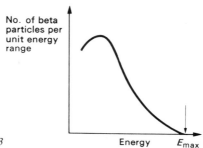

Fig. 22.3

(iii) Gamma rays. These fall into several distinct monoenergetic groups, giving a 'line' spectrum. The gamma rays from cobalt 60 (see below) have two different energies of 1.2 and 1.3 MeV.

(d) *Sources.* Those used for instructional purposes are usually supplied mounted in a holder with a 4 mm plug. The active material is sealed in metal foil which is protected by a wire gauze cover, Fig. 22.4a. When not in use they are stored in a small lead castle, Fig. 22.4b, in a wooden box.

The health hazard is negligible when using the weak closed sources listed shortly, provided they are (i) *always lifted with forceps*, (ii) *held so that the open window is directed away from the body* and (iii) *never brought close to the eyes for inspection.*

Approved *closed* sources have low activities, about 5 microcuries (5 μCi: see p. 538) and include

(i) Radium 226 for α, β and γ rays.

(ii) Americium 241 or plutonium 239 for α particles only (but the former also emits some low energy γ rays).

(iii) Strontium 90 for β particles only.

(iv) Cobalt 60 for γ rays. (An aluminium cover disc absorbs the beta particles also emitted.)

(v) Some other very weak sources (0.1 μCi) are not

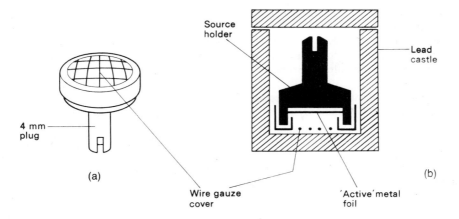

4 mm plug

Source holder

Lead castle

Wire gauze cover

'Active' metal foil

(a)

(b)

Fig. 22.4

completely enclosed but the radioactive material is secured to a support. Such sources are used in school cloud chambers.

Sources (*ii*), (*iii*) and (*iv*) are prepared in nuclear reactors.

Nuclear radiation detectors

In a nuclear radiation detector *energy* is transferred from the radiation to atoms of the detector and may cause

(*i*) ionization of a gas as in an ionization chamber, a Geiger-Müller tube, a cloud (or similar) chamber,

(*ii*) exposure of a photographic emulsion,

(*iii*) fluorescence of a phosphor as in a scintillation counter, or

(*iv*) mobile charge-carriers in a semiconducting solid state detector.

The radiation is thus detected by the effects it produces.

(*a*) *Ionization chamber.* In its simplest form this comprises two electrodes between which ion-pairs, i.e. electrons and positive ions, can be produced from neutral gas atoms and molecules by ionizing radiation from a source inside or outside the chamber. One electrode of the chamber is often a cylindrical can and the other a metal rod along the axis of the cylinder. Under the influence of an electric field between the electrodes, electrons move to the anode and positive gas ions to the cathode to form an ionization current. Some means of detecting the current is necessary. Fig. 22.5a shows the basic arrangement required.

The current depends on the nature of the radiation and the volume of the chamber. A 5 μCi alpha source creates a current of the order of 10^{-10} A in a small chamber; beta particles and particularly gamma rays

cause very much smaller currents. A school-type d.c. amplifier (p. 251) can detect the currents due to alpha particles but is insensitive to beta and gamma radiation.

To measure the ionization current the d.c. amplifier is first calibrated (p. 253) then arranged as in Fig. 22.5b. The supply voltage is increased until the output meter reading reaches a maximum value which is recorded. All ions produced are then being collected by the electrodes of the chamber and the ionization current has its saturation value *I*. This is a measure of the

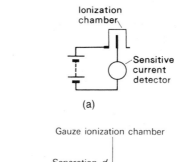

Ionization chamber

Sensitive current detector

(a)

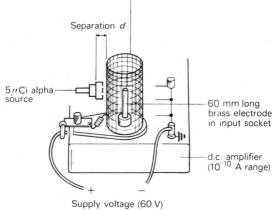

Gauze ionization chamber

Separation *d*

5 μCi alpha source

60 mm long brass electrode in input socket

d.c. amplifier (10^{-10} A range)

+   −

Supply voltage (60 V)

Fig. 22.5   (b)

intensity of the radiation from the 5 μCi alpha source (p. 474). It is calculated from $I = V/R$ where V is the p.d. across the input resistor R (10^{10} Ω) at saturation; V is obtained from the calibration measurement. A very rough estimate of the energy of an alpha particle may also be made (see Qn. 1, p. 556).

Other experiments with an ionization chamber and d.c. amplifier are given on pp. 539 and 541.

(*b*) *Geiger-Müller tube.* The G-M tube is a very sensitive type of ionization chamber which can detect single ionizing events. It consists of a cylindrical metal cathode (the wall of the tube) and a coaxial wire anode, containing argon at low pressure, Fig. 22.6*a*. The very thin mica end-window allows beta and gamma radiation to enter, as well as more energetic alpha particles; gamma rays can also penetrate the wall.

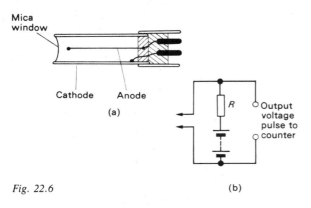

Mica window

Cathode   Anode

(a)

R Output voltage pulse to counter

Fig. 22.6   (b)

A p.d. of about 450 V is maintained between anode and cathode and, since the anode is very thin, an intense electric field is created near it. If an ionizing 'particle' passing through the tube produces an ion-pair from an argon atom, the resulting electron is rapidly accelerated towards the anode and when close to it has

sufficient energy itself to produce ion-pairs in encounters. The electrons thus freed produce additional ionization and an avalanche of electrons spreads along the whole length of the wire, which absorbs them to produce a large pulse of anode current. In this way, one electron freed in one ionizing event can lead to the release, in a few tenths of a microsecond, of as many as 10^8 electrons.

During the electron avalanche the comparatively heavy positive ion members of the ion-pairs have been almost stationary round the anode. After the avalanche has occurred they move towards the cathode under the action of the electric field, taking about 100 microseconds to reach it. They now have appreciable energy and would cause the emission of electrons from the cathode by bombardment. A second avalanche would follow, maintaining the discharge and creating confusion with the effect of a later ionizing particle entering the tube. The presence in the tube of a small amount of a *quenching agent* such as bromine tends to prevent this, since the positive ion energy is used to decompose the molecules of the quencher. In a halogen quenched tube these subsequently recombine and are available for further quenching.

A Geiger-Müller tube has a *dead time* of about 200 microseconds due to the time taken by the positive ions to travel towards the anode. Ionizing particles arriving within this period will not give separate pulses, i.e. are not resolved. If radioactive substances emitted particles at regular intervals a maximum of 5000 pulses per second could be detected, but this is not so and in practice the counting rate is less. Almost every beta particle entering a Geiger-Müller tube is detected. By contrast, the detection efficiency for gamma rays is less than 1 per cent. Gamma photons produce ion-pairs indirectly in the gas of the tube as a result of the secondary electrons they create when absorbed by the tube wall (cathode). The number of such electrons is

(a)

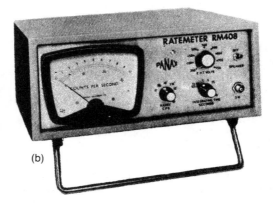

(b)

Fig. 22.7

small since gamma rays interact weakly with matter and this accounts for the low detection efficiency.

If a resistor R is connected in series with a G-M tube as in Fig. 22.6b, a voltage pulse (of about 1 V) is created which can be applied to an electronic counter such as a scaler or a ratemeter. A *scaler*, Fig. 22.7a, counts the pulses and indicates on a series of dials the total received in a certain time. The *ratemeter*, Fig. 22.7b, has a meter marked in 'counts per second' from which the average pulse rate can be read directly. It usually has a loudspeaker which gives a 'click' for every pulse. For quantitative work the scaler is more accurate.

The characteristic curve of a G-M tube shows how its response depends on the applied p.d. and allows the selection of an appropriate working p.d. The curve is obtained by placing a beta source at a short distance from the tube and noting the count-rates on a scaler (or ratemeter) as the p.d. is increased. A typical curve is given in Fig. 22.8. When the p.d. reaches the *threshold* value, the count-rate remains almost constant over a range called the *plateau* (of 100 V or more). Here a full avalanche is obtained along the entire length of the anode and all particles whatever their energy (and nature) produce the same output pulse. Beyond the plateau the count-rate for the same intensity of radiation increases rapidly with p.d. and a continuous discharge occurs which damages the tube.

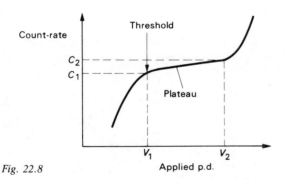

Fig. 22.8

the slope increases. The latter should be less than 0.15 per cent per volt.

Other experiments with a G-M tube are outlined on pp. 540 and 542–3.

(*c*) *Cloud chambers.* There are two types, and in both saturated vapour (alcohol) is made to condense on air ions created by the radiation and the resulting white line of tiny liquid drops shows up as a track in the chamber when suitably illuminated. Note that what we see is the track, not the ionizing radiation.

In the *expansion cloud chamber* the vapour is cooled by a rapid expansion and condensation occurs. This is achieved in the small chamber of Fig. 22.9 by withdrawing sharply the piston of a bicycle pump having the cup washer reversed so that it removes air. Ions are being produced all the time (from a very weak source inside the chamber) and would cause blurred tracks at the moment of expansion. A high p.d. (e.g. 1 kV) between the top and bottom of the chamber provides a clearing field which removes them. Consequently only the tracks of radiation which has just left the source are observed.

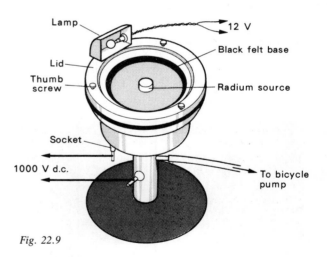

Fig. 22.9

Normally a tube is operated about the middle of the plateau where the count-rate is unaffected by p.d. fluctuations. The plateau usually has a slight upward slope, calculated from

$$\text{percentage slope} = \frac{C_2 - C_1}{V_2 - V_1} \times \frac{100}{C_M} \text{ per cent per volt}$$

where $C_M = (C_1 + C_2)/2$. As the condition of a tube deteriorates, the length of the plateau decreases and

In the *diffusion cloud chamber* tracks are produced continuously. The upper compartment in the simple chamber of Fig. 22.10 contains air which is at room temperature at the top and at about $-78\,°C$ at the bottom due to the layer of 'dry ice' (solid carbon dioxide) in the lower compartment. The felt ring at the top of the chamber is soaked with alcohol, which vaporizes in the warm upper region, diffuses downwards and is cooled. About 1 cm from the floor of the chamber, the air contains a layer of saturated alcohol

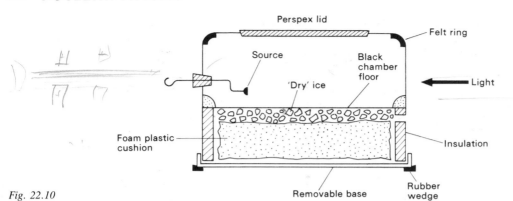

Fig. 22.10

vapour where conditions are suitable for the condensation of droplets on air ions produced along the path of the radiation from the source. Tracks are seen in this sensitive layer which are well-defined if an electric field is created by frequently rubbing the Perspex lid of the chamber with a cloth.

Alpha particles give bright, straight tracks like those in Fig. 22.11a. Very fast beta particles produce thin, straight tracks whilst those travelling more slowly give short, thicker, tortuous ones, Fig. 22.11b. Gamma rays (like X-rays) cause electrons to be ejected from air molecules and give tracks similar to X-rays; Fig. 22.11c is due to an intense beam of X-rays.

The *bubble chamber*, in which ionizing radiation leaves a trail of bubbles in liquid hydrogen, has re-placed the cloud chamber and is now used extensively in high-energy nuclear physics. The photographs which they give of atomic collisions yield information about the particles involved. Estimates of mass and speeds can be obtained assuming momentum and energy are conserved.

Radioactive decay

(*a*) *Disintegration theory*. The changes accompanying the emission of radiation from radioactive substances are unlike ordinary chemical changes in certain fundamental respects. They are spontaneous, they cannot be controlled and they are unaffected by chemical combination and physical conditions such as tempera-

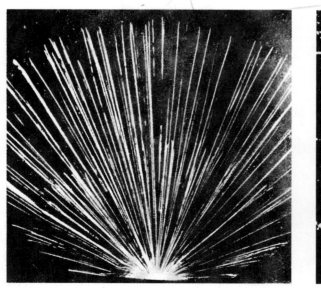

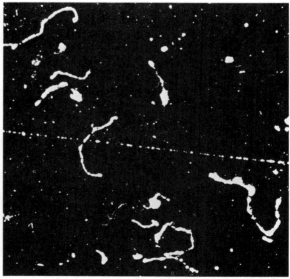

Fig. 22.11　　　　(a)　　　　　　　　　　　　　　(b)

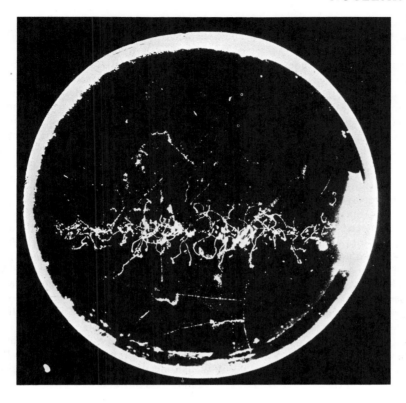

Fig. 12.11c

ture and pressure. Energy considerations too suggest that they are different. Thus the energy released by a radioactive substance emitting alpha particles is several million electron-volts per atom compared with a few electron-volts per atom in any chemical change.

We believe that radioactivity involves the nucleus of an atom—not its extranuclear electrons as do chemical changes—and is an attempt by an *unstable* nucleus to become more stable. The emission of an alpha or beta particle from the nucleus of a radioactive atom produces the nucleus of a different atom—called the *daughter* or *decay atom*—which may itself be unstable. The disintegration process proceeds at a definite rate through a certain number of stages until a stable end-product is formed. When first advanced by Rutherford and others the theory conflicted with existing ideas about the permanency of atoms but it is a view which accounts satisfactorily for the known facts.

An alpha particle is a helium nucleus consisting of 2 protons and 2 neutrons and when an atom decays by alpha emission, its mass number decreases by 4 and its atomic number by 2. It becomes the atom of an element two places lower in the periodic table. For example, when radium of mass number 226 and atomic number 88 emits an alpha particle, it decays to radon of mass

number 222 and atomic number 86. The change may be written

$$^{226}_{88}\text{Ra} \rightarrow {}^{222}_{86}\text{Rn} + {}^{4}_{2}\text{He}$$

Radon in turn disintegrates and after seven more disintegrations, a stable isotope of lead is formed.

When beta decay occurs a neutron changes into a proton and an electron. The proton remains in the nucleus and the electron is emitted as a beta particle. The new nucleus has the same mass number, but its atomic number increases by one since it has one more proton. It becomes the atom of an element one place higher in the periodic table. Radioactive carbon, called carbon 14, decays by beta emission to nitrogen.

$$^{14}_{6}\text{C} \rightarrow {}^{14}_{7}\text{N} + {}^{0}_{-1}\text{e}$$

Beta emitters have a higher proportion of neutrons; few occur naturally in appreciable quantities but many are obtained artificially by irradiating matter with neutrons.

The emission of gamma rays is explained by considering that nuclei (as well as atoms) have energy levels and if an alpha or beta particle is emitted, the nucleus is left in an excited state. A gamma ray photon is emitted when the nucleus returns to the ground state.

The existence of different nuclear energy levels would account for the 'line-type' energy spectrum of gamma rays (p. 532).

Naturally-occurring radioactive nuclides of high atomic number fall into three decay series, known as the *thorium* series, the *uranium* series and the *actinium* series. Radium is a member of the uranium series and at any time all the daughter nuclides will be present—which explains why a radium source emits alpha particles and gamma rays of several energies, as well as beta particles.

(*b*) *Decay law*. Radioactive decay is a completely haphazard or random process in which nuclei disintegrate quite independently. Since there is always a very large number of active nuclei in a given amount of radioactive material we can apply the methods of statistics to the process and obtain an expression for the certain fraction of the nuclei originally present that will have decayed on average in a given time interval.

We assume that the *rate of disintegration of a given nuclide at any time is directly proportional to the number of nuclei N of the nuclide present at that time;* that is, in calculus notation,

$$-\frac{dN}{dt} \propto N$$

The negative sign indicates that N decreases as t increases. The *radioactive decay constant λ* is defined as the constant of proportionality in this expression, giving

$$-\frac{dN}{dt} = \lambda N \qquad (1)$$

If there are N_0 undecayed nuclei at some time $t = 0$ and a smaller number N at a later time t, then integrating (1),

$$\int_{N_0}^{N} \frac{dN}{N} = -\lambda \int_0^t dt$$

$$\therefore \quad (\ln N)_{N_0}^{N} = -\lambda t \qquad (\ln = \log_e)$$

$$\ln(N - N_0) = \ln(N/N_0) = -\lambda t \qquad (2)$$

$$\therefore \quad N = N_0 e^{-\lambda t} \qquad (3)$$

This is the decay law and states that *a radioactive substance decays exponentially with time*—a fact confirmed by experiment. The law is a statistical one, it does not tell us when a particular nucleus will decay but only that after a certain time a certain fraction will have decayed. On the microscopic scale the process is purely random as is evident from the variations which occur when particles from a source are counted. On the macroscopic scale, however, where large numbers of particles are concerned, a definite law is followed.

The rate of decay of a source, i.e. the number of disintegrations per second, is called its *activity* and is often expressed in curies (Ci). One curie is the activity of a source which on average undergoes 3.70×10^{10} disintegrations per second. Sources used in school laboratories have activities of about 5 μCi which is very small compared with some medical sources.

From equation (1) we see that $\lambda = -dN/(N \cdot dt)$ and so the decay constant is the fraction of the total number of nuclei present which decays in unit time, provided the unit of time is small.

(*c*) *Half-life*. An alternative but more convenient term to the decay constant λ is the *half-life $t_\frac{1}{2}$*. This is the *time for the number of active nuclei present in a source at a given time to fall to half its value*. Whereas it is often difficult to know when a substance has lost practically all its activity, it is less difficult to find out how long it takes for the activity to fall to half the value it has at some instant.

Half-lives vary from millionths of a second to thousands of millions of years. Radium 226 has a half-life of 1622 years, therefore starting with 1 g of pure radium, $\frac{1}{2}$ g remains as radium after 1622 years, $\frac{1}{4}$ g after 3244 years and so on. An exponential decay curve, like that given on p. 250 for the discharge of a capacitor through a high resistor, is shown in Fig. 22.12 and illustrates the idea of half-lives again.

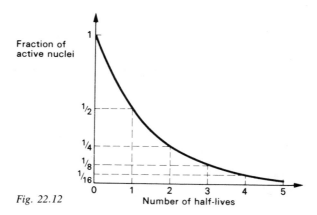

Fig. 22.12

The relationship between λ and $t_\frac{1}{2}$ can be derived from equation (3) for the decay law. Thus

$$N = N_0 e^{-\lambda t}$$

$$\therefore \quad \frac{N_0}{N} = e^{\lambda t}$$

When $N = N_0/2$, $t = t_{\frac{1}{2}}$

$$\therefore \quad 2 = e^{\lambda t_{\frac{1}{2}}}$$

Taking logs to base e, $\quad \ln 2 = \lambda t_{\frac{1}{2}}$

$$\therefore \quad t_{\frac{1}{2}} = \frac{0.693}{\lambda}$$

$t_{\frac{1}{2}}$ (and λ) is characteristic for each radionuclide and is an important means of identification.

Note. An analogue experiment using dice to illustrate the random decay law is outlined in Appendix 19.

(*d*) *Nuclear stability.* Whilst the chemical properties of an atom are governed entirely by the number of protons in the nucleus (i.e. the atomic number Z), the stability of an atom appears to depend on both the number of protons and the number of neutrons. In Fig. 22.13 the number of neutrons ($A-Z$, where A is the mass number) has been plotted against the number of protons for all known nuclides, stable and unstable, natural and man-made. A continuous line has been drawn approximately through the stable nuclides (only a few are labelled) and the shading on either side of this line shows the region of unstable nuclides.

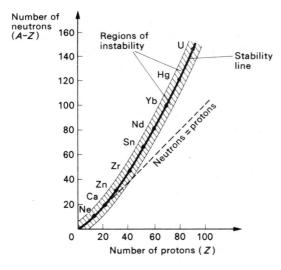

Fig. 22.13

For *stable* nuclides the following points emerge:
(*i*) The lightest nuclides have almost equal numbers of protons and neutrons.
(*ii*) The heavier nuclides require more neutrons than protons, the heaviest having about 50 per cent more.

(*iii*) Most nuclides have both an even number of protons and an even number of neutrons. The implication is that two protons and two neutrons, i.e. an alpha particle, form a particularly stable combination and in this connection, it is worth noting that oxygen ($^{16}_{8}O$), silicon ($^{28}_{14}Si$) and iron ($^{56}_{28}Fe$) together account for over three-quarters of the earth's crust.

For *unstable* nuclides the following points can be made:

(*i*) Disintegrations tend to produce new nuclides nearer the 'stability' line and continue until a stable nuclide is formed.

(*ii*) A nuclide above the line decays so as to give an increase of atomic number, i.e. by beta emission (in which a neutron changes to a proton and an electron). Its neutron-to-proton ratio is thereby increased.

(*iii*) A nuclide below the line disintegrates in such a way that its atomic number decreases and its neutron to proton ratio increases. In heavy nuclides this can occur by alpha emission.

Measuring half-lives

The next two experiments can be performed in a school laboratory.

(*a*) *Half-life of radon 220 ($^{220}_{86}Rn$).* This is an alpha-emitting radioactive gas, usually called *thoron*, which collects above its solid, parent nuclides in an enclosed space and is readily separated from them. If a polythene bottle containing thorium hydroxide is connected to an ionization chamber mounted on a d.c. amplifier, Fig. 22.14, a small quantity of thoron enters the chamber when the bottle is squeezed.

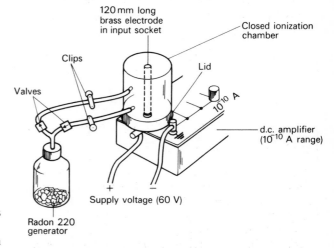

Fig. 22.14

As the thoron decays the ionization current decreases and is always a measure of the number of alpha particles present (i.e. the quantity of thoron remaining) so long as the p.d. across the chamber produces the saturation value of the current. If the time is determined for the reading of the output meter on the d.c. amplifier to fall to a half (or better still, a quarter) of some previous value, the half-life can be found. Alternatively the meter may be read every ten seconds and the half-life found from a graph of meter reading against time.

The chief constituent of the thoron source is thorium 232 and part of its decay series is shown below.

The protactinium 234 is extracted almost completely by an organic solvent such as amyl acetate, from an acidified aqueous solution of uranyl nitrate (proportions—1 g uranyl nitrate, 3 cm³ water, 7 cm³ concentrated hydrochloric acid). Only the high-energy beta particles from the decay of protactinium 234 are detected by a G-M tube.

Equal volumes of the organic and aqueous solutions are contained in a thin-walled, stoppered, polythene bottle so that each layer is as deep as the width of the G-M tube window. The bottle is well shaken and then arranged as in Fig. 22.15 with the G-M tube opposite the top half of the bottle. As soon as the layers have

$$\text{Thorium 232} \quad \text{Actinium 228} \quad \text{Radium 224} \quad \text{Polonium 216}$$
$$(^{232}_{90}\text{Th}) \quad (^{228}_{89}\text{Ac}) \quad (^{224}_{88}\text{Ra}) \quad (^{216}_{84}\text{Po})$$

$$1.4 \times 10^{10} \text{ years} \xrightarrow{\alpha} \xleftarrow{\beta} 6.7 \text{ years} \quad 1.1 \text{ hours} \xrightarrow{\beta} \xleftarrow{\alpha} 1.9 \text{ years} \quad 3.6 \text{ days} \xrightarrow{\alpha} \xleftarrow{\alpha} 52 \text{ seconds} \quad 0.16 \text{ second} \xrightarrow{\alpha} \xleftarrow{\beta} 10.6 \text{ hours}$$

$$\text{Radium 228} \quad \text{Thorium 228} \quad \text{RADON 220} \quad \text{Lead 212}$$
$$(^{228}_{88}\text{Ra}) \quad (^{228}_{90}\text{Th}) \quad (^{220}_{86}\text{Rn}) \quad (^{212}_{82}\text{Pb})$$

The decay products following radon 220 do not affect the result appreciably since their half-lives are so very different from that of radon 220.

If some thoron is puffed into a closed diffusion cloud chamber, the two tracks can be observed which are due to the two alpha particles emitted almost simultaneously (Why?) when radon 220 decays to polonium 216 and then to lead 212. V-shaped tracks are obtained.

(*b*) *Half-life of protactinium 234* ($^{234}_{91}\text{Pa}$). This is a beta-emitting daughter product in the decay of uranium 238; part of the series is given below.

separated, the scaler or ratemeter connected to the G-M tube is started and counts taken at ten-second intervals without stopping the counter. After allowing for the background count (p. 545), a graph of count-rate against time can be plotted and the half-life found.

The thorium from which the protactinium 234 has been extracted is left in the aqueous solution and immediately starts to generate the protactinium again as it decays. A second experiment may be carried out with the G-M tube opposite the lower half of the bottle to observe the growth of $^{234}_{91}\text{Pa}$.

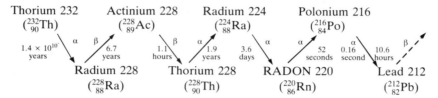

$$\text{Uranium} \quad \xrightarrow[\substack{4.5 \times 10^9 \\ \text{years}}]{\alpha} \quad \text{Thorium} \quad \xrightarrow[\substack{24 \\ \text{days}}]{\beta} \quad \overset{\text{low-energy}}{\text{PROTACTINIUM}} \quad \xrightarrow[\substack{72 \\ \text{seconds}}]{\overset{\text{high-energy}}{\beta}} \quad \text{Uranium}$$
$$238 \quad\quad 234 \quad\quad 234 \quad\quad 234$$
$$(^{238}_{92}\text{U}) \quad\quad (^{234}_{90}\text{Th}) \quad\quad (^{234}_{91}\text{Pa}) \quad\quad (^{234}_{92}\text{U})$$

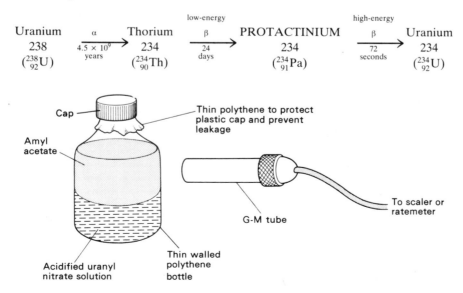

Cap

Thin polythene to protect plastic cap and prevent leakage

Amyl acetate

To scaler or ratemeter

G-M tube

Thin walled polythene bottle

Acidified uranyl nitrate solution

Fig. 22.15

Absorption of alpha, beta and gamma rays

(a) *Absorption processes.* In their passage through matter alpha and beta particles are 'absorbed' by losing kinetic energy in ionizing encounters with atoms of the absorbing medium, i.e. an electron is 'knocked out' of an atom to form an ion-pair. After travelling a certain distance, called the *range*, they have insufficient energy to produce any more ion-pairs and are then considered to have been absorbed. An alpha particle can also have an encounter with a nucleus (p. 476), and beta particles, because of their small mass, are readily 'back-scattered' by atoms of the medium to emerge from the incident surface.

A measure of the intensity of the ionization produced when a charged particle passes through a gas is given by the *specific ionization*. This is defined as the number of ion-pairs formed per centimetre of path. It increases with the size of the charge on the particle and decreases as the speed of the particle increases. A fast-moving particle spends less time near an atom of the gas through which it is travelling and so there is less chance of an ion-pair being formed. An alpha particle may produce 10^5 ion-pairs per centimetre in air at atmospheric pressure while a beta particle of similar energy produces about 10^3 on account of its higher speed (on average about ten times greater) and smaller charge. It should, however, be noted that for alpha and beta particles of the same energy the *total* number of ion-pairs formed would be of the same order since a beta particle travels about one hundred times farther in air than an alpha particle. For each ion-pair produced in the track of any type of charged particle in air the *average* energy loss is 34 eV.

Gamma rays are usually most strongly absorbed by elements of high atomic number such as lead, and the absorption process is complex and differs from that occurring with charged particles. Whereas the alpha or beta particle gradually loses kinetic energy by a series of ionizing encounters with electrons belonging to atoms of the absorber, a gamma ray photon may interact with either an electron or, if it has enough energy, with a nucleus in several ways. The energy given up by the photon produces one or more high-speed 'secondary' electrons and it is these which are responsible for the ionization created in a gas by gamma rays. They enable gamma rays to be detected by a G-M tube. The specific ionization of gamma rays therefore depends on the energy of the secondary electrons.

(b) *Range of alpha particles in air.* An ionization chamber with a d.c. amplifier can be used with the source (e.g. 5 μCi americium 241) supported just outside the wire gauze ionization chamber which is connected to the d.c. amplifier, Fig. 22.16a. The ionization current (saturation) in the chamber is large when the source is close to the gauze and decreases as d increases by, say, 5 mm steps. The corresponding readings of the d.c. amplifier output meter are noted and from a graph like that in Fig. 22.16b the range in air can be estimated.

(c) *Range of beta particles in aluminium.* Beta particles have a continuous energy spectrum and the number emerging from an absorber falls off gradually as the thickness of the absorber increases; this behaviour contrasts with the fairly sharp cut-off given by alpha

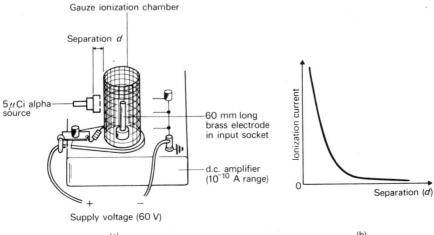

Gauze ionization chamber

Separation d

5 μCi alpha source

60 mm long brass electrode in input socket

d.c. amplifier (10^{-10} A range)

+   −
Supply voltage (60 V)

Ionization current

0

Separation (d)

Fig. 22.16   (a)   (b)

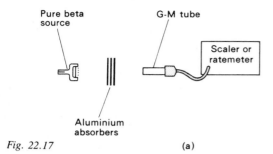

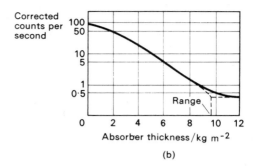

Fig. 22.17   (a)   (b)

particles. For practical purposes the range of beta particles is defined as the thickness of aluminium beyond which very few particles can be detected.

The range may be determined by inserting an increasing number of aluminium sheets between a pure beta source (e.g. 5 μCi strontium 90) and a G-M tube connected to a scaler or ratemeter, Fig. 22.17a. Count-rates are measured and an absorption curve is drawn.

A typical curve is given in Fig. 22.17b in which the wide range of count-rates is accommodated by plotting the logarithm of the count-rate. This is most conveniently done by plotting the count-rate (after subtracting the background count-rate) directly on the log scale of semi-log graph paper. The absorber thickness can be expressed either in millimetres of aluminium or, as is more common, in terms of the *surface density* of the absorber. This is the mass per unit area and equals the product of the density and the thickness of the absorber. For example, a sheet of aluminium 2.0 mm (2.0 $\times$ 10^{-3} m) thick of density 2.7 $\times$ 10^3 kg m^{-3} has a surface density of 5.4 kg m^{-2}. The range in kg m^{-2} is found by extrapolation from the absorption curve, as shown.

(*d*) *Inverse square law for gamma rays.* Gamma rays are highly penetrating on account of their small interaction with matter. In air they suffer very little absorption or scattering and, like other forms of electromagnetic radiation, their intensity falls off with distance according to the inverse square law. This states that the intensity of radiation I is inversely proportional to the square of the distance d from a *point* source. That is, $I = k/d^2$ where k is a constant.

The law may be investigated by placing a G-M tube at various distances from a pure gamma source (e.g. 5 μCi cobalt 60) and measuring the corresponding count-rates, Fig. 22.18a. If after correction for background, C is the count-rate, then C is directly proportional to I and we can write $C = k_1/d^2$, where k_1 is another constant. A graph of C against $1/d^2$ should be a straight line through the origin but small errors occur in

the measurement of d and become important as d decreases. Non-linearity of the graph results. The errors can be eliminated by an alternative procedure.

Let $d = D + x$, where D = distance measured from the source to any point on the G-M tube and x = an unknown correction term which gives the true distance d. The law may then be written

$$C = k_1/(D + x)^2$$

$$\therefore \quad D + x = (k_1/C)^{\frac{1}{2}}$$

Hence $$D = (k_1/C)^{\frac{1}{2}} - x$$

A graph of D against $1/C^{\frac{1}{2}}$ should be a straight line of slope $k_1^{\frac{1}{2}}$ and intercept $-x$ on the D-axis, as shown in Fig. 22.18b.

In practice, departure from the law arises because of (*i*) the finite size of the source and (*ii*) counting losses due to the G-M tube having a 'dead-time' when the source is close to the tube and count-rates high.

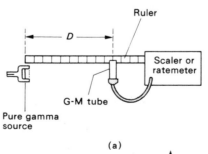

(a)

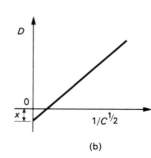

Fig. 22.18   (b)

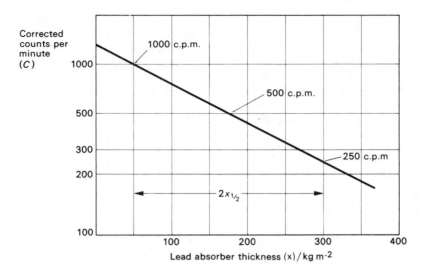

Fig. 22.19

(e) *Half-thickness of lead for gamma rays.* The 'half-thickness', denoted by $x_{\frac{1}{2}}$, is a convenient term for dealing with the absorption of gamma rays. It is defined as the thickness of absorber, usually lead, which reduces the intensity of the gamma radiation to half its incident value.

The half-thickness can be determined by inserting sheets of lead between a gamma source (e.g. 5 μCi cobalt 60) and a G-M tube and counter in a similar way to that described for beta absorption.

The count-rate C, corrected for background, is directly proportional to the intensity of the radiation, and it is again convenient to plot C directly on semi-log graph paper. The thickness of lead is usually expressed in the surface density unit of kg m^{-2}. An absorption curve for a parallel beam of monoenergetic gamma rays is shown in Fig. 22.19; by considering two (or more) half-thicknesses, i.e. the thickness required to halve the count-rate twice, a more accurate result is obtained for $x_{\frac{1}{4}}$.

The graph of log C against absorber thickness is seen to be a *straight line*. This implies that the intensity of the gamma radiation decreases *exponentially* with increasing absorber thickness. Thus if I_0 is the intensity of the gamma rays from the source in the absence of absorbing material and I is the intensity after passing through a thickness x of absorber, then

$$I = I_0 \, e^{-\mu x}$$

where μ is a constant, called the *linear absorption coefficient*. Taking count-rate as a measure of intensity

$$C = C_0 \, e^{-\mu x}$$

$$\therefore \quad \ln C = \ln C_0 - \mu x$$

and

$$\log C = \log C_0 - \frac{\mu}{2.303} \cdot x$$

The graph of log C against x should thus be a straight line if the absorption is exponential. This is only true for a parallel beam of monochromatic (i.e. of one wavelength and energy) gamma rays. μ, which is an alternative term to $x_{\frac{1}{2}}$, can be found from the slope of the line and is smaller for high energy (more penetrating) rays than for low energy rays.

Radioisotopes and their uses

(a) *Man-made radioactivity.* The first artificial radioactive substance was produced in 1934 by Irène Joliot-Curie (daughter of the discoverer of radium) and her husband. They bombarded aluminium with alpha particles and, as a result of a nuclear reaction (p. 545), obtained an unstable isotope of phosphorus:

$$^{27}_{13}\text{Al} + {}^{4}_{2}\text{He} \rightarrow {}^{30}_{15}\text{P} + {}^{1}_{0}\text{n} \qquad [{}^{27}_{13}\text{Al}(\alpha, \text{n}){}^{30}_{15}\text{P}]$$

The expression in brackets is the symbolic way of writing a nuclear reaction, i.e.

initial nuclide (incoming particles, outgoing particles)
final nuclide

Since then artificial radioisotopes (i.e. radioactive isotopes) of every element have been produced and today about 1500 are known. They are made by bombarding a stable element with neutrons in a nuclear reactor (p. 550) or with charged particles in a particle accelerator (p. 553). Their use in industry, research and medicine grows annually.

(b) *Some uses.* Some applications use the fact that

the extent to which radiation is absorbed when passing through matter depends on its thickness and density. For example, in the manufacture of paper the thickness can be checked by having a beta source below the paper and a G-M tube and counter above it. A thickness gauge of this type may be adapted for automatic control of the manufacturing process. Level indicators also depend on absorption and are used to check the filling of toothpaste tubes and packets of detergents.

Leaks can be detected in underground pipe-lines carrying water, oil, etc., by adding a little radioactive solution to the liquid being pumped. Temporary activity gathers in the soil around the leak which can be detected from the ground above, Fig. 22.20a. This technique is now used widely when studying river pollution, sand-wave movement in river estuaries and in the accurate measurement of fluid flow. In research into wear in machinery a small amount of radioactive iron is introduced into the bearings and the rate of wear found from the resulting radioactivity of the lubricating oil.

The use of radioactive 'tracers' in medicine, agriculture and biological research depends on the fact that the radioisotope of an element can take part in the same processes as its non-active isotope since it is chemically identical (and only slightly different physically). A scan of a normal brain is shown in Fig. 22.20b. It can be used to detect the concentration of a radioisotope given earlier to the patient and brain tumours can be located in this way. The use of radioactive phosphorus as a tracer in agriculture has provided information about the best type of phosphate fertilizer to supply to a particular crop and soil.

Gamma rays from high-activity cobalt 60 sources have many applications. In radiotherapy they are replacing X-rays from expensive X-ray machines in the treatment of cancer. The rapidly growing cells of the diseased tissue which cause cancer are even more affected by radiation than are healthy cells. Medical instruments and bandages are sterilized after packing by brief exposure to gamma rays. Food may be similarly treated and meat made to stay fresh for fifteen days instead of three. This is perfectly safe since no radioactivity is produced in the material irradiated by gamma rays.

(c) *Carbon 14 dating.* A natural radioisotope interesting from an archaeological point of view is carbon 14, formed in a nuclear reaction when neutrons ejected from nuclei in the atmosphere by cosmic rays collide with atmospheric nitrogen.

$$^{14}_{7}N + ^{1}_{0}n \rightarrow ^{14}_{6}C + ^{1}_{1}H \qquad [^{14}_{7}N\,(n,\,p)\,^{14}_{6}C]$$

Subsequently carbon 14 forms radioactive carbon dioxide and may be taken by plants and trees for the manufacture of carbohydrates by photosynthesis. The normal activity of living carbonaceous material is 15.3 counts per minute per gram of carbon, but when the organism dies (or when part of it ceases to have any interaction with the atmosphere, as in the heartwood of trees) then no fresh carbon is taken in and carbon 14 starts to decay by beta emission with a half-life of

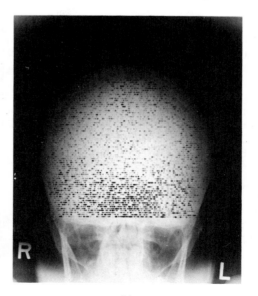

Fig. 22.20   (a)   (b)

5.70×10^3 years. By measuring the residual activity, the age of any ancient carbon-containing material such as wood, linen or charcoal may be estimated within the range 1000 to 50 000 years. This has been done with the Dead Sea Scrolls (about 2000 years old) and charcoal from Stonehenge (about 4000 years old).

The ages of radioactive rocks have been similarly estimated.

(d) *Transuranic elements*. Uranium has the highest atomic number (92) of any naturally-occurring element, but traces of elements with atomic numbers from 93 to 104 have been made artificially and, like other artificial isotopes made from stable atoms in particle accelerators or nuclear reactors, they are radioactive; they are called the *transuranic* elements.

Radiation hazards

The radiation hazards to human beings arise from (i) exposure of the body to external radiation and (ii) ingestion or inhalation of radioactive matter.

The effect of radiation depends on the nature of the radiation, the part of the body irradiated and the dose received. The hazard from alpha particles is slight (unless the source enters the body) since they cannot penetrate the outer layers of skin. Beta particles are more penetrating in general: most of their energy is absorbed by surface tissues and adequate protection is afforded by a sheet of Perspex or aluminium a few millimetres thick. Gamma rays present the main external radiation hazard since they penetrate deeply into the body and may require substantial lead or concrete shielding.

Radiation can cause immediate damage to tissue and, according to the dose, is accompanied by radiation burns (i.e redness of the skin followed by blistering and sores which are slow to heal), radiation sickness and, in extremely severe cases, by death. Delayed effects such as cancer, leukaemia and eye cataracts may appear many years later. Hereditary defects may also occur in succeeding generations due to genetic damage. The most susceptible parts are the reproductive organs, blood-forming organs such as the liver, and to a smaller extent the eyes. The hands, forearms, feet and ankles are less vulnerable. Damage to human cells is due to the creation of ions which upset or destroy them.

The *radiation dose* is the energy absorbed (in J kg^{-1}) by unit mass of the irradiated material. Equal doses of different ionizing radiations provide the same amount of energy in a given absorber but they do not have the same biological effect on the human body. Thus

$$\text{effective dose} = \text{total dose} \times \text{r.b.e.}$$

where r.b.e. is the *relative biological effectiveness*. For beta particles, X- and gamma rays, the r.b.e. is about 1; for alpha particles, protons and fast neutrons it is about 10.

We all unavoidably absorb background radiation due to cosmic rays, radioactive minerals, radon in the atmosphere, potassium 40 in the body, X-rays from television screens. The dose rate from one of the weak sources used for experimental work in school physics is very small. In industry and research strong sources have to be handled by a 'master-slave' manipulator and protection obtained for the operator from concrete and lead walls. The effective dose for radiation workers must not exceed a certain maximum in a certain period.

Nuclear reactions

Much information concerning the nucleus has been obtained by studying the effects of bombarding atomic nuclei with fast-moving particles. Three historic nuclear reactions will be considered.

(a) *Bombardment of nitrogen by alpha particles*. In 1919 Rutherford bombarded gases with alpha particles. His apparatus is shown in Fig. 22.21. One end of the metal cylinder was closed by thin silver foil capable of stopping most of the alpha particles from a radioactive source near the other end. Any which penetrated fell on a fluorescent screen and their scintillations were observed through a microscope. Various gases were admitted: with nitrogen, scintillations were obtained. Evidently more penetrating particles were produced and further experiments involving the deflection of these longer range particles in a magnetic field showed they were protons of high energy.

Two explanations seem plausible. Either an alpha particle 'chips' a proton off a nitrogen nucleus, in which case the alpha particle survives the encounter; or the alpha particle actually enters the nitrogen nucleus and the latter immediately ejects a proton. In either case the proton is energetic enough to cause a scintillation on the screen. Since only one alpha particle in about a million caused a disintegration the chance of identifying the residual nucleus seemed small.

In 1925, however, about 20 000 photographs of the tracks of alpha particles passing through nitrogen in a cloud chamber were taken. In eight of these, a forked track like that of Fig. 22.22 was obtained. The alpha source is at the bottom and the collision occurs near the top. The short thick track at one o'clock is due to the residual nucleus and the long thin one at eight o'clock is that of a proton. There is nothing to indicate the existence of the alpha particle after the disintegration.

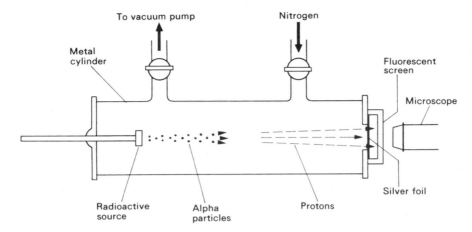

Fig. 22.21

Fig. 22.22

This nuclear reaction may therefore be represented by the equation (assuming neutrons and protons are conserved),

$$\underset{\text{nitrogen nucleus}}{^{14}_{7}\text{N}} + \underset{\text{alpha particle}}{^{4}_{2}\text{He}} \rightarrow \underset{\text{residual nucleus}}{^{17}_{8}\text{O}} + \underset{\text{proton}}{^{1}_{1}\text{H}} \qquad [^{14}_{7}\text{N}(\alpha, p)^{17}_{8}\text{O}]$$

The residual nucleus is an isotope of oxygen. This was the first occasion on which one element (nitrogen) was changed into another (oxygen) although the number of atoms transformed was small. The event is sometimes referred to as the 'splitting of the atom'. One other point requires comment.

If the inverse square law holds right up to the nucleus the repulsive electrostatic force would become infinitely large as the distance approaches zero. A charged particle would never enter a nucleus, however great its energy. Nuclear reactions do occur, however, as Rutherford demonstrated and so the repulsive force

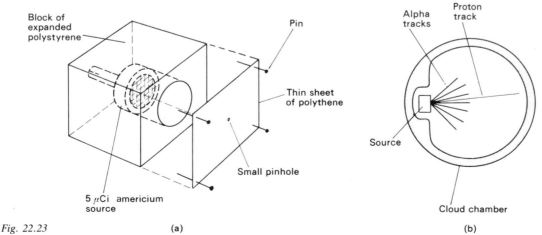

Fig. 22.23   (a)   (b)

must be replaced by an extremely powerful attractive force acting over a range of about 10^{-15} m. It is in fact this force which keeps nucleons together in a nucleus and is responsible for the existence of matter; it is called the *nuclear force*.

The tracks produced by protons in a diffusion cloud chamber (p. 535) may be observed by making a small pin-hole in a thin sheet of polythene and placing it over a 5 μCi americium source, Fig. 22.23a, in the chamber. Alpha particles emerging from the pinhole produce short tracks all about four centimetres long, but occasionally a much longer track due to the ejection of a proton from the hydrogen-rich polythene appears as a result of alpha particle bombardment, Fig. 22.23b.

(b) *Discovery of the neutron.* After the transmutation of nitrogen many other light elements were found to emit protons when bombarded by alpha particles. Beryllium, however, behaved unexpectedly and was found to emit radiation capable of penetrating many centimetres of lead. It could also cause protons of high energy to be shot out from materials such as paraffin wax which contain hydrogen, Fig. 22.24.

At first it was suggested that 'beryllium radiation' might be very energetic gamma rays until it was realized that if this were so then energy and momentum were not conserved in the collision producing it. However, in 1932 Chadwick suggested that there would be conservation if the radiation consisted of uncharged particles with a mass almost the same as a proton's. This neutral particle was called a *neutron* ($_0^1$n) and Chadwick was able to account for all the observed effects and established that the action of alpha particles on beryllium was represented by

$$_4^9\text{Be} + {}_2^4\text{He} \rightarrow {}_6^{12}\text{C} + {}_0^1\text{n} \qquad [{}_4^9\text{Be}\,(\alpha,\,\text{n})\,{}_6^{12}\text{C}]$$

Because of its lack of charge the neutron penetrates matter easily and leaves no tracks in a cloud chamber unless it has a collision with a nucleus, when charged particles can be released that do give tracks. Although a neutron decays outside the nucleus with a half-life of 13 minutes to a proton and an electron, inside it is perfectly stable and an entity in its own right.

In general, nuclear reactions induced by alpha particles cause the emission of either protons or neutrons depending on the energy of the alpha particles and the nucleus under attack.

(c) *Cockcroft and Walton's experiment.* Alpha particles from radioactive sources have limitations as atomic projectiles. This is due partly to the fact that, because their energy is limited, they cannot overcome the powerful repulsion of the large positive nuclear charge of a heavy atom. In addition only a very small proportion of particles are generally able to cause disintegration, and so radioactive sources of high activity which emit a large number of particles per second would be necessary. These are both difficult to obtain and to manipulate. Consequently, in the early 1930s attention was given to the problem of building machines, called *particle accelerators*, to accelerate charged particles such as protons to high speeds by means of p.d.s of hundreds of thousands of volts.

The first nuclear reaction to be induced by artificially-accelerated particles was achieved in 1932 by Cockcroft and Walton. They bombarded lithium with protons and obtained alpha particles. The reaction is represented by the equation

$$_3^7\text{Li} + {}_1^1\text{H} \rightarrow {}_2^4\text{He} + {}_2^4\text{He} \qquad [{}_3^7\text{Li}\,(\text{p},\,\alpha)\,{}_2^4\text{He}]$$

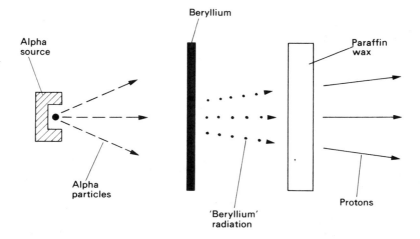

Fig. 22.24

Mass and energy

(a) *Einstein's relation.* In 1905, while developing his special theory of relativity, Einstein made the startling suggestion that energy and mass are equivalent. He predicted that if the energy of a body changes by an amount E, its mass changes by an amount m given by the equation

$$E = mc^2$$

where c is the speed of light.

Everyday examples of energy gain are much too small to produce detectable changes of mass. For example, when 1.0 kg of water absorbs 4.2×10^3 J to produce a temperature rise of 1 K, according to $E = mc^2$ the increase in mass of the water is

$$m = \frac{E}{c^2} = \frac{4.2 \times 10^3}{(3.0 \times 10^8)^2}$$

$$= 4.7 \times 10^{-14} \text{ kg}$$

The changes of mass accompanying energy changes in chemical reactions are not much greater and cannot be used to prove Einstein's equation.

However, radioactive decay, which is a spontaneous nuclear reaction, is more helpful. Thus for a radium atom, the combined mass of the alpha particle it emits and the radon atom to which it decays is, by atomic standards, appreciably less than the mass of the original radium atom. Atomic masses can now be measured to a very high degree of accuracy by mass spectrographs and the mass decrease m for the decay of one radium atom is 8.8×10^{-30} kg. The energy equivalent E is given by

$$E = mc^2 = 8.8 \times 10^{-30} \times (3.0 \times 10^8)^2 \text{ J}$$

$$= 7.9 \times 10^{-13} \text{ J}$$

But

$$1 \text{ eV} = 1.6 \times 10^{-19} \text{ J}$$

$$\therefore \quad E = \frac{7.9 \times 10^{-13}}{1.6 \times 10^{-19}} \text{ eV}$$

$$= 4.9 \times 10^6 \text{ eV}$$

$$= 4.9 \text{ MeV}$$

The alpha particle carries off 4.8 MeV of this energy; some appears as k.e. of the recoiling nucleus, and the rest is emitted soon afterwards as a γ-ray photon. Mass therefore appears as energy and the two can be regarded as equivalent. In nuclear physics mass is measured in *unified atomic mass units* (u), 1 u being one-twelfth of the mass of the carbon 12 atom and equals 1.66×10^{-27} kg. It can readily be shown using

$E = mc^2$, that

931 MeV has mass 1 u

A unit of energy may therefore be considered to be a unit of mass, and in tables of physical constants the masses of various atomic particles are often given in MeV as well as in kg and u. For example, the electron has a rest mass of about 0.5 MeV.

If the principle of conservation of energy is to hold for nuclear reactions it is clear that mass and energy must be regarded as equivalent.

The implication of $E = mc^2$ is that any reaction producing an appreciable mass decrease is a possible source of energy. Shortly we will consider two types of nuclear reaction in this category.

(b) *Binding energy.* The mass of a nucleus is found to be less than the sum of the masses of the constituent protons and neutrons. This is explained as being due to the binding of the nucleons together into a nucleus and the mass defect represents the energy which would be released in forming the nucleus from its component particles. The energy equivalent is called the *binding energy* of the nucleus; it would also be the energy needed to split the nucleus into its individual nucleons if this were possible.

Consider an example. The helium atom, ^4_2He, has an atomic mass of 4.0026 u; this includes the mass of its two electrons. Its constituents comprise two neutrons each of mass 1.0087 u and two hydrogen atoms (i.e. two protons and two electrons) each of mass 1.0078 u; the total mass of the particles is thus $(2 \times 1.0087 + 2 \times 1.0078) = 4.0330$ u. The *mass defect* is $(4.0330 - 4.0026) = 0.0304$ u and since 1 u is equivalent to 931 MeV, it follows that the binding energy of helium is 28.3 MeV.

The binding energy is derived in a similar manner for other nuclides and is found to increase as the mass number increases. For neon, $^{20}_{10}\text{Ne}$, it is 160 MeV. If the binding energy of a nucleus is divided by its mass number (i.e. the number of nucleons), the *binding energy per nucleon* is obtained. The graph of Fig. 22.25 shows how this quantity varies with mass number; in most cases it is about 8 MeV. Nuclides in the middle of the graph have the highest binding energy per nucleon and are thus the most stable since they need most energy to disintegrate. The smaller values for higher and lower mass numbers imply that potential sources of nuclear energy are reactions involving the disintegration of a heavy nucleus or the fusing of particles to form a nucleus of higher mass number. In both cases nuclei are produced having a greater binding energy per nucleon and there is consequently a mass transfer during their formation.

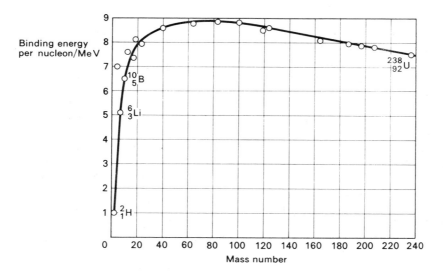

Fig. 22.25

Nuclear energy

(*a*) *Fission.* The discovery of the neutron provided nuclear physicists with an important new missile. Being uncharged, it is not repelled on approaching the large positive charge on the nucleus of a heavy atom as are protons and alpha particles. In 1939, following some work by Fermi in Italy and Hahn and Strassmann in Germany, it was found that the bombardment of uranium by neutrons may split the uranium nucleus into two large nuclei—for example, those of barium and krypton.

This nuclear reaction, called *nuclear fission*, differs from earlier nuclear reactions in three respects.

(*i*) The nucleus is deeply divided into two large *fission fragments* of roughly equal mass.

(*ii*) The *mass decrease* is appreciable.

(*iii*) Other neutrons, called *fission neutrons*, are emitted in the process.

Fission occurs in several heavy nuclides as well as in the two main isotopes of uranium, $^{238}_{92}U$ and $^{235}_{92}U$. The latter is the more useful and one reaction which occurs is

$$^{235}_{92}U + {}^{1}_{0}n \rightarrow {}^{144}_{56}Ba + {}^{90}_{36}Kr + 2\,{}^{1}_{0}n$$
$$[{}^{235}_{92}U\,(n, 2n)\,{}^{144}_{56}Ba,\,{}^{90}_{36}Kr]$$

The mass decrease in this reaction is about 200 MeV per fission, which is 45 million times greater per atom of fuel (uranium 235) than in a chemical reaction where the energy change is no more than a few electron-volts per atom. The uranium 235 nucleus is evidently a vast storehouse of energy. The fission energy appears mostly as kinetic energy of the fission fragments (e.g.

barium and krypton nuclei) which fly apart at great speed. The kinetic energy of the fission neutrons also makes a slight contribution. In addition, one or both of the large fragments are highly radioactive and a small amount of energy takes the form of beta and gamma radiation.

The production of fission neutrons (two in the above case) raises the possibility of having a *chain reaction* in which fission neutrons cause further fission of uranium 235 and the reaction keeps going, Fig. 22.26. In practice only a proportion of the fission neutrons is available for new fissions since some are lost by escaping from the surface of the uranium before colliding with another nucleus. The ratio of neutrons escaping to those causing fission decreases as the size of the piece of uranium 235 increases and there is a critical size (about the size of a cricket ball) which must be attained before a chain reaction can start.

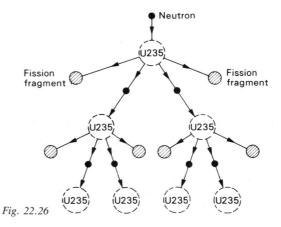

Fig. 22.26

In the *atomic bomb* an increasing uncontrolled chain reaction occurs in a very short time when two pieces of uranium 235 (or plutonium 239) are rapidly brought together to form a mass greater than the critical size. The reaction is started either by having a small neutron source in the bomb or by stray neutrons from the occasional spontaneous fission of a uranium 235 nucleus.

(*b*) *Nuclear reactors.* In a nuclear reactor the chain reaction is steady and controlled so that on average only one neutron from each fission produces another fission. The reaction rate is adjusted by inserting neutron-absorbing rods of boron steel into the uranium 235. The basic arrangement is shown in Fig. 22.27. The graphite core is called the *moderator* and is needed to maintain a chain reaction when the fuel is not fairly pure uranium 235.

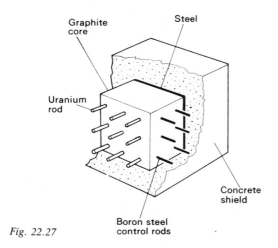

Fig. 22.27

Natural uranium contains over 99 per cent of uranium 238 and less than 1 per cent of uranium 235 and unfortunately the former captures the medium-speed fission neutrons without fissioning. The reaction which occurs is

$$^{238}_{92}U + ^1_0n \rightarrow ^{239}_{92}U \rightarrow ^{239}_{93}Np + ^0_{-1}e \quad [^{238}_{92}U(n, e)^{239}_{93}Np]$$

Neptunium 239 is the first transuranic element and decays, also by beta emission, to plutonium, the second transuranic element.

$$^{239}_{93}Np \rightarrow ^{239}_{94}Pu + ^0_{-1}e$$

Uranium 238 only fissions with very fast neutrons. On the other hand, uranium 235 (and plutonium 239)

fissions with *slow neutrons* and the job of the moderator is to slow down the fission neutrons very quickly so that most escape capture by uranium 238 and then cause the fission of uranium 235. A bombarding particle gives up most energy when it has an elastic collision with a particle of similar mass. For neutrons, hydrogen atoms would be most effective but unfortunately absorption occurs. However, deuterium (in heavy water) and carbon (as graphite) are both suitable.

In a nuclear power station a nuclear reactor provides the heat required to produce steam instead of a coal- or oil-burning furnace. In the reactor core the large fission fragments share their kinetic energy with surrounding atoms as a result of collisions and raise the internal energy of the system. In a gas-cooled reactor a current of high pressure carbon dioxide gas is pumped through the core of the reactor and transfers heat to the heat exchanger where water is converted to steam which is then used as in a conventional power station to drive a turbo-generator. A thick concrete shield gives protection from neutrons and gamma rays.

Nuclear reactors like that just described are called *thermal reactors* because fission is caused by slow neutrons with thermal energies, i.e. energies equal to the average kinetic energy of the surrounding atoms. The *fast breeder reactor* has a core made of highly enriched fuel (i.e. natural uranium from which uranium 238 has been extracted), fission occurs by *fast* neutrons and no moderator is required. The core is surrounded by a blanket of natural uranium which is converted to plutonium 239 by fission neutrons that escape from the core and would otherwise be lost. New fissionable material is thus 'bred' for the reactor. Fig 22.28*a* shows an external view of the Prototype Fast Reactor (PFR) at Dounreay in the north of Scotland; it is designed to operate a 250 MW electricity generator. Fig. 22.28*b* shows the reactor top and the charge machine console.

Important by-products of nuclear reactors are artificial radioisotopes; they are made when stable nuclides, inserted in the core of the reactor, are bombarded by neutrons.

(*c*) *Fusion.* The union of light nuclei into heavier nuclei can also lead to a transfer of mass and a consequent liberation of energy. Such a reaction has been achieved in the 'hydrogen bomb' and it is believed to be the principal source of the sun's energy.

A reaction with heavy hydrogen or deuterium which yields 3.3 MeV per fusion is

$$^2_1H + ^2_1H \rightarrow ^3_2He + ^1_0n \quad [^2_1H(^2_1H, n)^3_2He]$$

By comparison with the 200 MeV per fission of uranium 235 this seems small, but per unit mass of material

(a)

(b)

Fig. 22.28

it is not. Fusion of the two deuterium nuclei, i.e. deuterons, will only occur if they overcome their mutual electrostatic repulsion. This may happen if they collide at very high speed when, for example, they are raised to a very high temperature. If fusion occurs, enough energy is released to keep the reaction going and since heat is required, the process is called *thermonuclear fusion*.

Temperatures between 10^8 and 10^9 K are required and in the uncontrolled thermonuclear reaction occur-ring in the hydrogen bomb, the high initial temperature is obtained by using an atomic (fission) bomb to trigger off fusion. If a controlled fusion reaction can be achieved an almost unlimited supply of energy will become available from deuterium in the water of the oceans. The problem is, first, how to achieve the high temperature necessary and, second, how to keep the hot reacting gases (called the plasma) from touching the vessel holding it. Research continues on a world-wide scale into this difficult but challenging task. Fig.

Fig. 22.29

22.29 shows some of the equipment being used for research into fusion at the United Kingdom Atomic Energy Authority's laboratories at Culham.

Particle physics

Insight into the structure of the nucleus has been obtained by bombarding matter with high-speed particles, thereby causing nuclear reactions. As we have seen, alpha particles from radioactive sources were the earliest projectiles but these have limitations (p. 547), among them a maximum energy of 10 MeV. The various types of particle accelerator are designed to provide a copious but controlled supply of a variety of particles such as electrons, protons, deuterons, etc., of known, high energy. Before describing some of these in outline, another natural source of high-speed particles will be considered.

(*a*) *Cosmic rays.* Ionizing radiations from outer space, called *cosmic rays*, were first postulated to explain the gradual discharge of a well-insulated electroscope. The rays have been of great interest to physicists, not only because of the mystery of their origin but also on account of their energies which are far in excess of any that present-day accelerators can produce. Their use as projectiles for bombarding nuclei has led to the discovery of other particles such as the positron (a positive electron) and various mesons (short-lived and most with masses between that of the electron and the proton).

Those rays arriving at the outer limit of the earth's atmosphere are called *primary cosmic rays* and consist mostly of protons. Their mean energy is 10 GeV (10^{10} eV) but some particles have energies greater than 10^{10} GeV. Information about the primary rays has been obtained by sending stacks of special photographic plates to heights of 30 km using balloons, and more recently rockets and artificial satellites have been used. The high-energy primary rays react with atomic nuclei in the earth's upper atmosphere, causing nuclear reactions in which the nucleons themselves change and are not just rearranged within the nucleus as in lower energy reactions. As a result *secondary cosmic rays* are produced comprising electrons, mesons, gamma rays, positrons, neutrons and protons and these create most of our 'background' radiation.

(*b*) *Electrostatic accelerator.* A high p.d. is used to accelerate charged particles (protons, deuterons, alpha particles, etc.) between two electrodes in an evacuated tube. Cockcroft and Walton (p. 547) produced a p.d. of 2 MV from an a.c. input using a voltage multiplying circuit. A van de Graaff generator may also be used and a maximum p.d. of 14 MV obtained.

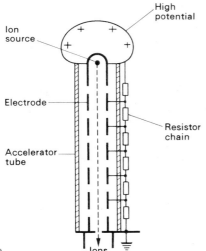

Fig. 22.30

Ions from an appropriate source enter the accelerator tube, Fig. 22.30, which contains a series of cylindrical electrodes at decreasing potentials. The ions are accelerated while travelling between the electrodes (since the electric field well inside a charged conductor is zero) and, because the field lines are curved there, the ions are focused into a narrow beam. They emerge through a fine slit at the bottom on to the target.

(*c*) *Linear accelerator.* It is possible to produce high-energy particles without especially large p.d.s by a method called *synchronous acceleration* and which is used in the linear accelerator. The electrodes are again a series of coaxial cylinders but they increase in length towards the target and alternate ones are connected to the same output terminal of a high-frequency alternating p.d. as in Fig. 22.31.

Acceleration occurs in the small gaps between the electrodes, the electric field inside an electrode being zero. Positive ions from the source which reach the gap between P and Q when Q is at a high negative potential relative to P will be greatly accelerated. If the length of Q is such that the time taken by the ion (travelling with constant velocity) to travel through Q is half the period of oscillation of the supply p.d. then the field between Q and R will produce further acceleration, i.e. R will be at a negative potential. If the peak value of the supply p.d. is 200 kV, the ions gain 200 keV of energy at each gap and because their speed increases as they travel along the tube, the electrodes must progressively increase in length. What is the advantage of using a very high frequency supply?

Electron linear accelerators working on the same principle are also in use.

(*d*) *Cyclotron.* The cyclotron also uses synchronous acceleration but a magnetic field makes the charged particles traverse a spiral of increasing radius rather than a long straight path. The machine consists of two semi-circular boxes, D_1 and D_2 in Fig. 22.32*a*, called 'dees' because of their shape, enclosed in a chamber C containing gas at low pressure. C is arranged between the poles of an electromagnet so that a nearly uniform

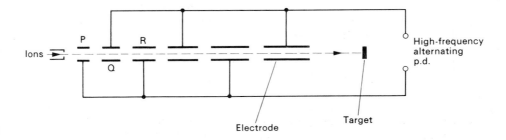

Fig. 22.31

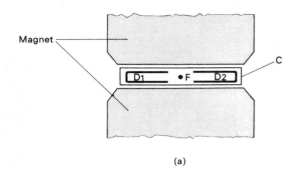

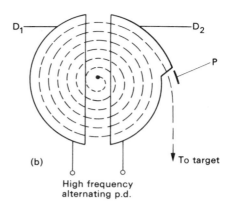

(a)

(b)

High frequency
alternating p.d.

Fig. 22.32

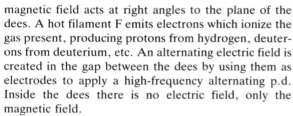

magnetic field acts at right angles to the plane of the dees. A hot filament F emits electrons which ionize the gas present, producing protons from hydrogen, deuterons from deuterium, etc. An alternating electric field is created in the gap between the dees by using them as electrodes to apply a high-frequency alternating p.d. Inside the dees there is no electric field, only the magnetic field.

Suppose that at a certain instant D_1 is positive and D_2 negative. A positively charged particle starting from F will be accelerated towards D_2 and when inside this dee it describes a semi-circular path at constant speed since it is under the influence of the magnetic field alone. The radius r of this path is given by

$$BQv = \frac{mv^2}{r}$$

where B is the magnetic flux density, Q the charge on the particle, v its speed inside D_2 and m its mass. Hence

$$r = \frac{mv}{BQ} \qquad (1)$$

If the frequency of the alternating p.d. is such that the particle reaches the gap again when D_1 is negative and D_2 positive, it accelerates across the gap and describes another semi-circle inside D_1 but of greater radius since its speed has increased. The particle thus gains kinetic energy and moves in a spiral of increasing radius, Fig. 22.32*b*, provided that the time of one complete half-oscillation of the p.d. equals the time for the particle to make one half-revolution. We shall now show that this condition generally holds.

Let T be the time for the particle to describe a semi-circle of radius r with speed v then

$$T = \frac{\pi r}{v}$$

Substituting for r from (1), we get

$$T = \frac{\pi m}{BQ}$$

T is therefore independent of v and r and constant if B, m and Q do not alter. Hence, for paths of larger radius the increased distance to be covered is exactly compensated by the increased speed of the particle. After about 100 revolutions, a plate P, at a high negative potential, draws the particles out of the dees before it bombards the target under study.

(*e*) *Synchrotron*. It is not possible to obtain protons with energies greater than about 20 MeV using a cyclotron. The limit arises when the speed of the particle is sufficiently great for its relativistic increase of mass to increase the time of revolution and upset the synchronization. The synchrotron, on the other hand, uses this effect.

The speed of the particle is first increased to be close to that of light, either by a subsidiary accelerator or in some other way. The particle then traverses the same circular path under the action of a magnetic field and its energy increased by applying synchronous electrical pulses from a high-frequency alternating p.d. connected to several cylindrical electrodes spaced round the track, Fig. 22.33. As a result, the relativistic mass of the particle increases but its speed remains more or less the same.

Synchrotrons have been built to accelerate both electrons and protons but the initial acceleration is more easily achieved for electrons. Fig. 22.34*a* shows part of the 30 GeV proton synchrotron at CERN (The European Organization for Nuclear Research) near Geneva, whose energy has now been increased to 300 GeV. The whole laboratory covers a large area, Fig. 22.34*b* and is devoted to pure research into the

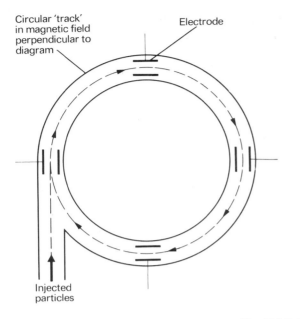

Circular 'track' in magnetic field perpendicular to diagram

Electrode

Injected particles

Fig. 22.33

Fig. 22.34(a)

Fig. 22.34(b)

fundamental structure of matter by groups of visiting scientists from the twelve European member countries.

(*f*) *Elementary particles*. In 1932 it seemed that all matter was built from three basic particles—the proton, the neutron and the electron. Today about 200 'parti-cles' have been produced in nuclear reactions brought about by particle accelerators. Many have lives of less than 10^{-10} s, their nature and function is one of the exciting mysteries of nuclear physics at the present time.

QUESTIONS

1. Estimate the energy in MeV of an alpha particle from a source of activity 1.0 μCi which creates a saturation current of 1.0×10^{-9} A in an ionization chamber. Assume $e = 1.6 \times 10^{-19}$ C, 1 Ci = 3.7×10^{10} disintegrations s^{-1} and 30 eV is needed to produce one ion pair.

2. Part of the uranium decay series is shown below.

$$\underset{92}{^{238}}U \overset{(1)}{\rightarrow} \underset{90}{^{234}}Th \overset{(2)}{\rightarrow} \underset{91}{^{234}}Pa \overset{(3)}{\rightarrow} \underset{92}{^{234}}U \overset{(4)}{\rightarrow} \underset{90}{^{230}}Th \overset{(5)}{\rightarrow} \underset{88}{^{226}}Ra$$

(a) What particle is emitted at each decay?
(b) How many pairs of isotopes are there?
(c) If the stable end-product of the complete uranium series is lead 206, how many alpha particles are emitted between radium 226 and the end of the series?

3. A radioisotope of silver has a half-life of 20 minutes. (a) How many half-lives does it have in one hour? (b) What fraction of the original mass would *remain* after one hour? (c) What fraction would have *decayed* after two hours?

4. Taking the half-life of radium 226 to be 1600 years
(a) what fraction of a given sample remains after 4800 years
(b) what fraction has decayed after 6400 years, and
(c) how many half-lives does it have in 9600 years?

5. State the law governing the rate of decay of a radioactive substance, and explain the terms decay constant (λ) and half-life (T). Show that these two quantities are related by the equation

$$\lambda T = \ln 2$$

Describe briefly how the decay law may be verified experimentally for a source of half-life of about one hour.
Two radioactive sources A and B initially contain equal numbers of radioactive atoms. Source A has a half-life of one hour, and source B a half-life of two hours. What is the ratio of the rate of disintegration of source A to that of source B (a) initially, (b) after two hours, and (c) after ten hours?

(O. and C.)

6. Discuss the assumption on which the law of radioactive decay is based.
What is meant by the *half-life* of a radioactive substance?
A small volume of a solution which contained a radioactive isotope of sodium had an activity of 12 000 disintegrations per minute when it was injected into the bloodstream of a patient. After 30 hours the activity of 1.0 cm^3 of the blood was found to be 0.50 disintegrations per minute. If the half-life of the sodium isotope is taken as 15 hours, estimate the volume of blood in the patient.

(J.M.B.)

7. Describe, with the aid of a diagram, the structure of a Geiger-Müller tube. Why does the tube contain a small quantity of a halogen or an organic gas?
Explain in what important respect a Geiger-Müller tube

suitable for detecting beta particles differs from a tube used for detecting gamma radiation. State, giving your reasons, whether either tube would be suitable for detecting alpha particles.
A Geiger-Müller tube in conjunction with a scaler was used to investigate the rate of decay of a radioactive isotope of protactinium. Counts were made over periods of ten seconds, each count starting 30 seconds after the previous count was completed. The results, corrected for background radiation, were recorded as follows:

Time interval from start:	0–10	40–50	80–90 seconds
Count:	3410	2310	1620

Time interval from start:	120–130	160–170	220–210 seconds
Count:	1110	770	505

Determine, graphically or otherwise, the half-life of the isotope.

(L.)

8. What is gamma-radiation? Explain *one* way in which it originates.
An experiment was conducted to investigate the absorption by aluminium of the radiation from a radioactive source by inserting aluminium plates of different thicknesses between the source and a Geiger tube connected to a ratemeter (or scaler). The observations are summarized in the following table:

Thickness of aluminium (cm)	Corrected mean count rate (min^{-1})
2.3	1326
6.9	802
11.4	496
16.0	300

Use these data to plot a graph and hence determine for this radiation in aluminium the *linear absorption coefficient*, μ (defined by $\mu = -\dfrac{dI}{I} \cdot \dfrac{1}{dx}$ where I is the intensity of the incident radiation and dI is the part of the incident radiation absorbed in thickness dx).
Draw a diagram to illustrate the arrangement of the apparatus used in the experiment and describe its preliminary adjustment.
What significance do you attach to the words *corrected* and *mean* underlined in the table?

(J.M.B.)

9. Write short notes on *atomic number, mass number, neutron*.
Discuss briefly the reaction represented by

$$\underset{4}{^{9}}Be + \underset{2}{^{4}}He = \underset{6}{^{12}}C + \underset{0}{^{1}}n.$$

What is the nature and possible origin of the particle $\underset{2}{^{4}}He$?

(L.)

10. A particle of mass m and speed v has a head-on collision with a stationary particle of mass M. Assuming the collision is elastic, derive an expression for the velocity of each particle after impact.

Hence determine what happens if the moving particle is an alpha particle and the stationary particle is (a) an electron, (b) a helium atom, and (c) a gold atom.

Under what conditions is there maximum energy transfer?

11. Two collisions between atomic particles are shown in the cloud chamber photographs of Fig. 22.35a and b. In both photographs most of the tracks are due to alpha particles travelling from left to right. In one case there is hydrogen in the chamber and in the other nitrogen and in each an elastic collision has occurred giving a 'split' track.

(a) Which photograph shows the nitrogen collision? Why?

(b) Copy the hydrogen collision tracks (three parts) and label each part to show the tracks of the alpha particle before and after the collision and of the hydrogen atom set into motion.

(c) Draw the collision tracks for an alpha particle having an oblique elastic collision with a helium atom. Mark the size of any angle which you think is important.

12. Compare nuclear fission and nuclear fusion. How many fissions must occur per second to generate a power of 1.00 MW if each fission of uranium 235 liberates 200 MeV. ($1 \text{ eV} = 1.60 \times 10^{-19}$ J.)

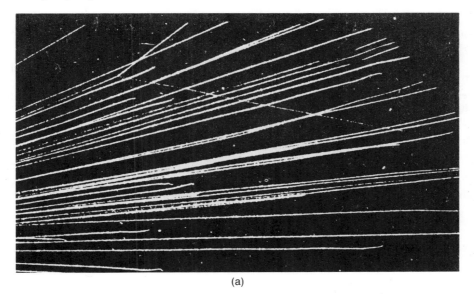

(a)

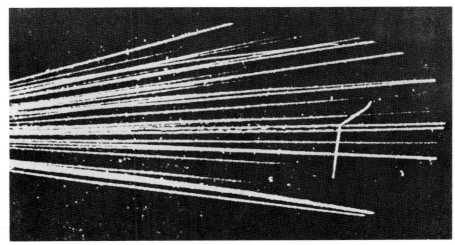

(b)

Fig. 22.35

Objective-type revision questions for Part 5

The first figure of a question number gives the relevant chapter, e.g. **19.2** is the second question for chapter 19.

Multiple choice

Select the response which you think is correct.

19.1. At pressure P and absolute temperature T a mass M of an ideal gas fills a closed container of volume V. An *additional* mass $2M$ of the same gas is introduced into the container and the volume is then reduced to $V/3$ and the temperature to $T/3$. The pressure of the gas will now be

   A $\dfrac{P}{3}$ **B** P **C** $3P$ **D** $9P$ **E** $27P$ (*J.M.B.*)

19.2. If, at a pressure of 10^5 Pa (N m^{-2}), the density of oxygen is 1.4 kg m^{-3}, it follows that the root-mean-square velocity of oxygen molecules in m s^{-1} is

   A 5   **B** 18   **C** 120   **D** 270   **E** 460   (*J.M.B.*)

19.3. An ideal gas at 300 K is adiabatically expanded to twice its original volume and then heated until the pressure is restored to its initial value. What is the final temperature?

   A 300 K   **B** 400 K   **C** 450 K   **D** 600 K
                                  (*J.M.B. Eng. Sc.*)

19.4. A Carnot engine operates between temperatures of 600 K and 300 K and accepts a heat input of 1000 joules. The work output is

   A 300 J   **B** 400 J   **C** 500 J   **D** 600 J
                                  (*J.M.B. Eng. Sc.*)

19.5. A heat pump is to be installed to heat a greenhouse; the heat is to be extracted from a neighbouring stream at a maximum rate of 20 kW. The maximum temperature of the greenhouse is 300 K.

When the maximum temperature difference between the greenhouse and the stream is 40 K assuming a Carnot cycle the input power required for the pump will be of the order of

   A 1 kW   **B** 3 kW   **C** 10 kW   **D** 30 kW
                                  (*J.M.B. Eng. Sc.*)

20.1. An electrostatic field E and a magnetic flux density B act over the same region, and an electron enters the region. Which one of the combinations of E and B in Fig. 24 can be made to cause the electron to pass undeflected?

                                  (*J.M.B. Eng. Sc.*)

20.2. Which one of the following phenomena cannot be explained by the wave theory of light?

   A Refraction   **B** Interference   **C** Diffraction
   D Polarization   **E** Photoelectric effect

20.3. A photocell is illuminated with ultraviolet. The intensity of the illumination is reduced resulting in

   A a reduction in the average kinetic energy of the electrons but no change in their rate of emission.
   B no change in either the rate at which electrons are emitted or in their average kinetic energy.
   C a reduction in the rate at which electrons are emitted but no change in their average kinetic energy.
   D a reduction in both the rate at which they are emitted and their average kinetic energy.   (*J.M.B. Eng. Sc.*)

20.4. Which one of the following statements is *not* correct? The value obtained by Thomson for the ratio of the charge to the mass of cathode rays

   A did not depend on the gas in the tube.
   B did not depend on the electrode material.
   C did not depend on the tube material.
   D did not depend on the accelerating voltage.
   E was lower than the value obtained for protons.

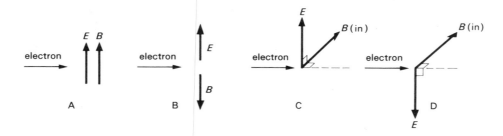

Fig. 24

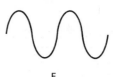

Fig. 25　　A　　　　　　B　　　　　C　　　　　D　　　　　　E

21.1. When a linear time base of 25 Hz is applied to the X-plates of an oscilloscope and a sinusoidal voltage of 50 Hz is applied to the Y-plate, the trace could be, Fig. 25.

22.1. Two radioactive elements X and Y have half-value periods ('half-lives') of 50 minutes and 100 minutes respectively. Samples of A and B initially contain equal numbers of atoms. After 200 minutes what is the value of the following fraction?

$$\frac{\text{number of atoms of X unchanged}}{\text{number of atoms of Y unchanged}}$$

is　　**A** 4　　**B** 2　　**C** 1　　**D** $\frac{1}{2}$　　**E** $\frac{1}{4}$　　　　(*J.M.B.*)

22.2. When a lithium nucleus ($^{7}_{3}$Li) is bombarded with certain particles, two alpha particles only are produced. The bombarding particles are

A electrons　　**B** protons　　**C** deuterons　　**D** neutrons
E photons

Multiple selection

In each question one or more of the responses may be correct. Choose one letter from the answer code given.

*Answer **A** if (i), (ii) and (iii) are correct*
*Answer **B** if only (i) and (ii) are correct*
*Answer **C** if only (ii) and (iii) are correct*
*Answer **D** if (i) only is correct*
*Answer **E** if (ii) only is correct*

20.5. If the accelerating voltage across an X-ray tube is doubled

(*i*) the wavelengths of the characteristic lines are halved.
(*ii*) the minimum wavelength of the X-rays is halved.
(*iii*) the X-rays are most probably more penetrating.

Appendix 1

Method of dimensions

(*a*) *Physical quantities*. These can be classified as *basic* quantities and *derived* quantities. Seven basic quantities are chosen for their convenience and are: *mass, length, time, electric current, temperature, luminous intensity* and *amount of substance*. All other quantities are derived from one or more of the basic quantities.

(*b*) *Dimensions*. The dimensions of a quantity show how it is related to the basic quantities. For example, the derived quantity *volume* has three dimensions in length; it is measured basically by multiplying three lengths together and this is shown by writing $[V] = [L^3]$. The square brackets round V indicate that we are dealing with the dimensions of V. *Density* is measured by dividing a mass by a volume and has dimensions $[ML^{-3}]$. The dimensions of *velocity* are $[LT^{-1}]$ and of *acceleration* $[LT^{-2}]$. Fundamentally *force* is measured by multiplying a mass by an acceleration ($F = ma$) and so has dimensions $[MLT^{-2}]$. Every derived quantity has dimensions.

What are they for *pressure*?

Some quantities are partly dimensional. For example, *frequency* is a number of oscillations per unit time and has dimension $[T^{-1}]$ since the number of oscillations part is dimensionless. Some quantities such as refractive index are dimensionless.

(*c*) *Dimensional analysis*. If an equation is correct the dimensions of the quantities on either side must be identical. This fact is used in the method of Dimensional analysis which enables predictions to be made about how quantities may be related. The method is particularly helpful when dealing with viscosity problems and is used to derive Poiseuille's formula on p. 207 and Stokes' law on p. 208.

Assumptions based on experiment or intuition have first to be made about what quantities could be involved and in general no more than *three* such dependent quantities can be considered. Neither does the method yield the value of any dimensionless constants, e.g. π.

560

Appendix 2

SI units

SI units (standing for *Système International d'Unités*) were adopted internationally in 1960.

(*a*) *Basic units.* The system has seven basic units, one for each of the basic quantities.

Basic quantity	Unit	
	Name	Symbol
mass	kilogram	kg
length	metre	m
time	second	s
electric current	ampere	A
temperature	kelvin	K
luminous intensity	candela	cd
amount of substance	mole	mol

Units must be easily reproducible and unvarying with time and so are often based on the properties of atoms. Thus the *metre* is now the length which equals 1 650 763.73 wavelengths in a vacuum of a specified radiation from a krypton-86 atom. Definitions of some of the other units are given when the quantity arises in the text.

(*b*) *Derived units.* These are obtained from the basic units by multiplication or division; no numerical factors are involved. Some derived units with complex names:

Derived quantity	Unit	
	Name	Symbol
volume	cubic metre	m^3
density	kilogram per cubic metre	$kg\ m^{-3}$
velocity	metre per second	$m\ s^{-1}$
acceleration	metre per second squared	$m\ s^{-2}$
momentum	kilogram metre per second	$kg\ m\ s^{-1}$

Some derived units are given special names due to their complexity when expressed in terms of the basic units.

Derived quantity	Unit	
	Name	Symbol
force	newton	N
pressure	pascal	Pa
energy, work	joule	J
power	watt	W
frequency	hertz	Hz
electric charge	coulomb	C
electric resistance	ohm	Ω
electromotive force	volt	V

When the unit is named after a person the *symbol* has a capital letter.

(*c*) *Standard prefixes.* Decimal multiples and submultiples are attached to units when appropriate. In general, prefixes involving powers which are multiples of three are preferred but others are used, e.g. 10^{-2} (centi).

Multiple	Prefix	Symbol	Submultiple	Prefix	Symbol
10^3	kilo	k	10^{-3}	milli	m
10^6	mega	M	10^{-6}	micro	μ
10^9	giga	G	10^{-9}	nano	n
10^{12}	tera	T	10^{-12}	pico	p

(*d*) *Coherence.* SI units are coherent. This means that there is only *one* unit for each quantity (ignoring multiples and submultiples) and if these are used for the quantities in an expression, the answer is obtained in the correct SI unit. For example if in $F = ma$, m is expressed in kg and a in $m\ s^{-2}$ then F will be automatically in newtons.

Appendix 3

Measurement of length

(*a*) *Vernier scale.* The simplest type enables a length to be measured to 0.01 cm. It comprises a small sliding scale which is 9 mm long and is divided into 10 equal divisions, Fig. A3.1*a*. Hence

1 vernier division = 9/10 mm = 0.9 mm = 0.09 cm

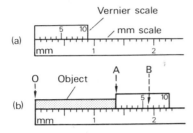

Fig. A3.1

One end of the length to be measured is made to coincide with the zero of the millimetre scale and the other end with the zero of the vernier scale. The length of the object in Fig. A3.1*b* is between 1.3 cm and 1.4 cm. The reading to the second place of decimals is obtained by finding the *vernier mark* which is exactly opposite (or nearest to) a mark on the millimetre scale. In this case it is the 6th mark and the length is 1.36 cm since

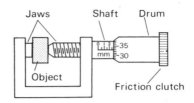

Fig. A3.2

OA = OB − AB

= (1.90 cm) − (6 vernier divisions)

= 1.90 − 6(0.09) = 1.90 − 0.54 cm

= 1.36 cm

Vernier scales are often used on calipers, barometers, travelling microscopes and spectrometers.

(*b*) *Micrometer screw gauge.* This measures very small objects to 0.001 cm. One revolution of the drum opens the accurately plane, parallel jaws by 1 division on the scale on the *shaft* of the gauge; this is usually $\frac{1}{2}$ mm, i.e. 0.05 cm. If the *drum* has a scale of 50 divisions round it, then rotation of the drum by 1 division opens the jaws by 0.05/50 = 0.001 cm, Fig. A3.2. A friction clutch ensures that the jaws always exert the same force when the object is gripped. The object shown has a length

= 2.5 mm on the shaft scale + 33 divisions on the drum scale

= 0.25 cm + 33(0.001) cm

= 0.283 cm.

Appendix 4

Graphs

When plotting a graph from experimental results, as much of the paper as possible should be used, points should be marked O or X and a smooth curve or straight line drawn so that the points are distributed equally on either side of it.

(*a*) *Straight line graph.* If a straight line is obtained its equation is of the form

$$y = mx + c$$

where m is the slope of the line (QR/PQ in Fig. A4.1*a*) and c is the intercept (OS) on the y-axis.

In the magnification method for finding the focal length of a converging lens (p. 107), if a graph is plotted of magnification m against image distance v, a straight line should be obtained of equation $m = v/f - 1$. The slope of the graph is $1/f$ and the intercept on the m-axis is -1.

If a graph passes through the origin O then $c = 0$ and $y = mx$, i.e. $y \propto x$ and so y is directly proportional to x. A straight line graph *not* passing through the origin only indicates that y is directly proportional to x *plus a constant*, rather than solely dependent on x.

If it is not possible to include on the graph the origins for one or both axes, c can be calculated by substituting the co-ordinates of a point on the graph, say (x_1, y_1) and the value of m, in the equation $y = mx + c$ and solving for c. Thus if $x = 2$, $y = 3$ and $m = +1$ then $3 = 1 \times 2 + c$, therefore $c = 1$ and the graph does not pass through the origin of the y-axis.

When possible, quantities are usually plotted which give a straight-line graph since this is easier to interpret. Thus if quantities T and l are related by the equation $aT = bl^2 + c$, (*i*) what would you plot to obtain a straight line and (*ii*) what would be the slope?

(*b*) *Log graph.* Sometimes we wish to find experimentally the relationship between two quantities x and y. If we assume that

$$y = kx^n$$

where k and n are constants, then taking logs

$$\log y = \log k + n \log x$$

This is of the form $y = c + mx$ and so a graph of $\log y$ against $\log x$ should be a straight line of slope n and intercept $\log k$ on the $\log y$ axis, Fig. A4.1*b*. Hence k and n can be found.

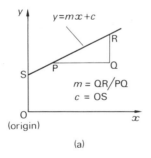

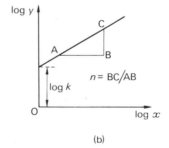

Fig. A4.1

(a)

(b)

Appendix 5

Treatment of errors

(a) *Types of error*. Experimental errors cause a measurement to differ from its true value and are of two main types.

(i) A *systematic error* may be due to an incorrectly calibrated scale on, for example, a ruler or ammeter. Repeating the observation does not help and the existence of the error may not be suspected until the final result is calculated and checked, say by a different experimental method. If the systematic error is small a measurement is accurate.

(ii) A *random error* arises in any measurement, usually when the observer has to estimate the last figure, possibly in an instrument which lacks sensitivity. Random errors are small for a good experimenter and taking the mean of a number of separate measurements reduces them in all cases. A measurement with a small random error is precise but it may not be accurate.

Fig. A5.1a shows random errors only in a meter reading, whilst in Fig. A5.1b there is a systematic error as well.

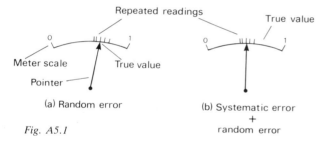

Repeated readings

True value

Meter scale   True value

Pointer

(a) Random error   (b) Systematic error
+
random error

Fig. A5.1

(b) *Estimating errors in single measurements*. In more advanced work, if systematic errors are not eliminated they can be corrected from the observations made. Here we shall assume they do not exist and then make a reasonable estimate of the likely random error. Two examples follow.

Using a metre rule the length of an object is measured as 2.3 cm. At very worst the answer might be 2.2 or 2.4 cm, i.e. the maximum error it is possible to make using a ruler marked in mm is ± 0.1 cm. The *possible error* (p.e.), is said to be ± 0.1 cm and the length is written (2.3 ± 0.1) cm. The *percentage possible error* (p.p.e.) is $(\pm 0.1 \times 100)/2.3 \simeq \pm 4\%$.

Using vernier calipers capable of measuring to 0.01 cm, the length of the same object might be read as (2.36 ± 0.01) cm. In this case the p.e. is ± 0.01 cm and the p.p.e. is $(\pm 0.01 \times 100)/2.36 \simeq \pm 0.4\%$.

If a large number of readings of one quantity are taken the mean value is likely to be close to the true value and statistical methods enable a *probable error* to be estimated. Here, we adopt the simpler procedure of estimating the maximum error likely, i.e. the *possible error*.

(c) *Combining errors*. The result of an experiment is usually calculated from an expression containing the different quantities measured. The combined effect of the errors in the various measurements has to be estimated. Three simple cases will be considered.

(i) *Sum*. Suppose the quantity Q we require, is related to quantities a and b which we have measured, by the equation

$$Q = a + b$$

Then total p.e. in Q = p.e. in a + p.e. in b. Thus if $a = 5.1 \pm 0.1$ cm and $b = 3.2 \pm 0.1$ cm then $Q = 8.3 \pm 0.2$ cm. That is, in the worst cases, if both a and b are read 0.1 cm too high $Q = (5.2 + 3.3) = 8.5$ cm, but if both are 0.1 cm too low then $Q = (5.0 + 3.1) = 8.1$ cm.

(ii) *Difference*. If $Q = a - b$, the same rule applies, i.e. the total p.e. in Q = p.e. in a + p.e. in b.

(iii) *Product and quotient*. If the individual measurements have to be multiplied or divided it can be shown that the *total percentage possible error equals the sum of the separate percentage possible errors*. For example, if a,b and c are measurements made and

$$Q = \frac{ab^2}{c^{1/2}}$$

then if the p.p.e. in a is $\pm 2\%$, that in b is $\pm 1\%$ and that in c is $\pm 2\%$, then the p.p.e. in b^2 is $2(\pm 1)\% = \pm 2\%$ and in $c^{1/2}$ is $\frac{1}{2}(\pm 2)\% = \pm 1\%$. Hence

total p.p.e. in Q = $\pm$(p.p.e. in a + p.p.e. in b^2
$$+ \text{ p.p.e. in } c^{1/2})$$

$$= \pm(2 + 2 + 1) = \pm 5\%$$

The answer for Q will therefore be accurate to 1 part in 20 and if the numerical result for Q is 1.8 then it is written

$$Q = 1.8 \pm \tfrac{1}{20} \times 1.8 = 1.8 \pm 0.1$$

It would not be justifiable to write $Q = 1.852$ since this would be claiming an accuracy of a few units in 1852. According to our estimate this accuracy is not possible with the apparatus used.

It is instructive to estimate whenever possible the total p.p.e. for an experiment; it indicates (i) the number of significant figures that can be given in the result, (ii) the limits within which the result lies and (iii) the measurements requiring particular care. There is little point in making one measurement to a very high degree of accuracy if it is not possible with the others; a chain is only as strong as the weakest link.

Appendix 6

Construction of model crystal for microwave analogue demonstration

The model has a face-centred cubic structure and is made from 190 5 cm diameter polystyrene balls glued together with Durofix in seven hexagonal close-packed layers, Fig. A6.1.

Start with *layer 4* which has 37 balls, Fig. A6.2*a* (the black dots represent the centres of the balls).

The black dots in Fig. A6.2*b* show the 36 balls in *layer 3* and their positions in the hollows of layer 4 which is shown by circles. Each ball should be glued to all those it touches.

The 27 balls of *layer 2* are the black dots in Fig. A6.2*c* (the circles are layer 3) and the layer 2 balls should be placed *over the hollows in layer 4* (not over the balls in layer 4) to give the

ABCABC stacking of an FCC crystal.

Fig. A6.2*d* shows the 19 balls of *layer 1* as black dots and layer 2 as circles.

Half the model is now made and when set, it can be turned over to build the other half.

In Fig. A6.2*e* the circles are layer 4 (now uppermost), the crosses represent layer 3. *Layer 5* is shown by 36 black dots.

The 23 black dots in Fig. A6.2*f* show *layer 6* and the circles layer 5. Care should again be taken to ensure that the balls in layer 6 are over hollows in layer 4.

Fig. A6.2*g* gives the position of the 12 balls in *layer 7*.

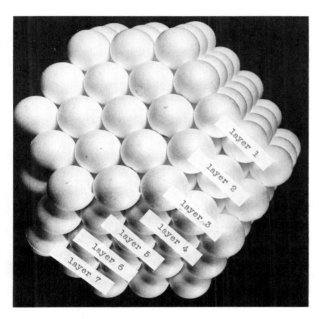

Fig. A6.1

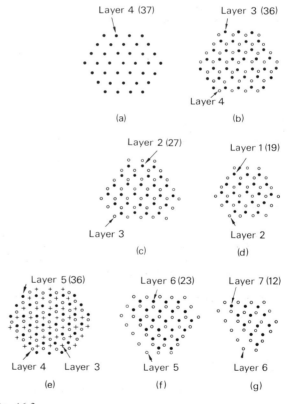

Fig. A6.2

Appendix 7

Speed of an electrical pulse along a cable

The drift speed of the electrons forming the current in a conductor is about 1 mm s^{-1} but the electric and magnetic fields which constitute the signal that sets them in motion almost simultaneously round the circuit, travel at very great speeds. Basically these fields are the same as those of an electromagnetic wave in free space.

Using a double beam C.R.O. on a fast time base speed (1 μs cm^{-1}), the time taken by an electrical pulse from a 200 kHz pulse generator to travel along a 200 m length of coaxial cable can be found, Fig. A7.1. The in-going pulse is applied to the Y$_1$ input (set at 0.2 V cm^{-1}) as well as to the near end of the cable and the outcoming pulse from the far end of the cable is applied to the Y$_2$ input (also on 0.2 V cm^{-1}). The distance between the pulses is measured and the time it represents calculated. The speed is roughly 2 × 10^8 m s^{-1}. If air or a vacuum replaced the polythene insulation between the central and outer conductors of the cable and through which the fields travel, the speed would be that of light (3 × 10^8 m s^{-1}). Fig. A7.2 shows the shapes of the electric and magnetic fields travelling along the cable; they are at right angles to each other and to the direction of the current.

Coaxial cable is used because it does not pick up unwanted interference if the outer conductor is connected to E on the C.R.O. at both ends. The 68 Ω resistors should be connected to the C.R.O. terminals directly. A full explanation of their action requires a more advanced treatment but without them the pulse would be reflected backwards and forwards along the cable, setting up a standing wave system on it. Instead, the resistors 'absorb' the pulses and ensure they are applied to the C.R.O.

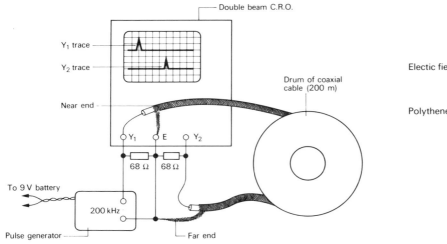

Fig. A7.1

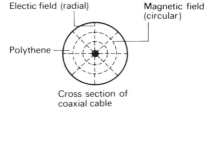

Fig. A7.2

Appendix 8

Construction and action of a flame probe

(*a*) *Construction.* Details are shown in Fig. A8.1. In use the flame should be made as small as possible by *slowly* reducing the gas supply (a screw clip helps).

(*b*) *Action.* The deflection of the electroscope is a measure of the potential at the point where the flame probe is situated. Roughly, the action is as follows. When brought near to a positively charged body the metal probe (i.e. the needle) has a negative charge induced in it whilst a positive charge appears on the electroscope movement. However, the flame is producing positive and negative ions; the former are attracted to the probe and neutralize its charge whilst the latter are repelled. The electroscope remains positively charged and is at the same potential as the uncharged probe since they are connected. The closer the probe is to the charged body the higher is its potential and the greater is the positive charge induced in the electroscope and so the greater the deflection.

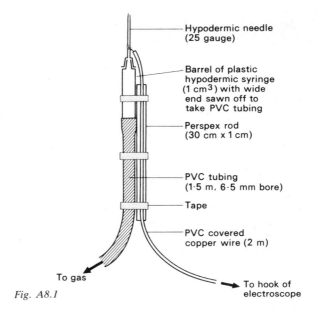

Fig. A8.1

Labels in figure:
- Hypodermic needle (25 gauge)
- Barrel of plastic hypodermic syringe (1 cm^3) with wide end sawn off to take PVC tubing
- Perspex rod (30 cm x 1 cm)
- PVC tubing (1·5 m, 6·5 mm bore)
- Tape
- PVC covered copper wire (2 m)
- To gas
- To hook of electroscope

Appendix 9

Capacitor experiments using a d.c. amplifier

A calibrated d.c. amplifier can replace the reed switch in certain capacitor experiments considered earlier.

(*a*) *Parallel-plate capacitor investigation* (also see p. 239). The arrangement is shown in Fig. A9.1, the capacitor being two thick metal plates, each of side 25 cm, kept apart by four small polythene spacers (5 mm × 5 mm) about 1.5 mm thick, at the corners.

The capacitor is charged to a known p.d. using a flying lead from a protective resistance (10 MΩ) in series with the output from an h.t. power supply. The flying lead is removed and the charge on the capacitor measured by touching the upper plate with the tip of the screened extension cable from the *calibrated* electrometer set on the charge range. The input capacitor of the electrometer should be discharged (by switching to 'short').

By varying the charging p.d. V in steps and noting the corresponding charges Q on the capacitor, it will be seen that $Q \propto V$.

If V is kept constant and the separation d of the plates changed by using more spacers at the corners, $Q \propto 1/d$ can be tested.

Finally if V and d are fixed and the area of overlap A of the plates is varied, a test of $Q \propto A$ is possible.

(*b*) *Measurement of ε_0 and ε_r* (also see p. 242). To find ε_0, the permittivity of free space, we use the expression for the capacitance of a parallel-plate capacitor, $C = A\varepsilon_0/d$ (p. 241) and experimental results from (*a*). Since $Q = VC$ we have

$$\varepsilon_0 = \frac{Qd}{AV}$$

A mean value of Q/V can be obtained from a graph of Q against V; d and A are readily measured.

The relative permittivity ε_r of, say, polythene is found by charging the parallel-plate capacitor to the same p.d. first with a sheet of polythene filling the space between the plates (giving capacitance C), then with the same thickness of vacuum (or air) between them (capacitance C_0) and measuring the charges (Q and Q_0 respectively) stored in each case, with the electrometer. The charges Q and Q_0 are in the ratio of the capacitances C and C_0, hence

$$\varepsilon_r = \frac{C}{C_0} = \frac{Q}{Q_0}$$

(*c*) *Decay curve for capacitor discharge* (also see p. 250). A capacitor-resistor combination is used, 10^{-8} F (0.01 μF) and 10^{10} Ω giving the convenient time constant of 100 s. The capacitor is first charged to 1 V to give a full-scale deflection on the meter, then the resistor is brought into circuit across it. Meter readings are taken every 10 seconds whilst the capacitor discharges through the resistor. A decay curve is plotted from the results.

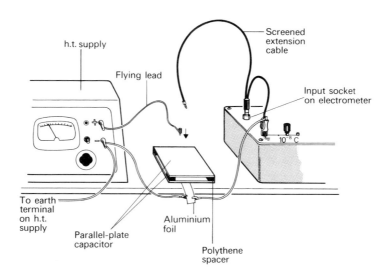

Fig. A9.1

Appendix 10

Calculation of B for an infinitely long straight wire using the Biot-Savart law

We require to find the flux density at P, perpendicular distance a in air from the long straight wire carrying current I.

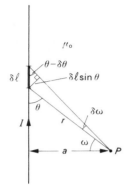

Fig. A10.1

The small element δl in Fig. A10.1 contributes flux density δB at P. By Biot and Savart,

$$\delta B = \frac{\mu_0 I \delta l \sin \theta}{4\pi r^2} = \frac{\mu_0 I}{4\pi} \cdot \frac{\delta \omega}{r} \qquad \text{(since } \delta l \sin \theta = r \cdot \delta \omega\text{)}$$

The total flux density B at P is obtained by integrating this expression over the whole length of wire between the limits $-\pi/2$ and $+\pi/2$, where these are the angles subtended at P by the ends of the wire. We have $\cos \omega = a/r$, hence

$$B = \int_{-\pi/2}^{+\pi/2} dB = \frac{\mu_0 I}{4\pi a} \int_{-\pi/2}^{+\pi/2} \cos \omega \, d\omega$$

$$= \frac{\mu_0 I}{4\pi a} \left(\sin \omega \right)_{-\pi/2}^{+\pi/2}$$

$$= \frac{\mu_0 I}{2\pi a}$$

Appendix 11

Demonstration of rectification and smoothing using a CRO

The circuit of Fig. A11.1a can be connected on a circuit board and the various waveforms studied on a CRO.

(a) With S_1 and S_2 closed, S_3 and S_4 open, the diode is short-circuited and the alternating input p.d. is developed across the load R.

(b) With S_2 closed, S_1, S_3 and S_4 open, half-wave rectifica-tion occurs but no smoothing.

(c) With S_2 and S_3 closed, S_1 and S_4 open, smoothing is produced by the reservoir capacitor C_1.

(d) With S_2, S_3 and S_4 closed, S_1 open, the capacitor-input filter LC_2 supplements the smoothing due to C_1.

(e) To show full-wave rectification and smoothing the 1N4001 diode is replaced by a bridge rectifier (e.g. BY 164), Fig. A11.1b.

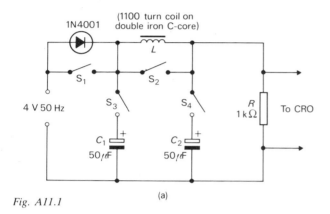

(a)

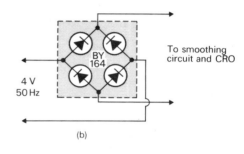

(b)

Fig. A11.1

Appendix 12

Speed of sound in a solid

Let an external force F be applied to the end of a solid rod of cross-sectional area A, setting it in motion with speed V and causing a compression pulse to travel along the rod with speed v. In time t the pulse covers a distance vt and this length of rod is compressed Vt, Fig. A12.1. Hence

$$\text{stress} = \frac{\text{force}}{\text{area}} = \frac{F}{A}$$

$$\text{strain} = \frac{\text{compression}}{\text{length compressed}} = \frac{Vt}{vt} = \frac{V}{v}$$

If the material has the Young modulus E then

$$E = \frac{\text{stress}}{\text{strain}} = \frac{Fv}{AV}$$

Therefore

$$F = \frac{EAV}{v}$$

But, *force $\times$ time* equals *change of momentum* and momentum of mass of length vt of rod set into motion with speed V is $(vtA\rho)V$ where ρ is the density of the material. Thus

$$Ft = \left(\frac{EAV}{v}\right)t = vtA\rho V$$

$$\therefore \quad v^2 = \frac{E}{\rho}$$

or,

$$v = \sqrt{\frac{E}{\rho}}$$

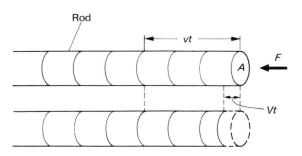

Fig. A12.1

Appendix 13

Ruling optical slits

Using a paint brush with fine bristles, one side of a microscope slide is coated with a smooth, thin paste of colloidal graphite (Aquadag) and water. The slide is allowed to dry, preferably overnight, and then inserted in the holder, Fig. A13.1. A blunt needle is held against the cross-piece and a slit ruled by drawing it across the slide so that it removes the graphite.

If a second slit is required, the screw on the end of the holder is turned, thus moving the slide along slightly. A separation of 0.5 mm (usually one revolution of the screw) between the slits is suitable.

In the double-slit experiment the closer the slits are together the greater is the fringe spacing and the easier are the fringes seen. Also, wider slits give brighter fringes but fewer are obtained.

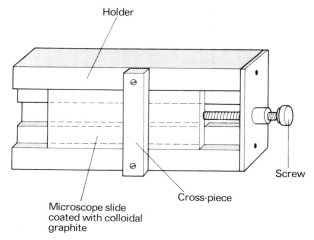

Fig. A13.1

Appendix 14

Bromine diffusion experiment (pp. 435 and 436)

Bromine liquid and vapour blister the skin and the vapour will cause a sore throat. They also attack most materials but not glass.

(*a*) *Precautions*. A 500 cm^3 beaker containing 0.88 concentrated ammonia solution diluted with its own volume of water should be to hand when preparing and performing the experiment. This solution must be poured at once on any bromine spilt on the bench or skin (harmless ammonium bromide is formed) but for the eyes use plenty of cold water.

(*b*) *Cleaning the apparatus*. Using rubber gloves the whole apparatus is placed in a plastic bucket half full of weak ammonia solution (200 cm^3 of 0.88 ammonia to half a bucket of water). The rubber stopper is eased out from the side tube *under the solution*. As the solution enters, white clouds of ammonium bromide are formed. The apparatus is manipulated until all signs of bromine have gone. The diffusion tube is emptied, washed out with hot water and dried in a current of air.

The rubber tubing from the tap is removed *under the solution* and the tap dismantled before washing, drying and lightly lubricating with Vaseline.

(*c*) *Use of rubber stoppers and rubber tubing*. The rubber tubing should be discarded after use *once*. The stopper on the tap may be used two or three times within a day or two; thereafter it hardens and must be replaced.

Appendix 15

The 'random walk' game (p. 436)

Six arrows each making 60° with its neighbour are drawn in pencil in the centre of a piece of isometric grid paper (i.e. graph paper ruled with 60° triangles) and labelled 1 to 6, Fig. A15.1.

A die is thrown and one step marked with a pen from the centre of the sheet in the direction given by the number on the die. The die is thrown again and a further step of one triangle-side taken *following on from the end of the first*. This is repeated for a total of 25 throws.

The distance in 'steps' from the starting point to the finishing point 'as the crow flies' is measured and compared with the theoretical value of $\sqrt{25}$.

If the game is played by the members of a class the scatter of results can be examined to see the large probable error in the result.

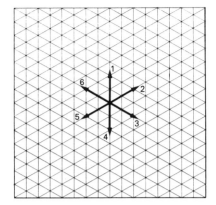

Fig. A15.1

Appendix 16

Evidence of energy levels from the absorption spectrum of iodine vapour

The experimental arrangement is shown in Fig. A16.1a, and what is seen in Fig. A.16.1b. The test-tube (hard glass) containing one or two small iodine crystals, after being warmed along its length and lightly corked, is heated at the bottom until the iodine vaporizes and colours the tube strongly. As it cools it is observed in front of the straight filament lamp through a fine diffraction grating (about 300 lines per millimetre).

The iodine molecules absorb from the light, those frequencies whose quanta have the correct amount of energy to enable them to jump from one energy level to another. Certain frequencies are therefore missing in the continuous spectrum of the light from the lamp and show up as dark lines. The presence of so many of these lines indicates that a whole 'ladder' of energy levels exists. In a gas or vapour the molecules are sufficiently far apart not to affect each other's energy levels to any extent.

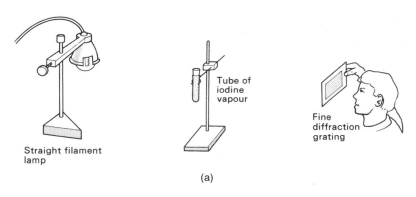

Straight filament lamp

Tube of iodine vapour

Fine diffraction grating

(a)

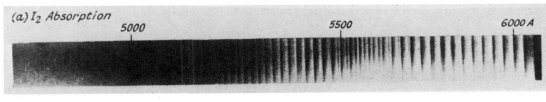

(a) I₂ Absorption

5000   5500   6000 A

(b)

Fig. A16.1

Appendix 17

Photoelectric effect and rough estimate of Planck's constant

The apparatus is shown in Fig. A17.1. Before the electrometer/d.c. amplifier is connected to the photoelectric unit any input resistor is removed so that it acts as a voltmeter whose resistance is as high as possible.

The photoelectric unit contains a photocell (see p. 467) with a photocathode of potassium (in a vacuum) which loses electrons when light falls on it. These travel to a collecting wire (of material that does not emit photoelectrons for light) which now becomes negatively charged and soon stops the arrival of further electrons. A steady p.d. is thus created between the cathode and the collecting wire and is measured by the electrometer connected across the photocell. If this is, say, 1.0 V then this must be the maximum k.e. of the electrons emitted by the cathode, otherwise they would continue to reach the collecting wire. The maximum k.e. of the photoelectrons is thus indicated by the electrometer reading (provided its input resistance is sufficiently high).

(a) *Effect of colour.* The spectrum is rotated *slowly* so that the colours from red to violet and beyond fall in turn on the slit in front of the photocell. The reading on the electrometer meter rises steadily indicating that the higher the frequency of the light, the greater the maximum k.e. of the photoelectrons.

The p.d. readings for red and violet light should be noted.

(b) *Effect of intensity.* With blue light on the slit a stop is placed in front of the lamp to halve (approx.) the light intensity. The electrometer reading should remain *almost* constant showing that the maximum k.e. of the photoelectrons is more or less independent of the brightness of the light. (What is the effect of decreased intensity?)

(c) *Planck's constant.* An estimate of h may be obtained if we assume that for red light $f = 4.5 \times 10^{14}$ Hz and for blue light $f = 6.5 \times 10^{14}$ Hz. If the corresponding p.d.s are, say, 0.25 V and 1.00 V then the maximum k.e.s are 0.25 eV (i.e. $0.25 \times 1.6 \times 10^{-19}$ J) and 1.00 eV (i.e. $1.00 \times 1.6 \times 10^{-19}$ J) respectively. Hence from Einstein's photoelectric equation (p. 466) we have,

$$hf - \Phi = \tfrac{1}{2}mv_{\max}^2$$

For blue light

$$h \times 6.5 \times 10^{14} - \Phi = 1.00 \times 1.6 \times 10^{-19}$$

For red light

$$h \times 4.5 \times 10^{14} - \Phi = 0.25 \times 1.6 \times 10^{-19}$$

Subtracting, $h = \dfrac{(1.00 - 0.25)\, 1.6 \times 10^{-19}}{(6.5 - 4.5)\, 10^{14}}$

$$= 6.0 \times 10^{-34} \text{ J s}$$

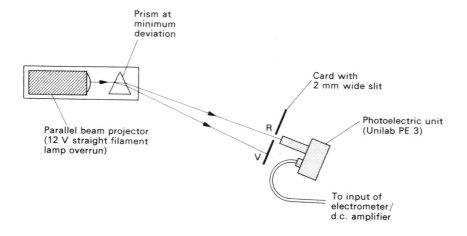

Fig. A17.1

Appendix 18

Electron collision experiments using a commercial xenon-filled thyratron valve (EN 91)

(a) *Excitation potential.* The circuit is shown in Fig. A18.1a. Electrons emitted by the cathode C of the thyratron are accelerated by the positive potential V_g on the grid G_1, and initially are able to reach the anode A since, for small values of V_g, any collisions they have with atoms of the xenon gas are elastic. The electron flow I_a through the thyratron is recorded by the light beam galvanometer on its most sensitive range and maintains A at a small negative potential with respect to G_1.

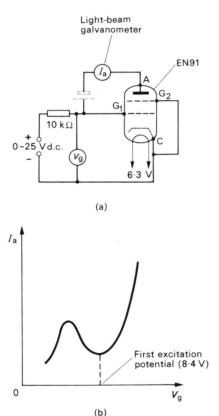

(a)

(b)

Fig. A18.1

As V_g is increased some electrons have just enough energy to cause excitation of xenon atoms and, having lost all their kinetic energy in the inelastic collision, are unable to overcome the retarding p.d. between A and G_1 causing I_a to fall. Further increase of V_g causes more electrons to lose their energy after excitation and I_a decreases further. When most of the electrons passing through G_1 produce excitation, I_a is a minimum and the corresponding value of V_g gives the *first*

excitation potential of xenon. Increasing V_g beyond this value enables electrons, even after they have had inelastic collisions, to overcome the retarding p.d., reach A and I_a rises again. The form of the I_a—V_g graph is shown in Fig. A18.1b.

The 10 kΩ resistor in the grid circuit prevents the current exceeding 10 mA (and destroying the thyratron) in the event of the gas ionizing during the experiment. With some thyratrons it may be necessary to *apply* a small retarding p.d. to prevent electrons reaching A. This is done by connecting a 1.5 V cell between A and G_1, in series with the galvanometer, as shown by the dotted symbol in Fig. A18.1a.

(b) *Ionization potential.* The thyratron, used as a diode with A joined to G_1, is connected as in Fig. A18.2a. When V equals the ionization potential of xenon, the current recorded by the milliammeter increases due to electrons from C having inelastic collisions with xenon atoms and ionizing them. The positive xenon ions created act as a new source of current. Fig. A18.2b shows the form of the current-p.d. graph.

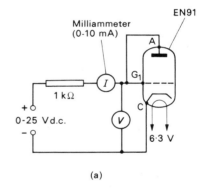

(a)

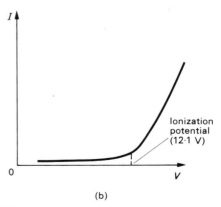

(b)

Fig. A18.2

577

Appendix 19

Radioactive decay analogue experiment using dice

Radioactive atoms are represented by small wooden cubes having one face marked in some way. If 100 of these are placed in a large can or jar and then thrown on to the bench so that they are in a single layer, the cubes with the marked face upper-most are considered to have 'decayed' and are removed and counted. The remaining 'undecayed' cubes are returned to the can, thrown again, the 'decayed' cubes removed and counted as before. The process is repeated at least 15 times until only a few cubes have not 'decayed'.

The whole procedure is repeated *another four times*, always starting with 100 cubes and making the same number of throws so that the effect is the same as if 500 cubes had been thrown initially. The results can be recorded as shown, the total number N of surviving cubes being obtained for each throw t.

A graph of N against t is plotted and a smooth curve drawn through the points. An estimate of the half-life for cube decay, i.e. the number of throws necessary for half the original number of cubes to decay, can be made.

Assuming the decay law is exponential we have

$$N = N_0 e^{-\lambda t} \qquad \text{(p. 538)}$$

Hence

$$\ln N = \ln N_0 - \lambda t \qquad (\ln = \log_e)$$

But $\ln N = 2.303 \log_{10} N$,

$$\therefore \quad \log_{10} N = \log_{10} N_0 - \frac{\lambda}{2.303} \cdot t$$

What should be the form of a graph of $\log_{10} N$ against t?

Number of throw	Number of 'decayed' cubes						Number of 'surviving' cubes					
t	(1)	(2)	(3)	(4)	(5)	Total	(1)	(2)	(3)	(4)	(5)	Total N
0 1 2	0	0	0	0	0	0	100	100	100	100	100	500

Appendix 20

Converting a single beam into a double beam CRO using an electronic beam splitter

The circuit is shown in Fig. A20.1. It consists of an astable multivibrator producing pulses, at about 2.5 kHz, that are fed into a transistor unit which 'squares' them before they are fed into the beam splitter unit. This switches the CRO beam to and fro rapidly between the two inputs to the unit. If the switching frequency is appreciably higher than the signal frequency, both traces seem almost continuous when a suitable time base speed is chosen.

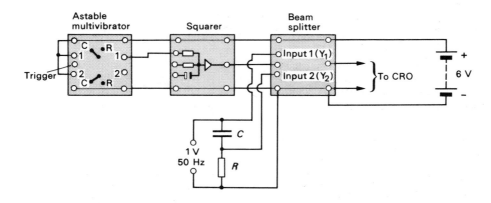

Fig. A20.1

Answers

Chapter 1. Structure of materials

1. (a) 4:3   (b) 16   (c) 16 g
2. (a) 14 g   (b) 70 g
3. (a) 1.5×10^{23} (b) 18×10^{23} (c) 3.0×10^{23}
4. (a) 4.0×10^{-23} g   (b) 1.2×10^{-21} g   (c) 3.0×10^{-23}
5. (a) 6.0×10^{23} (b) 2.0 g
   (c) 2.2×10^{4} cm^3 (d) 2.7×10^{17}
6. 6.1×10^{23} mol^{-1}
7. (a) 3.0×10^{-23} g   (b) 3.9×10^{-10} m (0.39 nm)
8. (a) 5.0×10^{3} J g^{-1} (b) 5.3×10^{-19} J/atom
9. (a) (i) 4   (ii) 6. Closest in hexagonal packing
   (b) (i) hexagonal   (ii) square
10. (a) $2\sqrt{2}\, r$ cm   (b) $16\sqrt{2}\, r^3$ cm^3 (c) $1/(16\sqrt{2}\, r^3)$
11. (a) 8   (b) 8   (c) 8/8 = 1   (d) 6   (e) 2   (f) 6/2 = 3
    (g) 1 + 3 = 4   (h) $1/(4\sqrt{2}\, r^3)$
12. 8.4×10^{22}
13. 1.3×10^{-10} m (0.13 nm)
14. If a is the separation between adjacent Na and Cl ions, a cube of side a and volume a^3 will contain 4 Na ions and 4 Cl ions, i.e. 4 NaCl molecules. But each corner (ion) of the cube is shared by 8 neighbouring cubes and so each volume a^3 contains ⅛ of 4 molecules of NaCl, i.e. ½ molecule of NaCl.

   $$\therefore \quad \text{Volume occupied by } \tfrac{1}{2} \text{ molecule} = a^3 = \frac{1}{2} \cdot \frac{M}{L\rho}$$

   where M = formula weight (molecular mass) and ρ = density of NaCl and L = the Avogadro constant.

   $$a = 2.82 \times 10^{-8} \text{ cm } (0.282 \text{ nm})$$

Chapter 2. Mechanical properties

1. (b) 4.0×10^{7} Pa
   (c) Yes
   (d) (i) ½   (ii) 50%
   (e) 2.002 m
2. (c) 1.0×10^{11} Pa; 0.040%; 4.0×10^{2} N; no; smaller.
4. (a) 1.1 (1.06) mm   (b) 1.0×10^{2} (101) N
6. (a) 1.86×10^{-3} m   (b) 7.75×10^{-2} J

Chapter 3. Electrical properties

1. 2.4×10^{2} C; 1.5×10^{21}
2. (a) 8.5×10^{28} (b) 8.5×10^{22} (c) 1.4×10^{4} C
   (d) 6.8×10^{3} s   (e) 0.15 mm s^{-1}
3. 8.0 A
4. (a) 8.0 V; 2.5 A   (b) 240 V
5. (a) 4.0 V   (b) 240 J
6. (a) 2 A   (b) 2 A   (c) 6 V   (d) 10 V   (e) 5 Ω
7. (a) $I = I_1 + I_2$ (b) 6 V   (c) 3 A, 2 A, 1 A   (d) 2 Ω
8. (a) 4 V; 2 V   (b) 4.8 V; 1.2 V
9. 150 kΩ, 25 kΩ
10. (a) 10 V   (b) 40 V
12. (i) 95 Ω in series   (ii) 0.050 Ω in parallel
13. 3:2
14. 4.20×10^{-3} °C^{-1}
15. (a) 1.00×10^{3} J s^{-1} (b) 798 °C
16. (a) 12 J   (b) 36 J   (c) 240 J
17. (a) 6 V   (b) (i) 2 J   (ii) 6 J   (c) 0.03 Ω   (d) 3 A
18. (b) 3×10^{4} J   (c) 960 Ω
19. Internal resistance 5 Ω
20. 2.0 (4) Ω; 0.50 A; 0.75 W
22. 4.9 Ω; 8.9 Ω
23. 0.0020 °C^{-1}; 50.2 cm
24. (i) 2.0 V   (ii) 2.7 Ω in parallel
25. 996 Ω; 0.2 cm
26. 2.00 Ω

Chapter 4. Thermal properties

2. (a) 63.16 °C   (b) 47.7 °C
3. 16.0 V
4. 1.70 J g^{-1} K^{-1}
5. 4.4 J g^{-1}K^{-1}; ±10%
6. 0.050 K s^{-1}; 6.8 revolutions per second
7. 26.0%
8. 3.3×10^{2} °C
9. True height is 76.46 *divisions* of scale. A division marked as 1 cm is only 1 cm at 0 °C. At 15 °C 1 division has length $1(1 + 1.9 \times 10^{-5} \times 15)$ cm, therefore 76.46 divisions at 15 °C have a true length of $76.46(1 + 1 \times 1.9 \times 10^{-5} \times 15)$, i.e. 76.48 cm
10. 1.2 N
11. (a) 1:15 (14.5)   (b) 66.7 °C
12. 1.9×10^{3} W
13. Copper 3.0(2.98) °C cm^{-1}; aluminium 5.5 °C cm^{-1}
14. 4.4×10^{-3} W cm^{-2}
15. 18 °C

Chapter 5. Optical properties

1. 15 cm; 5.0 cm
2. 20 cm; 10 cm behind mirror
3. Virtual, 13 (12.9) cm behind convex mirror
4. (a) 2.4 m   (b) 1.3×10^{-2} m
5. (a) 40 (.5)°   (b) 40 (.5°)   (c) 35 (.4)°
6. (i) ray refracted in water at an angle of 34 (.2)°
   (ii) critical angle for glass–water boundary = 63 (62.7)°, therefore ray totally internally reflected
7. Angle in liquid must exceed 67.9°
8. (a) 37 (.4)°
   (b) 28 (27.9)°
9. (a) 55.6°; 39.6° to 90.0°   (b) 60.3°
10. 48.8°
11. (a) 20 cm; +15 cm   (b) +4.0 cm
13. 20 cm from first position of screen
14. (a) +60 cm   (b) +2.4 m
15. +20 cm, +20 cm, −20 cm, −60 cm
16. Virtual, 39 cm from lens on same side as object
17. $f = -500$ cm gives far point at infinity.

The nearest distance an object can now be brought up to his eye is that object distance which has its virtual image at 60.0 cm, i.e. we have to find u when $v = -60.0$ cm and $f = -500$ cm. Using $1/f = 1/u + 1/v$ we get $u = 68.2$ cm. His range of vision is now infinity to 68.2 cm, i.e. when wearing concave spectacle lenses of $f = -500$ cm objects within this range can be seen.

18. $f = 66.7$ cm gives near point at 25.0 cm.

The new far point will be the object distance which gives a virtual image 200 cm from the spectacle lens, i.e. we have to find u when $v = -200$ cm and $f = 200/3$ cm. From $1/f = 1/v + 1/u$ we get $u = 50.0$ cm. The new range of vision is therefore 25.0 to 50.0 cm.

19. magnifying power = 5.0; magnification = 6.0

20. (b) 58 (.3)

21. Separation of lenses 25 cm; $f = 6.0$ cm

22. 112 (.4) cm; 14.7

23. 80 cm; 16; 5.3 cm; 13 (12.8) cm

24. Reduced $\frac{1}{4}$; 5 times larger

Chapter 6. Statics and dynamics

2. (a) 1.2×10^3 N   (b) 1.1×10^3 N; 74° to horizontal.

3. 4.3×10^2 N; 69° to the horizontal.

4. AB $\sqrt{3} \times 10^6$ N; BC 2×10^6 N; AC 10^6 N; CD $\sqrt{3} \times 10^6$ N. Minimum cross-section area for BC = $2(.5) \times 10^{-2}$ m^2

5. (i) 0.245 s   (ii) 12.2 m s^{-1} (iii) 12.5 m s^{-1}

6. 2.17×10^4 m; 3.13×10^3 m; 25.0 s.

7. $\sqrt{3}/3$; $\sqrt{3}\,g/3$ m s^{-2}; $2\sqrt{3}\,mg/3$ (m = mass of body)

9. $P/(5\,m)$; $P/5$; $2\,P/5$

10. 3.6×10^3 N; 5.4×10^4 W

11. $mg/2$

12. (a) 10/3 N   (b) 5/9 W   (c) 5/18 W

13. 8.00×10^3 N; 1.20×10^4 W

14. $2E/103$

Chapter 7. Circular motion and gravitation

1. (a) time $= \dfrac{\text{arc AB}}{\text{speed}} = \dfrac{\pi \times 5\ \text{m}}{11\ \text{m s}^{-1}} = \dfrac{22 \times 5\ \text{s}}{7 \times 11} = \dfrac{10\ \text{s}}{7}$

(b) average velocity $= \dfrac{\text{displacement}}{\text{time}} = \dfrac{10\ \text{m}}{10/7\ \text{s}}$

$= 7.0$ m s^{-1} *to the right*

The displacement is diameter AB to the right.

(c) average acceleration $= \dfrac{\text{change of velocity}}{\text{time}}$

$= \dfrac{22\ \text{m s}^{-1}\ \text{downward}}{10/7\ \text{s}}$

$= 154/10$ m s^{-2} *downward*

2. (a) 1.5π rad s^{-1} (b) 0.18π m s^{-1}

3. (a) 5.0 rad s^{-1}; 25 N

(b) 30 N: 20 N

4. (i) 4/3 N   (ii) $\sqrt{3}\pi/5$ s

5. 1.01×10^3 N; 1.70×10^2 N; 1.85×10^3 N

6. (a) 44 rad s^{-1} (b) 11 m s^{-1} (c) 22 m s^{-1} (d) 87 J

7. 1.9×10^{-2} N m

8. 15.8 s

9. (b) 4.0×10^{-4} kg m^2

10. 9.83 N

11. $v_B/v_C = y/x$

12. 24 hours

13. 7.71×10^{22} kg; 1.15×10^3 m

14. 3.83×10^5 km

15. (a) (i) 0.12 N   (ii) 0.10 N   (c) 7.0×10^5 J

Chapter 8. Mechanical oscillations

1. (a) 75 cm s^{-1} (b) 1.4×10^4 cm s^{-2}

2. 14 cm s^{-1}

3. 1.04

4. (a) 31 cm s^{-1} (b) 50 cm s^{-2} (c) 0.33 s

(d) 5.0×10^{-3} J at A; zero at limits of motion

(e) 5.0×10^{-3} J

5. An s.h.m. of period $\pi/\sqrt{50}$ s: maximum velocity = 0.71 m s^{-1}; maximum acceleration = 10 m s^{-2}

6. 24(.5) N kg^{-1} m^{-1}

7. 16 Hz

8. 1.3×10^2 cm s^{-1}; 1.6 cm

Chapter 9. Fluids at rest

1. 1.0×10^4 N; 5.0×10^2 N

2. (b) 9 N

3. Silver 4.90 g   Gold 30.3 g

4. 2.4 cm

5. 757.3 mm; 0.2736 mm; 760.3 mm

6. 118 (The general expression is $p_n = p[V/(V + v)]^n$ where p is the initial pressure of the air in the vessel, V is the volume of the vessel, v is the volume swept out by the piston of the pump per stroke and n is the number of strokes to achieve pressure p_n. The expression is obtained by repeated application of Boyle's law to the mass of air in the vessel at the start of each stroke.)

8. 7.1×10^{-2} N m^{-1}

9. 49 mm; the water does not overflow (and violate conservation of energy) but it remains at the top of the tube (i.e. $h = 30$ mm) with an angle of contact of 52.2°. The weight of the raised column of water is then supported by the vertical components of the surface tension forces.

10. 76.93 cm

Chapter 10. Fluids in motion

3. 8.7×10^{-4} m s^{-1}

4. 0.26 mm^2 (Air and petrol do not have the same velocities; Bernoulli's equation must be used)

5. Yes. (Hint: apply Bernoulli's equation using

$$\frac{\text{velocity of water in barrel}}{\text{velocity of water from jet}} = \frac{\text{area of jet}}{\text{area of barrel}}$$

to find velocity of water from jet. Find the time taken by a drop of water to travel 3.5 m (assuming its horizontal velocity is constant) and then calculate how far it falls in this time.

Chapter 11. Electric fields

1. (a) 8.9×10^4 V m^{-1} (b) 8.9×10^3 V
2. (a) 29 V (b) 4.6×10^{-18} J (c) 4.6×10^{-18} J
3. $_BE_A = 1.0 \times 10^3$ V m^{-1};
   $_CE_B = 2.0 \times 10^3$ V m^{-1}; $_CE_D = 3.0 \times 10^3$ V m^{-1}
4. (a) 2.9×10^{11} N C^{-1}
   (b) 4.3×10^6 N C^{-1}; ratio 0.68×10^5:1
   (c) 1.4×10^{11} N C^{-1}
   (d) 5.8×10^7 N C^{-1}; ratio 0.25×10^4:1
5. (i) 8.0×10^{-16} J (ii) 1.6×10^{-15} J
6. (i) 10^{-12} J (ii) 10^{-12} J (iii) 5×10^{-11} N
   (iv) 5×10^4 N C^{-1} (v) 5×10^4 V m^{-1}

Chapter 12. Capacitors

1. (a) 6 V; 18 μC; 3 μF
   (b) (i) 4 μC (ii) 4 μC (iii) 4 V (iv) 2 V
   (v) 2 μF (vi) 2/3 μF; 4 μC
2. (a) 10^{-12} F (1 pF) (b) 10^{-9} C
   (c) $V = Q_1/C_1 = Q_2/C_2 \therefore Q_1/Q_2 = C_1/C_2 =$
   $10^{-9}/10^{-12} = 10^3/1$
   (d) (i) 10^{-9} C (ii) zero (e) equally
3. (a) 28.6 V; 71.4 V (b) 143 μC
   (c) 2.05×10^{-3} J; 5.11×10^{-3} J
4. (a) (i) 3.7×10^{-8} C (ii) 9.3×10^{-8} C (b) 2.3 m
5. 1.07×10^{-4} m; 2.1×10^3 V
6. 50 J; 33 J
7. 3 banks of 3 capacitors in series
8. 1.1×10^{-9} F; 8.8×10^{-12} F m^{-1}; 1.8×10^9 Ω
9. (a) 1.2×10^{-5} C (b) 8.0×10^5 V (c) 20 cm
10. 11 s

Chapter 13. Magnetic fields

1. (b) downwards (c) upwards
2. (i) 1×10^{-2} N (ii) 0.5×10^{-2} N
3. 2.0×10^{-5} T; 2.8×10^{-5} T
4. 1.6 A $(5/\pi)$
5. 0.23 A
6. 3.6 A
7. 1.1 kV
8. $Be/(2\pi m)$; 2.8×10^7 rev s^{-1}
9. (a) No
   (c) $Ee = Bev \therefore E = Bv$. But $E = V_H/d \therefore V_H = Bdv$
   (d) 1 mm s^{-1}
10. 1.0×10^{-4} T; 1.0×10^{-3} N
11. 4.0×10^{-5} T; 4.0×10^{-3} N
12. 1.0×10^{-3} N m
13. 1.1×10^{-3} A
14. (a) 10 (b) $\frac{1}{4}$

Chapter 14. Electromagnetic induction

2. 1.5 mA
3. 0.38 V
4. 16 mV

5. (a) 50 Hz (b) 23.6 V
6. 0.31 mV
9. 237 V, 948 W; 195 V, 11.7×10^3 W; 329 r.p.m.
10. 4 A
11. 0.20 V
12. 4.0 A; 2.0 A s^{-1}
13. 0.50 T
14. 1.43 T

Chapter 15. Alternating current

1. (a) 12 V (b) 17 V (c) 2.8 cm
2. (a) 99 kV, 0 (b) 19.8 MW, 0; 141 kV
3. (a) 7.0×10^{-4} C (c) (i) Current a maximum when rate of change of charge is a maximum, i.e. when $Q = 0$ (ii) Current a minimum when rate of change of charge is zero, i.e. when $Q = \pm 7.0 \times 10^{-4}$ C
   (d) See Fig. 15.10b (e) 1.6×10^{-1} A
4. 6.37 μF
5. (a) 0 (b) 10^{-4} A (c) 2 V s^{-1}
6. (a) $V_R = 3.0$ V; $V_L = 4.0$ V (c) 37°
7. 120 Ω; 0.66 H; 15 μF
8. 225 Hz (a) $10^4/\sqrt{2}$ V (b) $10^4/\sqrt{2}$ V (c) 0
9. (a) 40 Ω (b) 10 H (c) 1.6×10^3 V
10. (a) 0.10 A, 40 V (b) 3.0 J s^{-1}
11. (i) 4.7 A, 89 W; $V_R = 19$ V; $V_L = 15$ V
12. 2 A; 40 W
13. Damped oscillations occur of frequency about 5 Hz

Chapter 16. Wave motion

1. 19°
2. 47°; 42° with the normal to the oil surface
3. (a) $\lambda/2 = S_2Q - S_1Q$; $\lambda = S_2R - S_1R$
   (b) The spacings PQ and QR (i) decrease (ii) increase
4. 4.7 m s^{-1}
5. 20.0 m to 17.0 mm
6. (a) 200 kHz (b) 60.0 cm; 59.4 cm (c) 5×10^{14} Hz
7. (a) In phase and polarized in the same direction
   (b) same signal
   (c) more than 100 m
8. $(7.16 \pm 0.21) \times 10^{-2}$ N m^{-2}
9. (a) 3.3×10^2 m s^{-1} (b) 6.6×10^{-4} m s^{-1}

Chapter 17. Sound

1. At the point C of zero intensity the path difference for sound from A and B is $(13.2 - 11.0)/0.40 = 5\frac{1}{2}$ wavelengths.

   If I_A and I_B are the intensities due to A and B respectively at C then $I_A = I_B$ but $I_A \propto P_A/AC^2$ and $I_B \propto P_B/BC^2$ (inverse square law holds) where P_A and P_B are the rates of emission of sound energy from A and B respectively.

   $$\therefore \quad \frac{P_A}{AC^2} = \frac{P_B}{BC^2} \quad \text{or} \quad \frac{P_A}{P_B} = \frac{AC^2}{BC^2} = \frac{13.2^2}{11.0^2} = 1.44.$$

ANSWERS 583

The sound intensity is therefore zero at C because the path difference is an odd number of half wavelengths and the distances are such that the intensities (and therefore the amplitudes) are equal.

2. Stationary wave system set up with wavelength of 3.0 m; nodes occur every 1.5 m. A beat note due to Doppler effect of 1.3 Hz.
3. 344 Hz
4. 215 Hz; 645 Hz; 1075 Hz
5. 306 m s^{-1}
6. 0.78 m; 2.8
8. 190 Hz; 202 N
9. Resonant vibration at 115 Hz
10. $\dfrac{10}{9}f$; $\dfrac{11}{10}f$
11. 320 Hz; 282 Hz; 1.0 s
12. 136 cm s^{-1}
13. 120 Hz
14. (a) 400 m s^{-1} (b) 408 m s^{-1}
15. Stationary wave pattern formed of wavelength $2 \times (5.66 - 2.99)/4$ mm = 1.34 mm; 1.34×10^3 m s^{-1}
16. 0.266 m
17. 280, 350, 420 or 490 Hz

Chapter 18. Physical optics

1. 3.0×10^8 m s^{-1}
2. (a) 2.0×10^8 m s^{-1} (b) 5.0×10^{14} Hz (c) 4.0×10^{-7} m
4. 5.0×10^{-7} m
6. (i) 0.200 mm (ii) 0.300 mm (iii) 0.600 mm
7. 1.5; to side of covered slit
8. 100
9. 0.20 mm
10. 5.87×10^{-7} m
11. 1.51 m; 1.33
12. 9°
13. 2.5×10^5 m^{-1}
14. 4.8×10^{-6} m
15. 20.1°
16. 0.950°; 3.32 mm
20. 9.0×10^2 °C
21. 0.96 °C s^{-1}
22. 16 (15.6) J s^{-1}
23. 1.8×10^3 K

Chapter 19. Kinetic theory: thermodynamics

1. 8.20 J mol^{-1} K^{-1}
2. (b) 2.66×10^{25}
3. 289 mm (*Hint.* Apply $p_1/T_1 = p_2/T_2$ to the *air* in the mixture of air and saturated water vapour.)
4. 6.8×10^2 mm
5. 6.63×10^2 m s^{-1}
6. 450 K; 2.40×10^3 m s^{-1}

8. (a) 1.2×10^{16} (b) 1.6×10^{-7} m (c) 1.9×10^3 m s^{-1}
9. (iv) 200 cmHg (v) −205 °C (68 K); 273 °C (546 K) (vi) 136 J at constant pressure
10. $\dfrac{3}{2}nRT$ (n = no. of moles) $= \dfrac{3}{2} \times \dfrac{1}{4} \times 8.3\,T = 3.1\,T$; $c_v = 3.1 \times 10^3$ J kg^{-1} K^{-1}; $c_p = 5.2 \times 10^3$ J kg^{-1} K^{-1}
11. 0.167×10^6 J
12. 5.3×10^2 J kg^{-1} K^{-1}; 1/2
13. 1.49 atmospheres
14. 0.53 atmosphere; −130 °C (143 K); 6.3 litres

Chapter 20. Atomic physics

1. 1.60×10^{-19} J; 3.20×10^{-14} J; 6.19×10^6 m s^{-1}
2. (a) 0.39×10^{-15} N (b) 4.2×10^{14} m s^{-2} (c) 5.6×10^6 m s^{-1}
3. 2.7×10^7 m s^{-1}; 3.2×10^{-4} W
4. (a) 2.65×10^7 m s^{-1} (b) 9.09×10^{-4} T
5. 2.01×10^3 V
6. 3; 19.6 m s^{-2}
7. (a) 4.1×10^{-7} m (b) 8.8×10^5 m s^{-1}
8. (a) 1.5 eV (b) 3.5×10^{14} Hz
9. 6.6×10^{-19} J
10. 5.1×10^{14} Hz
11. 8.5×10^7 m s^{-1}
12. (a) 6.3×10^{15} electron s^{-1} (b) 2.4×10^{-16} J
13. 1.2×10^{-11} m
16. A: 0.65 μm; B: 0.48 μm; C: 0.43 μm; D: 0.41 μm
17. (a) 4.9 eV; 6.7 eV; 8.8 eV; 10.4 eV
   (b) 10.4 eV
   (c) (i) 5 eV, (5 − 4.9) = 0.1 eV
      (ii) 10 eV, (10 − 4.9) = 5.1 eV, (10 − 6.7) = 3.3 eV
   (d) (i) absorbed and disappears
      (ii) scattered

Chapter 21. Electronics

2. (b) (i) 5 waves (ii) ½ wave (c) 50 Hz
3. 1 V; 0.7 V
8. 36 kΩ
9. (i) 5.4 V (ii) 54 μA (iii) 3.2 mA (iv) 3.2 V (v) 2.8 V
10. (b) $I_C = 2.5$ mA, $I_B = 30$ μA, $V_{CE} = 4.5$ V
   (c) 11.3 mW (d) (i) 4.5 V ± 2.0 V (ii) ± 2.0 V peak
   (e) ± 20 mV (f) 100 (g) 130 kΩ
13. (c) 10
16.

Input 1	Input 2	NOR output	OR output	AND output	NAND output
0	0	1	0	0	1
1	1	0	1	1	0
1	0	0	1	0	1
0	1	0	1	0	1

17.

R	A	G
1	0	0
1	1	0
0	0	1
0	1	0

Chapter 22. Nuclear physics

1. 5.1 MeV
2. (a) alpha, beta, beta, alpha, alpha (b) 2 (c) 5
3. (a) 3 (b) 1/8 (c) 63/64
4. (a) 1/8 (b) 15/16 (c) 6
5. (a) 2:1 (b) 1:1 (c) 1:16
6. 6.0×10^3 cm^3
7. 75 s
8. 0.11 cm^{-1}
10. $v_m = (m - M)v/(m + M)$; $v_M = 2mv/(m + M)$
 (a) alpha v; electron $2v$
 (b) alpha zero; helium atom v
 (c) alpha $- v$; gold atom zero
  When there is a head-on collision between a moving and a stationary particle of equal mass
11. (a) Fig. 22.35b (c) 90°
12. 3.13×10^{16}

Objective-type questions

1.1. B   **1.2.** A
2.1. C   **2.2.** C   **2.3.** B   **2.4.** A, E, D
3.1. A   **3.2.** D   **3.3.** C   **3.4.** E   **3.5.** D   **3.6.** B
3.7. C (this is a *balanced* Wheatstone bridge)
3.8. C   **3.9.** D   **3.10.** E
4.1. B   **4.2.** C   **4.3.** E   **4.4.** B
5.1. B   **5.2.** A   **5.3.** D   **5.4.** A
6.1. B   **6.2.** C   **6.3.** E   **6.4.** D   **6.5.** B   **6.6.** C
7.1. A   **7.2.** D   **7.3.** D   **7.4.** C
8.1. E   **8.2.** B   **8.3.** C
9.1. C
10.1. E
11.1. C   **11.2.** E   **11.3.** A
12.1. A   **12.2.** (a) E (b) D   **12.3.** C
13.1. D   **13.2.** C   **13.3.** D
14.1. C   **14.2.** D   **14.3.** C   **14.4.** C   **14.5.** C   **14.6.** A
15.1. C   **15.2.** A   **15.3.** C
16.1. D   **16.2.** E   **16.3.** A   **16.4.** A
17.1. D   **17.2.** A
18.1. D   **18.2.** B   **18.3.** C
19.1. C   **19.2.** E   **19.3.** D   **19.4.** C   **19.5.** B
20.1. C   **20.2.** E   **20.3.** C   **20.4.** E   **20.5.** C
21.1. E
22.1. E   **22.2.** B

Index